COMPREHENSIVE SERIES IN PHOTOCHEMISTRY & PHOTOBIOLOGY

Series Editors

Donat P. Häder
Professor of Botany

and

Giulio Jori
Professor of Chemistry

European Society for Photobiology

COMPREHENSIVE SERIES IN PHOTOCHEMISTRY & PHOTOBIOLOGY

Series Editors: Donat P. Häder and Giulio Jori

Titles in this Series

Volume 1 UV Effects in Aquatic Organisms and Ecosystems
Edited by E.W. Helbling and H. Zagarese

COMPREHENSIVE SERIES IN PHOTOCHEMISTRY
& PHOTOBIOLOGY – VOLUME 1

UV Effects in Aquatic Organisms and Ecosystems

Editors

E. Walter Helbling
Estación de Fotobiología Playa Unión
Playa Unión, Rawson, Chubut
Argentina

and

Horacio Zagarese
Centro Regional Universitario Bariloche
Universidad Nacional del Comahue
Bariloche
Argentina

RS•C
advancing the chemical sciences

ISBN 0-85404-301-2

A catalogue record for this book is available from the British Library

Published by The Royal Society of Chemistry,
Thomas Graham House, Science Park, Milton Road,
Cambridge CB4 0WF, UK

Registered Charity Number 207890

For further information see our web site at www.rsc.org

Typeset by Vision Typesetting, Manchester
Printed and bound by Bookcraft Ltd, UK

Preface for the ESP series in photochemical and photobiological sciences

"Its not the substance, it's the dose which makes something poisonous!" When Paracelsius, a German physician of the 14th century, made this statement he probably did not think about light as one of the most obvious environmental. But his statement applies as well to light. While we need light, for example for vitamin D production, too much light might cause skin cancer. The dose makes the difference. These diverse findings of light effects have attracted the attention of scientists for centuries. The photosciences represent a dynamic multidisciplinary field which includes such diverse subjects as behavioral responses of single cells, cures for certain types of cancer and protective potential of tanning lotions. It includes photobiology and photochemistry, photomedicine as well as the technology for light production, filtering and measurement. Light is a common theme in all these areas. In recent decades a more molecular centered approach has changed both the depth and the quality of the theoretical as well as the experimental foundation of photosciences.

An example of the relationship between global environment and the biosphere is the recent discovery of ozone depletion and the resulting increase in high energy ultraviolet radiation. The hazardous effects of high energy ultraviolet radiation on all living systems is now well established. This discovery of the result of ozone depletion put photosciences in the center of public interest with the result that in an unparalleled effort scientists and politicians worked closely together to come to international agreements to stop the pollution of the atmosphere.

The changed recreational behavior and the correlation with several diseases in which sunlight or artificial light sources play a major role in the causation of clinical conditions (e.g. porphyrias, polymorphic photodermatoses, *Xeroderma pigmentosum* and skin cancers) have been well documented. As a result, in some countries (i.e. Australia) public services inform people about the potential risk of extended periods of sun exposure. The problems are often aggravated by the phototoxic or photoallergic reactions produced by a variety of environmental pollutants, food additives or therapeutic and cosmetic drugs. On the other hand, if properly used, light-stimulated processes can induce important beneficial effects in biological systems, such as the elucidation of several aspects of cell structure and function. Novel developments are centered around photodiagnostic and phototherapeutic modalities for the treatment of cancer, artherosclerosis, several autoimmune diseases, neonatal jaundice and others. In addition, classic research areas like vision and photosynthesis are still very active. Some of these developments are unique to photobiology, since the peculiar physico-chemical properties of electronically excited biomolecules often lead to the promotion of reactions which are characterized by high levels of selectivity in space and time. Besides the biologically centered areas, technical developments have paved the way for the harnessing of solar energy to produce warm water and electricity or the development of environmentally friendly techniques for addressing problems of large social impact (e.g. the decontamination of polluted waters). While also in use in Western countries, these techniques are of great interest for developing countries.

The European Society for Photobiology (ESP) is an organization for developing and coordinating the very different fields of photosciences in terms of public knowledge and scientific interests. Due to the ever increasing demand for a comprehensive overview over the photosciences the ESP decided to initiate an encyclopedic series, the "Comprehensive Series in Photochemical and Photobiological Sciences". This series is intended to give an in-depth coverage of all the very different fields related to light effects. It will allow investigators, physicians, students, industry and laypersons to obtain an updated record of the state-of-the-art in specific fields, including a ready access to the recent literature. Most importantly, such reviews give a critical evaluation of the directions that the field is taking, outline hotly debated or innovative topics and even suggest a redirection if appropriate. It is our intention to produce the volumes at a sufficiently high rate to generate a timely coverage of both well established and emerging topics. As a rule, the individual volumes are commissioned; however, comments, suggestions or proposals for new subjects are welcome.

Donat-P. Häder and Giulio Jori
Spring 2002

Volume preface

The surge in the systematic study of UV effects on aquatic habitats is contemporary with the discovery of the ozone hole in the 1980s. Since then, and for the last two decades, the number of publications on UV related issues has grown virtually exponentially. Paralleling the explosive development of this new "field", a number of reviews have attempted to summarize the available knowledge in the primary literature. These works have evolved from environmental agencies' reports to symposia volumes to multi-authored edited books, many of which are excellent, some of which are reasonably comprehensive, and a few of which are quite recent. Thus, as soon as we were offered the opportunity of producing a new book on UV effects on aquatic ecosystems, we wondered how a new book on this subject could provide new insights or a different perspective, and perhaps the stimulus or inspiration for future research. In this book, we have attempted to bridge the gap between the environmental studies of UVR effects and the broader, traditional fields of ecology, oceanography and limnology. Our purpose has been to provide evidence to persuade a general ecologist that UV driven processes are relevant to aquatic ecosystems. But, at the same time, we wanted to adopt the point of view that UV is only one of several important ecological processes operating synchronously in the natural environment. If we have succeeded, the message from this book should be that the search for environmental UV effects must be framed within a wider ecological context.

What's in the book

The book is divided into five sections, which are intended to cover the most salient aspects of UV research. The introductory chapter provides an overview of the role of UVR in aquatic systems with a strong emphasis on the interaction between UVR and DOM. This interaction will be a recurrent subject in subsequent chapters, and reflects the contemporary perception of UV researchers about the key role played by DOM in controlling UV optics, and directly or indirectly regulating chemical and biological processes.

The Physics section provides information and discussions on global UVR climatology at the Earth's surface level and the factors controlling the transmittance of UVR through the atmosphere (Chapter 2), and the penetration into the water column (Chapter 3). The last chapter within this section (Chapter 4) introduces basic notions of mixing and advection and outlines the implications of vertical water motion for photochemical and photobiological processes.

The four chapters within the Chemistry section discuss the effects of UVR on biogeochemical cycles of various elements (Chapter 5), the photochemistry of DOM (Chapter 6), the photo-activated toxicity of several natural and anthropogenic substances (Chapter 7) and the environmental implications of photoinduced formation of reactive oxygen species (Chapter 8). Once again, the central role of DOM emerges as a unifying theme.

The fourth section focuses on individual and sub-individual effects and responses. The first chapter within this section (Chapter 9) reviews the effects of UVR on DNA, which has long been identified as one of the primary targets of UVR in biological systems. It is followed by a discussion of the main physiological photoprotective mechanisms in aquatic organisms (Chapter 10). Chapter 11 reviews the available literature on UVR effects on autotrophs, while Chapters 12 and 13 present two different and complementary perspectives on the effects of UVR on heterotrophs. This section ends with an extensive review on the role of sensory systems and behavioral responses to UVR (Chapter 14).

Three chapters within the last section address the effects of UVR from the community and ecosystem perspective that has been anticipated in the introductory section. Chapter 15 provides a thorough review of the effects of UVR on species interactions, including predation, competition, parasitism and diseases. Chapter 16 discusses the methods for reconstructing the radiation history of aquatic ecosystems and presents evidence for different UVR paleoclimates. Chapter 17 speculates on potential future UVR scenarios in a world that is experiencing several climatic changes from regional acidification to global warming and the also global depletion of stratospheric ozone.

Acknowledgements

Many people have contributed to make this book possible. First, we would like to thank all the authors for their time and commitment, and for their great disposition to help us out in every way. It has been a pleasure and a honour to work with them. We are deeply indebted to Ruben Sommaruga, who "volunteered" to work as Associate Editor for the chapters authored by either one of us. His help has been essential to assure the impartial review process of our chapters. Virginia Villafañe helped us at every stage of the editing process and her inputs, comments and criticisms have been fundamental.

The contribution of the external reviewers deserves a separate paragraph. We cannot overemphasize how much this book benefited from the comments, opinions and generous suggestions made by the many reviewers: María Marta Bianchi, J. Platt Bradbury, Howard Browman, Cynthia Carey, Ron Douglas,

Bruce Greenberg, David Hamilton, Bruce Hargreaves, Osmund Holm-Hansen, Ron Kiene, Peter Kiffney, Susanne Kratzer, George Losey, John Marra, Diane McKnight, Tim van Oijen, James T. Oris, Isabel Reche, David Schindler, Ralph Smith, Francesco Zaratti, and seven anonymous reviewers.

Finally, we would like to acknowledge the support of European Society of Photobiology, Consejo Nacional de Investigaciones Científicas y Técnicas (CONICET), Universidad Nacional del Comahue and Fundación Playa Unión.

E. Walter Helbling
Horacio Zagarese

Contributors

Anastazia T. Banaszak
Unidad Académica Puerto Morelos
ICML-UNAM
Apartado Postal 1152
Cancún
Quintana Roo 77500
México.

Mario Blumthaler
Institute for Medical Physics
Muellerstr. 44
A-6020 Innsbruck
Austria

Peter Boelen
Department of Marine Biology
Center for Ecological and
Evolutionary Studies
University of Groningen,
P.O. Box 14,
1790 AA Haren
The Netherlands

Anita G.J. Buma
Department of Marine Biology
Center for Ecological and
Evolutionary Studies
University of Groningen
P.O. Box 14
1790 AA Haren
The Netherlands

Stephen A. Diamond
U.S. Environmental Protection
Agency
Mid-Continent Ecology Division
6201 Congdon Boulevard
Duluth, MN 55804
USA

David Fabacher
USGS Columbia Environmental
Research Center
4200 New Haven Road
Columbia, MO 65201
USA

Félix L. Figueroa
Departamento de Ecología
Facultad de Ciencias
Universidad de Málaga
Campus Universitario de Teatinos
s/n
E-29071 Málaga
Spain

Bruce R. Hargreaves
Lehigh University
Department of Earth &
Environmental Sciences
31 Williams Drive
Bethlehem, PA 18015
USA

E. Walter Helbling
Estación de Fotobiología Playa Unión
and
Consejo Nacional de Investigaciones Científicas y Técnicas (CONICET)
Casilla de Correos No. 153
(9100) Trelew
Chubut
Argentina

Dag O. Hessen
Department of Biology
University of Oslo
P.O.Box 1027
Blindern
0316 Oslo
Norway

Dominic A. Hodgson
British Antarctic Survey
Natural Environment Research Council
High Cross
Madingley Road
Cambridge, CB3 0ET
United Kingdom

Wade H. Jeffrey
Center for Environmental Diagnostics and Bioremediation
University of West Florida
11000 University Parkway
Pensacola, FL 32514
USA

Sönke Johnsen
Biology Department
Woods Hole Oceanographic Institution
Woods Hole, MA 02543
USA

David J. Kieber
State University of New York
College of Environmental Science and Forestry
Chemistry Department
1 Forestry Drive
Syracuse, NY 13210
USA

Peter R. Leavitt
Limnology Laboratory
Dept. of Biology
University of Regina
Regina
Saskatchewan
Canada, S4S 0A2

Dina M. Leech
Department of Earth and Environmental Sciences
Lehigh University
Bethlehem, PA 18015
USA
Present address:
Biology Department
Box 90338
Duke University
Durham, NC 27708
USA

Edward E. Little
USGS Columbia Environmental Research Center
4200 New Haven Road
Columbia, Missouri 65201
USA

Donald P. Morris
Department of Earth & Environmental Sciences
Lehigh University
31 Williams Drive
Bethlehem, PA 18015
USA

Patrick J. Neale
Smithsonian Environmental
Research Center
P. O. Box 28
Edgewater MD 21037
USA

Christopher L. Osburn
Chemistry Division, Code 6115
Naval Research Laboratory
4555 Overlook Ave SW
Washington, DC 20375
USA

Barrie M. Peake
Chemistry Department
University of Otago
Union Place
Dunedin
New Zealand

Reinhard Pienitz
Paleolimnology-Paleoecology
Laboratory
Centre d'Études Nordiques
Université Laval, Québec
Québec
Canada, G1K 7P4

Norman M. Scully
Chemistry Department
Center for Marine Science
University of North Carolina at
Wilmington
One Marvin K. Moss Lane
Wilmington, NC 28409
USA

Ruben Sommaruga
Institute of Zoology and Limnology
University of Innsbruck
Technikerstr. 25
A-6020 Innsbruck
Austria

Kristina Sundbäck
Department of Marine Botany
Göteborg University
Box 461
SE-405 30
Sweden

Virginia E. Villafañe
Estación de Fotobiología Playa
Unión
and
Consejo Nacional de
Investigaciones Científicas y
Técnicas (CONICET)
Casilla de Correos No. 153
(9100) Trelew
Chubut
Argentina

Ann R. Webb
Department for Physics
University Manchester Institute for
Science and Technology
Manchester M60 1QD
United Kingdom

Robert G. Wetzel
Department of Environmental
Sciences and Engineering
School of Public Health
The University of North Carolina
Chapel Hill, NC 27599-7431
USA

Craig E. Williamson
Department of Earth and
Environmental Sciences
31 Williams Drive
Lehigh University
Bethlehem, PA 18015-3188
USA

Horacio E. Zagarese
Laboratorio de Fotobiología
Centro Regional Universitario
Bariloche
Universidad Nacional del Comahue
& (CONICET)
Unidad Postal Universidad
8400 Bariloche
Argentina

Richard G. Zepp
U.S. Environmental Protection
Agency
960 College Station Road
Athens, GA 30605-2700
USA

Contents

Individual and Sub-indiviudal Effects and Responses

Community and Ecosystem Perspectives

Abbreviations and symbols

Φ, quantum yield
Φ_a, apparent quantum yield
$\alpha^*_{cdom,\lambda}$, chromophoric dissolved organic matter-specific optical absorption
$\alpha^*_{Chl,\lambda}$, chlorophyll-specific spectral absorption factor
α^*_{DOC}, dissolved organic carbon-specific absorption
6-4 PP, (6-4) pyrimidone photoproduct
8-oxoG, 8-hydroxyguanosine
a_{CDOM}, absorption coefficient of CDOM
ADP, adenosine di-phosphate
AOP, apparent optical properties
ATP, adenosine tri-phosphate
BAF, bioaccumulation factor
BCF, bioconcentration factor
BED, biologically effective dose
BEE, biologically effective exposure
BEI, biologically effective irradiance
BLP, biologically-available photoproduct
BWF, biological weighting function
CA, carbonic anhydrase
CAT, catalase
CCN, cloud condensation nuclei
CDOM, chromophoric dissolved organic matter
chl-*a*, chlorophyll *a*
CO_2, carbon dioxide
CO, carbon monoxide
COS, carbonyl sulfide
CPD, cyclobutane pyrimidine dimer
CTM, chemical transport models
DIC, dissolved inorganic carbon
DMS, dimethyl sulfide
DMSP, dimethylsulfonium propionate

Dn, diadinoxanthin
DNA, desoxy ribonucleic acid
DOC, dissolved organic carbon
DOM, dissolved organic matter
DON, dissolved organic nitrogen
DOS, dissolved organic sulfur
Dt, diatoxanthin
DVM, diel vertical migration
E_d, downwelling irradiance
EEMS, excitation–emission matrix spectra
ENSO, El Niño-Southern Oscillation
E_{ph}, euphotic zone (1% of surface PAR)
EPR, electron paramagnetic resonance spectroscopy
E_u, upwelling irradiance
FAD, flavin adenine dinucleotide
FP, fluorescent pigments
FWHM, full-width at half maximum
G3PDH, glyceraldehyde -3-phosphate dehydrogenase
GCM, general circulation models
GST, glutathione transferase
HMW, high molecular weight
HNF, heterotrophic nanoflagellates
$HO^{\bullet}$, hydroxyl radical
HOMO, highest occupied molecular orbital
HPLC, high-performance liquid chromatography
IOP, inherent optical properties
IR, infrared
K_d, diffuse attenuation coefficient for downwelling irradiance
K_{ow}, organic–water partitioning coefficient
K_u, diffuse attenuation coefficient for upwelling irradiance
LMW, low molecular weight
LUMO, lowest unoccupied molecular orbital
MAA, mycosporine-like amino acid
MCH, melanin-concentration hormone
MDR, mean damage ratio
MPB, microphytobenthos
NCDOM, non-chromophoric dissolved organic matter
NO, nitric oxide
NR, nitrate reductase
PAH, polycyclic aromatic hydrocarbons
PAM, pulse amplitude modulated (fluorescence)
PAR, photosynthetically active radiation (400–700 nm)
PER, photoenzymatic repair
POC, particulate organic carbon
POM, particulate organic matter
PS II, photosystem II

PUFAs, polyunsaturated fatty acids
PWF, phototoxicity weighting function
QFT, quantitative filter technique
QSAR, quantitative structure–activity relationship
RAF, radiation amplification factor
ROS, reactive oxygen species
RNA, ribonucleic acid
RPA, relative photodynamic activity
RUBISCO, ribulose-1,5-biphosphate carboxylase/oxygenase
SBB, single strand breaks
SOD, superoxide dismutase
SPE, solid phase extraction
SWF, spectral weighting function
SZA, solar zenith angle
TT, thymine dimers
UML, upper mixed layer
UV-A, ultraviolet radiation (315–400 nm)
UV-B, ultraviolet radiation (280–315 nm)
UV-C, ultraviolet radiation (200–280 nm)
UV_{eff}, weighted UVR for photosynthesis inhibition
UVR, ultraviolet radiation
WSC, Weddell–Scotia confluence
z_a, depth (m) of 1% of surface irradiance.
Z_{Eph}, depth of the euphotic zone
$Z_{n\%}$, percent attenuation depth
Z_{UML}, depth of the upper mixed layer

Introduction

- Solar radiation as an ecosystem modulator

Chapter 1

Solar radiation as an ecosystem modulator

Robert G. Wetzel

Table of contents

Abstract

Solar radiation is the fundamental ecosystem modulator. Nearly all generation of organic matter is photosynthetic and as such the distribution of light in aquatic ecosystems is critical to regulation of major energetic inputs. However, simultaneously specific components of solar radiation, in particular the UV, function as both an accelerator of microbial degradation by enhancing bioavailability of complex organic substrates to microbes and by complete photolysis and oxidation of components of organic macromolecules to CO_2 and other inorganic forms of nutrients. Alterations in UV intensities impinging upon and within inland aquatic and coastal marine ecosystems by natural or anthropogenic causes will modify the rates of metabolism and biogeochemical processes associated with these macromolecules. This cascade of effects can greatly modify the functioning of natural ecosystems.

1.1 Introduction

In the subsequent chapters of this volume, detailed evaluations provide a summary of contemporary understanding of the properties of ultraviolet radiation (UVR) in aquatic ecosystems and its effects on aquatic organisms. Here I attempt to provide an overview of the coupling of these properties to emphasize how individual effects of UVR are integrated and, at the ecosystem level, provide a master level of regulation of ecosystem biogeochemical cycling, energy fluxes, productivity, and system evolution.

In regard to these detailed treatments of specific components of solar radiation and their effects, it is useful to emphasize several related universal characteristics of aquatic ecosystems. Namely, ecosystems are biological systems, ecosystems are biogeochemical systems, and the cycling of materials and energy in ecosystems is regulated by a highly variable set of intercoupled physical, chemical, and biological parameters. It is extraordinarily important to evaluate the influences and changes of UVR in the ecological contexts of a highly dynamic, changing environment – dynamic spatial and especially temporal scales. The question then is whether UV effects within the ecosystems are so variable that analyses are chaotic or whether certain stoichiometric analyses allow quantitative predictions of generic system responses to changes in UVR.

The approach taken is to first analyze our present understanding of how UVR influences ecosystem processes and how these processes are intercoupled with other related influences of those processes, such as climatic or atmospheric processes related to UVR. Finally, can one reasonably predict how ecosystems of different characteristics will respond to changes in atmospheric or aquatic conditions that alter UVR.

1.2 Size matters – radiation attenuation in relation to loadings of organic matter

Nearly all UV-C (<280 nm) is absorbed by the stratospheric gases and by the water of aquatic ecosystems. Although relatively little UV-B (280–320 nm) passes through the stratosphere (Chapter 2), UV-B is highly energetic and an important photactivating agent in waters. UV-A (320–400 nm) is less energetic than UV-B but is absorbed less readily and penetrates more deeply into water. The near UV light in the blue portion of the visible spectrum (400–500 nm) has recently been shown to be functionally similar to the adjacent UV-A radiation in many of the important photochemical reactions influenced by UVR and must be considered in any evaluation of composite effects.

Recent measurements in situ have demonstrated great variability in the penetration of UV-B and UV-A, but penetration has been found to be much greater than was believed previously (Chapters 3, 6, [1]). When referenced against pure water, the transmission of radiation is reduced drastically with increasing concentration of naturally occurring chromophoric dissolved organic compounds, particularly humic acids. UV-B attenuation depths ($z_a = 1\%$ of surface irradiance) range from a few centimeters to >10 m among a number of waters [2–6]. Much (>90%) of the among-habitat variation in diffuse attenuation coefficients (K_d) could be explained by differences in dissolved organic carbon (DOC) concentrations. Throughout the solar UV-B and UV-A range, K_d was well estimated with a univariate power model based on DOC concentration, particularly in waters of low to moderate phytoplanktonic productivity. The z_a is strongly dependent on DOC concentrations when below 2 mg C l^{-1}. In eutrophic lakes, densities of phytoplankton can begin to influence UV attenuation [7].

Only certain portions of the heterogeneous dissolved organic matter (DOM) absorb solar radiation. In inland waters, phenolic and other aromatic-based humic compounds (fulvic and humic acids), largely of terrestrial and higher aquatic plant origin, form a major component of dissolved organic acids and can constitute some 80% of the total DOM, 30–40% of which is composed of aromatic carbon compounds [8]. Humic substances are the largest component of chromophoric dissolved organic matter (CDOM). Of the soluble part of humic substances, heterogeneous fulvic acids have molecular weights from 500 to 1200 Da and contain many acidic functional groups, primarily carboxylic acids [9–11]. Humic acids are less hydrophilic than fulvic acids and are of greater molecular weight (mean ca. 4000–5000 Da) [12]. Humic substances dominate CDOM and are the most important component in the absorption of solar UV and blue radiation [4,13].

Concentrations of 4–8 mg organic acids liter^{-1} are common in surface waters and often exceed 50 mg l^{-1} in organic-rich waters, such as those of wetlands, flood plains of river ecosystems, and interstitial waters of hydrosoils [1]. Concentrations of both CDOM and humic substances commonly decrease along the gradient of fresh-to-coastal-to-oceanic waters.

Because the effects of UVR on aquatic ecosystems are so strongly influenced by concentrations of CDOM, factors that influence the loading rates of CDOM to aquatic ecosystems will influence strongly the selective distribution of UV and its effects on habitats and biota. Two aspects are particularly important in this regard. Firstly, the proportion of the DOM that is derived from higher plant tissues (terrestrial and wetland/littoral sources) that are dominated by chromophoric humic compounds vs. that derived form algae, which contain few fulvic and no humic constituents [10,14,15]. The DOM of streams and rivers is almost totally dominated by partial decomposition products of terrestrial and wetland higher plants. Similarly, small lakes receive a high proportion of DOM from terrestrial and wetland sources dominated by higher plant productivity and a high proportion of humic substance residues from partial degradation of structural tissue constituents, particularly lignocelluloses.

Secondly, the morphology of the receiving aquatic ecosystem is imperative because of the direct relationships between lake basin volume to water retention times, dilution of influent DOM, and mixing frequencies into photic zones. Most of the millions of lakes are small (< 10 km^2) and relatively shallow, usually < 10 m in depth [1,16]. As a result, the frequency of interaction of DOM-entrained water with solar radiation is often high both within stratified lakes and in shallow non-stratified lakes and ponds. Similarly, the frequency with which DOM in water of streams and rivers interacts with solar radiation is also high, particularly among larger stream orders ($>$3rd order) where the influence of shading from riparian tree canopy is small.

1.3 Precipitation matters – importance of frequency and intensity of influents

Because the penetration of UVR and its effects on ecosystem metabolism and functioning is so strongly influenced by DOM, the rates and timing of loading of DOM to receiving waters is important. Many studies have demonstrated the dominance of allochthonous inputs of terrestrial organic matter, in the form of detrital DOM and particulate organic matter (POM) for material and energy cycling in stream and river ecosystems. Much of that DOM is released from soils into groundwater and from anaerobic processes in adjoining wetlands [e.g., 1,17–19].

The DOM inputs from terrestrial organic matter to streams and lakes results from direct leaching from living vegetation and from soluble compounds carried in runoff from dead plant materials in various stages of decomposition. Very high concentrations of organic matter emanate from wetlands. Inputs of DOM are often directly correlated with precipitation, with high loading rates to receiving waters in the initial flushing stages of precipitation events. DOM loading then declines markedly in the later stages as dilution increases and eventually the discharge volume declines. Similarly, the DOM loading during the initial stages of snowmelt is much higher than subsequently. Although the total loading of DOM is high during these flushing events, dilution is also high. Some of the highest DOM concentrations and resulting UV attenuation occur during periods of low flow in rivers. In stratified lakes, the longer residence time allows for higher rates of photolysis of DOM in the photic zone. As in shallow, nonstratified lakes that mix frequently to the surface layers of high UV insolation, the concentrating effects of water residence time are countered by time available for UV alteration and microbial mineralization (Chapter 4).

The seasonal timing of the DOM loading also affects the effectiveness of UV photolysis and microbial utilization. Obviously, runoff loading events in cold, low light periods of the year will lead to less effective degradation and utilization of the organic compounds by biota of the ecosystem. These altered rates of UV-mediated metabolism will in turn affect rates of nutrient regeneration and subsequent productivity at many biotic levels.

As the DOM is delivered to marine coastal regions by rivers, reduction of transport rates occurs in the estuarine regions with complex hydrodynamic dispersion of water currents. The less dense saline water overlies that of the coastal waters and is exposed to solar photolysis with greater intensity and frequency than the underlying waters. The result is increased rates of partial and complete photolysis, largely by UV radiation, with higher mineralization rates of CDOM to CO_2 by enhanced microbial metabolism and by direct degradation to CO_2. As a result, a significant portion of the residual DOM is non-chromophoric (NCDOM). This relatively recalcitrant NCDOM, constituting perhaps 10–20% of the total DOM, tends to persist in marine environments with appreciable chemical stability and longevity (decades to centuries).

How the loading rates of allochthonous dissolved organic matter to freshwater ecosystems and to continental marine regions are and will be affected by climatic changes is unclear. There are indications among long-term data sets that DOC concentrations are declining gradually in lakes over several decades [e.g., 1(p. 779),20,21]. Particularly in oligotrophic lakes where DOC concentrations are often low, UVR penetrates to depths of several meters and can negatively influence organisms by genetic damage, diverting production to increased synthesis of protective pigments, or in high elevations or latitudes where higher plant source materials and DOM loading is low. Organisms in such lakes can be exposed to high intensities of UVR [22]. Even in lakes with higher concentrations of DOM, the long-term trends are often toward slowly decreasing concentrations of DOM [1].

There is little question that both temperature and carbon dioxide concentrations of the atmosphere are increasing. Rising temperature has also influenced precipitation patterns and has led to large regions in which rainfall and snow accumulations have been reduced [1]. Droughts are a cumulative result of numerous meteorological factors affecting precipitation, evapotranspiration, and other water losses. Droughts usually do not become severe until after long periods of deficient rainfall and unrestrained water use.

DOC in some lakes has declined appreciably over the last quarter century coincident with substantial warming [e.g., 1(p. 780),23]. Reduced precipitation and increased evapotranspiration in the drainage basin result in reduced stream flows and lower DOC loading to the streams and lakes. Transparency of lake water to UV photolysis increases under these conditions. Similar reductions in DOC have been observed in streams [19]. The decrease in annual DOC yields of streams occurs in spite of higher concentrations in storm flows following periods of prolonged drought [23,24].

1.4 Direct effects of UVR

Photosynthesis of algae is clearly inhibited by exposure to natural levels of UV-B and especially UV-A radiation. Physiological and genetic recovery occurs, and as a result a quasi-steady physiological state is found commonly between damage and recover processes [25,26, Chapters 9, 11, and 13]. Many species repair damage to photosystems and DNA during daily periods of darkness. Many species produce UV-absorbing compounds – mycosporine-like amino acids are an important and ubiquitous class of such compounds [27,28, Chapter 10]. Many species have biochemical defenses against toxic end products of UVR, such as radical scavenging by carotenoid pigments and superoxide dismutase (Chapter 15). Some species have limited abilities to avoid intense surface UV by migration to deeper areas.

UV radiation can impact zooplankton and fish directly in shallow water habitats by damage to DNA and generation of harmful photochemicals (free radicals, reactive oxygen species) [29,30, Chapter 8]. Although many animals can avoid UV-intense habitats, as well as develop photoprotective pigments (carotenoids, cuticular melanin), both of these strategies can alter their susceptibility to predation by other organisms, particularly fish.

1.5 Allochthonous vs. autochthonous organic matter – key UV-VIS mediated processes regulate heterotrophic utilization

Some 90 per cent or more of the total metabolism in aquatic ecosystems is microbial, accomplished by heterotrophic metabolism of bacteria, fungi, and many protists, all of a size less than 100 μm [1,31]. Therefore, the material and energy fluxes of aquatic ecosystems is totally dominated by metabolism of particulate detritus (non-living) and especially DOM from autochthonous and allochthonous sources. The pelagic open water is but a portion of the whole lake or river ecosystem. In relation to loading and fluxes of DOM, allochthonous and littoral sources are critical because of their chemical differences from that produced by algal photosynthesis.

The modes of senescence, death, and degradation rates of biota are also of considerable importance to rates and pathways of degradation and energetic utilization. For example, the continual slow senescence and release of DOM from a higher aquatic plant is very different from the relatively instantaneous biochemical death and release of DOM from a bacterium or alga. Non-predatory death and metabolism of non-living detrital POM and DOM by prokaryotic and protistan heterotrophs dominate in all aquatic ecosystems.

In providing a synthesis of the ramifications of UV on aquatic ecosystems, a key component is the simultaneous importance of DOM in regulating the distribution and attenuation of UVR as well as the effects that UV has both directly and indirectly on the metabolism, growth, reproductive, and production efficacy of biota. Because these effects of UV are so interactive and coupled, it is difficult to separate them without some redundancy. Several points can be characterized, however, in summary of some of the more detailed discussions in subsequent chapters.

Physical processes, such as partial or complete *photochemical modification of organic macromolecules*, can result in major alterations in biological availability of portions of complex, heterogeneous dissolved organic compounds. These photochemical processes can result in:

1.5.1 Alterations of enzymatic accessibility by the macromolecules

Polyphenolic organic acids, which occur in great abundance (commonly 4–8 mg l^{-1}) in many fresh waters, can complex with or induce precipitation of proteins by binding to one or more sites on the protein surface to yield a monolayer that is less hydrophilic than the protein itself [32,33]. This complexation, as well as cross-linking of polypeptide chains with polyphenolic humic substances can lead to enzymatic inhibition or reduction of activity [e.g., 33–35]. More aromatic and condensed humic acid molecules are more rigid and can distort bound enzymes to a greater extent than is the case with simpler compounds, such as fulvic acids [e.g., 36]. The inhibition of enzymes occurs in a classical noncompetitive manner, in which the inhibitor, polyphenol, and substrate bind simultaneously to the enzyme. Furthermore, dissolved humic substances can complex by peptidization and alter biological susceptibility to enzymatic hydrolysis. For example, membrane properties, such as lipid hydrophobicity, can be altered by humic substances and in turn affect enzyme hydrolysis rates and nutrient transport mechanisms [e.g., 37,38]. An important ecosystem aspect is that these protein or enzyme complexes can be stored in an inactivated state for long periods, transported within the ecosystems, and later reactivated by partial photolytic cleavage by UVR [1,34,35].

1.5.2 Photolysis of humic macromolecules

Partial photolysis of humic macromolecules, particularly with the generation of volatile fatty acids and related simple compounds that serve as excellent substrates for bacterial degradation [e.g., 39–42]. It is important to recognize that of the total photolytic irradiance, about a quarter of the partial photolysis of organic substrates results from UV-B, about half from UV-A, and about a quarter from the lower wavelengths (400–500 nm) of photosynthetically active radiation (PAR, 400–700 nm). Transmittance and photolytic activity from UV-B and UV-A is restricted largely to the surface waters. In contrast, PAR, although much weaker energetically than UV, penetrates into water to much greater depths. Although photolysis of organic compounds is appreciably less than that induced by UV at surface waters, the photolytic generation of simple substrates is appreciable by PAR as well as by UV [1,43,44]. Results of such studies are indicating that an appreciable portion of photolytic generation of some simple substrates is generated by PAR.

1.5.3 Photolysis of dissolved organic nitrogen and phosphorus compounds

Photolytic degradation of dissolved organic nitrogen and phosphorus compounds release inorganic nutrient compounds such as nitrite, ammonia, and phosphate, as well as CO and CO_2 [e.g., review of 41,45–47]. Stimulatory effects of increase nutrient availability by such processes clearly occurs [e.g., 47].

1.5.4 Complete photolysis of humic substances to CO and CO_2

Photochemical oxidation by solar radiation of natural dissolved organic compounds to both CO and dissolved inorganic carbon (CO_2 and HCO_3^-) has been known for some years [e.g., 48]. Depending on dissociation and saturation conditions, some excess CO_2 will evade to the atmosphere. Previous studies on the photolytic degradation of dissolved organic matter suggested that the dominant photolytic components of solar radiation were UV-B and UV-A, and that PAR was of little consequence. Many of these studies, however, were not performed under sterile conditions, and as a result findings were confounded by nearly instantaneous microbial utilization of organic compounds generated with rapid degradation and generation of CO_2. Moreover, many of the DOM sources of these studies had been exposed to natural radiation for long (e.g., weeks) and non-comparable periods of light. Contemporary research is indicating that although UV-B and UV-A are significant and can contribute to more than half of photochemical mineralization, PAR is also a major photolytic agent [43,49,50]. For example, from nearly 200 separate photolytic experiments on DOM from different waters and plants under different conditions, the UV-B portion of the spectrum was always most effective in complete photodegradation to CO_2, but UV-A was also highly effective with small differences from the photolytic capacities of UV-B [1]. PAR is also highly effective in photolytic degradation of DOM to CO_2 and frequently about a quarter to half of the collective photolysis can be attributed to the largely blue portion of the PAR spectrum. Bioavailability of CDOM may increase [40], remain unaltered, or decrease from photolysis [47,51,52]. Bioavailability is clearly related to the stages of photolysis and alteration of the dominant components of the heterogeneous natural aggregation of natural organic compounds.

Both partial photolysis to the generation of volatile fatty acids, and the complete photolysis with the generation of large quantities of CO_2 by PAR are important findings because of the much lower extinction rates of PAR in water in comparison to those of UVR. Photolytic processes, so important to nutrient cycling, are therefore not restricted to the uppermost strata of a few centimetres of aquatic ecosystems, but rather affect much of the seasonally-variable volume of the photic zone.

1.5.5 Less direct but important biogeochemical interactions of UVR

Biogeochemical interactions of UVR upon DOM in aquatic systems are also important, but poorly studied at the ecosystem level. Continued intensive study of natural dissolved organic substances in aquatic ecosystems is resulting in improved understanding of the many ways in which these diverse compounds, particularly humic compounds, can interact with other important metabolic components. Any of these processes will be altered by UV partial or complete photolysis of DOM. Examples are manifold:

(a) Interact with inorganic compounds, particularly in complexation reactions such as chelation [reviewed in 53]. Depending on the concentration ratios of the complexing DOM to inorganic elements, the mode of organic complexation, biological availability and, in some cases, elemental toxicity can be increased or decreased. All of these processes will be altered by UV photolysis of CDOM.

(b) Interact with other organic compounds, such as peptidization, and alter biological susceptibility to enzymatic hydrolysis. For example, membrane properties, such as lipid hydrophobicity, can be altered by humic substances and in turn affect enzyme hydrolysis rates and nutrient transport mechanisms [e.g., 37,38]. In a most interesting interaction, humic substances can complex with proteins, particularly enzymes both freely soluble and membrane-bound, with non-competitive inhibition [54,55]. Enzymes can be stored for long periods (days, weeks) in this complexed, inactive state, be redistributed in the ecosystem with water parcel movements, and reactivated by partial photolytic cleavage by UVR [31,34,35,43].

(c) Alter chemical properties such as redox and pH. A predominance of humic acids can result in an organic acidity that can influence, and at times exceed, inorganically derived acidity form natural or anthropogenic sources [reviewed in 1]. Exposure of natural dissolved organic matter to UV can form reduced reactive oxygen species, particularly hydrogen peroxide (H_2O_2) [56,57; Chapter 8]. H_2O_2 has a half-life of several hours in natural waters and can radically alter redox cycling of metals [58].

(d) Microbially reduced humic substances can, upon entering less reduced zones of sediments, serve as electron donors for the microbial reduction of several environmentally significant electron donors [59]. Once microbially reduced, humic substances can transfer electrons to various Fe(III) or Mn(IV) oxide forms abiotically and recycle the humic compounds to the oxidized form, which can then accept more electrons from the humic compound-reducing microorganisms. The interactions of UVR on these highly reactive processes in shallow waters, particularly littoral and wetland areas, are unclear.

(e) Change physical properties such as selective modifications of light penetration. The well-known selective attenuation of light by CDOM [cf. 1] can further modify biogeochemical cycling in numerous ways. Such modification of the light regimes can alter rates of photosynthesis, hormonal activities, and migratory distribution and reproductive behaviors. Absorption of UVR by humic substances can protect organisms from genetic damage as well as modify macromolecules and enhance bioavailability of organic substrates.

1.6 Recalcitrant organic matter, metabolic stability, and photolysis

The commonly observed incomplete photolysis of DOC is critical to accelerated utilization of these macromolecules, but is clearly not mandatory. Portions of the complex DOM pools, including fractions of humic and fulvic acid compounds, are degraded, but total degradation rates are clearly slow. Chemical organic recalcitrance of DOM is instrumental in providing a thermodynamic stability to metabolism within lake, reservoir, wetland-littoral land–water interface, and river ecosystems [1,31,54,60–62]. The chemical recalcitrance is a "brake" on ecosystem metabolism, and that brake is critical for maintenance of the integrated stability of heterotrophic utilization of synthesized or imported organic matter and energy. UVR can alter the effectiveness of that chemical recalcitrance "brake".

Most of the detrital organic pool, both in particulate and dissolved phases, of inland aquatic ecosystems consists of residual organic compounds of plant structural tissues. The more labile organic constituents of complex dissolved and particulate organic matter are commonly hydrolyzed and metabolized more rapidly than more recalcitrant organic compounds that are less accessible enzymatically. The result is a general increase in concentration of the more recalcitrant compounds, commonly exceeding 80% of the total, with slower rates of metabolism and turnover. These recalcitrant compounds, however, are metabolized at rates slowed and regulated in large part by their molecular complexity and bonding structure.

In every detailed *annual* organic carbon budget of lake and river ecosystems, organic matter generated by phytoplankton will not support all of the heterotrophic metabolism of the ecosystem. At least several fold support of the total metabolism is by organic subsidies from the land–water interface communities and allochthonous production. From the standpoint of metabolic stability, it is particularly important that most of the organic carbon is dissolved and relatively recalcitrant, widely distributed within the inland waters. The chemical recalcitrance of this dominating DOM ameliorates the violent metabolic and growth oscillations so characteristic of the pelagic biota components of the ecosystem when resources are available in abundance. In addition, much of the POM formed in the dominating land–water interface regions of lake, river, and estuarine ecosystem, is displaced to reducing, anoxic environments of the littoral and profundal sediments. The DOC, largely of higher plant origins, provides the stability and is the currency for the quantitatively more important detrital pathways in aquatic ecosystems. The same underpinnings of that stability prevail in terrestrial ecosystems and likely in coastal as well as much if not most of the marine ecosystem.

Detritus includes non-living particulate, colloidal, and dissolved organic matter, and metabolically size only affects rates of hydrolytic attack [31]. Inland aquatic ecosystems collect organic matter, particularly in dissolved forms, from terrestrial, wetland, and littoral sources in quantities that supplement if not exceed those produced autochthonously. Rates of utilization of that organic matter are slowed by a combination of chemical recalcitrance as well as displacement to anoxic environments. As a result, inland aquatic ecosystems are heterotrophic and functionally detrital bowls, not algal bowls.

The high organic matter production of terrestrial and particularly land–water interface regions (wetlands, littoral areas) commonly results in loading of excessive organic carbon, usually primarily in the form of dissolved organic compounds, to inland waters. A significant portion of that DOM is metabolized, sorbed and sedimented, or photolyzed while moving through lakes and rivers, but nonetheless a portion does reach coastal oceanic regions. The extent of this allochthonous loading to oceanic waters is unclear, although estimates are as high as 20% of the total oceanic DOC [63,64]. Because of long periods of exposure of much of this allochthonous DOM to photolytic degradative processes en route to the open ocean, its metabolic regulatory functions are clearly less than is the case in inland water ecosystems. More labile DOM products of algal photosynthesis dominate in the marine pelagic, and as a result undergo rapid utilization and exploitation until limiting conditions for sustained growth prevail. It is hypothesized that these conditions are appreciably less stable that those containing high concentrations of chromophoric and non-chromophoric DOM emanating largely from higher plant tissues. As a result, effects of altered rates of fluence of UVR in the oceanic pelagic impact the ecosystem by more direct means, such as direct damage to genetic constituents of the biota, rather than the major roles in altering the chemistry of organic macromolecules.

In long-term evolutionary scales, humans now have the abilities to intervene rapidly in this interdependent relationship and alter the stability of the rates of metabolism of organic matter. For example, reduction of ozone in the stratosphere and associated increased UV-B could lead to accelerated photolytic degradation of macromolecules of DOM to CO_2 by both abiotic and biotic pathways. In addition, the photolytic enhancement of substrates for bacterial metabolism by UV photolysis can result in accelerated rates of biogeochemical cycling of nutrients and stimulated productivity of the ecosystems. In addition to decreasing the metabolic stability of the lakes and streams, the enhanced microbial respiration will certainly lead to increased generation of CO_2 and evasion to the atmosphere.

References

1. R.G. Wetzel (2001). Limnology: Lake and River Ecosystems, Academic Press, San Diego.
2. J.T.O. Kirk (1994). Optics of UV-B radiation in natural waters. *Arch. Hydrobiol. Beih. Ergebn. Limnol.*, **43**, 1–16.
3. N.M. Scully, D.R.S. Lean (1994). The attenuation of ultraviolet radiation in temperate lakes. *Arch. Hydrobiol. Beih. Ergebn. Limnol.*, **43**, 35–44.
4. D.P. Morris, H. Zagarese, C.E. Williamson, E.G. Balseiro, B.R. Hargreaves, B. Modenutti, R. Moeller, C. Queimalinos (1995). The attenuation of solar UV radiation in lakes and the role of dissolved organic carbon. *Limnol. Oceanogr.*, **40**, 1381–1391.
5. C.E. Williamson, R.S. Stemberger, D.P. Morris, T.M. Frost, S.G. Paulsen (1996). Ultraviolet radiation in North American lakes: Attenuation estimates from DOC measurements and implications for plankton communities. *Limnol. Oceanogr.*, **41**, 1024–1034.
6. R. Sommaruga, R. Psenner (1997). Ultraviolet radiation in a high mountain lake of the Austrian Alps: Air and underwater measurements. *Photochem. Photobiol.*, **65**, 957–963.
7. Y. Hodoki, Y. Watanabe, (1998). Attenuation of solar ultraviolet radiation in eutrophic freshwater lakes and ponds. *Jpn. J. Limnol.*, **59**, 27–37.
8. R.L. Malcolm (1990). The uniqueness of humic substances in each of soil, stream and marine environments. *Anal. Chem. Acta*, **232**, 19–30.
9. E. Saski, A. Vähätalo, K. Salonen, M. Salkinoja-Salonen (1996). Mesocosm simulation on sediment formation induced by biologically treated bleached kraft pulp mill wastewater in freshwater recipients. In: M. Servos, K. Munkittrick, J. Carey, G. Kraak, (Eds)., *Environmental Fate and Effects of Pulp and Paper Mill Effluents.* (pp. 261–270.) St. Lucie Press, Delray Beach, FL.
10. D.M. McKnight, G.R. Aiken (1998). Sources and age of aquatic humus. In: D.O. Hessen, L.J. Tranvik (Eds)., *Aquatic Humic Substances: Ecology and Biogeochemistry.* (pp. 9–39). Springer-Verlag, New York.
11. J. Peuravuori, K. Pihlaja (1999). Characterization of aquatic humic substances. In: J. Keskitalo, P. Eloranta (Eds), (pp. 11–39). *Limnology of Humic Waters.* Backhuys Publishers, Leiden, The Netherlands.
12. S.E. Cabaniss, Q. Zhou, P.A. Maurice, Y.-P. Chin, G.R. Aiken (2000). A log-normal distribution model for the molecular weight of aquatic fulvic acids. *Environ. Sci. Technol.*, **34**, 1103–1109.
13. G.M. Ferrari, M.D. Dowell (1998). CDOM absorption characteristics with relation to fluorescence and salinity in coastal areas of the Southern Baltic Sea. *Estuarine, Coastal Shelf Sci.*, **47**, 91–105.
14. D.M. McKnight, G.R. Aiken, R.L. Smith (1991). Aquatic fulvic acids in microbially based ecosystems: Results from two Antarctic desert lakes. *Limnol. Oceanogr.*, **36**, 998–1006.
15. D.M. McKnight, E.D. Andrews, R.L. Smith, R. Dufford (1994). Aquatic fulvic acids in algal-rich Antarctic ponds. *Limnol. Oceanogr.*, **39**, 1972–1979.
16. R.G. Wetzel (1990). Land-water interfaces: Metabolic and limnological regulators. *Verhand. Internat. Verein. Limnol.*, **24**, 6–24.
17. C.N. Dahm, E.H. Trotter, J.R. Sedell (1987). Role of anaerobic zones and processes in stream ecosystem productivity. In: R.C. Averett, D.M. McKnight (Eds)., *Chemical*

Quality of Water and the Hydrologic Cycle, (pp. 157–178) Lewis Publishers, Chelsea, MI.

18. B.W. Eckhardt, T.R. Moore, (1990). Controls on dissolved organic carbon concentrations in streams, southern Québec. *Can. J. Fish. Aquat. Sci.*, **47**, 1537–1544.
19. P.J. Dillon, L.A. Molot (1997). Effect of landscape form on export of dissolved organic carbon, iron, and phosphorus from forested stream catchments. *Water Resour. Res.*, **11**, 2591–2600.
20. D.W. Schindler, T.W. Frost, K.H. Mills, P.S.S. Chang, I.J. Davies, D.L. Findlay, D.F. Malley, J.A. Shearer, M.J. Turner, P.I. Brezonik, A. Swenson (1991). Comparisons between experimentally- and atmospherically-acidified lakes during stress and recovery. *Proc. Royal Soc. Edinburgh*, **97B**, 193–226.
21. D.W. Schindler, P.J. Curtis (1997). The role of DOC in protecting freshwaters (*sic*) subjected to climatic warming and acidification from UV exposure. *Biogeochemistry*, **36**, 1–8.
22. D.M. McKnight, R. Harnish, R.L. Wershaw, J.S. Baron, S. Schiff (1997). Chemical characteristics of particulate, colloidal, and dissolved organic material in Loch Vale watershed. *Biogeochemistry*, **36**, 99–124.
23. D.W. Schindler, S.E. Bayley, P.J. Curtis, B.R. Parker, M.P. Stainton, C.A. Kelly (1992). Natural and man-caused factors affecting the abundance and cycling of dissolved organic substances in Precambrian Shield lakes. *Hydrobiologia*, **229**, 1–21.
24. M.J. Hinton, S.L. Schiff, M.C. English (1997). The significance of storms for the concentrations and export of dissolved organic carbon from two Precambrian Shield catchments. *Biogeochemistry*, **36**, 67–88.
25. J.J. Cullen, P.J. Neale (1994). Ultraviolet radiation, ozone depletion, and marine photosynthesis. *Photosyn. Res.*, **39**, 303–320.
26. D. Karentz, M.L. Bothwell, R.B. Coffin, A. Hanson, G.J. Herndl, S.S. Kilham, M.P. Lesser, M. Lindell, R.E. Moeller, P.J. Neale, R.W. Sanders, C.S. Weiler, R.G. Wetzel (1994). Impact of UV-B radiation on pelagic freshwater ecosystems. *Arch. Hydrobiol. Beih. Ergebn. Limnol.*, **43**, 31–69.
27. W.F. Vincent, S. Roy (1993). Solar ultraviolet-B radiation and aquatic primary production: Damage, protection, and recovery. *Environ. Rev.*, **1**, 1–12.
28. F. Xiong, J. Komenda, J. Kopecky, L. Nedbal (1997). Strategies of ultraviolet-protection in microscopic algae. *Physiol. Plant.*, **100**, 378–388.
29. D.O. Hessen (1994). *Daphnia* responses to UV-light. *Arch. Hydrobiol. Beih. Ergebn. Limnol.*, **43**, 185–195.
30. O. Siebeck, T.L. Vail, C.E. Williamson, R. Vetter, D. Hessen, H. Zagarese, E. Little, E. Balseiro, B. Modenutti, J. Seva, A. Shumate (1994). Impact of UV-B radiation on zooplankton and fish in pelagic freshwater ecosystems. *Arch. Hydrobiol. Beih. Ergebn. Limnol.*, **43**, 101–114.
31. R.G. Wetzel (1995). Death, detritus, and energy flow in aquatic ecosystems. *Freshwat. Biol.*, **33**, 83–89.
32. E. Haslam (1988). Plant polyphenols (syn. vegetable tannins) and chemical defense – a reappraisal . *J. Chem. Ecol.*, **14**, 1789–1805.
33. E. Haslam (1988). *Practical Polyphenolics*: *From Structure to Molecular Recognition and Physiological Action*, Cambridge University Press, Cambridge.
34. R.G. Wetzel (1991). Extracellular enzymatic interactions in aquatic ecosystems: Storage, redistribution, and interspecific communication. In: R.J. Chróst (Ed.), *Microbial Enzymes in Aquatic Environments*, (pp. 6–28). Springer-Verlag, New York
35. M.-J. Boavida, R.G. Wetzel (1998). Inhibition of phosphatase activity by dissolved

humic substances and hydrolytic reactivation by natural UV. *Freshwat. Biol.*, **40**, 285–293.

36. J.N. Ladd, J.H.A. Butler (1975). Humus-enzyme systems and synthetic, organic polymer-enzyme analogs. In: E.A. Paul, A.D. McLaren, (Eds), *Soil Biochemistry* (Vol. 4, pp. 142–194). M. Dekker, Inc., New York.
37. M.J. Lemke, P.F. Churchill, R.G. Wetzel (1995). Effect of substrate and cell surface hydrophobicity on phosphate utilization in bacteria. *Appl. Environ. Microbiol.*, **61**, 913–919.
38. M.J. Lemke, P.F. Churchill, R.G. Wetzel (1998). Humic acid interaction with extracellular layers of wetland bacteria. *Verh. Internat. Verein. Limnol.*, **26**, 1621–1624.
39. A.J. Stewart, R.G. Wetzel (1981). Dissolved humic materials: Photodegradation, sediment effects, and reactivity with phosphate and calcium carbonate precipitation. *Arch. Hydrobiol.*, **92**, 265–286.
40. R.G. Wetzel, P.G. Hatcher, T.S. Bianchi (1995). Natural photolysis by ultraviolet irradiance of recalcitrant dissolved organic matter to simple substrates for rapid bacterial metabolism. *Limnol. Oceanogr.*, **40**, 1369–1380.
41. M.A. Moran, R.G. Zepp (1997). Role of photoreactions in the formation of biologically labile compounds from dissolved organic matter. *Limnol. Oceanogr.*, **42**, 1307–1316.
42. M.A. Moran, J.S. Covert. Photochemically-mediated linkages between dissolved organic matter and bacterioplankton. In: S. Findlay, R. Sinsabaugh (Eds), *Integrating Approaches to Microbial-Dissolved Organic Matter Trophic Linkages*, Academic Press, San Diego, in press.
43. R.G. Wetzel (2002). Origins, fates and ramifications of natural organic compounds of wetlands. In: M.M. Holland, M.L. Warren, J.A. Stanturf (Eds.), *Sustainability of Wetlands and Water Resources*, Gen. Tech. Rep. SRS-50, Forest Service. (pp.183–189) U.S. Department of Agriculture, Ashevill, NC.
44. R.G. Wetzel, N.C. Tuchman, Effects of CO_2 enrichment on the production of plant degradation products and their natural photodegradation and biological utilization (2002). *Limnol. Oceanogr.*, **47**, in review.
45. B.A. Manny, M.C. Miller, R.G. Wetzel (1971). Ultraviolet combustion of dissolved organic nitrogen compounds in lake waters. *Limnol. Oceanogr.*, **16**, 71–85.
46. R.J. Kieber, A. Li, P.J. Seaton (1999). Production of nitrite from the photodegradation of dissolved organic matter in natural waters. *Environ. Sci. Technol.*, **33**, 993–998.
47. A.V. Vähätalo, K. Salonen, U. Münster, M. Järvinen, R.G. Wetzel (2002). Photochemical transformation of allochthonous organic matter provides bioavailable nutrients in a humic lake. *Arch. Hydrobiol.*, in press.
48. W.L. Miller, R.G. Zepp (1995). Photochemical production of dissolved inorganic carbon from terrestrial organic matter: Significance to the oceanic organic carbon cycle. *Geophys. Res. Lett.*, **22**, 417–420.
49. A.V. Vähätalo, M. Salkinoja-Salonen, P. Taalas, K. Salonen (2000). Spectrum of the quantum yield for photochemical mineralization of dissolved organic carbon in a humic lake. *Limnol. Oceanogr.*, **45**, 664–676.
50. R.G. Wetzel (2000). Natural photodegradation by UV-B of dissolved organic matter of different decomposing plant sources to readily degradable fatty acids. *Verhand. Int. Verein. Limnol.*, **27**, 2036–2043.
51. A.M. Anesio, C.M. Denward, L.J. Tranvik, W. Granéli (1999). Decreased bacterial growth on vascular plant detritus due to photochemical modification. *Aquat. Microb. Ecol.*, **17**, 159–165.

52. I. Obernosterer, B. Reitner, G.J. Herndl (1999). Contrasting effects of solar radiation on dissolved organic matter and its bioavailability to marine bacterioplankton. *Limnol. Oceanogr.*, **44**, 1645–1654.
53. E.M. Perdue (1998). Chemical composition, structure, and metal binding properties. In: D.O. Hessen, L.J. Tranvik (Eds), *Aquatic Humic Substances: Ecology and Biogeochemistry* (pp. 41–61). Springer-Verlag, Berlin.
54. R.G. Wetzel (1992). Gradient-dominated ecosystems: Sources and regulatory functions of dissolved organic matter in freshwater ecosystems. *Hydrobiologia*, **229**, 181–198.
55. R.G. Wetzel (1993). Humic compounds from wetlands: Complexation, inactivation, and reactivation of surface-bound and extracellular enzymes. *Verhand. Internat. Verein. Limnol.*, **25**, 122–128.
56. W.J. Cooper, R.G. Zika, R.G. Petasne, J.M.C. Plane (1988). Photochemical formation of H_2O_2 in natural waters exposed to sunlight. *Environ. Sci. Technol.*, **22**, 1156–1160.
57. N.M. Scully, D.R.S. Lean, D.J. McQueen, W.J. Cooper (1995). Photochemical formation of hydrogen peroxide in lakes: Effects of dissolved organic carbon and ultraviolet radiation. *Can. J. Fish. Aquat. Sci.*, **52**, 2675–2681.
58. J.W. Moffet, R.G. Zika (1987). Reaction kinetics of hydrogen peroxide with copper and iron in seawater. *Environ. Sci. Technol.*, **21**, 804–810.
59. D.R. Lovley, J.L. Fraga, J.D. Coates, E.L. Blunt-Harris (1999). Humics as an electron donor for anaerobic respiration. *Environ. Microbiol.*, **1**, 89–98.
60. R.G. Wetzel (1983). *Limnology*, (2nd ed.) Saunders College Publishing, Philadelphia,
61. R.G. Wetzel (1984). Detrital dissolved and particulate organic carbon functions in aquatic ecosystems. *Bull. Mar. Sci.*, **35**, 503–509.
62. R.G. Wetzel (2000). Freshwater ecology: Changes, requirements, future demands. *Limnology* **1**, 3–11.
63. M. Meybeck (1993). Riverine transport of atmospheric carbon: Sources, global typology and budget. *Water Air Soil Poll.*, **70**, 443–463.
64. M. Meybeck (1993). Natural sources of C, N. P and S. In: R. Wolast, F.T. Mackenzie, L. Chou (Eds), *Interactions of C, N. P and S Biogeochemical Cycles and Global Change* (pp. 163–193). Springer-Verlag, Berlin.

Physics

- UVR climatology
- Water column optics and penetration of UVR
- Modulation of UVR exposure and effects by vertical mixing and advection

Chapter 2

UVR climatology

Mario Blumthaler and Ann R. Webb

Table of contents

Abstract

Ultraviolet radiation at the Earth's surface is determined by the emission of the sun, and subsequent modification when passing through the atmosphere. The Earth–sun distance and the position of the observer on the Earth determine the power of the incoming radiation while the UV spectrum at the ground varies with time and place, because of the wavelength dependent attenuation processes within the atmosphere. The most important determinants are solar elevation, ozone and aerosol content, altitude and albedo, and cloudiness. Measurements with high spectral resolution allow detailed investigation of the effects of these parameters. Simpler broadband measurements provide information from a great number of locations. Model calculations can estimate irradiance levels well if the input parameters (state of atmosphere) are well known. Thus, estimates based on space-born measurements provide for world-wide distribution and temporal variation of UVR, but incomplete knowledge of some atmospheric parameters still limits the absolute accuracy.

As total ozone amount decreases, especially in mid- and high-latitudes, UV-B tends to increase. The interaction of all potential climate change influences on ozone makes predictions of future UV difficult, but best estimates do not expect recovery on a global scale earlier than within 10 to 20 years.

2.1 Introduction

The UVR reaching aquatic organisms in their natural habitat comes from the sun. Extra-terrestrial radiation is modified as it passes through the Earth's atmosphere and there are many factors that influence the radiation arriving at the surface of the Earth. These include the state of the atmosphere (clear, clean, cloudy, polluted), position on the Earth (latitude and altitude) and season (relative position of the sun to location on Earth). Further attenuation then occurs as the radiation passes through the water environment to reach aquatic organisms. These latter complications are dealt with in Chapter 3; here we deal only with the UVR incident at the ground, or water surface.

UVR covers the part of the electromagnetic spectrum at wavelengths below 400 nm, between X-rays and visible radiation. The UV is split, somewhat arbitrarily, into narrower wavebands with designations (from different branches of science) such as vacuum-, far-, near-UV. In considering UV at the surface of the Earth we are concerned with the longest wavelengths in the UV part of the spectrum, those between 280–400 nm, designated as UV-B and UV-A radiation (the more harmful UV-C (200–280 nm) and shorter wavelengths are completely attenuated by the atmosphere). The Commission Internationale d'Eclairage (CIE) define UV-B as 280–315 nm, and UV-A as 315–400 nm. However, UV-B can frequently be found described as 280–320 nm, for the pragmatic reason that 320 nm is about where the solar spectrum "flattens out", and where biological action spectra approach a region of zero or very small response. In reality these waveband distinctions are arbitrary boundaries in the continuous spectra of both solar radiation and biological reactions.

The first section of this chapter discusses the basic physics of radiation and radiative transfer in general. In the following sections, measurements of UVR are discussed, subdivided in ground-based and space-born methods. The instrumentation and the results with respect to the different parameters affecting UVR at the Earth's surface are presented. Finally, trends in solar UVR are analyzed. Starting with the observed long-term ozone changes, the resulting changes in UVR are discussed. Future levels of UVR refer to forecasting for a short time scale (days) as well as possible scenarios in the next decades.

2.2 Theory

2.2.1 Energy from the sun

Natural UVR originates with nuclear reactions in the interior of the sun. The energy generated in this way travels outwards through the gaseous body of the sun to a layer called the photosphere. The photosphere is the layer that emits the radiation we receive on Earth. It emits approximately like a blackbody at 5800 K, that is its emission is continuous across the electromagnetic spectrum and the spectral shape is determined by Planck's law. The temperature of the photosphere is such that the emission covers the spectral region from gamma rays to the near infra-red (about 4000 nm) (Figure 1). The wavelength of maximum emission is given by Wien's law

$$\lambda_{max} = 2897/T$$

and for the temperature of the sun this is 0.5 μm (500 nm), in the blue–green visible part of the spectrum. However, the shape of the Planck curve, the relative sensitivity of the human eye, and the spectrally dependent interactions of the radiation with the atmosphere (discussed below) lead to the yellow sun that we observe.

The total amount of solar radiation emitted by the sun is determined by the Stefan–Boltzmann law

$$E = \sigma T^4$$

where σ is the Stefan–Boltzmann constant of 5.67×10^{-8} W m^{-2} K^{-4}. However, incident energy at a distance from an emitting object is proportional to the square root of distance, hence the energy reaching the top of the Earth's atmosphere becomes

$$S_0 = \sigma T^4 \, r_s^2 / \, r_0^2$$

where S_0 is the solar constant, r_s is the solar radius, and r_0 is the average Earth–sun distance. The solar "constant", best estimated as 1370 W m^{-2}, varies on several time scales. Over the lifetime of the sun its temperature, and therefore both its total emission and spectral properties, have changed (it is estimated that emission has increased by 20–40% in 4.5×10^9 years). On a time scale that we can

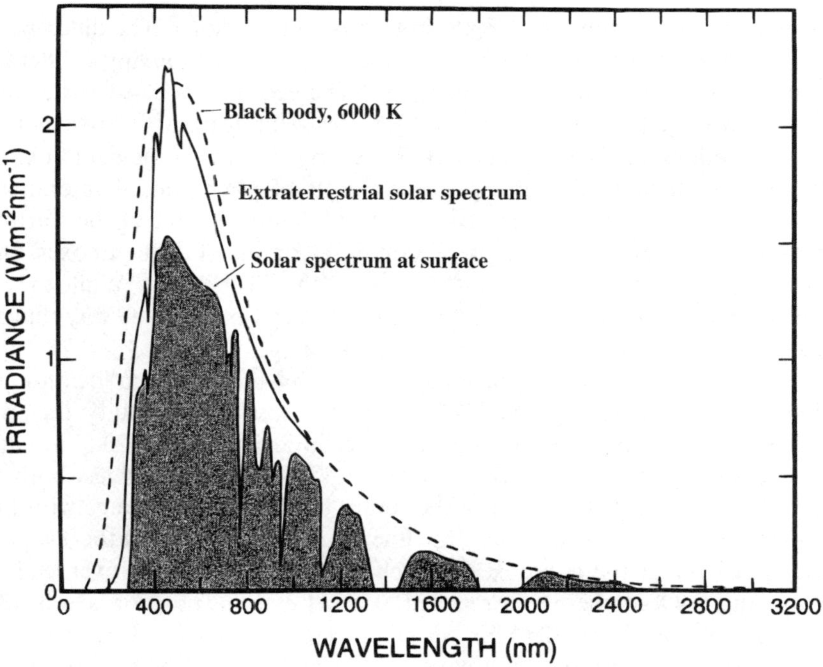

Figure 1. Spectra of a black body at 6000 K, the sun outside the atmosphere, and the sun at the Earth's surface.

comprehend, the activity of the sun, associated with the observable sunspots, varies in a broadly cyclic manner of 22 years duration. However, this includes a reversal of the sun's magnetic field and the cycle in sunspot numbers (our main concern) is 11 years. Active sunspots appear as dark patches on the face of the sun and their magnetic activity leads to solar flares – great eruptions of energy with enhanced ultraviolet and X-ray emission. These rather unpredictable emissions affect the solar constant, but in a wavelength dependent manner: the peak-to-peak change for a wavelength of 160 nm is about 10%, while for wavelengths greater than 300 nm it is less than 1%, and for the solar constant as a whole it is of order 0.1%. Finally, the 27-day rotation cycle of the sun leads to variation of several percent in the solar UV output, although again at longer wavelength (>250 nm) this variation is less than 1%.

2.2.2 Planetary motion and geometry

Shorter time-scale changes, and more immediately relevant, are due to the astronomical motions of the Earth and sun. The Earth's annual orbit around the sun is slightly elliptical and the Earth–sun distance varies, leading to small changes in the available energy throughout the year. The current eccentricity of the orbit means that the Earth is closest to the sun (perihelion) in the January, the Southern Hemisphere summer (Northern Hemisphere winter) and furthest from the sun (aphelion) in July, the Northern Hemisphere summer. The difference in Earth–sun distance between the two extremes is about 3.4%, giving a difference in extraterrestrial radiation of about 6.9%. The eccentricity itself varies on a 110 000 year cycle (becoming more and then less elliptical), with extreme positions that would give no more than a 0.17% change in Earth's incident flux. The position of perihelion also changes as a result of gravitational interactions (mainly with the planet Jupiter) that cause the elliptical orbit of the Earth to precess, which in turn leads to a precession of the timing of the equinoxes. Such changes occur over time periods of 18 800 years and 23 000 years. While they do not affect the total energy received by the earth they do affect the way that the energy is distributed over the surface of the planet.

The most noticeable change in solar energy received at a given location, the seasonal effect, is caused by the tilt of the Earth's axis. This obliquity, the angle between the earth's axis and the plane of the ecliptic, is currently 23.5° (it varies between 22 and 24.5° over a period of about 40 000 years). It affects both the length of daylight and the height of the sun in the sky, which change with time and location on the Earth's surface. In June the sun is overhead at the Tropic of Cancer (23.5°N), at the equinoxes (March and September) it is overhead the equator, and in December it has reached its other extreme position overhead at the Tropic of Capricorn (23.5°S). Since the Earth's axis is tilted there is a differential shading of latitudinal bands that changes with the position of the overhead sun (Figure 2) and day-length is approximately equal to the fraction of a latitude circle that is unshaded. At the equinoxes this is 12 hours everywhere, while the polar circles go from 24 hours darkness in their winter to 24 hours daylight in their respective summers. The sun's height in the sky is usually expressed in terms of the solar zenith angle, z. This is the angle between the local vertical and the position of the sun. The solar zenith angle is given by

$$\cos z = \sin \theta \sin \delta + \cos \theta \cos \delta \cos h$$

where θ is the latitude, δ is the solar declination (latitude where the sun is

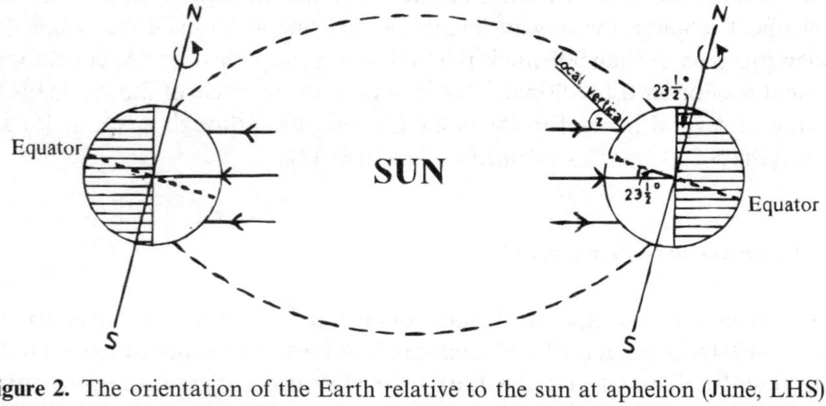

Figure 2. The orientation of the Earth relative to the sun at aphelion (June, LHS) and perihelion (December, RHS). Note that the sun is overhead at 23.50°N and S respectively, resulting in the polar regions experiencing 24 hours of either light or dark because of the tilt of the Earth's axis.

overhead at noon), and h is the hour angle. The hour angle is zero at local solar noon and increases by 15° ($\pi/12$) for every hour from noon. Note that local solar noon is a function of longitude and is not necessarily coincident with the local time zone (clock time). Local solar noon is further modified by the "equation of time", which gives a variation within the year by about ± 15 minutes, as a consequence of the elliptical Earth's orbit around the sun and of the tilt of the Earth's axis relative to the plane of the orbit. The sun rises and sets when $\cos z = 0$, leading to an expression for the half daylength H of

$$\cos H = -\tan\theta \tan\delta$$

The amount of incoming energy on a horizontal surface at the top of the atmosphere above a given location is then

$$E_0 = S_0 (r_0/r)^2 \cos z$$

where r is the instantaneous Earth–sun distance, and r_0 its mean value.

As it enters the atmosphere this radiation becomes subject to interactions with the atmospheric constituents. The atmosphere changes in density, composition and temperature as a function of height so the types of interaction and the wavelengths of radiation affected are also a function of height. At the surface we observe the net effect of attenuation throughout the depth of the atmosphere.

2.2.3 The atmosphere

The atmosphere is not a homogenous medium. At best it can be considered as a series of uniform horizontal layers, the simplification that is most often made when calculating radiative transfer through the atmosphere. In reality many of the atmospheric properties can change on a range of space and time scales. However, the physics can be discussed in terms of a 1-dimensional atmosphere of horizontal layers.

The vertical temperature and density structure of the atmosphere are shown in Figure 3, while the composition of the lower atmosphere is shown in Table 1. Number density, n (the number of gas molecules in unit volume) can be determined from the ideal gas law

$$n = P/kT$$

where P is pressure, T is absolute temperature and k is Boltzmann's constant (1.381×10^{-23} J K^{-1}). In the atmosphere pressure and number density both decrease with altitude (h) in an approximately exponential way. Under the hypothesis of constant temperature, to the pressure applies

$$P(h) = \mathrm{P}(0)\exp(-h/H)$$

where H is a scale height and is about 8 km in the lower regions of the atmosphere. Note that n also depends on temperature (see above) which is neither a constant nor a simple function of height for the whole atmosphere (Figure 3), so n does not have a purely exponential decrease with altitude.

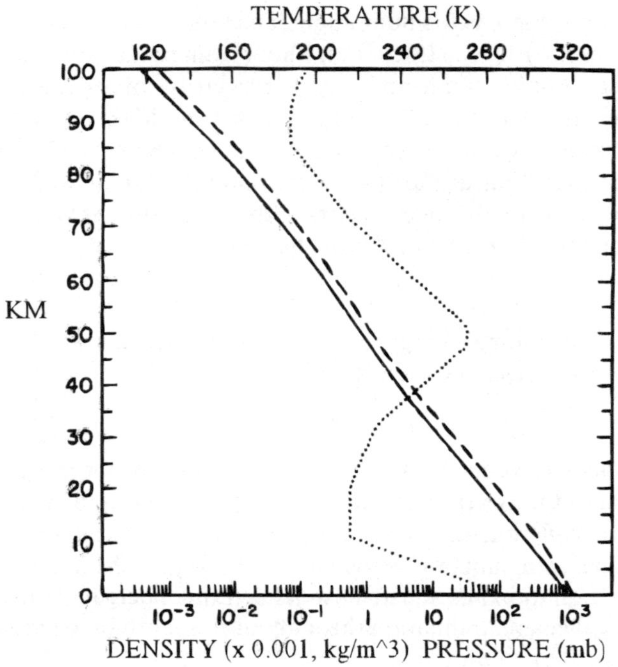

Figure 3. The vertical pressure, density and temperature of a standard atmosphere.

Table 1. Composition of the lower part of the atmosphere (without water vapor)

Gas	Percent of volume
Nitrogen	78.084
Oxygen	20.947
Argon	0.934
CO_2	0.0314 (variable)
Neon	0.001818
Helium	0.000524
Methane (CH_4)	0.0002 (variable)
Krypton	0.000114
H_2	0.00005
N_2O	0.00005
Xenon	0.0000087
Ozone	0.000007 (variable)

The regions of the atmosphere are defined by the vertical temperature profile. At the bottom is the troposphere where temperature decreases with height from the surface (which is warmed by the sun). The rate of change of temperature (the lapse rate) depends on the amount of moisture in the air since the latent heats of condensation and evaporation affect the heat of a rising or descending air parcel. For dry air the dry adiabatic lapse rate is -9.8 K km^{-1}, but a more typical value of the environmental lapse rate (for air containing some water vapor) is -6.5 K km^{-1}. The troposphere extends up to about 10 km, though this varies with location and season (it is higher in the Tropics than at the Poles), and above that the temperature becomes nearly constant for a short distance (the tropopause) before increasing with height through a region known as the stratosphere. This temperature inversion at the tropopause constrains much of the convective vertical motion to the lower atmosphere and reduces interaction between the troposphere and stratosphere. The heating in the stratosphere is a result of the absorption of sunlight (mainly UV) by the ozone at this level, with maximum ozone and so maximum lapse rate around 20–30 km. The top of the stratosphere is marked by the stratopause at about 50 km, and here the temperature profile reverses again and temperature drops with height in a region known as the mesosphere. Higher still are the thermosphere and ionosphere where air density is extremely low.

Over 99% of the atmosphere is composed of oxygen and nitrogen, in a constant ratio. The remaining one percent is made up of other gases, some of which are currently increasing (those with a source in human activity), and one (ozone) that has shown some depletion in the last decades, at some latitudes. The most variable constituents in the atmosphere are water vapor and ozone. Water vapor is found predominantly in the troposphere and is a strong absorber at infrared wavelengths. It varies widely in both time and space. Ozone is concentrated in the stratosphere, though it is also found in the troposphere ($<10\%$ of the column total), and absorbs strongly at UV wavelengths. Ozone has a seasonal variation (highest in the spring, lowest in the autumn) and also varies with latitude (Figure 4). The source region for stratospheric ozone is in the Tropics, but stratospheric transport moves the ozone efficiently to higher latitudes and the equatorial regions, together with the polar regions, have the lowest zonal values of ozone, while the maxima occur at about 60° latitude. Ozone also follows longer-term cycles associated with atmospheric motion (the quasi-biennial oscillation) and solar activity (the sunspot cycle). In the past two decades there has been an observable decrease in ozone over parts of the globe attributed to the destructive catalytic action of chlorofluorocarbons. This is most pronounced over Antarctica in the spring months where it is commonly known as the ozone hole. Similar but less dramatic depletion has been seen over the Arctic, while mid–high latitude sites have witnessed smaller but significant declines in ozone column. There has been no change in Tropical regions. Tropospheric ozone is generally highest in industrialized and polluted regions since it is a product of the photodissociation of nitrogen dioxide (NO_2). Ozone also reaches the troposphere by downward transport from the stratosphere during tropical convection and the passage of mid-latitude depressions.

2.2.4 Absorption and scattering

Radiation traveling through a medium, in this case the atmosphere, can be absorbed, scattered, or transmitted without interruption. The energy of absorbed radiation usually results in either heating or the breaking of a chemical bond in the absorbing molecule: for example the oxygen molecule in the stratosphere

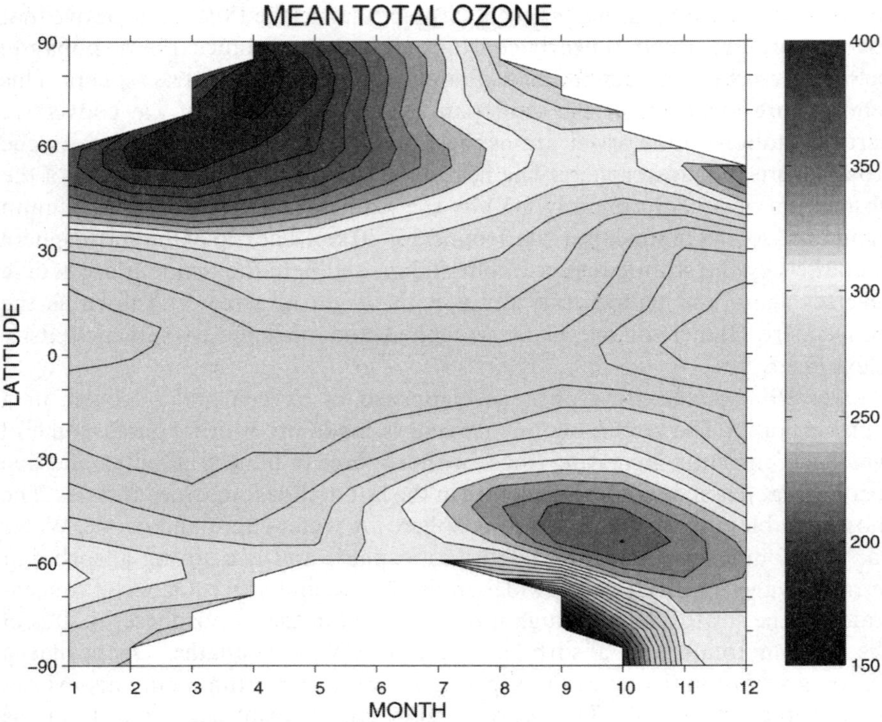

Figure 4. Monthly mean total ozone in 2000 in dependence on latitude, averaged over all longitudes. The white areas in the top corners and in the middle at the bottom correspond to lack of measurements due to polar night. [From EPTOMS, NASA (http://toms.gsfc.nasa.gov/ozone/ozoneother.html).]

absorbs energy from radiation of wavelengths <240 nm and is split into two oxygen atoms – part of the process of ozone formation. In terms of solar radiative transfer this energy is lost. Scattering on the other hand is the redirection of the energy (photon) in a different direction, and as no energy is lost the wavelength remains the same. The scattered radiation may return to space (backscattering) and so be lost to the system, or it may go on to reach the surface by a different route to the unscattered solar beam after one or more scattering events. This redirected radiation reaches the ground from all parts of the sky hemisphere and is known as diffuse radiation, while the photons that remain in the solar beam are direct radiation.

If a parallel beam of monochromatic radiation travels at an incident angle θ through a uniform, plan-parallel layer (thickness h) of simple absorbing gas, the transmitted radiation is given by the Beer–Lambert law

$$I = I_0 \exp\left[-\tau(\lambda, h)/\cos\theta\right]$$

where I_0 is the incident radiation and τ is the vertical optical depth given by

$$\tau(\lambda, h) = \sigma(\lambda) n h$$

with σ (λ) the absorption cross-section and n the number density of the gas molecules. The absorption cross-section depends on the gas, wavelength, often temperature and occasionally pressure. As n varies with altitude in the atmosphere the optical depth can be expressed in terms of the vertical column density N (molecules m^{-2}) where

$$\tau(\lambda) = \sigma(\lambda)N$$

For the atmosphere the angle θ is the solar zenith angle, for direct beam radiation. As the atmosphere is spherical and not plan-parallel, the path length of the radiation through the atmosphere has to be described by the relative atmospheric mass m, which can be approximated for zenith angles less than about 85° by

$$m = 1/\cos\theta$$

The larger the angle θ, the longer the distance that the radiation has to pass through the atmosphere, or one of its layers, and the greater the chance of attenuation.

Attenuation of a single beam of radiation by scattering can be described in a similar way, using the number densities of particles and gases that scatter, and the associated scattering cross-section. The scattering process becomes more complex than absorption because the radiation stays in the system and while it is removed from the beam in one direction the scattered radiation becomes a source for beams in other directions. One photon can also undergo scattering several times in what is known as multiple scattering. The interchangeable sinks and sources of directional radiation over the whole hemisphere and throughout the depth of the atmosphere result in the diffuse radiation at the ground.

There are two main scattering regimes in the atmosphere, known as Rayleigh scattering and Mie scattering, determined by the relative sizes of the scattering body and the wavelength of the radiation. When the scattering particles are small compared to the wavelength (radius $< 0.1\lambda$), as is the case for atmospheric gases, Rayleigh scattering occurs, predominantly in the lower atmosphere where number densities are greatest. The main characteristics of Rayleigh scattering are that the scattering is proportional to λ^{-4} and the angular redistribution of the scattered radiation is symmetrical, i.e. there is an equal probability of the scattered radiation propagating in a forward or backward direction (Figure 5). The wavelength dependency of the scattering leads to the blue color of the sky when the atmosphere is clean (most scattering is by gases), since blue light is scattered about 10 times more than red light. UVR, at even shorter wavelengths, undergoes additional Rayleigh scattering with the result that a large fraction of UVR will always be diffuse and the direct beam is comparatively poor in UV.

Mie scattering is relevant for larger particles such as aerosols and cloud particles (it is most effective when scattering particles and wavelength are of similar size). Mie scattering does not have the same strong wavelength dependency as Rayleigh scattering and can be described by the Angstrom formula,

$$\tau(\text{aerosol}) = \beta\lambda^{-\alpha}$$

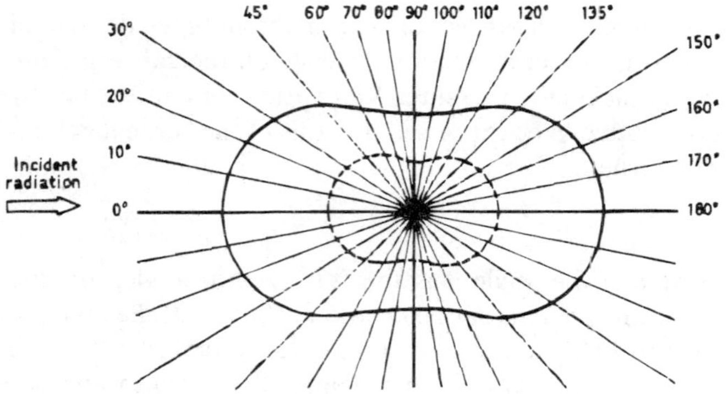

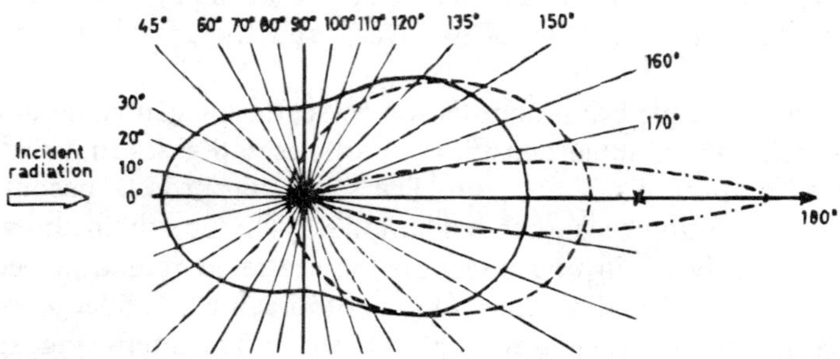

Figure 5. The intensity of radiation as a function of scattering angle for Rayleigh scattering (top) and Mie scattering (bottom) at a wavelength of 520 nm (green light). For Rayleigh scattering the particle size is radius 25 nm and the inner dashed line shows the reduced scattering of red light (wavelength 700 nm). In the bottom plot the Mie scattering is by 3 different particle sizes: 50 m (solid, $X=0.1$), 100 nm (dashed, $X=1$) and 500 nm (dash-dot, $X=4000$). The point X provides a relative scale of intensity for the different particle sizes, showing the greatly increased intensity of the forward scattered beam as the particle size increases.

that shows the wavelength dependency of the scattering is determined by α. The value of α depends on the type of aerosols involved in the scattering. For a continental aerosol it is typically about 1.3. As the size of the scattering particles decreases, α increases. The value of β characterizes the amount of aerosols, as $\tau(\text{aerosol})=\beta$ for a wavelength of 1 μm (1000 nm).

The direction of scattered radiation becomes increasingly concentrated in the forward direction as the wavelength approaches the size of the particle (Figure 5). For particle sizes larger than the wavelength of the radiation there is still a strong narrow lobe of forward scattered radiation, but other weaker lobes also occur at wider scattering angles. In a polluted atmosphere the sky will appear milky or pale blue/grey because there is no preferential wavelength scattering by the pollution and more of all colors of light is distributed across the sky.

The angular redistribution of light during a scattering event is described by the scattering phase function, $P(\Theta)$. This gives the probability that a photon from one specified 3-dimensional direction will be scattered into another specified direction. A simple way of specifying the overall directionality of the phase function is to look at the asymmetry factor, g. This has a value ranging from $+1$ (complete forward scattering) through zero (equal forward and backward, as for Rayleigh scattering) to -1 (complete backward scattering). Aerosols might have values of g between about 0.6 and 0.8 while large cloud droplets can reach 0.9.

Both scattering and absorption can occur on all particles and molecules and the relative probability of each is given by the single scattering albedo, ω_0, expressed as

$$\omega_0 = \sigma_s/(\sigma_s + \sigma_a)$$

where σ_s and σ_a are the scattering and absorption cross-sections. Pure absorption has a value for ω_0 of zero, while pure scattering has $\omega_0 = 1$. For most molecules the attenuation will be dominated by one or other of the processes for a given wavelength, and ω_0 can effectively be taken as either zero or 1. However, mixtures of gases with each process dominating for some of the gases in the mixture can lead to a range of values for the overall ω_0 in a layer of atmosphere. An example of this is the mixture of ozone and air in the stratosphere; ozone is a strong absorber at UV wavelengths while air scatters. Measurable amounts of both scattering and absorption can occur on cloud droplets and aerosol particles, though scattering tends to be stronger with values of ω_0 that can typically be 0.8–0.99 at UV wavelengths.

2.2.5 Determining the UV spectrum at the ground

The extraterrestrial solar radiation contains wavelengths of electromagnetic radiation far shorter than those experienced at the ground. These high energy photons are absorbed in the atmosphere: generally the higher the photon energy, the higher up in the atmosphere it is absorbed. At wavelengths shorter than 102 nm the photon energy is sufficient for the ionization of molecular oxygen and this part of the solar output does not penetrate below about 100 km in the atmosphere. In the thermosphere, dissociation of oxygen molecules to two atoms (one excited) occurs (at wavelengths <175 nm) in the Schumann–Runge continuum. Still above 60 km, further radiation is removed in the Schumann–Runge bands (vibrational excitation of oxygen). Below 60 km the main oxygen absorption process results in photodissociation of the molecule to two ground state atoms (wavelengths 200–242 nm), but this becomes a secondary process to absorption by ozone in the stratosphere. This leads to the heating and the temperature inversion observed above the tropopause and protects the Earth from the life-damaging UV-C and UV-B radiation.

2.2.5.1 Ozone absorption

Ozone has three major absorption bands for solar radiation named the Hartley, Huggins and Chappuis bands. It also has an absorption band at terrestrial infrared wavelengths (not relevant here). The strongest ozone absorption is in the

Hartley band (a decreasing continuum from 200 to 310 nm) and this merges with the spectrum of lines in the Huggins band (310–400 nm). The two UV absorption bands together prevent all UV-C and much of the UV-B radiation from propagating down below the stratosphere. The detectable short wavelength limit of the ground level solar spectrum depends on solar zenith angle (determining pathlength) and ozone column (density of absorbing molecules) but is about 290 nm. When the solar zenith angle is small and ozone is low (in the Tropics at midday) shorter wavelengths (by a few nm) may be detected; when the solar zenith angle and/or ozone are high (high latitudes, winter or early/late in the day) the short-wave limit will be significantly longer. The ground level solar spectrum then rises rapidly through the UV-B flattening out into the UV-A, thus mirroring the decreasing absorption cross-section of ozone. At wavelengths longer than 340 nm ozone essentially has no influence on the solar spectrum. The third, Chappuis band is a continuous absorption band starting in the visible (400–850 nm). It is a weaker absorption band than the other two, but is at the peak of the solar spectrum and becomes important in the troposphere. All three absorption bands are associated with the photodissociation of ozone, so due to the Chappuis band even ozone at the surface is rapidly dissociated.

The importance of ozone as a shield against damaging short-wave radiation reaching the Earth requires an understanding of its existence and the recently observed stratospheric depletion.

2.2.5.2 Stratospheric ozone chemistry

The amount of ozone in the stratosphere is determined by a dynamic balance between the processes of production and destruction. Ozone is formed by a two stage process that begins with photodissociation of oxygen by wavelengths <242 nm

$$O_2 + h\nu \rightarrow O + O$$
$$O + O_2 \rightarrow O_3$$

Ozone production is a maximum in the stratosphere because it requires both oxygen and the correct wavelengths of radiation. At higher altitudes there is little oxygen; at lower altitudes there is no radiation of wavelengths <242 nm since it has already been absorbed by oxygen and ozone at higher levels. Production is greatest in equatorial regions because this is where there is the most radiation.

Ozone loss occurs through a number of different reactions

$$O_3 + h\nu \rightarrow O + O_2 \ (\lambda < 800 \text{ nm})$$
$$O + O_3 \rightarrow 2O_2$$
$$O_3 + M \rightarrow O_2 + MO$$
$$O + MO \rightarrow O_2 + M$$

The catalytic cycles involving M leave M unchanged and able to go around the cycle many more times destroying further ozone molecules. The most important molecules that can take the place of M are OH, NO and Cl, though there are other examples and more complex chemical reactions that can also destroy ozone. The active species, M, are lost through conversion into less active species (called reservoir species) in other reactions. The reservoir species can be converted back into active species by photodissociation or reactions that occur on the surfaces of aerosol particles or stratospheric clouds heterogeneous reactions). Alternatively they are gradually transported to the troposphere and lost to the atmosphere, e.g. through rainout and deposition.

The production and destruction processes above occur naturally to produce the dynamic photochemical equilibrium of ozone distribution described earlier. This equilibrium has been upset by the addition of extra species of M to the stratosphere, most notably chlorine (Cl), from human sources.

2.2.5.3 Changes in stratospheric ozone

Unprecedented loss of stratospheric ozone was first observed over the Antarctic and reported in 1985 [1]. Since then each spring has seen the almost complete destruction of ozone in the lower stratosphere for the air within the polar vortex, and is correlated with high amounts of ozone-destroying compounds. The destruction is aided by the special conditions over Antarctica where the winter polar vortex is very strong and traps air in a huge mobile column that oscillates about covering most of Antarctica and sometimes drifting or extending over parts of South America. Active chlorine species originating from chlorofluorocarbons and previously photodissociated into reservoir species by the high energy photons in the stratosphere reside in the Antarctic stratosphere during the polar night. Temperatures become low enough for the formation of polar stratospheric clouds so that when solar radiation reaches the air in the vortex in the spring the ozone is trapped with ozone destroying species and the surfaces suitable for rapid heterogeneous reactions, and is rapidly destroyed. The sunlight also warms the atmosphere and eventually provides sufficient energy to breakdown the vortex and allow ozone rich air to enter the region from lower latitudes and ozone poor air to be transported away and mixed.

Ozone loss over the Arctic has been less dramatic than that over the Antarctic, mainly because the different distribution of land and sea in the Northern Hemisphere allows for only a weak vortex over the Arctic. There is more mixing of air with that from lower latitudes and temperatures do not become low enough for routine formation of polar stratospheric clouds. In years when the Arctic has been cold enough for cloud formation similar ozone destruction has been observed, but for less prolonged periods than over Antarctica. Trends in ozone over the rest of the globe have been small compared to those of the Antarctic, or even Arctic, and are quantified in section 2.4.1.

2.2.5.4 Tropospheric ozone chemistry

As in the stratosphere the tropospheric ozone is in a dynamic balance between sources and sinks. Ozone can be transported to the troposphere from the stratosphere or can be produced in situ. The main production follows the dissociation of NO_2

$$NO_2 + h\nu \rightarrow NO + O (\lambda < 400\ \text{nm})$$
$$O + O_2 \rightarrow O_3$$

However, the NO produced in the photodissociation of NO_2 destroys ozone

$$O_3 + NO \rightarrow O_2 + NO_2$$

Thus the net effect of dissociating nitrogen dioxide is neutral. Net production of tropospheric ozone occurs as a result of other reactions that convert NO into NO_2 without destroying ozone. There are many such reactions, most of which involve the photooxidation of chemicals like carbon monoxide, methane and other hydrocarbons. Since these are produced by traffic and industrial processes, ozone production is a feature of polluted regions, and ozone itself is considered a pollutant at low levels of the atmosphere where it is detrimental to human and other life forms. Sinks of ozone include photodissociation and reactions with OH and HO_2 (as in the stratosphere) and deposition.

2.2.5.5 Other attenuators

Despite its importance as an absorber of short wavelength UVR, ozone is not the only attenuator of UVR. Scattering by air molecules has already been discussed but there are other gaseous absorbers and other scattering bodies in the atmosphere. All of these tend to be very variable with both time and position.

The other main gaseous absorbers of UV are SO_2 and NO_2. However, they are only present as trace gases, and then only in significant amounts in fairly heavily polluted air. High values of the gases (that might have a noticeable effect on UV) are generally confined to the lower levels of the troposphere around large urban areas during high pollution episodes.

Aerosols (suspended particles) can be natural in origin or related to human activity such as combustion (of fossil fuels) or biomass burning. They can be, for example, sea salt, mineral dust, soot, dilute sulfuric acid droplets, and their existence will depend on the proximity of sources and suitable conditions (e.g., windspeed, humidity) for their formation and transport. Aerosols, which can both scatter and absorb, are most concentrated in the lower troposphere (planetary boundary layer) and decrease quickly with altitude. High altitude aerosols are usually insignificant in terms of UV transmission, except in unusual circumstances, such as immediately after a large volcanic eruption such as Mount Pinatubo in 1991.

Finally, clouds must be considered. They attenuate radiation in a manner that is not strongly wavelength dependent but is highly variable and generally unquantified because of the difficulty of defining the cloud situation. Attenuation will depend on both the microphysical properties of the cloud (affecting scattering and absorption properties) and the macrophysical arrangement of the clouds in the sky. Clouds can be composed of ice particles (upper troposphere) or water droplets (lower troposphere) or a mixture of the two, and they may contain impurities, while the drop size distributions cover a wide range. The macrophysical properties of the clouds, their size, depth and shape, are continually varying. Cloud observations usually report the fraction of the sky covered by cloud (in eights or tenths) and the visible cloud type. However, clouds may be in several layers with only the lower layer visible from the ground and it is very rare to have detailed macro and microphysical information about a cloud situation, which is still only valid for a brief time and a single cloud formation. In general clouds reduce radiation reaching the surface, but in certain conditions (e.g. cumulus clouds close to but not covering the sun) the forward scattering from the cloud sides can increase the ground irradiation above its clear sky value for short periods of time. This is well known for total solar radiation, and applies to UVR too. Finally, in broken cloud situations the apparent attenuation of UV is less than that of longer wavelengths because there is less UV in the direct beam due to the strong Rayleigh scattering. Thus, if most of the sky is clear and there is a cumulus cloud before the sun the UV "shade" will be less dark than the visible shade.

2.2.5.6 Final result at the surface

The discussion above has shown that the UVR at the bottom of the atmosphere is dependent on many factors, some of which are well quantified and others that are very poorly defined. Astronomical factors determining solar zenith angle and so the extraterrestrial radiation and the pathlength through the atmosphere are fixed and easily calculated if geographical position and time are known. The major absorber, ozone, is highly variable on a short time scale, but has seasonal and climatological norms known from long time series of measurements (since about 1920 [2]). However, ozone depletion must also now be considered especially at high latitudes [3]. The effects of a clean, clear atmosphere are well defined by Rayleigh scattering, but the real atmosphere contains clouds, aerosols and pollutant gases that are very variable in amount and can have significant effects on the incident UVR.

At the bottom of the atmosphere is the Earth's surface, which is neither flat nor uniform and has its own influences on the irradiance it receives. The two major properties of the surface that affect the UV irradiance are the elevation and the albedo (reflectivity).

Elevation is the height of the ground surface above sea level. Much of the surface of the globe is water, and so by definition at sea level, but the ground surface of the continents can rise several kilometers into the atmosphere. The effect of this is almost the same as rising to the same height in the atmosphere over the sea. The pressure and density of the air is lower and the radiation has had a shorter pathlength through the atmosphere, in particular it has not had to travel through the densest air, which generally contains the most aerosol, moisture and tropospheric ozone. Thus there has been less attenuation and the irradiance is higher at elevated sites than at nearby sites at sea level, in the same weather conditions. The decreased scattering at elevation can be observed in the deeper blue of the sky seen on high mountain tops, and this also holds in the UV, so there will be more radiation in the direct beam at elevation than at low levels. The effect of elevation on UV has a spectral dependence, since attenuation processes are also spectral and it is greatest at the shortest wavelengths.

Snow, which is a permanent feature at high mountain sites and polar regions, and a seasonal feature at high latitudes and moderate mountain regions, is highly reflective in the UV. Radiation that is reflected from the Earth's surface becomes another source of radiation in the atmosphere and is subject to the same

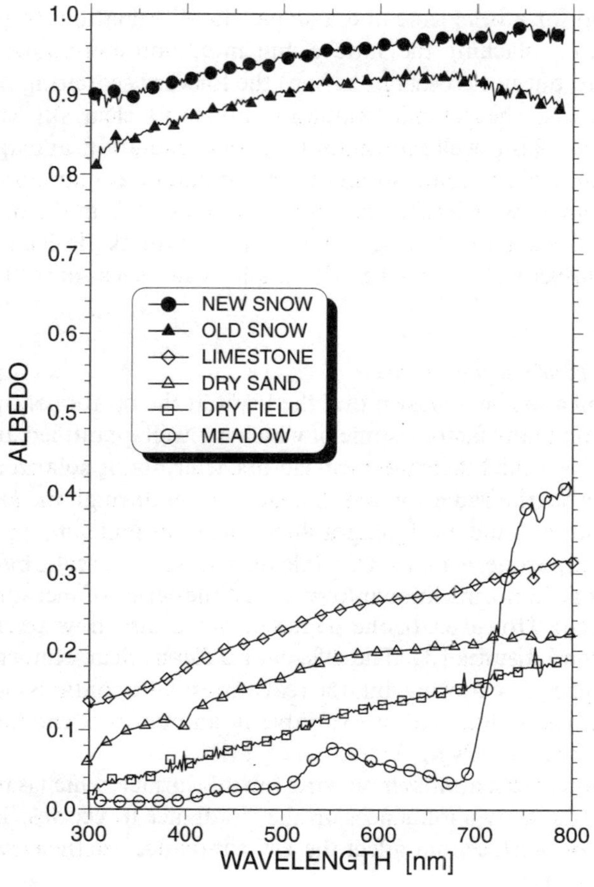

Figure 6. Spectral dependence of albedo for various types of surfaces.

radiation laws as any other scattered radiation. As it re-enters the atmosphere where it is most dense further scattering is likely and some of the radiation will return again to the surface, increasing the down-welling irradiance. The surface albedo (the ratio of reflected to incident radiation) is generally low for most natural surfaces in the UV: lower than the total solar albedo or visible albedo. The UV albedo tends to increase with wavelength but typical values for crops and vegetation are generally less than 5%, soils and rock can reach about 10% and water is 5–10% (Figure 6) [4–7]. Over low albedo surfaces the effect of the surface (reflected radiation) is only apparent at low altitudes: above about 500 m the wavelength dependency of the up-welling radiation reverses, increasing with decreasing wavelength and indicating the backscattering properties of the atmosphere rather than the surface [5]. Sand has a variable albedo, but can reach values of around 20%, similar to some man made surfaces e.g. cement, concrete and light colored asphalt. The highest albedo is found for snow and ice. Snow albedo can vary from about 20% (old, wet and dirty) to virtually 100% (fresh, clean and dry). Ice has an albedo of up to about 75%, again dependent on the state of the ice.

The effect of albedo on UVR is not dependent only on the albedo of the surface in the immediate vicinity of the location of interest, but on the albedo of the whole surrounding area, so unless the area is homogenous estimating the albedo, and so its influence, can be difficult.

2.3 Measurements

Measurements of solar UVR started in the first decades of the 20th century with chemical detectors [8], where the changing of the color of a solution was an indicator of UV-B irradiance. Physical measurements of UVR use photoelectric methods to give quantitative information about the intensity and spectral distribution of UVR. The first extensive data sets originate from the 1960's, from Davos (Switzerland) [9], but it was not until the 1990's that more of such high quality spectral measurements were made at further locations world-wide. Since the 1970's a greater number of UV measurements have been made with a different type of detector, sensitive to a broad wavelength range in the UV-B (and to a lesser extent in the UV-A), the so-called broadband erythemal detectors. They have been used in many parts of the world for monitoring over periods of several years, but systematic, long-term observations are very rare and again were mainly established in the 1990's in response to stratospheric ozone depletion. More recently, estimations of UVR based on measurements with instruments on satellite platforms allow more global coverage.

2.3.1 Ground-based measurements

2.3.1.1 Instrumentation

Different types of UV instruments are used for measuring solar UVR depending on the objectives of the measurements and on the required accuracy. Instrument types differ in the extent of spectral information that they provide between high (e.g., spectroradiometers), moderate (e.g., multifilter radiometers) or low (e.g., broadband radiometers, dosimeters) spectral resolution. In addition, time integration is usually done by frequent sampling, whereas with dosimeters a time integration is inherent in the measurement.

The entrance optics of all types of instrument are usually designed to measure global irradiance, that is radiation falling on a horizontal surface with a spatial weighting according to the cosine of the zenith angle of the radiance. This response is achieved by the so-called cosine-diffuser. Flat pieces of Teflon provide a first approximation to the cosine-law, but for higher quality measurements more care is necessary (i.e., by using specially shaped Teflon). As an alternative, integrating spheres are used in some cases as input optics. For measurements of actinic fluxes (the radiation on the surface of a small sphere) instead of irradiances, the input optic has to have a uniform sensitivity in all directions, which is realized by a 4π-head, or a 2π-head for one hemisphere only.

Spectral instruments: The highest amount of information about UVR can be gathered with spectral instruments. These instruments use a monochromator to disperse (separate) the incoming radiation into small spectral intervals, in most cases by using a diffraction grating, and then measure the signal at individual wavelengths with a photoelectric detector. The spectral resolution is determined by the geometry of the entrance- and exit-slits of the monochromator, which define the slit function. It has usually the shape of a triangle and is characterized by the full-width at half of the maximum signal (FWHM). Typically values for FWHM are between 0.2 and 2 nm.

As a consequence of the steep decline of the solar spectrum at UV-B wavelengths, high stray-light rejection is necessary to avoid erroneous signals from higher wavelengths (in the UV-A and visible), where the intensity of the solar radiation is several orders of magnitude higher than the UV-B. Therefore, sensitivity outside the ideal slit function should be as small as possible. This can be achieved best by using a double-monochromator, where a second monochromator follows at the exit slit of the first one, and also gives the additional advantage of higher spectral resolution.

The wavelength setting of a monochromator may be carried out by rotating the grating(s), which directs radiation of the individual wavelengths to the exit slit, where it is measured with a photomultiplier tube. This type is called a scanning spectroradiometer, and it is the usual practice for double monochromators. For single monochromators it is also possible to mount a diode array or a CCD-element at the exit slit, providing information on the intensity at individual wavelengths without any mechanically moving parts. Furthermore, the information about all wavelengths is taken simultaneously, whereas scanning radiometers usually need a time in the range of 1 to 10 minutes to measure a wavelength range of about 100 nm. However, for measurements of solar UV-B radiation the best results are currently achieved with scanning spectroradiometers with an attached photomultiplier, providing superior stray light rejection and a high dynamic range.

Spectroradiometers for field measurements of solar radiation are complex and sensitive instruments, usually they require temperature stabilization and they need frequent control of their calibration with respect to wavelength and to irradiance. Therefore maintenance of these instruments is a challenging task and needs significant manpower and experienced operators.

The calibration of the output signal of a spectroradiometer to the incoming radiation is usually done by comparison with a 1000 W quartz tungsten halogen lamp. Standard lamps of this type are calibrated by national calibration laboratories, giving the irradiance (W m^{-2}) at defined wavelengths and at a defined distance. However, even from the National Reference Laboratories world-wide, the uncertainty of the calibration of these lamps in the UV range is in the order of 2–4% (95% range), which ultimately limits the possible accuracy of solar UV measurements. Adding transfer uncertainties, leads to a best-achievable final uncertainty of solar UV measurements of about 5% (95% range) [10]. A level of agreement of about 5% has also been found between several UV-measurement groups in the most recent spectroradiometer intercomparisons, where up to 20 different spectroradiometers were simultaneously measuring solar UVR [11].

Broadband instruments: Broadband instruments measure solar irradiance in a specified wavelength range, typically 20 nm to 100 nm wide. This range is defined by the construction of the detector and it results from a combination of different optical elements such as filters and photoelectric sensors. The output signal of broadband instruments corresponds to the integral of the incident irradiance multiplied by the spectral response of the detector. Therefore, any information about the detailed spectral structure of the incident solar radiation is lost. On the other hand, the measurement is instantaneous and thus allows rapid changes in irradiance to be followed, due to fast moving clouds for example.

One of the most commonly used type of broadband detectors for measurement of solar UVR, the so-called Robertson–Berger type detector [12,13], has a spectral sensitivity which is adapted to the standardized erythema action spectrum of the human skin [14]. It has the maximum of its sensitivity around 297 nm, then sensitivity decreases steeply to about 320 nm and in the UV-A range the sensitivity is about 1000 times smaller than at the maximum in the UV-B. Thus these detectors give a direct measure for the biologically relevant irradiance. However, as no one available detector has a spectral sensitivity perfectly matched with the erythema action spectrum, corrections are necessary to get a standardized output from these instruments. These corrections depend on the variation of the solar spectrum, mainly correlated with solar zenith angle and with total atmospheric ozone content, and are specific to an individual instrument. This conversion from detector based units into absolutely defined erythemally weighted units may not be necessary if relative variations of UVR are observed over longer time scales at one station only and no absolute comparison is made with results from other detectors.

Currently a great number of detectors of this type is in use world wide, but the calibration to a common reference is often not possible. To help address this problem the World Meteorological Organization has organized two (1995, 1999) international intercomparisons of broadband detectors to encourage homogenization of the data between different sites [15,16]. Meanwhile, in the USA a central calibration facility has been established [17] to serve the different organizations that operate broadband instruments, thus helping to maintain a constant quality of data.

As the absolute calibration of broadband detectors is usually based on field intercomparisons with spectroradiometers and an additional conversion of raw data into standardized erythemally weighted units is necessary, the overall uncertainty of broadband detectors is higher than that for spectroradiometers. Under careful operation and frequent recalibration, an uncertainty of about 7–8% (95% range) might be achievable. This is the absolute uncertainty of an individual detector – if similar detectors are used in a network, then the relative uncertainty between these individual detectors might be, at best, in the order of 2–3% [18]. It is important to mention that although the price of broadband detectors is relatively low; these detectors also need a significant amount of quality control and quality assurance to give reliable results. Operating a network of broadband meters at a high level of quality is a challenging task. Long-term stability cannot be expected by itself, but it has to be verified by the operator. In addition it now appears that internal humidity of the detectors might significantly affect the response and therefore careful maintenance of the attached desiccant is important.

Moderate bandwidth instruments: Instruments that measure solar radiation with a bandwidth between about 2 and 20 nm are called moderate bandwidth instruments. In the most cases, interference filters are used in combination with photodiodes. For absolute calibration the spectral sensitivity of the detector has to be known with high accuracy over the whole wavelength range of solar radiation, in order to avoid distortions from secondary transmission regions far away from the central wavelength. Filters in the UV-B range around 300 nm are particularly sensitive to this problem, and the long-term stability of the transmission of the interference filters also has to be tested carefully. The small size of filters and diodes allows several channels to be combined in one instrument, thus offering simultaneous measurements with moderate bandwidths over a broad wavelength range, usually with 4–8 channels from the UV to the visible range. Therefore spectral effects of solar radiation (in the range of a few nm) can be investigated under all weather conditions.

Data from multi-filter, moderate bandwidth instruments are often post-processed in combination with radiative transfer modeling, which allows the reconstruction of the full solar spectrum. In a second step, the calculated solar spectrum can be weighted with any biological weighting function, i.e. for determination of erythemal doses [19].

Biological UV dosimeters: Dosimeters for UVR measure the integrated dose over the time of exposure. A well utilized dosimeter based on a polymer and with a response spectrum similar to that of erythema is polysulfone film, developed in the 1970's [20]. More recently biological UV dosimeters have been developed in the 1990's. These are based directly on a specific biological reaction (e.g., DNA damage), the inactivation of bacterial spores or bacteriophages, a photochemical reaction in the in situ photosynthesis of vitamin D, or on uracil molecules [21,22]. The biological material is exposed to solar UVR and its response is then measured (often by a slow analysis process some time after irradiation). In order to quantify the reading of these dosimeters, the spectral response function of the specific reaction has to be known, as well as exposure geometry, linearity and stability to environmental parameters. Usually the dynamic range of these detectors is rather small. On the other hand, for some biological applications it is an advantage to collect directly the dose (total over time) of the biologically relevant component of incident radiation fluxes. However, the conversion into absolute radiative quantities may still have uncertainties greater than 10% [23].

2.3.1.2 Results

The effects of the different parameters that determine solar UVR at the Earth's surface are illustrated with the results of UV measurements, demonstrating the great natural variability of the solar UVR.

Solar elevation: The dependence of solar UVR on solar elevation is shown in

Figure 7 for a cloudless day at Izaña Observatory, Canary Islands, Spain (latitude 28.3°N, longitude 16.5°W, altitude 2367 m above sea level). Solar elevation at noon on 18.07.1995 was 80.3°, total ozone 282 DU and aerosol optical depth 0.06 at 350 nm. Therefore the measured irradiance was close to the expected maximum for these solar elevations, due to the relatively low ozone content, the very low amount of aerosols and the high altitude. Higher values could only be expected if the surroundings had been covered with snow: in fact under the measurement conditions the albedo of the surrounding was very low. From Figure 7 it can be seen that for erythemally weighted irradiance the noon value was 350 mW m^{-2}, which corresponds to an UV-Index of 14 (1 UV-Index = 25 mW m^{-2}). In addition to global erythemal irradiance, the diurnal variation of spectral irradiance at 302 nm and at 320 nm is shown in relative units, normalized to the maximum of erythemal irradiance. This shows that for shorter wavelengths the change of irradiance with solar elevation is steeper than for longer wavelengths, and the variation of erythemal irradiance as a broadband integral with a central wavelength around 310 nm (dependent on solar elevation) lies in between. Therefore the range of variation of irradiance for solar elevation from 20° to 80° is about a factor 6 at 320 nm and a factor 70 at 302 nm, showing that at low solar elevations the relative contribution of shorter wavelengths is much lower than at higher solar elevations. The reason for this effect is the absorption in the ozone layer, which is more pronounced at low

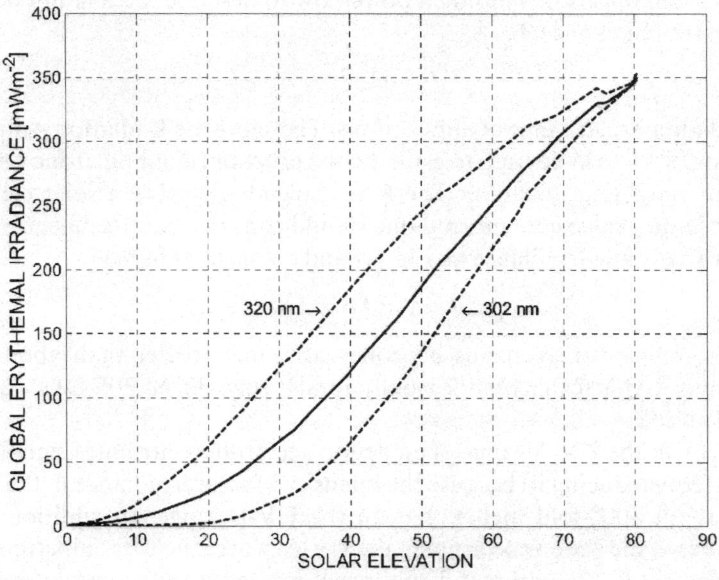

Figure 7. Dependence of global erythemal irradiance (solid curve) on solar elevation for a clear sky day (18 July, 1995) at Izaña Observatory (Canary Islands, Spain, 28.3°N, 16.5°W, 2367 m above sea level) with total ozone 282 DU and aerosol optical depth 0.06 at 350 nm. Spectral irradiances at 302 nm and 320 nm (dashed curves) on the same day are given in relative units, normalized to the maximum of erythemal irradiance.

solar elevations and more effective for short wavelengths in the UV-B.

From Figure 7 the seasonal variation of UVR at noon can also be estimated, remembering that noon solar elevation increases from winter solstice to summer solstice by about 47°. Annual maximum noon solar elevation ($se_{(max)}$) can be calculated for any latitude higher than 23.5° as ($se_{(max)}$) = 113.5° − latitude, while for latitudes lower than 23.5° the maximum noon solar elevation is of course 90°. As already mentioned, the absolute values of erythemal irradiance shown in Figure 7 correspond to a situation with extremely high irradiance; however, relative variations of irradiance with solar elevation can be estimated from these measurements.

The change in solar elevation can also be interpreted in terms of changing latitude. At 20° latitude northwards of the station Izaña, the noon intensity would be (under all other identical conditions) 100 mW m^{-2} less, which is about 30% smaller. Moving another 20° northwards would reduce erythemal irradiance again by about 125 mW m^{-2}, which means that irradiance at 68°N is about 50% less than at 48°N.

Ozone: The relation between changes of total ozone column and the corresponding changes in UVR is well established through measurements. That means that if ozone decreases then UVR increases, when all other parameters that influence UVR are constant. The shorter the wavelength in the UV-B range, the stronger is the increase due to ozone decrease, as a consequence of the spectral shape of the ozone absorption cross-section. For broadband spectral ranges, the simplification by using a power law to describe the relation between irradiance (I) and ozone (O),

$$I \propto O^{-\mathrm{RAF}}$$

is justified for a broad range of applications. Therefore the Radiation Amplification Factor (RAF) can be used to estimate the effect of changing ozone on UVR for various weighting functions. The RAF depends slightly on solar elevation and on absolute ozone content, and one should consider that the linear relation between variation of irradiance (ΔI in %) and ozone (ΔO in %)

$$\Delta I = -\mathrm{RAF} \times \Delta O$$

becomes significantly erroneous if ozone variations greater than about 20% occur. Some RAF's (calculated for daily totals, July, 30°N, 305 DU) [25] are given in Table 2.

Sensitivity in the UV-A range of an action spectrum contributes significantly to the RAF (by reducing it) because the intensity of solar irradiance in the UV-A range is about 1000-fold higher than in the UV-B range. In addition, small uncertainties in the action spectrum in the UV-A, where the determination of the action spectrum is often more difficult, result in significant uncertainties of the overall RAF.

Aerosols: The amount, type and optical characteristics of aerosols are usually not very well known when UV measurements are made. This is the cause of the greatest uncertainties in the comparison of UV measurements with results of radiative transfer models under cloudless skies. The determination of the aerosol

Table 2. Radiation amplification factors (RAF) for various action spectra, calculated for daily totals of solar radiation in July, 30°N, 305 DU ozone amount

Action spectrum	RAF
Skin erythema [14]	1.2
Skin cancer [26]	1.1
DNA damage [27]	2.1
Photokeratitis [28]	1.1
Cataract [28]	0.7
Phytoplankton motility [29]	1.5
J $O(^1D)$ Photolysis [30]	1.5

optical depth, which is the basic parameter used to quantify the actual aerosol amount, is usually carried out by sunphotometric measurements. With these instruments the intensity of direct solar irradiance, either spectral or in specific wavelength ranges, is measured and from the known extraterrestrial spectrum and the known attenuation of direct solar irradiance by molecules (Rayleigh-scattering) the aerosol optical depth is derived. If absorption by gases takes place too (in the UV by ozone), this quantity must also be known in order to derive aerosol optical depth. The vertical distribution of aerosols can be determined with lidar systems, which are becoming more portable and now feasible additions to measurement campaigns. The separation between tropospheric aerosols and stratospheric aerosols is important, because they affect solar irradiance differently.

The separation between scattering and absorbing component of the aerosols is characterized by the single scattering albedo. For in situ measurements (at the ground or from an airplane) instruments are available to provide this information. Attempts have recently been made to derive the single scattering albedo from radiance measurements of the scattered radiation of the sky in combination with radiative transfer calculations [31,32], which gives reasonable results for higher aerosol optical depth (greater about 0.4). In a similar way, the spatial characteristics of the forward scattering of the aerosols can be derived. Therefore measurements of clear sky radiance distributions at wavelengths in the UV range have the potential to provide valuable information about the atmosphere.

The most complete information about aerosols would be vertical profiles of their size distribution and their chemical composition, which would allow the complex refractive index to be determined. However, such detail is usually not available.

The regional and temporal distribution of aerosols is strongly variable, therefore measurements of their effect on solar irradiance are primarily local information. In general, in addition to a tropospheric and stratospheric background aerosol load local enhancements typically occur due to urban pollution or natural events (e.g., forest fires). An example of the effect of aerosols on UVR is shown in Figure 8, where the attenuation of UV-B irradiance by varying aerosols is shown [33]. The measurements near the city of Athens (Greece) in summer 1996 were first simulated with a radiative transfer model, where the aerosol optical depth was independently measured and the aerosol single scattering

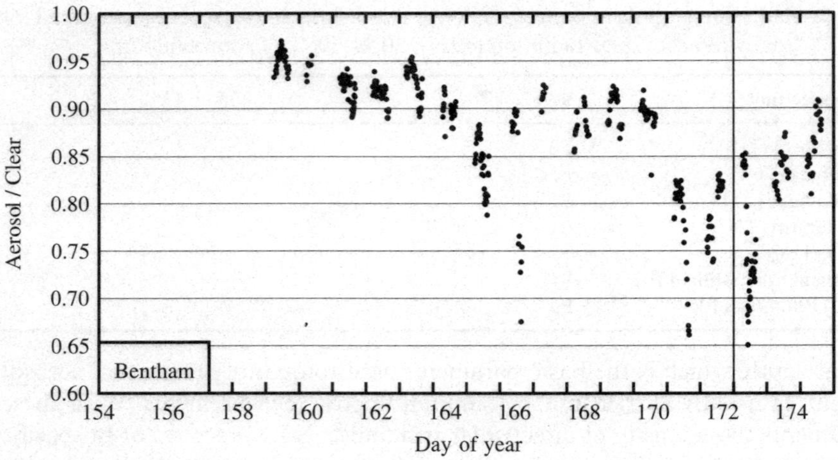

Figure 8. Attenuation of erythemal UVR (measured with a Bentham spectroradiometer) by aerosols on several days in summer 1996 near Athens, Greece. Ratio of model calculations including the actual aerosols relative to calculations without aerosols.

albedo was the only free parameter to fit the model to the measurements. In this case, a variation of single scattering albedo between 0.85 and 0.98 was found, consistent for two independent spectroradiometric measurements. Comparing the model result for calculated spectra with and without aerosols shows great temporal variability and reductions of up to 35% due to the aerosols. Only situations without any observable clouds are illustrated here.

Worldwide enhancements of aerosols in the stratosphere are observed after big volcanic eruptions (i.e., by the volcano Pinatubo in 1991), which led to a decrease in direct solar irradiance and an increase in diffuse irradiance. This effect could be measured especially well at a high mountain station, where the disturbance by urban aerosol pollution is very small. In this case, diffuse solar radiation was increased nearly twofold, while global (direct and diffuse) radiation was reduced by about 4% [34].

Albedo: Model calculations show for increasing albedo an increase of solar irradiance with decreasing wavelength in the UV-A and a maximal effect around 320 nm (Figure 9). In the UV-B range, the effect of increasing albedo is less pronounced due to absorption of the reflected radiation by tropospheric ozone. The amplification of global irradiance at 320nm for an increase of albedo by 0.1 is between 3% and 4%. Therefore a change of the terrain from snow free (average UV albedo 0.03) to a fresh snow covered terrain (UV albedo 0.9) will result in an increase at 320 nm of about 30%, and of about 18% at 400 nm. These numbers are valid for cloudless sky. If there are clouds, then the effect of surfaces with high albedo is enlarged due to multiple reflections between the cloud and the surface.

The calculations above are valid for the assumption of a homogeneous, horizontally unlimited surface with a specific albedo. Of course this is not true in reality. Recently, 3D radiative transfer models became available, using the

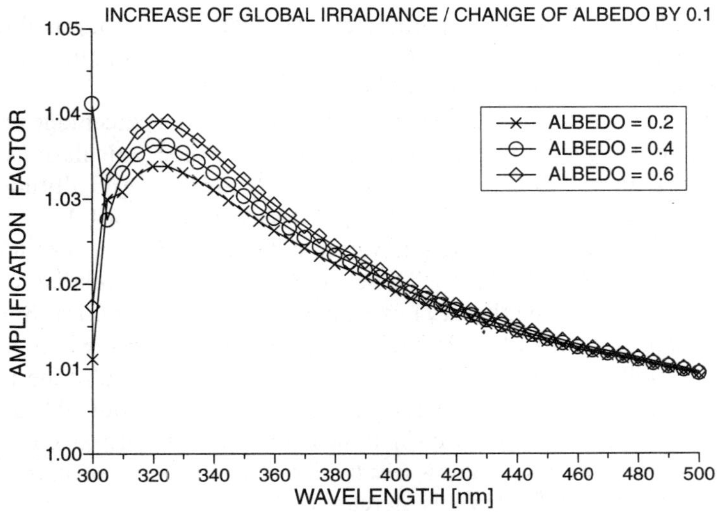

Figure 9. Increase of spectral global irradiance as a consequence of an increase of albedo by 0.1 for three different levels of albedo.

Monte Carlo method to calculate irradiance for any inhomogeneous surface with variable albedo. This can be applied to specific terrains to show the local variability [35]. For general application in the standard 1D radiative transfer models, an effective albedo is assumed. This is the value for a homogeneous surface that has the same effect on modeled irradiance as the real, inhomogeneous distribution. In Alpine areas with snow coverage, values of effective albedo in the range 0.3 to 0.8 have been derived, depending on local topography and ground vegetation.

3D models were also used to derive the radius of significance around a measurement site, i.e. the area over which the albedo still has an influence on local UVR. It was found that for clear sky conditions this radius can extend to 20–30 km [36], whereas it is significantly smaller in the presence of cloud.

So far the effect of changing albedo was discussed only on global irradiance, that means on horizontal detector surfaces. If the surface is tilted or if actinic detectors (2π or 4π) are used, then significantly higher effects due to high albedo have to be expected.

Altitude: The intensity of solar radiation increases with increasing altitude under cloudless conditions due to the smaller amount of scattering molecules and particles at higher altitudes. That means that direct irradiance is increasing strongly, whereas diffuse radiation is nearly constant or slightly decreasing with altitude. As the separation of global irradiance into direct and diffuse components is dependent on wavelength (with a higher diffuse component at shorter wavelengths), the increase of global irradiance with altitude is wavelength dependent. In addition, because of the wavelength dependence of the scattering coefficients, the increase with altitude is more pronounced at the shorter wavelengths, and the absorption by tropospheric ozone will also result in an enhanced increase of irradiance at shorter wavelengths in the UV-B range with increasing altitude. As so many parameters act together, it is not possible to state only one number for the increase of irradiance with altitude, but it is necessary to give a range of values, depending on the specific conditions.

Usually the term "altitude effect" is used to describe the percentage increase of irradiance for an increase in altitude by 1000 m. Model calculations with an atmosphere without any aerosols and without any ozone give an altitude effect of about 6% in the UV-A and 9% at 310 nm. Measurements of UVR at different altitudes in the Chilenean Andes have indicated altitude effects of 8% in the UV-A and 9% in the UV-B [37]. These values are related to relatively clean air at all altitudes with probably very low tropospheric ozone pollution. Significantly higher values for the altitude effect are found in the Alps, where a marked gradient in pollution from the lower altitude to the higher altitude is typical. Several studies [38,39] have shown values in the UV-A of about 10% and at 310 nm of about 15% to 20% (Figure 10). If the higher altitudes are covered by snow, while the lower altitudes are snow free (which is a typical situation in the Alps), then the altitude effect for UVR is increased by about 10%.

Clouds: The discussion of solar UV variability in preceeding paragraphs has assumed cloud free conditions. However, in most parts of the world this is the exception, usually there are either broken cloud fields or a more or less homogeneous cloud cover. In general, clouds reduce solar radiation, but the amount of reduction is extremely variable due to the variable nature of clouds. For the estimation of the effect of clouds on solar irradiance, the most important par-

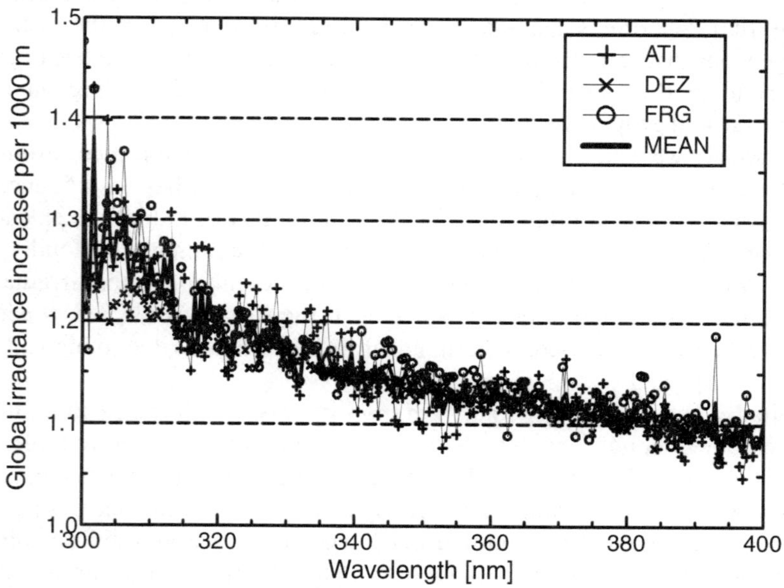

Figure 10. Increase of spectral global irradiance for an increase in altitude by 1000 m, based on measurements of three spectroradiometers (ATI, DEZ, FRG) at different altitudes (1200 m, 1750 m, 2964 m) relative to an instrument at a valley station (Garmisch-Partenkirchen, Germany, 730 m). The solid line is the average of the individual measurements.

ameter is not the amount of clouds, which cover the sky, but it is the coverage of the sun by clouds. Investigations have shown that even in situations when 70% or 80% of the sky are covered by clouds, the global solar irradiance is only marginally reduced, as long as the sun is not covered by clouds. The reduction by complete cloudiness, when the direct sun is not visible, can range up to more than 90%: average values are about 75% at sea level and about 50% at high altitudes, as there the optical thickness of the clouds is usually smaller.

Within clouds, solar radiation is mainly scattered on small water droplets. This scattering process is only slightly dependent on wavelength, therefore the color of clouds appears white or grey or dark, depending on the thickness, as long as the illuminating sun is "white" (when the solar elevation is low and the sun is reddish, then clouds also may appear reddish). However, the effect of clouds on global irradiance at the ground is dependent on wavelength: irradiance at shorter wavelengths is less attenuated than at longer wavelengths. The reason is the different distribution between direct and diffuse irradiance at different wavelengths. If a cloud is blocking the direct sun, then the reduction of visible global irradiance is much higher than that of UVR, where already a great part of global irradiance is diffuse. From radiation measurements together with detailed cloud observations at the High Alpine Research Station Jungfraujoch (3576 m above sea level, Switzerland) it is found that UV-A and erythemally weighted UVR are attenuated about in the same amount, whereas total solar radiation (300 nm to 3000 nm) is attenuated about 40% more [40] (Figure 11). From measurements of spectral irradiance above and below a cloud layer in Garmisch-Partenkirchen (Germany) again a spectral dependence of the effect of clouds was derived, where the transmission of the cloud was in the UV-B 57% and in the UV-A 45% [41].

The multiple scattering of photons in the cloud and between the cloud and the ground enlarges the average photon path length. Therefore any absorption by tropospheric ozone or aerosols is amplified and thus the spectral shape of the spectrum is significantly modified by clouds.

Global distribution: When comparing measurements of UVR at the Northern and Southern hemisphere, significantly higher values are found on the Southern hemisphere at the same solar elevations in the respective summer time. There are several reasons for this: firstly, the Earth–sun distance has its minimum in the Southern hemisphere summertime. Secondly, the average aerosol load in the planetary boundary layer and in the free troposphere is significantly lower in the Southern hemisphere. Thirdly, tropospheric ozone levels are lower for the Southern hemisphere. All three facts act together in increasing the level of UV-B irradiance on the Southern hemisphere in relation to the Northern. In a measurement campaign in 1990/1991 [42] it was shown that erythema doses in summertime at mid-latitudes (about 48°) in New Zealand were about double those in Germany in the corresponding summer. At that time, systematically lower total ozone content was also observed in the Southern summer relative to Northern summer.

Worldwide maxima of measured irradiances for erythemally weighted UV are found in the high altitude region of Northern Argentina, where altitudes around

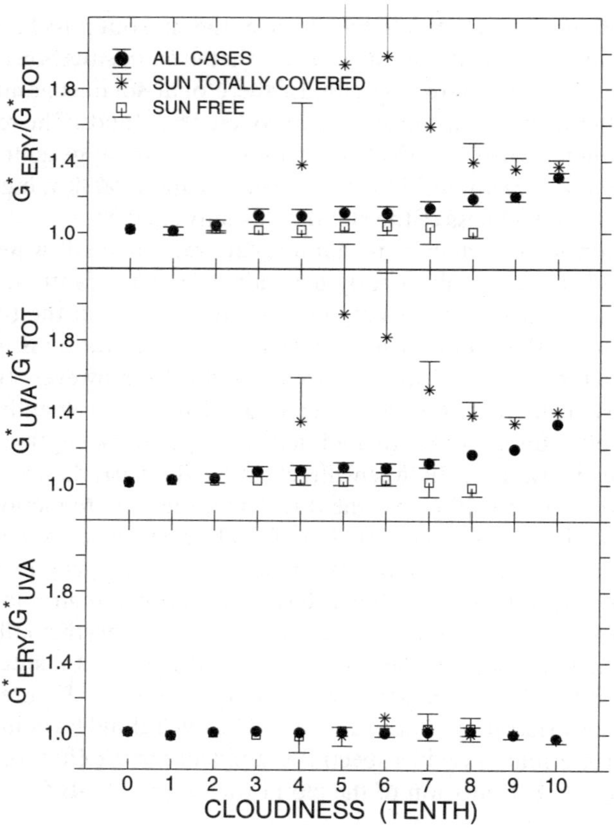

Figure 11. Ratio of global radiation fluxes (normalized to cloudless conditions) for erythemal radiation, UV-A radiation and total solar radiation in dependence on cloudiness, separated for cases when the sun is free or when the sun is totally covered by clouds. Bars indicate ± 3 standard deviation of the mean.

3500 m are regularly populated. In summertime, when the sun is close to the zenith, a maximal UV Index of 20 was measured at 22° South and 3500 m above sea level [43], at a time when the total ozone amount had relatively low values (236 DU, the annual average at this site is about 260 DU).

2.3.2 Space-born measurements

Since 1978 spectral measurements of sunlight backscattered from the Earth to the space have been made from satellites, starting with NASA's Nimbus-7 satellite with the TOMS (Total ozone mapping spectrometer) instrument. This spectrometer was designed to measure backscattered UVR at six wavelengths in the UV-B and UV-A and to derive total column ozone amount from these radiance data. This application is well established and the uncertainty is well determined [44]. The combination of the space-born measurements of backscattered sunlight at several wavelengths in the UV range with a radiative transfer model also allows an estimate of the UV spectrum at the Earth's surface [45]. UV estimates based on measurements of TOMS instruments on board of various satellites are available now for November 1978 to May 1993, August 1991 to December 1994 and August 1996 to present. Surface UVR has also been estimated from the Earth Radiation Satellite (ERS-2) with the GOME (Global Ozone Monitoring Experiment) instrument [46].

The great advantage of estimating surface UVR from space-born measurements is that results for the whole globe can be derived. Thus a global daily image of the geographic distribution of UV irradiance in the UV-B and UV-A range is calculated. However, from the principle of this method it is clear that not all relevant parameters for the radiative transfer calculations can be measured by the instrument on the satellite. When ozone is known from the well established ozone retrieval, then the remaining main parameters are ground albedo, aerosols and clouds. From the long-term satellite measurements, a global climatology for albedo has been derived, based on the minimum values of reflectivity observed, which is assumed to be the real ground reflectivity without any clouds. This is done for a wavelength in the UV-A range to avoid any interaction with the absorption by ozone. The only difficulty is the separation between cloud-free snow-covered terrain and snow-free terrain with clouds, which is the source of an additional uncertainty. The determination of tropospheric aerosols is possible for absorbing aerosols above about 1.5 km [47], whereas the estimation of the effect of aerosols in the boundary layer remains uncertain. Also, the interaction of changing ground albedo and changing amounts of aerosols close to the ground remains a problem.

The great natural variability of cloudiness in space and in time requires further consideration. As the satellite has an overpass of a given place at regular intervals, ranging from hours to days, the information about cloudiness is only a snapshot. Combining the results of several satellites may improve the temporal resolution, but temporal averaging will still be necessary. Therefore, the most valuable results of space-born retrievals are monthly averages of solar irradiance. Only in specific regions, where cloudiness is very low over time scales of days, can instantaneous values be derived with reasonable confidence. The regional distribution of cloudiness results in a similar problem as the temporal variation. The minimum spatial resolution of data derived from satellite measurements is given by the pixel size of the instrument on the satellite. This ranges from about 100×100 km^2 for the TOMS instrument down to about 1×1 km^2 for the cloud information from the AVHRR (Advanced Very High Resolution Radiometer) instrument of NOAA (National Oceanic and Atmospheric Administration). Clouds smaller than the pixel size of the instrument cannot be detected.

Comparisons between UV estimates derived from satellite data and ground-based measurements at a few stations in Europe have shown that for daily doses there is only a bias between data sets of about 5%, but the standard deviation is in the order of about 30% [48]. For monthly doses, the scatter becomes significantly smaller, but differences of up to 30% are observed, depending on the site of the comparison [49,50]. The differences seem to correlate with tropospheric extinction and with seasonal changes in regional snow cover.

2.4 Trends in solar UVR

The impact of the observed ozone decline on UV-B radiation at the Earth's surface is one of the key factors which motivated long-term measurements of UVR worldwide, in order to investigate if there is an associated increase in solar UV-B. However, cloudiness has a stronger modulating impact on UVR reaching the Earth's surface than ozone. Therefore statistical analyses have shown that several decades of continuous measurements with broadband detectors under all conditions of cloudiness are necessary to identify the actual changes due to ozone [51].

2.4.1 Long-term ozone changes

Based on measurements of the total column ozone content of the atmosphere from the ground as well as from satellites, a consistent picture of the current loss of stratospheric ozone can be derived. The most recent results are discussed in ref. [3]. Relative to the values in the 1970's, the ozone loss at the end of the 1990's is estimated to be about 50% in the Antarctic spring, where the "ozone hole" appears every year, and about 15% in the Arctic spring. In the mid-latitudes of the Southern hemisphere the loss is about 5% all the year round, while in the Northern hemisphere it is about 6% in winter/spring and about 3% in summer/fall. No significant trend in ozone has been found in the Equatorial regions. In the second half of the 1990's relatively little change in ozone has been observed in the mid-latitudes of both hemispheres.

2.4.2 Long-term UVR changes

Very few long-term UV measurement records are available, and most of these were obtained with Robertson–Berger type broadband detectors (see Section 2.3.1.1). The longest continuous measurements of solar UVR extend back to the middle of the 1970's [52]. In some cases the monitoring efforts were interrupted in the 1980's, i.e. with the network of 14 stations in the USA, operational from 1974 to 1985. A first analysis of these measurements had suggested a decrease in erythemally weighted UVR [53], recalculations have been introduced to consider specific calibration practices [54], resulting in no significant trend within these data. Annual measurement campaigns at Northern mid-latitudes (Switzerland) at a high mountain site largely unaffected by air pollution have shown a slight but significant increase of erythemally weighted UVR for cloud free days between 1981 and 1991. The change of $7 \pm 3\%$ per decade (1 standard deviation) is in agreement (within the limits of uncertainty) with the observed ozone changes in this area [55].

As absorption by ozone is strongly dependent on wavelength in the UV-B-range, but the effects of clouds, aerosols and albedo have only weak spectral effects, spectral measurements have the advantage of allowing the various attenuation effects to be separated, and thus allow consequences of ozone changes to be more clearly identified [10]. However, records of spectral UV measurements are relatively short, the longest extending back to 1990, using Brewer single monochromator spectroradiometers. In general, deducing long-term changes from relatively short time series can be very dependent on the time interval selected.

In Reading, UK, the trend in erythemal UVR, after a first order correction for clouds and aerosols, was +4.3% from 1993–1997, compared to an ozone change of −5.9% in the same period [56]. An updated analysis of these time series to the end of year 2000 shows no change in erythemal UV from 1993–2000 and a small increase in ozone, though the results are statistically insignificant. This illustrates the effect of changing time periods of analysis in a short data record. The past few years have seen near normal levels of ozone over Southern England, while in the early years of the record (1993–1996) the effects of Mount Pinatubo were still apparent and then there were several instances of very low ozone over Northern Europe. Measurements from other mid-latitude sites show similar results for similar periods of measurement. Spectral measurements in Thessaloniki, Greece, since 1990, indicate an increase of irradiance at 325 nm as well as at 305 nm on clear sky days. Assuming that the increase at 325 nm is a consequence of changing air pollution, then the irradiance at 305 nm can be corrected for this effect. The remaining increase at 305 nm is about 10% per decade, corresponding to an average decrease of total ozone by about 4.5% per decade at the same place [57]. In Toronto, Canada, integrated daily doses in summer time between 1989 and 1997 show an increase of about 8–10% per decade at 305 nm and no increase at 324 nm, which again is in good agreement with the ozone decline of about 4.3% per decade during that time in Toronto [3]. Spectral measurements at high Northern latitudes (Sodankylä, Finland, 67° North) show for the time period from 1990 to 1998 especially for springtime a significant increase in irradiance at 305 nm of about 50% per decade, whereas the change at 325 nm is about 10% per decade. In summer and autumn, the increase at 305 nm is about 20% per decade and no significant change is observed at 325 nm [58].

Besides ground-based measurements of UVR, space-born estimates of UV levels can be used in producing estimates of long-term UV variations. An analysis of zonally averaged global UV data, derived using TOMS data from 1979 to 1992, shows an increase of erythemally weighted UV doses at 55° South by about $5.5 \pm 3.5\%$ per decade (2 standard deviations) and at 55° North by about 4.5% per decade. In Equatorial regions between $\pm 35°$, no significant trend on the 95% level is found [59]. The largest increase of UVR has occurred in spring time at high latitudes, corresponding to the observed decrease in stratospheric ozone.

Reconstruction of previous levels of UVR and determination of long-term variations have also been modeled using long-term ozone data from ground-based measurements together with records of pyranometer data (total solar irradiance between 300 nm and 3000 nm) and cloud information from satellites or from synoptic data. A comparison of the monthly mean trends derived with this method, with broadband measurements in Northern Europe and with TOMS satellite data has shown good agreement [60], with trends of erythemally weighted UVR of 5–10% per decade in the last two decades. Similar agreement was found for other stations in Europe [61].

2.4.3 Future levels of UVR

2.4.3.1 Forecasting UVR

During the past few years, many countries have included a prediction of the UVR for the next one or two days in operational weather forecasting. The maximum value of erythemally weighted UVR, expressed as the UV index, is then broadcast to the public. The radiative transfer calculations are based on a prediction of total ozone amount and use parameters specific to the area of interest. This works relatively well as long as cloud-free situations are considered, showing the expected uncertainties due to uncertainties of ozone, aerosols and ground albedo [62]. If clouds are part of the expected weather then the success of the cloud forecast dominates the success of the UV forecast. In a comparison of measured and forecasted UV indices at several stations in Europe under all weather conditions, the agreement was better than ± 1 UV index value in about 60% of the cases [63], with very high deviations if cloudiness was wrongly forecasted.

2.4.3.2 Future UV scenarios

Predictions about UVR levels in future years or decades are of interest too. However, for realistic estimates using radiative transfer calculations it is necessary to know the future levels of cloudiness, ozone, aerosols and albedo. Chemical transport models (CTM) allow predictions of future ozone levels based on stratospheric chemical processes and on emission scenarios for the relevant halogen gases. They suggest that the maximum globally averaged ozone depletion would take place around year 2000 and a recovery of stratospheric ozone would occur around 2050 at the earliest, if the reduction of the halogen emissions strictly follows the Montreal protocol and all its amendments [3]. However, there is a strong feedback between ozone decline and stratospheric temperature, which in turn has a strong effect on stratospheric chemistry. Furthermore, general circulation models (GCM) suggest that the increasing atmospheric carbon dioxide content will lead to a cooling of the winter stratosphere [64]. Therefore it is necessary to study the whole climate-chemistry system with coupled GCM-CTMs in order to derive future ozone estimates. As a first approach to calculation of future UV levels, climatological means of cloudiness, aerosols and albedo can be used [65], which show peak values of UV irradiance occurring significantly later than in pure CTM calculations: they may take place around 2010–2020. Additional studies are needed to take into account also future changes in cloudiness and albedo as a consequence of climate change.

References

1. J.C. Farman, B.G. Gardine, J.D. Shanklin (1985). Large losses of total ozone in Antarctica reveal seasonal ClOxNOx interaction. *Nature*, **315**, 207–210.
2. J. Staehelin, A. Renaud, J. Bader, R. McPeters, P. Viatte, B. Hoeffer, V. Bugnion, M. Giroud, H. Schill (1998). Total ozone series at Arosa (Switzerland): Homogenization and data compression. *J. Geophys. Res.*, **103**, 5827–5841.
3. Scientific Assessment of Ozone Depletion: 1998, Global Ozone Research and Monitoring Project. World Meteorological Organization Report Nr. 44, Geneva (1999).
4. M. Blumthaler, W. Ambach (1988). Solar UVB–Albedo of various surfaces. *Photochem. Photobiol.*, **48**, 85–88.
5. A.R. Webb, I.M. Stromberg, H. Li, L.M. Bartlett (2000). Airborn spectral measurements of surface reflectivity at ultraviolet and visible wavelengths. *J. Geophys. Res.*, **105,** 4945–4948.
6. R. Mckenzie, M. Kotkamp, W. Ireland (1996). Upwelling UV Spectral Irradiances and Surface Albedo Measurements at Lauder, New-Zealand. *Geophys. Res. Lett.*, **23**, 1757–1760.
7. U. Feister, R. Grewe (1995). Spectral albedo measurements in the UV and visible region over different types of surfaces. *Photochem. Photobiol.*, **62**, 736–744.
8. L. Hill (1927). Measurement of the biologically active ultraviolet rays of sunlight. *Roy. Soc. Proc.* A, **114**, 268–276.
9. P. Bener. Approximate values of intensity of natural ultraviolet radiation for different amounts of atmospheric ozone, Final Technical Report, Contract DAJA 37-68-C-1017, Davos, Switzerland, 1972.
10. G. Bernhard, G. Seckmeyer (1999). The uncertainty of measurements of spectral solar UV irradiance. *J. Geophys. Res.*, **104**, 14321–14345.
11. A. Bais, B. Gardiner, H. Slaper, M. Blumthaler, G. Bernhard, R. McKenzie, A. Webb, G. Seckmeyer, B. Kjeldstad, T. Koskela, P. Kirsch, J. Grobner, J. Kerr, S. Kazadzis, K. Leszczynski, D. Wardle, W. Josefsson, C. Brogniez, D. Gillotay, H. Reinen, P. Weihs, T. Svenoe, P. Eriksen, F. Kuik, A. Redondas (2001). SUSPEN intercomparison of ultraviolet spectroradiometers. *J. Geophys. Res.*, **106**, 12509–12526.
12. D. Berger (1976). The sunburning ultraviolet meter: design and performance. *Photochem. Photobiol.*, **24**, 587–593.
13. M. Morys, D. Berger (1993). The accurate measurements of biologically effective ultraviolet radiation. *Proc. SPIE*, **2049**, 152–161.
14. A.F. McKinlay, B.L. Diffey (1987). A reference action spectrum for ultraviolet induced erythema in human skin. *CIE J.*, **6**, 17–22.
15. K. Leszczynski, K. Jokela, L. Ylianttila, R. Visuri, M. Blumthaler (1998). Erythemally weighted radiometers in solar UV monitoring: Results from the WMO/STUK intercomparison. *Photochem. Photobiol.*, **67**, 212–221.
16. A.F. Bais, C. Topaloglou, B., S. Kazadzis, M. Blumthaler, J. Schreder, A. Schmalwieser, D. Henriques, M. Janouch (1999). Report of the LAP/COST/WMO intercomparison of erythemal radiometers. *WMO/GAW report No. 141*, World Meteorological Organization, Geneva.

17. K. Lantz, P. Disterhoft, J. DeLuisi, D. Bigelow, J. Slusser (1999). Methodology for deriving clear-sky erythemal calibration factors for UV broadband radiometers of the U.S. Central UV Calibration Facility. *J. Atmos. Ocean. Tech.*, **16**, 1736–1752.
18. A. Cede, E. Luccini, L. Nunez, R. Piacentini, M. Blumthaler (2002). Calibration and uncertainty estimation of erythemal radiometers in the Argentine ultraviolet network. *Appl. Opt.*, in press.
19. A. Dahlback (1996). Measurements of biologically effective UV doses, total ozone abundances, and cloud effects with multichannel, moderate bandwidth filter instruments. *Appl. Opt.*, **35**, 6514–6521.
20. A. Davis, G.H.W. Deane, B.L. Diffey (1976). Possible dosimeter for ultraviolet radiation. *Nature*, **261**, 169–170.
21. G. Horneck (1995). Quantification of biological effectiveness of environmental UV-radiation. *J. Photochem. Photobiol. B: Biol.*, **31**, 43–49.
22. G. Ronto, S. Gaspar, P. Grof, A. Berces, Z. Gugolya (1994). Ultraviolet dosimetry in outdoor measurements based on bacteriophage T7 as a biosensor. *Photochem. Photobiol.*, **59**, 209–214.
23. N. Munakata, S. Kazadzis, A. Bais, K. Hieda, G. Rontó, P. Rettberg, G. Horneck (2000). Comparisons of Spore Dosimetry and Spectral Photometry of Solar-UV Radiation at Four Sites in Japan and Europe. *Photochem. Photobiol.*, **72**, 739–745.
24. International Commission on Non-Ionizing Radiation Protection (ICNIRP) (1995). Global Solar UV Index, WHO, WMO, UNEP, ICNIRP-1/95 Oberschleißheim.
25. S. Madronich, R.L. McKenzie, L.O. Björn, M.M. Caldwell (1998). Changes in biologically active ultraviolet radiation reaching the Earth's surface. *J. Photochem. Photobiol. B: Biol.*, **46**, 5–19.
26. F.R. de Gruijl, J.C. van der Leun (1994). Estimate of the wavelength dependency of ultraviolet carcinogenesis and its relevance to the risk assessment of a stratospheric ozone depletion. *Health Phys.*, **4**, 317–323.
27. R.B. Setlow (1974). The wavelengths in sunlight effective in producing skin cancer: a theoretical analysis. *Proc. Natl. Acad. Sci. USA*, **71**, 6666–6670.
28. D.G. Pitts, A.P. Cullen, P.D. Hacker (1977). Ocular effects of ultraviolet radiation from 295 to 365 nm. *Invest. Ophthalmol.*, **16**, 932–939.
29. D.P. Häder, R.C. Worrest (1991). Effects of enhanced solar ultraviolet radiation on aquatic ecosystems. *Photochem. Photobiol.*, **53**, 717–725.
30. W.P. de More, S.P. Sander, D.M. Golden, R.F. Hamspon, M.J. Kurylo, C.J. Howard, A.R. Ravishankara, C.E. Kolb, M.J. Molina (1997). *Chemical Kinetics and Photochemical Data for use in Stratospheric Modeling* (Evaluation number 12, JPL Publication 97–4,) PL, Pasadena, CA.
31. O. Dubovik, M.D. King (2000). A flexible inversion algorithm for retrieval of aerosol optical properties from sun and sky radiance measurements. *J. Geophys. Res.*, **105**, 20673–20696.
32. T. Nakajima, G. Tonna, R. Rao, P. Boi, Y. Kaufman, B. Holben (1996). Use of sky brightness measurements from ground for remote sensing of particulate polydispersions. *Appl. Opt.*, **35**, 2672–2686.
33. A. Kylling, A.F. Bais, M. Blumthaler, J. Schreder, C.S. Zerefos (1998). The effect of aerosols on solar UV irradiances during the PAUR campaign. *J. Geophys. Res.*, **103**, 26051–26060.
34. M. Blumthaler, W. Ambach (1994). Changes in solar radiation fluxes after the Pinatubo eruption. *Tellus*, **46B**, 76–78.
35. A. Kylling, T. Persen, B. Mayer, T. Svenoe (2000). Determination of an Effective

Spectral Surface Albedo From Ground Based Global and Direct UV Irradiance Measurements. *J. Geophys. Res.*, **105**, 4949–4959.

36. M. Degunther, R. Meerkötter, A. Albold, G. Seckmeyer (1998). Case study on the influence of in-homogeneous surface albedo on UV irradiance. *Geophys. Res. Lett.*, **25**, 3587–3590.
37. H. Piazena (1993). Vertical distribution of solar UV irradiation in the tropical Chilenian Andes. *Proc. of the First European Symposium on The effects of Environmental UV-B Radiation on Health and Ecosystems*, Munich, Germany ISBN 92-826-4121-X, 59–62.
38. M. Blumthaler, A.R. Webb, G. Seckmeyer, A.F. Bais, M. Huber, B.Mayer (1994). Simultaneous Spectroradiometry: A Study of Solar UV Irradiance at Two Altitudes. *Geophys. Res. Lett.*, **21**, 2805–2808.
39. J. Gröbner, A. Albold, M. Blumthaler, T. Cabot, A. De la Casinieri, J. Lenoble, T. Martin, D. Masserot, M. Müller, R. Philipona, T. Pichler, E. Pougatch, G. Rengarajan, D. Schmucki, G. Seckmeyer, C. Sergent, M. L. Touré, P. Weihs (2000). Variability of spectral solar ultraviolet irradiance in an Alpine environment. *J. Geophys. Res.*, **105**, 26991–27003.
40. M. Blumthaler, W. Ambach, M. Salzgeber (1994). Effects of cloudiness on global and diffuse UV irradiance in a high-mountain area. *Theor. Appl. Climatol.*, **50**, 23–30.
41. G. Seckmeyer, R. Erb, A. Albold (1996). Transmittance of a cloud is wavelength-dependent in the UV-range. *Geophys. Res., Lett.*, **23**, 2753–2755.
42. G. Seckmeyer, R.L. McKenzie (1992). Increased ultraviolet radiation in New Zealand (45°S) relative to Germany (48°N). *Nature*, **359**, 135–137.
43. A. Cede, E. Luccini, L. Nunez, R. Piacentini, M. Blumthaler (2002). Monitoring of erythemal irradiance in the Argentine ultraviolet network. *J. Geophys. Res.*, in press.
44. R.D. McPeters, G. Labow (1996). An assessment of the accuracy af 14.5 years of Nimbus 7 TOMS version 7 ozone data by comparison with the Dobson Network. *Geophys. Res. Lett.*, **23**, 3695–3698.
45. J. Herman, N. Krotkov, E. Celarier, D. Larko, G. Labow (1999). Distribution of UV radiation at the Earth's surface using TOMS-measured UV-backscattered radiances. *J. Geophys. Res.*, **104**, 12059–12076.
46. R. Meerkoetter, B. Wissinger, G. Seckmeyer (1997). Surface UV from ERS-2/GOME and NOAA/AVHR data: A case study. *Geophys. Res. Lett.*, **24**, 1939–1942.
47. N.A. Krotkov, P.K. Bhartia, J.R. Herman, V. Fioletov, J. Kerr (1998). Satellite estimation of spectral surface UV irradiance in the presence of tropospheric aerosols: 1, cloud-free case. *J. Geophys. Res.*, **103**, 8779–8793.
48. T. Martin, B.G. Gardiner, G. Seckmeyer (2001). Uncertainties in satellite-derived estimates of surface UV doses. *J. Geophys. Res.*, **15**, 27005–27011.
49. A. Arola, S. Kalliskota, P.N den Outer, K. Edvardsen, G. Hansen, T. Koskela, T. J. Martin, J. Matthijsen, R. Meerkoetter, P. Peeters, G. Seckmeyer, P. Simon, H. Slaper, P. Taalas, J. Verdebout (2001). Four UV mapping procedures using satellite data and their validation against ground-based UV measurements. *J. Geophys. Res.*, submitted.
50. R. McKenzie, G. Seckmeyer, A.F. Bais, J.B. Kerr, S. Madronich (2001). Satellite-retrievals of erythemal UV dose compared with ground-based measurements at Northern and Southern mid-latitudes. *J. Geophys. Res.*, **106**, 24051–24062.
51. E.C. Weatherhead, G.C. Reinsel, G.C. Taio, X. Meng, D. Choi, W. Cheang, T. Keller, J. DeLuisi, D.J. Wuebbles, J.B. Kerr, A.J. Miller, S.J. Oltmans, J.E. Frederick (1997). Factors affecting the detection of trends: statistical considerations and applications to environmental data, *J. Geophys. Res.*, **103**, 17149–17161.

52. J. Borkowski (2000). Homogenization of the Belsk UV-B series (1976–1997) and trend analysis. *J. Geophys. Res.*, **105**, 4873–4878.
53. J. Scotto, G. Cotton, F. Urbach, D. Berger, T. Fears (1988). Biologically effective ultraviolet radiation-surface measurements in the United States 1974–1985. *Science*, **239**, 762–764.
54. E.C. Weatherhead, G.C. Taio, G.C. Reinsel, J.E. Frederick, J.J. DeLuisi, D. Choi, W.-K. Tam (1997). Analysis of long-term behaviour of ultraviolet radiation measured by Robertson–Berger meters at 14 sites in the United States. *J. Geophys. Res.*, **102**, 8737–8754.
55. M. Blumthaler, W. Ambach (1990). Indication of increasing solar ultraviolet-B radiation flux in Alpine regions. *Science*, **248**, 206–208.
56. L.M. Bartlett, A. Webb (2000). Changes in ultraviolet radiation in the 1990s: spectral measurements from Reading, England. *J. Geophys. Res.*, **105**, 4889–4893.
57. C.S. Zerefos, C. Meleti, D.S. Balis, K. Tourpali, A.F. Bais (1998). Quasi-Biennial and longer-term changes in clear-sky UV-B solar irradiance. *Geophys. Res. Let.*, **25**, 4345–4348.
58. K. Masson, E. Kyrö (2001). *A Decade of Spectral UV-B Measurements at Sodankylä, Finland*, (EGS Newsletter 203). European Geophysical Society XXXVI General Assembly 2001.
59. J. Herman, P.K. Bhartia, J. Ziemke, Z. Ahmad, D. Larko (1996). UV-B radiation increases (1979–1992) from decreases in total ozone. *Geophys. Res. Lett.* **23**, 2117–2120.
60. J. Kaurola, P. Taalas, T. Koskela, J. Borkowski, W. Josefsson (2000). Long-term variations of UV-B doses at three stations in northern Europe. *J. Geophys. Res.* **105**, 20813–20820.
61. P. den Outer, H. Slaper, J. Matthijsen, H.A. Reinen, R. Tax (2000). Variability of ground level UV: model and measurement. *Radiation Protection Dosim.* **91**, 105–110.
62. H. De Backer, P. Koepke, A. Bais, X. de Cabo,T. Frei, D. Gillotay, C. Haite, A. Heikkilä, A. Kazantzidis, T. Koskela, E. Kyrö, B. Lapeta, J. Lorente, K. Masson, B. Mayer, H. Plets, A. Redondas, A. Renaud, G. Schauberger, A. Schmalwieser, H. Schwander, K. Vanicek (2001). Comparison of measured and modeled UV indices. *Met. Appl.*, **8**, 267–277.
63. P. Koepke, H. De Backer, P. Ericsen, U. Feister, D. Grifoni, T. Koskela, A. Lehmann, Z. Litynska, A. Oppenrieder, A. Schmalwieser, H. Staiger, K. Vanicek (2001). An overview of the results from the comparison of UV-Index forecasted and measured at all atmospheric conditions including clouds. *International Radiation Symposium*, St. Petersburg, Russia, in press.
64. J. Austin, N. Butchart, K.P. Shine (1992). Possibility of an Arctic ozone hole in a doubled-CO2 climate. *Nature* **360**, 221–225.
65. P. Taalas, J. Kaurola, A. Kylling, D. Shindell, R. Sausen, M. Dameris, V. Grewe, J. Herman, J. Damski, B. Steil (2000). The impact of greenhouse gases and halogenated species on future solar UV radiation doses. *Geophys. Res. Lett.* **27**, 1127–1130.

Chapter 3

Water column optics and penetration of UVR

Bruce R. Hargreaves

Table of contents

Abstract

Penetration of ultraviolet radiation (UVR) into natural waters depends on the concentration and optical qualities of dissolved organic matter (DOM), phytoplankton, other suspended particles, and on the optical properties of pure water. Optical classification schemes developed for lakes or ocean regions, and relevant to UVR penetration, indicate which of these factors contribute to underwater attenuation of solar radiation. In most cases the best predictor of UV attenuation is optical absorption by chromophoric DOM (CDOM). In natural waters in which UVR penetration is greatest (Crater Lake, USA, and Lake Vanda, Antarctica), CDOM and phytoplankton are so scarce near the surface that UV attenuation by water molecules is significant. Recent evidence from these lakes supports a downward revision of older estimates for UV attenuation by pure water. Variations in atmospheric scattering, sun angle, and depth can cause diffuse attenuation measurements to deviate from the Beer–Lambert Law, but a simple correction is available. Bio-optical models to predict the penetration of UVR into natural waters from measurements will require better understanding of specific absorption of CDOM and phytoplankton, of the reactivity of CDOM, and of the linkage between the microbial community and autochthonous production of CDOM. Spatial and temporal patterns relating UV attenuation with dissolved organic carbon (DOC) concentration and optical quality in aquatic ecosystems appear to be driven by the rates of DOC flux and photobleaching, and by hydrologic properties (residence time and evaporation rate).

3.1 Introduction and background

The penetration of UVR into natural waters leads to exposure of organisms and nonliving matter to energetic photons. The response of organisms and nonliving matter to UVR exposure may change the UV transparency of the aquatic environment. When the intensity (irradiance) of UVR underwater is sufficient to cause biotic damage then a stressful range of depths exists in the water column. Ecological consequences of this stress will depend on the depth of penetration and spectral shifts in the underwater irradiance relative to the depth of the mixed layer ([1], Chapter 4). Spectral shifts may influence vision and behavior of aquatic organisms, especially those with UVR receptors (Chapter 14). They may also influence survival and productivity of aquatic organisms because of the interplay between damaging UV-B wavelengths and the sometimes-beneficial UV-A wavelengths ([2], Chapters 11–13). Ecologists can benefit from improved understanding of patterns and determinants of the intensity and spectral quality of underwater UVR, especially in view of future changes in global climate and in stratospheric ozone. While column ozone and clouds have a strong influence on damaging UVR that reaches the Earth's surface, other factors, especially the concentration and optical qualities of DOM, are more important to the penetration of UVR in the water column.

The topic of underwater optics has been extensively treated with a marine emphasis by Preisendörfer [3], Jerlov [4], and Mobley [5], and with added coverage of freshwater systems and photosynthesis by Kirk [6]. Reviews of solar radiation penetration into natural waters began with Smith and Tyler [7], and with emphasis on UVR and varying emphasis on marine and freshwater environments include Smith and Baker [8], Baker and Smith [9], Kirk [10], Booth and Morrow [11], and Whitehead et al. [12]. Xenopoulos and Schindler [13] recently reviewed underwater UVR in the context of terrestrial ecosystems and climate change. While marine and freshwater systems are often treated separately in the optical literature, here their optical properties will be compared to reveal common features as well as information gaps and different approaches used by oceanographers and limnologists.

Solar radiation is typically measured underwater as irradiance, the energy striking a unit of surface area (e.g., W m^{-2}), and is further characterized by its wavelength (units, nanometers, nm). Spectral irradiance is reported as energy integrated over a waveband, which may be narrow (e.g., 1 nm) or broad (e.g., UV-A, 320–400 nm; UV-B, 280–320 nm). The solar UVR spectrum, about 10% of the incoming energy reaching the Earth's surface, includes UV-A and UV-B wavebands. Photosynthetic organisms use wavelengths starting at about 400 nm and extending to 700 nm (Photosynthetically Active Radiation or PAR) in the process of photosynthesis. Roughly half of the incoming solar energy is represented by infrared wavelengths from 700 to 2000 nm. Ozone in the atmosphere and DOM in natural waters strongly absorb UV-B wavelengths. Water molecules in the atmosphere and in aquatic systems strongly absorb far-red and infrared wavelengths.

UVR will pass through the air–water interface if it is not reflected. Reflection depends on the angle of incidence and follows Fresnel's Law [6]. Penetration of irradiance coming directly from the sun depends on the solar zenith angle (SZA) for a flat (calm) surface and decreases as SZA rises (i.e. for decreasing sun elevation). Windy conditions can increase the penetration of direct solar irradiance for high SZA but may have the opposite effect at low SZA. Penetration of the diffuse component of sunlight has been modeled [14,15]. The proportions of direct and diffuse solar radiation vary with wavelength, atmospheric conditions, and sun angle (some examples are shown in Figure 4A below). For UV-B wavelengths the distribution is often similar to that for all wavelengths under an overcast sky or when the sun is near the horizon, when models predict 96% transmission, [14]. The effects of surface waves and variations in direct versus diffuse radiation combine to complicate the interpretation of optical field measurements at shallow depths unless temporal and spatial averaging are adopted at appropriate scales [16].

Underwater light is similar in many coastal and freshwater environments. Figure 1 shows an example of underwater UVR and PAR spectra at several depths in a typical coastal ocean site (San Diego, California; unpublished data provided by J.H. Morrow). Water absorbs strongly in the red and longer wavelengths (>600 nm) with the result that the right side of the curves shows rapid attenuation with depth. Photosynthetic pigments in phytoplankton (such as chlorophyll *a*) absorb blue (450 nm) wavelengths most strongly but phyto-

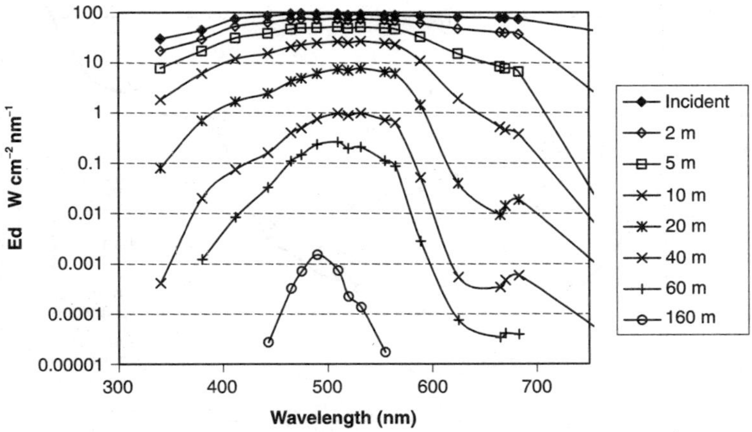

Figure 1. Typical coastal ocean underwater spectra of downwelling cosine irradiance with moderate levels of CDOM and algae. San Diego coastal waters, 5 miles offshore (5 January 2000), using Biospherical Instruments PRR-800 multichannel reflectance profiler (J.H. Morrow, unpublished data). This spectrum is plotted on a log scale to show similar percentage changes over the range of intensity that spans seven decades of magnitude. Except for the peak around 685 nm (caused by algal fluorescence), deep irradiance beyond 520 nm is likely to be caused by Raman scattering from shorter wavelengths.

plankton also absorb at shorter wavelengths. For the component of DOM called CDOM, absorption increases exponentially from the mid-visible wavelengths into the UVR range. The red peak (centered at 685 nm) that is evident at greater depths (most noticeable in the 20 m curve of Figure 1) is fluorescence emitted by photosynthetic cells – a small fraction of the visible light they absorb. The characteristic spectrum of underwater light (note the blue–green peak at 500 nm in the 160 m curve of Figure 1) is caused by combining the strong red absorption of water, the blue absorption by photosynthetic cells, and the violet and UVR absorption by DOM. The color of light emerging from deep natural waters is a product of this selective absorption of the medium and backscattering in the upward direction that increases at shorter wavelengths.

UV transparency of natural waters can be described empirically by two measures that are wavelength-specific and inter-related: the *downwelling diffuse attenuation coefficient*, K_d, and the *percent attenuation depth*, $Z_{n\%}$. A downwelling diffuse attenuation coefficient is nominally proportional to the concentration of substances in the water that absorb or scatter UVR [17,42]. It is typically calculated for specific wavelengths (λ) from measurements of downwelling irradiance ($E_{d,\lambda}$) by fitting the following equation (in units of m^{-1}) [8] to irradiance versus depth data:

$$E_{d(Z,\lambda)} = E_{d(0-,\lambda)}\, e^{(-K_{d,\lambda} * Z)} \quad (1)$$

where Z is geometric depth measured in vertical metres from the mean surface, $E_{d,0-}$ represents downwelling irradiance just below the water surface, and $E_{d(Z,\lambda)}$ is the downwelling irradiance at depth Z (m) and wavelength λ (nm). Figure 2

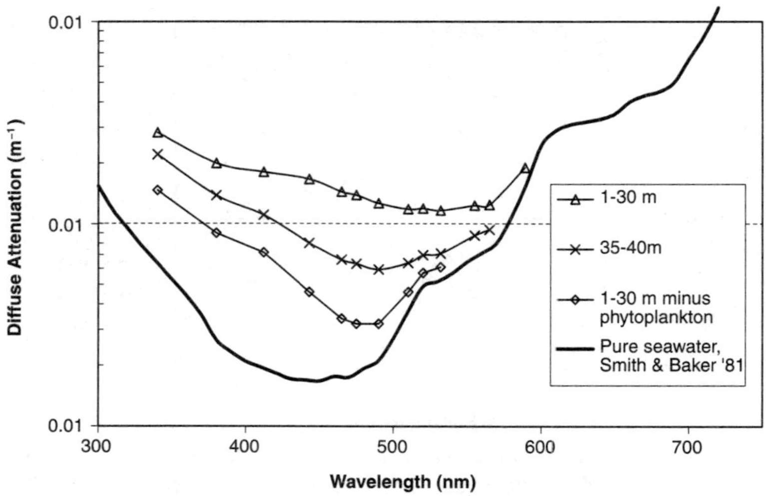

Figure 2. Spectral diffuse attenuation of downwelling irradiance from Figure 1 compared with K_w for pure seawater estimated by Smith and Baker [18]. The phytoplankton concentration (based on chlorophyll *a* fluorescence) was highest in the upper 30 m. The curve labeled "K_d 1–30 m minus phytoplankton" was calculated by regression of spectral K_d against chlorophyll fluorescence for a range of depths, a method that also removes effects of scattering and absorption (including that of CDOM) that covary with phytoplankton fluorescence.

shows diffuse attenuation spectra computed from irradiance data shown in the previous figure. In this example the upper mixed layer of water shows the highest attenuation because of the higher concentration of phytoplankton there. Compared to surface waters, the spectral attenuation at 35–40 m depth is reduced in the blue and UV wavelengths because the chlorophyll concentration is 42% lower than at shallower depths. To show the residual effects on spectral attenuation caused by other substances (DOM and suspended non-algal particles), adjusted K_d values were calculated (by regression of K_d versus algal biomass as estimated from chlorophyll fluorescence) and are plotted as "1–30 m minus phytoplankton" in Figure 2. An estimate of attenuation by pure seawater, K_w, from Smith and Baker [18] is presented as a contrast to "K_d minus phytoplankton" and suggests the magnitude of attenuation by these other substances. Subtracting K_w from "1–30 m minus phytoplankton" yields an exponential curve (exponent $= -0.015$) typical of CDOM (Figure 6 below).

Rearranging equation (1) and substituting the symbol ζ (Greek "z") for $\text{Ln}(E_{d,0-}/E_{d,z})$, yields Kirk's [6] general equation for the depth (in meters) at which irradiance for a specific wavelength is reduced from 100% just below the surface to n% in a uniformly mixed water column:

$$Z_{n\%,\lambda} = \zeta / K_{d,\lambda} \tag{2}$$

Kirk [6] calls ζ the "optical depth" while Mobley [5] uses "optical depth" differently. Using Kirk's definition for ζ, and using f for the fraction of surface irradiance reaching $Z_{n\%}$, the general solution for optical depth is $\zeta = \text{Ln}(f^{-1})$ with

$\zeta = 1$, 2.3, and 4.6 corresponding to $(f \times 100) = 37\%$, 10%, and 1% respectively. Calculations of $Z_{10\%}$ and $Z_{1\%}$ have been used increasingly in the UV literature [e.g., 10,12,19], but infer significance primarily from the PAR waveband where $Z_{1\%}$ is considered the bottom and $Z_{10\%}$ the midpoint of the euphotic zone for photosynthesis [6]. No general name for these measures is generally accepted, although Williamson [20,21] used Z_a (which he referred to as "the attenuation depth") for the specific case of the 1% attenuation depth. This author suggests that $Z_{n\%,\lambda}$ be referred to as the "% attenuation depth" to serve as a general term for the depth at which irradiance is reduced to n% of the value just below the surface. Equation (2) can be reduced to:

$$Z_{37\%,\lambda} = 1/K_{d,\lambda} \tag{3}$$

Lacking biological or physical significance for a particular percentage within UV wavelengths, the 37% attenuation depth has the advantage that it is less likely to extend below the mixed layer in the water column. Because K_d is typically determined from mixed layer measurements, the 1% and 10% attenuation depths are more likely to misrepresent the penetration of UVR (specifically, whenever the computed value of $Z_{n\%}$ extends below the mixed layer). This is because K_d often changes below the mixed layer, either increasing (e.g., Figures 5A and 12, also [22–24]) or decreasing (e.g Figure 2, also [19]) compared to surface K_d values. Figure 3 (adapted from Whitehead et al. [12] using data from Table 1) shows $Z_{37\%}$ attenuation depths for 320 nm UVR for the mixed layer at a number of freshwater and marine sites. Also included in Figure 3 are estimates for pure water calculated by Smith and Baker [18] and a new estimate for pure water (described in Section 3.3.4).

The diffuse attenuation coefficient (K_d) is one of several "apparent optical properties" (AOPs) of natural waters described by Preisendorfer [25]. Unlike inherent optical properties (IOPs) described below, AOP's depend on the quality of incident light as well as the optical qualities of the water. In spite of this apparent limitation (and in part because the differences between AOP's and IOP's were said to be small in many instances [26]), a case was argued for the standard use of K_d to characterize natural waters for purposes of optical comparisons and bio-optical models [27,28]. Gordon [17,29] provided a practical means to adjust K_d measurements to remove much of its dependence on the ambient light field. In particular, Gordon [17] established that, after adjustment (described below), $K_{d,0}$ averaged from surface to $Z_{10\%}$ is proportional to the summed concentrations of constituent optical compounds.

In contrast to AOP's, "inherent optical properties" (IOP's) depend solely on the water and its optically active constituents. The IOP's include the beam absorption coefficient "α", beam scattering coefficient "b", and beam attenuation coefficient "c", which are related as follows:

$$c = \alpha + b \tag{4}$$

The absorption coefficient "α" is the sum of absorption by constituent components (including the solvent, water) and is proportional to the concentration of absorbing substances. Similarly, the scattering coefficient is the sum of constitu-

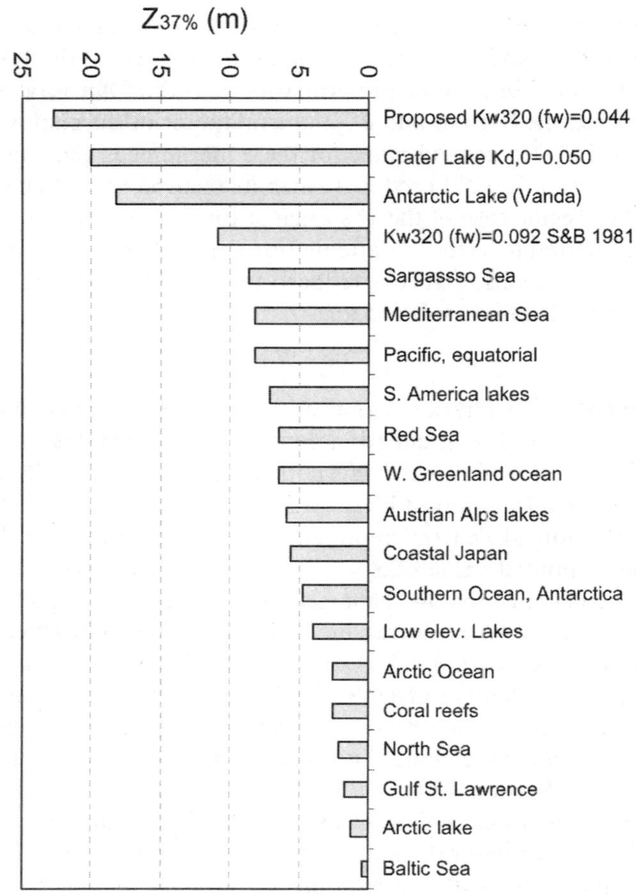

Figure 3. Maximum 37% attenuation depths for 320 nm (depth where irradiance is attenuated to 37% of the value just beneath the surface, computed as $1/K_{d320}$). Values have been calculated for the lowest K_d's reported for each category from Table 1.

ent component scattering and is proportional to the concentration of scattering substances. These three IOPS's are expressed in units of m^{-1}.

A property of the light field that relates AOPs with IOPs is the mean cosine ($\bar{\mu}$). It summarizes the angular distribution of photons according to equation (5). Direct measurement of $\bar{\mu}$ combines downwelling irradiance (E_d, with cosine response to solar zenith angle, maximal to vertical light from above and no response to horizontal light or light from below), upwelling irradiance (E_u), and scalar irradiance (E_o, equal response to light from any direction):

$$\bar{\mu} = (E_d - E_u)/E_o \tag{5}$$

Underwater $\bar{\mu}$ varies with sky and water conditions, the angle of the sun from vertical (solar zenith angle), wavelength, and depth [17]. From equation (5) one can establish a theoretical value (assuming no scattering) of 1.0 for a collimated beam from above, −1.0 for a beam from below, and zero for a completely diffuse light field. Direct measurements of scalar irradiance have not been reported using commercial underwater UV instruments although Danish scientists have reported upwelling and downwelling scalar irradiance and corresponding values for $\bar{\mu}_d$ and $\bar{\mu}_u$ [117]. Commercial underwater radiometers are available for determining spectral reflectance ratios for visible and UV wavelengths (E_d, E_u, and also upwelling radiance L_u). Stramska et al. [30] have proposed and tested a method to calculate $\bar{\mu}$ as well as α and b_b (the backscattered portion of b) from field measurements of E_d, E_u, and L_u in the wavelengths from 400–560 nm. This approach is promising for determining $\bar{\mu}$ over UV wavelengths but it will require validation beyond the currently specified range of wavelengths.

The mean cosine relates "α" and K_d in natural waters when these optical properties describe a narrow waveband. An exact relationship valid for all depths in the absence of any "internal light source" [17,31] is

$$K_d = (1 - R)\alpha/\bar{\mu} + RK_u \tag{6}$$

where R is irradiance reflectance (E_u/E_d) and K_u is the diffuse attenuation coefficient for upwelling irradiance. Internal light sources include Raman scattering and fluorescence emitted from DOM or chlorophyll after absorbing light at shorter wavelengths. When $E_u \ll E_d$, and thus when R becomes very small, equation (6) becomes

$$K_d \approx \alpha/\bar{\mu} \tag{7}$$

where $\bar{\mu}$ is less than 1 and thus K_d is greater than "α" to account for the longer mean path traveled by either diffuse or off-vertical sunlight for each vertical metre in the water column. The exact relationship in this form represents Gershun's Law [32]:

$$K_E = \alpha/\bar{\mu} \tag{8}$$

where K_E is the attenuation coefficient for the net downward irradiance, $E_d - E_u$. From the modeling work of Gordon [17], the mean cosine for downwelling irradiance ($\bar{\mu}_d$) also relates $K_{d,0}$ (measured just below the surface) to IOPs at that depth in a useful empirical equation:

$$K_{d,0} = 1.0395(\alpha + b_b)/\bar{\mu}_{d,0} \tag{9}$$

In equation (9) $\bar{\mu}_{d,0}$ is the fraction of downwelling scalar irradiance contributed by downwelling cosine irradiance ($E_d/E_{o,d}$), measured just below the surface, and b_b is the backscattered fraction of b. This relationship was developed for Case 1 waters (described in Section 2) and should be tested for validity in non-Case 1 waters. Other empirical relationships for predicting K_d from IOPs are described in Kirk [6,10].

Underwater spectral UVR measurements have sometimes been summarized by integration over a broad waveband (e.g., for UV-B and UV-A bands in [33] and [34]). The response of a broadband attenuation measurement, whether calculated from a detailed spectrum or measured with a broadband sensor (e.g., PAR), is subject to errors when used to characterize the optical properties of the water. These errors occur in attempting to characterize a uniform section of the water column (where IOPs are constant) because the effective K_d and effective "central wavelength" for the waveband will shift with depth and with the magnitude of attenuation [11,35]. In the UV-B waveband, for example, a simple spectral attenuation model can demonstrate that the wideband K_{dUVB} calculated for a specific depth deviates from the surface K_{dUVB} (full solar spectrum) by 12–19% over a range of K_d (K_{d320}=0.1–22 m^{-1}) and depths ($Z_{1\%}$–$Z_{37\%}$). For K_{dUVR} the effect is even greater; using published coastal ocean spectral data Booth and Morrow [11] calculated that K_{dUVR} would change 36% with depth (from 0.32 m^{-1} at the surface to 0.25 m^{-1} at 15 m depth, with an asymptote of 0.21 m^{-1} at much greater depth). Instruments with sensor bandwidth of ≤8 nm appear to perform well throughout the UV-A and UV-B ranges [36]. Spectral modeling has confirmed this: errors caused by spectral shift for the 8–10 nm bandwidth sensors of a widely-used UV radiometer (the PUV-500 from Biospherical Instruments, Inc.) are in the range of 1% from the surface down to $Z_{10\%}$ and less than 5% down to the limit of detection [35]. If data reduction from full spectral data is required, a better strategy than broadband integration is to present attenuation or irradiance values for several narrow wavebands within the UV-B and UV-A wavelengths.

3.2 Optical classification of natural waters

Natural waters differ optically from one another in color, transparency, and composition. Oceanographers and limnologists have developed different types of optical classification schemes to account for one or more of these attributes, and these are reviewed by Mobley [5] and Kirk [6]. Jerlov [4] was first to establish the concept of optical classifications for regions of the ocean, and described classes of open-ocean and coastal water based on transparency. Morel and Prieur [37] divided ocean waters according to optically dominant constituents: in Case 1 waters, phytoplankton and their products dominate; in Case 2 waters, the dominant constituents are either mineral particles or dissolved organic matter not associated with phytoplankton. Kirk [38] classified inland waters according to optical constituents (**W**, **G**, **A**, and **T** are used alone or in combinations to indicate the role of **W**ater, **G**ilvin=CDOM, **A**lgae=Phytoplankton, and **T**ripton=inorganic particles). Prieur and Sathyendranath [39] proposed a similar optical classification system for seawater. Kirk's scheme is used below in the description of different optical constituents of natural waters.

The terms DOM, DOC, and CDOM can be confusing and are sometimes used interchangeably. DOM describes the uncharacterized dissolved organic matter in natural waters. While DOM concentration could be quantified on a dry-mass basis (g m^{-3}) and values are found in the older literature, current analytical techniques (e.g., high temperature oxidation [40]), are calibrated in terms of the concentration of carbon atoms. The term DOC is now used when specific concentrations are reported. A molar or mass-based carbon concentration (e.g., g C m^{-3}) is thus the preferred unit of measure for either DOM or DOC. In contrast to these measures, CDOM is a generic description of the "optical concentration" of DOM, or the concentration of colored substances such as humic and fulvic acids. CDOM is measured as a spectral absorption coefficient ($\alpha_{cdom,\lambda}$) with units of m^{-1}.

Table 1 summarizes data on UV optical properties of natural waters from different regions. Section A describes marine sites while Section B describes freshwater sites. Each section presents UV-A and UV-B values, and entries are nominally sorted by value of attenuation or absorption coefficients. Attenuation, absorption and scattering coefficients have been converted, where feasible, to either 380 or 320 nm for easier comparison. The values of $K_{d,380}$ in the marine environment range from 0.03–0.8 m^{-1}; in freshwater the range is 0.02–32 m^{-1}. The values for $K_{d,320}$ in the marine environment range from 0.07–37 m^{-1}; in freshwater the range is 0.05–165 m^{-1}. The lowest values are found in open ocean (Sargasso Sea and eastern Mediterranean) and deep lakes (Crater Lake and L. Vanda). These low values occur in environments where the "hydraulic residence time" is long and where the water is isolated from terrestrial sources of DOM and nutrients either by distance from land, high elevation, or high latitude. In most cases this UV-transparent water is also exposed to high levels of UVR. The values for K_{dUV} in the region of the Baltic Sea and North Sea [41] and other coastal areas with large rivers [42] tend to vary with salinity in response to mixing of high-CDOM waters discharged by major rivers with low-CDOM waters from the open ocean. UV attenuation has been reported rarely in turbid systems (but see [49]). High values in Table 1 reflect either high DOM loading or evapoconcentration. UVR in lakes and estuaries where erosion or bottom resuspension contribute to extremely high turbidity can be assumed to exhibit rapid attenuation with depth.

3.3 Constituents controlling UV attenuation in natural waters: bio-optical models

Bio-optical models have been developed to predict spectral attenuation as a function of conveniently measured parameters and are discussed extensively in Mobley [5] and Kirk [6]. Since the pioneering work of Smith and Baker [42], bio-optical models typically break down diffuse attenuation into optical constituents of natural waters. In an approach covering UV-B and UV-A wavelengths summarized by Baker and Smith [9,43], these components are represented as partial attenuation coefficients (λ for each term not shown for simplicity):

$$K_{d\,Total} = K_{d\,Water} + K_{d\,CDOM} + K_{d\,Phyto} \tag{10}$$

to which $K_{d\,Tripton}$ may be added for waters where attenuation is caused by nonliving particles.

These components can be individually computed from measurements at a site after proper "calibration" to establish "specific attenuation" factors. For example, from a series of sites that differ in DOC concentration and having low or constant levels of phytoplankton pigments, $K_{d\,total}$ and [DOC] are measured

Table 1. UV optical properties, ranked in order of UV transparency

Site	WL (nm)	$c = \alpha + b$	b_w	α_w (m^{-1})	α_{cdom} (m^{-1})	α_p (m^{-1})	K_d (m^{-1})	Source
A. Ocean and coastal measurements								
UV-A ocean data								
Pure seawater (see Freshwater table for α_w)								
Pure seawater	380		0.0094					Morel 1974 in [6]
Clearest natural waters	380	[0.031]		0.022			0.027	Smith & Baker [18]
Sargasso Sea	380						0.044–0.045	Tyler & Smith [110]
East Mediterranean	375						0.05	Jerlov [95]
Gulf of Mexico USA	380						0.06	Smith & Baker [8]
Eastern Pacific (Mexico)	380						0.075	Tyler & Smith [110]
West Mediterranean	375						0.02–0.08	Højerslev [65]
Antarctica (61°S)	380						0.08–0.15	Helbling et al. [44]
Coastal Japan (3/98–3/99)	380				0.15[a]	0.16[a]	0.09–0.59, 0.24[a]	Kuwahara et al. [108]
Gulf of California (Mexico)	380						0.11–0.28	Tyler & Smith, [110]
Arctic polynya	380						0.18–0.50	Belzile et al. [48]
Tropical, near coral reefs	380						0.18–0.76	Dunne & Brown [111]
Gulf of St. Lawrence	380						0.1–0.8	Kuhn et al. [57]
Gulf coast, Florida USA	380						0.28	Smith & Baker [8]
Continental Slope	365	0.10			0.02			Clark & James '39 in [4]
Bermuda	380	0.20			0.03			Ivanov et al. '61 in [4]
North Sea	375				0.09–0.15			Højerslev [65]
	375				0.16–0.53			Stedmon et al. [77]
Kattegat	380	0.54			0.11	0.3		Jerlov '55a in [4]
	375				0.30–0.68			Højerslev [65]
	375				0.46–0.92			Stedmon et al. [77]
South Baltic Sea	375				0.71–0.90			Stedmon et al. [77]
Baltic Sea	375				0.89–1.04			Højerslev [65]
Delaware Bay mouth USA	380				0.6–1.2			Vodacek et al. [23]
Skagerrak	380				0.1	0.3		Malmberg '64 in [4]
	375				0.12–0.31			Højerslev [65]
	375				0.09–0.42			Stedmon et al. [77]

UV-B ocean data								
Pure seawater	320		0.0200					Morel 1974, in [6]
Pure seawater	320	[0.104]		0.084			0.094	Smith & Baker [18]
Sargasso Sea	320						0.069	Højerslev 1985 in Aas et al. [117]
	310						0.15	Højerslev [65]
East Mediterranean	310						0.15	Jerlov [95]
Central Equatorial Pacific	310						0.15	Smith & Baker [8]
West Mediterranean	310						0.16–0.43	Smith & Baker [8]
Gulf of Mexico, USA	310						0.18	Smith & Baker [8]
Red Sea (Gulf of Aqaba)	310						0.19	Iluz '93 in [10]
Western Greenland	310						0.19–0.21	Højerslev [65]
Orkney – Shetland	310						0.39	Højerslev [65]
Antarctica (61°S)	320						0.21–0.23	Helbling et al. [44]
Arctic polynya	320						0.39–0.83	Belzile et al. [48]
Coastal Japan (3/98–3/99)	320				0.44[a]	0.30[a]	0.18–0.98, 0.52[a]	Kuwahara et al. [108]
Tropical, near coral reefs	320						0.39–1.5	Dunne & Brown [111]
Gulf of St. Lawrence	310						0.68–2.0	Kuhn et al. [57]
Gulf coast, Florida USA	310						0.8	Smith & Baker [8]
German Bight, North Sea	310				0.07–1.49		0.53–5.0	Højerslev [65]
Skagerrak	310						0.59–1.20	Højerslev [65]
Kattegat	310						1.2–2.4	Højerslev [65]
Baltic	310						3.0–3.5	Højerslev [65]
Delaware Bay mouth USA 320					1.3–2.8			Vodacek et al. [23]
Yellow Sea	310						32–37	Højerslev '88 in [10]
B. Lake measurements								
UV-A lake data								
(Pure water)	375	0.045		[0.038]				Clark & James '39 in [4]
(Pure water)	380		0.0072					Morel '74 in [3], see also [58]
(Pure water)	380	[0.017]		0.010				Sogandares & Fry [112]
(Pure water)	380	[0.017]		0.010				Pope & Fry [101]
(Pure water)	380						0.018	Hargreaves (unpub. Crater Lake 8/01)
(Pure water)	380	[0.029]		0.022			0.025	Smith & Baker [18]
Crater Lake, Oregon USA	380					0.004–0.009	0.022	Hargreaves (unpub. Crater Lake, 0–20 m 8/01, divided by D_0)

Table 1. (*cont.*)

Site	WL (nm)	$c=\alpha+b$	b_w	α_w (m^{-1})	α_{cdom} (m^{-1})	α_p (m^{-1})	K_d (m^{-1})	Source
B. Lake measurements (*cont.*)								
Crater Lake, Oregon USA	380						0.016–0.33	Hargreaves, Larsen, Girdner (unpub.) Crater Lake, 0–15 m, Jun-Jul '96–'99
Crater Lake, Oregon USA	380						0.022	Tyler & Smith [110]
Lake Vanda, Antarctica	380						0.023	Vincent et al. [68]
12 Lakes, S. Argentina	380				0.05–1.9		0.07–3.0	Morris et al. [60]
Lake Tahoe, USA	380						0.08	Smith et al. [99]
High lake, Austria	380						0.08–0.14	Sommaruga & Psenner [63]
14 Lakes, NE USA	380				0.23–44		0.16–32	Morris et al. [60]
L. Giles, PA USA	380				0.07		0.16	Morris et al. [60]
4 Arctic lakes, Canada	380						0.41–2.8	Laurion et al. [61]
L. Biwa, Japan	380						0.5–7.5	Belzile et al. [49]
18 Subarctic lakes, Canada	380						0.67–16	Laurion et al. [61]
7 Lakes, Canada	380						0.48–9.1	Scully & Lean [33]
20 Lakes, Colorado USA	380				0.51–4.7		1.3–17	Morris et al. [60]
San Vincente Reservoir, San Diego, CA USA	380						2.69	Tyler & Smith [110]
L. Lacawac, PA USA	380				1.45		3.22	Morris et al. [60]
13 Lakes, Alaska USA	380				0.90–7.0		3.7–13	Morris et al. [60]
UV-B lake data								
(Pure water)	313	0.041		[0.027]				Boivin et al. [118]
(Pure water)	320		0.0153					Morel 1974 in [6], see also [58]
(Pure water)	320	0.022		[0.007]				Quickenden & Irvin [113][b]
(Pure water)	320						0.04	Hargreaves, unpub. (Crater L. 8/01)
(Pure water)	320	[0.099]		0.084			0.092	Smith & Baker [18]
Crater Lake, OR, USA	320					0.006-0.017	0.050	Hargreaves, unpub. (0–20 m, 8/01, divided by $D_0=1.13$)

Lake Vanda, Antarctica	320			0.055	Vincent et al. [68]
Crater Lake, OR USA	320			0.051–0.71	Hargreaves et al. (unpub.) 0–15 m, Jun–Jul '96–99
12 Lakes, S. Argentina	320	0.14–6.5[a]		0.14–7.7[a]	Morris et al. [60]
High elev. lake, Norway	320			0.17	Hessen [114]
High elev. lake, Austria	320			0.17–0.26	Sommaruga & Psenner [63], seasonal
11 Lakes (rocky) Alps & Pyr 320		0.21–1.8[a]		0.17–2.5[a]	Laurion et al. [62]
Lake Giles, PA USA	320			0.25–1.6	Hargreaves & Moeller, unpub.[c]
	320	0.23		0.32	Morris et al. [60]
	320		0.1–0.7	0.55–1.3	Ayoub et al. [46]
14 Lakes, NE USA	320	0.23–165[a]		0.32–67[a]	Morris et al. [60]
10 Lakes (trees) Alps & Pyr	320	0.48–4.6		0.60–5.7	Laurion et al. [62]
4 Arctic lakes, Canada				0.75–7.9	Laurion et al. [61]
5 meadow lakes, Alps & Pyr	320	0.74–3.1		0.87–4.3	Laurion et al. [62]
L. Biwa, Japan	320	0.85–6.9		1.1–14	Belzile et al. [49]
7 Lakes, Canada	320			1.1–21.6	Scully & Lean [33]
18 Subarctic lakes, Canada				1.7–41	Laurion et al. [61]
20 Lakes, Colorado USA	320	2.0–17		2.8–37	Morris et al. [60]
13 Lakes, Alaska USA	320	5.8–27		7.1–48	Morris et al. [60]
Lake Lacawac, PA USA	320			6.0–19	Hargreaves & Moeller, unpub.[c]
	320	5.8		7.8	Morris et al. [60]
	320		1.2–10	10–16	Ayoub et al. [46]

[a] Error in either α_{CDOM}, α_p, or K_d likely because $\alpha_{CDOM} + \alpha_p + \alpha_w$ exceeds K_d.
[b]After conversion of decadic beam attenuation value into $\log_e$ (value × 2.303), the spectral attenuation (280–320 nm) was exponentially regressed against wavelength to estimate c_{w320}, then α_{w320} was computed by subtracting scattering coefficient for pure water (b_w).
[c] Seasonal range of K_{d320}, 1993–2001.

(the latter from water samples). Then a regression of $K_{d\ total}$ versus [DOC] is computed to yield the DOC-specific diffuse attenuation factor (K^*_{DOC}) from the regression slope. This approach can also be used to estimate other specific attenuation factors (K^*_{Chl}, $K^*_{Tripton}$) and a similar approach could estimate specific absorption factors as well. In early attempts to estimate K^*_{CDOM} and K_w, Smith and Baker [18] subtracted modeled phytoplankton attenuation spectra from $K_{d,total}$ calculated from natural waters low in CDOM to estimate K_w. Baker and Smith [43] subtracted modeled phytoplankton attenuation spectra (developed as described above) and K_w from $K_{d,Total}$ for coastal waters to estimate $K_{d,CDOM}$. Various non-linear models have been developed to relate absorption and attenuation of algal cultures and natural phytoplankton to chlorophyll *a* concentration [5]. There is some information on UV attenuation by algae [44–51] but numerous studies have focused only on visible wavelengths [30,38,52–56].

The use of partial attenuation coefficients in bio-optical models has been criticized because the measured value of K_d does not depend solely on the natural waters [31]. AOP's such as K_d vary with sun angle, sky conditions, and depth, although some [26] have argued that the effects are small if the sun elevation is reasonably high and thus K_d can be used as a quasi-inherent property. Gordon [17] has established through optical modeling that K_d can be converted into a quasi-inherent optical property, for near-surface conditions at least, if it is first adjusted to remove atmospheric effects using equation (11). The procedure is to measure K_d near the surface or average K_d over 0–10% attenuation depth, and then multiply K_d by $\bar{\mu}_{d,0}$. The adjusted K_d can be expected to respond proportionally to changes in concentration of absorbing substances. Gordon's method for determining $1/\bar{\mu}_{d,0}$ (for which he used the symbol D_0) applies to calm (flat) surfaces of case 1 waters [17]:

$$1/\bar{\mu}_{d,0} = D_0 = f_{direct}/\cos(\theta_w) + 1.197 f_{diffuse} \tag{11}$$

where f_{direct} and $f_{diffuse}$ are the fractions of incident irradiance in the direct rays from the sun and in the indirect skylight respectively. The value of $\cos(\theta_w)$, the angle from the zenith of direct sunlight just beneath the surface after direct sunlight has been refracted from its incident angle ($\cos(\theta_\alpha)$) by passing through the horizontal water surface, can be determined from Snell's Law [6]. For light passing from air to water, $\cos(\theta_w) = \cos(\theta_a)/(n_a/n_w)$, where n_a/n_w is the ratio of refractive indices for air and water (nominally 1.33 in the visible wavelengths but is within 0.3% of 1.345 over the wavelengths 300 nm to 400 nm). Gordon's equation is almost identical to a "mean pathlength" equation derived independently by Zepp and Cline [15] to determine the amount of light absorbed in a vertical metre of water column based on laboratory measurements of absorption coefficients and modeled or measured incident sunlight. Gordon [17] suggests simple adjustments of D_0 for windy conditions that cause surface waves. He recommends using either $K_{d,0}/D_0$ or $K_{d,0\text{–}10\%}/D_0$ whenever the objective is to compare inherent optical properties of natural waters. This approach has been mentioned rarely in UVR studies (but see [57]). It will be used later in this chapter to estimate spectral attenuation by pure water.

The values for f_{direct} and $f_{diffuse}$ vary with atmospheric conditions (e.g., aerosols), sun angle, and wavelength. They can be determined with simple field measurements [17]: a vertically oriented radiometer with cosine sensor records full sun and sky, then is partially shaded to block only the direct irradiance from the sun. The ratio $E_{d,shaded}/E_{d,full}$ is $f_{diffuse}$, while f_{direct} is $(1-f_{diffuse})$. Figure 4A shows summer values for $f_{diffuse}$ for extremely clear air at Crater Lake, Oregon (1882 m elevation) and the hazier air over Lake Lacawac, Pennsylvania (400 m elev.) on clear "blue sky" summer days. Under overcast conditions and low solar elevation the value of $f_{diffuse}$ approaches 100% and D_0 (Figure 4B) approaches 1.2. At high solar elevation the longer UV wavelengths become more direct, more so

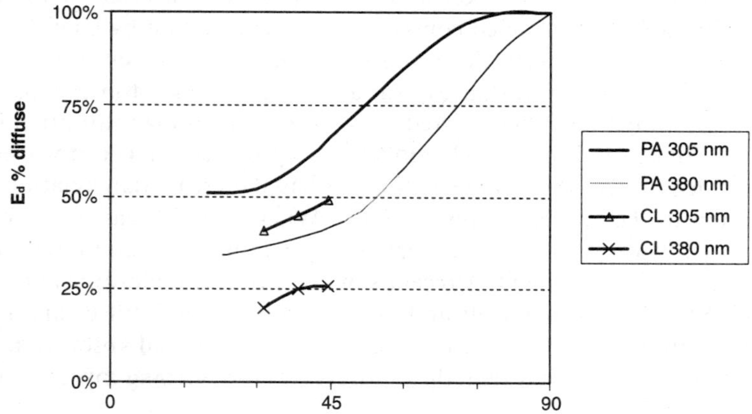

(A)

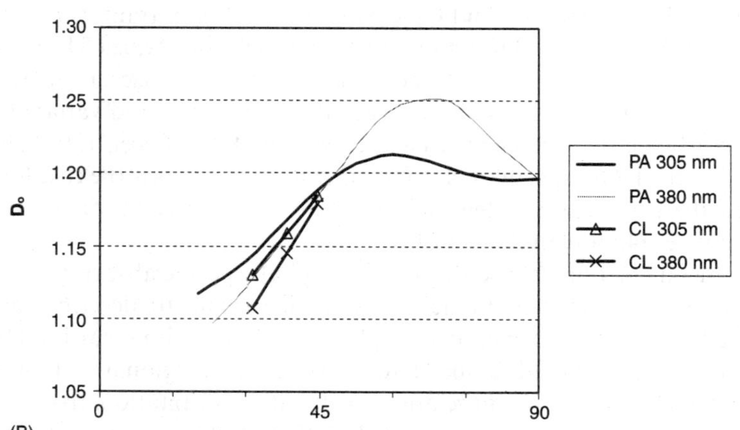

(B)

Figure 4. Examples of direct and diffuse solar irradiance and a correction factor for diffuse path length in K_d measurements (Hargreaves, unpublished). (A) Diffuse fraction of irradiance as a function of solar zenith angle during summer, 1996, L. Lacawac, Pennsylvania (41.3°N) and August 2001, Crater Lake, Oregon (42.9°N). (B) Calculated correction [17] to remove effects of irradiance field from near-surface diffuse attenuation (K_d) measurements, based on data in part (A).

in the clean dry air over Crater Lake than over L. Lacawac in the N.E. USA. For this range of conditions D_0 varies from 1.09 to 1.26. The value of $f_{diffuse}$ at 320 nm during summer in the Gulf of St. Lawrence (latitude 47–50°N, solar zenith angle 24–54°) ranged from 48–72% [57]. If K_d measurements are made with SZA $\leq 50°$, the effect of incident diffuse and direct light on K_d will always be less than 20%. At the Latitude (41.3°N) of L. Lacawac, the SZA will be less than 50° if measurements are made within 3 hours of solar noon between the dates of 17 April and 26 August. Before 26 February and after 14 October the SZA will be greater than 50° even at solar noon.

The first spectral models, developed for marine systems by Prieur and Sathyendranath [39] and Baker and Smith [43], emphasized phytoplankton optics, although they included components for attenuation by CDOM, phytoplankton, and water. From these and subsequent studies (reviewed by Morel [58]) has emerged the importance of variation in phytoplankton attenuation per unit of chlorophyll, hereafter referred to as "specific phytoplankton attenuation". While the Baker and Smith [43] model was optimized for UV-B wavelengths and included a CDOM component, this was based on scant data relating optical absorption to the concentration of DOM. When CDOM was detected in the open ocean it was assumed to covary with phytoplankton and often modeled without direct measurement. A recent summary of ocean bio-optical models in Morel [58] does not specifically address UV wavelengths. Yentsch and Phinney [59] mention that the blue and UV regions of algal spectral absorption are the most variable, especially when algae produce UV-absorbing mycosporine-like amino acids (MAA's).

Freshwater optical models relevant to UV attenuation have been developed [33,60,61]. These generally find CDOM is the most important factor, although absorption by phytoplankton and detrital particles [46,49,62,63] or scattering by suspended solids [49] has also been important in some cases. From recent UV investigations in freshwaters has emerged the importance and variability of the CDOM absorption coefficient scaled per unit of DOC concentration, hereafter referred to the DOC-specific absorption factor ($\alpha^*_{\lambda DOC}$). As is the case for marine systems, there are relatively few measurements of UV attenuation or absorption by freshwater phytoplankton [46,49].

Despite an emerging consensus on chlorophyll-specific absorption for visible wavelengths [58], current models appear inadequate to describe the highly variable UV attenuation exhibited by phytoplankton. With regard to DOC and CDOM, several studies (described later) have revealed regional patterns relating K_d and CDOM to [DOC] in relation to climate, precipitation, river discharge, and watershed properties. Integration of patterns and processes to explain UVR penetration into aquatic systems has been lacking.

3.3.1 Role of CDOM (Kirk type G waters)

CDOM was central to the earliest optical models for seawater (Jerlov, 1968 [4]). Interest in CDOM in freshwater goes back to the early 1900's (Shapiro, 1957

[64] cites a 1908 paper on humic substances precipitated from Finnish lakes). It has long been recognized that the penetration of ultraviolet radiation (UVR) depends on concentrations and optical qualities of dissolved organic matter (DOM) [4,65–67]. In non-turbid waters where UV attenuation is high, absorption by CDOM easily surpasses attenuation by other components. Somewhat surprising is the recent finding that CDOM tends to be the most predictive optical component of UV attenuation even in low DOC systems. UV bio-optical models developed recently for lakes reflect a dominating effect of DOC and CDOM, with only a small or immeasurable contribution by phytoplankton in surface waters [33,60,62,68].

Lakes matching Kirk's "Type G" typically have moderate to high levels of DOC. Figure 5 shows an example the close relationship between $\alpha_{cdom380}$ measured in a spectrophotometer and $K_{d,380}$ calculated from field measurements of underwater irradiance at 380 nm in a moderately clear lake (L. Giles, PA, listed in Table 1). The two signals change together and increase below the mixed layer (9.5 m on this date in early September). Attenuation and α_{cdom} decrease substantially in the mixed layer of this lake during summer months when rates of UVR photobleaching of CDOM exceed the rates of CDOM production and import [22].

In the older literature CDOM has been called "yellow substance", "gelbstoff", and "gilvin'. It is considered to be a mixture of compounds chemically characterized as humic and fulvic acids [6,10,69,70]. Figure 6 shows typical absorption spectra for the CDOM in water from two mid-latitude lakes that has passed through a fine glass fiber filter. The samples are from the depth of the mixed layer of two lakes surrounded by mixed conifer–deciduous forest: L. Giles (watershed soils well-drained) and L. Lacawac (bordered 50% by a sphagnum bog). The values come from absorbance ($A_{sample,\lambda}$) recorded in a spectrophotometer using a quartz cuvette, corrected by subtracting (optically or numerically) the value for highly purified water, $A_{water,\lambda}$, and the cuvette to compute $A_{cdom,\lambda}$. We assume negligible absorbance by inorganic dissolved matter such as ferrous iron, nitrite, or sulfate ions. An adjustment is often made to correct for instrument baseline drift and optical scattering within the cuvette that would otherwise cause errors in estimating $A_{cdom,\lambda}$. A long reference wavelength (λ_{base}) should be chosen for the correction of offset (e.g., >650 nm) where absorption by CDOM is assumed nil; 700 nm or 775 nm are particularly useful in avoiding the strong temperature effects on the absorption of pure water that reach a peak at about 750 nm [71,72] but a longer wavelength (up to 900 nm) may be needed for highly concentrated CDOM. Suggestions for subtracting a spectral scattering term [67,73] from measured A_{λ} are derived from empirical models showing scattering from small particles in nominally-filtered natural waters varies in proportion to λ^{-1}. Depending on the sample filtration and optical configuration of the instrument and cuvette, spectral scattering may affect the measurement; the suggested correction, $A_{cdom,\lambda} = A_{\lambda cdom\_raw} - A_{base}\,(\lambda_{base}/\lambda)$, has not been rigorously tested. $A_{cdom,\lambda}$ is then converted into a (Napierian) absorption coefficient, $\alpha_{cdom,\,\lambda}$ (units m^{-1}):

$$\alpha_{cdom,\,\lambda} = \mathrm{Ln}(10) A_{cdom,\,\lambda}/\ell \qquad (12)$$

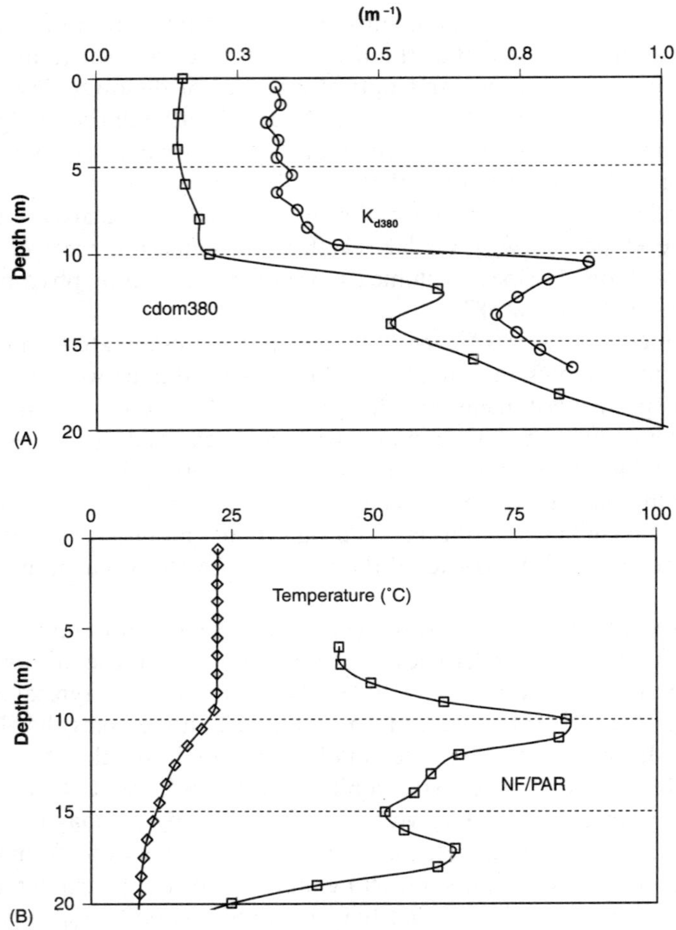

Figure 5. (A) UV-A CDOM absorption and diffuse attenuation coefficients (380 nm) for L. Giles (2 September 1999) determined using binned data from PUV-501 profiling radiometer and laboratory analysis of GF/F filtered water samples in 10 cm quartz cuvettes and Shimadzu UV160U spectrophotometer. (Hargreaves, unpublished). (B) Supplementary data from the PUV-501 profiling radiometer: water temperature and chlorophyll index (upwelling natural 685 nm fluorescence, NF, divided by downwelling PAR). In (B) The thermally mixed zone above 9.5 m corresponds to the optically mixed zone in (A). Algal biomass increases with two peaks near 10 m and 17 m that correspond to optical changes in (A) (Hargreaves, unpublished).

where ℓ is the cuvette path length (in meters). In the chemical literature the decadic absorption coefficient ($a = A/\ell$) is sometimes reported. Although the units are identical (m^{-1}), decadic units must be multiplied by 2.303 (the natural logarithm of 10) to be numerically equivalent to the standard (Napierian) exponential units from equation (12).

CDOM has an absorption spectrum that is nominally exponential in shape [4] and has been frequently characterized by the two exponential parameters,

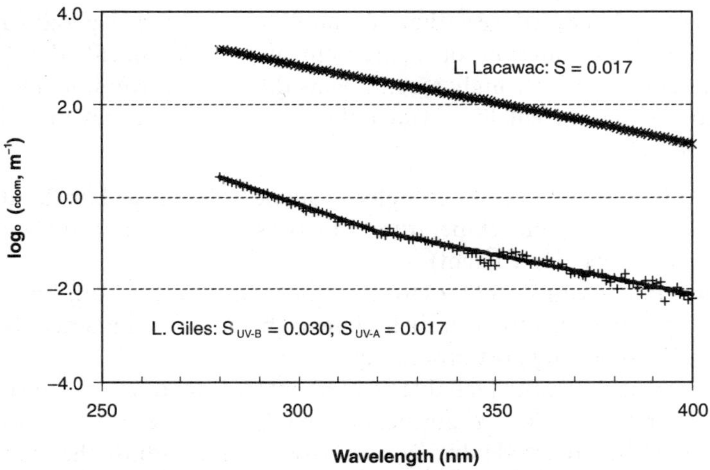

Figure 6. Spectral slope of CDOM from two lakes (Hargreaves, unpublished). S (nm^{-1}) is an exponential parameter from the relationship $\alpha_{cdom,\lambda} = ae^{-Sl}$. The value of S can be computed as the absolute value of the slope when Ln($\alpha_{cdom,l}$) is plotted against wavelength over the UV and blue range. Such plots tend to be linear over UV wavelengths when DOC is high (upper curve) but can sometimes be separated into a steeper UV-B slope (280–320 nm) and shallower UV-A slope (320–380 nm) when substantial photobleaching has occurred (lower curve). These lake samples are from the upper mixed layer, June 2001 (particles removed with GF/F filter, Shimadzu UV-1601 spectrophotometer, 10 cm quartz cuvette, low DOC deionized water spectrum subtracted; small glitch at 345–350 nm in lower curve is caused by spectrophotometer imperfection).

"spectral slope" (S) and reference absorption ($\alpha_{cdom, \lambda_{ref}}$):

$$\alpha_{cdom, \lambda} = \alpha_{cdom, \lambda_{ref}} e^{-S(\lambda - \lambda ref)} \tag{13}$$

where ($\lambda - \lambda_{ref}$) is the difference between the desired wavelength and the reference wavelength over the range 350–700 nm [67]. The value of S (units, nm^{-1}) is typically computed from a linear regression of Ln(α_{cdom}) versus wavelength. The waveband used in numerous published reports has varied but frequently covers the range from UV-B through 700 nm [74]. Bricaud et al. [67] likely chose 350 nm as their lower limit after observing nonlinear regions at shorter wavelengths in their published spectra of open ocean CDOM. When α_{cdom} is high, spectra tend to be exponential from below 300 nm well into the visible spectrum (upper curve in Figure 6). At lower levels of α_{cdom}, typically following substantial exposure to sunlight, the spectra become more irregular below 350 nm. Under these circumstances, separate values for S may be calculated for the UV-B (280–320 nm) and UV-A (320–340 nm) ranges of the spectrum ([75] and K. Mopper, personal communication), as shown in the lower curve of Figure 6.

Several authors [74,76,77] have recently suggested computing S using a nonlinear regression technique that gives less weight to longer (and noisier) wavelengths. They assume that S is uniform throughout the range of wavelengths included in the nonlinear regress (otherwise the nonlinear approach would bias S toward the slope at the shortest wavelengths where α_{cdom} has the greatest value) and they suggest that the nonlinear technique avoids a bias caused by log-transformation of instrument noise present at the longer wavelengths. This author strongly recommends the more conventional log–linear regression with a caveat to consider the following guidelines in order to compute S accurately:

- The spectrophotometer must be completely stable (e.g., warmed up for at least an hour at a stable room temperature; this is especially important with the diode array variety of instrument).
- A single carefully-cleaned quartz cuvette should be used for both blank and sample scans (referenced to air in the reference beam) with numerical correction for the blank during post processing.
- If ultrapure water is not available (stored water can develop substantial absorbance and many water purification methods leave a UV-absorbing residue) it may be preferable during post-processing to adjust the measured blanks recorded in the field with a file recorded earlier with the best quality water using the same instrument and cuvette.
- The initial selection of wavelength range for S should consider the shape of the spectrum (mentioned above).
- The baseline should be carefully adjusted to zero during post-processing at a non-absorbing waveband (e.g., 775–800 nm); this should be accompanied by visual inspection of a linear graph of α_{cdom} versus wavelength (with scales expanded to show detail, e.g., ± 0.05 m^{-1} for the range 600–800 nm).
- The longer wavelength should be revised if necessary so as to avoid wavelengths near the instrument limit of detection (typically $A = \pm 0.001$ after subtraction of the blank and A_{base}) where noise can return negative α_{cdom} values. The effect of the baseline adjustment (described above) is to ensure that the noise is symmetrical with respect to zero, but if half the noise values are negative (and thus automatically excluded from the regression), the value of S will be underestimated to an extent that depends on how many of these "noise" data are included.

A similar exponential treatment has also been applied to spectral modeling of UV diffuse attenuation coefficients for natural waters [57,61,76] but this seems ill-advised unless the absorption spectrum of phytoplankton or other particles is insignificant or has been observed to follow the same exponential pattern as CDOM that may be present.

DOM molecules are the chemical basis for CDOM optical absorption (α_{cdom}, m^{-1}), but because of molecular variations in DOM and its chemical environment, DOC-specific absorption, (α^*_{DOC}) and DOC-specific attenuation ($K_d{}^*_{DOC}$) vary in natural waters. Both have units of m^{-1}[g m^{-3}]$^{-1}$, typically simplified to m^2 g $^{-1}$. The optical properties of CDOM are known to vary with the source, including type of watershed vegetation and in situ production [6,74,78–80] and modification in the water column [22,23,74,81].

Although variation in CDOM specific absorption has been recognized for some time there is disagreement among researchers on patterns of variation with DOC concentration. Currently the scaling factor of choice is DOC concentration, rather than DOM concentration, because of standardization in methods for measuring the carbon content of DOM [40]. A linear relationship between α^{*}_{CDOM} and DOC has been assumed to date [23,43] in marine systems but given the pattern of variation in DOC-specific attenuation (described later), a non-linear model is proposed for DOC-specific CDOM absorption:

$$\alpha_{CDOM,\lambda} = \alpha^{*}_{DOC,\lambda}\ DOC^{x} \tag{14}$$

Values for $\alpha^{*}_{DOC,320}$ computed from available data (Table 2) range from 0.3 to 3.2 $m^2\ g^{-1}$ for both marine and freshwater sites (converted to 320 nm from other wavelengths as needed using reported S values and equation (13)). Data from a study of 61 lakes [60] reveal that DOC-specific absorption increases together with DOC concentration with an exponent of 1.12 and $\alpha^{*}_{DOC,320} = 1.2\ m^{-1}$ (Table 2). CDOM from surface waters in the Gulf of Mexico was concentrated and separated into fulvic and humic fractions [69] to indicate their relative contributions to absorption. Fulvic acids have a much lower specific absorption than humic acids. While shifting proportions of the fulvic and humic fractions in

Table 2. Variations in DOC-specific absorption of CDOM (α_1), using $\alpha_{cdom320} = \alpha_1\ DOC^{x}$ where units are m^{-1} for $\alpha_{cdom320}$ and $g\ m^{-3}$ for DOC concentration

*α_1	*α_5	†x	r^2	DOC	Region	Data from
Coastal marine DOC (extracted from Gulf of Mexico surface water)						
0.06	0.3	(1)			extracted fulvic acids	Carder et al. [69]
0.5	2.5	(1)			extracted humics	Carder et al. [69]
Coastal marine CDOM						
0.3	1.4	(1)		1.6	Japanese coastal, 13 months	Kuwahara et al. [108]
0.7	3.4	(1)		0.3–3.8	Danish coastal	Stedmon et al. [77] [a]
1.3	6.5	(1)			Danish coastal	Nyquist in Højerslev [65] [b]
1.9	9.5	(1)		6.2	estuarine salt marsh	Miller & Moran [109]
Coastal water receiving Delaware River discharge, comparing seasons						
2.1	10.6	(1)		0.8–1.7	Spring, water column mixed	Vodacek et al. [23]
0.6	2.8	(1)		1.3–1.5	August, surface layer	Vodacek et al. [23]
Mid-latitude lakes, comparing seasons						
1.8	9.0	(1)		0.7	Spring, L. Giles	Morris & Hargreaves [22]
0.3	1.3	(1)		1.0	Summer surface, L. Giles	Morris & Hargreaves [22]
3.2	16.0	(1)		4.4	Spring, L. Lacawac	Morris & Hargreaves [22]
1.4	7.0	(1)		5.7	Summer surface, L. Lacawac	Morris & Hargreaves [22]
Mid-latitude lake surveys						
1.2	7.4	1.12	0.79	1–24	61 Lakes, mid-latitudes	Morris et al. [60]
0.7*	10.6	1.70	0.91	4–22	30 Lakes, Northern USA	Reche et al. [115] [c]
0.8	10.8	1.58	0.90	0.1–15	85 Adirondack lakes	Bukaveckas & Robbins-Forbes [24] [c]

* α_1 is equivalent to α^{*}_{DOC320}; α_5 is $\alpha_{cdom320}$ computed for $DOC = 5\ g\ m^{-3}$

†(1) indicated for x where a proportional scaling pattern for DOC has been assumed.

[a] TOC used to calculate specific absorption.

[b] DOM $g\ m^{-3}$ instead of DOC $g\ m^{-3}$ [c]

α_1 and α_2 adjusted from 440, 340 or 300 nm to 320 nm using equation (13) and $S = 0.01565$.

the CDOM source may cause some of the variation in specific absorption of natural CDOM, other factors include changes in pH, ionic composition, and photobleaching. While the absorption by humic acids is stable over a wide range of pH (6–11), that of fulvic acids is not [82]. Stewart and Wetzel [83] studied humic substances in experimental leachate of decaying plants and DOM from 55 lakes of southwestern Michigan. They concluded that calcium concentration affects average molecular size and this in turn affects DOC-specific absorption. Vodacek et al. [23] attributed the decline in specific absorption for CDOM in the coastal plume from the Delaware River to photobleaching in the surface mixed layer. Specific absorption at 320 nm was reduced from 2.1 $m^2 g^{-1}$ during winter conditions of high river flow and low irradiance to 0.6 m^2 g^{-1} in offshore stratified surface waters during high summer irradiance (August). Morris and Hargreaves [22] observed similar declines from spring to summer in K_{d320} and $\alpha_{cdom,320}$ for several lakes differing in their CDOM source (one surrounded by a sphagnum bog, the other by well-drained soil). They established a major causal role for photobleaching through experimental exposure of particle-free lake water to different wavebands of the solar spectrum (Tables 2 and 3 and Figure 5A). In some cases a sampling artifact appears to interfere with measurements of DOC-specific absorption. In the mountain lake study by Laurion et al. [62], surface $\alpha_{cdom,320}$ was generally reduced compared to deeper in the water column, a pattern that may have been caused by photobleaching or surface inhibition of phytoplankton. However, measured $\alpha_{cdom,320}$ was greater than K_{d320} for 73% of lakes with low DOC and rocky watersheds and 21% of lakes with higher DOC and forested or meadow-covered watersheds. The authors suggested that UV-screening pigments (MAAs discussed in the next section) known to be present in the phytoplankton may have leaked out of cells during filtration. This problem might partially explain a similar anomaly in several of the 61 lakes sampled by Morris et al. [60].

CDOM exhibits fluorescence by emitting blue light after absorbing UVR. The maximum fluorescence response per unit of absorbed energy occurs when coastal CDGM is excited at 380 nm [84]. Although CDOM fluorescence is sometimes well-correlated with UV attenuation [61] and CDOM absorption [23], it has also been a somewhat variable predictor of variations in UV attenuation or absorption in other cases when the CDOM source varies [62,83]. DOC-specific fluorescence appears to vary both among and within lakes. As in the case of CDOM absorption, variations in fluorescence properties of DOC are likely to reflect differences in source as wells as a history of photochemical and biological processing. CDOM from terrestrial and marine sources can be distinguished from each other using three-dimensional excitation–emission fluorescence spectra [85,86]. McKnight et al. [80] showed that for excitation at 370 nm the CDOM emission peak of an acidified filtered water sample would occur at 442–448 nm for microbially-derived fulvic acids and at 457–461 nm for plant-derived (terrestrial) fulvic acids. Their fluorescence index (Em450 : Em500) based on these differences yielded 1.9 for microbial-derived DOM and 1.4 for terrestrial-derived DOM. This index is reported to be affected by environmental acidification, which changes DOC composition, but not by photobleaching [87].

Table 3. Relationship between $K_{d,\,CDOM}$ (m^{-1}) and DOC ($g\ m^{-3}$) using $K_{d320} - K_{w320} = k_1\,DOC^x$; Published K_{d320} and DOC data were used with $K_{w320} = 0.04$ to fit k_1 and x (least squares regression) where $k_1 = K_{CDOM,\,320}$ at DOC = 1 $g\ m^{-3}$ and $k_5 = K_{CDOM,\,320}$ at DOC = 5 $g\ m^{-3}$

k_1	k_5	x	r^2	DOC	Region	Data from
Coastal & Marine						
1.3	7	(1)			Ocean (based on DOM)	Højerslev [65]
0.8[a]	4[a]	(1)		2.5	St. Lawrence Estuary, Stn 24	Kuhn & Browman [57]
0.4		(1)		1.7	Coastal Japan, 8 m. w/rain	Kuwahara et al. [108]
0.2		(1)		1.5	Coastal Japan, 5 m. dry	Kuwahara et al. [108]
43 Canadian prairie lakes, ponds, wetlands, including saline systems						
6.7[a]	18[a]	0.61	0.41	24–80	52°N Wetlands, ponds	Arts et al. [34][b]
1.4[a]	5	0.76	0.50	4–156	52°N Lakes	Arts et al. [34][b]
Freshwater: photobleaching effects in surface waters of lakes						
2.0		(1)	0.7		41°N, L. Giles, spring	Morris & Hargreaves [22]
0.3		(1)	1		41°N, L. Giles, summer	Morris & Hargreaves [22]
	19	(1)	4.4		41°N, L. Lacawac, spring	Morris & Hargreaves [22]
	9	(1)	5.7		41°N, L. Lacawac, summer	Morris & Hargreaves [22]
Freshwater: lakes differing in land cover and altitude						
0.6	3[a]	0.89	0.54	0.2–1	Alps & Pyr., rocky	Laurion et al. [62][c]
1.4	7[a]	0.81	0.68	0.4–4	Alps & Pyr., trees, meadows	Laurion et al. [62][c]
1.0	8[a]	1.33	0.81	0.2–4	Alps & Pyrenees, combined	Laurion et al. [62][c]
Freshwater: high latitude lakes						
0.3	10	2.08	0.93	0.3-11	Arctic, subArctic, Antarctic	Vincent et al. [68]
0.3	13[a]	2.43	0.99	0.3–1	Antarctic	Vincent et al. [68]
0.4[a]	10	2.06	0.86	2–11	Sub-Arctic Canada	Laurion et al. [61]
0.7[a]	10	1.62	0.78	4–11	Alaska, USA	Morris et al. [60][d]
0.8	6	1.28	0.97	1–5	Arctic Canada	Laurion et al. [61]
Freshwater: mid-latitude lakes						
0.6	7	1.62	0.91	0.5–8	41–51°N. USA & Canada	Scully & Lean [33][d]
0.6	7	1.57	0.90	1–24	41° N Pennsylvania,USA	Morris et al. [60][d]
1.7	13[a]	1.24	0.78	0.4–3	40°S, Argentina	Morris et al. [60][d]
3.1	12	0.83	0.60	0.8–10	Colorado, USA	Morris et al. [60]
1.5	10	1.20	0.84	0.4–24	Average, mid-latitudes	Morris et al. [60][d]

[a] k_1 or k_5 extrapolated beyond measured range of DOC.
[b] Waveband used for K_d was UV-B (280–320 nm) instead of narrow-band 320 nm.
[c] Excluding lakes when $C/L > 50$ (C = catchment area, L = lake area).
[d] Excluding data when DOC low relative to phytoplankton (DOC/chl < 1400, units, $g\ m^{-3}$).

Figure 7 shows the emission spectrum of CDOM fluorescence for samples excited at 365 nm from two lakes having DOC in the range of 1–5 mg l^{-1} (Hargreaves, unpublished). The small peak is Raman scattering by water molecules (centered at 417 nm, a shift in wavenumber of $-3400\ cm^{-1}$ from the excitation wavenumber, where wavenumber is 10^7 divided by wavelength in nm). The Raman water peak can be used to provide scale calibration of fluorescence emission spectra [88,89]. The broad peak in Figure 7 is contributed predominantly by the fulvic acid fraction of DOM [80]. The peak wavelength and

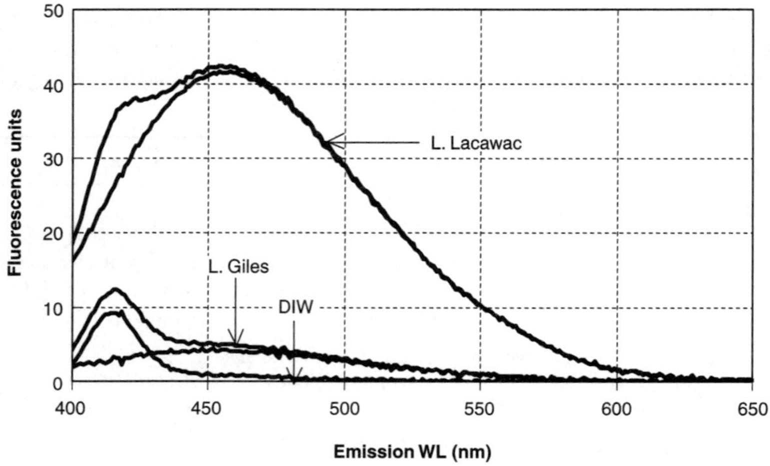

Figure 7. CDOM fluorescence of water from two lakes (Hargreaves, unpublished): emission scans for excitation at 370 nm (Shimadzu 551 fluorometer), before and after subtraction of water blank. Samples: deionized water (DIW), L. Giles water (ca. 1 g m^{-3} DOC), L. Lacawac water (ca. 5 g m^{-3} DOC) The Raman scattering peak at 417 nm represents a shift in wavenumber by 3400 cm^{-1} from the excitation wavenumber. The broad peak is contributed predominantly by the fulvic acid fraction of DOM. The peak wavelength and fluorescence index ratio for these samples (L. Giles, 452 nm peak and ratio = 1.5; L. Lacawac, 455 nm peak and ratio = 1.4) suggest a slight difference in CDOM source [80].

fluorescence index ratio for these samples (L. Giles, 452 nm peak and ratio = 1.5; L. Lacawac, 455 nm peak and ratio = 1.4) suggest a slight difference in CDOM source.

Values for $K_d{}^*{}_{DOC,320}$ ranging from 0.3 to 3.8 (Table 3) have been calculated from published UV attenuation data for both marine and freshwater sites (converted into 320 nm as needed using reported S values and equation (14)). The attenuation of pure water was subtracted ($K_{w,320}$ discussed below) and sites with high chlorophyll relative to DOC (DOC/chl < 1400; units g m^{-3}) were excluded where noted. Although some K_d's may be elevated by phytoplankton and other particles that attenuate underwater irradiance, the predominant source of variation in $K_d{}^*{}_{DOC,320}$ is DOC quality. While Scully and Lean [33] reported no effect of phytoplankton in their optical model (chlorophyll ranged from 1.3–33 mg m^{-3}), a reassessment shows that when lakes with a low ratio of DOC to algal chlorophyll (DOC/Chl < 1400; units g m^{-3}) were excluded, there was an improved r^2 for the regression of K_d versus DOC and a reduced K^*_{DOC}. The regressions of Morris et al. [60] were also improved by reanalysis in which lakes with low DOC/Chl ratios were excluded, although only three lakes (out of 64 sampled) had chlorophyll levels exceeding 5 mg m^{-3}.

A linear model for scaling $K_d{}^*{}_{DOC,320}$ to DOC concentration was adopted by Baker and Smith [43], who used a constant value of $K_d{}^*{}_{DOC,320} = 1.3$ in their bio-optical model but cited a range of values from as low as $K^*_{DOC,320} = 0.75$ for clear Sargasso Sea water to as high as $K^*_{DOC,320} = 6.2$ for a coastal site. A linear relationship between K^*_{DOC} and DOC has been assumed in several studies [61,62]. A power relationship between K^*_{cdom} and DOC (similar to that describe for CDOM absorption above) has been assumed by others:

$$K_{d,cdom,320} = K^*_{DOC,320}\, DOC^x \tag{15}$$

where $K^*_{DOC,320}$ is the DOC-specific attenuation of CDOM at 320 nm.

The power model for relating UV attenuation to DOC was used by Scully and Lean [33], Morris et al. [60] and Vincent et al. [68]. Vincent et al. [68] found an unusually strong relationship between UV attenuation depths and DOC for high-latitude lakes (replotted in Figure 8 as $K_{d320} - K_{w320}$ versus DOC), a relationship with an exponent much greater than one. Arts et al. [34] used both linear and power models (but preferred the power relationship) to relate wide-band UV-B attenuation to DOC in prairie lakes of Canada. As reported by Arts et al. [34] a reasonable fit was obtained with a power model, but when all ponds and wetlands are grouped (including three with high salinity), they have exponents less than one, similar to the fresh and saline lakes (Table 3). While attenuation for the UV-B waveband (derived from detailed spectral irradiance) is too broad to serve as a rigorous attenuation coefficient, it is an index of K_{d320} that is largely a function of DOC concentration and quality. Although Arts et al. [34] concluded that UV-B irradiance penetrates more deeply into saline water bodies that it does into freshwater systems of similar DOC concentration, this reanalysis of their data supports a somewhat contrary conclusion (Figure 9). Saline systems tend to have higher [DOC] than freshwater systems (probably because evaporation causes DOC to become more concentrated). For similar [DOC], the greater penetration is actually observed in lakes (especially the large,

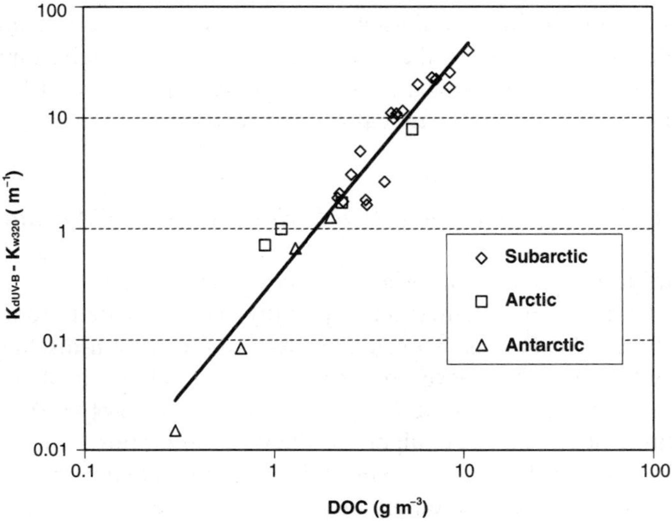

Figure 8. Lake data computed from Vincent et al. [68] to show the power relationship between $(K_{d320} - K_{w320})$ and DOC concentration for a range of high latitude lakes. The equation for all sites combined is $(K_{d320} - K_{w320}) = 0.34\text{DOC}^{2.08}$ ($r^2 = 0.93$).

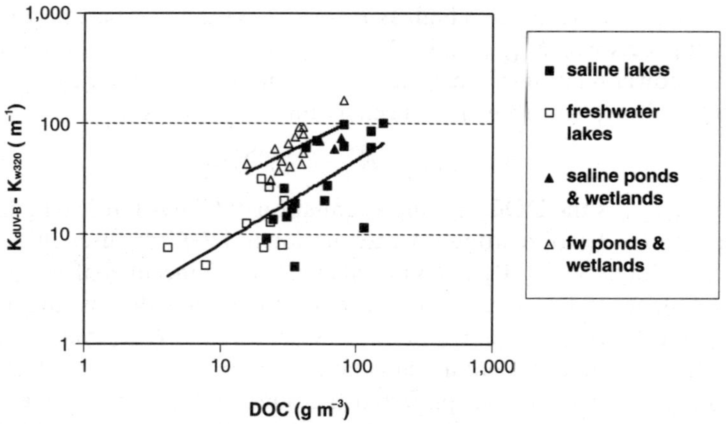

Figure 9. Attenuation of UV-B irradiance in saline prairie lakes, ponds, and wetlands of Canada (52°N) from Arts et al. [34]. Saline systems tend to have higher [DOC] than freshwater systems (probably because evaporation causes DOC to become more concentrated). For similar [DOC], greater penetration of UVR is observed in lakes (especially the large, deep ones) compared to small and shallow ponds and wetlands. The equation for all ponds and wetlands (triangle symbols) is $K_{dUV\text{-}B} = 6.7\ DOC^{0.61}$ ($r^2 = 0.41$); for freshwater ponds and wetlands (open triangles) $K_{dUV\text{-}B} = 2.5\ DOC^{0.91}$ ($r^2 = 0.56$); for all lakes (squares), $K_{dUV\text{-}B} = 1.4\ DOC^{0.76}$ ($r^2 = 0.50$).

deep ones) compared to small and shallow ponds and wetlands. Arts et al. [34] noted this pattern and hypothesized that attenuation is lower per unit of DOC in deep lakes because their greater residence time allows for more complete photobleaching. Waiser and Robarts [90], studying one of these large lakes, found lower [DOC] but higher UV-B attenuation per unit of DOC in a major stream feeding the lake compared to the lake water column. Although salinity covaries with DOC in xeric regions such as this, it is a weak predictor of UV-B attenuation because DOC-specific attenuation also varies.

3.3.2 Role of phytoplankton and CDOM (Kirk type A and GA natural waters)

Phytoplankton can contribute significantly to UV attenuation in waters with moderate to low UV attenuation, especially when isolated from watershed sources of CDOM. In Case 1 oceanic waters where coastal discharge is not a source of CDOM [52] the levels of locally-produced DOC, algal pigments, and the associated microbial community are assumed to co-vary [43,58] but perhaps with a time delay between algal production and appearance of CDOM [67,69,85,91]. It can be difficult to establish the indirect contribution of phytoplankton to water column optics by their release of DOM which increases α_{cdom}. For example, the spatial correlation between a peak in α_{cdom} just below the mixed layer of a lake (Figure 5A) and a peak in abundance of phytoplankton at the same depth (Figure 5B) are suggestive of a causal relationship, but not definitive.

Studies of optical properties of photosynthetic organisms are numerous, but relatively little has been published on the role of UV attenuation by phytoplankton. Blough and Del Vecchio [74] summarize spatial and temporal relationships between CDOM, DOC and phytoplankton in coastal waters. Belzile et al. [49] report data from L. Biwa, Japan, in which both α_{cdom} and [chl *a*] (over the range 1.5–7.5 mg m^{-3}) are highly correlated with K_{d320} and K_{d380}. Twardowski and Donaghay [116] inferred from optical measurements at a coastal site the direct production of CDOM from a thin layer of phytoplankton in the water column.

A key parameter in bio-optical models is the chlorophyll-specific spectral absorption factor ($\alpha^*_{Chl,\lambda}$). Chlorophyll concentration is most often measured optically after extraction from phytoplankton. In vivo methods involving measurement of fluorescence can provide a convenient index of biomass but are confounded with acclimation and species effects on calibration parameters. The Quantitative Filter Technique (QFT) is widely used to provide a measure of particulate absorption over the waveband (typically 400–700 nm) of photosynthetically active radiation (PAR). The method, pioneered by Yentsch [53], involves concentrating particles onto a filter and then measuring absorption on the filter in a spectrophotometer. Many workers have contributed to refining this method; Kishino et al. [54] added an option to estimate separately the contribution of photosynthetic pigments and detritus by extraction with hot methanol, while Mitchell [55,56] established a standard technique and then extended it to different instruments and filter types. Numerous modifications to the QFT have been proposed (most recently by Roesler [92], Lohrenz [51] and Tassan et al. [93]) because of complications with calibration and the effects of loading and particle type. The QFT is the basis for a recent review by Morel [58] summarizing systematic variation in $\alpha^*_{Chl,PAR}$ of marine phytoplankton: the values are lowest in eutrophic waters and highest in oligotrophic waters. Over the chlorophyll range from 0.03 to 30 mg m^{-3} the value of α^*_{Chl} decreases by a factor of 10. He attributes roughly half of this variation to changes in cell size, and half to changes in accessory pigments, but does not address the issue of UV absorption. A new technique involving water column profiles made with an in situ absorption instrument (WET Labs, Inc. AC-9) is revolutionizing the characterization of α_{phyto} in visible wavelengths by allowing in vivo IOP measurements of phytoplankton and associated optical constituents at nine wavelengths (e.g., [49]). When a pair of instruments is lowered together, one can record absorption and beam attenuation at nine visible wavelengths for whole water while simultaneously recording the same signals for particle-free water [30].

The QFT can be used to measure UV absorption of particles, including phytoplankton cells, if the samples are analyzed shortly after filtering [47]. Figure 10 shows summer near-surface particulate absorption spectra for three lakes with different chlorophyll *a* and DOC concentrations (Hargreaves, unpublished). Comparing the three lakes, Crater Lake near-surface waters have the lowest concentrations of DOC (<0.2 g m^{-3}, [94]) and chlorophyll *a* (<0.2 mg m^{-3}, Emmanuel Boss, personal communication). Lake Giles is intermediate (1.1 g m^{-3} and 0.6 mg m^{-3}), and L. Lacawac has the highest concentrations

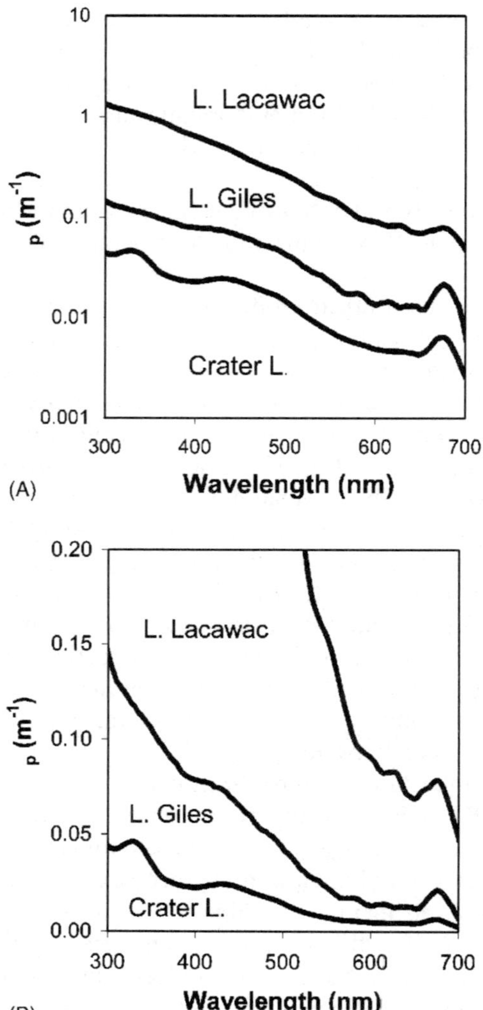

Figure 10. Particulate absorption for three lakes with different chlorophyll *a* and DOC concentrations (Hargreaves, unpublished). Crater L conditions: August 2001, Depth 25 m, $K_{d320} = 0.066$. Typical late summer near surface DOC < 0.2 g m^{-3} and chl-*a* <0.2 mg m^{-3} ([94], and Emmanuel Boss, personal communication). L. Giles conditions: June 2001, depth 0–6 m; DOC = 1.1 g m^{-3}, chl-*a* = 0.6 mg m^{-3}, $\alpha_{cdom320} = 0.48$ m^{-1}, $K_{d320} = 0.68$ m^{-1}. L. Lacawac conditions: June 2001, Depth, 0–2 m; DOC = 4.7 g m^{-3}, chl-*a* = 2.7 mg m^{-3}, $\alpha_{cdom320} = 12.3$ m^{-1}, $K_{d320} = 14.9$ m^{-1}. QFT method from Mitchell [56] as adapted by Roesler [92] using GF/F filters supported on quartz discs in a Shimadzu UV-1601 spectrophotometer. (A) Log scale. (B) Linear scale. The spectra for L. Lacawac ($\alpha_{p320} = 1.2$ m^{-1}, $\alpha_{p675} = 0.08$ m^{-1}) and L. Giles ($\alpha_{p320} = 0.12$ m^{-1}, $\alpha_{p675} = 0.02$ m^{-1}) show the blue and red peaks for chlorophyll superimposed on a CDOM-like exponential pattern of particulate absorption. Crater L. ($\alpha_{p320} = 0.045$ m^{-1}, $\alpha_{p675} = 0.006$ m^{-1}) shows a spectrum of algal pigments with a peak at 330 nm in addition to typical blue and red chlorophyll peaks. Crater L. chlorophyll *a* at 25 m depth calculated from the 675 nm peak (proportional to the other lakes) is 0.18–0.20 mg m^{-3}.

(4.7 g m^{-3} and 2.7 mg m^{-3}). In L. Giles and L. Lacawac the value for chlorophyll-specific absorption at 675 nm ($\alpha^{*}_{chl,675}$) is 0.03 from these data. The spectra for L. Lacawac ($\alpha_{p320}=1.2\ m^{-1}$, $\alpha_{p675}=0.08\ m^{-1}$) and L. Giles ($\alpha_{p320}=0.12\ m^{-1}$, $\alpha_{p675}=0.02\ m^{-1}$) show the blue and red peaks for chlorophyll superimposed on a CDOM-like exponential pattern of particulate absorption. Crater L. ($\alpha_{p320}=0.045\ m^{-1}$, $\alpha_{p675}=0.006\ m^{-1}$) shows a spectrum of algal pigments with a peak at 330 nm in addition to typical blue and red chlorophyll peaks.

Figure 11 shows underwater spectral irradiance in Crater Lake, OR (Hargreaves, unpublished). Data were collected during continuous lowering of two radiometers from 13:20–14:30 local time, 20 August 2001 (solar zenith angle 31–36°, clear sky) using a PRR-800 multichannel reflectance profiling radiometer to 153 m and a PUV-2500 profiling UV radiometer to 60 m. These filter-band radiometers have a sensor bandwidth of 8–10 nm and can rapidly and accurately track downwelling cosine irradiance (and upwelling radiance, in the PRR-800) over more than 7 decades of intensity. (Biospherical Instruments, Inc.). The PRR-800 was similar to the unit used for Figures. 1 and 2, but was equipped to sense a slightly different set of sensor wavebands (both PRR-800's included 340 and 380 nm channels). The PUV-2500 has sensors for downwelling cosine irradiance centered at 305, 313, 320, 340, and 380 nm and PAR and for upwelling 683 nm radiance (not shown). The hybrid spectrum in Figure 11 combines

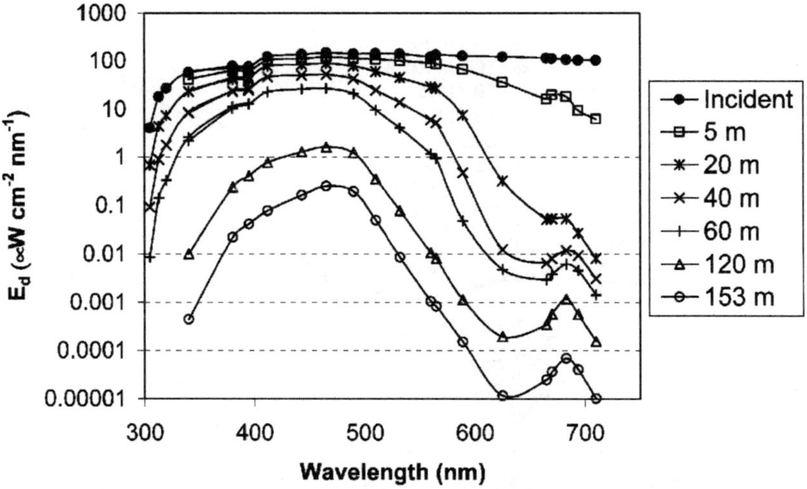

Figure 11. Crater Lake, Oregon: Underwater spectra of downwelling cosine irradiance with very low CDOM and algae concentrations, especially in the surface waters above 20 m. Data collected during continuous lowering from 13:20–14:30 local time, 20 August 2001 (solar zenith angle 31–36°, clear sky) using a PRR-800 multichannel reflectance profiling radiometer to 153 m and a PUV-2500 profiling UV radiometer to 60 m (8–10 nm bandwidth sensor response, Biospherical Instruments, Inc.). This hybrid spectrum using depth-binned data (with overlap at 340, 380, and 395 nm) is plotted on a log scale to show a range of irradiance spanning seven decades of magnitude. Except for the peak around 685 nm (caused by algal fluorescence), deep irradiance at WL $>$ 520 nm likely was caused by Raman scattering from shorter wavelengths.

depth-binned data from both instruments for 20, 40, and 60 m (overlap from the two instruments is visible at 340, 380, and 395 nm) on a log scale to show a range of irradiance spanning seven decades of magnitude. Except for the peak around 685 nm (caused by algal fluorescence), deep irradiance at WL >520 nm likely was caused by Raman scattering from shorter wavelengths. At these longer wavelengths accurate estimates of K_d are restricted to high-irradiance surface waters. Compared to Figure 1, the shorter wavelengths penetrate much more deeply in Crater Lake (in the coastal water at 40 m the irradiance at 340 nm is similar to that at 153 m in Crater L.). As was the case in Figure 1, the red peak caused by algal fluorescence is visible at all depths but the relative magnitude of the red peaks are different.

Figure 12 shows spectral K_d for Crater Lake calculated at 10 nm intervals from underwater spectral irradiance data collected at nearly the same time as the data in Figure 11, in this case using a LI-COR spectroradiometer (model LI-1800UW, 8 nm bandwidth). Spectral values for $K_{w,\lambda}$ (freshwater) estimated by Smith and Baker [18] are also included for comparison, and two new estimates of $K_{w,\lambda}$ that will be discussed in section 3.3.4. LI-COR $K_{d,\lambda}$ values for Crater Lake have been replicated using PRR-800, PUV-501, and PUV-2500 instruments (Biospherical Instruments, Inc.) on several dates. The accuracy of the LI-COR and PUV-501 instruments for measuring UV diffuse attenuation has

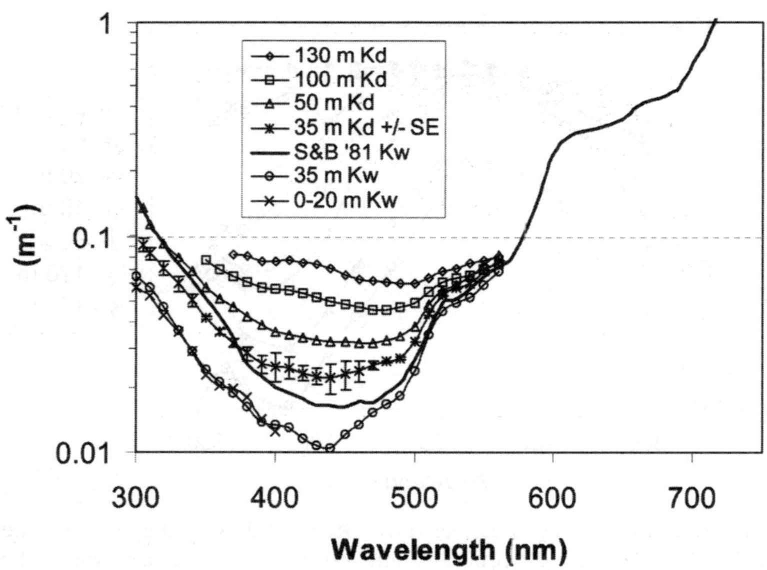

Figure 12. Spectral diffuse attenuation coefficients ($K_{d,\lambda}$) calculated at 10 nm intervals for Crater Lake Oregon, from downwelling irradiance scans at fixed depths, 12:00–13:00 local time, 20 August 2001 (SZA = 31°, clear sky) using a LI-COR LI-1800UW spectral radiometer (8 nm bandwidth single monochromator). $K_{w,\lambda}$ for freshwater estimated by Smith and Baker [18] is also compared to two new estimates computed by subtracting particulate absorption (measured similarly to spectra in Figure 10) from $K_{d,\lambda}$ averaged over 0–20 m and 30–40 m, adjusted for sky and sun angle (Figure 4B).

also been demonstrated in a field comparison [36]. Details of Crater Lake UV attenuation data will be published separately (B. Hargreaves, G. Larsen, J. Morrow, and others).

In Crater Lake during August, phytoplankton are extremely scarce in the upper 15 m (chlorophyll concentration ca. 0.1–0.2 mg m^{-3} [94 and Emmanuel Boss, personal communication]), a depth that also represents $Z_{37\%}$ for 320 nm. Typically the Crater Lake chlorophyll concentration in August increases with depth to a maximum concentration near 130 m (ca. 0.6 mg m^{-3} [94 and Emmanuel Boss, personal communication]). As the phytoplankton pigment increases with depth, the phytoplankton optical effects (direct and via CDOM) on UV and PAR diffuse attenuation increase as well (Figure 12). By comparing the curves for K_{d320} and K_{w320} in Figure 12 it is apparent that phytoplankton and associated CDOM absorption are contributing about 50% of the diffuse 320 nm attenuation at 50 m ($K_{d320}=0.092$ m^{-1}, $K_{w320}=0.045$ m^{-1}). Ayoub et al. [46] used the QFT to determine relative importance and seasonal variations for UV absorption by phytoplankton in the two lakes represented in Figure 10. The contribution of particulates was also high for oligotrophic L. Giles (particles contributed 20–55% of total absorption at 320 nm) but was low for the humic L. Lacawac (8–18% of total absorption, with particles dominated by CDOM-like detritus rather than phytoplankton). Other cases of within-lake variation in UV attenuation attributed to phytoplankton include high mountain lakes [62,63] and Lake Biwa, Japan [49]. More work will be required to determine if depth-variation in phytoplankton biomass and in chlorophyll-specific absorption represent the same biotic mechanisms and range of values reported for horizontal trophic gradients in ocean waters [58] and among lakes.

Absorption in the UV-A wavelengths includes accessory pigments associated with photosynthesis while UV-B absorption by living cells is caused in part by proteins and nucleic acids. Compounds that absorb with various peaks in the UV range accumulate in some organisms subject to UVR exposure and may serve as UV-B screening compounds: myco-sporine-like amino acids (MAA's) in algae and invertebrates and scytonemin and its derivatives in cyanobacteria [45,50, Chapter 10]. Chlorophyll-specific absorption is known to change with depth in the water column, but while the specific absorption of photosynthetic pigments tends to increase in dim light as depth increases, the response of UV-screening pigments may be opposite. Helbling et al. [44] adapted the QFT to measure UV and visible chlorophyll-specific absorption by marine phytoplankton (α^*_{Chl}) in marine waters of Antarctica and observed a UV-protective pigment whenever algae were exposed to at least 1% of incident UV-B (320 nm) irradiance. They observed a UV absorption peak between 310 and 330 nm for algae within the upper mixed layer as deep as 20 m (roughly the $Z_{1\%}$ attenuation depth for 320 nm) and down to 90 m when thermal stratification was weak enough to allow deep mixing. The value for α^*_{Chl} at 327 nm varied inversely with the depth of the mixed layer for a large number of samples, especially when diatoms were dominant, suggesting that some algae produce a UV screening pigment in proportion to their UV-B exposure. A similar UV-B particulate absorption peak was observed in Crater Lake, Oregon above the $Z_{1\%,320}$ depth

(Hargreaves, unpublished). In this case the peak was evident at 50 m (Figure 10), but not at 100 m, when the mixing depth was approximately 10 m and $Z_{1\%, 320}$ was 60 m (Figure 11).

3.3.3 Ultra-low attenuation (Kirk type W, WA, or WG natural waters)

Early studies of UV attenuation identified several ocean regions as unusually transparent. Table 1 shows three regions (Sargasso Sea, East Mediterranean, and central equatorial Pacific) with $K_d \leq 0.15$ in the range of 310–320 nm, including the earliest radiometer measurement of underwater UV-B attenuation [95]. Crater Lake Oregon is the only freshwater site to receive early attention for its "near distilled water" transparency near the surface. Crater Lake spectra for visible (Tyler [96]) and UV-A (Smith and Tyler [97]) wavelengths were recorded by the first underwater scanning spectroradiometer (Tyler and Smith [98]). An improved instrument recorded UV-A and visible spectral irradiance underwater in Crater Lake during July 1969 [99]. The authors commented on the similarities of Crater Lake water optical properties to those of pure water. There are no published data for Crater Lake using later versions of the spectroradiometer [8,100] that included the capacity to record UV-B wavelengths.

In the 1990's new commercial UV instruments made possible a large number of lake measurements, with the result that high mountain lakes in Austria, North America, and South America have been identified with $K_{d320} \leq 0.17$ (Table 1). Vincent et al. [68] measured UV attenuation in several Antarctic lakes and reported record low values for the depth range 10–20 m below the ice-covered surface of Lake Vanda ($K_{d320} = 0.055$), values smaller than the attenuation estimated for pure water reported by Smith and Baker [18]. Recent UVR measurements at Crater Lake OR revealed similar low values in surface waters (K_{d320} from 0.050 to 0.071, Table 1). As indicated by equation (10), K_d values include contributions from water, phytoplankton, and DOC. These recent low values for two lakes suggest that CDOM and phytoplankton concentrations are very low. The implication that the Smith and Baker's [18] values for K_w are too high in the UV wavelengths will be addressed in the next section. In both Crater Lake and Lake Vanda, UV attenuation changes with depth and is minimal near the surface. From the limited information available it appears that both phytoplankton and CDOM contribute to this change in UV attenuation with depth, as discussed in the previous section.

At Crater Lake, DOC levels have been reported to be less than 16 μM (equivalent to 0.2 mg l^{-1}) with algal biomass in surface waters <0.2 mg m^{-3} chlorophyll [94 and Emmanuel Boss, personal communication]. At Lake Vanda, DOC in the UV-transparent surface water is reported to be 0.3 g m^{-3} with algal biomass <0.1 mg m^{-3} chlorophyll [68]. The role of CDOM at such low attenuation levels is difficult to measure by spectrophotometer and even the concentration of DOC is at the level of "blank" values for high temperature oxidation instruments [40]. If one assumes that the surface value for DOC in Crater Lake is actually 0.2 g m^{-3} at the time of low K_d measurements, the absorption by CDOM can be estimated using the DOC specific absorption relationship published for other sites. Marine DOC concentrated from Gulf of Mexico and Mississippi River plume water by Carder et al. [69] would give $\alpha_{cdom320}=0.02\ m^{-1}$ if all the DOC consisted of marine-like fulvic acids (equivalent to $\alpha_{cdom,412}\approx 0.001$), or $\alpha_{cdom320}=0.11\ m^{-1}$ if it were all marine-like humic acids. The lower estimate, $0.02\ m^{-1}$ at 320 nm, could be a component of the minimum K_{d320} value in Crater Lake of $0.05\ m^{-1}$, depending on the absorption and scattering attributed to water molecules and phytoplankton. Using the reported DOC value of $0.3\ g\ m^{-3}$ for Lake Vanda surface waters, the predicted K_{d320} is $0.035\ m^{-1}$ (again, assuming marine-like fulvic acids), compared to measured $K_{d320}=0.055\ m^{-1}$. K_{d320} can also be estimated using a regression of K_{d320} versus DOC from a series of high latitude lakes (Figure 8) that included Lake Vanda and other Antarctic lakes ([68] and Table 3). Predicted K_{d320} ranges from 0.060 to $0.090\ m^{-1}$ for lakes with DOC in the range 0.2–0.3 $mg\ l^{-1}$ (after adding $K_{w320}=0.04$). To avoid introducing a latitudinal bias (caused by sun angle differences) future relationships of this sort should be adjusted using equation (11).

3.3.4 *Attenuation and absorption by pure water*

The extremely low UV attenuation in surface waters of Crater Lake provides an opportunity to improve the Smith and Baker [18] "upper bounds estimate" of UV attenuation in pure water, K_w. The recent near-surface measurements of spectral K_d (Figure 12) and spectral absorption of particles (Figure 10) can be combined to make a new "upper bounds estimate" of K_w. Spectral $K_{w,\lambda}$ values averaged over depths of 0–20 m and 30–40 m are shown in Figure 12, along with the freshwater K_w spectrum from Smith and Baker [18]. The new "K_w" estimates have been computed by first adjusting K_d's with equation (11) for diffuse and direct sunlight (see Figure 4B) and then subtracting particle absorption appropriate to the depth. Because of uncertainties with the QFT calibration for this lake, the particulate absorption values were adjusted downward from the values calculated by the Roesler method [92] until the two estimates of spectral K_w (0–20 m and 30–40 m) converged. The final adjustment was to 37% of the original particulate absorption coefficient values. A justification for this approach is that unpublished measurements of α_{cdom} made several weeks later by Emmanuel Boss (personal communication) using a Wetlabs AC-9 in situ absorption meter showed $\alpha_{cdom440}$ essentially uniform in the top 40 m and just above the limit of detection for the instrument. Another supporting argument is that the same adjustment improves an estimate for chlorophyll *a* concentration at 25 m depth calculated from the particulate absorption peak at 675 nm (Figure 10). By using $\alpha^{*}_{chl,675}=0.012$ derived from September data (2001 AC-9 particle absorption and extracted chlorophyll *a* concentration; Emmanuel Boss, personal communication), the uncorrected estimate for chlorophyll *a* concentration ($0.5\ mg\ m^{-3}$) was reduced to a reasonable value within the normal range for that depth in Crater Lake ($0.18\ mg\ m^{-3}$). Tassan et al. [93] have suggested that small phytoplankton (such as those abundant in the surface of Crater Lake) cause the Roesler method [92] to overestimate particulate absorption. The resulting "K_w" estimates may still include scattering effects of particles present plus absorption by any cdom present that was not correlated with particle absorption. For UV wavelengths the Crater Lake "high" estimates for K_{w320} from 0–20 m and 30–40 m range from 0.043 to 0.047 m^{-1}. These are substantially lower than the Smith and Baker [18] value of $K_{w320}=0.09\ m^{-1}$. Note that these freshwater values of $K_{w,\lambda}$ from Smith and Baker [18] include slightly smaller backscatter terms than the seawater $K_{w,\lambda}$ values used in Figure 2.

A laboratory study of spectral absorption by pure water [101], which extended only into the UV-A wavelengths, reported pure water absorption as $\alpha_{w380}=0.011\ m^{-1}$. This value can be converted into K_{w380} using equation (9) (with $b_{bw380}=0.007$ from Mobley [5]) to yield $K_{w380}=0.017\ m^{-1}$, compared to the Crater Lake estimate, $K_{w380}=0.018\ m^{-1}$ (using near surface K_{d380} averaged from several instruments, August 2001). Both values are lower than the Smith and Baker [18] estimate, $K_{w380}=0.026$. Further refinements in estimates for spectral K_w and α_w for UV wavelengths could result from improved measurements of spectra for K_d, particle absorption, CDOM absorption, and particle scattering in the remarkably clear waters of Crater Lake.

3.4 Predicting levels of UV-attenuating constituents

A range of bio-optical models (described in Section 3.3) can be used to estimate UV attenuation in natural waters when the concentrations and optical properties of algal pigment and DOC are known. How can these concentrations and optical properties be predicted from regional-scale features of the environment? Predicting typical concentrations and optical properties for either algal pigments or DOC in an aquatic system will require knowledge about sources and sinks, mechanisms for changing optical properties, and how each of these responds to the environment, a subject that will be only briefly reviewed here. Only passing mention will be made here of the role of inorganic particles, which can be extremely important as UV attenuators in some aquatic systems [6, 10, 49]. Such systems are shallow or subject to erosion: streams and rivers with disturbed or non-vegetated watersheds, and shallow lakes or coastal systems where the high energy keeps particles suspended in the water column. One special case is the precipitation of carbonates that imparts a whitish color (observed in several lakes by Laurion et al. [62]). Another special case is the meltwater from glaciers, where interactions with the underlying rock surface create small particles that give the water a milky appearance.

Phytoplankton abundance responds to nutrients, light, and grazing pressure. The primary autochthonous source of CDOM is often assumed to be photosynthetic organisms, but heterotrophic bacteria may play an important role by processing the relatively UV-transparent photosynthate and releasing modified compounds that absorb UVR more strongly [91]. Spatial and temporal linkage between CDOM and the microbial community appears complex and difficult to predict at present, in part because the microbial community also attenuates UVR. Predictions about microbial UV attenuation are rather difficult at present because there are so few measurements and reliability of current techniques (e.g. QFT) is still being debated. The adaptive physiological and evolutionary responses of phytoplankton to UVR are likely to differ from their adjustments of photosynthetic pigments to light and depth. Here again there is too little information to make predictions except that one should not be surprised to find UVR screening pigments in phytoplankton exposed to high UVR (perhaps when 320 nm irradiance exceeds 1% of surface levels?).

The balance between sources and sinks for DOC will determine its concentration in natural waters and thus establish a major factor in UV attenuation. The levels and optical properties of CDOM in aquatic systems will depend in part on whether the source is within the water column (autochthonous) or coming from elsewhere (allochthonous). Allochthonous sources include watershed runoff (e.g. from soil or wetlands, especially water-saturated soils but also from man-modified surfaces which may be enriched with petroleum hydrocarbons), and wastewater discharge. In estuaries the large CDOM load typically carried by rivers is both diluted (and to a lesser extent, precipitated) as it mixes with brackish water. Allochthonous sources are affected by precipitation, evaporation, soil hydraulic residence time, and temperature, which influence production of CDOM in saturated soils and transfer, dilution, and concentration of this material in receiving waters. In a comparative study of 337 lakes from northern USA and Canada, Rasmussen et al. [102] found a strong positive correlation between CDOM and the ratio of drainage area to lake area, and a negative correlation between CDOM and average slope of the drainage landscape. CDOM in coastal regions is strongly correlated with river discharge and inversely correlated with salinity [74]. CDOM in two Australian reservoirs was highly correlated with season, increasing during rainy periods and declining during dry, sunny periods in response to microbial and photochemical degradation [103]. The two reservoirs differed in average level of CDOM, with 97% of the variability in average annual CDOM accounted for by difference in hydraulic residence time. Similarly, Arts et al. [34] suggested that long residence time in prairie lakes (Canada) causes the DOC to become more UV transparent. Climate can affect both the watershed yield of DOC to lakes and coastal oceans and hydraulic residence time by influencing snowmelt, precipitation, evaporation, and watershed soil properties [13]. A period of warm and dry years resulted in a decline in DOC of Canadian shield lakes [104]. Temporal variation in DOC and CDOM absorption in lakes of N. Michigan (USA) was correlated with ice-out date and spring-summer precipitation [119].

At regional and larger scales several end members and other patterns of UV attenuation have emerged. The clearest lakes and ocean regions are those most isolated from terrestrial sources of CDOM and nutrients. A barren watershed containing little or no vegetation surrounds Crater Lake and Lake Vanda, and the permanently unmixed bottom waters likely serve as a trap for nutrients present in sinking biomass. The open ocean may appear to be equally isolated, but the connection with a deep-circulating source of CDOM and nutrients may provide a low baseline level of UV attenuating substances that increases in the presence of deep mixing or upwelling. High elevation lakes tend to have lower levels of DOC [20] along with less vegetation in the watershed [62]. Mountain lakes of the Alps and Pyrenees [62] show an expected decrease in watershed vegetation with elevation, but when meadows and forest are compared at the same elevation the meadow-dominated watersheds have higher DOC levels. Lakes in contact with wetlands tend to have high levels of DOC (reviewed in [24]),but even higher levels occur in arid closed basin lakes [34].

Variations in DOC quality (DOC – specific attenuation and absorption) warrant further discussion. These variations can be divided into two related categories: changes in specific attenuation correlated with DOC concentrations, and changes in specific attenuation correlated with lake and watershed characteristics. The immediate basis for the power relationships between K_{d320} and [DOC] (Table 3) comes directly from the power relationship between a_{d320} and [DOC] (Table 2) for the data of Morris et al. [60]. But why should DOC quality change together with DOC concentration over the scale of lakes in this study? What determines the exponent in the power model relating specific absorption and DOC concentration?

Land cover in lake watersheds, climate, and hydraulic residence time influence the spatial pattern of DOC quality. Data from Laurion et al. [62] show that lakes with high-elevation watersheds had lower DOC concentrations than those at lower elevations in the same region. Lakes at similar elevations had lower DOC-specific attenuation if the watershed was rocky compared to watersheds with meadow or forested land cover. These data also showed that, for lake watersheds with similar land cover, DOC-specific attenuation was elevated when catchment area was more than 50 times larger than lake area, a characteristic associated with short residence times. For all the lake surveys in Table 2 and the majority in Table 3 the power equation exponent is greater than 1. These cover mountain lakes and wet areas at mid to high latitudes. The exceptions, where DOC-specific absorption varies inversely with DOC concentration, include ponds and lakes in the arid prairies of central Canada.

Temporal variation in DOC quality provides clues to explain the spatial variation. The decline of DOC-specific absorption in stratified surface waters [22, 23] and lake versus feeder stream [90] are attributed to cumulative photobleaching of the DOC pool. While photobleached DOC is in some cases subject to enhanced microbial utilization [105], the old and previously bleached DOC of saline prairie lakes is metabolized very slowly [90]. It is self-evident that photobleached DOC will not be dominant in the DOC pool while the rate of influx of non-bleached DOC is high. If hydraulic residence time is short then even low rates of DOC influx or production will be adequate to prevent accumulation of photobleached DOC.

3.4.1 A conceptual model for UV-DOC relationships

In most comparative studies to date the concentration and optical qualities of

DOC are the best predictors of UV attenuation. A conceptual model is proposed here to predict spatial and temporal patterns of UVR attenuation, including the relationship of K_{dUV} to [DOC], and α_{cdom} to [DOC], for all non-turbid aquatic systems. The DOC concentration will tend to be high in aquatic systems where influx or production is high relative to the rate of dilution or flushing by direct precipitation, snowmelt, or other source of low-DOC water. The [DOC] will also be high when a system with long hydraulic residence time experiences evaporation rates that exceed the rates of DOC degradation (microbial and photochemical). DOC concentration will be low in aquatic systems where DOC influx or production is low relative to rates of degradation or flushing with low-DOC water. DOC optical quality reflects the biotic source and the extent of photobleaching. For systems with short hydraulic residence times the DOC-specific absorption will be higher for sources from higher plant and lower for sources from the microbial community. Systems with long hydraulic residence times will tend to have low DOC-specific absorption when the long term rate of influx or production of DOC is slow compared to the rate of photobleaching by sunlight. DOC-specific absorption will become low as well in water that is seasonally isolated by density differences in a thin surface layer exposed to prolonged sunlight. Combining the processes influencing DOC concentration and specific absorption results in two hypotheses explaining the power relationships between K_d and [DOC] in equation (15). When DOC is not concentrated by water loss, the power exponent tends to exceed 1.0 because the systems with low [DOC] have experienced more cumulative photobleaching. In arid or frozen regions where cumulative water loss occurs in systems with long hydraulic residence time, the power exponent tends to be less than 1.0 because systems with high [DOC] have experienced more cumulative photobleaching.

3.5 Future directions

Fundamental to progress in understanding UV attenuation by natural waters is an improved understanding of absorption by pure water. Crater Lake and others like it offer the possibility to achieve optical closure for IOPs and AOPs with appropriate instrumentation. Advanced spectral absorption metres (e.g., the ICAM, [101,107]) should be extended into UV-B wavelengths and applied to testing ultra-pure water in the laboratory as well as the absorption of phytoplankton and other suspended particles at low concentrations. Temperature and salinity effects should be established for absorption by water in UV wavelengths.

Optical methods can be improved in large part by extending existing methods developed for visible wavelengths. A multispectral scalar UV sensor combined with existing uplooking and downlooking cosine sensors would improve our ability to measure $\bar{\mu}$ and thus to relate IOP's with AOP's. Lacking this, published algorithms for estimating $\bar{\mu}$ from multispectral reflectance meters should be extended into the UV wavelengths. Future underwater measurements should be more consistently combined with determination of sun angle (requires geographic coordinates and time of day) and the diffuse fraction for solar UV irradiance in order that attenuation calculations can be standardized for sun and sky effects. CDOM measurements should be combined more often with measurement of [DOC] and with CDOM spectral fluorescence using the new method of McKnight et al. [80] to help characterize DOM optical qualities. At very low levels of CDOM the use of a proxy such as fluorescence will be needed for field measurements; even better would be an instrument like the Wetlabs AC-9 that operates in UV wavebands. CDOM calculations should consistently include corrections for scattering (where needed) and baseline offset, and regression techniques should avoid the noise region of the measuring instrument (where negative absorption coefficients can create a bias in calculated spectral slope). Spectral slope should be calculated for UV-B and UV-A wavebands. CDOM values should be checked against matched diffuse attenuation measurements to detect (and correct) errors where $K_d \leq (\alpha_{cdom} + \alpha_{phyto} + \alpha_{water})$. Measurement of UV absorption and attenuation by particles should be improved, including better methods for calibration of QFT using field samples. Phytoplankton pigment should be measured in combination with field optical measurements to develop trends in chlorophyll-specific absorption in the water column.

More field measurements of UV absorption by phytoplankton are needed and these should be scaled to chlorophyll concentration and evaluated for spatial and temporal patterns across gradients (trophic, UV, nutrient, grazing). The effect of depth and mixing should be evaluated for effects on UV exposure and on the response of cells to adjust UV-absorbing and photosynthetic pigments.

Sources and sinks of CDOM should be evaluated in lakes and ocean regions to achieve closure for carbon and optical budgets. Spatial correlation between microbial communities and CDOM is now possible with field absorption instruments but more work is also needed on microbial processing of DOC, especially in low-DOC systems. Effects of climate change on DOC and UV attenuation should be explored for different geographic regions, and should take into account changes in DOC-specific absorption as well as DOC concentration. Patterns of DOC-specific CDOM absorption and other DOC characteristics should be determined for a larger range of saline environments, including saline lakes, coastal and the open ocean, with comparisons between surface and deep water. More work is needed on the roles for watersheds and hydrology (including concentration of [DOC] by regulating its influx rate or by evaporative loss of water) in determining DOC optical qualities. The counterpart for oceans is to better determine basin-scale changes in CDOM quality as a function of water column biotic processes, photochemical processes, mixing, and other sources and sinks [74].

Acknowledgements

Support to the author for some of the published and unpublished data presented here has been provided by the Keck Foundation, the National Science Foundation, Lacawac Sanctuary, and Blooming Grove Hunting and Fishing Club. This manuscript was completed while the author was on sabbatical leave from Lehigh

University and supported in the laboratory of Horacio Zagarese, Univ. Comahue, Bariloche Argentina, by IAI, (Inter-American Institute for Global Change Research). Some unpublished data and instrument use at Crater Lake, OR have been provided by J.H. Morrow (Biospherical Instruments), Gary Larson (US Geologic Survey, Corvallis, OR) and Scott Girdner (National Park Service, Crater Lake). Helpful comments and unpublished data were also provided by Emmanuel Boss (Oregon State University). Other helpful comments were provided by an anonymous reviewer.

References

1. C.E. Williamson (1995). What role does UV-B radiation play in freshwater ecosystems? *Limnol. Oceanogr.*, **40**, 386–392.
2. C.E. Williamson, P.J. Neale, G. Grad, H.J. De Lange, B.R. Hargreaves (2001). Beneficial and detrimental effects of UV radiation on aquatic organisms: implications of variation in spectral composition *Ecol. Appl.* **11**, 1843–1857.
3. W. Preisendorfer (1976). *Hydrologic Optics* (vol. 1–6). U.S. Dept. of Commerce, NOAA, Honolulu, Hawaii.
4. G. Jerlov (1968). *Optical Oceanography*. Elsevier Publishing Co., Amsterdam.
5. C.D. Mobley (1994). *Light and Water: Radiative Transfer in Natural Waters*. Academic Press, San Diego.
6. J.T.O. Kirk (1994). *Light and Photosynthesis in Aquatic Ecosystems*, (2nd ed.). Cambridge University Press, Cambridge, 509 pp.
7. R.C. Smith, J.E. Tyler (1976). Transmission of solar radiation into natural waters. In: K.C. Smith (Ed.), *Photochemical and photobiological reviews*, (Volume 1, pp. 117–155). Plenum, New York.
8. R.C. Smith, K.S. Baker (1979). Penetration of UV-B and biologically effective doesrates in natural waters. *Photochem. Photobiol.*, **29**, 311–323.
9. K. S. Baker, R.C. Smith (1982). Spectral Irradiance penetration in natural waters. In: J. Calkins (Ed.), *The Role of Solar Ultraviolet Radiation in Marine Ecosystems, NATO Conference Series IV, Marine Sciences* (Vol. 7, pp. 233–246). Plenum Press, New York.
10. J.T.O. Kirk (1994). Optics of UV-B radiation in natural waters. *Arch. Hydrobiol. Beih., Ergbn. Limnol.*, **43**, pp. 226.
11. C.R. Booth, J.H. Morrow (1997). The Penetration of UV into Natural Waters, *Photochem. Photobiol.*, **65**, 254–257.
12. R.F. Whitehead, S.F. de Mora, S. Demers (2000). Enhanced UV radiation – a new problem for the marine environment. In: S. de Mora, S. Demers, M. Vernet (Eds), *The effects of UV Radiation in the Marine Environment*, (pp. 1–34). Cambridge University Press, Cambridge UK.
13. M.A. Xenopoulos, D.W. Schindler (2001). Physical factors determining ultraviolet radiation flux into ecosystems. In: C.S. Cockell and A.R. Blaustein (Eds), *Ecosystems, Evolution, and Ultraviolet Radiation*, (pp. 36–62). Springer, New York.
14. R.C. Smith, K.S. Baker (1986). Analysis of ocean optical data II. *Proc. SPIE, Ocean Optics VIII*, **636**, 95–107.
15. R.G. Zepp, D.M. Cline (1977). Rates of direct photolysis in the aquatic environment. *Environ. Sci. Technol.*, **11**, 359–366.
16. J.R. Zaneveld, E. Boss, A. Barnard (2001). Influence of surface waves on measured and

modeled irradiance profiles *Appl. Opt.*, **40**, 1442–1449.

17. H.R. Gordon (1989). Can the Beer-Lambert law be applied to the diffuse attenuation coefficient of ocean water? *Limnol. Oceanogr.*, **34**, 1389–1409.
18. R.C. Smith, K.S. Baker (1981). Optical Properties of the clearest natural waters (200–800 nm). *Appl. Opt.*, **20**, 177–184.
19. V.S. Kuwahara (2000). Variability in the relative penetration of ultraviolet radiation to photosynthetically available radiation in temperate coastal waters, Japan. *J. Oceanogr.*, **56**, 399–408.
20. C.E. Williamson, R.S. Stemberger, D.P. Morris, T.M. Frost, S.G. Paulsen (1996). Ultraviolet radiation in North American lakes: Attenuation estimates from DOC measurements and implications for plankton communities. *Limnol. Oceanogr.*, **41**, 1024–1034.
21. C.E. Williamson (1996). Effects of UV radiation on freshwater ecosystems. *Int. J. Environ. Stud.*, **51**, 245–256.
22. D.P. Morris, B.R. Hargreaves (1997). The role of photochemical degradation of dissolved organic carbon in regulating the UV transparency of three lakes on the Pocono Plateau. *Limnol. Oceanogr.*, **42**, 239–249.
23. A. Vodacek, N.V. Blough, M.D DeGrandpre, E.T. Peltzer, R.K. Nelson (1997). Seasonal variation of CDOM and DOC in the Middle Atlantic Bight: Terrestrial inputs and photooxidation, *Limnol. Oceanogr.*, **42**, 674–686.
24. P.A. Bukaveckas, M. Robbins-Forbes (2000). Role of dissolved organic carbon in the attenuation of photosynthetically active and ultraviolet radiation in Adirondack lakes. *Freshwat. Biol.*, **43**, 339–354.
25. R.W. Preisendorfer (1961). Application of radiative transfer theory to light measurements in the sea. *Union geod. Geophys. Inst. Monogr.*, **10**, 11–30, cited in [3].
26. K.S. Baker, R.C. Smith (1979). Quasi-inherent characteristics of the diffuse attenuation coefficient for irradiance, *Ocean Optics VI, SPIE*, **208**, 60–63.
27. V.N. Pelevin, V.A. Rutkovskaya (1977). On the optical classification of ocean waters from the spectral attenuation of solar radiation. *Oceanology*, **17**, 28–32.
28. R.C. Smith, K.S. Baker (1978). The bio-optical state of ocean waters and remote sensing. *Limnol. Oceanogr.*, **23**, 247–259.
29. H.R. Gordon (1976). Radiative transfer in the ocean: a method for determination of absorption and scattering properties. *Appl. Opt.*, **15**, 2611–2613.
30. M. Stramska, D. Stramski, B.G. Mitchell, C.D. Mobley (2000). Estimation of the absorption and backscattering coefficients from in-water radiometric measurements. *Limnol. Oceanogr.*, **45**, 628–641.
31. R.H. Stavn (1988). Lambert-Beer law in ocean waters: Optical properties of water and of dissolved/suspended material, optical energy budgets. *Appl. Opt.*, **27**, 222–231.
32. A.A. Gershun (1936). Fundamental ideas of the theory of a light field (vector methods of photometric calculations). (in Russian). *Izvestiya Akad. Nauk SSSR*, pp. 417–430, cited in Mobley [13].
33. N.M. Scully, D.R.S. Lean (1994). The attenuation of ultraviolet radiation in temperate lakes. *Arch. Hydrobiol. Ergebn. Limnol.*, **43**, 135–144.
34. M.T. Arts, R.D. Robarts, F. Kasai, M.J. Waiser, V.P. Tumber, A.J. Plante, H. Rai, H.J. de Lange (2000). The attenuation of ultraviolet radiation in high dissolved organic carbon waters of wetlands and lakes on the northern Great Plains. *Limnol. Oceanogr.*, **45**, 292–299.
35. K.W. Patterson, R.C. Smith, C.R. Booth (1997). A method for removing a majority of the error in PUV attenuation coefficients due to spectral drift in response with depth in the water column. *Ocean Optics XIII, SPIE*, **2963**, 737–742.

36. J.T.O. Kirk, B.R. Hargreaves, D.P. Morris, R.B. Coffin, B. David, D. Frederickson, D. Karentz, D.R.S. Lean, M.P. Lesser, S. Madronich, J.H. Morrow, N.B. Nelson, N.M. Scully (1994). Measurements of UV-B radiation in two freshwater lakes: an instrument intercomparison. *Arch. Hydrobiol. Beih Ergebn. Limnol.*, **43**, 71–99.
37. A. Morel, L. Prieur (1977). Analysis of variations in ocean colour. *Limnol. Oceanogr.*, **22**, 709–722.
38. J.T.O. Kirk (1980). Spectral absorption properties of natural waters: contribution of the soluble and particulate fractions to light absorption in some inland waters of southeastern Australia. *Aust. J. Mar. Freshwater Res.*, **31**, 287–296.
39. L. Prieur, S. Sathyendranath (1981). An optical classification of coastal and oceanic waters based on the specific spectral absorption curves of phytoplankton pigments, dissolved organic matter, and other particulate materials. *Limnol. Oceanogr.*, **26**, 671–689.
40. J.H. Sharp, R. Benner, L. Bennett, C.A. Carson, R. Dow, S.E. Fitzwater (1993). Re-evaluation of high temperature combustion and chemical oxidation measurements of dissolved organic carbon in seawater. *Limnol. Oceanogr.*, **38**, 1174–1782.
41. E. Aas, N.K. Højerslev (2001). Attenuation of ultraviolet irradiance in North European coastal waters. *Oceanologia*, **43**, 139–168.
42. R.C. Smith and K.S. Baker (1978). Optical classification of natural waters. *Limnol. Oceanogr.*, **23**, 260–267.
43. K.S. Baker, R.C. Smith (1982). Bio-optical classification and model of natural waters. 2 *Limnol. Oceanogr.*, **27**, 500–509.
44. E. W. Helbling, V. Villafane, O. Holm-Hansen (1994). Effects of ultraviolet radiation on Antarctic marine phytoplankton photosynthesis with particular attention to the influence of mixing. In: C.S. Weiler, P. A. Penhale (Eds). *Ultraviolet Radiation In Antarctica: Measurements and Biological Effects, Antarctic Research Series.* American Geophysical Union, Washington, D.C. (Vol. 62, pp. 207–227).
45. D. Karentz, S. McEuen, M.C. Land, W.C. Dunlap (1991). Survey of mycosporine-like amino acid compounds in Antarctic marine organisms: Potential protection from ultraviolet exposure. *Mar. Biol.*, **108**, 157–166.
46. L. Ayoub, B.R. Hargreaves, D.P. Morris (1997). UVR attenuation in lakes: Relative contribution of dissolved and particulate material. *Ocean Optics XIII, SPIE*, **2963**, 338–352.
47. H.M. Sosik (1999). Storage of marine particulate samples for light-absorption measurements. *Limnol. Oceanogr.*, **44**, 1139–1141.
48. C. Belzile, S.C. Johannessen, M. Gosselin, S. Demers, W.L. Miller (2000). Ultraviolet attenuation by dissolved and particulate constituents of first-year ice during late spring in an Arctic polynya. *Limnol. Oceanogr.*, **45**, 1265–1273.
49. C. Belzile, W.F. Vincent, M. Kumagai (2002). Contribution of absorption and scattering to the attenuation of UV and photosynthetically available radiation in Lake Biwa, *Limnol. Oceanogr.*, **47**, 95–107.
50. S.W. Jeffrey, H.S. MacTavish, W.C. Dunlap, M. Vesk, K. Groenewoud (1999). Occurrence of UVA- and UVB-absorbing compounds in 152 species (206 strains) of marine microalgae. *Mar. Ecol. Progr. Ser.*, **189**, 35–51.
51. S.E. Lohrenz (2000). A novel theoretical approach to correct for pathlength amplification and variable sampling loading in measurements of particulate spectral absorption by the quantitative filter technique. *J. Plankton. Res.*, **22**, 639–657.
52. A. Morel (1980). In-water and remote measurement of ocean colour. *Boundary-layer Meteorol.*, **18**, 177–201.
53. C.S. Yentsch (1962). Measurement of visible light absorption by particulate matter in

the ocean. *Limnol. Oceanogr.*, **7**, 207–217.
54. M. Kishino, N. Takahashi, N. Okami, S. Ichimura (1985). Estimation of spectral absorption coefficients of phytoplankton in the sea. *Bull. Mar. Sci.*, **37**, 634–642.
55. B.G. Mitchell D.A. Kiefer (1984). Determination of absorption and fluorescence excitation spectra for phytoplankton. In: O. Holm-Hansen, L. Bolis, R. Gilles (Eds). *Marine Phytoplankton and Productivity* (pp. 157–169). Springer-Verlag, Berlin.
56. B.G. Mitchell (1990). Algorithms for determining the absorption coefficient of aquatic particulates using the quantitative filter technique (QFT). *Ocean Optics X, SPIE*, **1302**, 137–148.
57. P.S. Kuhn, H.I. Browman, B. MacArthur, J.-F. St.-Pierre (1999). Penetration of ultraviolet radiation in the waters of the estuary and Gulf of St. Lawrence. *Limnol. Oceanogr.* **44**, 710–716.
58. A. Morel (1997). Optical properties of oceanic Case 1 waters, revisited. *Ocean Sciences XIII, SPIE*, **2963**, 108–114.
59. C.S. Yentsch, D.A. Phinney (1989). A bridge between ocean optics and microbial ecology. *Limnol. Oceanogr.*, **34**, 1694–1705.
60. D.P. Morris, H. Zagarese, C.E. Williamson, E.G. Balseiro, B.R. Hargreaves, B. Modenutti, R. Moeller, C. Queimalinos (1995). The attenuation of solar UV radiation in lakes and the role of dissolved organic carbon. *Limnol. Oceanogr.*, **40**, 1381–1391.
61. I. Laurion, W.F. Vincent, D.R.S. Lean (1997). Underwater Ultraviolet radiation: development of spectral models for northern high latitude lakes. *Photochem. Photobiol.*, **65**, 107–114.
62. I. Laurion, M. Ventura, J. Catalan, R. Psenner, R. Sommaruga (2000). Attenuation of ultraviolet radiation in mountain lakes: Factors controlling the among- and within-lake variability. *Limnol. Oceanogr.*, **45**, 1274–1288.
63. R. Sommaruga, R. Psenner (1997). Ultraviolet radiation in a high-mountain lake of the Austrian Alps: Air and underwater measurements. *Photochem. Photobiol.*, **65**, 957–963.
64. J. Shapiro (1957). Chemical and biological studies on the yellow organic acids of lake water, *Limnol. Oceanogr.*, **2**, 161–179.
65. N.K. Højerslev (1982). Yellow substance in the sea. In: J. Calkins (Ed.). *The Role of solar ultraviolet radiation in marine ecosystems* (pp. 263–279). Plenum Press, New York.
66. J.T.O. Kirk (1976). Yellow substance (gelbstoff) and its contribution to the attenuation of photosynthetically active radiation in some inland and coastal southeastern Australian waters. *Aust. J. Mar. Freshwater Res.*, **27**, 61–71.
67. A. Bricaud, A. Morel, L. Prieur (1981). Absorption by dissolved organic matter in the sea (yellow substance) in the UV and visible domains. *Limnol. Oceanogr.*, **26**, 43–53.
68. W.F. Vincent, R. Rae, I. Laurion, L. Howard-Williams, J. Priscu (1998). Transparency of Antarctic ice-covered lakes to solar UV radiation. *Limnol. Oceanogr.*, **43**, 618–624.
69. K.L. Carder, R.G. Stewart, G.R. Harvey, P.B. Ortner (1989). Marine humic and fulvic acids: effect on ocean color and fluorometric detection. *Limnol. Oceanogr.*, **34**, 68–81.
70. D.M. McKnight, R. Harnish, R.L. Wershaw, J.S. Baron, S. Schiff (1997). Chemical characteristics of particulate, colloidal, and dissolved organic material in Loch Vale watershed, Rocky Mountain National Park. *Biogeochemistry*, **36**, 99–124.
71. W.S. Pegau, J.R.V. Zaneveld (1993). Temperature-dependent absorption of water in the red and near-infrared portions of the spectrum. *Limnol. Oceanogr.*, **38**, 188–192.
72. W.S. Pegau, D. Gray, J.R.V. Zaneveld (1997). Absorption and attenuation of visible and near-infrared light in water: dependence on temperature and salinity. *Appl. Opt.*, **36**, 6035–6046.

73. R.J. Davies-Colley, W.N. Vant (1987). Absorption of light by yellow substance in freshwater lakes. *Limnol. Oceanogr.*, **32**, 416–425.
74. N.V. Blough, R. Del Vecchio (2002). Chromophoric dissolved organic matter (CDOM) in the coastal environment. In: D.A. Hansell and C.A. Carlson (Eds). *Biogeochemistry of Marine Dissolved Organic Matter* (pp. 509–546), Academic Press.
75. B.R. Hargreaves, C.L. Osburn, D.P. Morris, R.E. Moeller (2001). Spectral slope response to solar and biotic processes: Comparison of model with estimates from CDOM absorption and UV attenuation measurements. *Aquatic Sciences Meeting of ASLO*, Albuquerque, New Mexico, USA, February (abstract).
76. S. Markager, W.F. Vincent (2000). Spectral light attenuation and the absorption of UV and blue light in natural waters. *Limnol. Oceanogr.*, **45**, 642–650.
77. C.A. Stedmon, S. Markager, H. Kaas (2000). Optical properties and signatures of chormophoric dissolved organic matter (CDOM) in Danish coastal waters. *Estuarine, Coastal Shelf Sci.*, **51**, 267–278.
78. R.G. Wetzel (1992). Gradient-dominated ecosystems: sources and regulatory functions of dissolved organic matter in freshwater ecosystems. *Hydrobiologia*, **29**, 181–198.
79. C.E. Williamson, D.P. Morris, M.L. Pace, O.G. Olson (1999). Dissolved organic carbon and nutrients as regulators of lake ecosystems: Resurrection of a more integrated paradigm. *Limnol. Oceanogr.*, **44**, 795–803.
80. D.M McKnight, E.W. Boyer, P.K. Westerhoff, P.T. Doran, T. Kulbe, D.T. Anderson (2001). Spectrofluorometric characterization of dissolved organic matter for indication of precursor organic material and aromaticity. *Limnol. Oceanogr.*, **46**, 38–48.
81. H. DeHaan (1993). Solar UV-light penetration and photodegradation of humic substances in peaty lake water. *Limnol. Oceanogr.*, **38**, 1072–1076.
82. R.G. Zepp, P.F. Schlotzhauer (1981). Comparison of photochemical behavior of various humic substances in water: 3. Spectroscopic properties of humic substances. *Chemosphere*, **10**, 479–486.
83. A.J. Stewart, R.G. Wetzel (1981). Asymmetrical relationships between absorbance, fluorescence, and dissolved organic carbon. *Limnol. Oceanogr.*, **26**, 590–597.
84. S.A. Green, N.V. Blough (1994). Optical absorption and fluorescence properties of chromophoric dissolved organic matter in natural waters. *Limnol. Oceanogr.*, **39**, 1903–1916.
85. C.E. Del Castillo, F. Gilbes, P.G. Coble, F.E. Müller-Karger (2000). On the dispersal of riverine colored dissolved organic matter over the West Florida Shelf. *Limnol. Oceanogr.*, **45**, 1425–1432.
86. P.G. Coble (1996). Characterization of marine and terrestrial DOM in seawater using excitation-emission matrix spectroscopy. *Mar. Chem.*, **52**, 325–346.
87. W.F. Donahue, D.W. Schindler, S.J. Page, M.P. Stainton (1998). Acid-induced changes in DOC quality in an experimental whole-lake manipulation. *Environ. Sci. Technol.*, **32**, 2954–2960.
88. F.E. Hoge, A. Vodacek, N.V. Blough (1993). Inherent optical properties of the ocean: Retrieval of the absorption coefficient of chromophoric dissolved organic matter from fluorescence measurements. *Limnol. Oceanogr.*, **38**, 1394–1402.
89. S. Determann, R. Reuter, P. Wagner, R. Willkomm (1994). Fluorescent matter in the eastern Atlantic Ocean. Part 1: method of measurement and near-surface distribution. *Deep-Sea Res.*, **41**, 659–675.
90. M.J. Waiser, R.D. Robarts (2000). Changes in composition and reactivity of allochthonous DOM in a prairie saline lake. *Limnol. Oceanogr.*, **45**, 763–774.
91. N.B. Nelson, D.A. Siegel, A.F. Michaels (1998). Seasonal dynamics of colored dissol-

ved material in the Sargasso Sea, *Deep Sea Res, Part I*, **45,** 931–957.
92. C.S. Roesler (1998). Theoretical and experimental approaches to improve the accuracy of particulate absorption coefficients derived from the quantitative filter technique. *Limnol. Oceanogr.*, **43**, 1649–1660.
93. S. Tassan, G.M. Ferrari, A. Bricaud, M. Babin (2000). Variability of the amplification factor of light absorption by filter-retained aquatic particles in the coastal environment. *J. Plankton. Res.*, **22,** 659–668.
94. E. Urbach, K.L. Vergin, L. Young, A. Morse, G.L. Larson, S.J. Giovannoni (2001). Unusual bacterioplankton community structure in ultra-oligotrophic Crater Lake. *Limnol. Oceanogr.*, **46**, 557–572.
95. N.G. Jerlov (1950). Ultra-violet Radiation in the Sea. *Nature*, **166**, 111–112.
96. J.E. Tyler (1965). In situ spectroscopy in ocean and lake waters. *J. Opt. Soc. Am.*, **55**, 800–805.
97. R.C. Smith, J.E. Tyler (1967). Optical properties of clear natural water. *J. Opt. Soc. Am.*, **57**, 589–595.
98. J.E. Tyler, R.C. Smith (1966). Submersible spectroradiometer. *J. Opt. Soc. Am.*, **56**, 1390–1396.
99. R.C. Smith, J.E. Tyler, C.R. Goldman (1973). Optical properties and color of Lake Tahoe and Crater Lake. *Limnol. Oceanogr.*, **18**, 189–199.
100. R.C. Smith, B.B. Prezelin, K.S. Baker, R.R. Bidigare, N.P. Boucher, T. Coley, D. Karentz, S. MacIntyre, H.A. Matlick, D. Menzies, M. Ondrusek, Z. Wan, K.J. Waters (1992). Ozone depletion: ultraviolet radiation and phytoplankton biology in Antarctic waters. *Science*, **255**, 952–959.
101. R.M. Pope, E.S. Fry (1997). Absorption spectrum (380–700 nm) of pure water. II. Integrating cavity measurements. *Appl. Opt.*, **36**, 8710–8723.
102. J.B. Rasmussen, L. Godbout, M. Schallenberg (1989). The humic content of lake water and its relationship to watershed and lake morphometry. *Limnol. Oceanogr.*, **34**, 1336–1343.
103. S.A. Townsend, J.T. Luong-Van, K.T. Boland (1996). Retention time as a primary determinant of colour and light attenuation in two tropical Australian reservoirs. *Freshwat. Biol.*, **36**, 57–69.
104. D.W. Schindler, S.E. Bayley, B.R. Parker, K.G. Beaty, D.R. Cruikshank, E.J. Fee, E.U. Schindler, M.P. Stainton (1996). The effects of climatic warming on the properties of boreal lakes and streams at the Experimental Lakes Area, northwestern Ontario. *Limnol. Oceanogr.*, **41**, 1004–1017.
105. M.A. Moran, W.M. Sheldon, Jr., R.G. Zepp (2000). Carbon loss and optical property changes during long-term photochemical and biological degradation of estuarine dissolved organic matter. *Limnol. Oceanogr.*, **45**, 1254–1264.
106. R.E.H. Smith, J.A. Furgal, M.N. Charlton, B.M. Greenberg, V. Hiriart, C. Marwood (1999). Attenuation of ultraviolet radiation in a large lake with low dissolved organic matter concentrations. *Can. J. Fish. Aquat. Sci.*, **56**, 1351–1361.
107. R.M. Pope, A.D. Weidemann, E.S. Fry (2000). Integrating cavity absorption metre measurements of dissolved and suspended particles in ocean water. *Dynamics Atmos. Oceans*, **31**, 307–320.
108. V.S. Kuwahara, H. Ogawa, T. Toda, T. Kikuchi, S. Taguchi (2000). Variability of bio-optical factors influencing the seasonal attenuation of ultraviolet radiation in temperate coastal waters of Japan, *Photochem. Photobiol.*, **72**, 193–199.
109. W.L. Miller, M.A. Moran (1997). Interaction of photochemical and microbial processes in the degradation of refractory dissolved organic matter from a coastal marine environment. *Limnol. Oceanogr.*, **42**, 1317–1324.

110. J.E. Tyler, R.C. Smith, (1970). *Measurements of Spectral Irradiance Underwater*, Gordon & Breach, New York.
111. R.P. Dunne, B.E. Brown (1996). Penetration of solar UVB radiation in shallow tropical waters and its potential biological effects on coral reefs; results from the central Indian Ocean and Andaman Sea. *Mar. Ecol. Prog. Ser.*, **144**, 109–118.
112. F.M. Sogandares, E.S. Fry (1997). Absorption spectrum (340–640 nm) of pure water. I. Photothermal measurements. *Appl. Opt.*, **36**, 8699–8709.
113. T.I. Quickenden, J.A. Irvin (1980). The ultraviolet absorption spectrum of liquid water. *J. Chem. Phys.*, **72**, 4416–4428.
114. D.O. Hessen (1993). DNA-damage and pigmentation in alpine and arctic zooplankton as bioindicators of UV-radiation. *Verh. Internat. Verein. Limnol.*, **25**, 482–486 (cited in [10]).
115. I. Reche, M.L. Pace, J.J. Cole (1999). Relationship of trophic and chemical conditions to photobleaching of dissolved organic matter in lake ecosystems. *Biogeochemistry*, **44**, 259–280.
116. M.S. Twardowski, P.L. Donaghay (2001). Separating in situ and terrigenous sources of absorption by dissolved materials in coastal waters. *J. Geophys. Res.*, **106(C2)**, 2545–2560.
117. E. Aas, J. Høkedal, N.K. Højerslev, R. Sandvik, E. Sakshaug (2002). Spectral Properties of UV Attenuation in Arctic Marine Waters, pp. 23–56 in: *UV Radiation and Arctic Ecosystems, Ecological Studies* 153, Springer Verlag, Berlin 321 pp.
118. L.P. Boivin, W.F. Davidson, R.S. Storey, D. Sinclair, E.D. Earle (1986). Determination of the attenuation coefficients of visible and ultraviolet radiation in heavy water. *Appl. Opt.*, **25**, 877–882.

113. [illegible]

Chapter 4

Modulation of UVR exposure and effects by vertical mixing and advection

Patrick J. Neale, E. Walter Helbling and Horacio E. Zagarese

Table of contents

Abstract

Vertical mixing processes determine the depth range of water that exchanges through the relatively shallow photoactive zone where most UVR effects occur as well as the residence time of constituents (molecules and organisms) in this zone. Depth of mixing varies through the global ocean and among lakes as a function of latitude, season and various regional effects. The rate of vertical mixing is a function of convective and turbulent energy relative to water column stability. Mixing rates can be broadly characterized through scaling relationships however estimates of residence times in the photoactive zone vary from a few minutes to longer than a day so specific estimates are needed to assess how UVR exposure is modulated in any particular system.

There are few studies of UVR effects in the context of variable vertical mixing, these have used both experimental and modeling approaches. The results suggest that UVR effects are modulated by vertical mixing to the extent that responses are dependent on the duration and irradiance of exposure. Vertical mixing can actually enhance UVR effects integrated over the water column when responses depend on cumulative exposure and UVR effectiveness declines with exposure duration (photobleaching of dissolved organic matter, photoinhibition of Antarctic phytoplankton, mortality of ichthyoplankton). In contrast, if repair is present, but modest, vertical mixing can moderate effects (e.g., increase survival of zooplankton). If damage and repair are balanced so that responses reach a steady state over typical exposure times, vertical mixing can have little effect on water column responses.

Our present understanding of mixing and UVR effects has been limited by both the availability of physical measurements and the oversimplified representation of mixing processes in experiments and analyses. This is expected to change in the future as it becomes more practical to incorporate mixing measurements into field work, and as experimental exposures and mixing models become more sophisticated and allow a better approximation of actual water column conditions.

4.1 Introduction

UVR effects in aquatic ecosystems are distinguished by strong vertical gradients (see Chapter 3), so that most effects occur in a "photoactive" zone near the surface. Vertical mixing mediates UVR effects by determining how much of a water body (ocean, lake, etc.) exchanges through this photoactive zone. Processes influencing this exchange are mainly those that determine (1) the depth of the upper mixed layer (UML) and (2) the rate of transport within the UML. In this chapter, we will catalog and briefly discuss primary physical processes influencing each of these UML characteristics. For more detail on these processes, the reader is referred to several recent general reviews on mixing in the upper ocean [1,2] and in lakes [3,4]. We then consider how changes in mixed layer depth and rate of mixing can influence the chemical and biological effects of UVR exposure.

4.2. Mixing processes

4.2.1 UML depth

The sun is the primary source of both the stirring energy that drives vertical mixing and heating that dampens vertical mixing. Variations in solar energy over the Earth (see Chapter 2) ultimately result in variations in mixing between different parts of the ocean and lakes in different latitudes. Heating of surface waters is a direct effect of the sun, which stratifies waters and dampens mixing. Surface forcing by winds is an indirect effect, caused by the gradients in solar heating of the atmosphere between the poles and the equator. At large scales in the ocean (10^4–10^3 km), prevailing wind fields set up gyres defining oceanic provinces with characteristic UML depths based on convergent or divergent flow within the gyre [2]. At intermediate scales (10^3–10^2 km), the horizontal and vertical boundaries of currents associated with gyres and along the continents are shear zones that produce turbulence and eddies which also influence surface mixing.

Given that both surface UVR exposure and vertical mixing are products of the global distribution and seasonality of solar irradiance, it is reasonable to expect these factors to co-vary in some systematic way. Overall, subpolar waters are cooler, fresher, and have strong winter heat losses that favor the development of deep mixed layers which, at least occasionally, transport water to the surface from deeper, UVR protected, layers. On the other hand, tropical and subtropical waters experience strong surface heating, producing warm and salty upper layers that persist for long periods and isolate deep waters from the surface. The overall salinity gradient also affects the strength (i.e., density gradient) of the pycnocline and the probability of a deep mixing event, and this differs between ocean basins ([2], Figure 1). Lakes follow similar trends, though both UML depths and UVR penetration are more variable but generally shallower. Usually, tropical lakes tend to have stable, shallow stratification, whereas seasonal deep mixing is more common in temperate and polar lakes. As a result, where UVR exposure is the highest, i.e. at low latitudes, average exposure over the UML (as a fraction of incident UVR) also tends to be high. Of course, the depth of the photoactive layer will depend on the water transparency (Chapter 3), but significant exposure ($>10\%$ of incident UV-B) rarely occurs deeper than 10 m and in turbid environments is usually <1 m [5]. In this case the UML sustains a much greater magnitude of chemical and biological effects of UVR than deeper waters. The surface layer of lakes during temperate summer, especially smaller lakes with a surrounding windbreak of trees (e.g., in forested regions of North America and Europe), is a similarly isolated zone of UVR action [6,7]. In contrast, at higher latitudes, or during the temperate zone winter, mixing extends deeper and will periodically flush the near surface zone with water from depths with very low UVR exposure. In this situation, surface UVR is lower, because of the latitude or season, but the constituents of these waters may be more labile to UVR effects since they have received little previous exposure.

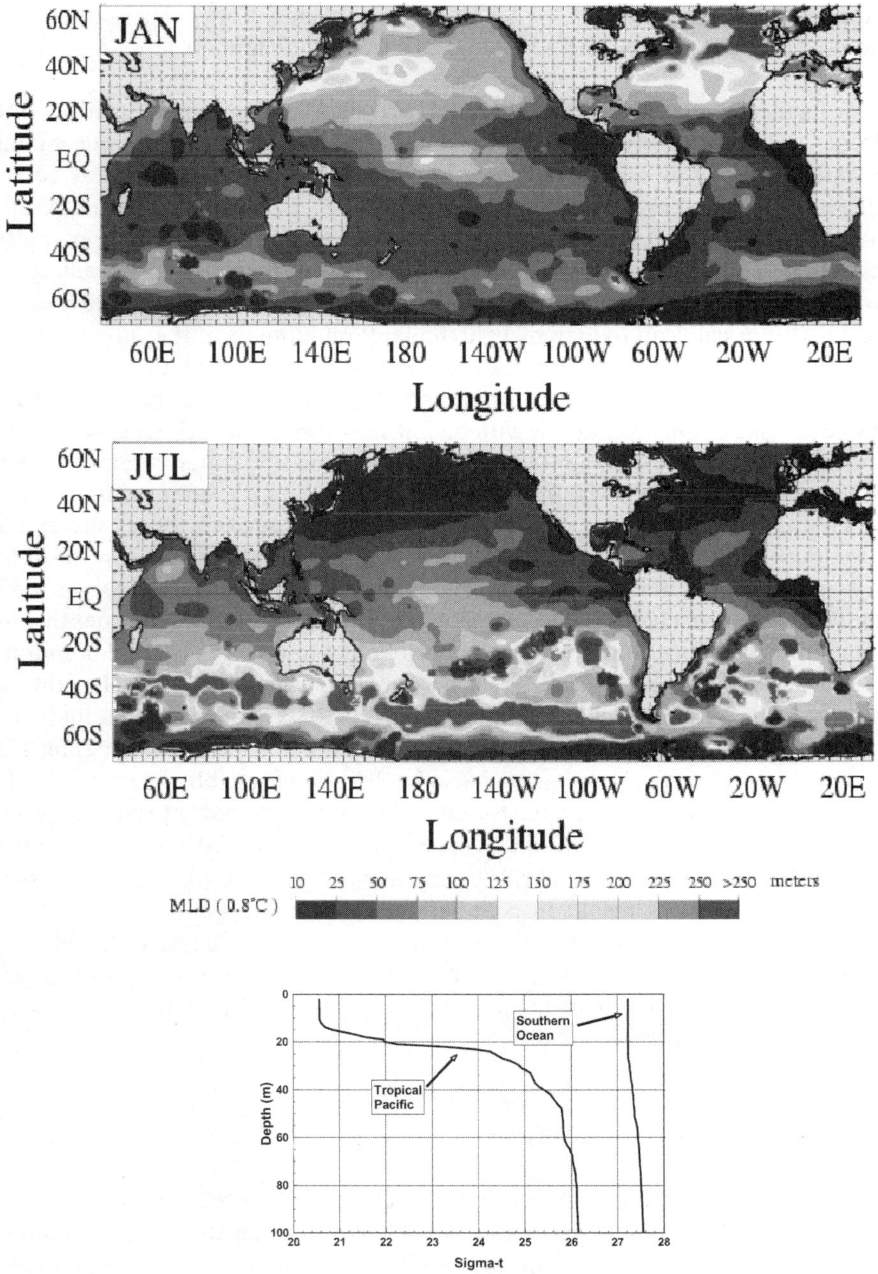

Figure 1. Maps of mixed layer depth in the global ocean as monthly averages for January (upper panel) and July (middle panel). The lower panel shows typical profiles of sigma-t (a measure of density) for a polar (Southern Ocean) versus tropical (Pacific) area of the ocean. The maps use global ocean temperature and salinity data sets compiled by the U.S. National Oceanic Atmospheric Administration as processed by Kara et al. [94].

Implications of the depth of the UML are discussed in more detail below in the context of specific chemical and biological effects of UVR in the aquatic environment.

A further source of variation in the UML depth occurs in association with fronts and eddies that form along the boundaries between major oceanic mixing regimes. Depending on whether they are cyclonic or anticyclonic, these eddies increase upwelling or downwelling, respectively [2]. Coastal upwelling, typically associated with increased productivity due to enhanced nutrient supply, also causes shallow UMLs (Figure 1). Along the equator, there also is upwelling, as well as complex patterns of convergence and divergence. Certain combinations of conditions can lead to extremely high plankton biomass at the surface [8].

In the coastal ocean, estuaries and shallow lakes, the depth of the water column is such that the bottom topography might limit the extent of the UML. For example, in some situations with high wind stress, such as those encountered during spring in the Patagonia region of Argentina, the water column is completely mixed and the UML therefore comprises the entire water column.

The patterns of vertical mixing that have been described reflect the global distribution of ocean–atmosphere energy exchange. It should be kept in mind that global climate change is affecting this distribution and changes in these patterns are expected. Indeed, climate induced changes in UML depth and strength of stratification have already been linked to changes in the community composition of plankton in the N. Pacific gyre [9]. Changes in vertical mixing as a consequence of global climate change are likely to have much greater influence on UVR effects on aquatic systems than ozone depletion per se (see Chapter 17).

The depth of vertical mixing may also be directly affected by increased UVR. In many systems (particularly freshwaters), the major component responsible for determining UVR penetration, chromophoric dissolved organic matter (CDOM), is also the primary absorber of visible radiation. Absorption of solar radiation by CDOM results in near-surface heating and shallow stratification [10,11]. However, CDOM absorbance is not constant, due to photobleaching by UVR [12]. As CDOM bleaches, there is deeper penetration of solar radiation and less pronounced surface heating, allowing deeper mixed layers to develop ([13] also see Chapters 3 and 6).

4.2.2 Time scales of vertical mixing

In addition to controlling the vertical structure of the water column, vertical mixing also determines the variability of UVR exposure, i.e. through determining the residence time of constituents (molecules and organisms) in the photoactive zone. Residence times depend on the particular processes contributing to vertical transport, and these can include convective or shear–stress driven turbulence, Langmuir circulation and breaking of internal gravity waves in the pycnocline [2]. Denman and Gargett [14] reviewed measurements of wind speed (U m s^{-1}), the buoyancy frequency (N, an index of stratification, units s^{-1}) and the rate of dissipation of turbulent kinetic energy (ε, m^2 s^{-3}) from several marine and freshwater mixed layers (for more background on N and ε see reviews [4,15]). Based on this data, they estimated the times of transport over a vertical displacement of 5 m (a nominal depth for the photoactive zone in the ocean) using scaling relationships. The estimated transport time near the surface varied from a few minutes (strong winds, >8 m s^{-1}) to a hundred hours (low winds <5 m s^{-1}). Deeper in the mixed layer the estimated 5 m displacement time varied from nearly a day to weeks over the same range of winds. Although vertical mixing in lakes and the ocean is an active research area, there have been few further estimates of vertical transport rates near the surface since Denman and Gargett's 1983 review. A study of vertical mixing in a turbid shallow lake (photoactive zone probably <50 cm) estimated mixing times over the upper 50 cm of 1–2 minutes during typical midday-afternoon breezes [16].

Rapid transport into and out of the photoactive zone results in extremely variable UVR exposure for organisms such as phytoplankton that are entrained in these flow features. As an example, we show exposure time-courses during eddy circulation in moderately clear lake which were simulated by rotation of bottles over a 0–4 m depth interval every 8 minutes (Figure 2). Photosynthetic active radiation (PAR) remains saturating for photosynthesis during the whole rotation; however, effective UVR for inhibition of photosynthesis (UV_{eff}) varies between insignificant at maximum depth to very strong at the surface. Such circular trajectories are about the only type of motion that has been practical to use in field studies. In contrast, the most common approach in numerical models is to generate constituent trajectories using a random walk approximation of vertical displacement [17–19]. In reality, the water motion is a composite of random displacements (sometimes anisotropic) and advection by large eddies [2]. Detailed studies of these processes, particularly turbulence, have been enabled by development of instruments that measure small-scale variations in physical properties [4,15]. The microstructure profiler measures mm–cm scale fluctuations in density (temperature) or velocity, from which ε can be estimated [4,20]. Scaling expressions similar to those described by Denman and Gargett [14] are then calculated, e.g. that the largest overturning scale of turbulence having dissipation rate, ε, in a background stratification measured by the buoyancy frequency, N, is given by the Ozmidov scale, $L_O \equiv (\varepsilon/N^3)^{1/2}$. A complementary approach is to estimate large-eddy scales using Thorpe scales [21]. This requires measuring cm–m scale density profiles with a free-fall instrument and sorting the vertical density profiles to stability (i.e., monotonically increasing density with depth). The (minimal) distance that must be moved to achieve stability becomes a measure of the local Thorpe displacement. The Thorpe scale, L_T, is defined as a root mean square value of Thorpe displacements, sometimes over fixed depth intervals, but more reliably over individual "overturns" [22]. Oceanic measurements to date have found $L_T = c_{OT}\ L_O$, with $0.5 < c_{OT} < 2$ [23–26]. While it is generally accepted that in stratified water columns the relevant time scale is the buoyancy period $T_b = 2\pi/N$, for studies of UVR effects, it is also necessary to determine local large-eddy length scales in the vertical dimension of the light gradient.

Scaling relationships have been developed from theories and observations of

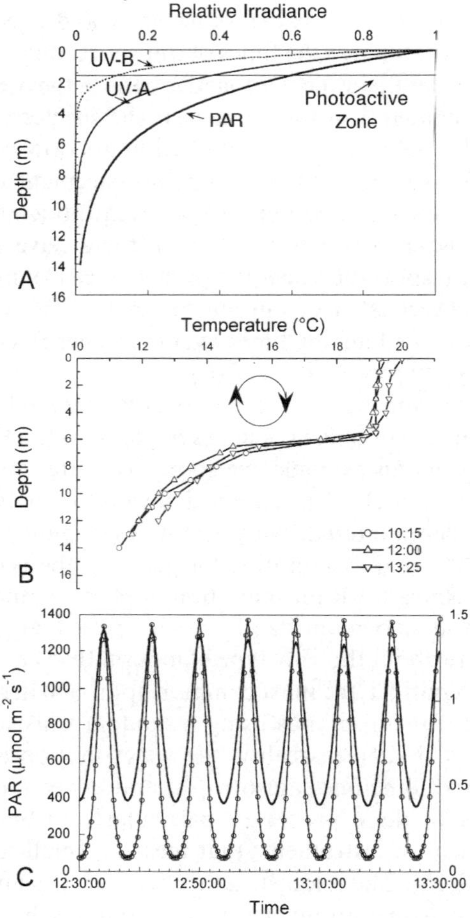

Figure 2. Time variable exposure to UVR and PAR in a simulated mixed layer of a temperate lake. Data is from Lake Lucerne, September 15, 1999 (adapted from [79]). (A) Profiles of UV-B (290–320 nm), UV-A (320–400 nm) and PAR (400–700 nm) as irradiance relative to the surface. A line shows the bottom of the "photoactive zone" as defined by 10% of surface UV-B (1.7 m in this case). (B) Temperature profiles showing an approximately 5 m thick upper mixed layer. Experimental mixing was conducted over the upper 4 m (circle). (C) Exposures for PAR (thick line) and UV_{eff} (circles, UVR weighted for inhibition of photosynthesis) obtained when samples were mixed with a 8 min rotation time. Photosynthesis was estimated to be about 60% lower at the surface compared to the maximum depth [79].

turbulence and large eddy scales for several cases. If turbulent intensity is zero, the vertical position of neutral particles in the upper mixed layer (UML) is affected by the vertical motions associated with the surface and internal wave fields, including near the bottom of the mixed layer where internal waves dissipate with buoyancy frequency near that of the pycnocline immediately below. In both cases, vertical displacements can be estimated with some accuracy [27]. If turbulence is so intense that the UML is completely mixed, reasonable scalings exist for surface shear–stress and/or convection driven motions [14,28], as well as for Langmuir circulation [29]. No good scaling exists for the more complex case in which turbulence of intermediate intensity proves insufficient to mix the upper layer thoroughly in the presence of stabilizing influences like solar heating and/or surface freshwater input from melting ice, the latter being important in polar seas. Direct measurements are the best approach under these conditions. Because vertical motions resulting from mixing may be highly intermittent in such cases, they offer a particular challenge for studies of the biological effects of UVR, as the balance between damage and repair may shift depending on the time and length scales involved.

Both the depth and rate of mixing reflect a dynamic balance over all time scales between kinetic energy transfer and dissipation, as well as heat gain and loss. The daily time scale is particularly important for UVR effects. Mixing processes operating on the daily scale, and the diel cycle of turbulent mixing and stratification, have been extensively studied in the ocean [30–32] and lakes [16, 33–36]. An illustration of the interplay between these processes is provided in the observations made by the serial release over a daily cycle of near neutrally buoyant floats that are acoustically tracked ([Figure 3, D'Asaro and Dairiki unpublished, as cited in 2]). The floats can be viewed as showing the depth–time variation of neutrally buoyant constituents (molecules or non-motile organisms). These observations show a typical cycle of nocturnal convective mixing, followed by heat gain during the morning leading to a shallow diurnal thermocline. As surface irradiance declines, wind stress and cooling again dominate and deep mixing resumes. This is a dramatic illustration of how vertical mixing processes control the time scale of UVR exposure – at times the tracers remain confined near the surfaces, in other instances there are rapid (time scale minutes) ascents (or descents) between the surface and 30 m. In another field deployment of floats in the Pacific ocean off of Vancouver during January, the principal source of energy for vertical mixing was the wind and the 5 m transport times of the floats varied between 17 minutes during low winds to 3 minutes during high winds [37]. Similar time scales were obtained during strong convective mixing [38].

4.3 Interactions between vertical mixing and UVR effects

The previous section showed how mixing processes determine the way UML constituents (molecules or organisms) enter the photoactive zone where they may participate in a UVR-mediated process. Under strong stratification, such transport is very limited, so UVR effects will only involve those constituents already present in the active zone. Such extreme stratification can be episodically important in systems where diurnal thermoclines form, but more typically the UML extends below the photoactive zone so that constituents in waters below the photoactive zone will also participate in UVR photochemistry and photobiology. The mixing characteristics of the UML influence photoprocesses in at least two fundamental ways. First, the depth of the UML influences the average UVR exposure that occurs; second, the rate of vertical transport, either due to advection or turbulent exchange, determines the residence time in and out of the photoactive zone (Figure 2). If all biological and chemical effects were linear functions of cumulative exposure over all time scales, then the timing and sequence of exposure would not matter and knowledge of the vertical distribution of irradiance (Chapter 3) would be sufficient to assess UVR effects in aquatic ecosystems. However, most UVR effects are non-linear at least over part of the environmentally relevant range of time scales. Another way of stating that effects are not proportional to cumulative exposure is that there is a failure of reciprocity. When reciprocity fails, effects are dependent on the duration and irradiance of exposure, so that residence time and average irradiance make a difference. Detailed examples of how these dependencies arise are considered in the next sections.

4.3.1 Photochemistry

The effects of mixing on photochemical reactions are not well known. As a corollary of the first law of photochemistry (see Chapters 6 and 8), primary photochemical reaction rates should be directly proportional to the rate of light absorption. However, the previous statement applies only to completely mixed, optically thin water bodies, or where the reactant is the sole absorber [39]. In contrast, in optically thick water columns with competition amongst chromophores for photons, differences in mixing rates can affect photochemical reaction rates. Under such circumstances, fast turnover would tend to release the CDOM pool from self-shading, which should translate into higher photoreaction rates [40].

A few studies have addressed the effect of mixing on photodegradation of natural organic matter from a mass transport perspective [40–42]. Both mesocosm experiments [40] and models [41,42] have shown that mixing rates can exert a strong influence on the rates and distribution of photoprocesses in the water column. In general, lack of, or slow mixing confines photoreactions to the very top of the water column in water bodies where there is strong light attenuation of wavelengths involved in the photoreactions and the rates of these primary photoreactions decrease exponentially with depth [41]. With more thorough mixing, fresh photochemical reactants are continually added to the photoactive layer, which results in higher rates of photolysis for the entire water column [43]. In addition, highly reactive species produced photochemically may be transported to deeper waters by vertical mixing. For example, reactive oxygen species (ROS, Chapter 8), produced near the surface may be transported by mixing and oxidize organic molecules several meters below the water surface.

Conversely, lack of mixing (i.e., stratification) may also be important because it isolates a certain CDOM pool within the UML during the stratified period. CDOM in the UML (epilimnion in lakes) becomes photobleached during seasonal exposure to UV, causing differences in CDOM absorbance between surface layer and deeper waters (hypolimnion in lakes). For example, studies in two Pocono lakes [12] found differences in photobleaching and the spectral weighting function for UVR bleaching of CDOM between epilimnetic and hypolimnetic samples. This suggests that the timing and extent of stratification may affect the annual production of photoproducts, including carbon gases, such as CO_2 and CO.

The previous paragraphs illustrate how mixing can affect photochemical reactions by resupplying surface waters with fresh, unbleached chromophores and reducing surface concentrations of photoproducts [42]. But mixing may have additional effects that are as yet unexplored. Organic molecules are complex structures having several photoactive sites (see Chapter 6). Absorbed energy decays through a variety of pathways, which may or may not involve molecular rearrangements and fragmentation [44]. In cases where the molecular structure is altered, the resulting new structure may have different absorption and photoreactivity characteristics than its parent molecule. Thus, in addition to affecting the mass transport along the water column, vertical circulation may influence the sequence of reactions undergone by organic molecules. To our knowledge, there is only one study that addressed the effect of fluctuating levels of solar radiation, on a time scale of minutes, on photodegradation of natural organic matter [45]. That study showed differences in (i) photobleaching, (ii) nutrient release, and (iii) subsequent use of CDOM by algae and bacteria, between bottles incubated at fixed depths and bottles rotating within the water column. Although far from definitive, such results suggest that vertical mixing should be explicitly considered in photochemical studies of natural waters.

This section has called attention to some ways that vertical mixing complicates the photochemistry of natural waters. On the other hand, if the rates of CDOM absorption and photochemistry can be quantified, then the steady state profile of a photochemical product (i.e. dissolved hydrogen peroxide) can be used to infer vertical mixing rates. This was possible in freshwater systems (Canadian Lakes and the St. Lawrence River) that accumulate higher levels of peroxide due to their CDOM content [36,46]. A similar attempt to model the depth–time variation of hydrogen peroxide in the ocean (where CDOM is much lower) was only partially successful in reproducing the observed distribution [47,48].

4.3.2 Photobiology – phytoplankton and bacterioplankton

Phytoplankton and bacterioplankton are small (0.2–100 μm) unicellular organisms that have no, or weak, motility. Some bacterioplankton exhibit chemokinetic motility, but velocities are small (on the order of 0.001–0.01 cm s^{-1}[e.g., 49]) compared to typical vertical transport rates (on the order of 1–10 cm s^{-1}). Phytoplankton have varying properties of buoyancy and motility that can interact with the mixing regime and affect residence time in the photoactive zone (examples [16,18]). Indeed, it has been argued that the diverse life forms of algae reflect the diversity of mixing regimes in aquatic environments [50]. Under strong mixing (cf. Figure 3), variations of cellular physical properties and motility may have only subtle effects on vertical transport rates but significant effects on vertical distribution, e.g. positive or negative buoyancy will result in a

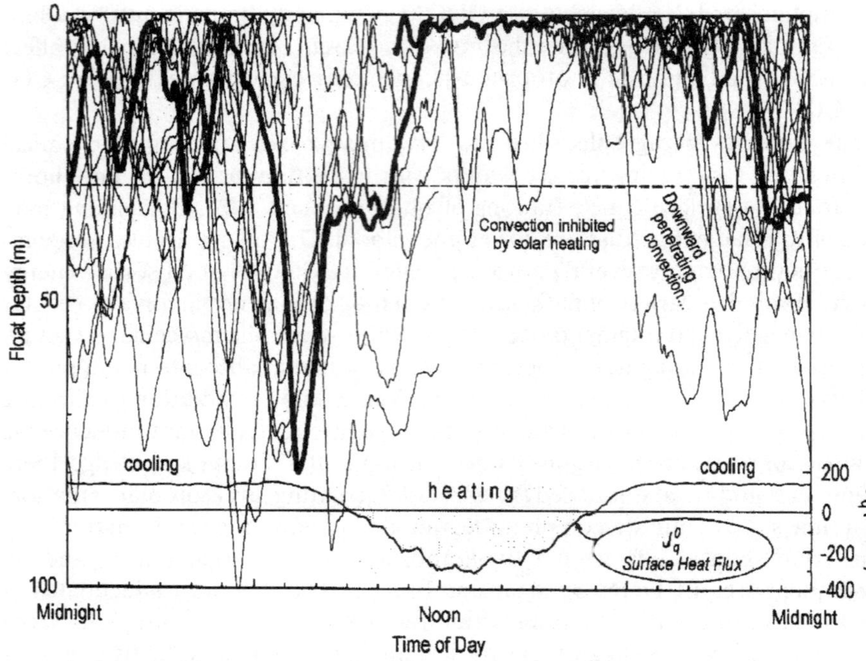

Figure 3. Schematic of the time vs. depth distribution of floats released at various times during a diel cycle of nocturnal mixing and diurnal stratification in the Labrador Sea [E. D'Asaro and G. Dairiki unpublished data as shown in 2]. The heavy line (a composite of three separate float deployments) depicts a possible depth history of a non-motile plankton over the 24 h period. Note that trajectories are terminated once they enter the stable diurnal thermocline since they actually start to ascend due to very slight positive buoyancy of the floats.

distribution with a larger proportion of plankton in the upper or lower (respectively) part of the mixed layer.

Photosynthesis by phytoplankton is an important process in aquatic environments and is well known to be time dependent. Accounting for the effects of vertical mixing as a determinant of light history is a long standing issue in the study of phytoplankton productivity (see reviews by [2,14]). Vertical mixing can influence productivity through several types of time-dependent responses [51]; however, integral photosynthesis is most often affected when the assemblage is sensitive to inhibition by near-surface irradiance [52].

Phytoplankton respond to light variation over time-scales of seconds to many days [51], with average light exposure over a generation time (ca. 1–3 days) having a strong effect on capacity to resist photoinhibition (see Chapter 10). Induction of these defense mechanisms depends on the balance between PAR and the energetic demands of phytoplankton metabolism. If average light intensity is low, then phytoplankton emphasize efficiency of photosynthesis through accumulation of light-harvesting pigments and have generally weak capacity to defend against high-light exposure (PAR and UVR). When such low-light acclimated phytoplankton are transported near the surface, they can be severely affected by high-light exposure. On the other hand, when light availability is high relative to energetic demands, phytoplankton emphasize protective and defensive mechanisms more than photosynthetic efficiency, accumulating protective pigments like xanthophylls and UVR-screening compounds, and maintaining efficient mechanisms to repair photodamage.

The main determinants of average exposure in a mixed layer are transparency of the water and the depth of the UML. Thus, in water bodies of relatively stable optical characteristics, sensitivity to UVR would be expected to increase with depth of mixing, but defense mechanisms should be inversely related to mixing depth. This hypothesis is supported by a few field studies. Helbling et al. [53] found an increased inhibition of photosynthesis in natural populations of Antarctic phytoplankton with increased depth of the UML. Whereas samples coming from an UML <25 m showed no significant inhibition, the inhibition due to UVR was about 40% and 75% for samples from mixed layers depths of 35 m and more than 100 m, respectively. In their study, photosynthetic inhibition also decreased from the Antarctic to the Equator, and part of this was attributed to the shallowing of the UML towards tropical areas (see also Figure 1). Vernet et al. [54] found that concentration of UVR screening compounds (one type of acclimation to high-light exposure) was inversely related to UML optical depth (and therefore directly related to average exposure) for waters near the Antarctic Peninsula. Neale et al. [55] reported that the sensitivity to UVR for phytoplankton in the Weddell–Scotia Confluence (Southern Ocean) was related to surface layer density, which was used as an indicator of the overall light history. Higher surface density reflects a greater contribution of deep water and thus deeper mixing. Depth of mixing appears to be an important factor explaining seasonal variation in sensitivity in the North American Great Lakes [56].

A number of studies have suggested that rapid vertical mixing can actually enhance production if the time-scale of surface exposure is short compared to the induction time of photoinhibition [57–60]. On the other hand, under the low vertical mixing of diurnal thermoclines, photoinhibition is quite pronounced [61,62]. Effects on other indicators of photoacclimation also seem to be greatest when the time scales of mixing and photoresponse are comparable [cf. 63,64]. This has led to the commonly expressed view that photoinhibition (by PAR or UVR) as measured in long-term incubations does not apply in situ under the prevalent conditions of vertical mixing and thus can be ignored [e.g. 65].

Early studies of productivity and vertical mixing did not specifically consider responses to UVR, despite some early, order-of-magnitude calculations that suggested it could be important [66]. Certainly, during the presence of diurnal thermoclines, the surface community receives extended UVR exposure and severe inhibition of photosynthesis can occur [36,67,68]. However, another consideration is overall effect of UVR exposure on water column (vertically integrated) production. Helbling et al. [69] were the first to consider the relationship between vertical mixing and UVR effects on integrated production using an experimental approach. Screens were rotated over quartz tubes providing a UVR and PAR gradient to simulate mixing and the results were compared to average rates obtained under static conditions (no screen rotation). Incubations were conducted using phytoplankton samples from near Elephant Island in the Southern Ocean during which the screens covering the tubes were rotated at a rate within the range of estimates for transport in the UML so that UVR varied between 100% and 3% of incident irradiance. Average inhibition of photosynthesis increased when Antarctic phytoplankton were incubated in the variable light regime (rotated screens) compared to a series of control bottles that received a fixed percentage of incident radiation (Figure 4). At low mean irradiances in the UML the phytoplankton that were static (i.e., fixed irradiances) had lower carbon fixation than the samples that were in a simulated UML. At higher irradiances, however, samples that were "mixing" had a rate of photosynthesis that was inhibited (lower) compared to the average of the static treatments.

That vertical mixing could actually enhance inhibition was a surprising result, considering earlier experiments that suggested that vertical mixing decreased the effect of photoinhibition by PAR. However, experiments with phytoplankton from the Weddell–Scotia Confluence (WSC, near Elephant Island) showed that the kinetics of photosynthetic response to UVR were quite different from those assumed in earlier studies. The decrease in photosynthesis occurred rapidly after the onset of exposure, and recovery, after exposure to near-surface irradiance, was absent or slow [55]. The absence of recovery meant that reciprocity of exposure applied over short-time-scales but near-surface inhibition was strong enough that responses were non-linear for exposures exceeding about 30 minutes

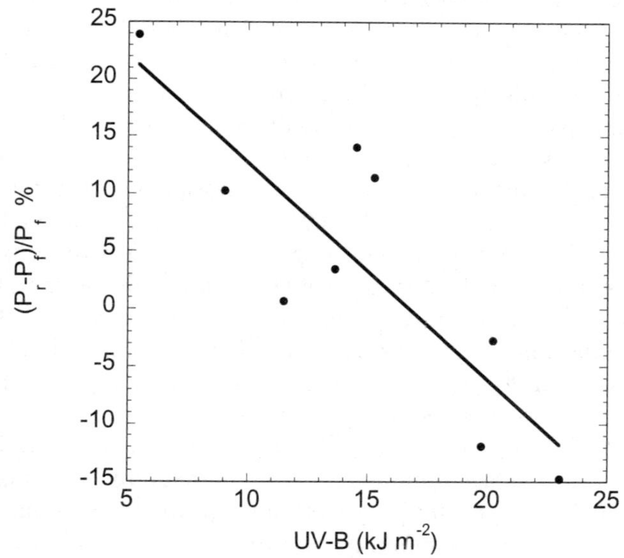

Figure 4. Integrated primary production in an incubator with 12 light levels (100% to 3% of incident) when samples are rotated (P_r) to simulate vertical movement in an UML as a percent of static (fixed) samples (P_f). Percent change is given as a function of total UV-B (290–320 nm, J m^{-2}) over the incubation period. Total UV-B was estimated from measured irradiance at 320 nm (E_{320}) by applying the ratio of measured to modeled E_{320} to a full UVR spectrum calculated by a radiative transfer model for clear sky conditions. [Adapted from Helbling et al. 69].

[55]. Thus, high-rates of photosynthesis were not maintained during brief exposures to near-surface irradiance as had been found for PAR effects. The consequence of these different kinetics were examined in a numerical model of varying UVR exposure in the mixed layer of the WSC [70]. Similar to the experimental results of Helbling et al. [69], vertical mixing enhanced inhibition of integrated water column production for moderate depth mixed layers ($z_{mix} < 40$ m) (Figure 5). Under static conditions, inhibition by UVR is severe but effects are limited to the near surface. Photosynthesis decreases with time according to an exponential [or survival curve, 71] relationship [55]. The implication of these kinetics is that UVR has the greatest absolute effect at the beginning of the exposure period, i.e. 75% of the inhibition occurs during the first half of the day. Thus, in static layers the decrease in productivity is small after the first hour or two of exposure in the morning. Vertical mixing results in a flux of phytoplankton from depths where exposure is low to the high exposure surface layer. Under these conditions, there will always be a relatively unexposed component of the surface phytoplankton that will experience large, UVR induced, decreases in photosynthetic rate. Since recovery is low (or nonexistent) when previously exposed phytoplankton are transported away from the surface layer, effects accumulate and integrated inhibition is higher than under strongly stratified conditions. This situation has analogies with the mixing enhancement of CDOM bleaching as discussed in the previous section (4.3.1).

The rate of mixing not only affects the overall impact of UVR, but also the response to ozone depletion (Figure 5). Stratospheric ozone depletion, which is

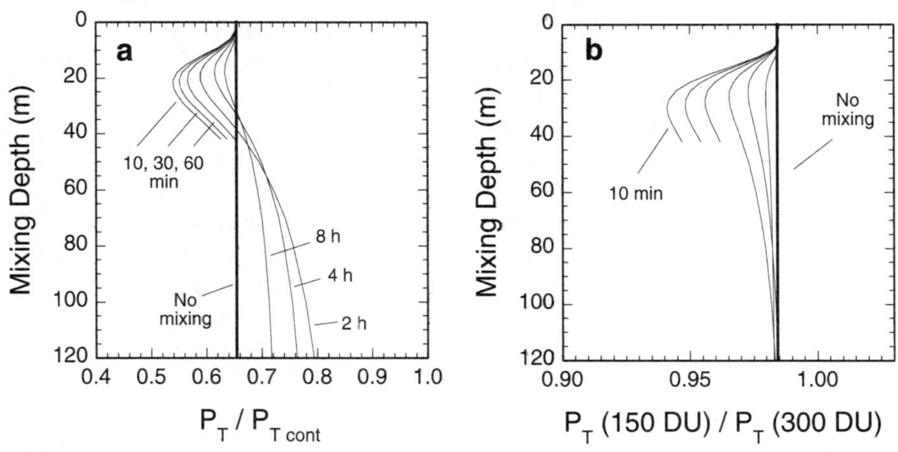

Figure 5. Results of a model for inhibition of photosynthesis in the surface layer of the Weddell–Scotia Confluence of the Southern Ocean (modified from Neale et al. [70]). (a) Interactive effects of mixing depth and mixing time scale on daily water column productivity relative to the uninhibited rate (curves labeled with t_{mix} = time scale of mixing). Langmuir circulation can mix the upper water column rapidly, so mixing times ≤ 1 h are modeled for mixing depths ≤ 42 m. (b) Interactive effects of O_3 depletion (150 DU vs 300 DU) and mixing time as a function of mixing depth: proportional change in P_T for the same curves as in (a).

most severe over the polar regions, has a spectrally specific effect on the solar UVR spectrum in which UV-B (280–315 nm) radiation is particularly enhanced (Chapter 2). The calculated effect of the Antarctic "ozone hole" (i.e. 50% ozone depletion) on water column phytoplankton production was around 1.5% reduction for the static case, but as much as 8% reduction in the mixing scenarios [70]. Again, this appears to be a consequence of the non-linear response of photosynthesis to cumulative exposure. Attenuation of UV-B with increasing depth is always more rapid than attenuation of UV-A (in the WSC, as elsewhere, Figure 2 also see Chapter 3). As a result the enhancement of exposure of organisms to UVR by ozone depletion is always greater near the surface. This increment in exposure results in more loss of photosynthesis for the relatively unexposed phytoplankton transported to the surface in the mixed case relative to the static case, where the enhanced UV-B exposure occurs mainly to the near-surface phytoplankton that have low photosynthetic rates even under normal UV-B conditions.

Recent experimentation conducted in temperate Patagonian marine waters [72] also indicated different responses of phytoplankton to UVR during simulated mixing. Vertical movement (i.e., change in irradiance) and mixing within the euphotic zone depth (E_{ph}) were simulated using a moving system in combination with neutral density screens, and compared with a fixed system. The rates of production of winter and summer samples were compared under the same simulated UML (UML/ $E_{ph} < 0.5$). In terms of UVR-dependent reduction of carbon fixation rates under a variable ("mixed") irradiance field, species adapted to low irradiance levels (i.e. winter samples, generally dominated by microplankton) were more sensitive than summer samples (dominated by nanoplankton). In addition, when summer samples were exposed to the irradiances encountered in shallow UMLs (UML/ $E_{ph} < 0.5$), vertical mixing enhanced the inhibition by UVR. However, when a deep simulated UML condition (UML/ $E_{ph} > 0.8$) was imposed on the samples, the phytoplankton that were in the moving system had a higher integrated carbon fixation rate than samples in the fixed system. In these studies, the kinetics of inhibition were dependent on the composition of the phytoplankton population; there was a tendency for samples dominated by nanoplankton to have relatively slow kinetics of inhibition. This was also observed in a study conducted in various Andean lakes [73], where samples dominated by smaller cells had slower inhibition rates as compared to larger cells.

The kinetics of photosynthetic response are clearly the key in predicting how vertical mixing will affect the impact of UVR exposure, as seen for the cases discussed so far. Contrasting kinetics have been observed for phytoplankton assemblages in other temperate environments. In these assemblages, recovery appears to be much more active and thus reciprocity is not observed [74–77]. While inhibition is still rapid, a steady-state rate is obtained after a short transition. This steady-state is the result of an equilibrium between damage and repair processes. When exposure decreases, photosynthetic activity is restored. When the time-scale of mixing is sufficiently long that steady-state is established, UVR inhibition of photosynthesis is a function of irradiance alone and thus should be independent of vertical mixing rate. However, the interaction of vertical mixing and UVR inhibition of photosynthesis has received little experimental examination in assemblages with active repair. An example of an assemblage with active repair is the phytoplankton in the surface layer of a temperate Swiss lake (Vierwaldstättersee or Lake Lucerne) [78]. The effect of fluctuating UVR on photosynthesis was measured in bottles that were circulated by a rotating lift system through a surface mixed layer at relatively fast rates (complete bottle rotation every 4–20 minutes, exposure conditions shown in Figure 2) [79]. The observed effect of UVR could be accounted for using a steady-state (irradiance dependent) relationship developed using laboratory incubations (Figure 6). Since the model of UVR effects is not time-dependent, this indicated that the effect of vertical mixing, if present, was too small to cause a deviation from the profile predicted under static conditions. While more studies are needed of how mixing interacts with UVR effects on phytoplankton production, the first indications are that results depend strongly on the kinetics of the UVR response.

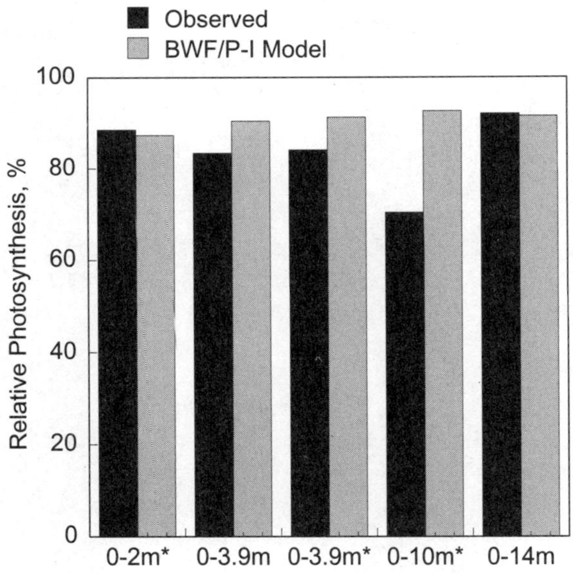

Figure 6. Effects of UVR on photosynthesis (total C assimilation) of phytoplankton moved through different mixing depths, presented as per cent photosynthesis in quartz (UVR transparent) relative to glass (partial UVR exclusion) bottles. Measured rates are for bottles that were circulated over the indicated depth ranges at the rate of once per 4 min (0–2 m), once per 8 min (0–3.9 m) and once per 20 min (0–10 and 0–14 m) for a 4 h midday incubation period. The modeled rates are the average of the steady-state (irradiance based) photosynthesis predicted using a biological weighting function and photosynthesis irradiance (BWF/P-I) curve applied to in situ irradiance estimated from recorded surface irradiance, depth of the bottles and measured vertical extinction coefficient. Model and measurements agree within measurement variability (ca. 10%) except for the 0–10 m incubation. Experiments were conducted in Lake Lucerne on September 13, 1999 (no asterisks) and September 15, 1999 (asterisks, see exposure data in Figure 2). [Modified from Köhler et al. 79.]

Production by bacterioplankton is also inhibited by UVR exposure (see review, [80]), but the relationship of inhibition to the depth and rate of vertical mixing has not been studied. However, the vertical distribution of damage to bacterial DNA, an indicator of the UVR effects, has been studied for water columns with different mixing regimes. The most common form of UVR-mediated damage is the formation of cyclobutane pyrimidine dimers (CPDs) (see Chapter 9). There are dramatic differences in the vertical profile of CPDs depending on whether calm conditions or strongly mixed conditions prevail (Figure 7). Boelen et al. [81] observed an in situ diel cycle of accumulation and decrease of near surface CPDs, while in parallel surface incubations CPDs increased mid-day but were not repaired in the afternoon. This suggested that vertical mixing was necessary to exchange severely damaged organisms (due to surface UVR exposure) out of the near-surface layer.

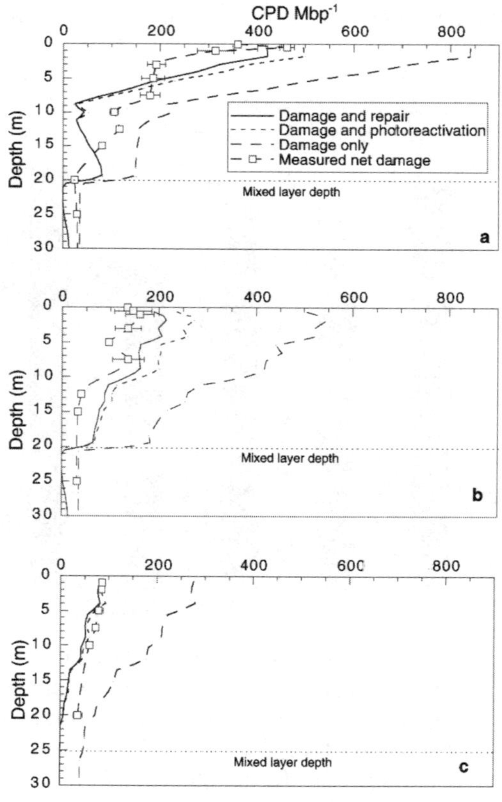

Figure 7. Measured and modeled profiles of cyclobutane pyrimidine dimers (CPD) for water columns with varying intensities of vertical mixing (a) Gulf of Mexico, 8 September 1994, mixed layer 20 m, wind speed low (2 m s^{-1}) (b) Gulf of Mexico, 7 September 1994, mixed layer 20 m, wind speed high (8 m s^{-1}) (c) Gerlache Strait, Antarctica, 6 October 1996, mixed layer 25 m, wind speed high (8 m s^{-1}). Modeled profiles are for damage (CPD formation) only, damage plus photoreactivation only, and damage and all repair mechanisms (photoreactivation and excision repair). [From Jeffrey et al. and Huot et al. 82.]

Vertical profiles of bacterial CPDs reflect the in situ dynamics of stratification and mixing, and thus present an opportunity to test our understanding of how the processes of vertical mixing, damage and repair interact. Such a test was done by Huot et al. who constructed a mathematical model of the depth profile of DNA damage in the UML [82]. In this model, the initiation of DNA damage (CPDs) was linearly related to exposure, and DNA repair was linearly related to accumulated CPDs at rates developed from laboratory studies. However, photoreactivation (PR) had to be arbitrarily reduced to 10% of the response predicted from laboratory measurements of repair in *Escherichia coli* in order for the model predictions to fit observed profiles of DNA damage (Figure 7). Under these assumptions, mixing rates affected the vertical distribution DNA dimers but had little effect on average dimer concentration [82,83]. The authors recognized that more data on DNA repair in natural assemblages of bacteria is needed before realistic modeling can be done. Both rates and linearity need to be examined. Observations that damage is rapidly repaired when total CPDs are low, but very slowly repaired when CPDs are high ([81], Jeffrey et al. unpublished), suggest that the relative effectiveness of PR decreases when damage exceeds a threshold. If this is the case, average damage within the water column will be higher when mixing slows to the point that the rate of CPD accumulation in the photoactive zone exceeds repair capacity.

4.3.3 Aquatic biota – zooplankton and fish

Another important biological effect of UVR is increased mortality in multicellular aquatic organisms, particularly zooplankton and ichthyoplankton. As for phytoplankton, sensitivity of zooplankton species to UVR varies (Chapter 12), and differences in sensitivity seem to be mostly related to variations in the ability to repair UVR-induced DNA damage (see Chapter 9) and the presence of UVR screening compounds (Chapter 10). However, an important distinction needs to be made between non-motile or slow-moving organisms (e.g., fish eggs) and active swimmers, e.g. crustacean zooplankton and fish larvae. The former may be viewed as passive constituents, responding to the average UVR over the whole mixed layer, but responses for the latter category will be more complicated. Most zooplankton can migrate vertically over long distances, and it is now becoming apparent that they can adjust their vertical position in response to UVR, and perhaps even to UV-B [84–86, Chapter 14]. These zooplankton may be able to seek a deep refuge from UVR effects depending on whether they inhabit shallow or deep environments.

In shallow environments the water vertical velocity induced by turbulent mixing may exceed the maximum swimming speed of organisms at any depth in the water column. Under calm conditions, the organisms may or may not be able to reach a UVR refuge in deep water (depending on water transparency and swimming speed), but under turbulent conditions the organisms may not avoid being cycled throughout the water column (cf. Figure 3). The latter would be the only case for strong swimmers in which it would make sense to relate average survival over the whole water column to the rate of vertical mixing. Under the above scenario (shallow and windy), the average exposure corresponds to the mean biologically effective irradiance over the water column and the biological effects depends on whether the organisms are able to repair UVR damage. When repair is absent or low, reciprocity applies though there is a non-linear relationship between the mortality and exposure, either as a simple survival curve [as for phytoplankton, see 87,88] or a somewhat more complicated (but also non-linear) logistic curve [89]. When repair is present, reciprocity is no longer obeyed and the actual effect will depend on irradiance and period of exposure.

This type of shallow environment is common in the Patagonian steppe and the Pampean regions of Argentina. A study of zooplankton mortality in lakes in these areas found that the presence or absence of reciprocity strongly affected whether data from fixed depth incubations could be used to predict the survivorship in samples undergoing simulated mixing (Figure 8). A single logistic relationship with cumulative exposure was successful in predicting survivorship of a copepod, *Boeckella gracilipes*, which lacked photorecovery, whether the copepod was exposed in fixed or rotating (varying UVR) frame [89,90]. In contrast, another species, *Ceriodaphnia dubia*, which is able to repair DNA damage by photoreactivation, did not obey the same exposure–response relationship in fixed vs. variable (rotated) exposures. Indeed, survivorship was higher in the rotated samples compared to fixed samples receiving the same cumulative exposure [90]. These results show that presence or absence of reciprocity needs to be taken into account in predicting zooplankton survivorship under mixed conditions.

Unlike the previous work on phytoplankton productivity these zooplankton studies did not directly address the issue of whether average survivorship over the water column is affected by the rate of vertical mixing. This question was examined in more detail for a pair of slow moving zooplankton for which responses to UVR appeared to obey reciprocity: eggs of the Atlantic Cod (Gadus morhua) and embryos of a marine copepod (Calanus finmarchicus). For these zooplankton, survivorship was modeled using an exponential function of cumulative dose, similar to the inhibition of photosynthesis in Weddell–Scotia Confluence (WSC) phytoplankton [87,88]. A similar result was obtained as for the WSC phytoplankton (see section 4.3.2): When averaged over the water column, survivorship was higher for low mixing or static cases compared to rapid mixing [83]. Similar results (increased average survivorship in low mixing) could be expected for other zooplankton which lack repair mechanisms (e.g. *Boeckella gracilipes*). However, when repair is active, vertical mixing could increase average survival in a surface layer, as is suggested by the *Ceriodaphnia dubia* case already discussed (rotated samples had better survival than fixed samples receiving the same dose).

In deep environments, strong swimmers are able, in principle, to escape from the turbulent upper layer, and avoid damaging UVR exposure by adjusting their vertical position. Consequently, the net UVR-induced mortality is probably minimal. In this case, the costs, if any, are likely to be indirect, i.e., as a result of the trade-off between UVR exposure and risk of predation in surface waters

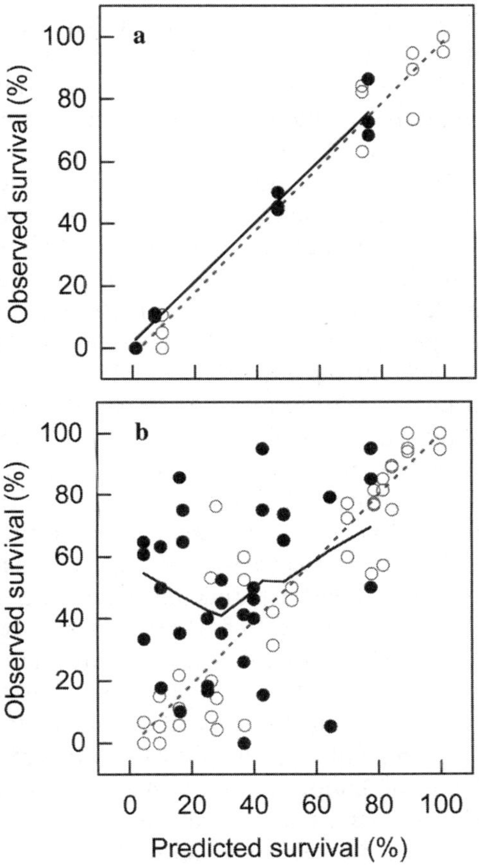

Figure 8. Relationship between observed and predicted survival for (A) a copepod that lacks recovery, *Boeckella gracilipes*. Predicted survival agrees with observation in both static (dashed line, open circles) and rotated (solid line, filled circles) incubations. (B) Another zooplankton species that has photorepair, *Ceriodaphnia dubia*. Observed survival in rotating (solid line, filled circles) is higher than predicted based on survival under static conditions (dashed line, open circles). [From Zagarese et al. [90], reprinted with permission of Oxford University Press.]

versus suboptimal temperature and food availability at depth. This trade-off is explored in more detail in Chapter 12.

4.4 Summary

This chapter has covered some aspects of the physics of the upper mixed layer that modulate how molecules and organisms in aquatic ecosystems are exposed to UVR. The basic implications of a vertically mixed environment have been realized by aquatic UVR researchers for some time. The contents of the near-surface "photoactive" zone are in continual motion and, over some time scale, exposed constituents at the surface will exchange with relatively unexposed constituents at depth. The modulation of UVR exposure is greater than for visible (PAR) part of the spectrum because of the stronger vertical gradients in the UV. Moving beyond this qualitative understanding to quantitative descriptions of mixing effects, however, has been slow.

To some extent, progress has been limited by the availability of measurements on exchange processes. Until recently, temperature microstructure measurements were the primary approach to quantify near-surface turbulence. The instruments needed to do this were expensive and difficult to operate. The situation is now considerably improved. Microstructure sensors more suited to field use are commercially available, and are more "user-friendly". Alternative methods to observe or infer mixing processes have also been perfected, including free-fall CTDs, acoustic doppler sensors and acoustically monitored floats. As these techniques are refined and deployment, operation, and analysis become more routine, it will become increasingly practical to incorporate a "mixing" component into field studies of UVR effects.

Even with good measurements, another technical problem in understanding how mixing modifies UVR effects is translating a given mixing regime into an exposure regime. Up to the present, experiments on mixing and UVR have used simplified schemes to simulate mixing. In field experiments, the exposure usually consists of deterministic cycles as would occur in a circular, vertically rotating eddy. Numerical models, on the other hand, have mainly used a random walk approach which emphasizes turbulent diffusion. Actual mixing is more complicated in that it is a combination of deterministic and random motions, with the contribution of each type of motion varying with circumstances. At present, we don't know if simplified versions of vertical mixing used in experiments or models result in the losing some important aspect of the exposure regime which affects overall response. We can attempt to use more realistic schemes in future studies. For example, with microprocessor controls it is now easier to structure complex sequences of light exposures (e.g. [91]). Another approach that has been developed (but not yet applied in UVR studies) is to measure photosynthesis in a sample attached to a Lagrangian float [92]. There are now sophisticated numerical models of mixing which capture the full range of motions [93] and increases in computing power make it increasingly possible to run such models on a desktop computer. Implementation of full numerical mixing models with accurate estimates of the spectral and temporal dependence of UVR responses will be a significant step towards accurate assessment of UVR effects in the aquatic environment.

The few studies that have been done of mixing and UVR effects on molecules and non-motile or slow moving organisms have confirmed the expectation that mixing strongly affects the vertical distribution of such effects as photobleaching, photoinhibition and DNA damage. The deeper the mixed layer and the more vigorous the mixing, the more products of UVR photochemistry or UVR affected plankton will be found at depths below the photoactive zone. However, the extent to which vertical mixing just changes the distribution of UVR effects or, in addition, affects the integral damage (and/or repair) over the water column is a more open question. In the examples considered here, including UVR effects on CDOM, phytoplankton photosynthesis, bacterial DNA damage and zooplankton mortality, there are some cases in which mixing effects on the integral (or average) response were significant and other cases in which it was not. The primary requirement for defining when mixing will be important (or not) is a good understanding of the kinetics of UVR damage and repair. A recommendation to those in the UVR research community who would like to see progress in understanding interactions with mixing processes would be to include a kinetics component in any study of UVR exposure and response. With increased attention to kinetics and more widespread measurements of mixing processes, we expect that the next few years will see rapid progress in our understanding of this topic.

Acknowledgements

We thank Ann Gargett and Gustavo Buscaglia for their comments on the manuscript and Eric D'Asaro for comments on mixing processes. This work was supported in part by International Foundation for Science (grant H/2325–2), InterAmerican Institute for Global Change Research (CRN–016), U.S. National Science Foundation (OPP–9615342, OCE–9812036, DEB–9973938), Consejo Nacional de Investigaciones Científicas y Técnicas (CONICET PIP 0457/98), Fundación Antorchas (No. 13887–83 and No. 13955–3), and Agencia Nacional de Promoción Científica y Tecnológica (PROALAR, No. 104).

References

1. J.N. Moum, W.D. Smyth (2001). Upper ocean mixing processes. In: J. Steele, S. Thorpe, K. Turekian (Eds), *Encyclopedia of Ocean Sciences* (pp. 3093–3100). Academic Press, NY.
2. A.E. Gargett, J. Marra (2001). Effects of upper ocean physical processes – turbulence, advection, and air–sea interaction – on oceanic primary production. In: A.R. Robinson, J.J. McCarthy, B.J. Rothschild (Eds), *The Sea: Biological–Physical Interactions in the Ocean* (pp. 19–49). Wiley, New York.
3. D.M. Imboden, A. Wüest (1995). Mixing mechanisms in lakes. In: A. Lerman, D.M. Imboden, J.R. Gat (Eds), *Physics and Chemistry of Lakes* (pp. 83–138). Springer-Verlag, Berlin.
4. J. Imberger (1994). Transport processes in lakes: a review. In: R. Margalef (Ed.), *Limnology Now: A Paradigm of Planetary Problems* (pp. 99–193). Elsevier, Amsterdam.
5. A. Vasilkov, N. Krotkov, J. Herman, C. McClain, K.Arrigo, W.Robinson (2001). Global mapping of underwater UV fluxes and DNA-weighted exposures using TOMS and SeaWifs data products. *J. Geophys. Res.*, **106**, 27205–27219.
6. D.P. Morris, B.R. Hargreaves (1997). The role of photochemical degradation of dissolved organic carbon in regulating the UV transparency of three lakes on the Pocono Plateau. *Limnol. Oceanogr.*, **42**, 239–249.

7. R. Sommaruga, R. Psenner (1997). Ultraviolet radiation in a high mountain lake of the Austrian Alps: Air and underwater measurements. *Photochem. Photobiol.*, **65**, 957–963.
8. J.A. Yoder, J. Aiken, R.N. Swift, F.E. Hoge, P.M. Stegmann (1993). Spatial variability in near-surface chlorophyll *a* fluorescence measured by the Airborne Oceanographic Lidar (AOL). *Deep-Sea Res.*, **40**, 37–53.
9. D. Karl (1999). Sea of change: Biogeochemical variability in the North Pacific subtropical gyre. *Ecosystems*, **2**, 181–214.
10. E.J. Fee, R.E. Hecky, S.E.M. Kasian, D.R. Cruikshank (1996). Effects of lake size, water clarity, and climatic variability on mixing depths in Canadian shield lakes. *Limnol. Oceanogr.*, **41**, 912–920.
11. A. Pérez-Fuentetaja, D.J. Dillon, N.D. Yan, D.J. McQueen (1999). Significance of dissolved organic carbon in the prediction of themocline depth in small Canadian shield lakes. *Aquat. Ecol.*, **33**, 127–133.
12. C.L. Osburn, H.E. Zagarese, W. Cravero, D.P. Morris, B.R. Hargreaves (2001). Calculation of spectral weighting functions for the solar photobleaching of chromophoric dissolved organic matter in temperate lakes. *Limnol. Oceanogr.*, **46**, 1455–1467.
13. C.E. Williamson, D.P. Morris, M.L. Pace (1999). Dissolved organic carbon and nutrients as regulators of lake ecosystems: Resurrection of a more integrated paradigm. *Limnol. Oceanogr.*, **44**, 795–803.
14. K.L. Denman, A.E. Gargett (1983). Time and space scales of vertical mixing and advection of phytoplankton in the upper ocean. *Limnol. Oceanogr.*, **28**, 801–815.
15. A.E. Gargett (1997). "Theories" and techniques for observing turbulence in the ocean euphotic zone. *Sci. Mar.*, **61**, 25–45.
16. S. MacIntyre (1993). Vertical mixing in a shallow, eutrophic lake: possible consequences for the light climate of phytoplankton. *Limnol. Oceanogr.*, **38**, 798–817.
17. J.D. Woods, R. Onken (1982). Diurnal variation and primary production in the ocean – preliminary results of a Lagrangian ensemble model. *J. Plankton Res.*, **4**, 735–756.
18. H. Yamazaki, D. Kamykowski (1991). The vertical trajectories of motile phytoplankton in a wind-mixed water column. *Deep-Sea Res.*, **38**, 219–241.
19. A.W. Visser (1997). Using random walk models to simulate the vertical distribution of particles in a turbulent water column. *Mar. Ecol. Progr. Ser.*, **158**, 275–281.
20. N.S. Oakey (2001). Turbulence Sensors. In: J.H. Steele, S.A. Thorpe, K.K. Turekian (Eds). *Encyclopedia of Ocean Sciences*, (pp. 3063–3069). Academic Press, San Diego.
21. S.A. Thorpe (1977). Turbulence and mixing in a Scottish loch. *Phil. Trans. Roy. Soc. A*, **286**, 125–181.
22. J.N. Moum (1996). Energy-containing scales of turbulence in the ocean thermocline. *J. Geophys. Res.*, **101**, 14095–14109.
23. W.R. Crawford (1986). A comparison of length scales and decay times of turbulence in stably stratified flows. *J. Phys. Oceanogr.*, **16**, 1847–1854.
24. T.M. Dillon (1982). Vertical overturns: A comparison of Thorpe and Ozmidov length scales. *J. Geophys. Res.*, **87**, 9601–9613.
25. H. Peters, M.C. Gregg, J.M. Toole (1988). On the parameterization of equatorial turbulence. *J. Geophys. Res.*, **93**, 1199–1218.
26. J.C. Wesson, M.C. Gregg (1994). Mixing at the Camarinal Sill in the Strait of Gibraltar. *J. Geophys. Res.*, **99**, 9847–9878.
27. P.H. LeBlond, L.A. Mysak (1978). *Waves in the Ocean*, Elsevier, Amsterdam.
28. C.P. Lombardo, M.C. Gregg (1989). Similarity scaling of viscous and thermal dissipation in a convecting surface boundary layer. *J. Geophys. Res.*, **94**, 6273–6284.
29. M. Li, C. Garrett (1997). Mixed layer deepening due to Langmuir circulation. *J. Phys.*

Oceanogr., **27**, 121–132.
30. K.L. Denman, A.E. Gargett (1988). Multiple thermoclines are barriers to vertical exchange in the subarctic Pacific during SUPER, May 1984. *J. Mar. Res.*, **46**, 77–103.
31. J. Moum, D. Caldwell, C. Paulson (1989). Mixing in the equatorial surface layer: the diurnal thermocline. *J. Geophys. Res.*, **94**, 2005.
32. K.E. Brainerd, M.C. Gregg (1993). Diurnal restratification and turbulence in the oceanic surface mixed layer. 2. Modeling. *J. Geophys. Res.*, **98**, 22657–22664.
33. T. Powell, M.H. Kirkish, P.J. Neale, P.J. Richerson (1984). The diurnal cycle of stratification in Lake Titicaca: Eddy diffusion. *Verh. Int. Verein. Limnol.*, **22**, 1237–1243.
34. J. Imberger (1985). The diurnal mixed layer. *Limnol. Oceanogr.*, **30**, 912–920.
35. M.A. Xenopoulos, D.W. Schindler (2001). The environmental control of near-surface thermoclines in boreal lakes. *Ecosystems*, **4**, 699–707.
36. N.M. Scully, W.F. Vincent, D.R.S. Lean (2000). Exposure to ultraviolet radiation in aquatic ecosystems: estimates of mixing rate in Lake Ontario and the St. Lawrence River. *Can. J. Fish. Aquat. Sci.*, **57**, 43–51.
37. E.A. D'Asaro (2001). Turbulent vertical kinetic energy in the ocean mixed layer. *J. Phys. Oceanogr.*, **31**, 3530–3537.
38. E. Steffen, E.A. D'Asaro (2001). Deep convection in the Labrador Sea as observed by Lagrangian Floats. *J. Phys. Oceanogr.*, **32**, 475–492.
39. R.G. Zepp, D.M. Cline (1977). Rates of direct photolysis in aquatic environment. *Environ. Sci. Technol.*, **11**, 359–366.
40. R.F. Whitehead, S. de Mora, S. Demers, M. Gosselin, P. Monfort, B. Mostajir (2000). Interactions of ultraviolet-B radiation, mixing, and biological activity on photobleaching of natural chromophoric dissolved organic matter: A mesocosm study. *Limnol. Oceanogr.*, **45**, 278–291.
41. J.M.C. Plane, R.G. Zika, R.G. Zepp, L.A. Burns (1987). Photochemical modeling applied to natural waters. In: R.G. Zika, W.J. Cooper (Eds). *Photochemistry of Environmental Aquatic Systems* (pp. 250–267). American Chemical Society.
42. S.C. Doney, R.G. Najjar, S. Stewart (1995). Photochemistry, mixing, and diurnal cycles in the upper ocean. *J. Mar. Res.*, **53**, 341–369.
43. R. Hedlund, C. Youngson (1972). Rates of photodecomposition of Picloram in aqueous systems. In: S. Faust (Ed.), *Fate of Organic Pesticides in the Aquatic Environment* (pp. 159–172). American Chemical Society Washington DC.
44. K. Mopper, D.J. Kieber (2000). Marine photochemistry and its impact on carbon cycling. In: S. de Mora, S. Demers, M. Vernet (Eds), *The Effects of UV Radiation in the Marine Environment* (pp. 101–129). Cambridge University Press, New York.
45. H.E. Zagarese, M. Diaz, F. Pedrozo, M. Ferraro, W. Cravero, B. Tartarotti (2001). Photodegradation of natural organic matter exposed to fluctuating levels of solar radiation. *J. Photochem. Photobiol. B*, **61**, 35–45.
46. N.M. Scully, W.F. Vincent, D.R.S. Lean, S. MacIntyre (1998). Hydrogen peroxide as a natural tracer of mixing in surface layers. *Aquat. Sci.*, **60**, 169–186.
47. R.J. Sikorski, R.G. Zika (1993). Modeling mixed-layer photochemistry of H_2O_2: Optical and chemical modeling of production. *J. Geophys. Res.*, **98**, 2315–2328.
48. R.J. Sikorski, R.G. Zika (1993). Modeling mixed-layer photochemistry of H_2O_2: Physical and chemical modeling of distribution. *J. Geophys. Res.*, **98**, 2329–2340.
49. J.G. Mitchell, L. Pearson, A. Bonazinga, S. Dillon, H. Khouri, R. Paxinos (1995). Long lag times and high velocities in the motility of natural assemblages of marine bacteria. *Appl. Environ. Microbiol.*, **61**, 877–882.
50. R. Margalef (1978). Life forms of phytoplankton as survival alternatives in an un-

stable environment. *Oceanol. Acta.*, **1**, 493–509.
51. H.L. MacIntyre, T.M. Kana, R.J. Geider (2000). The effect of water motion on short-term rates of photosynthesis by marine phytoplankton. *Trends Plant Sci.*, **5**, 12–17.
52. C.L. Gallegos, T. Platt (1985). Vertical advection of phytoplankton and productivity estimates: a dimensional analysis. *Mar. Ecol. Prog. Ser.*, **26**, 125–134.
53. E.W. Helbling, V. Villafañe, M. Ferrario, O. Holm-Hansen (1992). Impact of natural ultraviolet radiation on rates of photosynthesis and on specific marine phytoplankton species. *Mar. Ecol. Progr. Ser.*, **80**, 89–100.
54. M. Vernet, E.A. Brody, O. Holm-Hansen, B.G. Mitchell (1994). The response of Antarctic phytoplankton to ultraviolet radiation: absorption, photosynthesis, and taxonomic composition. In: C.S. Weiler, P.A. Penhale (Eds). *Ultraviolet Radiation in Antarctica: Measurements and Biological Effects* (pp. 143–158). American Geophysical Union, Washington, D.C.
55. P.J. Neale, J.J. Cullen, R.F. Davis (1998). Inhibition of marine photosynthesis by ultraviolet radiation: Variable sensitivity of phytoplankton in the Weddell–Scotia Sea during the austral spring. *Limnol. Oceanogr.*, **43**, 433–448.
56. R.E.H. Smith, J.A. Furgal, D.R.S. Lean (1998). The short-term effects of solar ultraviolet radiation on phytoplankton photosynthesis and photosynthate allocation under contrasting mixing regimes. *J. Gt. Lakes Res.*, **24**, 427–441.
57. J. Marra (1978). Effect of short-term variation in light intensity on photosynthesis of a marine phytoplankter: a laboratory simulation study. *Mar. Biol.*, **46**, 191–202.
58. J. Marra (1978). Phytoplankton photosynthetic response to vertical movement in a mixed layer. *Mar. Biol.*, **46**, 203–208.
59. P.J. Neale, J.F. Talling, S.I. Heaney, C.S. Reynolds, J.W.G. Lund (1991). Long time series from the English Lake District: Irradiance-dependent phytoplankton dynamics during the spring maximum. *Limnol. Oceanogr.*, **36**, 751–760.
60. P.J.S. Franks, J. Marra (1994). A simple new formulation for phytoplankton photoresponse and an application in a wind-driven mixed-layer model. *Mar. Ecol. Progr. Ser.*, **111**, 145–153.
61. W.F. Vincent, P.J. Neale, P.J. Richerson (1984). Photoinhibition: algal responses to bright light during diel stratification and mixing in a tropical alpine lake. *J. Phycol.*, **20**, 201–211.
62. P.J. Neale, P.J. Richerson (1987). Photoinhibition and the diurnal variation of phytoplankton photosynthesis – I. Development of a photosynthesis-irradiance model from studies of in situ responses. *J. Plankton Res.*, **9,** 167–193.
63. R. Lande, M.R. Lewis (1989). Models of photoadaptation and photosynthesis by algal cells in a turbulent mixed layer. *Deep-Sea Res.*, **36**, 1161–1175.
64. J.A. Dusenberry (2000). Steady-state single cell model simulations of photoacclimation in a vertically mixed layer: implications for biological tracer studies and primary productivity. *J. Mar. Syst.*, **24**, 201–220.
65. E.A. Laws, M.R. Landry, R.T. Barber, L. Campbell, M.-L. Dickson, J. Marra (2000). Carbon cycling in primary production bottle incubations: inferences from grazing experiments and photosynthetic studies using 14C and 18O in the Arabian Sea. *Deep-Sea Res.*, **47**, 1339–1352.
66. G. Kullenberg (1982). Note on the role of vertical mixing in relation to effects of UV radiation on the marine environment. In: J. Calkins (Ed.). *The Role of Solar UV Radiation on the Marine Ecosystems* (pp. 283–292). Plenum Press, N.Y.
67. V. Milot-Roy, W.F. Vincent (1994). UV radiation effects on photosynthesis: The importance of near-surface thermoclines in a subarctic lake. *Arch. Hydrobiol. Beih.*

Ergebn. Limnol., **43**, 171–184.

68. V.E. Villafañe, M. Andrade, V. Lairana, F. Zaratti, E.W. Helbling (1999). Inhibition of phytoplankton photosynthesis by solar ultraviolet radiation: studies in Lake Titicaca, Bolivia. *Freshwat. Biol.*, **42**, 215–224.
69. E.W. Helbling, V. Villafañe, O. Holm-Hansen (1994). Effects of ultraviolet radiation on Antarctic marine phytoplankton photosynthesis with particular attention to the influence of mixing. In: C.S. Weile, P.A. Penhale (Eds). *Ultraviolet Radiation in Antarctica: Measurements and Biological Effects* (pp. 207–227). American Geophysical Union, Washington, D.C.
70. P.J. Neale, R.F. Davis, J.J. Cullen (1998). Interactive effects of ozone depletion and vertical mixing on photosynthesis of Antarctic phytoplankton. *Nature*, **392**, 585–589.
71. W. Harm (1980). *Biological effects of ultraviolet radiation* (I.U.P.A.B. Biophysics series, Vol. 1), Cambridge University Press, Cambridge.
72. E.S. Barbieri, V.E. Villafañe, E.W. Helbling. Experimental assessment of UVR effects upon temperate marine phytoplankton when exosed to variable radiation regimes. *Limnol. Oceanogr.* Submitted.
73. E.W. Helbling, V.E. Villafañe, E.S. Barbieri (2001). Sensitivity of winter phytoplankton communities from Andean lakes to artificial ultraviolet-B radriation. *Rev. Chil. Hist. Nat.*, **74**, 391–400.
74. M.P. Lesser, J.J. Cullen, P.J. Neale (1994). Carbon uptake in a marine diatom during acute exposure to ultraviolet B radiation: relative importance of damage and repair. *J. Phycol.*, **30**, 183–192.
75. J.J. Cullen, M.P. Lesser (1991). Inhibition of photosynthesis by ultraviolet radiation as a function of dose and dosage rate: Results for a marine diatom. *Mar. Biol.*, **111**, 183–190.
76. P.J. Neale, J.J. Fritz, R.F. Davis (2001). Effects of UV on photosynthesis of Antarctic phytoplankton: Models and application to coastal and pelagic assemblages. *Rev. Chil. Hist. Nat.*, **74**, 283–292.
77. P.J. Neale (2001). Effects of ultraviolet radiation on estuarine phytoplankton production: Impact of variations in exposure and sensitivity to inhibition. *J. Photochem. Photobiol. B*, **62**, 1–8.
78. P.J. Neale, E. Litchman, C. Sobrino, C. Callieri, G. Morabito, V. Montecino, Y. Huot, P. Bossard, D. Steiner, C. Lehmann (2001). Quantifying the response of phytoplankton photosynthesis to ultraviolet radiation: Biological weighting functions versus in situ measurements in two Swiss lakes. *Aquat. Sci.*, **63**, 265–285.
79. J. Köhler, M. Schmitt, H. Krumbeck, M. Kapfer, E. Litchman, P.J. Neale (2001). Effects of UV on carbon assimilation of phytoplankton in a mixed water column. *Aquat. Sci.*, **63**, 294–309.
80. W.H. Jeffrey, J.P. Kase, S.W. Wilhelm (2000). UV radiation effects on heterotrophic bacterioplankton and viruses in marine ecosystems. In: S.J. de Mora, S. Demers, M. Vernet (Eds). *The Effects of UV Radiation on Marine Ecosystems* (pp. 206–236). Cambridge University Press, Cambridge.
81. P. Boelen, M.J.W. Veldhuis, A.G.J. Buma (2001). Accumulation and removal of UVBR-induced DNA damage in marine tropical plankton subjected to mixed and simulated non-mixed conditions. *Aquat. Microb. Ecol.*, **24**, 265–274.
82. Y. Huot, W.H. Jeffrey, R.F. Davis, J.J. Cullen (2000). Damage to DNA in bacterioplankton: A model of damage by ultraviolet radiation and its repair as influenced by vertical mixing. *Photochem. Photobiol.*, **72**, 62–74.
83. P.S. Kuhn, H.I. Browman, R.F. Davis, J.J. Cullen, B. McArthur (2000). Modeling the effects of ultraviolet radiation on embryos of *Calanus finmarchicus* and Atlantic cod

(*Gadus morhua*) in a mixing environment. *Limnol. Oceanogr.*, **45**, 1797–1806.
84. C. Alonso (2001). *The role of Ultraviolet Radiation as a Determinant of the Vertical Distribution of Planktonic Crustaceans in Three Andean Lakes* (*in Spanish*), Thesis, Centro Regional Universitario Bariloche, Universidad Nacional del Comahue, Bariloche, Argentina.
85. D.M. Leech, C.W. Williamson (2001). In situ exposure to ultraviolet radiation alters the depth distribution of *Daphnia. Limnol. Oceanogr.*, **46**, 416–420.
86. U.C. Storz, R.J. Paul (1998). Phototaxis in water fleas (*Daphnia magna*) is differently influenced by visible and UV light. *J. Comp. Physiol. A.*, **183**, 709–717.
87. J.H.M. Kouwenberg, H.I. Browman, J.J. Cullen, R.F. Davis, J.-F. St-Pierre, J.A. Runge (1999). Biological weighting of ultraviolet (280–400 nm) induced mortality in marine zooplankton and fish. I. Atlantic cod (*Gadus morhua*) eggs. *Mar. Biol.*, **134**, 269–284.
88. J.H.M. Kouwenberg, H.I. Browman, J.J. Cullen, R.F. Davis, J.-F. St-Pierre, J.A. Runge (1999). Biological weighting of ultraviolet (280–400 nm) induced mortality in marine zooplankton and fish. II. *Calanus finmarchicus* (Copepoda) eggs. *Mar. Biol.*, **134**, 285–293.
89. H.E. Zagarese, W. Cravero, P. Gonzalez, F. Pedrozo (1998). Copepod mortality induced by fluctuating levels of natural ultraviolet radiation simulating vertical mixing. *Limnol. Oceanogr.*, **43**, 169–174.
90. H.E. Zagarese, B. Tartarotti, W. Cravero, P. Gonzalez (1998). UV damage in shallow lakes: implication of water mixing. *J. Plankton Res.*, **20**, 1423–1433.
91. F. Gervais, T. Hintze, H. Behrendt (1999). An incubator for the simulation of a fluctuating light climate in studies of planktonic primary productivity. *Int. Rev. Hydrobiol.*, **84**, 49–60.
92. R.E. Reed, G.J. Kirkpatrick, D. Kamykowski (1994). Short-period photophysiological responses of *Thalassiosira pseudonana* during photoacclimation to near-surface irradiance. In: S.G. Ackleson (Ed.). *Ocean Optics XIII* (pp. 514–519). SPIE-The International Society for Optical Engineering, Bellingham, WA.
93. W.G. Large, J.C. McWilliams, S.C. Doney (1994). Oceanic vertical mixing – a review and a model with a nonlocal boundary-layer parameterization. *Rev. Geophys.*, **32**, 363–403.
94. A.B. Kara, P.A. Rochford, H.E. Hurlburt (2000). An optimal definition for ocean mixed layer depth. *J. Geophys. Res.*, **105**, 16803–16821.
95. W.H. Jeffrey, R.J. Pledger, P. Aas, S. Hager, R.B. Coffin, R. Von Haven, D.L. Mitchell (1996). Diel and depth profiles of DNA photodamge in bacterioplankton exposed to ambient solar ultraviolet radiation. *Mar. Ecol. Progr. Ser.*, **137**, 283–291.

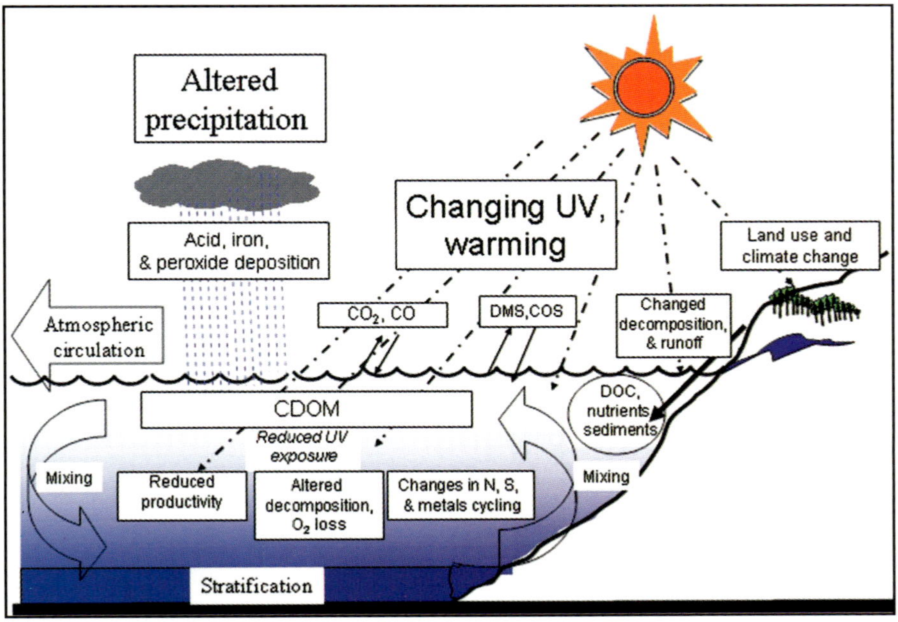

Chemistry

- Solar UVR and aquatic carbon, nitrogen, sulfur and metals cycles
- Photochemistry of chromophoric dissolved organic matter in natural waters
- Photoactivated toxicity in aquatic environments
- Reactive oxygen species in aquatic ecosystems

Chapter 5

Solar UVR and aquatic carbon, nitrogen, sulfur and metals cycles

Richard G. Zepp

Table of contents

Abstract

Solar UVR at the Earth's surface (280–400 nm) has wide-ranging impacts on biological and chemical processes that affect the cycling of elements in aquatic environments. This chapter uses recent field and laboratory observations along with models to assess these impacts on carbon, nitrogen, sulfur and metals cycles. Much emphasis is placed on the interactions of UVR with carbon capture and storage, decomposition, and trace gas exchange. UV exposure generally inhibits phytoplankton photosynthesis and also affects microbial processes both through direct inhibition of bacterial activity as well as through effects on the biological availability of carbon and nitrogen substrates. One important aspect of UV interactions with carbon cycling involves the formation and decomposition of UV-absorbing organic matter, principally chromophoric dissolved organic matter (CDOM). CDOM controls UV exposure in the sea and in many freshwater environments. It can be directly photodecomposed to dissolved inorganic carbon (DIC), carbon monoxide, and various carbonyl-containing compounds. UV can potentially affect nitrogen and sulfur cycling in a variety of ways such as alterations in nitrogen fixation, effects on the biological availability of dissolved organic nitrogen (DON), UV photoinhibition of phytoplankton, bacterioplankton and zooplankton that affect sources and sinks of dimethyl sulfide (DMS) and UV-initiated photoreactions that oxidize DMS and produce carbonyl sulfide. Metal cycling also interacts in many ways with UVR via photoinhibition of microbial redox cycling, direct photoreactions of dissolved metal complexes and metal oxides and indirect reactions that are mediated by photochemically-produced reactive oxygen species (ROS). Photoreactions can affect the biological availability of essential trace nutrients such as iron and manganese, transforming the metals from complexes that are not readily assimilated into free metal ions or metal hydroxides that are available.

5.1 Introduction

Biogeochemical cycles describe the complex interaction of biological, chemical, and physical processes that control the exchange and recycling of matter and energy at and near the Earth's surface. Considerable recent research has demonstrated that the cycles are sensitive to solar UVR in a variety of ways. A primary objective of this paper is to present and analyze interactions between UV and aquatic biogeochemistry, taking into account the fact that other co-occurring global environmental changes can influence the UV effects. Interest in this topic has been stimulated in part by declines in stratospheric ozone over the past two decades that have resulted in increases in solar UV-B radiation (280–315 nm) reaching the Earth's surface. Current projections indicate that return of the ozone layer thickness to pre-1980s levels may not occur for another 50 years.

Aquatic biogeochemical cycles are sensitive to other factors that affect underwater UV exposure such as UVR attenuation in the water, mixing and stratification. Underwater UV exposure is thus sensitive to global changes in climate, land use and other human activities that affect aquatic transport, composition and optical properties. These effects are changing the UV-A (315–400 nm) as well as the UV-B spectral region so the discussions here include the biogeochemical effects that are influenced by the entire solar UV spectrum.

Some of the many interactions between solar UV and aquatic biogeochemical cycles are illustrated in Figure 1. Biological processes are directly affected through increased damage to DNA, and damage to other molecules, including proteins, RNA, membrane-associated molecules, and chlorophyll. Resulting effects include changes in photosynthesis and respiration by phytoplankton and alterations in microbial respiration, biological availability of carbon and nitrogen, sulfur cycling, metal redox reactions and organic pollutant dissipation. Geochemical processes that are strongly influenced by UVR include: formation of greenhouse and chemically important trace gases (carbon dioxide (CO_2), carbon monoxide (CO), nitric oxide (NO), DMS, carbonyl sulfide (COS)); con-

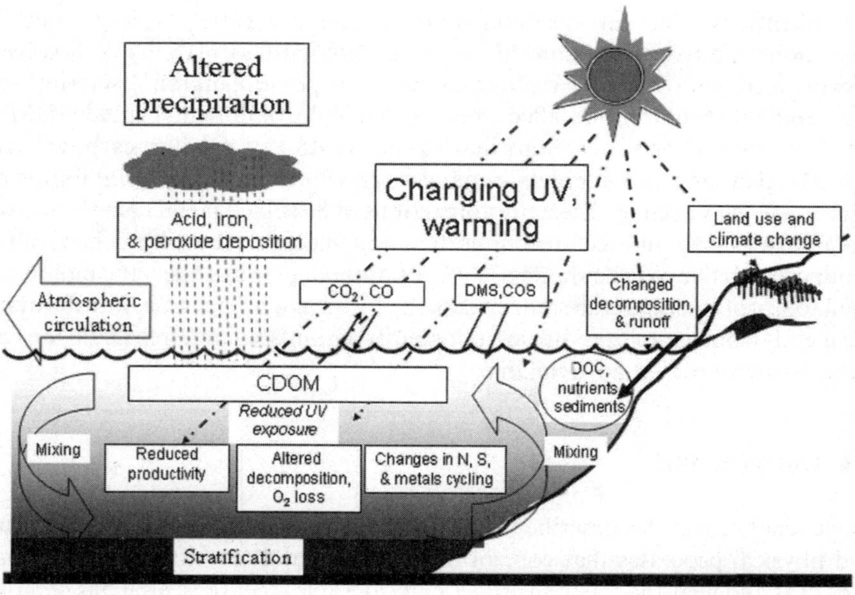

Figure 1. Schematic illustrating factors that influence the effects of solar UVR on aquatic ecosystems. Aquatic biogeochemical cycles are affected by increased UVR caused by stratospheric ozone depletion and its interaction with other co-occurring environmental changes such as global warming and land use change. Changes in precipitation chemistry are caused by UV-induced photoreactions in cloud droplets and changing climate affects the frequency and intensity of precipitation which in turn affects UV penetration and UV-induced processes in freshwaters and the sea (see text for additional explanation). Key: ROS reactive oxygen species, principally hydrogen peroxide; DMS dimethyl sulfide; COS, carbonyl sulfide; CO carbon monoxide; CO_2 carbon dioxide; CDOM chromophoric dissolved organic matter, the primary UV absorbing substance in aquatic environments; DOC dissolved organic carbon.

version of organically bound nitrogen into biologically available inorganic N; conversion of refractory organic matter into lower molecular weight, biologically available organic compounds; photoinduced changes in the optical properties of DOM; and changes in the redox state of the upper ocean and freshwaters through formation of peroxides and changes in transition metal speciation.

The cycles of various elements are discussed separately in this chapter, but it should be emphasized that biogeochemical cycles are tightly interwoven and are subject to significant feedback interactions. This paper is not intended to provide a detailed review of the burgeoning research in this area, but rather to provide the reader with a glimpse of the exciting studies that are taking place and the types of measurement approaches and concepts that are being applied. Wherever possible, recent reviews are cited to provide more background to this chapter. Excellent overviews of global biogeochemical cycles are available [1,2], and several useful reviews related to topics considered here have recently been published [3–12].

5.2 Carbon cycle

Although particulates that are predominantly clay mineral in content are important attenuators of UVR in turbulent streams and rivers, dissolved and particulate organic substances largely control the penetration of UV into most lakes and the sea. Hence, this discussion of the carbon cycle begins with a discussion of the interactions of UV with CDOM, with emphasis on the optical properties of aquatic ecosystems and penetration of solar UVR into the water (see also Chapters 3 and 6).

5.2.1 UV–CDOM interactions and aquatic optical properties

5.2.1.1 CDOM sources and characteristics

Because CDOM is the most important UV-absorbing dissolved organic constituent in aquatic systems, it plays an important role in the interactions between UVR and aquatic carbon cycling. As used in this chapter, CDOM includes hydrophobic colored organic matter referred to as "humic substances" [7,13]. These substances are chemically complex and poorly characterized mixtures of anionic organic substances known to contain phenolic and carboxyl groups [13–18]. Terrestrially-derived CDOM originates through the decomposition of dead plant material and it is introduced into water via leaching and runoff from land. CDOM also can be produced through the decay of algal detritus; this source of CDOM has been referred to as "microbially-derived" [15,16]. There is some evidence that CDOM in the ocean can also be produced by photoreactions of triglycerides and fatty acids [19] in regions such as the sea-surface microlayer [5].

There are distinguishable differences between the optical and physicochemical properties of humic substances that are related to the molecular properties of these substances [13–18]. CDOM accounts for only a small fraction of the DOC in the open ocean and it has a significantly-lower aromatic content and specific absorption coefficients than terrestrially-derived CDOM [13]. These differences can be ascribed to variability in sources [13,15,16,18] and/or transformations [19–35] of the CDOM.

Light absorption by CDOM typically decreases in an approximately exponential fashion with increasing wavelength [13]. The absorption coefficient $a(\lambda)$ is defined as:

$$a(\lambda) = 2.303A(\lambda)/\,l \tag{1}$$

where $A(\lambda)$ is the absorbance of a water sample or aqueous solution of isolated CDOM at wavelength λ in a cell of pathlength l. The relationship between $a(\lambda)$ and wavelength (in nm) can be represented by an equation of the following form

$$a(\lambda) = a(\lambda_o)\,e^{S(\lambda_o - \lambda)} \tag{2}$$

where S, the spectral slope coefficient, is computed by a non-linear least-squares fitting routine, and $a(\lambda_o)$ is the absorption coefficient at a reference wavelength λ_o. S is a parameter that characterizes how rapidly the absorption decreases with increasing wavelength. The specific absorption coefficient, $a(\lambda)^*$, is defined by the relation:

$$a(\lambda)^* = a(\lambda)/[\mathrm{DOC}] \tag{3}$$

where [DOC] is the DOC concentration expressed as mg C L^{-1}.

The coefficients, $a(\lambda)^*$ and S, vary both spatially and temporally [13,18] although the variation is not large for CDOM in freshwaters and coastal waters. Generally, values of $a(\lambda)^*$ in the UV-B region have been observed to be lower for very clear oligotrophic seawaters than for coastal seawaters and freshwaters strongly influenced by terrestrial input. S for the UV region generally increases with decreasing $a(\lambda)^*$, ranging from as low as 0.012–0.013 nm^{-1} for some highly absorbing coastal waters to over 0.02 nm^{-1} for weakly absorbing oligotrophic waters [13]. Sharp changes in S can occur in coastal regions along seaward transects [36–38] where oceanic (microbially-derived?) sources of CDOM are becoming dominant along the transect, perhaps as terrestrially-derived CDOM is removed by photochemical and biological processes.

5.2.1.2 CDOM effects on UV penetration

Detailed concepts and elegant numerical models have been developed to describe the transmission of solar radiation in aquatic ecosystems and several thorough reviews of this literature have appeared [39–41]. Solar irradiance at a given wavelength decreases approximately exponentially as it penetrates into a region of uniform composition in a natural water body. The slope of natural logarithmic plots of irradiance versus depth has been referred to as the "diffuse attenuation coefficient [$K_d(\lambda)$]." The diffuse attenuation coefficient for downwelling irradiance at a given wavelength can be modeled as a function of the composition of seawater, focusing on absorption by water, chlorophyll-like pigments, and dissolved organic carbon (DOC) [40,42,43]. In certain freshwaters, non-living particulate matter, which includes detritus and suspended sediments made primarily of clay minerals, can make a significant contribution to UV and visible light attenuation [41,44,45]. Microorganisms that are often exposed to UV-B radiation can develop cellular UV-protective substances such as mycosporine-like amino acids (MAAs, see Chapter 10) that absorb in the UV region. Such organisms or detritus derived from them can contribute significantly to UV attenuation in ecosystems that have low concentrations of DOC [41,46].

Studies of a wide range of freshwater and marine environments have shown that DOC, and, to a lesser extent in the open ocean, organic colloids, play an important role in the attenuation of solar UVR [13,45–54]. Not all of the DOC is responsible for such attenuation; CDOM is the UV-absorbing component. CDOM strongly absorbs UV-B and UV-A radiation and its absorption is sufficiently great in the visible (400–700 nm) region that it affects the remote sensing of ocean color [13,50–52,54]. One important feature of the absorption spectra of CDOM is the exponential increase in absorption with decreasing wavelength in the UV region. This feature enables CDOM to effectively protect aquatic ecosystems from harmful UV-B radiation while permitting beneficial photorepairing (UV-A) and PAR to be much more efficiently transmitted into the water. Efforts to model the spectral effects of CDOM on primary production [55] and microbial processes [56] have been reported.

Changes in CDOM concentrations in freshwaters and the sea cause significant changes in UV penetration. These changes occur seasonally [13,46,48–54,57] and they also likely are linked to climate [50, Chapter 17], acid deposition [50] and land use changes.

5.2.1.3 Effects of UV on CDOM optical properties

Microorganisms do not readily decompose the polymeric substances that make up CDOM, but CDOM transformation is accelerated when it is exposed to solar UVR. In this section, the effects of such transformations on CDOM optical properties are discussed. Also, see Section 5.2.3 and Chapter 6 for discussion of other aspects of the effects of UV on CDOM decomposition. It has been known for years that absorption of solar UVR by terrestrially-derived humic substances and CDOM results in a reduction in its light absorption and fluorescence (i.e. photobleaching) [3–8,13,19–35,52–54,57,58]. The CDOM in water samples, freshwater and coastal regions impacted by riverine inputs is readily bleached on exposure to natural solar radiation with concurrent increases in the spectral slope coefficient [18,35]. A study of freshwater lakes in the USA concluded that acid-neutralizing capacity has particularly important effects on photobleaching rates [28], but other studies in South America and Antarctica have found no effects of alkalinity on the rates [Zagarese, personal communication, 2002]. Factors such as oxygen concentration [59] and temperature [60] also affect photobleaching efficiency.

Measuring "apparent quantum yield" spectra can quantitate wavelength effects on photoreactions. The term "apparent quantum yield" denotes the moles of photoproduct formed per absorbed Einstein (mole photons) of radiation at a

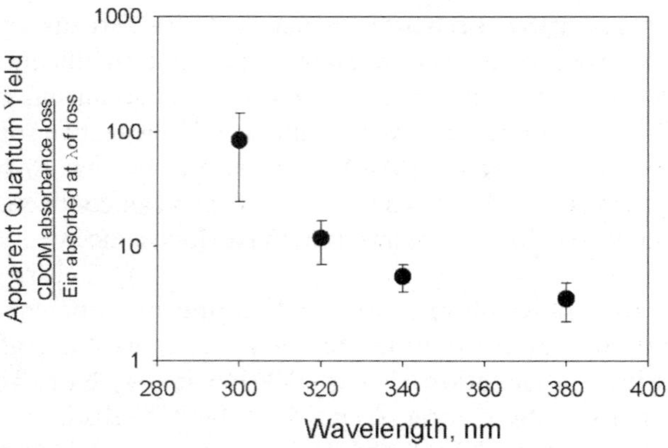

Figure 2. Apparent quantum yield spectra for the photobleaching of CDOM derived from the St. Lawrence Estuary. [Adapted from Whitehead et al. [29], Figure 4, p. 285, Copyright 2000, The American Society for Limnology and Oceanography, Inc.]

certain wavelength. Apparent quantum yields provide a useful unitless gauge for comparison of photoreactions of DOM from different natural waters. Recently published quantum yield spectra for CDOM photobleaching are shown in Figure 2, where the wavelength dependence of CDOM absorbance loss per absorbed Einstein is plotted. These and other quantum yield spectra indicate that the efficiency for the photobleaching of CDOM absorption and fluorescence is greatest in the UV region [23,29,35,61].

The excitation–emission matrix spectra (EEMS) of CDOM fluorescence also are altered on exposure to solar radiation, generally with reductions in fluorescence that approximately parallel absorption losses [36–38]. Hypsochromic shifts (shifts to shorter wavelengths) occur in both excitation and emission maxima on irradiation [36–38]. Interactions between photochemical and microbial degradation [38] are involved.

The photobleaching of CDOM involves two general classes of photoreactions that occur in aquatic environments: direct and indirect (photosensitized) [3,62,63, Chapter 8]. Direct photoreactions involve light absorption by the photoreactive constituent(s) of CDOM to produce reactive excited states. On the other hand, photoreactions also can involve indirect photoprocesses that also are initiated through light absorption by the CDOM. Indirect photoreactions occur through the intermediacy of various short-lived reactive transients, such as excited states, or species produced there from, that diffuse through the system and then react with the CDOM, sulfur compounds (see Section 4) or other constituents of the water. There is abundant evidence from continuous or laser flash photolysis experiments that various reactive transients are produced on irradiation of CDOM [62, Chapter 8]. Molecular oxygen often is involved in the formation of the reactive transients. CDOM photoreactions potentially can proceed by either direct or indirect pathways or by both pathways. The finding that superoxide reacts rapidly with CDOM [64] suggests that indirect photoreactions involving this transient may contribute to CDOM photochemistry. Reactions of CDOM with hydroxyl radicals account for a small fraction of the photobleaching [65]. A recent study by Del Vecchio and Blough [35] has shown that photobleaching of CDOM with monochromatic radiation occurs most rapidly at the irradiation wavelength, indicating that direct photolysis is at least partly involved in the photobleaching. With complex substances such as CDOM, it is conceivable that photoreactions may involve trapping of reactive transients before they can diffuse away into bulk aqueous solution.

5.2.2 UV interactions with aquatic carbon capture and storage

Phytoplankton communities are primarily responsible for the production of biomass in large lakes and the ocean. Submerged and partially-submerged aquatic vegetation play a central role in creation of biomass in many freshwater systems [66]. Freshwaters also receive inputs of allochthonous (externally produced) organic matter derived from terrestrial plants and soils, and rivers transport large amounts ($\sim$400 teragrams (Tg) C yr^{-1}; 1 Tg$=10^{12}$ g) of terrestrial organic matter into coastal waters [67–71]. The effects of UV-B on terrestrial plant productivity [72] and submerged plants [45, Chapter 11] has been discussed elsewhere.

5.2.2.1 Phytoplankton photosynthesis

Given the vital role of phytoplankton biology in aquatic carbon capture, understanding UV interactions with phytoplankton is of critical importance. Phytoplankton in the sea carry out about half of the photosynthesis on Earth. Phytoplankton photosynthesis reduces the partial pressure of carbon dioxide in the upper ocean and thereby promotes the absorption of CO_2 from the atmosphere. The organic carbon produced by photosynthesis forms the base of the marine food web. About 25% of the organic carbon produced by phytoplankton photosynthesis is exported from the upper ocean into intermediate and deep water [73]. On a global basis this "new production" is estimated to be about 11 to 16 Gt C per year [1 gigaton (Gt) of carbon equals 10^{15} g] and it is believed that most of this organic carbon is remineralized in the top km of the sea [73]. This mechanism for movement of carbon from the surface to deep ocean has been referred to as the "biological pump" and it has been shown that this process keeps atmospheric CO_2 concentrations about 150 to 200 ppmv lower than if there were no marine phytoplankton [73].

Here I briefly consider aspects of the direct effects of UV on phytoplankton that are relevant to carbon cycling. For more detailed considerations of UV interactions with phytoplankton see the chapters by Villafañe et al. (Chapter 11) and Neale et al. (Chapter 4) in this book. Indirect effects include changes in trophic level interactions as well as in the biological availability of micronutrients such as iron or in interactions with damaging reactive oxygen species. The latter are discussed elsewhere in the chapter as are the role of phytoplankton in the effects of UV on sulfur cycling.

A variety of studies have demonstrated that exposure to UV can directly inhibit photosynthesis in phytoplankton [11,12,45,74]. Studies in several different locations have shown that reductions in current levels of solar UV-B result in enhanced primary production, and Antarctic experiments under the ozone hole demonstrated that primary production is inhibited by enhanced UV-B. For example, investigations of depth-integrated in situ phytoplankton productivity in austral spring of 1990 [75] indicated that productivity was reduced 6 to 12% inside the ozone hole in the Bellinghausen Sea compared with productivity outside the hole. On an annual basis, this range corresponds to an estimated annual productivity loss of 7 to 14 teragrams, which is 2 to 4% of production in the Antarctic marginal ice zone.

Action spectra describe the wavelength dependency of radiation in producing some biological or chemical response [8,12,76,77]. Action spectra can be used to estimate the biological impacts of UV changes that result with changes in the ozone layer as well as changes in location, time-of-day, season, and depth. The term "biological weighting function" (BWF) has been used to distinguish a type of action spectrum measured using polychromatic UV and visible radiation with a series of cutoff filters [12], as originally described by Rundel [78]. Unlike action spectra measured using monochromatic radiation [76], the Rundel approach helps take into account the fact that there are interactions between various part of the spectrum, such as photorepair of UV-B damage by UV-A radiation. The evaluation of action spectra for UV inhibition of phytoplankton photosynthesis also must take into account the dependence of photosynthesis on exposure, in particular whether reciprocity applies. The term "reciprocity" applies to systems in which biological or chemical responses to UV depend on cumulative exposure alone, independent of the duration of exposure or the irradiance [11,12,79]. Reciprocity does not apply to phytoplankton that rapidly repair UV damage. Instead, a steady state that reflects a balance between damage and repair is attained with continuous UV exposure [12,80,81]. This steady state can be described as a function of weighted irradiance. Elegant procedures for modeling these effects have been developed over the past decade [11,12,80,81]. Using these procedures, a recent study has shown that seasonally-averaged action spectra for phytoplankton inhibition by UV in a mid-latitude estuary (Rhode River in Maryland, USA) are remarkably similar to action spectra for photoinhibition of Antarctic phytoplankton (Figure 3) [80]. Interestingly, the action spectra observed in the mid-latitude studies did not exhibit large seasonal changes as was expected, but they did indicate major short-term changes that may be attributable to factors such as changes in species composition, nutrient availability, temperature, and light acclimation.

In addition to its direct inhibitory effects on production, UV may also be involved in indirect effects on oceanic new production by its effects on the biological availability and chemical reactivity of micronutrients, iron, manganese, and copper, in particular. UVR also strongly affects the production of ROS and such effects can be mediated through the reactions of ROS with trace metals such as copper and iron. The specific effects are discussed in more detail below.

5.2.2.2 Other UV interactions with carbon capture and storage

Ozone depletion may influence the ability of the ocean to take up atmospheric

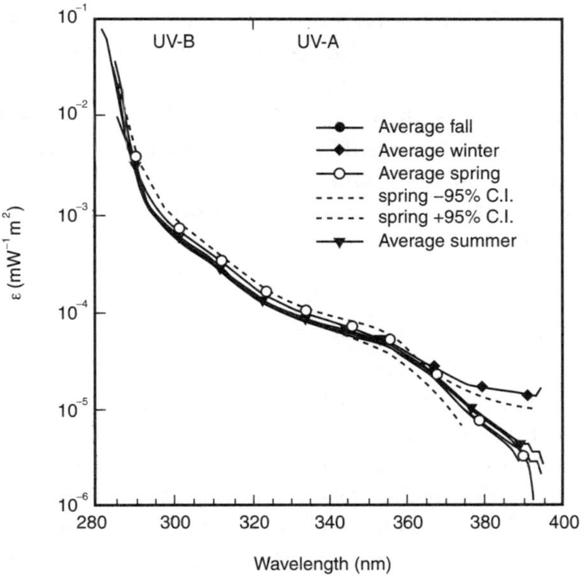

Figure 3. Comparison of biological weighting functions for UV inhibition of photosynthesis for Rhode River (mid-latitude site, North America) and Antarctic phytoplankton. [Reprinted with permission from Banaszak and Neale [80], Figure 2, p. 597, Copyright 2001, The American Society for Limnology and Oceanography, Inc.]

CO_2, but the net impact of a reduction in primary production on the ocean sink for atmospheric CO_2 is uncertain. In addition to the factors influencing phytoplankton photosynthesis that were discussed earlier, aquatic circulation (Chapter 4), microbial cycling and photodegradation, macro- and micronutrient availability and other factors affect net carbon storage. Moreover, other indirect effects involving trophic level interactions may also affect ecosystem productivity [82]. For example, the vertical migration of zooplankton has been recently shown to be sensitive to UVR [83,84]. This finding has important implications for the flux of carbon through the microbial food web, which involves transfer of biomass from the primary producers to metazoa and bacteria. Sulfur cycling also may be affected by UV-induced changes in zooplankton grazing. Thus, the net impact on carbon capture is clearly not a linear function of increased UV exposure.

Moreover, it is likely that interactions with other global-scale environmental changes will affect the biological pump and other aspects of global biogeochemical cycles. For example, changes in atmospheric circulation associated with climate change likely will affect aquatic mixing dynamics and thus the impact of UV on photosynthesis as well as decomposition (Figure 1). Moreover, recent remote sensing observations indicate that changes in thermocline depth in the tropical Pacific Ocean occur during El Niño/La Niña events that strongly influence new production and carbon export from the upper to deep ocean [85]. These changes can affect the impact of UV on phytoplankton photosynthesis and microbial decomposition as well as the air–sea exchange of gases. The unprecedented strength of the 1982/83 and 1997/98 El Niño events has indicated that they are becoming strongly influenced by anthropogenic activities and thus that strong El Niños may become more frequent in the future.

In the case of freshwaters, the past effects of climate change on UV exposure have impacted sedimentary records in a remarkable way. Analysis of fossil diatom assemblages in Canadian subarctic lake sediments has provided evidence of the interactive impacts of climate change and solar UVR on CDOM concentrations during the Holocene [86].

5.2.3 UV effects on decomposition

In Section 2.1.3 the effects of solar UVR on the optical properties of CDOM were discussed. Here the broader effects of UV on decomposition are considered. Detailed considerations of the effects of UVR on chemical and biological decomposition were initiated during the 1980s [63,87]. These early studies revealed that DOC plays a central, multifaceted role in aquatic photochemistry and photobiology. DOC photoreacts to produce atmospherically-important trace gases and biologically-available carbon- and nitrogen-containing compounds, and to initiate free radical and photosensitized reactions that affect aquatic composition.

UV effects on decomposition of aquatic organic matter are caused by inhibition of microbial activity, by direct photodegradation of the CDOM and particulate organic carbon (POC) to CO_2 and other gases, and by UV-induced photodegradation of the persistent, polymeric components of the DOC to readily decomposable compounds. These effects are discussed in the following sections and then, to complete this section on carbon cycling, modeling and experimental techniques are described and then used to evaluate the role of UV-induced decomposition of organic matter in selected freshwater and marine environments.

5.2.3.1 Microbial photoinhibition

Bacterial activity is inhibited by UV-B radiation [10,56,88] and direct DNA damage (pyrimidine dimerization) has been demonstrated in field studies [89,90]. The greatest damage is observed in poorly-mixed, stratified waters. However, observations in lakes, coastal waters, and the Gulf of Mexico and modeling studies showed that the reduction in microbial activity is attenuated with increased winds and surface layer mixing and the activity is rapidly restored in the dark (within a few hours) via repair and regrowth [56,89,90]. A modeling study concluded that changes in UVR caused by ozone depletion can have a more serious net impact on bacterial activity than UV increases attributable to decreased CDOM concentrations [56]. The radiation amplification factor (RAF) for UV damage to bacterioplankton is close to that computed for generalized DNA damage (see Section 8) and thus DNA damage for these microorganisms must generally be more susceptible to ozone depletion than phytoplankton photosynthesis. However, like phytoplankton, bacterioplankton also can repair

DNA damage by photoreactivation and excision repair. The model indicated that damage caused by decreased CDOM is partly offset by increased photoreactivation that is related to the modeled increase in repairing UV-A radiation caused by CDOM depletion [56]. It should be noted, however, that these results are a sensitive function of the assumed action spectra for damage and photoreactivation as well as the spectral slope coefficient for CDOM. The latter is considerably higher in seawater than the value of 0.014 that was assumed in the modeling study [13,14] and thus the study likely underestimated the dependence of changing CDOM concentrations on net UV damage. These studies indicate that adverse effects of UVR on microbial activity can change the timing and location of microbial decomposition of labile organic matter in the upper ocean. The amount and distribution of marine viruses also are affected by UV-B radiation in the sea [91,92]. Viruses can influence microbial diversity and activity, including decomposition. Light-induced repair of sunlight-damaged viruses, probably by photoreactivation, can be effected in the presence of bacteria [92].

As is the case for decomposers on terrestrial plant litter [6,72], UV exposure also affects bacterial and fungal growth on aquatic macrophyte detritus [93]. The effects are evidenced in part by changes in the attached microbial communities, which, for example, became dominated by bacteria in irradiated microcosms compared to shaded systems. Enzymatic activity of the microorganisms also was changed in UV-B irradiated systems, where significantly higher beta-glucosidase activity was observed [93].

5.2.3.2 Direct photodecomposition

In addition to its effects on microbial activity, solar UVR has direct effects on decomposition. A variety of recent studies have provided evidence that CDOM undergoes a complex array of other photoreactions that can involve a decrease in average molecular weight accompanied by cleavage to a variety of photoproducts [3–5,8,24,25,68,70,74,94–118], changes in isotopic content [60,119] and consumption of oxygen [59,68,106,116]. These reactions include the direct photochemical mineralization of the CDOM to carbon monoxide and dissolved inorganic carbon (DIC). Of these various direct pathways, the photoproduction of DIC is most efficient.

It has long been known that intense short-wavelength UVR can mineralize DOC [3,7,8]. Mineralizations achieved under such extreme conditions, however, are irrelevant to natural conditions. Only very recently have several reports indicated that DOC in freshwaters and seawater can be directly mineralized on exposure to sunlight [3,8,59,70,106,110–115,118]. Miles and Brezonik [106] were first to report this reaction in a natural freshwater system. They presented evidence that this process included photoreactions of DOC–iron complexes. More recent studies have provided a more detailed understanding of photoreactions involving iron in the natural photooxidation of DOC [59,107–109]. Appreciable iron concentrations are sometimes found in high-DOC, acidic freshwaters and iron can be introduced into the sea via riverine inputs, wet deposition, and deposition of Aeolian dust. It seems likely that future research will demonstrate an important role for iron in enhancing photochemical mineralization of

DOC in natural waters. However, other pathways for DIC photoproduction that do not involve iron must also be available [59,70]. Whatever the mechanism for DIC photoproduction, recent studies have shown that this process potentially could account for the global annual production of anywhere from 1 to 12 Gt of CO_2 C in the ocean [3,118].

In addition to effects on DOM, UV exposure also impacts the decomposition of POC [120]. Photoproduction of DIC has been observed from the sterilized detritus of several aquatic macrophytes in both air and immersed in water [120]. The highest production rates were observed in water. Although UVR was most effective at inducing detritus decomposition, visible light also played a role.

Carbon monoxide (CO) is also formed in aquatic environments from the photochemical degradation of DOM [3,4,8,22,94–105]. Strong gradients of CO have been observed in the lowest 10 metres of the atmosphere over the Atlantic Ocean [97]. The samples nearest the ocean surface were some 50 ppb higher than at the 10-metre altitude-sampling inlet. This implies that the ocean is a source of CO to the atmosphere and that this source can increase the atmospheric concentration. CO is reactive in the troposphere and thus its emissions from the ocean may influence the hydroxyl radical (OH) and ozone concentrations in the marine atmospheric boundary layer that is remote from strong continental influences.

Although the sea is thought to be a net source of CO, this source has been subject to a wide range of estimates. The most recent estimate has come from Zafiriou et al. [105] who, based on modeled results derived using CO quantum yields that were measured using Pacific water samples during 1994, concluded that the global open ocean photochemical source of CO is approximately 50 ± 10 Tg CO carbon per year. An approximate estimate of the coastal ocean source was about 10 Tg CO carbon annually. Most of this CO production was estimated to be consumed by microorganisms rather than escaping to the atmosphere. The microbial sink estimates were based on a series of incubations that quantified CO loss in freshly collected seawater samples. Previous, much higher estimates of CO photoproduction in the sea were based primarily on CO photoproduction from terrestrially-derived CDOM [8,105]. As shown in Figure

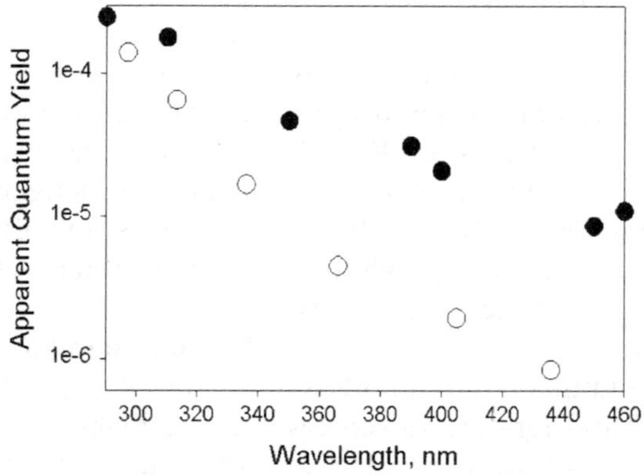

Figure 4. Comparison of apparent quantum yield spectra for production of CO from terrestrially-derived CDOM (●) [59,103] and from CDOM in the open ocean (○) [105].

4, apparent quantum yields for terrestrially-derived CDOM are much larger than those observed with open ocean water, especially in the UV-A and visible spectral regions. This difference in quantum yields, which largely accounts for the lower estimated fluxes in the open ocean, suggests that the photoreactivity of bluewater, algal-derived CDOM may be quite different from that of terrestrially-derived CDOM. The open ocean CO fluxes estimated by Zafiriou et al. [105] agree approximately with earlier estimates of Bates et al. [104], but are at the lower end of global flux estimates based on extensive studies during the 1980s of CO emissions in the Atlantic Ocean by Conrad and co-workers [100] and of CO emissions in the Pacific by Gammon and Kelley [102]. The differences in these estimates may reflect the fact that CO concentrations exhibit great spatial and temporal variability. But they may also reflect periodic large-scale changes in the nature of the CDOM in the upper ocean, related to El Niño/La Niña events [46,54,85]. For example, the higher CO fluxes reported by Gammon and Kelley [102] were based on Pacific observations during the 1987–1988 El Niño event, whereas the estimates of Zafiriou et al. [105] were based on observations during 1994 when El Niño conditions were not prevalent.

5.2.3.3 UV effects on lability of microbial substrates

UVR can potentially affect carbon cycling through modification of the biological availability of microbial substrates and microbial activity [4,6,8,9,38,67,112,116,121–138]. This effect, referred to here as photochemically-altered microbial degradation, is well documented in the case of terrestrially-derived DOC, where stimulation of microbial activity is usually observed. The relative importance of this pathway, compared to direct photodegradation, seems to strongly depend on the DOC source. Initial comparisons using early quantitative studies of identifiable biologically labile photoproducts indicated that other carbon photoproducts such as DIC are produced at rates many-fold higher than biologically-available photoproducts (BLPs) [8,9]. The photoproduction of BLPs has been quantified using microbial growth indicators (e.g., uptake of tritiated leucine), or cumulative bacterial oxygen consumption during post irradiation–incubation as an index (e.g., respiratory activity per absorbed photon). Recent studies, however, indicate that the quantum yields for formation of BLPs in coastal waters of the Southeastern United States are of the same order of magnitude as that for DIC photoproduction [67,125]. Likewise, direct and photochemically-stimulated microbial decomposition of the DOC from the Adriatic Sea and coastal North Sea were estimated to be approximately equivalent [128].

Not all results are consistent with a stimulating effect, however [128,132–137]. Surface water DOM in the open ocean [131,132] and a subtropical seagrass meadow [133] were not sources of biologically labile photoproducts. In the case of the seagrass meadow, BLPs were produced by exposure to solar radiation but the effect was attributed to algal exudates [133] not DOC photodecomposition. Photoreactions can reduce the microbial availability of certain organic substrates such as peptone and algal exudates [134–136], possibly via light-induced cross-linking between the CDOM and algal exudates [135]. Decreased bacterial activity also was observed on the leachate from UV-exposed detritus from vascular plants and this effect was attributed to decreased availability of the DOM and possibly the release of inhibitory substances from the detritus [137].

5.2.3.4 Modeling UV-induced decomposition

Large-scale models provide a useful technique for estimating global-scale fluxes through carbon pools and how environmental changes such as ozone depletion affect the fluxes. Models of the effects of UVR on decomposition and trace gas production require equations based on field or laboratory measurements under varying natural conditions and/or experimental manipulations that relate rates of UV-induced processes to changing environmental parameters. Time series observations of various indicators of UV effects such as atmospheric and aquatic concentrations of trace gases, aquatic CDOM concentrations and UV absorption coefficients, and isotopic and lignin composition of DOC can provide large-scale integrators of changes in biogeochemical cycling in aquatic ecosystems and serve as checks on fluxes inferred from models.

Various approaches have been used to model the rates of photoreactions in aquatic environments [3–6,8–10,67,68,139–141]. To illustrate the utility of the modeling approach, the wavelength dependence for CO photoproduction fluxes simulated for a mid-latitude location is shown in Figure 5. Quantum yield spectra that are shown in Figure 4 were used in these calculations. The equations and assumptions involved in these calculations are discussed in more detail in

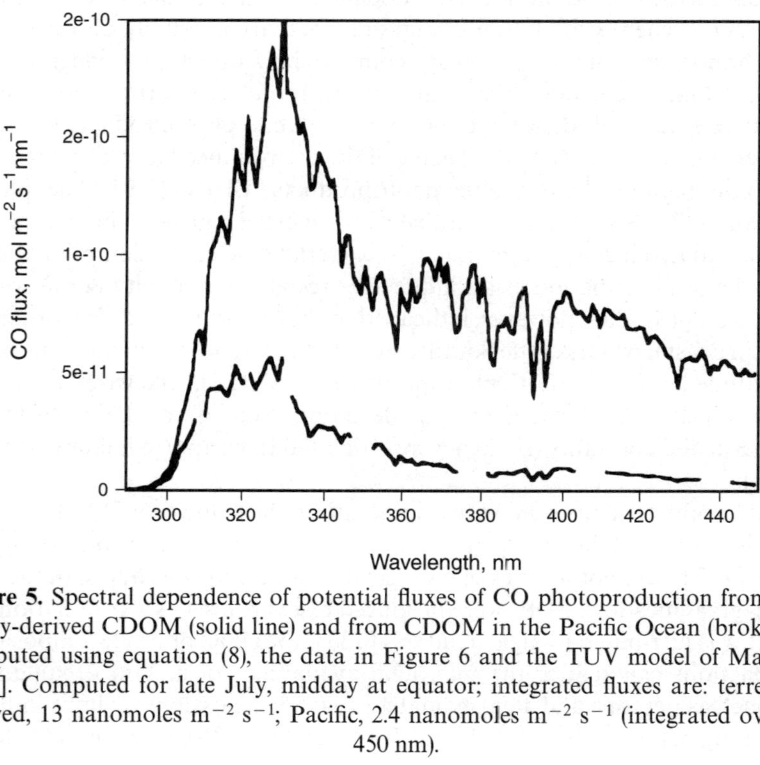

Figure 5. Spectral dependence of potential fluxes of CO photoproduction from terrestrially-derived CDOM (solid line) and from CDOM in the Pacific Ocean (broken line), computed using equation (8), the data in Figure 6 and the TUV model of Madronich [235]. Computed for late July, midday at equator; integrated fluxes are: terrestrially-derived, 13 nanomoles m^{-2} s^{-1}; Pacific, 2.4 nanomoles m^{-2} s^{-1} (integrated over 290–450 nm).

the Appendix. The results demonstrate the much larger area under the plot for terrestrially-derived CDOM, and thus much higher depth-integrated photoproduction, compared to that for open ocean CDOM. Longer wavelength UV is much more heavily involved in photoproduction of CO from the terrestrial organic matter. The difference may be due to an inherent difference in the source of the CDOM (derived from microbial processing of dead organic matter from terrestrial plants vs. phytoplankton) and/or to extensive photooxidation of the open ocean CDOM, decomposing the more reactive components and leaving less reactive substances in the residual CDOM.

5.2.3.5 UV-induced oxidation in coastal areas

Marine scientists have long puzzled over the fate of riverine organic matter on entry to the ocean [69,71]. Isotopic studies indicate that the DOC in the open ocean is primarily of marine origin [142,143], although some terrestrial character would have been expected. Models have been used to evaluate the potential loss of organic matter in coastal regions.

Quantum yield spectra determined for BLP production in coastal waters can be used in conjunction with simulated solar spectral irradiance to estimate potential seasonal and annual BLP fluxes as a function of latitude (Figure 6) [67]. Potential annual consumption of terrestrially-derived organic matter in coastal areas can then be calculated as the cross-product of the annual fluxes and coastal ocean areas for various 10° latitude bands. Using this approach it was estimated that most coastal BLP formation potentially occurs in the Northern Hemisphere, with substantial contributions from high-latitude coastal regions [67]. The estimated potential annual production of biologically labile photoproducts from coastal regions worldwide is 253×10^{12} g C, a value that approximately corresponds to the annual global input of riverine DOM (220×10^{12} g C [69]. Using quantum yield spectra for photochemical oxygen demand as a

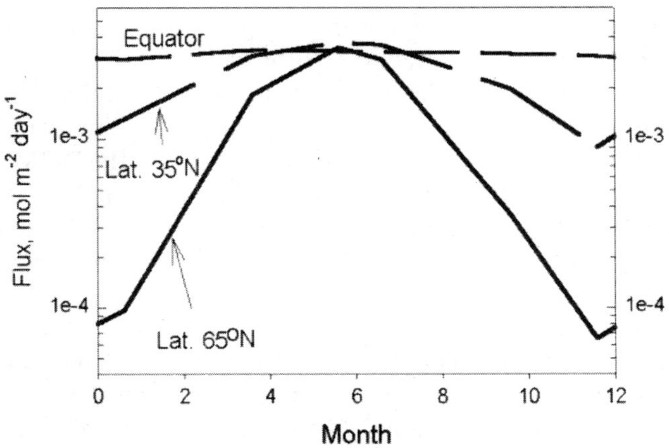

Figure 6. Seasonal changes estimated for potential biologically labile photoproduct production from terrestrially-derived CDOM at various latitudes in the Northern Hemisphere [67].

means to estimate efficiencies for direct photooxidation of CDOM, it was computed that, on an annual basis, the direct photooxidation of CDOM would potentially consume about 3.5×10^{14} moles O_2 yr^{-1} in the global ocean [68]. Assuming that coastal waters make up about 7% of the ocean's area and that 0.5 moles of O_2 are consumed per altered DOC carbon, this estimate corresponds to a potential consumption of 600×10^{12} g C per year in coastal regions.

These modeling results indicate that direct photooxidation combined with photochemically-stimulated microbial degradation can potentially consume about 850×10^{12} g C per year in coastal areas. The results indicate that even high-latitude coastal DOC is subject to major UV-induced oxidation in the latitudes of its input to the ocean. These likely are overestimates of the role of photochemistry in oxidizing terrestrially-derived organic matter. As noted earlier, these considerations apply only to the photoreactive component of the DOM (i.e., the CDOM). As noted above, riverine inputs of DOM appear to include a substantial non-reactive component. Estimates for the open ocean are clouded at this point by the previously-discussed findings that bluewater CDOM may be less photoreactive than terrestrially-derived CDOM and much less efficient in the net photoproduction of BLPs.

Other recent observations are consistent with the modeling results. The $\delta^{13}C$ isotopic composition of DOC in the open ocean, −20‰ to −21‰, is about the same is that found in marine plankton, but is significantly (5% to 7%) more positive than that of freshwater (terrestrially-derived) DOC [142,143]. Thus, organic carbon created originally by phytoplankton appears to be the primary source of the DOC in the open ocean, even though riverine inputs of terrestrially-derived DOC should be evident. These comparisons suggest that a large sink for the terrestrially-derived DOC must exist in the ocean or perhaps in estuaries [69,71]. If photooxidation plays an important role in this sink activity, as predicted by the models, then UV exposure of freshwater DOC should increase its $\delta^{13}C$. Recent studies have shown that this is indeed the case. Exposure of terrestrially-derived CDOM in water from coastal estuaries [60] or in mid-latitude lakes [119] resulted in significant changes in stable isotope composition: the $\delta^{13}C$ of DIC that was produced was isotopically "light" relative to the initial DOC, leaving a residual fraction of DOC that was isotopically "heavy" (Figure 7). It is likely that this effect involves photochemically-stimulated microbial degradation. Seasonal changes in the isotopic signature of the DOM in the upper layers of lakes also have been attributed to photooxidation [119]. These results are consistent with the model predictions and suggest that UV-induced photooxidation plays a role in the observed changes in DOC isotopic composition that occur in estuaries [144].

In addition to changes in stable isotope composition, photoreactions also cause decreases in lignin content of DOC [21,145,146]. Such decreases, especially loss (compared to freshwater DOC) of photochemically-sensitive molecular indicators such as syringyl phenols [142], also are observed in coastal DOC samples, indicating that UV-induced oxidation of part of the terrestrial DOC is rapid.

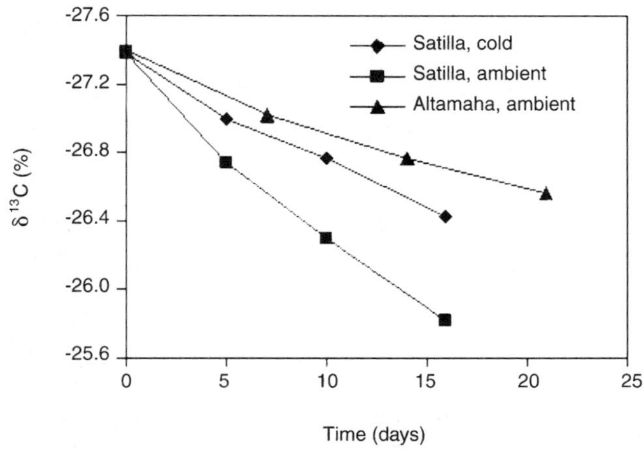

Figure 7. Changes in the stable isotopic composition of CDOM in a coastal river water sample on exposure to natural solar radiation . The change is consistent with observed changes in DOC isotopic composition that occur when terrestrially-derived DOM moves from land into the sea.

5.3 Nitrogen cycle

UVR can affect nitrogen cycling in several ways: (1) through effects on nitrogen-related enzymatic activity by microorganisms such as photoinhibition of nitrogen fixation by prokaryotes, principally cyanobacteria and, indirectly, through effects on the biological availability of essential trace elements, such as iron that stimulate the growth of nitrogen fixers; (2) through enhanced decomposition of persistent DON to biologically labile nitrogenous photoproducts. In addition, UV absorption by inorganic nitrogen species such as nitrate and nitrite produce ROS in aquatic environments, including the highly reactive hydroxyl radical [Chapter 8].

5.3.1 Nitrogen fixation

All organisms require nitrogen and, although its most common form in the environment, N_2 gas, is quite abundant, it can only be used by microorganisms that can "fix" nitrogen by using the enzyme nitrogenase to reduce N_2 to ammonium (NH_4^+). Nitrogenase is rapidly inhibited by exposure to UV-B radiation in vivo [45,147]. Moreover, the growth and survival of most N-fixing cyanobacteria that have been studied can be decreased by only a few hours of UV exposure [45,148]. However, few data are available on the effects of UV on open ocean nitrogen fixers. Damaging effects of UV include those that were discussed earlier, e.g. DNA and photosystem II damage [45]. Presumably, UV exposure can also affect the biological availability of certain organic substrates that affect the growth rates of cyanobacteria, but little attention has been paid to such effects thus far. Nitrogen is a limiting nutrient in remote parts of the open ocean and thus impacts of UV radiation on important nitrogen fixing cyanobacteria such as *Trichodesmium* [148] or the newly discovered N-fixing marine nanoplankton [149], although not proven thus far, could be quite ecologically significant.

Furthermore, iron plays a role in stimulating the growth of *Trichodesmium* [150]. In a later part of this chapter, the possible role of UV in iron cycling and biological availability is discussed in more detail. It has been proposed that the iron deposited in the sea via long range transport of Saharan dust can trigger rapid growth of this organism and that related increases in biologically-available nitrogen can trigger growth in toxic organisms such as *Gymnodinium breve* [148,151]. Riverine inputs and wet deposition of iron also can be important sources (Figure 1). Intense precipitation events can flush large amounts of riverine iron into coastal areas. Iron limitation of *Trichodesmium* may be even more pronounced than was originally believed, because much of the iron believed to be 'dissolved' in the upper ocean, is in fact in colloidal form and thus presumably less biologically-available [152].

5.3.2 Photoreactions of persistent DON

Biologically labile nitrogen compounds such as nitrate, ammonium and amino acids are rapidly recycled by the biota in aquatic systems, while N-containing substances whose structures are too complex or randomized to be readily assimilated accumulate in the water column. In lakes, estuaries, and parts of coastal regions, these persistent classes of DON are predominantly terrestrially-derived. In the open ocean, persistent DON is derived from a combination of bacterial and perhaps photochemical transformation of algal-derived organic matter. In oligotrophic waters with limited N fixation or external inputs of labile N, the labile compounds drop almost to immeasurable levels in the photic zone while the persistent DON accumulates [153]. Interactions of UVR and DON provide a pathway for the conversion of persistent DON into compounds that are more easily assimilated by aquatic microorganisms, such as ammonium [9,20,154–158], nitrite [159] and dissolved primary amines [9,20,154,155,160].

Ammonium photoproduction generally is the most rapid photoreaction of DON. Net production or, in some cases, consumption [156] of ammonium appears to be the result of a competition between production, possibly via hydrolysis of reactive organonitrogen intermediates such as Schiff bases [154], and reactions of ammonium with humic substances and their photoproducts [156]. Environmental conditions such as system pH and temperature affect the conversion of the intermediates into ammonium [59,154]. Other factors such as the degree of pre-exposure to solar radiation do not affect ammonium photoproduction, as evidenced by the low production rates observed in irradiated groundwater samples [156]. Under N-limiting conditions, the release of nitrogenous photoproducts from DOM photodegradation was found to significantly increase rates of bacterial growth, and it occurred most efficiently on exposure to

UV-B radiation [154]. Modeling results suggest that photochemically-produced labile nitrogen compounds can be an important source of biologically available nitrogen in coastal regions [154]. Such photoreactions also may be an increasingly important source of labile N in the upper open ocean, where future intensification of stratification in response to global warming may reduce upwelled sources of labile N [73]. However, few data are available on the photoreactions of open-ocean DON.

5.3.3 Nitrous oxide, nitric oxide, nitrate and nitrite

Nitrous oxide is a greenhouse gas with a poorly characterized, but possibly significant ocean source [161]. N_2O is produced in the ocean by nitrifying and denitrifying bacteria via processes that are inhibited by high dissolved oxygen levels and by light. The main production of N_2O occurs in the deep ocean where the nitrifiers are well protected from solar UVR. Thus, changes in UV fluxes at the ocean surface would only indirectly change the N_2O fluxes through, for example, alterations in N_2 fixation or labile N photoproduction.

UV-B radiation is mainly responsible for the photodegradation of nitrate in water [162–164]. This photoreaction is likely to be too slow (near-surface turnover in sunlight of about 0.008 day^{-1} in the tropics and 0.005 day^{-1} in mid-latitudes [163]) in comparison with the rapid biological uptake of nitrate to have a major effect on the environmental lifetime of nitrate in most aquatic environments, certainly in N-limited regions. Nonetheless, nitrate is an important source of reactive oxygen species in certain freshwaters [163,165]; among other species, hydroxyl radicals are produced in this photoreaction. Nitrite is also degraded by solar UVR to form nitric oxide (NO), hydroxyl radicals, and other products [162], but this reaction is mainly induced by UV-A radiation and thus its photolysis lifetime under solar radiation is considerably shorter than that of nitrate [162,166,167]. The rapid photolysis of nitrite constrains its buildup as a photoproduct from DOM photodegradation [159]. Photochemically produced NO may be an important source of this chemically-reactive gas in the remote marine boundary layer [167], but the buildup in its concentration is constrained by the co-occurrence of photochemically-produced superoxide ions in the upper ocean that rapidly react with NO [5,168, Chapter 8]. Jankowski et al. [162] have taken advantage of the sensitivity of nitrate and nitrite photoreactions to solar UV by using these substances in solar UV actinometers (i.e., photochemical systems for measuring UV irradiance).

5.4 Sulfur cycle

Atmospheric sulfur plays an important role in the radiative balance of the atmosphere [169–174]. Anthropogenic sources are dominant in highly-industrialized regions, such as those in the mid-latitudes of the Northern Hemisphere, and are well defined. Natural sources and sinks of sulfur gases are much less well defined, but have received greater scrutiny in recent years due to their potential involvement in the regulation of climate in remote parts of the ocean. The major source of natural sulfur gases is the sea [170,173]. Of particular interest are DMS and COS. Both of these compounds are formed predominantly in aerobic marine environments, i.e. the upper layers of the ocean, and their sources and sinks are affected by solar UVR.

5.4.1 Dimethyl sulfide

DMS reactions in the troposphere are believed to lead to enhanced reflectivity of marine clouds [171] and thus DMS emissions may have a cooling influence on the atmosphere. One of the best demonstrations of the link between the natural atmospheric sulfur cycle and the physical climate system are the observations that link the satellite derived stratus cloud optical depth and observed DMS derived cloud condensation nuclei (CCN) concentrations at Cape Grim, Australia [175]. Statistical evidence indicates that the optical depth of the clouds is correlated with the number of CCN in the atmosphere. Thus, any UV-related changes at the surface of the ocean that result in the alteration in DMS flux to the atmosphere and the subsequent formation of CCN would also alter the atmospheric radiation budget for the affected region.

DMS is the predominant volatile sulfur compound in the open ocean. It is derived from the transformation of dimethylsulfonium propionate (DMSP), an organosulfur compound that is synthesized by marine algae. Most marine phytoplankton synthesize DMSP, but their DMSP content is highly variable [176]. Concentrations of DMSP are highest in dinoflagellates and coccolithophores, algal classes that represent only a small fraction of the algae in the sea. Within major blooms of these algae there are readily discernable relationships between DMSP concentrations and algal biomass. High concentrations of DMS and DMSP occur in the Southern Ocean during early to mid spring as the sea ice melts [177], a time in which that region also experiences intense UV exposure due to ozone depletion. To first order, one might expect that DMS concentrations will decrease as surface UV increases, because DMS production is closely related to phytoplankton photosynthesis, which is inhibited by increased UV. Such decreases were observed under ozone-depleted regions of the Southern Ocean during 1993 [178]. However, this oversimplified view must be tempered by the fact that DMS concentrations are affected by a variety of biological and chemical processes that could be affected by UV exposure. For example, other research has shown that DMS emissions increase when the phytoplankton are stressed by zooplankton grazing [179]. Therefore, the effects of increased UV on zooplankton discussed earlier could have a positive effect on DMS release. Moreover, recent studies indicate that changes in mixing depth also can influence DMS emissions by controlling the conversion efficiency of DMSP into DMS [180,181]. Decreases in mixing depth are proposed to reduce assimilatory metabolism of DMSP via UV-induced photoinhibition (or perhaps nutrient limitation) thus enhancing the conversion of DMSP into DMS.

In addition to possible effects of solar UVR on the biological production of DMS, DMS emissions to the atmosphere also are strongly affected by biological and photochemical sinks of DMS in the upper ocean. DMS photoreactions accounted for 7% to 40% of the total turnover of DMS in the surface mixed layer of the equatorial Pacific Ocean [182]. The photoreaction involved conversion into dimethyl sulfoxide, but the yield of the conversion was only 14%, much less than was expected. DMS absorbs little or no light at wavelengths >295 nm, so its photolysis under solar radiation in pure water is very slow. Apparently, natural photosensitizers in the seawater initiated this indirect photooxidation (see Section 5.2.1.3). Although earlier results indicated that this reaction may be induced by visible light [182], more recent research using Sargasso Sea water indicates that solar UV is mainly responsible [183].

5.4.2 Carbonyl sulfide

COS is the most concentrated sulfur gas in the troposphere [170,173] and it is believed to play a role in the maintenance of the stratospheric sulfate layer [184,185], although this role may be more limited than was originally believed [185].

COS is produced in surface seawater by the photochemical degradation of DOM and it is degraded mainly by hydrolysis [94,141,186–192]. Comparison of the quantum yield spectra for COS from various regions of the sea again show that, as in the case of CO (Figure 4), COS is produced most efficiently in the UV region and that there appears to be a major difference between the quantum yield spectra in coastal regions and in the open ocean (Figure 8). Quantum yields are generally higher in coastal areas [189,193], especially in the UV-A region, and so are observed COS concentrations and fluxes [186]. Potential annual photo-production of COS in the open ocean and coastal regions at various 10 degree latitude bands is computed in Figure 9 by cross-multiplying the fluxes, computed as described in section 5.2.3.5 and the Appendix, and the areas of the open ocean and coastal ocean. Pos et al. [191] provided evidence that the photoproduction of COS and CO in the sea may be linked by competitive reactions involving free radical species, possibly helping to explain the similarity in the fractional change in quantum yields with increasing wavelength for CO and COS (see Figures 4 and 8).

In an impressive effort to model global air–sea fluxes of COS based on known information about its sources and sinks, Preiswerk and Najjar [141] estimated that the open ocean was a net source of COS (2.1 Gmol yr^{-1}), taking into account the possibility that dark production occurs. This is close to another global estimate of Ulshöfer and Andreae [192]. The estimates of Preiswerk and Najjar [141] did not include coastal contributions. Early observations and global estimates of Andreae and Ferek [186] suggested that COS emissions predominately come from such regions. Although the computed production comparisons indicate that this likely is not the case (Figure 9), it appears that substantial amounts of COS are produced in coastal regions, a large fraction of

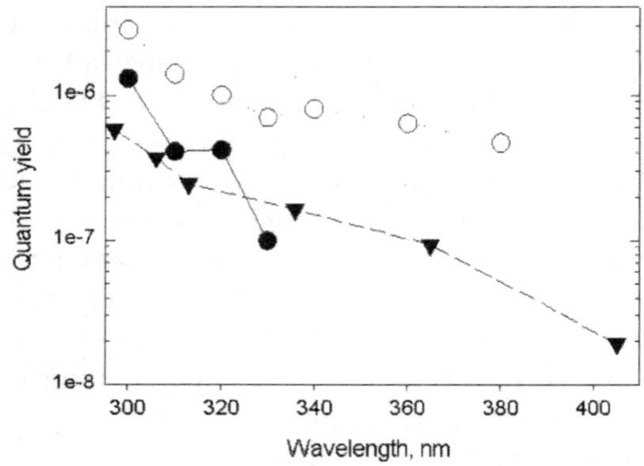

Figure 8. Quantum yield spectra for photoproduction of carbonyl sulfide in the ocean. Circles: Coastal regions in North Sea (filled) and Gulf of Mexico (open) [193]; triangles: Average in predominantly open Pacific Ocean during 1993–1994 [189]. Not shown are additional data of Weiss et al. [189] which indicate that coastal quantum yields are significantly higher.

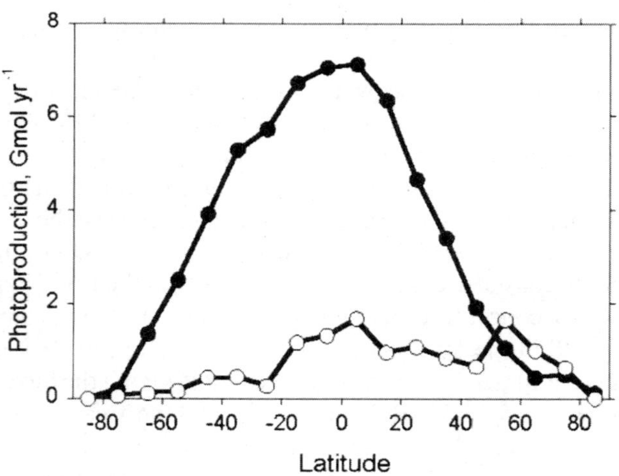

Figure 9. Latitudinal variations in potential production of carbonyl sulfide in the ocean: filled circles: open ocean estimated using apparent quantum yields for Pacific Ocean provided by Weiss et al. [189] and open ocean areas [8, 173]; open circles: coastal photoproduction estimated using average coastal AQYs determined by Zepp and Andreae [193] and coastal ocean areas [67, 173]. Integrated production was 58.3 Gmol yr^{-1} for the open ocean and 12.8 Gmol yr^{-1} for coastal regions, uncorrected for cloud cover. Light attenuation by clouds reduces UVR by approximately 30–40% on average [103]. A substantial fraction of the COS is hydrolyzed before it can escape to the atmosphere [141, 195].

which occurs in cold Northern latitudes where hydrolysis rates are much slower.

Factors that result in the higher quantum yields and concentrations of COS in coastal areas are not well understood, although it seems likely that the different chemical composition of CDOM and the higher concentrations of organosulfur compounds near the continents are mainly responsible. The observed regional differences in absorption spectrum and photoreactivity of CDOM may be due to UV-induced photobleaching (see section 5.2.1.3). Moreover, differences in the nature and concentration of dissolved organic sulfur (DOS) may account for the higher photoproduction quantum efficiencies in coastal regions. Zepp and Andreae [193] presented evidence that COS formation can involve photosensitized reactions whose rates depend upon both CDOM as well as DOS concentrations (see section 5.2.1.3). Thus, higher COS quantum yields in coastal areas potentially could be attributable to higher reactive DOS concentrations. Research on the distributions of organosulfur compounds in coastal areas [194] and open ocean [Cutter, peronal communication] is just starting to appear. Thiol concentrations in coastal North Sea waters were found to correlate with chlorophyll and appeared not to have significant riverine inputs [194]. On the other hand, Uher and Andreae [188] found that the photoproduction rates of COS in the North Sea correlated with absorption coefficients in the near UV (i.e., with CDOM concentrations), indicating that reactive DOS and CDOM must co-vary in this region. Whether this is the case in other parts of the sea is unknown.

COS may also influence metal cycling in the ocean. The hydrolysis of COS in the upper ocean results in the production of sulfide [195], a ligand that can reduce the biological lability of metals by chelation or formation of insoluble metal sulfides. Certain metal sulfide complexes photoreact efficiently when exposed to solar UVR. Other interactions of UV with metals cycling are considered in the next section.

5.5 Metals cycling

Complexation of metals has important interactive effects on biological availability and photochemical reactivity [117,152,165,196–225]. Iron, copper and manganese are essential micronutrients whose free metal ion concentrations in water, and hence biological availabilities, are affected by complexation or, in the case of metal oxides, by redox transformations. Complexation reduces biological availability by reducing free metal ion concentrations and dissolved iron [202] and copper [212,213] are quantitatively complexed by organic ligands. Solar UVR can interact with these processes by inducing direct photoreactions of the complexes, by enhancing redox reactions between the ligands and metal oxides, or by indirect photoreactions in which photochemically-produced ROS react with the complexes. Because both iron and manganese are limiting nutrients in parts of the sea [153,197], these photoreactions may play an important role in global biogeochemical cycles. To illustrate these possibilities, photoreactions involving iron, copper and manganese are discussed here.

5.5.1 Iron

In oxygenated water at circumneutral pH, Fe(III) (Fe^{3+} and complexes thereof) is extremely insoluble. The nature of the association between iron and DOM is imprecisely known [153], but interactions of the Fe(III) with DOM are known to increase its solubility and long-term retention in the water column. In addition to formation of soluble complexes, Fe(III) apparently can also associate with colloidal iron oxides and hydroxides thus producing high concentrations of filterable (0.4 μm) iron that reach over 10 micromolar in some coastal estuaries [59]. Whatever the exact nature of the iron complexes, field studies have provided ample evidence that their photoreduction occurs rapidly in most oxygenated natural waters [59,117,198,203–205,207–209,223,226]. Photoreaction of Fe(III) results in production of Fe(II) (Fe^{2+} and its complexes) plus oxidized ligand. The Fe(II) produced in such photoreactions is stable in acidic environments [203], but it is rapidly oxidized in circumneutral or basic natural waters, including seawater [153,207,224]. Rapid hydrolysis of the resulting Fe(III) then occurs, followed by slow formation of iron hydroxides and eventually unreactive iron (hydroxy)oxides. This sequence of photochemically-related events can transform Fe(III) into colloids, as demonstrated in a study of the photoreactions of the Fe in a coastal river [59]. Photoreactions and subsequent colloid formation may explain the high concentrations of colloidal iron that has recently been identified in the surface ocean [152]. The net effect of the UV-induced photoreactions, however, is to enhance steady-state concentrations of biologically available (and chemically reactive), soluble species of Fe(III) and Fe (II).

A recent study has clearly demonstrated the importance of Fe(III) photoreactions in controlling the biological uptake of iron in seawater [206]. Certain soluble complexes of iron with various strong organic ligands such as siderophores (i.e., microbial iron(III) binding ligands) are photoreactive and photoreactions of these iron complexes enhance the biological availability of the iron [207] (Figure 10).

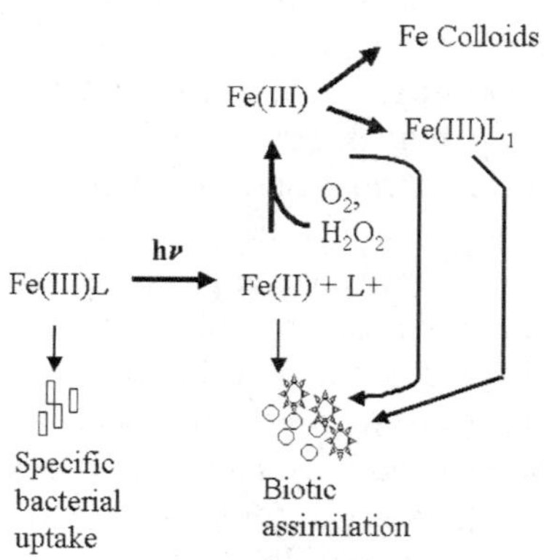

Figure 10. Schematic summary of siderophore-mediated photochemical cycling of iron in seawater [206]. Fe(III)L represents a photoreactive iron(III)–siderophore complex, L+ is the oxidized ligand photoproduct, and L_1 is another chelating ligand. The Fe(II) and its initial oxidation product, Fe(III) are readily assimilated by marine organisms but Fe(III) also can be readily hydrolyzed to Fe hydroxides which can then slowly polymerize to form unreactive, non-bioavailable iron(hydroxy)oxides. [Adapted from Barbeau et al. [206], Figure 4, p. 411, Copyright 2001, *Nature*.]

Moreover, iron also participates in the CDOM photoreactions and ROS reactions that were previously discussed. That iron plays a significant role in the photochemistry of coastal rivers was demonstrated by the inhibiting effects on DIC photoproduction of the addition of ligands (fluoride and desferal) that formed unreactive complexes with the iron [59] (Figure 11). Complexes of Fe(III) with organic carboxylate moieties in CDOM or with acids such as oxalic acid in cloudwater are involved in a complex array of UV-initiated reactions that affect ROS dynamics (Figure 12). For example, iron appears to play a role in the photochemical production of OH radicals in natural waters through reactions between Fe(II) and hydrogen peroxide (H_2O_2), two reactive compounds that are produced by UV-induced photoreactions of CDOM [165,227]. Light-induced photoreactions of iron complexes also contribute to production of Fe(II) in atmospheric cloud droplets, thus enhancing the biological activity of the iron in wet deposition to the sea [117,153,225,228].

Several possible mechanisms are available for UV-induced photoreactions of iron complexes. First, direct photoreactions involving ligand-to-metal charge transfer are likely to be one of the most important mechanisms for photoreaction [117,198,224]. Second, iron complexes can be reduced by photochemically-produced superoxide [207–209]. Superoxide ions are formed via the photoreduction of molecular oxygen by CDOM and it is one of the most concentrated radicals in seawater and is the precursor to hydrogen peroxide [Chapter 8]. Superoxide-induced reduction of Fe(III) is an important mechanism in certain lakes [207]. However, the fact that Fe(II) photoproduction can be more rapid in oxygen-free water than in air-saturated water in acidic estuaries [59] or model systems with well-defined organic acid complexes of Fe(III) [117] indicates that direct photolysis of Fe(III) is likely to be a dominant mechanism for Fe(II) photoproduction in many aquatic systems.

5.5.2 Copper

Copper is an essential trace element that is widely distributed in freshwaters and the sea. Human activities can release large amounts of copper into aquatic environments and there is concern about its potentially toxic effects on aquatic organisms. In the aquatic environment, copper is present predominantly as Cu(II) (Cu^{2+} and its complexes), a major fraction of which is complexed by organic substances of biological origin [153,210–213]. Terrestrial humic substances quantitatively complex copper in coastal waters [213], whereas strong ligands produced by marine organisms such as cyanobacteria, likely in response to Cu stress, chelate Cu in the open ocean [153,212]. Certain organocopper complexes are known to photoreact efficiently on absorption of UVR [214,215] and surface maxima in vertical profiles of Cu(I) (Cu^{+} and its complexes) in the upper layers of the Atlantic Ocean are consistent with a photochemical mechanism for Cu(II) reduction [210,211]. Natural organic substances such as amino acids and amines form complexes with copper on the surfaces of microorganisms via inner sphere types of coordination, and direct photolysis of such complexes

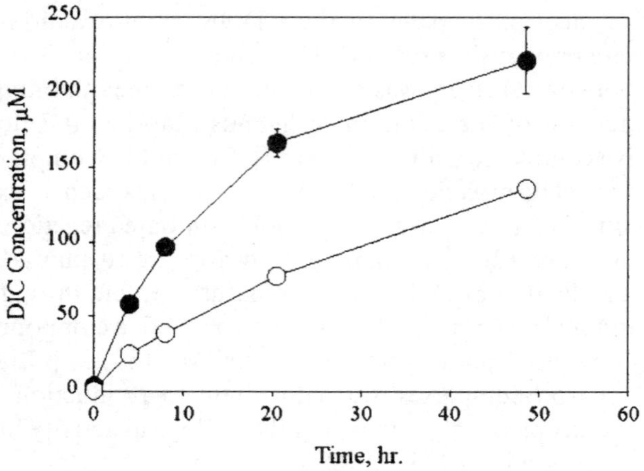

Figure 11. Retarding effects of added desferal (0.3 mM) on the photochemical production of DIC in a coastal river of the southeastern United States [59]. DIC concentration without (●) and with (○) added desferal. The effect is attributed to formation of unreactive Fe(III)-desferal complexes.

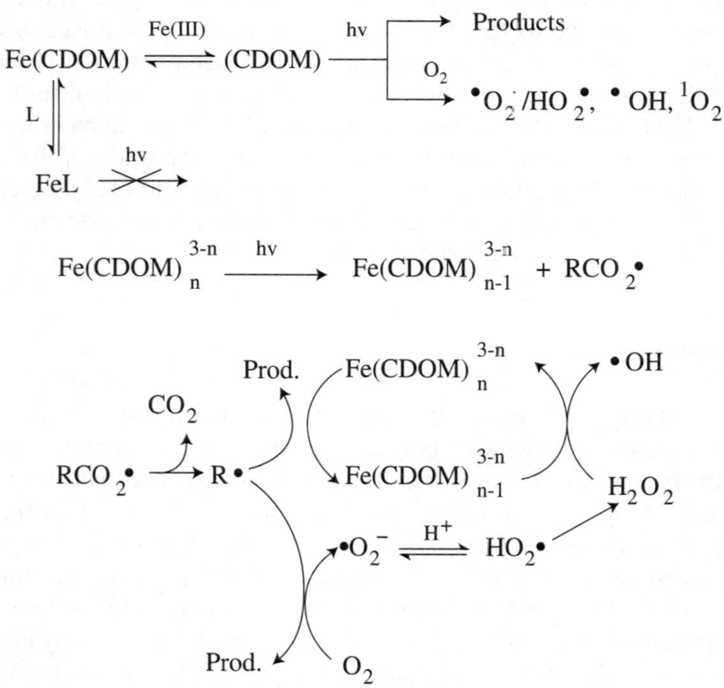

Figure 12. UV-initiated reactions involving iron carboxylate complexes and ROS. Such reactions play an important role in controlling ROS and biologically available iron concentrations in surface waters and in condensed phases of the troposphere [59,117,207]

by solar UVR may contribute to the sunlight-induced reduction of Cu(II) in natural waters [216]. Copper has a high affinity for aquatic biota and these photoreactions on cell surfaces may contribute to the biological damage caused by copper in polluted ecosystems [216]. Photoreactions of Cu(II)–humic complexes in polluted freshwaters also could produce biologically harmful concentrations of Cu^+ and Cu^{2+} by stripping away the ligands that help minimize its free ion concentration.

Like iron complexes, copper complexes have been shown to be an important sink for photochemically-generated superoxide in seawater and, based on the high reactivity of Cu(II) complexed by cyanobacterial-derived ligands, it is likely that redox reactions with superoxide significantly influence Cu redox speciation in the ocean [221,222]. These reactions also have important effects on the steady state concentrations of superoxide in seawater, reducing the concentration by at least an order of magnitude compared to previous estimates that ignored the reactions with copper complexes [222].

5.5.3 Manganese

Like iron and copper, manganese is an essential trace element [153] that also is involved with several biologically significant redox processes in the environment. Manganese concentrations in natural waters range from <1 nM in sea water up to several μM in polluted natural waters. Manganese occurs in three oxidation states in the natural environment. In fresh waters, Mn(II) occurs mainly as the dissolved, uncomplexed species, but may adsorb to particulates. In sea water, chloride and sulfate complexes become important species, but the free Mn^{2+} still predominates. Manganese(III) and (IV) occur as insoluble oxides, referred to here as MnOx. Manganese(III) disproportionates to Mn(II) and Mn(IV) in both acid and alkaline solutions; at neutral pH it may be stable as the oxide of mixed oxidation state Mn_3O_4, or in colloidal suspension as MnOOH (manganite).

The interconversion of the various oxidation states of Mn in natural waters is influenced by UVR through its effects on reactions involving ROS [Chapter 8] and natural phenols, photoinduced charge transfer reactions, and microbial processes. The oxidation of Mn^{2+} is slow at pH <8.5 in the absence of a catalyst. The oxidation of Mn(II) is faster on metal oxide surfaces than in homogeneous solution in the pH range of 8 to 9 [217], and its oxidation also can be biologically mediated in the environment [153]. In comparison to bacteria-free waters, the oxidation rate of Mn(II) in seawater is increased dramatically by catalysis on bacterial surfaces. However, even with such catalysis, its half-life still is of the order of weeks to months in open ocean waters [153].

In the absence of a catalyst, Mn(IV), Mn(III) and Mn(II) are interconverted by ROS such as H_2O_2 and O_2^- through a complex set of redox reactions [218–220,229]. Manganese oxides also are reduced by humic substances, probably through reactions with phenolic moieties [231] and the reaction rate is enhanced by light [196,226,230–234]. As a consequence of this reduction of MnOx, there is a surface maximum of soluble Mn(II) in the open ocean that helps to increase its availability to phytoplankton. In addition, biologically-labile low-molecular-weight organic products are produced from the oxidation of the humic substances [196]. The high Mn(II)/MnOx ratios in the upper layer of the sea have been attributed to both UV photoinhibition of Mn(II) oxidizing micro-organisms and photoreduction of the manganese oxides [232]. The photoreduction of MnOx by adsorbed aquatic humic substances is not greatly affected by removal of dioxygen, indicating that reduction primarily occurs via charge transfer from excited states of the sorbed humic substances on the oxide surface [231]. Although little is known about the nature of the oxidized substances resulting from these reactions, it is likely that the initial products are free radicals such as substituted phenoxyl radicals that interact with dioxygen to produce ROS.

These redox reactions of Mn have important effects on its oxidation state, solubility and biological availability in natural waters. As noted above, Mn(IV) and Mn(III) exist as insoluble oxides and Mn(IV) is the most thermodynamically stable form of manganese. Thus, as expected, Mn(IV) is the major form of Mn at great depths in the sea. In surface layers of the sea, however, almost all of the Mn exists as Mn(II).

5.6 Radiation amplification factors

Action spectra can be applied to estimate the biological and chemical impacts of ozone depletion and related UV increases. The fractional increases in biologically-active UVR are amplified compared to the changes in total ozone in the atmosphere. The degree of amplification, or "radiation amplification factor (RAF)", is defined by equation (4):

$$(UV_{biogeo})_2/(UV_{biogeo})_1 = [(O3)_1/(O_3)_2]^{RAF} \quad (4)$$

where $(UV_{biogeo})_2$ and $(UV_{biogeo})_1$ are the "weighted" irradiances that correspond, respectively, to total ozone amounts (O3)1 and (O3)2. The weighted irradiance is simply the integrated cross-product of the solar spectral irradiance and the action spectrum [3,8,11,12,80,235]. For small changes in total ozone, the RAF corresponds approximately to the % increase in UV_{biogeo} that would occur with a 1% decrease in ozone. Modeling studies using the Antarctic action spectra for phytoplankton photoinhibition have indicated that RAFs fall in the 0.2 to 0.8 range, which signifies a moderate sensitivity to ozone depletion [235]. Based on the close similarity between action spectra for Antarctic and Northern mid-latitude sites shown in Figure 3, it is likely that the mid-latitude RAFs are also in this range. By comparison, the RAF for generalized DNA damage, which applies to UV damage to bacterioplankton (see section 5.2.3.1), is close to 2 [235]. The lower sensitivity of phytoplankton photosynthesis to ozone depletion can generally be attributed to factors such as the occurrence of mycosporine-like amino acids or carotenoids in cells that help shield the DNA from UV-B damage, as well as repair mechanisms that involve longer-wavelength UV-A radiation that is little affected by ozone depletion [8,11,12,88]. RAFs for various photoreactions of CDOM fall in the 0.2 to 1.1 range [235], indicating that they are approximately as sensitive to changes in the stratospheric ozone layer as the effects on health, plants, and tropospheric photolysis.

5.7 Summary

The potential multi-faceted impacts of solar UVR on biogeochemical cycles have been assessed in this chapter using a combination of observations and modeling evaluations. In addition to the well-known effects of stratospheric ozone depletion on UVR, recent research indicates that other co-occurring global environmental changes in climate, land use and deposition are affecting the interactions of UV with element cycling. For example, it is now well-documented that CDOM, concentrations of which are sensitive to climate and land use changes, plays a key role in the interactions of UVR with aquatic biogeochemistry. CDOM controls the penetration of UV into many freshwater systems and the ocean and seasonal changes in CDOM concentrations linked to factors such as precipitation changes and photochemical and biological decomposition have important effects on aquatic UV exposure. Large scale changes in oceanic CDOM may be driven by El Niño-Southern Oscillation (ENSO) events that periodically affect oceanic mixed layer depth, upwelling, and mixing dynamics. Significant differences have been observed in the spectral, chemical, physical, and photochemical properties of the CDOM in freshwaters and the sea. These differences, which usually become apparent in transects along estuarine and coastal environments, can be attributed to differences in sources, as well as in UV-induced direct and photochemically-altered microbial decomposition.

UV exposure generally inhibits phytoplankton photosynthesis and recent results indicate that, on the average, BWFs for such inhibition are similar for mid-latitude and Antarctic phytoplankton. UV also indirectly affects phytoplankton photosynthesis through its effects on the biological availability of iron and other trace metal nutrients.

UV exposure affects microbial decomposition processes both through direct inhibition of bacterial activity as well as through effects on the biological availability of carbon and nitrogen substrates. Models of UV interactions with phytoplankton and bacteria indicate that factors such as vertical mixing dynamics and mixed layer depth have important effects on damage and repair. The net effects of UV exposure on the biological availability of CDOM are dependent on its source. Most current evidence indicates that UV exposure stimulates the lability of terrestrially-derived CDOM but reduces the availability of algal-derived organic matter.

CDOM also can be directly photodecomposed by solar UVR to DIC, carbon monoxide, and various carbonyl-containing compounds. This direct photodecomposition is accompanied by oxygen consumption although oxygen appears not to be directly involved in some of the photoreactions. Terrestrially-derived and open ocean CDOM have dissimilar quantum yield spectra for CO photoproduction, indicating that open ocean CDOM may produce CO less efficiently. Simulations using mathematical models indicate that UV induced decomposition potentially can consume all of the riverine CDOM that enters coastal regions. Observations of changes in UV-sensitive isotopic and lignin content are consistent with the simulations.

UVR can potentially affect nitrogen cycling through inhibition of nitrogen fixation and indirectly through changes in the biological availability of iron, an element that stimulates the growth of certain nitrogen-fixing cyanobacteria. UV also initiates the conversion of persistent DON into nitrogenous compounds that are readily assimilated by aquatic organisms.

Sulfur cycling is affected in a variety of ways, including UV photoinhibition of organisms such as bacterioplankton and zooplankton that affect sources and sinks of DMS and UV-initiated CDOM-sensitized photoreactions that oxidize DMS and produce carbonyl sulfide. Metal cycling also interacts in many ways with UVR via direct photoreactions of dissolved complexes and of metal oxides and indirect reactions that are mediated by photochemically-produced ROS. Photoreactions can affect the biological availability of essential trace nutrients such as iron and manganese, transforming the metals from complexes that are not readily assimilated into free metal ions or metal hydroxides that are available. Such photoreactions can enhance the toxicity of metals such as copper and can initiate metal redox reactions that transform non-reactive ROS such as superoxide into potent oxidants such as hydroxyl radicals.

5.8 Appendix: modeling rates of photoreactions

Aquatic photoreactions can be modeled using kinetic and spectroscopic properties derived from laboratory and field studies as well as empirical relationships derived from field observations [141]. Here the former approach is discussed. The rate at a given depth z, rate(z), can be computed using equation (5):

$$\text{rate}(z) = \int E_o(z, \lambda) f_p(\lambda) a(\lambda)\, \Phi_a(\lambda)\, d\lambda \qquad (5)$$

where $E_o(z, \lambda)$ is the scalar irradiance at z and wavelength λ, $a(\lambda)$ is the absorption coefficient, $f_p(\lambda)$ is the fraction of absorbed radiation that is absorbed by the photoreactive chromophores (usually assumed to be unity in the absence of more detailed information about system composition), $\Phi_a(\lambda)$ is the apparent quantum yield for the photoproduction of DIC, CO, or other photoproducts (see below), and the integration is across the range of photochemically-effective wavelengths. Note that photon scalar irradiance (e.g., photons $cm^{-2}\ s^{-1}\ nm^{-1}$) is used for simulations involving quantum yields in contrast to energy-based irradiance (e.g., watts $cm^{-2}\ nm^{-1}$) that usually is used for estimating the dose rates employing BWFs. Also, although many recent modeling efforts have assumed that all the absorbed radiation in the UV and blue region is photoactive (and hence that f_p is unity in this spectral region), this may not be the case for long-wavelength UV-A and blue radiation at the very low concentrations and absorbance of CDOM in bluewater regions of the ocean. Various recent studies have been conducted to provide data concerning the wavelengths responsible for photoproduction of DOC photoproducts in natural waters. As in the case of biological endpoints, such data are required in models that extrapolate measured photoproduct fluxes to unmeasured situations and estimate the changes in photoproduction in response to changes in atmospheric composition such as ozone depletion.

The data obtained in these studies have been presented as both apparent quantum yields and action spectra. Quantum yields are a useful indicator of the efficiency of a photoreaction, i.e. the fraction of absorbed radiation that results in chemical change. The term 'apparent' is used to signify that the nature of the light absorbing chromophores in the system has not been identified and likely could be a mixture. Apparent quantum yields are usually determined using monochromatic radiation. The quantum yield, $\Phi_a(\lambda)$, which is the fraction of absorbed radiation that results in formation (unitless) of the photoproduct, is defined by the following equation:

$$\Phi_a(\lambda) = \frac{\text{Rate}(\lambda)}{I_\lambda \ F_\lambda} \tag{6}$$

where Rate(λ) is the observed formation rate of photoproduct (e.g. in molecules $cm^{-3}\ s^{-1}$), F_λ is the fraction of light absorbed at wavelength λ, and I_λ is the number of photons that enter the photoreaction cell per unit volume and unit time (units of photons $cm^{-3}\ s^{-1}$). The latter is determined by chemical actinometry or by some physical light detecting device such as a calibrated thermopile or spectroradiometer. Quantum yields can also be determined using the Rundel approach that was discussed earlier for UV effects on aquatic photosynthesis. The Rundel approach involves measurement of both rates, absorption spectra and spectral irradiance of the light source with a series of filters in place that block out parts of the UVR. A light source with spectral irradiance close to that of solar radiation at the Earth's surface is used for the experiments. Using a modified version of the rate equation (equation 5) in conjunction with the production rates, measured spectral irradiance, and spectral absorptivity a best-fit solution for $\Phi_a(\lambda)$ can be obtained with a MATLAB© program based on Rundel's statistical approach for the optimization of action spectra [67,78,115]. Usually, the data are fitted assuming that the quantum yields decrease exponentially with increasing wavelength but other types of fitting equations, e.g. linear, also can be readily employed. The primary strength of the Rundel approach is conservation of effort and time; the monochromatic approach requires a large number of experiments whereas a quantum yield spectrum can be determined with one experiment using the Rundel approach. On the other hand, the required ad hoc assumption of a fitting equation to estimate the spectral dependence of the quantum yields can lead to imprecise estimates of quantum yields in spectral regions beyond where the wavelength cutoffs are applied (e.g. the visible region) or where the spectral irradiance of the light source is relatively low and rapidly changing with wavelength (the UV-B region). It is reasonable to assume that photoreactions are not affected by interactions between various parts of the spectrum, such as photorepair in the case of aquatic organisms, so both the monochromatic and Rundel approaches should yield similar quantum yield spectra. A recent study has indicated that the quantum yield spectra for CO determined by the monochromatic approach is nearly the same as that determined using the Rundel approach [67].

It can be shown that the quantum yield is related to the response function by the following relationship:

$$X_P(\lambda) = (a_\lambda)\,[\Phi_a(\lambda)] \tag{7}$$

where a_λ is the mean absorption coefficient (equation 1) of the water sample (e.g., in units of m^{-1}) during the irradiation period. When defined using these units, the response function has units of m^{-1}. Plots of $X_P(\lambda)$ versus wavelength represent action spectra for photoreactions. The photoproduction rate at wavelength λ is the cross product of the irradiance and the response function.

Although it has been generally assumed that aquatic photoreactions obey reciprocity, evidence in support of this assumption is mixed and can depend on the measured endpoint used to follow changes in irradiated system composition Fractional loss of the UV-absorbing CDOM component of DOC per unit time [38], as well as apparent quantum yields for photoproduct formation such as CO production [103], have been found to be conversion independent (and thus obey reciprocity). On the other hand, some photoreactions do not obey reciprocity because apparent quantum yields for photoreactions of CDOM are conversion dependent, i.e. dependent on the extent of photoreaction. For example, $\Phi_a(\lambda)$ for DIC photoproduction [70] and photochemical oxygen demand [68] can be conversion dependent. Conversion dependence of quantum yields can result from changes in CDOM composition that, for example, involves depletion or creation of more photoreactive chromophores as photoreaction proceeds. Other changes such as a shift in mechanism from predominantly indirect (photosensitized) to direct decomposition or a buildup or decrease in excited state quenchers (such as molecular oxygen) also can contribute to conversion dependence. A general consequence of conversion dependence is that predictions based on first order models are likely to overestimate the photochemical removal rates.

Conversion dependence of $\Phi_a(\lambda)$ is not the only cause of apparent deviation from first order kinetics in photoreactions of DOC. If total DOC loss (not CDOM) is used to follow photoreaction kinetics, first order kinetics does not provide an adequate fit over long irradiation periods. For example, Moran et al. [38] reported that the amount of sunlight required to bring about an equivalent proportional loss of the DOC pool in water samples from coastal rivers in the Southeastern United States increased as photodegradation progressed, even when delayed DOC mineralization that occurred via enhanced bacterial activity was considered. The total DOC pool may have included a component in the original material (about 65% in this case) that was especially refractory to photochemical degradation, perhaps because it only weakly absorbed solar UVR. Alternatively, UVR may have induced photoreactions that converted the original material into refractory compounds, possibly simply by photobleaching the reactive chromophores or, alternatively, by other structural modifications that reduced photoreaction quantum yields.

Models can also be used to provide estimates of depth-integrated production of photoproducts in aquatic environments. Under well-mixed conditions in a water column of depth z, the average irradiance at wavelength λ can be computed using equation 8:

$$E_{av}(\lambda, z) = E_o(\lambda,0)\frac{(1 - e^{-K_d(\lambda)z})}{K_d(\lambda)z} \tag{8}$$

where $E_o(\lambda,0)$ is the irradiance just below the water surface and $K_d(\lambda)$ is the diffuse attenuation coefficient. Under conditions in which all the radiation is absorbed this reduces to $(E_o(\lambda,0) / K_d z)$, the average rate becomes light-limited and inversely proportional to depth, and photoproduction can be described as depth-integrated "fluxes". These "fluxes," which are expressed, for example, in units of moles photoproduct per unit area and time are obtained by integrating the cross product of the net downwelling spectral irradiance just under the water surface $[\vec{E}(0,\lambda)]$ [39,41] and the apparent quantum yields $\Phi_a(\lambda)$ over the range of photoactive wavelengths (i.e., 280–450 nm) (equation 9).

$$\text{Flux} = \int f_p(\lambda)\vec{E}(0,\lambda)\ \Phi_a\ (\lambda)\ d\ \lambda \tag{9}$$

Because the upwelling irradiance is generally much lower than downwelling irradiance, the net downwelling irradiance approximately equals the downwelling irradiance [39,41]. This equation only provides a rough estimate of depth-integrated photodecomposition, because mixing effects, poor knowledge of $f_p(\lambda)$, lack of reciprocity, and other factors can render this approach inapplicable. Nonetheless, such flux estimates can provide a useful initial assessment of the potential impact of UVR on various chemical and biological processes.

Acknowledgements

I thank the reviewers for their helpful comments and N. Blough, R. Del Vecchio, O. Zafiriou, P. Neale, D. Kieber, and B. Peake for providing pre-publication copies of manuscripts that were cited. This work was supported in part by a grant from the Office of Naval Research (N00014-98-F-0202). This paper has been reviewed in accordance with the U.S. Environmental Protection Agency's peer and administrative review policies and approved for publication. Mention of trade names or commercial products does not constitute an endorsement or recommendation for use by the U.S. EPA.

References

1. M.C. Jacobson, R.J. Charlson, H. Ordhe, G. Orians (2000). *Earth System Science: From Biogeochemical Cycles to Global Change*. Academic Press, San Diego.
2. W.H. Schlesinger (1997). *Biogeochemistry: An Analysis of Global Change*. Academic Press, San Diego.
3. K. Mopper, D.J. Kieber (2000). Marine photochemistry and its impact on carbon

cycling. In: S. de Mora, S. Demers, M. Vernet (Eds), *The Effects of UV Radiation in the Marine Environment* (pp. 101–129). Cambridge University Press, Cambridge.

4. W.L. Miller (1999). Effects of UV radiation on aquatic humus: Photochemical principles and experimental considerations. In: D. O.Hessen, L.Tranvik (Eds), *Aquatic Humic Substances*, (pp. 125–143). Springer-Verlag, Berlin.
5. N.V. Blough (1997). Photochemistry in the sea-surface microlayer. In: P. Liss, R. Duce (Eds), *The Sea Surface and Global Change* (pp. 383–424). Cambridge University Press, Cambridge.
6. R.G. Zepp, T.V. Callaghan, D.J. Erickson (1998). Effects of enhanced solar ultraviolet radiation on biogeochemical cycles. *J. Photochem. Photobiol. B: Biol.*, **46**, 69–82.
7. R.G. Zepp (1988). Environmental photoprocesses involving natural organic matter. In: F.H. Frimmel, R.F. Christman (Eds), *Humic Substances and their Role in the Environment* (pp. 193–214). Wiley, New York.
8. M.A. Moran, R.G. Zepp (1997). Role of photoreactions in the formation of biologically labile compounds from dissolved organic matter. *Limnol. Oceanogr.*, **42**, 1307–1316.
9. D.J. Kieber (2000). Photochemical production of biological substrates. In: S. de Mora, S. Demers, M. Vernet (Eds), *The Effects of UV Radiation in the Marine Environment* (pp. 130–148). Cambridge University Press, Cambridge.
10. M.A. Moran, R.G. Zepp (2000). UV radiation effects on microbes and microbial processes. In: D. Kirchman (Ed.), *Microbial ecology of the oceans* (pp. 201–228). Wiley, New York.
11. J.J. Cullen, P.J. Neale (1994). Ultraviolet radiation, ozone depletion, and marine photosynthesis. *Photosynth. Res.*, **39**, 303–320.
12. P.J. Neale (2000). Spectral weighting functions for quantifying effects of UV solar radiation in marine ecosystems. In: S. de Mora, S. Demers, M. Vernet (Eds), *The Effects of UV Radiation in the Marine Environment* (pp. 72 –100). Cambridge University Press, Cambridge.
13. N. Blough, S. Green (1994). Spectroscopic characterization and remote sensing of non-living organic matter. In: R.G. Zepp, C. Sonntag (Eds), *Role of non-living organic matter in the Earth's carbon cycle* (pp. 42–57). Wiley, New York.
14. E.M. Thurman (1985). *Organic Geochemistry of Natural Waters.* Nijhoff/Junk, Boston.
15. D.M. McKnight, E.D. Andrews, S.A. Spaulding, G.R. Aiken (1994). Aquatic fulvic acids in algal-rich Antarctic ponds. *Limnol. Oceanogr.*, **36**, 998–1006.
16. D.M. McKnight, E.W. Boyer, P.K. Westerhoff, P.T. Doran, T. Kulbe, D.T. Andersen (2000). Spectrofluorometric characterization of dissolved organic matter for indicator of precursor organic material and aromaticity. *Limnol. Oceanogr.*, **46**, 38–48.
17. Y. Chin, G.R. Aiken, E. O'Loughlin (1994). Molecular weight, polydispersivity, and spectroscopic properties of aquatic humic substances. *Environ. Sci. Technol.*, **28**, 1853–1858.
18. N.V. Blough, R. Del Vecchio (2002). Distribution and dynamics of chromophoric dissolved organic matter (CDOM) in the coastal environment. In: D. Hansell, C. Carlson (Eds.), *Biogeochemistry of Marine Dissolved Organic Matter* (pp. 509–546). Academic Press.
19. R.J. Kieber, L.H. Hydro, P.J. Seaton (1997). Photooxidation of triglycerides and fatty acids in seawater: Implication toward the formation of marine humic substances. *Limnol. Oceanogr.*, **42**, 1454–1462.
20. J.A. Amador, M. Alexander, R.G. Zika (1989). Sequential photochemical and microbial degradation of organic molecules bound to humic acid. *Appl. Environ. Microbiol.*,

55, 2843–2849.

21. S. Opsahl, R. Benner (1998). Photochemical reactivity of dissolved lignin in river and ocean waters. *Limnol. Oceanogr.*, **43**, 1297–1304.
22. R.J. Kieber, X. Zhou, K. Mopper (1990). Formation of carbonyl compounds from UV-induced photodegradation of humic substances in natural waters: fate of riverine carbon in the sea. *Limnol. Oceanogr.*, **35**, 1503–1515.
23. A.M. Kouassi, R.G. Zika (1992). Light-induced destruction of the absorbance property of dissolved organic matter in seawater. *Toxicol. Environ. Chem.*, **35**, 195–211.
24. F.H. Frimmel (1994). Photochemical aspects related to humic substances. *Environ. Int.*, **20**, 373–385.
25. F.H. Frimmel (1998). Impact of light on the properties of aquatic natural organic matter. *Environ. Int.*, **24**, 559–571.
26. D.P. Morris, B.R. Hargreaves (1997). The role of photochemical degradation of dissolved organic matter in regulating UV transparency of three lakes on the Pocono Plateau. *Limnol. Oceanogr.*, **42**, 239–249.
27. C.L. Osburn, H.E. Zagarese, D.P. Morris, B.R. Hargreaves, W.E. Cravero (2001). Calculation of spectral weighting functions for the solar photobleaching of chromophoric dissolved organic matter in temperate lakes. *Limnol. Oceanogr.*, **46**, 1455–1467.
28. I. Reche, M.L. Pace, J.J. Cole (1999). Relationship of trophic and chemical conditions to photobleaching of dissolved organic matter in lake ecosystems. *Biogeochemistry*, **44**, 259–280.
29. R.F. Whitehead, S. de Mora, S. Demers, M. Gosselin, P. Monfort, B. Mostajir (2000). Interactions of ultraviolet-B radiation, mixing, and biological activity on photobleaching of natural chromophoric dissolved organic matter: A mesocosm study. *Limnol. Oceanogr.*, **45**, 278–291.
30. A.J. Stewart, R.G. Wetzel (1981). Dissolved humic materials: photodegradation, sediment effects, and reactivity with phosphate and calcium carbonate precipitation. *Arch. Hydrobiol.*, **92**, 265–286.
31. D. Strome, M.C. Miller (1978). Photolytic changes in dissolved humic substances. *Verh. Int. Verein. Limnol.*, **20**, 1248–1254.
32. D. Kotsias, M. Herrmann, A. Zsolnay, R. Bayerle-Pfnur, H. Parlar, F. Korte (1987). Photochemical aging of humic substances. *Chemosphere*, **16**, 1463–1468.
33. D. Hongve (1994). Sunlight degradation of aquatic humic substances. *Acta Hydrochim. Hydrobiol.*, **3**, 117–120.
34. E.T. Gjessing, T. Gjerdahl (1970). Influence of ultra-violet radiation on aquatic humus. *Vatten*, **26**, 144–145.
35. R. Del Vecchio, N.V. Blough, Photobleaching of chromophoric dissolved organic matter in natural waters: Kinetics and modeling. *Mar. Chem.*, submitted.
36. C.E. Del Castillo, P.G. Coble, J.M. Morell, J. M.Lopez (1999). Analysis of the optical properties of the Orinoco River plume by absorption and fluorescence spectroscopy. *Mar. Chem.*, **66**, 35–51.
37. C.E. Del Castillo, F. Gilbes, P.G. Coble, F.E. Muller-Karger (2000). On the dispersal of riverine colored dissolved organic matter over the West Florida shelf. *Limnol. Oceanogr.*, **45**, 1425–1432.
38. M.A. Moran, W.M. Sheldon, R.G. Zepp (2000). Carbon loss and optical property changes during long-term photochemical and biological degradation of estuarine dissolved organic matter. *Limnol. Oceanogr.*, **45**, 1254–1264.
39. C.D. Mobley (1994). *Light and Water: Radiative Transfer in Natural Waters*. Academic Press, San Diego.
40. C.D. Mobley, B. Gentili, H.R. Gordon, Z. Jin, G.W. Kattawar, A. Morel, P. Reiners-

man, K. Stamnes, R.H. Stavn (1993). Comparison of numerical models for computing underwater light fields. *Appl. Opt.*, **32**, 7484–7504.
41. J.T.O. Kirk (1994). *Light and Photosynthesis in Aquatic Ecosystems.* Cambridge University Press, Cambridge.
42. K.S. Baker, R.C Smith (1982). Spectral irradiance penetration in natural waters. In: J. Calkins (Ed.), *The Role of Solar Ultraviolet Radiation in Marine Ecosystems* (pp. 79–91). Plenum Press, New York.
43. H.R. Gordon (1989). Can the Lambert-Beer law be applied to the diffuse attenuation coefficient of ocean water? *Limnol. Oceanogr.*, **34**, 1389–1409.
44. G.C. Miller, R.G. Zepp (1979). Effects of suspended sediments on photolysis rates of dissolved pollutants. *Water Res.*, **13**, 453–459.
45. D.-P. Häder, H.D. Kumar, R.C. Smith, R.C. Worrest. (1998). Effects on aquatic ecosystems. *J. Photochem. Photobiol. B: Biol.*, **46**, 53–68.
46. D.A. Siegel, A.F. Michaels (1996). Quantification of non-algal light attenuation in the Sargasso Sea: Implications for biogeochemistry and remote sensing. *Deep-Sea Res.*, **43**, 321–346.
47. I. Laurion, M. Ventura, J. Catalan, R. Psenner, R. Sommaruga (2000). Attenuation of ultraviolet radiation in mountain lakes: Factors controlling the among- and within-lake variability. *Limnol. Oceanogr.*, **45**, 1274–1288.
48. D.P. Morris, H. Zagarese, C.E. Williamson, E.G. Balseiro, B.R. Hargreaves, B. Modenutti, R. Moeller, C. Queimalinos (1995). The attenuation of solar UV radiation in lakes and the role of dissolved organic carbon. *Limnol. Oceanogr.*, **40**, 1381–1391.
49. C.E. Williamson, R.S. Stemberger, D.P. Morris, T.M. Frost, S.G. Paulsen (1996). Ultraviolet radiation in North American lakes: Attenuation estimates from DOC measurements and implications for plankton communities. *Limnol. Oceanogr.*, **41**, 1024–1034.
50. D.W. Schindler, J.P. Curtis, B.R. Parker, M.P. Stainton (1996). Consequences of climate warming and lake acidification for UV-B penetration in North American boreal lakes. *Nature*, **379**, 705–708.
51. A. Vodacek, N.V. Blough, M.D. DeGrandpre, E.T. Peltzer, R.K. Nelson (1997). Seasonal variation of CDOM and DOC in the Middle Atlantic Bight: Terrestrial inputs and photooxidation. *Limnol. Oceanogr.*, **42**, 674–686.
52. S. Markager, W.F. Vincent (2000). Spectral light attenuation of UV and blue light in natural waters. *Limnol. Oceanogr.*, **45**, 642–650.
53. N.K. Højerslev (1982). Yellow substance in the sea. In: J. Calkins (Ed.), *The Role of Solar Ultraviolet Radiation in Marine Ecosystems* (pp. 263–281). Plenum Press, New York.
54. N.B. Nelson, D.A. Siegel, A.F. Michaels (1998). Seasonal dynamics of colored dissolved organic matter in the Sargasso Sea (Part I), *Deep-Sea Res.*, **45**, 931–957.
55. K.R. Arrigo, C.W. Brown (1996). Impact of chromophoric dissolved organic matter on UV inhibition of primary productivity in the sea. *Mar. Ecol. Prog. Ser.*, **140**, 204–216.
56. Y. Huot, W.H. Jeffrey, R.F. Davis, J.J. Cullen (2000). Damage to DNA in bacterioplankton: A model of damage by ultraviolet radiation and its repair as influenced by vertical mixing. *Photochem. Photobiol.*, **72**, 62–74.
57. N.V. Blough, R. Del Vecchio (2002). Distribution and dynamics of chromophoric dissolved organic matter (CDOM) in the coastal environment. In: D. Hansell C. Carlson (Eds), *Biogeochemistry of Marine Dissolved Organic Matter* (pp. 509–546). Academic Press.
58. H. De Haan (1993). Solar UV-light penetration and photodegradation of humic

substances in peaty lake water. *Limnol. Oceanogr.*, **38**, 1072–1076.
59. H. Gao, R.G. Zepp (1998). Factors influencing photoreactions of dissolved organic matter in a coastal river of the southeastern United States. *Environ. Sci. Technol.*, **32**, 2940–2946.
60. S. Opsahl, R.G. Zepp (2001). Photochemically-induced alteration of stable carbon isotope ratios (delta^{13}C) in terrigenous dissolved organic carbon. *Geophys. Res. Lett.*, **28**, 2417–2420.
61. A.M. Kouassi, R.G. Zika, J.M.C. Plane (1990). Photochemical modeling of marine humus fluorescence in the ocean. *Neth. J. Sea Res.*, **27**, 33–41.
62. N. Blough, R.G. Zepp (1995). Reactive oxygen species in natural waters. In: C.S. Foote, J.S. Valentine, A. Greenberg, J.F. Liebman (Eds), *Active Oxygen in Chemistry* (pp. 280–333). Chapman & Hall, New York.
63. O.C. Zafiriou, J. Joussot-Dubien, R.G. Zepp, R.G. Zika (1984). Photochemistry of natural waters. *Environ. Sci. Technol.*, **18**, 358–371.
64. J.V. Goldstone, B.M. Voelker (2000). Chemistry of superoxide radical in seawater: CDOM associated sink of superoxide in coastal waters. *Environ. Sci. Technol.*, **34**, 1043–1048.
65. J.V. Goldstone, M.J. Pullin, S. Bertilsson, B.M. Voelker (2002). Reactions of hydroxyl radical with humic substances: Bleaching, mineralization, and production of bioavailable carbon substrates. *Environ. Sci. Technol.*, **36**, 364–372.
66. R.G. Wetzel (1992). Gradient-dominated ecosystems: sources and regulatory functions of dissolved organic matter in freshwater ecosystems. *Hydrobiologia*, **229**, 181–198.
67. W.L. Miller, M.A. Moran, W.M. Sheldon, R.G. Zepp, S. Opsahl (2002). Determination of apparent quantum yield spectra for the formation of biologically labile photoproducts. *Limnol. Oceanogr.*, **47**, 343–352.
68. S.S. Andrews, S. Caron, O.C. Zafiriou (2000). Photochemical oxygen demand in marine waters: A major sink for colored dissolved organic matter? *Limnol. Oceanogr.*, **45**, 267–277.
69. J.I. Hedges (1992). Global biogeochemical cycles: progress and problems. *Mar. Chem.*, **39**, 67–93.
70. W.L. Miller, R.G. Zepp (1995). Photochemical production of dissolved inorganic carbon from terrestrial organic matter: significance to the oceanic organic carbon cycle. *Geophys. Res. Lett.*, **22**, 417–420.
71. W. Ludwig (2001). The age of river carbon. *Nature*, **409**, 466.
72. M.M. Caldwell, L.O. Björn, J.F. Bornman, S.D. Flint, G. Kulandaivelu, A.H. Teramura, M. Tevini (1998). Effects of increased solar ultraviolet radiation on terrestrial ecosystems. *J. Photochem. Photobiol. B: Biol.*, **46**, 40–52.
73. P. Falkowski, R.J. Scholes, E. Boyle, J. Canadell, D. Canfield, J. Elser, N. Gruber, K. Hibbard, P. Högberg, S. Linder, F.T. Mackenzie, B.I. Moore, T. Pedersen, Y. Rosenthal, S. Seitzinger, V. Smetacek, W. Steffen (2000). The global carbon cycle: A test of our knowledge of Earth as a system. *Science*, **290**, 291–296.
74. R. Zepp, T.V. Callaghan, D.J. Erickson (1998). Effects of enhanced solar ultraviolet radiation on biogeochemical cycles. In: J.C. van der Leun, X. Tang, M. Tevini (Eds), *Environmental Effects of Ozone Depletion – 1998 Assessment* (pp. 113–136). United Nations Environment Programme (UNEP), Nairobi, Kemya.
75. R.C. Smith, B.B. Prezelin, K.S. Baker, R.R. Bidigare, N.P. Boucher, T. Coley, D. Karentz, S. MacIntyre, H.A. Matlick, D. Menzies, M. Onderusek, Z. Wan, K.J. Waters (1992). Ozone depletion: Ultraviolet radiation and phytoplankton biology in Antarctic waters. *Science*, **255**, 952–959.

76. T.P. Coohil (1991). Photobiology school. Action spectra again? *Photochem. Photobiol.*, **54**, 859–870.
77. P.J. Neale, D.J. Kieber (2000). Assessing biological and chemical effects of UV in the marine environment: Spectral weighting functions. In: R.E. Hester, R.M. Harrison (Eds), *Causes and Environmental Implications of Increased UV-B Radiation.* (Issues in Environmental Science and Technology, No. 14, pp. 61–83). The Royal Society of Chemistry, Cambridge.
78. R.D. Rundel (1986). Action spectra and estimation of biologically effective UV radiation. *Physiol. Plant.*, **58**, 360–366.
79. P.J. Neale, J.J. Cullen, R.F. Davis (1998). Inhibition of marine photosynthesis by ultraviolet radiation: Variable sensitivity of phytoplankton in the Weddel–Scotia confluence during the austral spring. *Limnol. Oceanogr.*, **43**, 433–448.
80. A.T. Banaszak, P.J. Neale (2001). Ultraviolet radiation sensitivity of photosynthesis in phytoplankton from an estuarine environment. *Limnol. Oceanogr.*, **46**, 592–603.
81. M.P. Lesser, P.J. Neale, J.J. Cullen (1996). Acclimation of Antarctic phytoplankton: Ultraviolet absorbing compounds and carbon fixation. *Mol. Mar. Biol. Biotechnol.*, **5**, 314–325.
82. M.L. Bothwell, D.M.J. Sherbot, C.M. Pollock (1994). Ecosystem response to solar ultraviolet radiation: Influence of trophic level interactions. *Science*, **265**, 97–100.
83. S.C. Rhode, M. Pawlowski, R. Tollrian (2001). The impact of ultraviolet radiation on the vertical distribution of zooplankton of the genus Daphnia. *Nature*, **412**, 69–72.
84. D.M. Leech, C.E. Williamson (2001). In situ exposure to ultraviolet radiation alters the depth distribution of Daphnia. *Limnol. Oceanogr.*, **46**, 416–420.
85. D. Turk, M.J. McPhaden, A.J. Busalacchi, M.R. Lewis (2001). Remotely sensed biological production in the Equatorial Pacific. *Science*, **293**, 471–474.
86. R. Pienitz, W.F. Vincent (2000). Effect of climate change relative to ozone depletion on UV exposure in subarctic lakes. *Nature*, **404**, 484–487.
87. N.V. Blough, R.G. Zepp (1990). Introduction, in *Effects of Solar Ultraviolet Radiation on Biogeochemical Dynamics in Aquatic Environments*, N.V. Blough, R.G. Zepp (Eds), Technical Report No. WHOI-90-09, Woods Hole Oceanographic Institution, Woods Hole, MA, USA, 1990, pp. 1–2.
88. W.F. Vincent, P.J. Neale (2000). Mechanisms of UV damage to aquatic organisms. In: S. de Mora, S. Demers, M. Vernet (Eds), *The Effects of UV Radiation in the Marine Environment* (pp. 149–176). Cambridge University,Press.
89. W.H. Jeffrey, P. Aas, M.M. Lyons, R.B. Coffin, R.J. Pledger, D.L. Mitchell (1996). Ambient solar-radiation induced photodamage in marine bacterioplankton. *Photochem. Photobiol.*, **64**, 419–427.
90. W.H. Jeffrey, R.J. Pledger, P. Aas, S. Hager, R.B. Coffin, R. Vonhaven, D.L. Mitchell (1996). Diel and depth profiles of DNA photodamage in bacterioplankton exposed to ambient solar ultraviolet radiation. *Mar. Ecol. Prog. Ser.*, **137**, 283–291.
91. M.G. Weinbauer, S.W. Wilhelm, C.A. Suttle, D.R. Garza (1997). Photoreactivation of ultraviolet-radiation induced DNA damage is a major process controlling viral infectivity in the sea. *Appl. Environ. Microbiol.*, **63**, 2200–2205.
92. S.W. Wilhelm, M.G. Weinbauer, C.A. Suttle, W.H. Jeffrey (1998). The role of sunlight in the removal and repair of viruses in the sea. *Limnol. Oceanogr.*, **43**, 586–592.
93. C.M.T. Denward, H. Edling, L.J. Tranvik (1999). Effects of solar radiation on bacterial and fungal density on aquatic plant detritus. *Freshwat. Biol.*, **41**, 575–582.
94. D.J. Erickson, B.E. Eaton (1993). Global biogeochemical cycling estimates with CZCS satellite data and general circulation models. *Geophys. Res. Lett.*, **20**, 683–686.
95. D.J. Erickson (1989). Ocean to atmosphere carbon monoxide flux: Global inventory

and climate implications. *Global Biogeochem. Cycles*, **3**, 304–314.
96. K. Mopper, X. Zhou, R.J. Kieber, D.J. Kieber, R.J. Sikorski, R.D. Jones (1991). Photochemical degradation of dissolved organic carbon and its impact on the oceanic carbon cycle. *Nature*, **353**, 60–62.
97. M. Springer-Young, D.J. Erickson, T.P. Carsey (1996). Carbon monoxide gradients in the marine boundary layer of the North Atlantic Ocean. *J. Geophys. Res.-Atmos.*, **101**, 4479–4484.
98. J.W. Swinnerton, V.J. Linnenbom, R.A. Lamontagne (1970). The ocean. A natural source of carbon monoxide. *Science*, **167**, 984–986.
99. V.J. Linnenbom, J.W. Swinnerton, R.A. Lamontagne (1973). The ocean as a source of carbon monoxide. *J. Geophys. Res.-Atmos.*, **78**, 5333–5340.
100. R. Conrad, W. Seiler, G. Bunse, H. Giehl (1982). Carbon monoxide in seawater (Atlantic Ocean). *J. Geophys. Res.-Atmos.*, **87**, 8839–8852.
101. R. Conrad, W. Seiler (1988). Influence of the surface microlayer on the flux of nonconservative trace gases (CO, H_2, CH_4, N_2O) across the ocean-atmosphere interface. *J. Atmos. Chem.*, **6**, 83–94.
102. R.H. Gammon, K.C. Kelley (1990). Photochemical production of carbon monoxide in surface waters of the Pacific and Indian Oceans. In: R.G. Zepp, N.V. Blough (Eds), *Effects of Solar Ultraviolet Radiation on Biogeochemical Dynamics in Aquatic Environments* (pp. 58–60). Woods Hole Oceanographic Institution, Woods Hole, MA.
103. R.L. Valentine, R.G. Zepp (1993). Formation of carbon monoxide from the photodegradation of terrestrial dissolved organic carbon in natural waters. *Environ. Sci. Technol.*, **27** 409–412.
104. T.S. Bates, K.C. Kelly, J.E. Johnson, R.H. Gammon (1995). Regional and seasonal variation in the flux of oceanic carbon monoxide to the atmosphere. *J. Geophys. Res.*, **100**, 23093–23101.
105. O.C. Zafiriou, S.A. Andrews, W. Wang. Concordant estimates of oceanic carbon monoxide source and sink processes in the Pacific yield a balanced global "bluewater" CO budget, *Global Biogeochem. Cycles*, in press.
106. C.J. Miles, P.L. Brezonik (1981). Oxygen consumption in humic-colored waters by a photochemical ferrous-ferric catalytic cycle. *Environ. Sci. Technol.*, **15**, 1089–1095.
107. B.C. Faust, R.G. Zepp (1993). Photochemistry of aqueous iron(III)-polycarboxylate complexes: Roles in the chemistry of atmospheric and surface waters. *Environ. Sci. Technol.*, **27**, 2511–2517.
108. B.M. Voelker, B. Sulzberger (1996). Effects of fulvic acid on Fe(II) oxidation by hydrogen peroxide. *Environ. Sci. Technol.*, **30**, 1106–1114.
109. B.M. Voelker, F.M.N. Morel, B. Sulzberger (1997). Iron redox cycling in surface waters: Effects of humic substances and light. *Environ. Sci. Technol.*, **31**, 1004–1011.
110. M. Kulovaara, P. Backlund (1993). Effects of simulated sunlight on aquatic humic matter. *Vatten*, **49**, 100–103.
111. K. Salonen, A.Vähätalo (1994). Photochemical mineralization of dissolved organic matter in lake Skjervatjern. *Environ. Int.*, **20**, 307–312.
112. R.M.W. Amon, R. Benner (1996). Photochemical and microbial consumption of dissolved organic carbon and dissolved oxygen in the Amazon River system. *Geochim. Cosmochim. Acta*, **60**, 1783–1792.
113. W. Granéli, M. Lindell, L. Tranvikt (1996). Photooxidative production of dissolved inorganic carbon in lakes of different humic content. *Limnol. Oceanogr.*, **41**, 698–706.
114. A.V. Vähätalo, M.S. Salonen, P. Taalas, K. Salonen (2001). Spectrum of the quantum yield for photochemical mineralization of dissolved organic carbon in a humic lake. *Limnol. Oceanogr.*, **45**, 664–676.

115. S.C. Johannessen, W.L. Miller. Quantum yield for the photochemical production of dissolved inorganic carbon in the ocean. *Mar. Chem.*, in press.
116. B. Reitner, G.J. Herndl, A. Herzig (1997). Role of ultraviolet-B radiation on photochemical and microbial oxygen consumption in a humic rich shallow lake. *Limnol. Oceanogr.*, **42**, 950–960.
117. B.C. Faust, R.G. Zepp (1993). Photochemistry of aqueous iron(III)-polycarboxylate complexes: Roles in the chemistry of atmospheric and surface waters. *Environ. Sci. Technol.*, **27**, 2517–2522.
118. S.C.H. Johannessen (2000). A photochemical sink for dissolved organic carbon in the ocean (Ph.D. Thesis, p. 175). Dalhousie University.
119. C.L. Osburn, D.P. Morris, K.A. Thorn, R.E. Moeller (2001). Chemical and optical changes in freshwater dissolved organic matter exposed to solar radiation. *Biogeochemistry*, **54**, 251–278.
120. A.M. Anesio, L.J. Tranvik, W. Graneli (1999). Production of inorganic carbon from aquatic macrophytes by solar radiation. *Ecology*, **80**, 1852–1859.
121. D.J. Kieber, J. McDaniel, K. Mopper (1989). Photochemical source of biological substrates in sea water: implications for carbon cycling. *Nature*, **341**, 637–639.
122. J.J. Lindell, W. Granéli, L.J. Tranvik (1996). Effects of sunlight on bacterial growth in lakes of different humic content. *Aquat. Microb. Ecol.*, **11**, 138–141.
123. R.G. Wetzel P.G. Hatcher, T.S. Bianchi (1995). Natural photolysis by ultraviolet irradiance of recalcitrant dissolved organic matter to simple substrates for rapid bacterial metabolism. *Limnol. Oceanogr.*, **40**, 1369–1380.
124. G. Mullerniklas, A. Heissenberger, S. Puskaric, G.J. Herndl (1995). Ultraviolet-B radiation and bacterial metabolism in coastal waters. *Aquat. Microb. Ecol.*, **9**, 111–116.
125. W.L. Miller, M.A. Moran (1997). Interaction of photochemical and microbial processes in the degradation of refractory dissolved organic matter from a coastal marine environment. *Limnol. Oceanogr.*, **42**, 1317–1324.
126. G.J. Herndl, A. Brugger, S. Hager, E. Kaiser, I. Obernosterer, B. Reitner, D. Slezak (1997). Role of ultraviolet-B radiation on bacterioplankton and the availability of dissolved organic matter. *Plant Ecol.*, **128**, 42–51.
127. E. Kaiser, G.J. Herndl (1997). Rapid recovery of marine bacterioplankton activity after inhibition by UV radiation in coastal waters. *Appl. Environ. Microbiol.*, **63**, 4026–4031.
128. I. Obernosterer, G.J. Herndl (2000). Differences in the optical and biological reactivity of the humic and nonhumic dissolved organic carbon component in two contrasting coastal marine environments. *Limnol. Oceanogr.*, **45**, 1120–1129.
129. S. Bertilsson, L.J. Tranvik (1998). Photochemically produced carboxylic acids as substrates for freshwater bacterioplankton. *Limnol. Oceanogr.*, **43**, 885–895.
130. S. Bertilsson, L.J. Tranvik (2000). Photochemical transformation of dissolved organic matter in lakes. *Limnol. Oceanogr.*, **45**, 753–762.
131. I. Obernosterer, B. Reitner, G.J. Herndl (1999). Contrasting effects of solar radiation on dissolved organic matter and its bioavailability to marine bacterioplankton. *Limnol. Oceanogr.*, **44**, 1645–1654.
132. R. Benner, B. Biddanda (1998). Photochemical transformations of surface and deep marine dissolved organic matter: Effects on bacterial growth. *Limnol. Oceanogr.*, **43**, 1373–1378.
133. S. Ziegler, R. Benner (2000). Effects of solar radiation on dissolved organic matter in a subtropical seagrass meadow. *Limnol. Oceanogr.*, **45**, 257–266.
134. T. Naganuma, T. Konishi, T. Inoue, T. Nakane, S. Sukizaki (1996). Photodegrada-

tion or photoalteration? Microbial assay of dissolved organic matter. *Mar. Ecol. Prog. Ser.*, **135**, 309–310.
135. L.J. Tranvik, S. Kokalj (1998). Decreased biodegradability of algal DOC due to interactive effects of UV radiation and humic matter. *Aquat. Microb. Ecol.*, **14**, 301–307.
136. D.N. Thomas, R.J. Lara (1995). Photodegradation of algal derived dissolved organic carbon. *Mar. Ecol. Prog. Ser.*, **116**, 309–310.
137. A.M. Anesio, C.M.T. Denward, L.J. Tranvik, W. Graneli (1999). Decreased bacterial growth on vascular plant detritus due to photochemical modification. *Aquat. Microb. Ecol.*, **17**, 159–165.
138. J.J. Lindell, W. Granéli, L.J. Tranvik (1995). Enhanced bacterial growth in response to photochemical transformation of dissolved organic matter. *Limnol. Oceanogr.*, **40**, 195–199.
139. R.G. Zepp, D.M. Cline (1977). Rates of direct photolysis in the aquatic environment. *Environ. Sci. Technol.*, **11**, 359–366.
140. R.G. Zepp (1982). Experimental approaches to environmental photochemistry. In: O. Hutzinger (Ed.), *The Handbook of Environmental Chemistry* (pp. 19–41). Springer-Verlag, Berlin Heidelberg.
141. D. Preiswerk, R.G. Najjar (2000). A global, open-ocean model of carbonyl sulfide and its air-sea flux. *Global Biogeochem. Cycles*, **14**, 585–598.
142. P.M. Williams, E.R.M. Druffel (1987). Radiocarbon in dissolved organic matter in the central North Pacific Ocean. *Nature*, **330**, 246–248.
143. P.M. Williams, L.I. Gordon (1970). Carbon-13: carbon-12 ratios in dissolved and particulate organic matter in the sea. *Deep-Sea Res.*, **17**, 19–27.
144. P.A. Raymond, J.E. Bauer (2001). DOC cycling in a temperate estuary: A mass balance approach using 14C and 13C isotopes. *Limnol. Oceanogr.*, **46**, 655–667.
145. S. Opsahl, R. Benner (1997). Distribution and cycling of terrigenous dissolved organic matter in the ocean. *Nature*, **386**, 480–482.
146. R. Benner, S. Opsahl (2001). Molecular indicators of the sources and transformations of dissolved organic matter in the Mississippi River plume. *Org. Geochem.*, **32**, 597–611.
147. A. Kumar, R.P. Sinha, D.-P. Häder (1996). Effect of UV-B on enzymes of nitrogen metabolism in the cyanobacterium Nostoc calcicola. *J. Plant Physiol.*, **148**, 86–91.
148. J.A. Fuhrman, D.G. Capone (2001). Nifty nanoplankton. *Nature*, **412**, 593–594.
149. J.P. Zehr, J.B. Waterbury, P.J. Turner, J.P. Montoya, E. Omoregie, G.F. Steward, A. Hansen, N.-D.M. Karl (2001). Unicellular cyanobacteria fix N_2 in the subtropical North Pacific Ocean. *Nature*, **412**, 635–638.
150. I. Berman-Frank, J.T. Cullen, Y. Shaked, R.M. Sherrell, P.G. Falkowski (2001). Iron availability, cellular iron quotas, and nitrogen fixation in *Trichodesmium*. *Limnol. Oceanogr.*, **46**, 1249–1260.
151. J.M. Lenes, B.P. Darrow, C. Cattrall, C.A. Heil, M. Callahan, G.A. Vargo, R.H. Byrne, J.M. Prospero, D.E. Bates, K.A. Fanning, J.J. Walsh (2001). Iron fertilization and the *Trichodesmium* response on the West Florida shelf. *Limnol. Oceanogr.*, **46**, 1261–1277.
152. J. Wu, E. Boyle, W. Sunda, S.-L-S Wen (2001). Soluble and colloidal iron in the oligotrophic North Atlantic and North Pacific. *Science*, **293**, 847–5–849.
153. W.G. Sunda (1994). The influence of nonliving organic matter on the availability and cycling of plant nutrients in seawater. In: R.G. Zepp, C. Sonntag (Eds), *Role of Non Living Organic Matter in the Earth's Carbon Cycle* (pp. 191–207). Wiley, New York.
154. K.L. Bushaw, R.G. Zepp, M.A. Tarr, D. Schulz-Jander, R.A. Bourbonniere, R.E.

Hodson, W.L. Miller, D.A. Bronk, M.A. Moran (1996). Photochemical release of biologically available nitrogen from dissolved organic matter. *Nature*, **381**, 404–407.
155. K.L. Bushaw-Newton, M.A. Moran (1999). Photochemical formation of biologically-available nitrogen from dissolved humic substances in coastal marine systems. *Aquat. Microb. Ecol.*, **18**, 185–292.
156. D.J. Koopmans, D.A. Bronk. Photochemical production of inorganic nitrogen from dissolved organic nitrogen in waters of two estuaries and adjacent surficial groundwaters. *Aquat. Microb. Ecol.*, in press.
157. W.S. Gardner, J.F. Cavaletto, H.A. Bootsma, P.J. Lavrentyev, F. Troncone (1998). Nitrogen cycling rates and light effects in tropical Lake Maracaibo, Venezuela. *Limnol. Oceanogr.*, **43**, 1814–1825.
158. W.W. Wang, M.A. Tarr, T.S. Bianchi, E. Engelhaupt (2000). Ammonium photoproduction from aquatic humic and colloidal matter. *Aquat. Geochem.*, **6**, 275–292.
159. R.J. Kieber, A. Li, P.J. Seaton (1999). Production of nitrite from the photodegradation of dissolved organic matter in natural waters. *Environ. Sci. Technol.*, **33**, 993–998.
160. N.O.G. Jørgensen, L. Tranvik, H. Edling, W. Granéli, M. Lindell. (1998). Effects of sunlight on occurrence and bacterial turnover of specific carbon and nitrogen compounds in lake water. *FEMS Microbiol. Ecol.*, **25**, 217–227.
161. IPCC (1997). *IPCC Guidelines for National Greenhouse Gas Inventories.* Intergovernmental Panel for Climate Change, Paris, France.
162. J.J. Jankowski, D.J. Kieber, K. Mopper (1999). Nitrate and nitrite actinometers. *Photochem. Photobiol.*, **70**, 319–328.
163. R.G. Zepp, J Hoigné, H. Bader (1987). Nitrate-induced photooxidation of trace organic chemicals in water. *Environ. Sci. Technol.*, **21**, 443–450.
164. O.C. Zafiriou, M.B. True (1979). Nitrate photolysis in seawater by sunlight. *Mar. Chem.*, **8**, 336–42.
165. P.P. Vaughan, N.V. Blough (1998). Photochemical formation of hydroxyl radical by constituents of natural waters. *Environ. Sci. Technol.*, **32**, 2947–2953.
166. O.C. Zafiriou, R. Bonneau (1987). Wavelength-dependent quantum yield of OH radical formation from photolysis of nitrite ion in water. *Photochem. Photobiol.*, **45**, 723–727.
167. O.C. Zafiriou, M. McFarland (1981). Nitric oxide from nitrite photolysis in the Central Equatorial Pacific. *J. Geophys. Res.*, **86**, 3173–3182.
168. E. Micinski L.A. Ball, O.C. Zafiriou (1993). Photochemical oxygen activation: Superoxide radical detection and production rates in the Eastern Caribbean. *J. Geophys. Res.*, **98**, 2299–2306.
169. R.J. Ferek, R.B. Chatfield, M.O. Andreae (1986). Vertical distribution of dimethyl sulfide in the marine atmosphere. *Nature*, **320**, 514–516.
170. M.O. Andreae (1986). The ocean as a source of atmospheric sulfur compounds. In: P. Buat-Menard (Ed.), *The Role of Air–Sea Exchange in Geochemical Cycling* (pp. 331–362). D. Reidel, Dordrecht.
171. R.J. Charlson, J.E. Lovelock, M.O. Andreae, S.G. Warren (1987). Oceanic phytoplankton, atmospheric sulfur, cloud albedo and climate. *Nature*, **326**, 655–661.
172. F. Yin, D. Grosjean, J.H. Seinfeld (1990). Photooxidation of dimethyl sulfide and dimethyl disulfide: Mechanism development. *J. Atmos. Chem.*, **11**, 309–364.
173. T.S. Bates, B.K. Lamb, A. Guenther, J. Dignon, R.E. Stoiber (1992). Sulfur emissions to the atmosphere from natural sources. *J. Atmos. Chem.*, **14**, 315–337.
174. A.J. Kettle, M.O. Andreae et. al. (1999). A global database of sea surface dimethylsulfide (DMS) measurements and a procedure to predict sea surface DMS as a

function of latitude, longitude and month. *Global Biogeochem. Cycles*, **13**, 399–444.
175. R. Boers, P. Ayers, J.L. Gras (1994). Coherence between seasonal variation in satellite-derived cloud optical depth and boundary layer CCN concentrations at a mid-latitude Southern Hemisphere station. *Tellus Ser. B. Chem. Phys. Meteorol*, **46**, 123–131.
176. M.D. Keller, W.K. Bellows, R.R.L. Guillard (1988). A survey of dimethyl sulfide production in 12 classes of marine phytoplankton. In: E. Saltzman, W. Cooper (Eds), *Biogenic Sulfur in the Environment*, (pp. 167–182). American Chemical Society, Washington, D.C.
177. M.A.J. Curran, G.B. Jones (2000). Dimethyl sulfide in the Southern Ocean: Seasonality and flux. *J. Geophys. Res.*, **105**, 20451–20459.
178. D.J. Erickson, R.G. Zepp, E. Atlas (2000). Ozone depletion and the air-sea exchange of greenhouse and chemically reactive trace gases. *Chemosphere-Global Change Sci.*, **2**, 137–149.
179. J.W. Dacey, S.G. Wakeham (1986). Oceanic dimethylsulfide: Production during zooplankton grazing on phytoplankton. *Science*, **233**, 1314–1316.
180. R.P. Kiene (1999). Sulphur in the mix. *Nature*, **402**, 363–365.
181. R. Simó, C. Pedrós-Alió (1999). Role of vertical mixing in controlling the oceanic production of dimethyl sulfide. *Nature*, **402**, 396–398.
182. D.J. Kieber, J. Jiao, R.P. Kiene, T.S. Bates (1996). Impact of dimethylsulfide photochemistry on methyl sulfur cycling in the equatorial Pacific Ocean. *J. Geophys. Res.*, **101**, 3715–3722.
183. D.A. Toole, D.J. Kieber, R.P. Kiene, D.A. Siegel (2002). Quantum yield of dimethyl sulfide photo-oxidation in the Sargasso Sea. *Eos Trans. AGA*, **83**, OS184.
184. P.J. Crutzen (1976). The possible importance of COS for the sulfate layer of the stratosphere. *Geophys. Res. Lett.*, **3**, 73–76.
185. M. Chin, D.D. Davis (1995). A reanalysis of carbonyl sulfide as a source of stratospheric background sulfur aerosols. *J. Geophys. Res.*, **100**, 8993–9005.
186. M.O. Andreae, R.J. Ferek (1992). Photochemical production of carbonyl sulfide in seawater and its emission to the atmosphere. *Global Biogeochem. Cycles*, **6**, 175–183.
187. R. Zepp, T.V. Callaghan, D. Erikson (1994). Effects of increased solar untraviolet rdiation on biogeochemical cycles. In: J. an der Leun, M. Tevini, A. Teramura (Eds), *Environmental Effects of Ozone Depletion*, (pp. 79–93). United Nations Environment Programme Publication, Nairobi, Kenya.
188. G. Uher, M.O. Andreae (1997). Photochemical production of carbonyl sulfide in North Sea water: A process study. *Limnol. Oceanogr.*, **42**, 432–442.
189. P.S. Weiss, S.S. Andrews, J.E. Johnson, O.C. Zafiriou (1995). Photoproduction of carbonyl sulfide in South Pacific Ocean waters as a function of irradiation wavelength. *Geophys. Res. Lett.*, **22**, 215–218.
190. P.S. Weiss, J.E. Johnson, R.H. Gammon, T.S. Bates (1995). Re-evaluation of the open ocean source of carbonyl sulfide to the atmosphere. *J. Geophys. Res.*, **100**, 23083–23092.
191. W.H. Pos, D.H. Riemer, R.G. Zika (1998). Carbonyl sulfide (OCS) and carbon monoxide (CO) in natural waters: Evidence of a coupled production pathway. *Mar. Chem.*, **62**, 89–101.
192. V.S. Ulshöfer, M.O. Andreae (1998). Carbonyl sulfide (COS) in the surface ocean and the atmospheric COS budget. *Aquat. Geochem.*, **3**, 283–303.
193. R.G. Zepp, M.O. Andreae (1994). Factors affecting the photochemical formation of carbonyl sulfide in seawater. *Geophys. Res. Lett.*, **21**, 2812–2816.
194. R. Al-Farawati, C.M.G. van den Berg (2001). Thiols in coastal waters of the Western

North Sea and English Channel. *Environ. Sci. Technol.*, **35**, 1902–1911.
195. S. Elliot, E. Lu, F.S. Rowland (1989). Rates and mechanisms for hydrolysis of carbonyl sulfide in natural waters. *Environ. Sci. Technol.*, **23**, 458–461.
196. W.G. Sunda, D.J. Kieber (1994). Oxidation of humic substances by manganese oxides yields low-molecular-weight organic substrates. *Nature*, **367**, 62–64.
197. S.W. Chisholm (2000). Stirring times in the Southern Ocean. *Nature*, **407**, 685–687.
198. K. Barbeau, J.W. Moffett (2000). Laboratory and field studies of colloidal iron oxide dissolution as mediated by phagotrophy and photolysis. *Limnol. Oceanogr.*, **45**, 827–835.
199. W.G. Sunda, S A. Huntsman (1988). Effect of sunlight on redox cycles of manganese in the southwestern Sargasso sea. *Deep-Sea Res.*, **35**, 1297–1317.
200. M.L. Wells, N.M. Price, K.W. Bruland (1995). Iron chemistry in seawater and its relationship to phytoplankton: A workshop report. *Mar. Chem.*, **48**, 157–182.
201. N.M. Price, F.M.M. Morel (1998). Biological cycling of iron in the ocean. *Met. Ions Biol. Syst.*, **35**, 1–36.
202. E.L. Rue, K.W. Bruland (1995). Complexation of Fe(III) by natural organic ligands in the Central North Pacific as determined by a new competitive ligand equilibration/adsorptive cathode stripping voltammetric method. *Mar. Chem.*, **50**, 117–138.
203. D.M. McKnight, B.A. Kimball, K.E. Bencala (1988). Iron photoreduction and oxidation in an acidic mountain stream. *Science*, **240**, 637–641.
204. B. Sulzberger, J.L. Schnoor, R. Giovanall, J.G. Hering, J. Zobrist (1990). Biogeochemistry of iron in an acidic lake. *Aquat. Sci.*, **52**, 57–74.
205. M.L. Wells, L.M. Mayer (1991). The photoconversion of colloidal iron hydroxides in seawater. *Deep-Sea Res.*, **38**, 1379–1395.
206. K. Barbeau, E.L. Rue, K.W. Bruland, A. Butler (2001). Photochemical cycling of iron in the surface ocean mediated by microbial iron(III)-binding ligands. *Nature*, **413**, 409–413.
207. L. Emmenegger, R. Schwarzenbach, L. Sigg, B. Sulzberger (2000). Light-induced redox cycling of iron in circumneutral lakes. *Limnol. Oceanogr.*, **46**, 49–61.
208. W.L. Miller, D.W. King, J. Lin, D.R. Kester (1995). Photochemical redox cycling of iron in coastal seawater. *Mar. Chem.*, **50**, 63–77.
209. B.M. Voelker, D.L. Sedlak (1995). Iron reduction by photoproduced superoxide in seawater. *Mar. Chem.*, **50**, 93–102.
210. J.W. Moffett, R.G. Zika (1983). Oxidation kinetics of copper(I) in seawater: Implications for its existence in the marine environment. *Mar. Chem.*, **13**, 235–251.
211. J.W. Moffett, R.G. Zika (1987). Reaction kinetics of hydrogen peroxide with copper and iron in seawater. *Environ. Sci. Technol.*, **21**, 801–810.
212. P.L. Croot J.W. Moffett, L.E. Brand (2000). Photoproduction of extracellular Cu complexing ligands by eucaryotic phytoplankton in response to Cu stress. *Limnol. Oceanogr.*, **45**, 619–627.
213. M.G. Kogut, B.M. Voelker (2001). Strong copper-binding behavior of terrestrial humic substancres in seawater. *Environ. Sci. Technol.*, **35**, 1149–1156.
214. V. Balzani, V. Carrassiti (1970). *Photochemistry of Coordination Compounds.* Academic Press, London.
215. G. Ferraudi, S. Muralidharan (1981). Photochemical properties of copper complexes. *Coord. Chem. Rev.*, **36**, 45–88.
216. K. Hayase, R.G. Zepp (1991). Photolysis of copper(II)-amino acid complexes in water. *Environ. Sci. Technol.*, **25**, 1273–1279.
217. S.H.R. Davies, J.J. Morgan (1989). Manganese(II) oxidation kinetics on oxide surfaces. *J. Colloid Interface Sci.*, **129**, 63–77.
218. C. Lume-Pereira, S. Baral, A. Henglein, E. Janata (1985). Chemistry of colloidal

manganese dioxide. *J. Phys. Chem.*, **89**, 5772–5778.

219. S. Baral, C. Lume-Pereira, E. Janata, A. Henglein (1985). Chemistry of colloidal manganese dioxide. 1. Reaction with O and H_2O_2 (pulse radiolysis and stop flow studies). *J. Phys. Chem.*, **89**, 5779–5783.
220. A. Henglein (1989). Small-particle research: Physicochemical properties of extremely small colloidal metal and semiconductor particles. *Chem. Rev.*, **89**, 4342–4347.
221. O.C. Zafiriou, B.M. Voelker, D.L. Sedlak (1998). Chemistry of the superoxide ion (O_2^-) in seawater: Reactions with inorganic copper complexes. *J. Phys. Chem.*, **102**, 5693–5700.
222. B.M. Voelker, D.L. Sedlak, O.C. Zafiriou (2000). Chemistry of superoxide radicals (O_2^-) in seawater: Reactions with organic Cu complexes. *Environ. Sci. Technol.*, **34**, 1036–1042.
223. K.S. Johnson, K.H. Coale, V.A. Elrod, N.W. Tindale (1994). Iron photochemistry in the equatorial Pacific. *Mar. Chem.*, **46**, 319–334.
224. T.D. Waite, F.M.M. Morel (1984). Photoreductive dissolution of colloidal iron oxides in natural waters. *Environ. Sci. Technol.*, **18**, 860–868.
225. R.J. Kieber, K. Williams, J.D. Willey, S. Skrabal, G.B. Avery (2001). Iron speciation in coastal rainwater: Concentration and deposition to seawater. *Mar. Chem.*, **73**, 83–95.
226. T.D. Waite, I.C. Wrigley, R. Szymczak (1988). Photo-assisted dissolution of a colloidal manganese oxide in the presence of fulvic acid. *Environ. Sci. Technol.*, **22**, 778–7863.
227. R.G. Zepp, B.C. Faust, J. Hoigné (1992). Hydroxyl radical formation in aqueous reactions (pH 3–8) of iron(II) with hydrogen peroxide: the photo-Fenton reaction. *Envir. Sci. Technol.*, **26**, 313–321.
228. Y. Zuo, J. Hoigné (1992). Formation of hydrogen peroxide and depletion of oxalic acid in atmospheric water by photolysis of iron(III)-oxalato complexes. *Environ. Sci. Technol.*, **26**, 1014–1022.
229. A.T. Stone (1987). Reductive dissolution of manganese(III) and manganese(IV) oxides by substituted phenols. *Environ. Sci. Technol.*, **21**, 979–988.
230. W.G. Sunda, S.A. Huntsman, G.R. Harvey (1983). Photoreduction of manganese oxides in seawater and its geochemical and biological implications. *Nature*, **301**, 234–236.
231. D.J. Bertino, R.G. Zepp (1991). Effects of solar radiation on manganese oxide reactions with selected organic compounds. *Environ. Sci. Technol.*, **25**, 1267–1273.
232. W.G. Sunda, S.A. Huntsman (1990). Effects of sunlight and anthropogenic alterations in atmospheric solar attenuation on manganese redox cycles in surface seawater. In: N.V. Blough and R.G. Zepp (Eds), *Effects of Solar Ultraviolet Radiation on Biogeochemical Dynamics in Aquatic Environments* (pp, 104–107). Technical Report No. WHOI-90-09. Woods Hole Oceanographic Institution. Woods Hole, MA.
233. G.P. Klinkhammer, C.S. Chin, C. Wilson, M.D. Rudnicki, C.R. German (1997). Distributions of dissolved manganese and fluorescent dissolved organic matter in the Columbia River estuary and plume as determined by in-situ measurement. *Mar. Chem.*, **56**, 1–14.
234. T. Jickells, T. Church, A. Veron, R. Arimoto (1994). Atmospheric inputs of manganese and aluminum to the Sargasso Sea and their relation to surface-water concentrations. *Mar. Chem.*, **46**, 283–292.
235. S. Madronich, R.L. McKenzie, L.O. Björn, M.M. Caldwell (1998). Changes in biologically active ultraviolet radiation reaching the Earth's surface. *J. Photochem. Photobiol. B: Biol.*, **46**, 1–27.

Chapter 6

Photochemistry of chromophoric dissolved organic matter in natural waters

Christopher L. Osburn and Donald P. Morris

Table of contents

Abstract

The changes in the optics and chemistry of chromophoric dissolved organic matter (CDOM) caused by exposure to solar radiation (especially UVR) is an essential component of evaluating the effect of UVR on aquatic ecosystems. We briefly review photochemical concepts relevant to CDOM (light absorption in natural waters and photochemical reactions) and then describe methods for characterization of CDOM before considering the photochemical changes in CDOM and measurement of CDOM photoproducts. Experimental considerations necessary for working with polychromatic solar radiation are reviewed and briefly compared to monochromatic studies to provide a mechanism for predictive modeling of optical changes to CDOM based on measurements of solar radiation. CDOM sources, sinks, and transformations are discussed in context of UVR exposure in ecosystems and the cycling of carbon. Finally, we consider future directions of CDOM to include a more sophisticated connection between CDOM photodegradation and carbon cycling in aquatic ecosystems.

6.1 Introduction

Several reports have demonstrated that dissolved organic matter (DOM, principally dissolved organic carbon – DOC) is largely responsible for controlling the penetration of UVR in aquatic ecosystems [1–7, see also Chapters 1 and 3]. In addition, evidence suggests a strong coupling between the optical properties of natural waters and carbon cycling [8–12]. This understanding adds a decidedly ecological role to the photochemical reactions that DOM may undergo in the presence of natural solar radiation, by influencing the cycling of carbon. In this review, we will discuss "ecological photochemistry," or the photochemistry of DOM, especially the radiation-absorbing (*chromophoric*) fraction of DOM, termed CDOM. Our goal is to introduce the reader to the CDOM photochemistry that influences the optical properties of natural waters, by discussing the primary chemical reactions that influence the optical and chemical properties of DOM. While this will not be an exhaustive review of photochemical concepts related to natural waters (the reader is referred to refs. [12–16]), our review should provide some general reference to those approaching the topic for the first time, as well as providing a current "state-of-the-art" for those familiar with the photochemistry of CDOM. Moreover, we hope to introduce some basic concepts in predictive modeling of CDOM photobleaching in natural waters and suggest directions for future research.

6.1.1 The importance of photochemistry in the cycling of DOM

DOM is ubiquitous in natural waters, representing a substantial fraction of the total reduced carbon pool. Estimates of the total DOM concentration (as dissolved organic carbon, DOC) in lakes range from 1 to 10 mg C l^{-1}, while freshwater swamps, marshes, and bogs may have DOC concentrations ranging from 10 to 60 mg C l^{-1} [17]. Seawater and groundwater have substantially less DOC concentrations, ranging between 0.5 and 0.7 mg C l^{-1} on average [17]. A large source of the DOM found in natural waters originates from the degradation of terrestrial biomass and is present in the form of dissolved humic substances, predominantly humic and fulvic acids. In general, this material is transported to freshwaters by runoff or groundwater intrusion and to marine waters by riverine discharge. In fact, the global loading of terrestrial DOM is sufficient to overturn the oceanic pool of DOC in a relatively short time [18,19]. However, it has been demonstrated that terrestrial DOM is a very small fraction of the total marine DOM pool and that its turnover may be shorter than marine DOM synthesized in situ [20–22]. Explanations for this paradigm include removal mechanisms such as DOM flocculation [23], microbial utilization [21,24,25], and photodegradation [26–31].

6.1.2 Definitions and terms

Throughout this chapter, several terms relating to the photochemistry of DOM will be used and we provide a short glossary of terms here. *Photobleaching* is the loss of absorbance by CDOM in natural waters, also termed *fading* [32]. Photobleaching is an optical term, which is more appropriately connected with the chromophores found in CDOM, and, from here forward, we use "CDOM" to describe the optically-active component of bulk DOM that undergoes photochemical reaction and photophysical reaction (e.g., fluorescence). *Photodegradation* refers to the process of breaking down DOM to smaller compounds, which usually results in smaller molecular weight DOM products that may be rapidly consumed by bacteria [33–38]. The actual cleavage of chemical bonds by a photon of light energy during photodegradation is termed *photolysis*. *Photomineralization* is the actual oxidation of certain moieties to dissolved inorganic C (DIC), usually in the form of CO_2 but also in the form of CO and COS. *Photooxidation* is a catchall term that, in current usage, suggests a mix of the aforementioned specific processes. IUPAC recommends avoiding this term and using the more specific terms defined above and we will follow that convention.

6.2 The nature of light and its absorption in natural waters

The most essential step in photochemistry is the absorption of light by chemical species. All photochemistry is driven by the molecular excitation that occurs from the absorption of light. Light exhibits both wave and particle properties that impart the energy available for chemical and physical reactions. Wave theory can be used to describe the propagation of light through various media, where, if optically different, the light might be refracted and/or reflected (e.g., transmission of light through the air–water interface). Diffraction of light is important for monochromatic studies of photochemistry and will be briefly discussed in a later section. Polarization of light will not be discussed.

While wave theory is useful, it is an incomplete description of light. The absorption and emission properties of light are best described in terms of light particles, or photons. Planck's research on black body radiation led to the concept of a quantum of light energy, or a photon. The energy, E, of any photon may be calculated using Planck's relationship, where the photon's energy is inversely proportional to its wavelength:

$$E = h\nu = hc/\lambda \tag{1}$$

where h (6.63×10^{-24} J s) is a proportionality constant and c is the speed of light (3×10^{8} m s^{-1}). This relationship is important, because it allows us to quantify light energy reaching the Earth's surface and the surface of natural waters in terms of its wavelength. Often, the concept of the Einstein is employed in describing quanta of light, where 1 Einstein $= 6.023 \times 10^{23}$ photons, or a mole of photons. Many current light and UVR meters (e.g., Biospherical Instruments PUV and GUV radiometers) report radiation measurements in energy terms of watts (W) per squared area per discrete wavelength: μW cm^{-2} nm^{-1}. For our modeling example in section 7, we use units of energy in J m^{-2} nm^{-1}, which are the SI units for radiant exposure.

6.2.1 *The laws of photochemistry*

As mentioned above, it is the absorption of light that drives photochemical reaction. The 1st law of photochemistry formulated by Grotthus and Draper, plainly states that "*only the light which is absorbed by a molecule can be effective in producing photochemical change in the molecule.*" This is inherent in photochemistry, and perhaps taken for granted, but requires that we concern ourselves with accurate measurements of the light absorbed by the molecule(s) in natural waters. Therefore, we must not only measure the incident light at an aquatic surface, but also measure light propagation down through the water column (Figure 1). When light traveling through one transparent medium (e.g., the atmosphere) encounters a second medium in which the velocity of light is different (e.g., water) two different phenomena modify the light beam. First, a portion of the light beam is *reflected* at an angle (θ_a) equal to the angle of incidence. Second, the portion of the light beam transmitted into the new medium changes direction at the interface between the two media and is *refracted*. The angle of refraction (θ_w) is related to the angle of incidence by the different velocities of light in the two media, and the ratio of the velocity in medium 1 to the velocity in medium 2 is the *refractive index*. For most natural waters, this ratio is roughly 1.33, with subtle deviations due to the effects of temperature, light wavelength, and salt. Kirk [39] gives a thorough and rigorous discussion of the underwater light field, and the reader is directed to that reference, and also to Chapter 3.

The 2nd law of photochemistry, called the Stark–Einstein Law, states that "the absorption of light by a molecule is a one-quantum process, so that the sum of

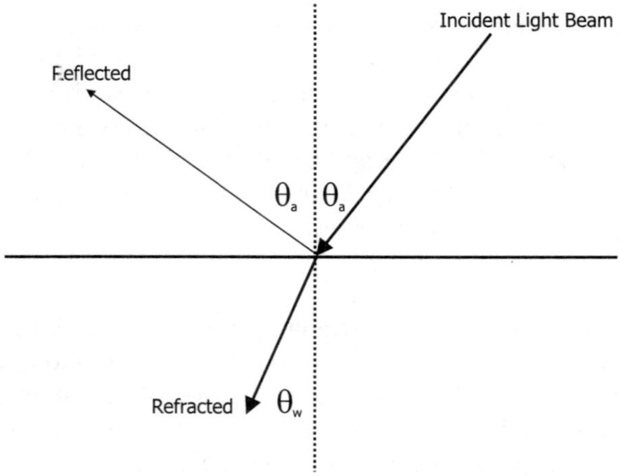

Figure 1. The change in angle of light beam intensity across the air–water interface, after Kirk [39].

the primary process quantum yields must be unity." Thus, a molecule that absorbs light and becomes electronically excited does so from only one photon, though bi-photonic absorption has been observed with laser light. The "primary" processes refer to the direct photochemical and photophysical reactions (e.g., dissociation, fluorescence, intersystem crossing) that occur due to light absorption. The quantum yield (Φ) is the amount of product (number of molecules) formed per unit time divided by the quanta of light absorbed per unit volume per unit time. In ecological photochemistry, this means that all of the photophysical and photochemical processes directly resulting from the absorption of a light photon must have individual quantum yields that add up to 1. We distinguish primary processes from "secondary" processes, which are the subsequent photochemical reactions that occur after a molecule has absorbed light energy. We describe these secondary processes as photosensitized reactions, because they occur due to the excess energy possessed by an excited chemical species. Photosensitized processes may be numerous (and actually induce further reactions), thus we typically observe quantum yields for individual processes that are quite small, and the overall quantum yield for a process such as photodegradation of CDOM is less than 1. The reason is that photodegradation of CDOM involves photosensitized processes that are not directly a result of the absorption of light energy.

We also point out the distinction between true and *apparent* quantum yields. An apparent quantum yield (Φ_a) is more appropriately used when working with CDOM, because we often do not know the true molar concentration of light-absorbing DOM molecules. That is, no molar basis exists for measuring the absorbance of CDOM, compared to measuring the absorbance for a single chemical of known molar concentration.

6.2.2 Absorbance of light

When dissolved species (denoted as S) absorb light photons, the outermost electron orbitals gain energy and electrons are elevated from their lowest energy state (the ground state, S_0) to a higher energy state (the excited state, denoted S*). Most ground state molecules are *singlet* ($^1S_0$), meaning that they have paired electrons resulting in a total electron quantum spin of zero; thus a single spin state. The exception is molecular dioxygen (O_2), which is a ground state *triplet* molecule (T_1), meaning that the molecule has unpaired electrons in its lowest energy state and may have three possible spin states (+1, 0, −1). The excited state that is initially produced by singlet molecules is also a singlet state ($^1S_1$) where the subscript "1" refers to the relative energy level above the ground state. The higher excited energy states ($^1S_2$, $^1S_3$, etc.) are very transient and usually decay rapidly to the $^1S_1$ state. Triplet states are generally longer-lived than singlet states, and we observe most photochemical reactions in CDOM from this state.

Excitation in natural water photochemistry typically involves the promotion of an electron from an n or π orbital (the bonding orbitals common in aromatic and carbonyl compounds) to a higher-energy anti-bonding orbital (π^*), and is referred to as n–π^* or π–π^* transition. In this excited state, S* has an excess of energy and the electronic orbital transitions impart dissimilar chemical reactivity to the excited molecule relative to its reactivity in the ground state. Therefore, several physical and chemical reactions may occur to release this excess energy and return the species to its ground state. Photophysical pathways are most common; in fact, most electron excitation results in the release of energy through various photophysical pathways that do not involve chemical reaction. For natural water photochemistry, the most common photophysical pathways are internal conversion, intersystem crossing, fluorescence, phosphorescence, and vibrational relaxation (cf. refs. [12,13,16,40]). CDOM fluorescence is a well-known phenomenon and has been studied extensively [41–46].

The presence of multiple types of chemical bonds in CDOM dictates its overall absorbance. Because CDOM is a heterogeneous mixture of perhaps hundreds or thousands of different compounds it is impossible to identify which of them is most responsible for the CDOM absorbance. However, several investigators have begun to use spectrophotometry and mass spectroscopy to identify individual chromophores [47–49]. Table 1 describes the maximum absorbance of certain molecular bonds and phenolic compounds that are likely to be present in CDOM derived from terrestrial sources (e.g., lignin) [12,50,51].

When chemical change does arise from molecular excitation by light absorption, it is usually due to the excitation to a triplet state rather than a singlet state. This may result from the longer lifetime of excited triplets (average lifetime of 10^{-3} s) versus excited singlets (average lifetime $<10^{-10}$ s). Also, a molecule in either the excited singlet or triplet state may transfer its energy to a receptor molecule (R) which becomes excited (R*) and may then undergo chemical reaction or return to its ground state through one of the previously mentioned photophysical pathways. An important example is the transfer of energy from an excited species to molecular oxygen (O_2) that is itself very reactive due to its two

Table 1 Maximum UV absorbance of and maximum wavelength to break bonds common to CDOM (superscripts indicate reference cited; n.r., not reported)

Chromophore	Maximum UVR absorbance (nm)	Maximum wavelength (nm) to break bond
C–C[a]	<180	346
C=C[a]	180	196
C–H[a]	<180	290
Phenols[b]	350	n.r
Aldehydes[b]	325	n.r.
Ketones[b]	350	n.r
Sinapic acid[c]	302, 327	n.r.
Protocatechuic acid[d]	258, 292	n.r

[a]Ref. [51].
[b]Ref. [12].
[c]Precursor of lignin, ref. [50].
[d]Lignin derivative, ref. [50].

unpaired electrons in the ground state (recall that the ground state is a triplet). The transfer of energy creates singlet oxygen ($^1O_2$), which is an effective quencher of triplet excited states and also very reactive (see Chapter 8).

6.2.3 Description of CDOM absorbance

CDOM is usually described in terms of its absorbance over the environmentally relevant wavelength range of 280 to 700 nm, encompassing the UV and the visible portions of the solar spectrum. Absorbance, measured by a spectrophotometer, is the log base 10 ratio of the light intensity, I_0, incident on the sample to the light intensity, I, transmitted by the system:

$$\text{Absorbance} = \log_{10}(I_0/I) \tag{2}$$

Units of absorbance are reported as absorption coefficients (a, in m^{-1}), reflecting the conversion of the raw absorbance $A_{CDOM}(\lambda)$ of a water sample measured by a spectrophotometer into its optical density (OD, or $a[\lambda]$). The absorption coefficient of CDOM is thus:

$$a_{CDOM}(\lambda) = A_{CDOM}(\lambda)2.303/l, \tag{3}$$

where l is the pathlength of the cuvette, and the 2.303 value converts the absorbance from base e into base 10 logarithms. The inclusion of the pathlength allows for the variable pathlength of cuvettes used to measure absorbance (typically 1, 5, and 10 cm). Examples of CDOM absorbance from several sources are shown in Figure 2. In most cases, absorption coefficients increase proportional to DOC concentration, though some saline lakes have very high DOC concentrations and very low absorption coefficients [Morris, unpublished data]. DOC-specific absorbance is another measurement that incorporates the DOC concentration in the optical measurement, and approximates the molar absorptivity commonly used to describe the spectroscopy of other discrete chemical

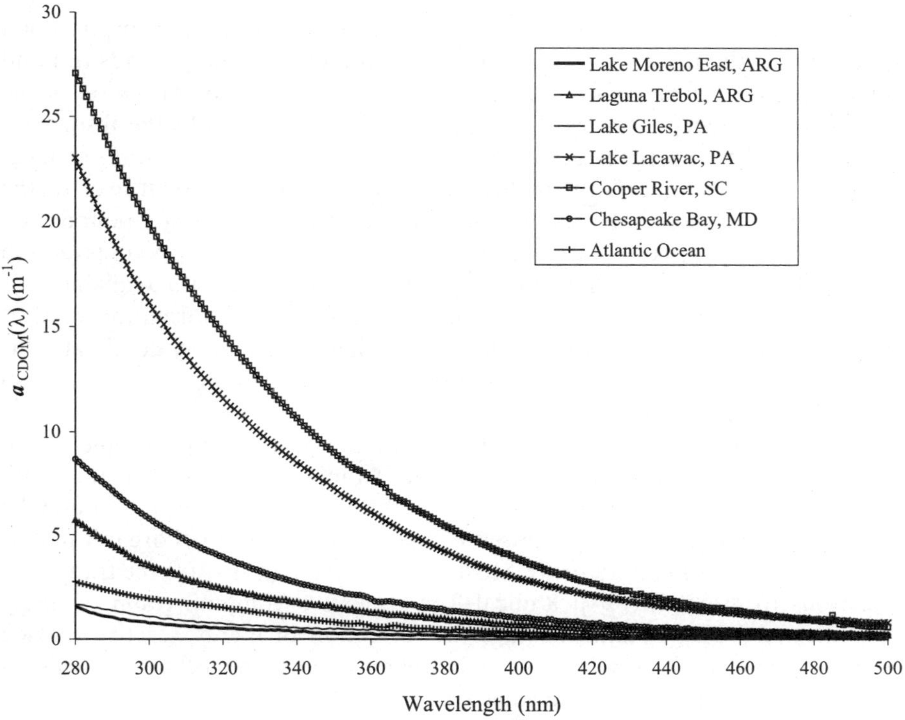

Figure 2. Absorbance spectra of multiple sources of CDOM exhibit wide variation.

species. The approximation is due to the heterogeneous nature of DOC. Some moieties of DOC may not absorb light and therefore would *not* be CDOM (Section 6.4). If a large fraction of the DOC is not CDOM, then DOC-specific absorbance is limited in its usefulness.

6.3 Direct and indirect photochemical reactions

The absorption of photon energy by CDOM can lead to several types of photophysical and photochemical reactions, and we emphasize that light absorption may be the first of many steps that can ultimately lead to the chemical changes we observe in CDOM. Most CDOM photochemistry involves the excitation of humic substances, which have a large degree of double bond character (C=C and C=O) that readily absorb sunlight energy. However, beyond direct chemical reaction from absorbing photon energy, excited species may participate in a number of indirect chemical reactions.

6.3.1 Direct photochemical reactions

Direct, or primary, photochemical reactions are the immediate chemical changes to CDOM such as isomerization, bond cleavage, and photolysis. Thus, we refer to the chemical rearrangements and reactions that result from absorption of a photon directly. In isomerization, the absorption of light energy leads to bond breakage and rearrangement, leading to a change in the conformation of the molecule. In direct photolysis, the excited state formed directly by the absorption of light energy undergoes bond cleavage leading to the degradation of large CDOM polymer units into smaller ones. Conceptually, this can lead to the liberation of smaller aliphatic compounds from larger aromatic compounds but, as Kieber and Mopper point out [52], even smaller organic compounds (α-keto acids) can also undergo photolysis Photochemical reaction with available oxygen may lead to photochemical decarboxylation and the formation of CO_2 [53,54]. Iron–CDOM complexes that absorb light energy may accelerate this process, whereby the oxidation of organic matter proceeds by a ligand–metal charge transfer [54–56].

Furthermore, humic substances in CDOM can gain energy (become *photosensitized*) as a result of initial absorption of radiation by another molecular entity (termed the *photosensitizer*). The energized humic species (HS*) may be involved in charge transfer processes leading to the oxidation of organic matter and/or the formation of radicals. Most of these reactions involve the transfer of electrons to dioxygen [14]. Radical formation is a common result of direct photolysis of CDOM and includes the formation of highly reactive oxygen species such as H_2O_2, singlet oxygen, inorganic and organic peroxy radicals (as discussed in Chapter 8), and solvated electrons [12,13].

6.3.2 Indirect photochemical reactions

Reactive species formed by the CDOM absorption of UVR can then undergo indirect, or secondary, photochemical reactions. In fact, the many possible reactions caused by photosensitized transient intermediates probably account for most of the photodegradation of CDOM that we observe. The many different photoprocesses involved in CDOM photochemistry make for a very complex pathway of reactions beginning with the initial absorption of light energy and ending with the final products of these multiple reactions.

Indirect, or secondary, photochemical reactions include the chemical changes brought on by photosensitizers, the molecules excited by the initial absorption of light. Sensitizers can include humic substances or other dissolved organic and inorganic species such as transition metals and nitrite and nitrate ions. Photosensitized organic matter has a short life span and tends to transfer its energy to a receptor molecule – usually dioxygen, forming singlet oxygen. Any photosensitized reaction involves the transfer of energy, hydrogen atoms, protons, or electrons [57,58], and the results of these charge transfers by intermediates are the underlying reactions that cause bond breakage or oxidation of the CDOM. Thus, indirect photochemistry has an important affect on the photochemical degradation of CDOM.

6.4 Characterization of CDOM

Because CDOM is an optical description of the photoreactive component of bulk DOM, several methods exist to characterize CDOM physically and chemically. These properties exhibit wide variations among CDOM source (e.g., freshwater vs. marine CDOM). At present, no universally accepted method for isolating CDOM from the bulk fraction exists, and here we briefly review several methods for its isolation. Each method has its benefits and limitations and we urge readers to carefully consider the procedure used to isolate CDOM for measurement or use in experiments.

6.4.1 Physical characterization of CDOM

Physically, CDOM is operationally defined as that material which passes through a 0.2 μm filter. However, many studies have shown that the range of CDOM size in natural waters is variable. The distinction of high molecular weight (HMW) DOM from low molecular weight (LMW) is determined using ultrafiltration with a 1000 Dalton (1 kDa) cutoff filter, and some studies have partitioned DOM among several size classes [24,59–63]. Though ultrafiltration may be used to represent CDOM, Osburn and Boyd have unpublished results that show recoveries of UV absorbance at 320 nm vary with both the source of DOM and the method of CDOM extraction (tangential flow filtration and solid phase extraction, Figure 3). This is not surprising since fulvic acids, which have conjugated double bonds and absorb readily in the UV region of the solar spectrum, are less than 1 kDa. Reverse osmosis is another physical method for isolating organic solutes from natural waters, but this method also concentrates salts and may actually polymerize DOM compounds, thus altering the original DOM material. Further, use of reverse osmosis techniques make additional chemical analysis of DOM with high salt content analytically difficult, especially if lyophilization is involved as the high salt content may preclude complete drying of the sample [Osburn, unpublished results].

6.4.2 Chemical characterization of CDOM

Chemical characterization of DOM has been studied extensively, providing a wealth of information regarding its chemical properties [17,20,25,60,64–89]. CDOM is usually characterized as aquatic humic substances, such as humic and fulvic acids, owing to the presence of multiple double bonds in aromatic, aldehyde, and ketone groups. Isolation of humic substances involves their separation by adsorption on macroporous resins (e.g., XAD-8 or XAD-4) and elution at various pH [90]. Humic acids are soluble above a pH of 2, while fulvic acids are soluble at any pH. Solid phase extraction (SPE) onto C_{18} resin is also employed to isolate CDOM [44,73]. Amador and coworkers have shown that

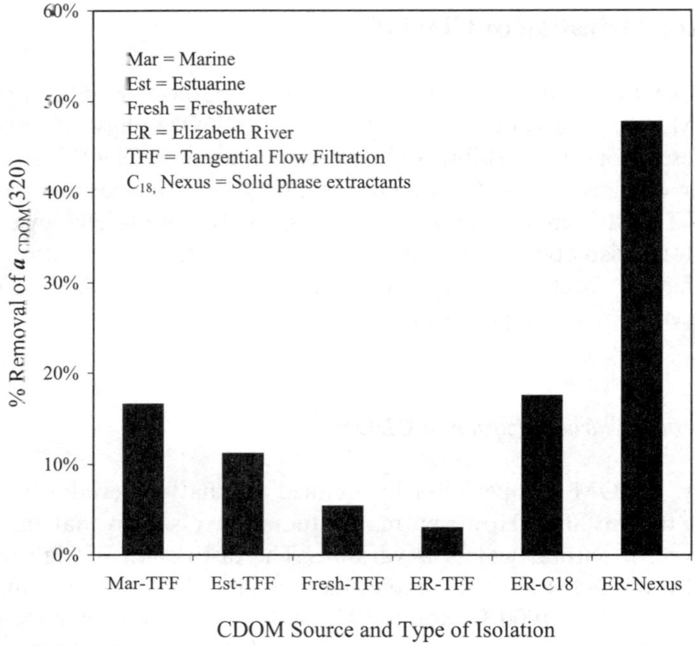

Figure 3. Comparison of CDOM recovery from tangential flow filtration (TFF) and solid phase extraction using two commercially-available resins (AbsElut and Nexus). Percent removal of initial CDOM absorbance at 320 nm.

SPE provides the best retention of marine CDOM based on recovery of chromophoric properties resembling the original marine DOM. There exists some debate on the extraction efficiencies of either technique and we refer the reader to current discussions on the subject [82,91–93].

Naturally, these parameters vary with the source of the CDOM. For example, freshwater (terrestrial or allochthonous) CDOM is higher in molecular weight and aromaticity than marine DOM, and freshwater CDOM typically has a lower C/N ratio as marine DOM. Carbon stable isotope evidence shows that marine DOM is on average −23%, reflecting a marine plankton source, whereas freshwater DOM is lighter, around −28%. The lighter isotopic signature reflects the contribution of terrestrial land plants. Additionally, Opsahl and Benner [31] have demonstrated the photoreactivity of lignin-derived phenols, suggesting their use as a proxy for terrestrial CDOM. Lignin originates primarily in terrestrial land plants and thus provides a good indicator of CDOM source. These analyses may be difficult in mixing environments, such as estuaries. Thus, one must consider the source of CDOM when interpreting the effects of solar radiation on CDOM chemical properties (see section 6.7).

6.5 Photochemical changes in DOM

Chemical changes in CDOM have been measured on several scales. Most common is measurement of bulk DOC by high temperature combustion or wet oxidation [94]. Other bulk measurements include carbon gases in the form of CO and CO_2. However, in recent years, compound-specific and molecular analyses have provided more detailed information on the products of CDOM photodegradation.

6.5.1 Measurement of organic photoproducts

LMW organic compounds produced by irradiation of DOM cover a wide range of compound classes, but are generally carbonyl compounds such as aldehydes and carboxylic acids, compounds readily available to aquatic microbes [26–28,30,33,35,38,95–111]. These compounds appear to arise from the degradation, or fragmentation, of larger humic structures into its component molecules by either direct photolysis or indirect secondary reactions discussed in Section 6.3. Measurement of these photoproducts usually involves gas chromatography/mass spectrometry of derivatized compounds or capillary ion electrophoresis (e.g., [37,38,52,112–114]).

Indirect evidence of organic photoproduct generation comes from structural analyses of DOM. Several reports have used ^{13}C NMR in the solid and solution state to show that the abundance of aromatic (ring containing) groups in the bulk DOM and humic and fulvic isolates is reduced after exposure to UVR [115,116]. Unpublished results by Thorn and Younger show reductions in both aromatic and carboxyl C groups of the Nordic fulvic acid after UV irradiation, corresponding to a 35% loss of dissolved absorbance at 465 nm. Osburn et al. [115] were able to suggested that these results are observable on a seasonal basis in a humic lake (Figure 4).

6.5.2 Measurement of inorganic photoproducts

The production of inorganic carbon compounds (DIC, primarily CO_2 and CO) has been widely reported from various CDOM sources [29,56,104,117–123]. Allard [124] has observed the photoproduction of DIC as carbonate ion from capillary electrophoresis. Production of DIC in natural waters was shown to have a strong dependence on wavelength band [120,123], pH [29,125], and cumulative dose [118]. CO_2 concentrations in fresh waters are easily measured by acidifying the sample with concentrated phosphoric acid and then measuring the evolved CO_2 either by gas chromatography or with a nondispersive infrared detector. However, Kieber (personal communication) notes the difficulty with measuring photoproduced CO_2 in marine samples or samples with high DIC background concentrations, in which case the DIC must be stripped out of the

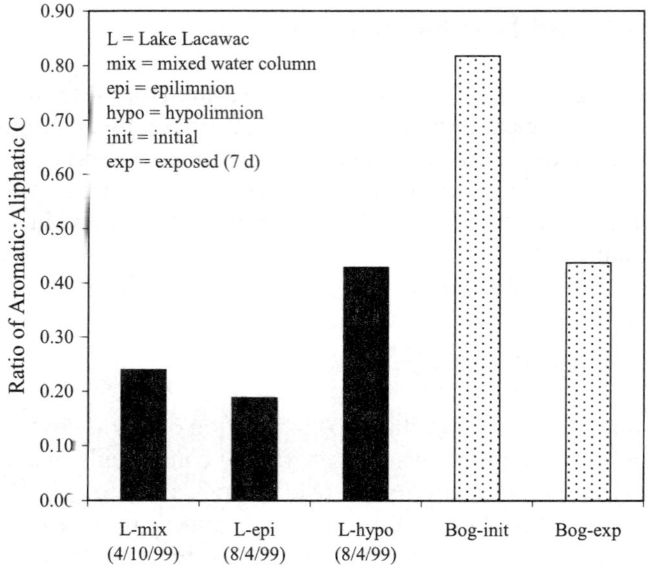

Figure 4. Change in the ratio of aromatic-to-aliphatic C in freshwater DOM exposed to solar radiation. Dark bars are 0.2 μm filtered and reverse osmosis concentrated samples collected from Lake Lacawac, Pennsylvania, USA during 1999. Stippled bars are results from sunlight exposure of bog DOM during 1998. From Osburn [158].

sample prior to irradiation. The reason is that the amount of DIC photo-produced is too small to measure above background DIC concentrations. CO concentrations are measured by gas chromatography fitted with either a methanizing flame ionization detector [29,117] or HgO/UV detector [126]. Photoproduction of CO_2 is at least an order of magnitude greater than CO [29,104].

In addition, photoproduction of inorganic nutrients such as phosphate [127] and ammonium [107,128] have been reported. Phosphate was shown to be bound to an iron–humic complex in a Midwestern US bog lake, and released upon irradiation with UVR. Ammonium was similarly released from irradiation of humic substances isolated from Skidaway and Satilla River estuaries. This result has opened up a new dimension to the biogeochemical cycling of nitrogen that may be mediated by UVR. Kieber et al. [129] have shown that humic substances isolated from various substances may produce nitrite upon photodegradation, though the rates of production were much less than those reported for ammonium production from ref. [16] (4 nM h^{-1} vs. 50 nM h^{-1}). Thus, UVR effects on CDOM may liberate inorganic nutrients that become active in the biogeochemical cycling of natural waters.

6.5.3 Recent approaches to measuring photochemical changes to CDOM

Other chemical methods have recently been employed to examine the effects of UVR on CDOM. Opsahl and Benner [31] have studied lignin-derived phenols of both HMW and LMW DOM in the Mississippi River plume. They showed that 75% of total dissolved lignin in riverine HMW DOM was lost during 28 d exposure to solar radiation; 80% of the remaining fraction of dissolved lignin was present as LMW material and the remaining result was less susceptible to further photochemical degradation. HMW lignin from the equatorial Pacific Ocean was found to be resistant to photodegradation. Interestingly, the photodegraded riverine dissolved lignin was similar to the marine dissolved lignin, suggesting that DOM photochemistry is an important factor in the composition of marine DOM. Furthermore, they found that the ratio of vanillic acid to vanillin reflects photochemical alteration and this may be a useful tracer for chemical changes in CDOM at the molecular level. Osburn et al. [115] have reported a correlation between the loss of DOC concentration, the loss of dissolved absorbance, and an enrichment in the carbon stable isotope value of DOM after sunlight exposure of 0.2 μm-filtered DOM from a *Sphagnum* bog. They were also able to show experimentally that the aromaticity of decreased by 47% after the exposure to sunlight. The aromatic loss corresponded to an increase in the stable carbon isotopic value of the DOM from −28% to −27% and a decrease in DOC concentration of 16% (Figure 5). These chemical changes suggest a removal of C as CO_2 and a change in the composition of the DOM to smaller molecular weight compounds. Similarly, Opsahl and Zepp [130] have also reported an increase in stable carbon isotopic values of riverine water exposed to sunlight. They also show loss of DOC and dissolved absorbance concurrent with isotopic enrichment; furthermore, they observed substantial reductions in lignin phenol concentration after sunlight exposure (>65%).

Vahatalo and coworkers [121] have provided direct evidence that lignin is photochemically reactive. Using synthetic lignin, radiolabeled with ^{14}C on the aromatic C ring only, they determined that approximately 20% of the ring-labeled C was mineralized to CO_2. Simultaneous exposure of DOC from lake water produced slightly higher (2–3%) results. These results provide convincing evidence that the aromatic fraction of CDOM is largely responsible for its photoreactivity.

These examples show the utility of using ^{13}C NMR spectroscopy, dissolved lignin, and stable isotopes as molecular tracers of photodegradation of CDOM. Other tools of mass spectroscopy and compound separation and identification should provide additional information on changes to DOM as it is photodegraded. Recently, several groups have presented spectroscopic and spectrophotometric methods for identification of chromophores [47–49,131].

6.6 Experimental and modeling considerations for working with CDOM photochemistry

Miller [32] has provided a concise consideration of issues when designing photochemical experiments using CDOM, and the careful measurements that must be made. Here, we will provide a general design concept for an experimental exposure of CDOM to a polychromatic light source (e.g., solar radiation) and

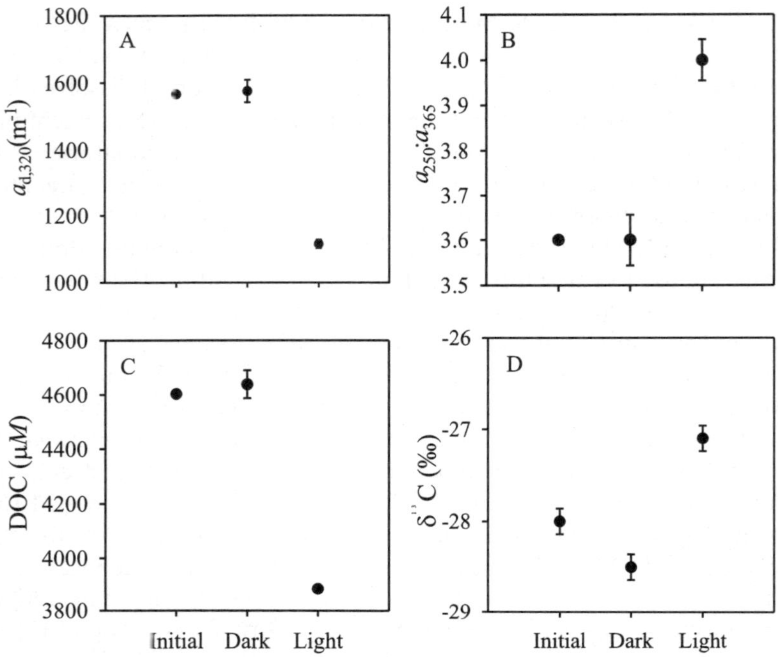

Figure 5. Changes in bog water DOM after exposure to solar radiation. (A) Loss of CDOM absorbance at 320 nm, (B) decrease in molecular weight (indicated by increase in ratio), (C) decrease in DOC concentration, and (D) increase in δ^{13}C of lyophilized DOM. After Osburn [158].

then discuss the methods for constructing a predictive model based on the results. We use as an example the process of photobleaching and the measurement of dissolved absorbance. While not comprehensive, this section should provide for new readers a general approach for designing and interpreting photochemical experiments with CDOM.

6.6.1 *Experimental methods*

The experimental design for measuring changes in bulk parameters of CDOM such as dissolved absorbance is straightforward. First, water is collected and then filtered to remove particles and bacteria, usually through a pre-cleaned (baked at 450°C then rinsed with ultrapure deionized and distilled water) glass fiber filter with a pore size of ~0.7 μm. Next, the filtrate is mechanically sterilized by passage through a 0.22 μm filter and carefully transferred to clean (acid-washed and baked) quartz vessels, usually test tubes or round-bottom flasks, reserving an aliquot of the sample for analysis of initial parameters. The quartz should be at least 99% pure to ensure transmission of all environmentally relevant UVR through the walls of the quartz vessel. In some cases, a bactericide may be used to suppress microbial growth, though many microbial inhibitors (NaN_3, $HgCl_2$) absorb UVR near the UV-B range, near 275 nm [132]. The vessels are then placed in a water bath and exposed to solar or artificial radiation. The water bath buffers the temperature of the CDOM solution that can vary by tens of degrees Celsius from early morning to evening for solar radiation exposures. Similarly, artificial light sources can also generate high temperatures around the samples. The exposure time may vary depending on the goals of the experiment. Some samples may be collected periodically during the exposure to generate a time-series of photochemical changes, which is useful for determining rate constants.

6.6.2 Reporting photobleaching results

Several methods exist to report the optical changes in CDOM after the photobleaching experiment. Loss of absorbance and fluorescence are the two most common parameters to report. The loss of absorbance is determined by subtraction of final minus initial absorbance spectra measured with a UV–Vis spectrophotometer. A sample spectrum of absorbance loss per nm for photobleached CDOM is shown in Figure 6. This figure illustrates that absorbance exhibits wide variation across the UV spectrum and between different sources of CDOM. Often, loss of absorbance is reported at specific wavelengths, with wavelengths in the UV-B or UV-A region being most common (e.g., 320 or 350 nm). It has been

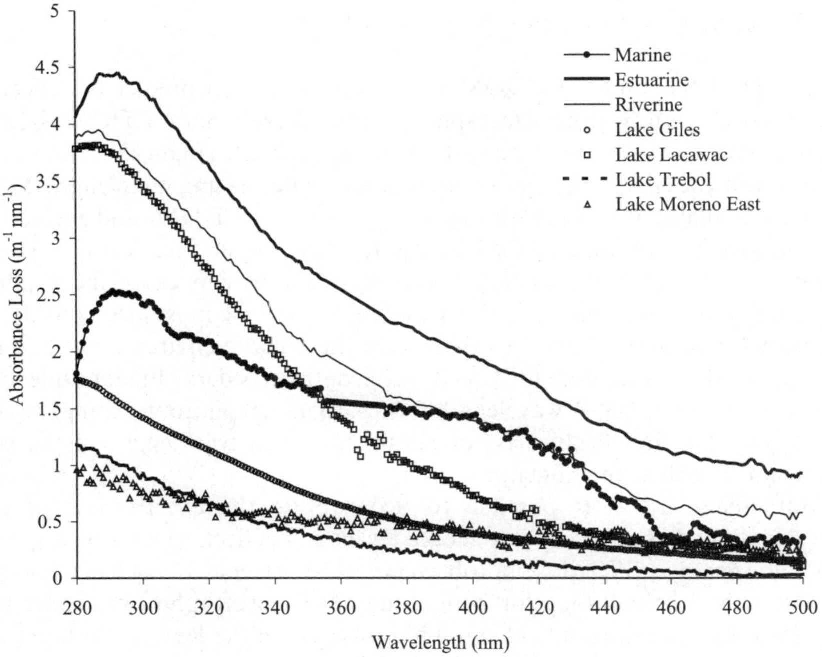

Figure 6. Loss of absorbance per nm in several different types of CDOM.

popular to report absorbance loss at 250 nm, which is an indicator of aromatic content. Similarly, loss of DOM fluorescence may be measured using excitation at one wavelength and emission at another wavelength. Recently, published reports of excitation–emission matrices have provided a three-dimensional picture of CDOM photochemical changes [41,122,133–135]. CDOM fluorescence is particularly useful because the results may be normalized to a standard, such as quinine sulfate. Currently, no accepted standard for CDOM absorbance exists and reporting loss of absorbance as absolute values makes for difficult comparisons among CDOM sources that vary in CDOM quantity and quality. Moreover, dissolved absorbance may be affected by dissolved species other than DOM. Thus, some workers have reported loss of absorbance normalized to the DOC concentration of the water sample, which approximates the molar absorptivity used in classic photochemistry [8,15,116,124,136–139].

Osburn et al. [132] have used the integrated photobleaching (loss of absorbance over the range of 280 to 500 nm) divided by the total absorbed energy (also from 280 to 500 nm) to calculate the photoreactivity of various CDOM sources. This parameter is descriptive of the capacity for the CDOM to lose absorbance, and normalizes different CDOM types to their absorbed energy, which is a function of CDOM quality and quantity. This is similar to the apparent quantum yield calculated for CDOM by Whitehead et al. [140]. Similar integrated photobleaching calculations have been used for CDOM photobleaching studies from estuarine [122] and lake [123] sources.

6.6.3 Modeling photochemical changes in CDOM

One goal of the study of CDOM photochemistry is to predict the effects of enhanced UV influx, from stratospheric ozone depletion, on CDOM. Because photon energy is inversely related to its wavelength (Planck's relationship, section 6.2), the energy per photon increases with decreasing wavelength (Figure 7). This is analogous to the absorption spectrum of CDOM, and means that wavelengths are variable in their ability to bring about chemical change. We might expect that higher-energy UV-B wavelengths can cause more photobleaching than lower-energy UV-A wavelengths, but we must also consider the quantity (or *fluence*) of photons that reach the aquatic environment. Figure 7 also shows that the fluence for UV-B wavelengths is orders of magnitude lower than the fluence of UV-A wavelengths. Thus, to model photobleaching, we need to account for the effectiveness of photons at each wavelength in the solar spectrum, as well as their fluence.

Mathematically, it is possible to deconvolute the effectiveness of each wavelength and assign each wavelength a weight; in effect, generating a *spectral weighting function* (SWF) for photobleaching. Osburn et al. [132] have described in detail the methodology for computing SWFs using photobleaching data obtained from multiple optical cutoff filters based on the Rundel method [141] and comments by Cullen and Neale [142]. The cutoff filters successively manipulate solar spectra by successively removing more UVR and creating an array of

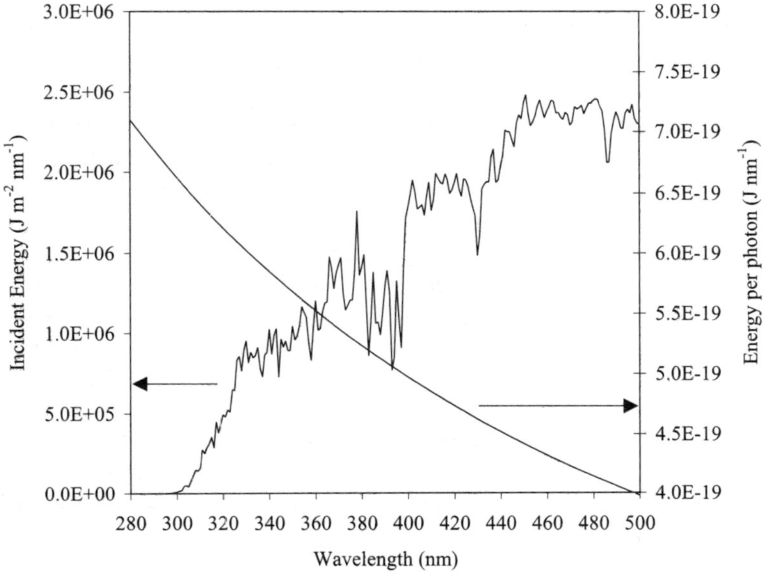

Figure 7. A comparison between a typical solar spectrum generated for temperate regions (Washington, DC, USA) and the energy per photon estimated by the Planck equation.

spectral treatments. A simple exponential was assumed for the shape of the SWF (based on the dissolved absorbance of CDOM and equation 1) and an iterative nonlinear regression of the iterative photobleaching in each spectral treatment versus the cumulative absorbed energy in each treatment was run to optimize the fit of the SWF to the observed data. The equation for the SWF was:

$$W(\lambda) = W_{300} \exp(-S_w[\lambda - 300]) \tag{4}$$

where W_{300} is the weight at 300 nm, and S_w is the slope of the exponential. The seed value for W_{300} was estimated from a regression of the differential photobleaching between adjacent optical cutoff filters versus the differential absorbed energy in adjacent treatments. The seed value for S_w was estimated from the slope of ln $a_{CDOM}(\lambda)$ versus wavelength (S, the spectral slope). The values for an average SWF (based on multiple experiments of surface water CDOM from several lakes) were: $W_{300} = -0.0103 \pm 0.006$ and $S_w = 4.34 \pm 1.79 \times 10^{-6}$).

This function can then be used to predict the amount of photobleaching that has occurred by multiplying the SWF by a measured irradiance spectrum. For example, we used the summary SWF of Osburn et al. [132] to predict daily changes in dissolved absorbance for humic Lake Lacawac, northeastern Pennsylvania, USA. First, daily measurements of ground level incident energy at four wavelengths (305, 320, 340, and 380 nm) were used to construct daily solar spectra for northeastern Pennsylvania, USA, during the summer of 1998. The measurements were recorded on a Biospherical Instruments GUV-521 radiometer and the total incident energy on the surface of Lake Lacawac was modeled to generate daily solar spectra at 1 nm intervals from 280 to 500 nm.

Modeled attenuation coefficients for the lake were used to modify the surface solar spectrum at 10 cm interval of depth throughout the mixed layer. Thus, they computed spectral energy at depth and could then predict the photobleaching at depth using the SWF. The daily spectra were multiplied by the spectral weights from the SWF to calculate daily photobleaching at the 10 cm intervals in the mixed layer of Lake Lacawac. The sum of the predicted photobleaching at depth was subtracted from an initial daily dissolved absorbance value, which equaled a daily change in dissolved absorbance for the mixed layer of Lake Lacawac. The model was run successively for that was repeated for each day of the study period. Thus, the model predicted daily changes in integrated dissolved absorbance caused solely by photobleaching. Figure 8 shows the measured changes in dissolved absorbance and the predicted changes using the SWF. The close match between modeled and measured dissolved absorbance suggest that CDOM photobleaching by solar UVR is the process that controls dissolved absorbance, and thus transparency, in lakes. Deviations of the model from observed change in the dissolved absorbance of the lake are attributed to recharge of fresh CDOM from various sources (precipitation, runoff, and advection).

6.6.4 Use of polychromatic vs. monochromatic radiation

The use of polychromatic radiation (*e. g.*, solar radiation) more closely resembles

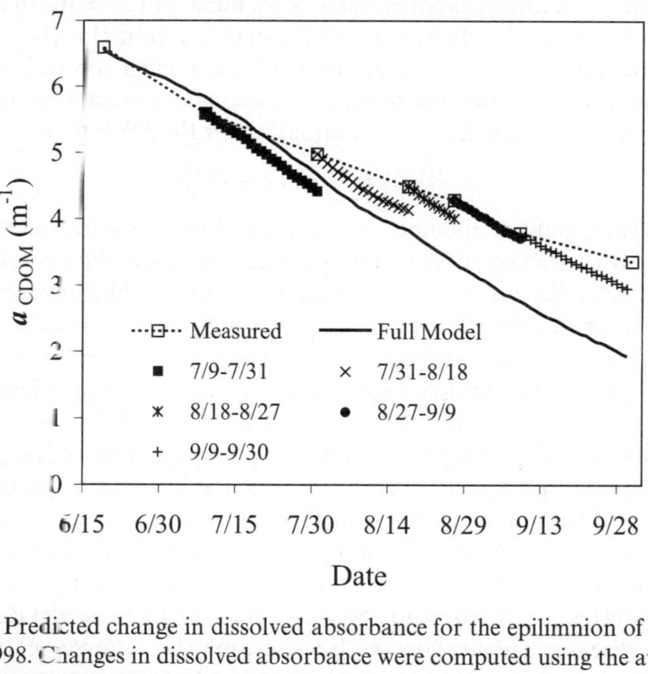

Figure 8. Predicted change in dissolved absorbance for the epilimnion of Lake Lacawac during 1998. Changes in dissolved absorbance were computed using the average SWF of Osburn [158]. Solid line is the model run from Day 1 to Day 104. Other symbols represent the model run for intervals in between dissolved absorbance measurements made during the modeling period (open squares).

the natural environment, and is different from classic photochemistry, which often uses monochromatic radiation. In this experimental setting, a monochromator is used to separate wavelengths generated from a light source. Thus, in an experiment, a sample is exposed to only one wavelength at a time and the effect of that one wavelength is recorded. This approach allows the researcher to easily calculate an effect per wavelength (*i.e.*, an action spectrum). If the radiant energy at each wavelength is known, the researcher can calculate a quantum yield for the photoreaction. However, the monochromatic approach implicitly assumes that the measured effect is only due to energy at the wavelength of irradiation. This makes application of action spectra to the natural environment difficult.

With polychromatic radiation, the effect is less obvious and necessitates the manipulation of the energy spectrum (e.g., with optical cutoff filters) to measure the effect and then to deconvolute the weighted effect at each wavelength. While the use of optical cutoff filters to modify polychromatic radiation more closely simulating the natural environment, their use introduces error to the calculations and may reduce the sensitivity of the analysis [141]. However, several lines of evidence suggest that with polychromatic solar radiation, multiple wavelength reactions contribute to the photodegradation of CDOM. Both Osburn et al. [132] and Whitehead et al. [140] have measured photobleaching at wavelengths that were excluded by optical filters. They suggest the interactive effect of photons from multiple wavelengths caused photobleaching at any one wavelength. The mechanism that drives this phenomenon may be that a chromophore absorbs over a range of wavelengths or a change in the relative abundance of chromophores (absorbing and different and multiple wavelengths) in the bulk DOM. This effect would further complicate the application of action spectra for CDOM photobleaching to measurements of polychromatic solar spectra.

6.7 CDOM dynamics in natural waters: sources, sinks, and transformations

We have seen that when CDOM absorbs UVR, several types of chemical reactions may occur. These reactions may be variable depending on the aquatic medium, which is also subject to wide variation. Aquatic ecosystems are dynamic systems, constantly influenced by biological activity, chemical reactions, and physical processes.

6.7.1 Chemical transformations of CDOM by UVR

As the chemical composition of the water changes, the amount of light energy available and the reactants available also changes. For example, terrestrial CDOM transported from a river into an estuary encounters a distinct change in the ionic composition of water due to increased salinity. This change in ionic composition may alter the solubility or conformation of certain C moieties and thus influence the optical properties of the CDOM. Osburn et al. [143] have shown that CDOM added from riverine and estuarine waters is more photoreactive in higher salinity CDOM-free permeates (Figure 9). Availability of O_2 can affect the degree of photodegradation. Allard et al. [124] showed that rate and degree of photodegradation for a fulvic acid solution was reduced when exposed under nitrogen gas. Prior exposure to UVR may also reduce the photoreactivity of CDOM, reducing the efficiency of photochemical reactions. Miller and Zepp [29] and Miller and Moran [104] both reported correlation between higher production of CO_2 and CO and higher absorption coefficient loss at 350 nm for humic acids isolated from a salt marsh. Thus the photochemical reactions are strongly influenced by the chemistry of the aquatic environment.

Sources of CDOM to aquatic ecosystems fall into two broad categories. Allochthonous sources of CDOM are derived from carbon sources outside of the aquatic ecosystem. This term is generally synonymous with terrestrial organic matter sources and the products of terrestrial organic matter biodegradation (e.g., lignin, tannins, and flavonoids). These compounds can also be characterized as humic substances, owing to the presence of many aromatic moieties. Autochthonous sources of CDOM come from the organic matter produced within a particular aquatic ecosystem. This material is largely derived from algae, though macrophytes can contribute CDOM in freshwater ecosystems. Several workers have shown that autochthonous sources of CDOM exist, primarily due to the humification of compounds released by algal senescence [66,144–148]. The mechanism of this humification is unclear, but appears to be caused by oxidative linkage of fatty acids either by microbial activity or sunlight [149]. It is also

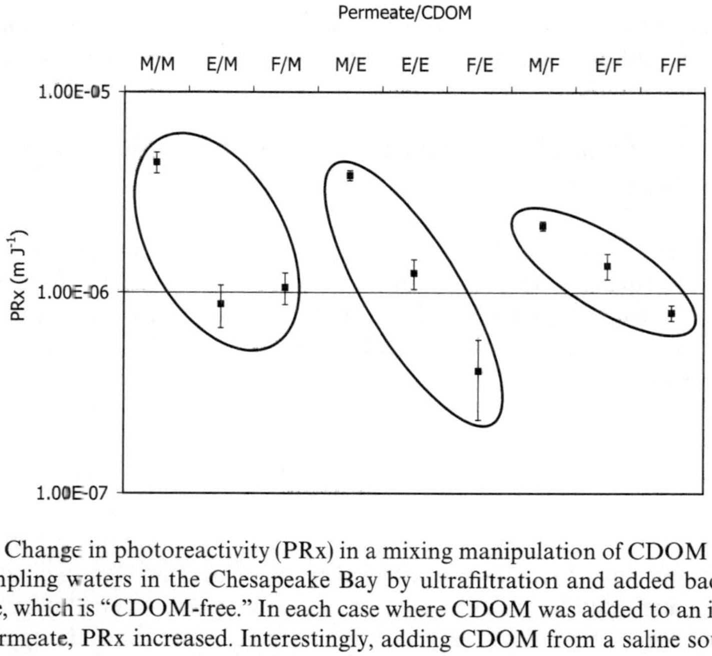

Figure 9. Change in photoreactivity (PRx) in a mixing manipulation of CDOM removed from sampling waters in the Chesapeake Bay by ultrafiltration and added back to the permeate, which is "CDOM-free." In each case where CDOM was added to an increased saline permeate, PRx increased. Interestingly, adding CDOM from a saline source to a fresher permeate (e.g., estuarine [E] CDOM in fresh water [F] permeate) decreased PRx.

possible that mineral surfaces may cause this oxidative process.

Reche and coworkers [125] have investigated the effect of water chemistry and trophic status on the photobleaching of CDOM in lake ecosystems. They examined about thirty lakes in the United States and found a strong and significant correlation ($r^2=0.94$) between acid-neutralizing capacity and loss of dissolved absorbance at 440 nm. Conductivity showed a less strong correlation ($r^2=0.74$), as did cation concentration (Ca^+ and Mg^{2++}, $r^2=0.62$), but both were still significant. Trophic status was estimated with chlorophyll *a* concentration, and showed a rather weak correlation ($r^2=0.15$). They conclude that high ionic composition in lakes likely increased photobleaching efficiency by changing the conformation of chromophoric moieties.

6.7.2 Ecological implications of CDOM photochemistry

Ultimately for ecologists, the photochemical reactions involving CDOM are of interest for multiple reasons, most dealing with the exposure of organisms to UV-B radiation. Many of these reasons are dealt with in other chapters of this book, and here we briefly speculate on a sample of direct and indirect connections between CDOM photochemistry and aquatic ecology. From the ecosystem perspective, CDOM photodegradation to smaller biolabile compounds might strongly influence carbon transfer among trophic levels by providing bacteria with carbon sources. Thus bacterial stimulation by photoproduced C (perhaps evidenced optically by an increase in the $a_{CDOM}(250){:}a_{CDOM}(365)$ ratio) might enhance its movement through the "microbial loop" [33–38,95,103,106,108,128,146,150,151]. Alternatively, some reports have suggested that CDOM photochemistry renders C unavailable to bacteria [109,147]. Because many reports show that CDOM photomineralization produces DIC [29,56,118–121,123,152,153], it is likely that this CDOM is utilized by primary producers (or by bacteria in the case of CO, [11,35,154]) at some point. This speculation has not been researched thoroughly, and DIC photoproduction is inextricably linked to other chemical factors in aquatic ecosystems (e.g., pH, iron, and conductivity; [38,54,119,125,152,155–157]). Its importance to carbon cycling remains unknown.

From the physiological perspective, the link between ecology and CDOM photochemistry is important through both direct and indirect associations. A direct effect is the potential for production of high reactive oxygen species (see Chapter 8) which can damage cellular membranes, and to a minor degree the photo-activated toxicity of organic compounds (Chapter 7). However, these effects are likely quite small compared to the indirect ecological effect of CDOM photochemistry: the change in the underwater radiation field caused by CDOM photobleaching. We see that CDOM photobleaching alters its spectral properties, most evident by changes in the spectral slope (S), and the net effect of CDOM photobleaching might be enhanced UV-B flux relative to UV-A flux down through the water column. This would potentially expose organisms to higher amount of damaging UV-B radiation, even if vertical mixing mediates these changes through reduced time at the surface (see Chapter 4), thus reducing the roles of photorepair (Chapters 10 and 13) or influencing behavioral responses (Chapter 14). A reduction in overall photosynthetic capacity of aquatic primary producers may also occur in an enhanced UV environment (Chapter 11). Because CDOM is colored, a contrasting ecological effect could be shading, whereby CDOM competes with primary produces for available photosynthetically active photons. Perhaps to some degree, CDOM photobleaching stimulates primary production in strongly colored waters such as wetlands and humic lakes.

It is clear from the above summary that CDOM plays a central role in the interaction of UVR and aquatic ecosystem function, and that the effects and implications of CDOM photochemistry are not always straightforward. Moreover, the effects become exacerbated by climate change and anthropogenic modification of aquatic ecosystems (e.g., acid deposition; see Chapter 17).

6.8 Conclusion

This review has highlighted some of the recent results of investigations into the ecological photochemistry of CDOM. While much information continues to accumulate on this subject, several issues remain. One issue is the relative contribution of CDOM photomineralization to atmospheric CO_2 flux out of natural waters. Another issue is the role of CDOM photodegradation in the transport of terrestrial C to the coastal ocean and its effectiveness relative to microbial degradation. Furthermore, it is unknown how the importance of CDOM photodegradation in controlling water column transparency varies among different types of natural waters and with latitude. While action spectra and SWFs are useful, we do not know if a general model can be used for all types of CDOM – we need more information on factors that affect spectral weights. For example, does prior solar radiation exposure affect the calculation of spectral weights [132]? How do spectral weights calculated for marine vs. fresh water differ? Answers to these questions are required to accurately model CDOM changes in the natural environment. Although variation exists in the approaches used to isolate CDOM and model its photochemistry, this exciting field continues to grow as a large component of biogeochemical study. Further research should vastly improve our understanding of the role of CDOM in biogeochemical cycling and water column transparency.

Acknowledgements

We thank Barrie Peake and an anonymous reviewer for helpful comments. Tom Boyd and Rick Coffin provided data and logistical support. This work was supported in part by NSF-DEB 9629639 (Lehigh Univeristy) and by the ONR Work Unit number N0001401WX20072 (Naval Research Laboratory).

References

1. N.M. Scully, D.R.S. Lean (1994). The attenuation of ultraviolet radiation in temperate lakes. In: C.E. Williamson, H.E. Zagarese (Eds), *The Impact of UV-B Radiation on Pelagic Freshwater Ecosystems* (*Arch. Hydrobiol., Beih. Ergebn. Limnol.*, **43**, 135–144). Stuttgart.
2. D.P. Morris, H.E. Zagarese, C.E. Williamson, E.G. Balseiro, B.R. Hargreaves, B. Modenutti, R.E. Moeller, C. Queimalinos (1995). The attenutation of solar UV radiation in lakes and the role of dissolved organic carbon. *Limnol. Oceanogr.*, **40**, 1381–1391.
3. N.D. Yan, W. Keller, N.M. Scully, D.R.S. Lean, P.J. Dillon (1996). Increased UV-B penetration in a lake owing to drought-induced acidification. *Nature*, **381**, 141–143.
4. I. Laurion, W.F. Vincent, D.R.S. Lean (1997). Underwater ultraviolet radiation: Development of spectral models for northern high latitude lakes. *Photochem. Photobiol.*, **65**, 107–114.
5. I. Laurion, M. Ventura, J. Catalan, R. Psenner, R. Sommaruga (2000). Attenuation of ultraviolet radiation in mountain lakes: Factors controlling the among- and within-lake variability. *Limnol. Oceanogr.*, **45**, 1274–1288.
6. J.A.E. Gibson, W.F. Vincent, B. Nieke, R. Pienitz (2000). Control of biological exposure to UV radiation in the Arctic Ocean: Comparison of the roles of ozone and riverine dissolved organic matter. *Arctic*, **53**, 372–382.
7. N.M. Scully, W.F. Vincent, D.R.S. Lean (2000). Exposure to ultraviolet radiation in aquatic ecosystems: estimates of mixing rate in Lake Ontario and the St. Lawrence River. *Can. J. Fish. Aquat. Sci.*, **57**, 43–51.
8. D.P. Morris, B.R. Hargreaves (1997). The role of photochemical degradation of dissolved organic carbon in regulating the UV transparency of three lakes on the Pocono Plateau. *Limnol. Oceanogr.*, **42**, 239–249.
9. C.E. Williamson, R.S. Stemberger, D.P. Morris, T.M. Frost, S.G. Paulsen (1996). Ultraviolet radiation in North American lakes: Attenuation estimates from DOC measurements and implications for plankton communities. *Limnol. Oceanogr.*, **41**, 1024–1034.
10. C.E. Williamson, D.P. Morris, M.L. Pace, A.G. Olson (1999). Dissolved organic carbon and nutrients as regulators of lake ecosystems: Resurrection of a more integrated paradigm. *Limnol. Oceanogr.*, **44**, 795–803.
11. R.G. Zepp, T. Callaghan, D. Erickson (1995). Effects of increased solar ultraviolet-radiation on biogeochemical cycles. *Ambio*, **24**, 181–187.
12. R.G. Zika (1980). Marine organic photochemistry. In: E.K. Duursma, R. Dawson (Eds), *Marine Organic Chemistry* (pp. 299–325). Elsevier, Amsterdam.
13. O.C. Zafiriou, J. Joussotdubien, R.G. Zepp, R.G. Zika (1984). Photochemistry of natural waters. *Environ. Sci. Technol.*, **18**, A358–A371.
14. R.G. Zepp (1988). Environmental photoprocesses involving natural organic matter. In: F.H. Frimmel, R.F. Christman (Eds), *Humic Substances and their Role in the Environment* (pp. 193–214). Wiley, Chichester.
15. F.H. Frimmel (1994). Photochemical aspects related to humic substances. *Environ. Int.*, **20**, 373–385.
16. W.L. Miller (1994). Recent advances in the photochemistry of natural dissolved organic matter. In: D. Crosby, G.R. Helz, R.G. Zepp (Eds), *Aquatic and Surface Photochemistry* (pp. 111–127). Lewis, Boca Raton.
17. E. Thurman (1985). *Organic Geochemistry of Natural Waters.* Martinus Nijhoff/Dr W. Junk, Dordrecht.

18. J.E. Bauer, P.M. Williams, E.R.M. Druffel (1992). C-14 Activity of dissolved organic-carbon fractions in the North-Central Pacific and Sargasso Sea. *Nature*, **357**, 667–670.
19. E.R.M. Druffel, P.M. Williams, J.E. Bauer, J.R. Ertel (1992). Cycling of dissolved and particulate organic-matter in the open ocean. *J. Geophys. Res.-Oceans*, **97**, 15639–15659.
20. K.J. Meyers-Schulte, J.I. Hedges (1986). Molecular evidence for a terrestrial component of organic matter dissolved in ocean water. *Nature*, **321**, 61–63.
21. J.I. Hedges, R.G. Keil, R. Benner (1997). What happens to terrestrial organic matter in the ocean?, *Org. Geochem.*, **27**, 195–212.
22. S. Opsahl, R. Benner (1997). Distribution and cycling of terrigenous dissolved organic matter in the ocean. *Nature*, **386**, 480–482.
23. R.F.C. Mantoura, E.M.S. Woodward (1983). Conservative behaviour of riverine dissolved organic carbon in the Severn Estuary: chemical and geochemical implications. *Geochim. Cosmochim. Acta*, **47**, 1293–1309.
24. R.M.W. Amon, R. Benner (1994). Rapid cycling of high molecular weight dissolved organic matter in the ocean. *Nature*, **369**, 549–552.
25. C.S. Hopkinson, I. Buffam, J. Hobbie, J. Vallino, M. Perdue, B. Eversmeyer, F. Prahl, J. Covert, R. Hodson, M.A. Moran, E. Smith, J. Baross, B. Crump, S. Findlay, K. Foreman (1998). Terrestrial inputs of organic matter to coastal ecosystems: An intercomparison of chemical characteristics and bioavailability. *Biogeochemistry*, **43**, 211–234.
26. R.M.W. Amon, R. Benner (1996). Photochemical and microbial consumption of dissolved organic carbon and dissolved oxygen in the Amazon River system. *Geochim. Cosmochim. Acta*, **60**, 1783–1792.
27. D.J. Kieber, J. McDaniel, K. Mopper (1989). Photochemical source of biological substrates in sea water: Implications for carbon cycling. *Nature*, **341**, 637–639.
28. R.J. Kieber, X.L. Zhou, K. Mopper (1990). Formation of carbonyl compounds from UV induced photodegradation of humic substances in natural waters: Fate of riverine carbon in the sea. *Limnol. Oceanogr.*, **35**, 1503–1515.
29. W.L. Miller, R.G. Zepp (1995). Photochemical production of dissolved inorganic carbon from terrestrial organic matter: Significance to the oceanic organic carbon cycle. *Geophys. Res. Lett.*, **22**, 417–420.
30. K. Mopper, X.L. Zhou, R.J. Kieber, D.J. Kieber, R.J. Sikorski, R.D. Jones (1991). Photochemical degradation of dissolved organic carbon and its impact on the oceanic carbon cycle. *Nature*, **353**, 60–62.
31. S. Opsahl, R. Benner (1998). Photochemical reactivity of dissolved lignin in river and ocean waters. *Limnol. Oceanogr.*, **43**, 1297–1304.
32. W.L. Miller (1998). Effects of UV radiation on aquatic humus: Photochemical principles and experimental considerations. In: D.O. Hessen, L.J. Tranvik (Eds), *Aquatic humic substances: Ecology and Biogeochemistry* (pp. 125–141). Springer, Berlin.
33. M.J. Lindell, W. Graneli, L.J. Tranvik (1995). Enhanced bacterial growth in response to photochemical transformation of dissolved organic matter. *Limnol. Oceanogr.*, **40**, 195–199.
34. R.G. Wetzel, P.G. Hatcher, T.S. Bianchi (1995). Natural photolysis by ultraviolet irradiance of recalcitrant dissolved organic matter to simple substrates for rapid bacterial metabolism. *Limnol. Oceanogr.*, **40**, 1369–1380.
35. M.A. Moran, R.G. Zepp (1997). Role of photoreactions in the formation of biologically labile compounds from dissolved organic matter. *Limnol. Oceanogr.*, **42**, 1307–1316.

36. N. Bano, M.A. Moran, R.E. Hodson (1998). Photochemical formation of labile organic matter from two components of dissolved organic carbon in a freshwater wetland. *Aquat. Microb. Ecol.*, **16**, 95–102.
37. S. Bertilsson, L.J. Tranvik (1998). Photochemically produced carboxylic acids as substrates for freshwater bacterioplankton. *Limnol. Oceanogr.*, **43**, 885–895.
38. S. Bertilsson, L.J. Tranvik (2000). Photochemical transformation of dissolved organic matter in lakes. *Limnol. Oceanogr.*, **45**, 753–762.
39. J.T.O. Kirk (1994). *Light and Photosynthesis in Aquatic Ecosystems.* Cambridge University Press, Cambridge.
40. J.G. Calvert, J.N. Pitts (1966). *Photochemistry*, Wiley, New York.
41. P.G. Coble, S.A. Green, N.V. Blough, R.B. Gagosian (1990). Characterization of dissolved organic matter in the Black Sea by fluorescence spectroscopy. *Nature*, **348**, 432–435.
42. R.F. Chen, J.L. Bada (1992). The fluorescence of dissolved organic matter in seawater. *Mar. Chem.*, **37**, 191–221.
43. A. Vodacek (1992). An explanation of the spectral variation in fresh water CDOM fluorescence. *Limnol. Oceanogr.*, **37**, 1808–1813.
44. S.A. Green, N.V. Blough (1994). Optical absorption and fluorescence properties of chromophoric dissolved organic matter in natural waters. *Limnol. Oceanogr.*, **39**, 1903–1916.
45. F.E. Hoge, M.E. Williams, R.N. Swift, J.K. Yungel, A. Vodacek (1995). Satellite retrieval of the absorption coefficient of chromophoric dissolved organic matter in continental margins. *J. Geophys. Res.-Oceans*, **100**, 24847–24854.
46. P.G. Coble (1996). Characterization of marine and terrestrial DOM in seawater using excitation emission matrix spectroscopy. **51**, 325–346.
47. R. Del Vecchio, N.V. Blough (2002). On the nature of the constituents underlying the absorption spectra of humic substances and chromophoric dissolved organic matter (CDOM). *Eos*, **83**, OS32R-10.
48. D. Repeta, S. John (2002). Chemical characterization of colored dissolved organic matter in seawater. *Eos*, **83**, OS22J–04.
49. E.R. Stabenau, C.A. Moore, R.G. Zika (2002). Application of LC/MSn to the study of DOM mediated optical properties: South Florida Coastal Zone 2001, *Eos*, **83**, OS22J-05.
50. P.G. Waterman, S. Mole (1994). *Analysis of Phenolic Plant Metabolites.* Blackwell Scientific, London.
51. R.B. Roy (1996). Review of the UV digestion procedures for the automated analysis of waste effluents. *Am. Environ. Lab.*, **8**, 27–29.
52. D.J. Kieber, K. Mopper (1987). Photochemical formation of glyoxylic and pyruvic acids in seawater. *Mar. Chem.*, **21**, 135–149.
53. Y. Chen, S.U. Khan, M. Schnitzer (1978). Ultraviolet irradiation of dilute fulvic acid solution. *Soil Sci. Soc. Am. J.*, **42**, 292–296.
54. C.J. Miles, P.L. Brezonik (1981). Oxygen-consumption in humic-colored waters by a photochemical ferrous-ferric catalytic cycle. *Environ. Sci. Technol.*, **15**, 1089–1095.
55. B.M. Voelker, F.M.M. Morel, B. Sulzberger (1997). Iron redox cycling in surface waters: Effects of humic substances and light. *Environ. Sci. Technol.*, **31**, 1004–1011.
56. H.Z. Gao, R.G. Zepp (1998). Factors influencing photoreactions of dissolved organic matter in a coastal river of the southeastern United States. *Environ. Sci. Technol.*, **32**, 2940–2946.
57. R.A. Larson, E.J. Weber (1994). Environmental photochemistry. In: *Reaction Mechanisms in Environmental Chemistry* (pp. 359–413). Lewis, Boca Raton.

58. P.L. Brezonik (1994). *Chemical Kinetics and Process Dynamics in AquaticsSystems.* Lewis, Boca Raton.
59. R.M.W. Amon, R. Benner (1996). Bacterial utilization of different size classes of dissolved organic matter. *Limnol. Oceanogr.*, **41**, 41–51.
60. R. Benner, J.D. Pakulski, M. McCarthy, J.I. Hedges, P.G. Hatcher (1992). Bulk chemical characteristics of dissolved organic matter in the ocean. *Science*, **255**, 1561–1564.
61. R. Benner, B. Biddanda, B. Black, M. McCarthy (1997). Abundance, size distribution, and stable carbon and nitrogen isotopic compositions of marine organic matter isolated by tangential-flow ultrafiltration. *Mar. Chem.*, **57**, 243–263.
62. L. Guo, P.H. Santschi, L.A. Cifuentes, S.E. Trumbore, J. Southon (1996). Cycling of high-molecular-weight dissolved organic matter in the Middle Atlantic Bight as revealed by carbon isotopic (-^{13}C and -^{14}C) signatures. *Limnol. Oceanogr.*, **41**, 1242–1252.
63. P.H. Santschi, L. Guo, M. Baskaran, S. Trumbore, J. Southon, T.S. Bianchi, B. Honeyman, L.A. Cifuentes (1995). Isotopic evidence for the contemporary origin of high-molecular weight organic matter in oceanic environments. *Geochim. Cosmochim. Acta*, **59**, 625–631.
64. A. Nissenbaum, I.R. Kaplan (1972). Chemical and isotopic evidence for the in situ origin of marine humic substances. *Limnol. Oceanogr.*, **17**, 570–582.
65. M.A. Wilson, P.F. Barron, A.H. Gillam (1981). The structure of freshwater humic substances as revealed by ^{13}C-NMR spectroscopy. *Geochim. Cosmochim. Acta*, **45**, 1743–1750.
66. G.R. Harvey, D.A. Boran, L.A. Chesal, J.M. Tokar (1983). The structure of marine fulvic and humic acids. *Mar. Chem.*, **12**, 119–132.
67. M.A. Wilson, A.H. Gillam, P. Collin (1983). Analysis of the structure of dissolved marine humic substances and their phytoplanktonic precursors by ^{1}H and ^{13}C nuclear magnetic resonance. *Chem. Geol.*, **40**, 187–201.
68. J.R. Ertel, J.I. Hedges, E.M. Perdue (1984). Lignin signature of aquatic humic substances. *Science*, **223**, 485–487.
69. J.R. Ertel, J.I. Hedges, A.H. Devol, J.E. Richey, M.D.G. Ribeiro (1986). Dissolved humic substances of the Amazon River system. *Limnol. Oceanogr.*, **31**, 739–754.
70. K.A. Thorn (1987). Structural characteristics of the IHSS Suwannee River fulvic and humic acids determined by solution state C-13 NMR spectroscopy. *Adv. Humic Substances Res.*, 175–183.
71. W. Michaelis, H.H. Richnow, A. Jenisch (1989). Structural studies of marine and riverine humic matter by chemical degradation. *Sci. Total Environ.*, **81–82**, 41–50.
72. L. Wassenaar, R. Aravena, P. Fritz (1989). The geochemistry and evolution of natural organic solutes in groundwater, *Radiocarbon*, **31**, 865–876.
73. J.A. Amador, P.J. Milne, C.A. Moore, R.G. Zika (1990). Extraction of chromophoric humic substances from seawater. *Mar. Chem.*, **29**, 1–17.
74. R. Francois (1990). Marine sedimentary humic substances – structure, genesis, and properties *Rev. Aquat. Sci.*, **3**, 41–80.
75. J. Baron, D. McKnight, A.S. Denning (1991). Sources of dissolved and particulate organic material in Loch Vale watershed, Rocky Mountain National Park, Colorado, USA. *Biogeochemistry*, **15**, 89–110.
76. J.I. Hedges P.G. Hatcher, J.R. Ertel, K.J. Meyersschulte (1992). A comparison of dissolved humic substances from seawater with Amazon River counterparts by 13C-NMR Spectrometry. *Geochim. Cosmochim. Acta*, **56**, 1753–1757.
77. M. McCarthy, J. Hedges, R. Benner (1996). Major biochemical composition of

dissolved high molecular weight organic matter in seawater. *Mar. Chem.*, **55**, 281–297.
78. A. Afcharian, Y. Levi, L. Kiene, P. Scribe (1997). Fractionation of dissolved organic matter from surface waters using macroporous resins. *Water Res.*, **31**, 2989–2996.
79. L.D. Guo, P.H. Santschi (1997). Composition and cycling of colloids in marine environments. *Rev. Geophys.*, **35**, 17–40.
80. D.M. McKnight, R. Harnish, R.L. Wershaw, J.S. Baron, S. Schiff (1997). Chemical characteristics of particulate, colloidal, and dissolved organic material in Loch Vale Watershed, Rocky Mountain National Park. *Biogeochemistry*, **36**, 99–124.
81. J. Peuravuori, K. Pihlaja (1997). Pyrolysis electron impact mass spectrometry in studying aquatic humic substances. *Anal. Chim. Acta*, **350**, 241–247.
82. J. Peuravuori, K. Pihlaja (1997). Isolation and characterization of natural organic matter from lake water: Comparison of isolation with solid adsorption and tangential membrane filtration. *Environ. Int.*, **23**, 441–451.
83. J. Peuravuori, K. Pihlaja, N. Valimaki (1997). Isolation and characterization of natural organic matter from lake water: Two different adsorption chromatographic methods. *Environ. Int.*, **23**, 453–464.
84. F.H. Frimmel (1998). Characterization of natural organic matter as major constituents in aquatic systems. *J. Cont. Hydrol.*, **35**, 201–216.
85. J. Peuravuori, K. Pihlaja (1998). Multimethod characterization of lake aquatic humic matter isolated with sorbing solid and tangential membrane filtration. *Anal. Chim. Acta*, **364**, 203–221.
86. J. Peuravuori, N. Paaso, K. Pihlaja (1999). Characterization of lake-aquatic humic matter isolated with two different sorbing solid techniques: pyrolysis electron impact mass spectrometry. *Anal. Chim. Acta*, **391**, 331–344.
87. O. Kracht, G. Gleixner (2000). Isotope analysis of pyrolysis products from Sphagnum peat and dissolved organic matter from bog water. *Org. Geochem.*, **31**, 645–654.
88. T. Lehtonen, J. Peuravuori, K. Pihlaja (2000). Characterisation of lake-aquatic humic matter isolated with two different sorbing solid techniques: tetramethylammonium hydroxide treatment and pyrolysis-gas chromatography/mass spectrometry. *Anal. Chim. Acta*, **424**, 91–103.
89. P. Louchouarn, S. Opsahl, R. Benner (2000). Isolation and quantification of dissolved lignin from natural waters using solid-phase extraction and GC/MS. *Anal. Chem.*, **72**, 2780–2787.
90. G.R. Aiken, D.M. McKnight, K.A. Thorn, E.M. Thurman (1992). Isolation of hydrophilic organic-acids from water using nonionic macroporous resins. *Org. Geochem.*, **18**, 567–573.
91. V.I. Esteves, N.M.A. Cordeiro, A.C. Duarte (1995). Variation on the adsorption efficiency of humic substances from estuarine waters using XAD resins. *Mar. Chem.*, **51**, 61–66.
92. K. Hautala, J. Peuravuori, K. Pihlaja (2000). Measurement of aquatic humus content by spectroscopic analyses. *Water Res.*, **34**, 246–258.
93. V. Roubeuf, S. Mounier, J.Y. Benaim (2000). Solid phase extraction applied to natural waters: efficiency and selectivity. *Org. Geochem.*, **31**, 127–131.
94. R. Benner, J.I. Hedges (1993). A test of the accuracy of fresh water DOC measurements by high-temperature catalytic oxidation and UV-promoted persulfate oxidation. *Mar. Chem.*, **41**, 161–165.
95. D.J. Strome, M.C. Miller (1978). Photolytic changes in dissolved humic substances. *Verh. Int. Verein. Limnol.*, **20**, 1248–1254.
96. A. Geller (1983). Degradability of dissolved organic lake water compounds in cultures of natural bacterial communities. *Arch. Hydrobiol.*, **99**, 60–79.

97. S.A. Visser (1984). Seasonal changes in the concentration and color of humic substances in some aquatic environments. *Freshwater Biol.*, **14**, 79–87.
98. A. Geller (1986). Comparison of mechanisms enhancing biodegradability of refractory lake water constituents. *Limnol. Oceanogr.*, **31**, 755–764.
99. V. Ittekkot, B. Haake (1990). The terrestrial link in the removal of organic carbon in the sea. In: V. Ittekkot (Ed.), *Facets of Modern Biogeochemistry* (pp. 318–325). Springer-Verlag, New York.
100. J.A. Amador, M. Alexander, R.G. Zika (1991). Degradation of aromatic compounds bound to humic acid by the combined action of sunlight and microorganisms. *Environ. Toxicol. Chem.*, **10**, 475–482.
101. M.A. Moran, R.E. Hodson (1994). Support of bacterioplankton production by dissolved humic substances from three marine environments. *Mar. Ecol. Prog. Ser.*, **110**, 241–247.
102. M. Lindell, H. Edling (1996). Influence of light on bacterioplankton in a tropical lake. *Hydrobiologia*, **323**, 67–73.
103. M.J. Lindell, H.W. Graneli, L.J. Tranvik (1996). Effects of sunlight on bacterial growth in lakes of different humic content. *Aquat. Microb. Ecol.*, **11**, 135–141.
104. W.L. Miller, M.A. Moran (1997). Interaction of photochemical and microbial processes in the degradation of refractory dissolved organic matter from a coastal marine environment. *Limnol. Oceanogr.*, **42**, 1317–1324.
105. R. Benner, B. Biddanda (1998). Photochemical transformations of surface and deep marine dissolved organic matter: Effects on bacterial growth. *Limnol. Oceanogr.*, **43**, 1373–1378.
106. I. Reche, M.L. Pace, J.J. Cole (1998). Interactions of photobleaching and inorganic nutrients in determining bacterial growth on colored dissolved organic carbon. *Microb. Ecol.*, **36**, 270–280.
107. S. Bertilsson, R. Stepanauskas, R. Cuadros-Hansson, W. Graneli, J. Wikner, L. Tranvik (1999). Photochemically induced changes in bioavailable carbon and nitrogen pools in a boreal watershed. *Aquat. Microb. Ecol.*, **19**, 47–56.
108. M.A. Moran, W.M. Sheldon, J.E. Sheldon (1999). Biodegradation of riverine dissolved organic carbon in five estuaries of the southeastern United States. *Estuaries*, **22**, 55–64.
109. I. Obernosterer, B. Reitner, G.J. Herndl (1999). Contrasting effects of solar radiation on dissolved organic matter and its bioavailability to marine bacterioplankton. *Limnol. Oceanogr.*, **44**, 1645–1654.
110. P.A. Raymond, J.E. Bauer (2000). Bacterial consumption of DOC during transport through a temperate estuary. *Aquat. Microb. Ecoli* **22**, 1–12.
111. T.N. Wiegner, S.P. Seitzinger (2001). Photochemical and microbial degradation of external dissolved organic matter inputs to rivers. *Aquat. Microb. Ecol.*, **24**, 27–40.
112. K. Mopper, W.L. Stahovec (1986). Sources and sinks of low molecular weight organic carbonyl compounds in seawater. *Mar. Chem.*, **19**, 305–321.
113. J.-F. Rontani (1991). Identification by GC/MS of acidic compounds produced during the photosenitized oxidation of normal and isoprenoid alkanes in seawater. *Int. J. Environ. Anal. Chem.*, **45**, 1–9.
114. N. Corin, P. Backlund, M. Kulovaara (1996). Degradation products formed during UV-irradiation of humic waters. *Chemosphere*, **33**, 245–255.
115. C.L. Osburn, D.P. Morris, K.A. Thorn, R.E. Moeller (2001). Chemical and optical changes in freshwater dissolved organic matter exposed to solar radiation. *Biogeochemistry.*, **54**, 251–178.
116. M. Kulovaara, N. Corin, P. Backlund, J. Tervo (1996). Impact of UV254-radiation

on aquatic humic substances. *Chemosphere*, **33**, 783–790.

117. R.L. Valentine, R.G. Zepp (1993). Formation of carbon monoxide from the photodegradation of terrestrial dissolved organic carbon in natural waters. *Environ. Sci. Technol.*, **27**, 409–412.
118. K. Salonen, A. Vahatalo (1994). Photochemical mineralization of dissolved organic matter in Lake Skjervatjern. *Environ. Int.*, **20**, 307–312.
119. W. Graneli, M. Lindell, L. Tranvik (1996). Photo-oxidative production of dissolved inorganic carbon in lakes of different humic content. *Limnol. Oceanogr.*, **41**, 698–706.
120. W. Graneli, M. Lindell, B.M. De Faria, F.D. Esteves (1998). Photoproduction of dissolved inorganic carbon in temperate and tropical lakes – dependence on wavelength band and dissolved organic carbon concentration. *Biogeochemistry.*, **43**, 175–195.
121. A.V. Vahatalo, K. Salonen, M. Salkinoja-Salonen, A. Hatakka (1999). Photochemical mineralization of synthetic lignin in lake water indicates enhanced turnover of aromatic organic matter under solar radiation. *Biodegradation*, **10**, 415–420.
122. M.A. Moran, W.M. Sheldon, R.G. Zepp (2000). Carbon loss and optical property changes during long-term photochemical and biological degradation of estuarine dissolved organic matter. *Limnol. Oceanogr.*, **45**, 1254–1264.
123. A.V. Vahatalo, M. Salkinoja-Salonen, P. Taalas, K. Salonen (2000). Spectrum of the quantum yield for photochemical mineralization of dissolved organic carbon in a humic lake. *Limnol. Oceanogr.*, **45**, 664–676.
124. B. Allard, H. Boren, C. Pettersson, G. Zhang (1994). Degradation of humic substances by UV irradiation. *Environ. Int.*, **20**, 97–101.
125. I. Reche, M.L. Pace, J.J. Cole (1999). Relationship of trophic and chemical conditions to photobleaching of dissolved organic matter in lake ecosystems. *Biogeochemistry.*, **44**, 259–280.
126. R.A. Bourbonniere, W.L. Miller, R.G. Zepp (1997). Distribution, flux, and photochemical production of carbon monoxide in a boreal beaver impoundment. *J. Geophys. Res.-Atmos.*, **102** , 29321–29329.
127. D.A. Francko, R.T. Heath (1982). UV-sensitive complex phosphorous: Association with dissolved humic material and iron in a bog lake. *Limnol. Oceanogr.*, **27**, 564–569.
128. K.L. Bushaw-Newton, M.A. Moran (1999). Photochemical formation of biologically available nitrogen from dissolved humic substances in coastal marine systems. *Aquat. Microb. Ecol.*, **18**, 285–292.
129. R.J. Kieber, A. Li, P.J. Seaton (1999). Production of nitrite from the photodegradation of dissolved organic matter in natural waters. *Environ. Sci. Technol.*, **33**, 993–998.
130. S.A. Opsahl, R.G. Zepp (2001). Photochemically-induced alteration of stable carbon isotope ratios (δ^{13}C) in terrigenous dissolved organic carbon. *Geophys. Res. Lett.*, **28**, 2417–2421.
131. E. Zanardi-Lamardo, C.D. Clark, R.G. Zika (2001). Frit inlet frit outlet flow field-flow fractionation: methodology for colored dissolved organic material in natural waters. *Anal. Chim. Acta*, **443**, 171–181.
132. C.L. Osburn, H.E. Zagarese, D.P. Morris, B.R. Hargreaves, W. Cravero (2001). Calculation of spectral weighting functions for the solar photobleaching of chromophoric dissolved organic matter in temperate lakes. *Limnol. Oceanogr.*, **46**, 1455–1467.
133. P.G. Coble, C.E. Del Castillo, B. Avril (1998). Distribution and optical properties of CDOM in the Arabian Sea during the 1995 Southwest Monsoon, **45**, 2195–2223.

134. C.E. Del Castillo, P.G. Coble, J.M. Morell, J.M. Lopez, J.E. Corredor (1999). Analysis of the optical properties of the Orinoco River plume by absorption and fluorescence spectroscopy. *Mar. Chem.*, **66**, 35–51.
135. C.E. Del Castillo, F. Gilbes, P.G. Coble, F.E. Muller-Karger (2000). On the dispersal of riverine colored dissolved organic matter over the West Florida Shelf. *Limnol. Oceanogr.*, **45**, 1425–1432.
136. P. Backlund (1992). Degradation of aquatic humic material by ultraviolet light. *Chemosphere*, **25**, 1869–1878.
137. H. DeHaan (1993). Solar UV light penetration and photodegradation of humic substances in peaty lake water. *Limnol. Oceanogr.*, **38**, 1072–1076.
138. S. Bertilsson, S. Bergh (1999). Photochemical reactivity of XAD-4 and XAD-8 adsorbable dissolved organic compounds from humic waters. *Chemosphere*, **39**, 2289–2300.
139. I. Reche, M.L. Pace, J.J. Cole (2000). Modeled effects of dissolved organic carbon and solar spectra on photobleaching in lake ecosystems. *Ecosystems*, **3**, 419–432.
140. R.F. Whitehead, S. de Mora, S. Demers, M. Gosselin, P. Monfort, B. Mostajir (2000). Interactions of ultraviolet-B radiation, mixing, and biological activity on photobleaching of natural chromophoric dissolved organic matter: A mesocosm study. *Limnol. Oceanogr.*, **45**, 278–291.
141. R.D. Rundel (1983). Action spectra and estimation of biologically effective UV radiation. *Physiol. Plant.*, **58**, 360–366.
142. J.J. Cullen, P.J. Neale (1997). Biological weighting functions for describing the effects of ultraviolet radiation in aquatic ecosystems. In: D.-P. Hader (Ed.), *The Effects of Ozone Depletion on Aquatic Ecosystems* (pp. 97–117). R. G. Landes.
143. C.L. Osburn, R.B. Coffin, T.J. Boyd (2002). Observed variation in the photoreactivity of CDOM from freshwater, estuarine, and marine sources in the Chesapeake Bay. *Eos*, **83**, OS21B-09.
144. G.R. Harvey, D.A. Boran, S.R. Piotrowicz, C.P. Weisel (1984). Synthesis of marine humic substances from unsaturated lipids. *Nature*, **309**, 244–246.
145. L.J. Tranvik (1993). Microbial transformation of labile dissolved organic matter into humic-like matter in seawater, **12**, 177–183.
146. N.O.G. Jorgensen, L. Tranvik, H. Edling, W. Graneli, M. Lindell (1998). Effects of sunlight on occurrence and bacterial turnover of specific carbon and nitrogen compounds in lake water. *Fems Microbiol. Ecol.*, **25**, 217–227.
147. L. Tranvik, S. Kokalj (1998). Decreased biodegradability of algal DOC due to interactive effects of UV radiation and humic matter. *Aquat. Microb. Ecol.*, **14**, 301–307.
148. A.M. Anesio, C.M.T. Denward, L.J. Tranvik, W. Graneli (1999). Decreased bacterial growth on vascular plant detritus due to photochemical modification. *Aquat. Microb. Ecol.*, **17**, 159–165.
149. R.J. Kieber, L.H. Hydro, P.J. Seaton (1997). Photooxidation of triglycerides and fatty acids in seawater: Implication toward the formation of marine humic substances. *Limnol. Oceanogr.*, **42**, 1454–1462.
150. R.J. Chrost, M.A. Faust (1999). Consequences of solar radiation on bacterial secondary production and growth rates in subtropical coastal water (Atlantic Coral Reef off Belize, Central America). *Aquat. Microb. Ecol.*, **20**, 39–48.
151. S. Ziegler, R. Benner (2000). Effects of solar radiation on dissolved organic matter cycling in a subtropical seagrass meadow. *Limnol. Oceanogr.*, **45**, 257–266.
152. M.J. Lindell, H. Graneli, S. Bertilsson (2000). Seasonal photoreactivity of dissolved

organic matter from lakes with contrasting humic content. *Can. J. Fisheries Aquat. Sci.*, **57**, 875–885.
153. S.C. Johannessen, W.L. Miller (2001). Quantum yield for the photochemical production of dissolved inorganic carbon in seawater. *Mar. Chem.*, **76**, 271–283.
154. R.G. Zepp, T.V. Callaghan, D.J. Erickson (1998). Effects of enhanced solar ultraviolet radiation on biogeochemical cycles. *J. Photochem. Photobiol. B-Biol.*, **46**, 69–82.
155. L.A. Molot, P.J. Dillon (1997). Photolytic regulation of dissolved organic carbon in northern lakes. *Global Biogeochem. Cycles*, **11**, 357–365.
156. I. Reche, E. Pulido-Villena, J.M. Conde-Porcuna, P. Carrillo (2001). Photoreactivity of dissolved organic matter from high-mountain lakes of Sierra Nevada, Spain. *Arct. Antarct. Alpine Res.*, **33**, 426–434.
157. C. Gunning, L.A. Molot, P.J. Dillon (2001). Enhanced photo-oxidation of dissolved organic carbon in acidic fresh waters. *Biogeochem.*, **52**, 339–354.
158. C.L. Osburn (2000). *Photochemical Changes in the Dissolved Organic Matter of Temperate Lakes: Implications for Organic Carbon Cycling and Lake Transparency* (100 pp.). Lehigh University, Bethleham, PA.

Chapter 7

Photoactivated toxicity in aquatic environments

Stephen A. Diamond

Table of contents

Abstract

Most aquatic organisms have evolved mechanisms to minimize damage by UVR. Many terrestrial species have additionally had to adapt to plant compounds (e.g., furanocoumarins) that are extremely toxic when activated by UVR. Over evolutionary time, it is unlikely that these compounds have been present in aquatic systems at concentrations sufficient to trigger adaptive responses. Within the last century, however, release of anthropogenic contaminants, particularly polycyclic aromatic hydrocarbons (PAHs), has greatly increased the potential for photoactivated toxicity in aquatic environments. Most phototoxic compounds exert toxicity via a photosensitizing process that produces (within tissues) ROS that ultimately damage biomacromolecules. Some phototoxic compounds, under certain conditions, may exert toxicity after they have been photochemically modified in the external environment. Both mechanisms require sufficient doses of chemical and UVR, particularly UV-A (315 to 400 nm). Assessment of the potential for phototoxic damage in aquatic systems requires thorough analysis of both of these elements, as well as species and lifestage-specific vulnerabilities. Because photoactivated contaminants are present in high concentrations at fairly isolated areas that may function as sinks, and are presently still being introduced into aquatic systems, consideration of phototoxicity will continue to be a significant ecological concern.

7.1 Introduction

Solar radiation is essential for life on Earth. The visible wavelength range, or photosynthetically active radiation (400 to 700 nm), provides the energy necessary for photosynthesis, which maintains oxygen in the atmosphere at levels sufficient for respiration and initiates carbon cycling in the earth biosphere. All wavelengths of solar radiation add heat to the Earth and its atmosphere and maintain temperatures within the ranges required by its biota. The shorter, ultraviolet wavelength ranges of solar radiation interact with molecular oxygen, primarily in the stratosphere, to produce ozone, which in turn protects biological systems in the troposphere from the very toxic UV-B wavelengths (see Chapter 2).

Solar radiation is also very toxic to biological processes. As is discussed thoroughly in these texts UVR can disrupt normal DNA replication and translation processes (e.g., [1–3], Chapter 9, 10), alter the structure and function of other biomacromolecules (e.g., [4–7]), inhibit photosynthesis (e.g., [8–10], Chapter 11), produce alterations in epidermal and other tissues (e.g., [11,12], Chapter 13). And, at the population level, UVR can alter plankton assemblages (e.g., [13–15]), and potentially affect the distribution and survival of other aquatic taxa [13]. Although exacerbated by recent anthropogenic impacts on stratospheric ozone, these processes are natural. Over millions of years of evolution species have developed mechanisms to avoid their damaging effects, at least at natural intensity and dose levels [1,9,13].

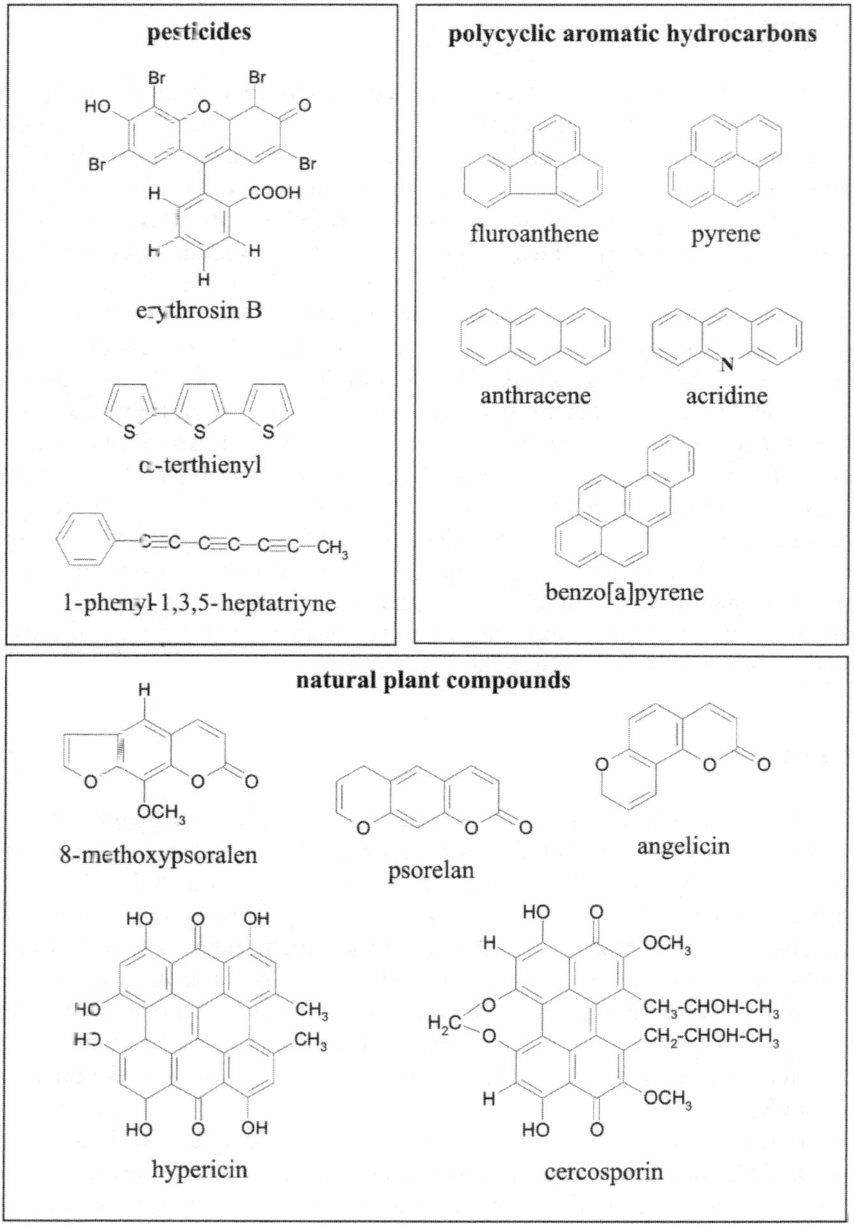

Figure 1. Examples of phototoxic compounds.

Solar radiation can also be harmful to biota via less direct mechanisms; specifically by dramatically increasing the toxicity of many natural and anthropogenic organic compounds [16–18] (see Figure 1). In fact, many species of plants and animals have evolved mechanisms that take advantage of photoactivated toxicity to defend against predators, foragers, and infectious agents [19].

These defense mechanisms involve production of compounds that, once ingested, adsorbed, or absorbed by predators, act as sensitizers of cellular and tissue damage in the threatening species. The numerous defensive compounds produced include acetylenes, benzopyrans and furans, furanocoumarins, and other classes of compounds [18,19]. The evolutionary processes underlying the history of plant photo-defense and pest adaptation is well understood and generally regarded as achieving some level of constantly-shifting balance. The same cannot be said for the phototoxic effects of anthropogenic compounds, including the fairly short-lived pesticide α-terthienyl, organic dyes, and PAHs.

The PAHs are of particular concern in aquatic environments. They have accumulated in many locations to concentrations in sediment and water that are well above those required to cause significant phototoxicity when tested in the laboratory. The PAHs also, in nearly every contaminated site, consist of hundreds of unsubstituted and variously substituted compounds that differ in their capacity for photoactivation, uptake, degradation, and environmental modification, including potential photo-modification to more toxic products [16,20–24]. The complexity of these mixtures makes predictions of phototoxicity risk a uniquely site-specific task, relative to other contaminants. PAHs are of ongoing concern because they still are introduced into surface waters by urban and industrial runoff, petroleum releases, and aerial deposition [22,25]. Most PAHs present in aquatic systems are also relatively recalcitrant to environmental degradation, and are bioaccumulative, having $\log K_{ow}$ values ranging from 2 to 7. The breadth of the PAH contamination problem is clear from the fact that PAHs are significant contributors to the contamination at over 60% of the United States Environmental Protection Agency's National Priorities List of SuperFund cleanup sites [http://www.atsdr.cdc.gov/tfacts69.html].

7.2 Terminology

The increase in toxicity of compounds in the presence of UV or visible radiation has been variously termed photoactivation, photoinduction, photosensitization, photodynamic action, phototoxicity, or combinations of these terms [16,17,26]. Largely due to the history of the science, the most consistent terminology is that photodynamic photosensitization, or photosensitized photodynamic action, refers very specifically to conditions where a sensitizing chemical is present, and toxic action requires the presence of oxygen [27]. A sensitizer is any chemical that responds to photons, and acts as a receptor for the transfer of that energy into a chemical or biological system. Not all photosensitized toxicity necessarily involves photodynamic action. While empirical evidence abounds for oxygen-associated photosensitization, there is also a strong theoretical basis for direct interaction of excited state sensitizers with biomacromolecules (e.g., intercalation of PAH with DNA [28]). There is also a strong theoretical basis, as well as some empirical evidence, for photon-mediated enhanced toxicity resulting from modification of parent compounds to more toxic products via photodegradation processes (which may involve oxygen) [21,23,29,30]. It has been theorized also that metals in aquatic systems that are maintained in relatively non-toxic, ligand-bound states, may be released from those complexes upon irradiation by solar radiation. In fact, this is a photosensitized reaction; as it is the ligand itself (e.g., dissolved organic carbons) that is the receptor of the photon energy which releases metal from the ligand–metal complex. Some photosensitized toxicity mechanisms do not directly photochemically involve the compounds that trigger the effects. For example, the disease porphyria, the accumulation of elevated levels of heme precursors, can be triggered by pesticides and pharmaceutical compounds that are not sensitizers themselves [8]. In this case, it is the heme precursors that act as photosensitizers once they achieve sufficient concentrations in irradiated tissues, primarily the epidermis.

Rather than attempt to incorporate complex terminology that would be precisely descriptive of each mechanism, I will use the very general terms photoactivated toxicity or phototoxicity throughout this chapter. Only where particular mechanisms have been identified, or are of importance to the discussion, will the more specific terms just discussed be applied, generally where photodynamic photosensitization or photomodified toxicity has been demonstrated.

7.3 Historical perspective

The fact that the chemical activity of many compounds is greatly increased in the presence of solar radiation has been recognized for millennia. Arguably, the first recorded documentation of photoactivation can be found in Egyptian and Indian writings dating as far back in history as 2000 BC, where application of the sap of plants such as false Bishop's Weed (*Ammi majus*, Umbelliferae), *Psorlea coryligolia*, and others, followed by immediate exposure to solar radiation, was recommended as a treatment for vitiligo [31,32]. The active ingredient present in the sap of these plants used to treat vitiligo was identified by Fahmy and Abu-Shady [33] as 8-methoxypsoralen. Since that time, numerous additional psoralens and other allelochemicals have been identified in a broad range of plant species, including figs, lemons, limes [34], and certain oranges (*Citrus bergamia*) [35], celery [36], and other species too numerous to list [37]. In many cases, the discovery and isolation of these compounds has led to their use in a variety of phototherapies for the treatment of psoriasis, eczema, some forms of cancer, and other afflictions [32].

As well as these beneficial (either for the plant or for humankind) uses of photoactivation of plant compounds, there are equally numerous and varied examples of harmful phototoxic responses to plant compounds. Many plants potentially ingested by livestock can initiate photosensitizing responses [38]. In particular, plants that produce sufficient concentrations of furanocoumarins, hypericin, and cercosporin have been documented to affect livestock in this manner [39]. A specific example of non-target species photosensitization by plant compounds is the dramatic erythemic response of agricultural workers exposed to the sap of celery (*Apium graveolens*) plants during harvest [40].

Extreme cases were first reported in celery strains bred specifically for their resistance to fungus (pink-rot disease). The compounds produced trigger fatal phototoxicity in pest species, as well as severe photodermatitis in field workers [41]. It has also been well documented that celery plants stressed by fungal invasion produce elevated levels of these compounds (3 to 30 times higher than normal) [42]. Cases of photodermatitis triggered in humans by ingestion of celery (e.g., [36]) and other plants [43] are also occasionally reported.

The number of photoactive compounds present in the environment has expanded considerably due to the activities of man; Santamaria and Prino [18] listed 380 compounds, including many that were anthropogenic, over three decades ago (examples of phototoxic compounds are illustrated in Figure 1). Several insecticides, either synthesized or refined from naturally-occurring compounds, have been produced and used in commercial agriculture e.g., organic dyes including erythrosin-B, registered for control of houseflies in chicken farming [44], and alpha-terthienyl, used extensively in mosquito control [45]. Additional photoactive compounds produced for purposes other than their phototoxic potential include many pharmaceuticals (e.g., tetracycline antibiotics and phenothiazine tranquilizers), organic dyes, and PAHs, which are present at some level in all petroleum products, coal gas, creosote, soot, and numerous other anthropogenic chemicals [27].

These early historical examples of phototoxicity were of terrestrial origin and occurrence. The potential for photoactivated toxicity in aquatic systems was not considered until the 1900s, probably because exposure to natural compounds described and prescribed by ancient herbalists and naturalists occurred terrestrially, and did not have an aquatic counterpart. Most natural compounds do not accumulate in aquatic systems because they are either rapidly degraded once free of plant tissues, do not enter aquatic systems, or are rapidly diluted if they do. The potential for photoactivated toxicity increased dramatically as PAHs, photoactivated pesticides, and other potential sensitizers were released into the environment and, in some cases, accumulated in high concentrations in aquatic systems.

That phototoxicity might be a concern in aquatic systems was first suggested by the work of Jodlbauer and Tappeiner [46] who demonstrated that anthracene was phototoxic to *Paramecia* (reference by Santamaria and Prino [18], see Table 1). Later, Mottram and Doniach [47,48] and Doniach [49] studied the photoactivation of several additional PAHs in *Paramecia*, with the specific goal of comparing the compound's potential for both carcinogenicity and phototoxicity. It is interesting to note that these researchers were the first to incorporate various controls to demonstrate that the compounds were not toxic in the absence of UVR, that irradiation of the exposure media prior to exposure did not increase toxicity, and that increasing the duration of the uptake period prior to UV exposure increased toxicity. These latter two points illustrated that, at these exposure levels, photoactivated toxicity occurred within the organism, rather than in the exposure matrix. Prior to the 1970s, most research on phototoxicity in aquatic organisms focused on the relationship between carcinogenicity and phototoxicity and was driven by the assumption that testing for the latter effect was much simpler than testing for the former, and could be used for rapid screening. *Paramecia* remained the model organisms in studies that both elucidated this relationship, and expanded the list of compounds known to be phototoxic [50–56]. In work discussed at length later, Morgan and Warshawsky [57] demonstrated that several PAHs were phototoxic to brine shrimp (*Artemia salina*).

The specific organisms used in these studies were selected for practical experimental expediency in addressing critical medical questions, rather than because they were aquatic, or because their response to exposure might represent potential effects in aquatic systems. The realization that phototoxicity could be of significant concern specifically in aquatic systems was made serendipitously by Bowling et al. [58] during anthracene fate and effect studies with fathead minnows (*Pimephales promelas*) conducted in natural radiation in 1980 and 1981. This work clearly demonstrated that solar radiation exposure dramatically increased the toxicity of anthracene, most notably at concentrations well below those that had caused mortality in laboratory exposures (in the absence of UVR). These exposures were conducted in outdoor troughs maintained at the Savannah River site (Aiken, SC). This system provided an elegant way for the authors to largely disprove the hypothesis that photomodification of anthracene dissolved in exposure water could cause toxicity. Fish held in shade downstream of unshaded sections were not affected, whereas fish held in full solar radiation downstream of shaded sections were. Also, when fish were allowed to depurate anthracene prior to solar radiation exposure, they were not affected. Although not definitive, these results strongly suggested that the primary mechanism of toxicity was excitation of chemical present in tissues, rather than photo-modification of external compound. Also, this was the first report of phototoxicity in fish or other aquatic vertebrates.

Subsequent to Bowling et al.'s [58] work, researchers have quantified or characterized the toxicity of numerous PAH and other mostly anthropogenic compounds, elucidated the chemical mechanisms underlying photactivated toxicity, and addressed several of the components necessary to begin ecological risk assessment for these effects in nature. In addition to the PAHs, the pure compounds α-terthienyl (e.g., [59,60]) and phenylheptatriene [61], 2,4,6-trinitrotoluene, dinitrotoluenes, diaminotoluene (and several of their metabolites) [62–64], and carbaryl [65] (although Wernersson [66] found no activation) have been demonstrated to be phototoxic in aquatic environments. In addition, complex mixtures (containing primarily PAHs) present in petroleum products [67–69] and various sediments [70–73] have been shown to be phototoxic. The dependence of oxygen on the phototoxicity process has been demonstrated numerous times (e.g., [61,74–77]) and factors (in addition to oxygen) that might ameliorate phototoxicity, including dissolved organic carbon [78,79], β-carotene [80], and turbidity [73], have been studied, and the sensitivity of early lifestages has been investigated [81–83].

In general, the research to date has defined the potential for phototoxicity of specific compounds to individual species. Several authors have discussed the importance of considering the environmental and ecological factors that mitigate these conditions, including photoperiod, UV dose, and the spectral characteristics of solar radiation in specific habitats [75,84–91] and potential adaptation in exposed populations [92,93].

7.4 Mechanisms of action

The chemical/physical process common to all photoactivated toxicity events is the absorption of photon energy (Figure 2A), generally in the 280 to 400 nm wavelength range, by a sensitizer molecule (e.g., PAH) [94,95]. The energy absorbed results in promotion of electrons from their ground-state orbitals to

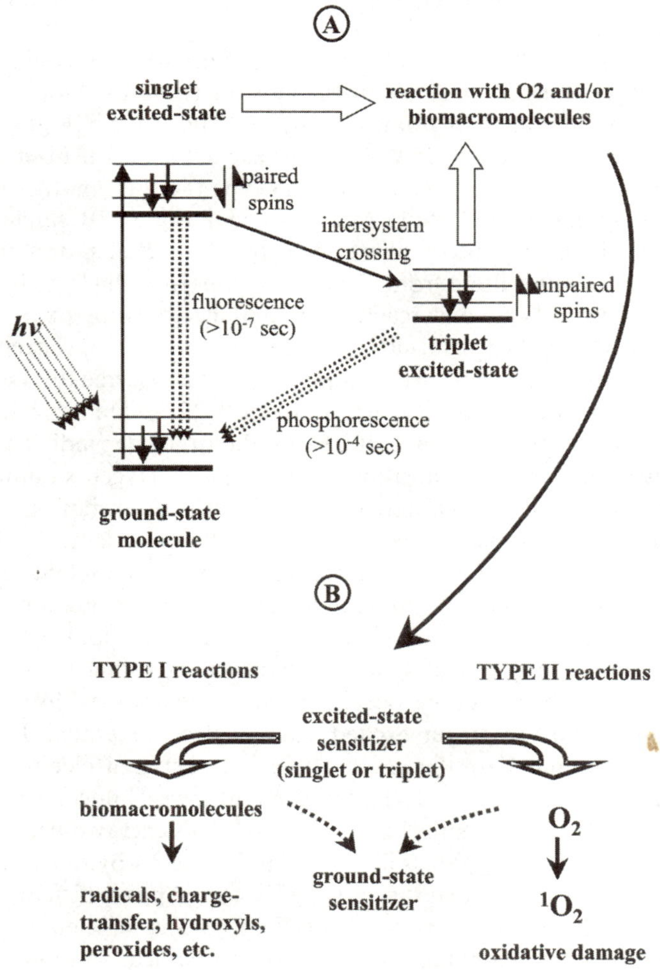

Figure 2. Photochemical processes involved in photoactivated toxicity. Generation of excited states and pathways for their decay are illustrated in (A). Possible mechanisms underlying toxic activity are illustrated in (B).

excited-state orbitals. The excess energy is then dissipated in radiationless transitions from excited to ground rotational and vibrational electron levels, radiative transitions from excited-state orbitals to ground-state orbitals (fluorescence or phosphorescence, depending on whether triplet state intermediates occur via intersystem crossing), or direct energy transfer from the sensitizer to other molecules present in the biological matrix. Biologically damaging events occur during these sensitizer transitions from excited to ground state. Although not generally considered to be a major component of environmental aquatic phototoxicity, there exists the potential for solar radiation activated modification of non-toxic parent compounds to more toxic photodegradation products (e.g., quinones and other oxygenation products). In this chapter, unless otherwise noted, discussion will be limited to photosensitization reactions, rather than photomodification reactions.

The photosensitization process follows two possible pathways, conventionally referred to as Type I and Type II reactions following the terminology suggested by Gollnick [96] and refined by Foote [97,98], and illustrated in Figure 2B. The two pathways are differentiated by whether the excited-state sensitizer molecule transfers energy directly to molecular oxygen (Type II) or to another molecule within the biological matrix (Type I). As illustrated in Figure 2B, singlet oxygen is the primary damaging intermediate in Type II reactions, and may also contribute to damage resulting from Type I reactions as well. Type I reactions produce molecular radicals and reactive oxygen species (superoxide radicals, peroxides, hydroxyl radicals) formed during interaction of the excited-state sensitizer and other constituents of the biological matrix (referred to as solvent or substrate in non-biological photochemical systems). These are frequently competing reactions, the predominance of one or the other depending upon the specific sensitizer and its concentrations, the extent of oxygen saturation, the nature of the biological matrix, and the wavelengths of excitation radiation present in the system [74,98]. For example, given sufficient quantities of oxygen (which favors a Type II pathway), a matrix of greater lipid content relative to higher water content is likely to produce higher substrate oxidation rates because singlet oxygen lifetimes are significantly longer in lipid-rich reaction systems. Additionally, experimental evidence indicates that singlet oxygen quantum yields are greater for reactions occurring within membranes [99].

The probability that sensitizer excited states will be generated during UV irradiation is determined by the energy difference between electrons in the highest occupied molecular orbital (HOMO) and the lowest unoccupied molecular orbital (LUMO) in the sensitizer molecule [95]. This energy difference, often termed the HOMO–LUMO gap, is illustrated in Figure 2A by the dark lines in the ground and excited-state diagrams. In PAHs, this gap is sufficiently narrow so that the relatively low energy present in UV photons is sufficient to promote electrons from occupied, bonding π orbitals to unoccupied, antibonding π orbitals (π to π^* transitions), thus initiating the photoactivation process. This relatively narrow HOMO–LUMO gap is characteristic of aromatic systems owing to the extensive π conjugation extent in the benzene constituents. In many heterocyclic and substituted PAHs, electron transitions from non-bonding 2p orbital to σ^* or π^* orbitals may also contribute to their potential phototoxicity. The HOMO–LUMO gap has been used in several QSAR (quantitative structure–activity relationship) analyses (discussed later) to predict whether specific PAHs are potentially photoactivated or non-photoactivated.

There are several aspects of photoactivated toxicity yet to be thoroughly elucidated. For most compounds, it has not been established clearly whether initial singlet excited states decay to excited triplet states prior to energy transfer to oxygen or substrate. This is not a trivial question, as the probability of singlet oxygen production increases by orders of magnitude when the sensitizer passes through a triplet excited state. This increased probability is due to the large difference in lifetimes for singlet (ns) and triplet (μs to ms) excited states. For most compounds, it also has yet to be established whether photoreactions occur primarily via Type I or Type II reactions.

Regardless of the exact excited-state processes involved, photoactivated damage to biological systems appears to proceed primarily via oxidative damage. Studies at the molecular level, as well as empirical whole-organism or whole-tissue studies, indicate that exclusion of oxygen from the experimental system or the presence of singlet oxygen quenchers (e.g., β-carotene, α-tocopherol, etc.) greatly, or entirely, eliminates photoactivated damage. As well as singlet excited-state oxygen, other reactive species may also be produced via Type I pathways where sufficient oxygen is present. Choi and Oris [104] recently demonstrated very clearly that simultaneous exposure of fish liver microsomes to PAH and UVR resulted in oxidative stress, specifically lipid peroxidation resulting from formation of superoxide anion. Other highly toxic species include oxygen-free radical, hydroxyl radicals, and peroxides, all of which have been demonstrated to disrupt cellular membranes, amino acids, DNA, and other cellular and tissue components [101].

7.5 Predicting phototoxicity

As was stated earlier, PAHs are among the more problematic contaminants, relative to potential environmental phototoxicity, because they generally occur in contaminated aquatic systems as complex mixtures. Thousands of unsubstituted and variously substituted PAHs have been identified as contributing to environmental contamination [22,102]. As it is untenable to test each of these possible compounds to determine the extent of their photoactivated toxicity, a significant effort has been made by various researchers to develop predictive models to address this issue. An interesting aspect of phototoxic chemicals is the clear relationship between their phototoxic potency and their carcinogenic potential. Some of the earliest efforts to predict phototoxicity potential were based on these relationships, although much effort was directed at accomplishing the opposite, that is predicting carcinogenicity based on phototoxic potential. The elucidation of the carcinogenicity–phototoxicity relationship facilitated development of QSAR models. Based on this relationship, it was considered parsimonious (and ultimately correct) to use the molecular parameters proven to be significant in carcinogenicity QSARs when developing phototoxicity QSARs.

The earliest studies suggesting a carcinogenicity–phototoxicity relationship were conducted by Mottram and Doniach [47,48], Doniach [49], and Calcutt [50] and suggested that many chemicals capable of causing tumors were also phototoxic. It should be noted though that these authors all suggested that, for some of these compounds (e.g., the arsenicals), the toxicity in the presence of light was mechanistically different than for the classic photosensitizers such as benzo[*a*]pyrene. This is an important distinction for the development of QSARs, as phototoxicity resulting from photodynamic action is likely to have a very different molecular basis relative to other modes of action (e.g., the photodermatitis associated with porphyria mentioned earlier).

Epstein et al. [56] tested the carcinogenicity and phototoxicity of 157 compounds spanning a broad structural range and including substituted and unsubstituted linear and cyclic compounds. The relationship between carcinogenicity and phototoxicity was apparent. Most notable was the fact that a larger portion of compounds that exhibited high phototoxic potency were carcinogens, compared to non-carcinogens. Rather than being diagnostic, these results suggested that potential for carcinogenicity and phototoxicity may have similar molecular foundations. Santamaria [103] also compared carcinogenicity and phototoxicity potentials of 35 PAHs in isolated mitochondria, and reported a significant correlation. Epstein [56] suggested, tested, and rejected the hypothesis that formation of charge-transfer complexes was a common molecular activity responsible for both carcinogenicity and for phototoxicity, a finding that caused other researchers to investigate alternative molecular explanations for both.

Morgan et al. [57] and Morgan and Warshawsky [102] linked several molecular parameters to carcinogenicity, and, based on the relationship between carcinogenicity and phototoxicity, suggested that these same parameters would be predictive of phototoxic potential. They determined energy levels for lowest singlet and triplets states, singlet–triplet splitting energy, and phosphorescence lifetimes for 18 carcinogens and 31 non-carcinogens. Of the parameters examined, excited singlet-state energy was highly significantly correlated, and singlet–triplet splitting energies significantly correlated with carcinogenicity, but triplet-state energy and phosphorescence lifetime were not. Compounds with singlet energies within the range of 297 to 310 kJ mol^{-1} were 22.8 times more likely to be carcinogenic. When compounds were plotted against singlet-state energy and singlet–triplet-state energy, carcinogens were clustered into a clearly defined ellipse, indicating that these two parameters could successfully discriminate between the carcinogenic and non-carcinogenic compounds examined.

Morgan and Warshawsky [57] related carcinogenicity, and indirectly the parameters used to predict it, to phototoxic potential by completing phototoxicity assays with brine shrimp (*Artemia salina*) nauplii. The greatest contribution of this work was the incorporation of quantum yields to accessing phototoxic potential, or potency. The authors accomplished this by testing each compound with essentially identical organisms and light levels. The data were then used to develop relative photodynamic activity (RPA) value, which could then be compared to carcinogenicity potential for these same compounds. RPA was calculated by first characterizing the relationship between the average number of nauplii immobilized (ANI) and several exposure parameters:

$$\text{ANI} = (I_0 l \alpha \phi_I)(2.303 \varepsilon C t) + B$$

where I_0 = light intensity, l = nauplii path-length, α = proportion of PAH absorbed by nauplii, ϕ_I = quantum yield for immobilization, ε = molar absorptivity of the PAH, C = exposure PAH concentration, t = duration of light exposure, and B = an integration constant. The RPA was then determined for each PAH tested:

$$\text{RPA} = \frac{I_0 l \alpha \phi_I}{I_0 l^r \phi_I^r} = \frac{\alpha \phi_I}{\alpha^r \phi_I^r}$$

where ϕ_I and ϕ_r^I are quantum yields for the each PAH and the reference PAH (benz[*c*]acridine), respectively, and α and α^r are molar absorptivities for each PAH and the reference PAH, respectively. A graphical representation of RPA estimation is shown in Figure 3. One limitation of this approach is that variability in the bioconcentration factors (BCF) for the examined compounds could significantly affect RPA, but was not accounted for. Thus Morgan and Warshawsky's [57] estimation of quantum efficiency was based on whole organism response, rather than a photochemical property specific to the compound itself. Regardless of this deficiency, their results were adequate to demonstrate a strong relationship between carcinogenic potential and phototoxicity, and to suggest

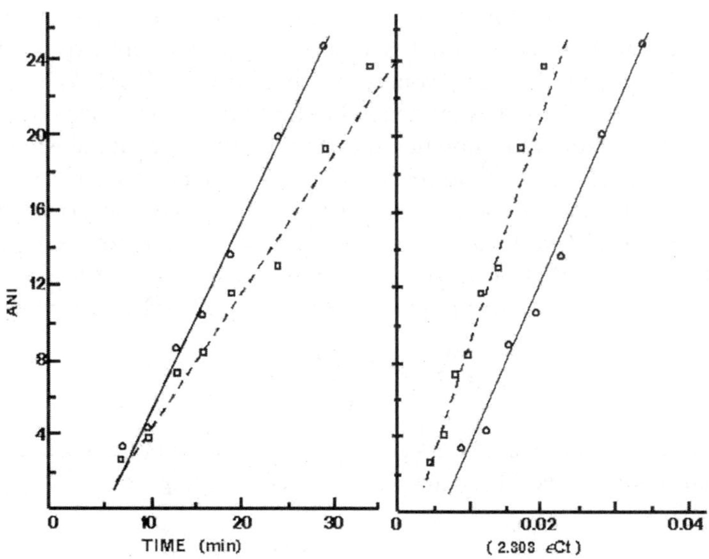

Figure 3. (A) Average number of nauplii immobilized (ANI) as a function of time. (B) Average number of nauplii immobilized as a function of (2.3034 ∊ *Ct*). Benz[*c*]acridine, □—, 85.7 nM; benzo[*a*]pyrene, ○—, 22 nM. [Data from [57], with permission of the publisher]

that the molecular parameters related to the former would also be related to the latter.

The predictive and QSAR models that were developed to predict phototoxicity have been based on the molecular characteristics that describe the probability that electrons will reach excited states when they interact with photons, and the pathway of decay to the ground state condition that is most likely to occur. As described previously, this latter question addresses the disposition of excess energy, and whether the decay processes are of sufficient duration to allow for energy, or electron, transfer among substrate or oxygen molecules. The characteristics examined in these modeling efforts include energy of lowest singlet and triplet states, HOMO–LUMO gap, energy of singlet–triplet interconversion, molecular connectivity, phosphorescence lifetime, and various parameters that describe molecular conformation and stability.

Newsted and Giesy [103] extended the work of Morgan and Warshawsky [102] by comparing the results of toxicity tests (*Daphnia magna* LT50) of 20 PAHs with several of their molecular parameters to determine which would best predict phototoxicity. These parameters included lowest-energy singlet and triplet states, singlet–triplet splitting energy, phosphorescence lifetime, and first- and second-order connectivity. These parameters, except for molecular connectivity, have been discussed in the section describing photosensitization mechanisms. Molecular connectivity describes molecular structure based on physical three-dimensional placement of skeletal atoms and their valence electrons, and molecular density, and has been correlated with bioconcentration potential and toxicity [106]. Following the assumption accepted by most phototoxicity researchers, that PAHs in organism tissues are responsible for toxicity (as opposed to those in the water column), Newsted and Giesy [103] attempted to conduct assays at equivalent molar tissue concentrations for each PAH. Where these attempts failed, they adjusted estimated lethal times based on tissue concentrations, resulting in estimates of potency they termed the median adjusted lethal time. Their modeling effort involved estimation of potency (ϕ for each PAH, quantification of the photon energy reaching the tissue PAHs, the molar absorptivity of each PAH, and values for each of the molecular parameters. Ultimately, an RPA, analogous to that of Morgan and Warshawsky [102], was calculated for each PAH as follows: First, a potency value for each PAH was calculated:

$$\phi = \frac{\mathrm{d}[\%\text{mortality}]/\mathrm{d}t}{I_\mathrm{a}}$$

where t = time and I_a = the rate of UVR absorbtion (quanta per minute) expressing I_a in quantifiable units. This equation was then rearranged:

$$\frac{\mathrm{d}(\%\text{mortality})}{\mathrm{d}t} = \frac{\sum_{\lambda}^{n} [(I_\lambda T_\lambda)(\varepsilon_\lambda b C_a)]}{n}\phi = \bar{A}\phi$$

where, λ is the integrated waveband (e.g., UV-A 315 to 336 nm), I_λ = the intensity of the waveband, T_λ, b, and C_a = optical transmittance, path length, and molar

PAH concentration in the organism, $\bar{A}$ = average quanta absorbed, n = number of λ summed. Integration of this equation yields:

$$d(\%\text{mortality}) = \bar{A}\phi t + B$$

which expresses the linear relationship of % mortality and exposure duration, with $B = y$-intercept, and slope = $\bar{A}\phi$t. Finally, RPA was calculated as:

$$\text{RPA} = \frac{\phi'}{\phi}$$

where ϕ' and ϕ are the potency for each PAH and for the reference PAH, benzo[*b*]anthracene, respectively.

Using linear regression analysis of the relationship between RPA and the various molecular parameters, the authors found that phosphorescence lifetime explained the greatest proportion of the variation, and that all other parameters, when added to the model, increased the regression R^2, but did not reduce residual variance. Curve–linear modeling produced a parabolic relationship with triplet-state energy providing the best fit (Figure 4). Finally, the authors successfully used principal-component analysis to cluster the 20 PAHs into three groups defined as very toxic, moderately toxic, and nontoxic. Discriminant analysis of these results indicated that phosphorescence lifetime and lowest triplet state energy were the parameters best able to reliably achieve these groupings.

Mekenyan et al. [107] and Newsted and Giesy [103] reanalyzed these data in an effort to develop a QSAR model that would predict phototoxicity using calculated ground-state molecular parameters, rather than excited-state parameters which are more difficult to compute. This approach was intended to predict toxicity where empirical data, e.g., measured triplet-state energy, were not available. The basis of Mekenyan et al.'s [107] modeling effort was the observation that the parabolic relationship revealed in Newsted and Giesy's [105] analysis might indicate multiple, competing processes, including internal factors (molecular parameters) and external factors (chemical and UV dosimetry). The molecular, internal parameters chosen for consideration included HOMO–LUMO gap and molecular stability.

Mekenyan et al.'s [108] results indicated that the HOMO–LUMO gap was a suitable ground state predictor of PAH phototoxicity and accurately placed the PAHs examined (the 20 studied by Newsted and Giesy [103]) into toxic or non-toxic groups (Figure 5). Adding to the significance of this work was the consistent relationship between the calculated HOMO–LUMO gap and the singlet and triplet state energies of the PAHs examined. This is significant, as the HOMO–LUMO gap is directly related to the probability that electrons will achieve the excited states necessary to initiate toxicity, but is not functionally related to the fate of that excess energy (a direct determinant of ultimate phototoxic potency), as were the parameters used by Newsted and Giesy [103]. As pointed out by Mekenyan et al. [108], the consistent relationship between these two parameters enables their QSAR approach to be fully consistent with

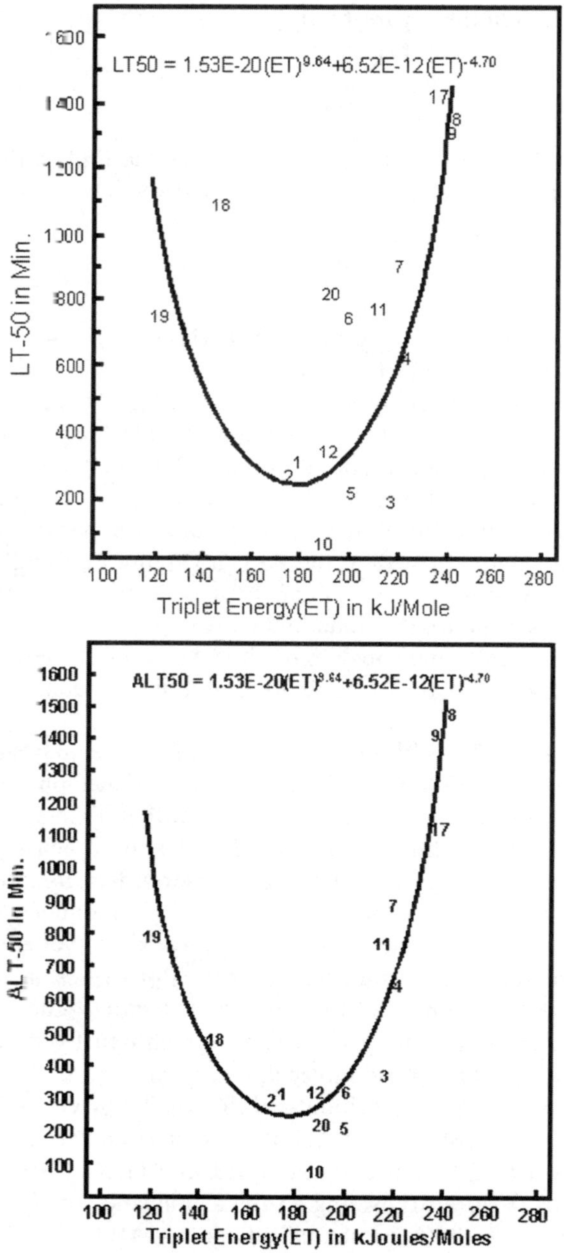

Figure 4. Top: Median lethal time (LT50) as a function of lowest triplet energy (ET). Bottom: Adjusted median lethal time (ALT50) as a function of lowest triplet energy. The polycyclic aromatic hydrocarbons are identified by number. [Data from [103], with permission of the publisher]

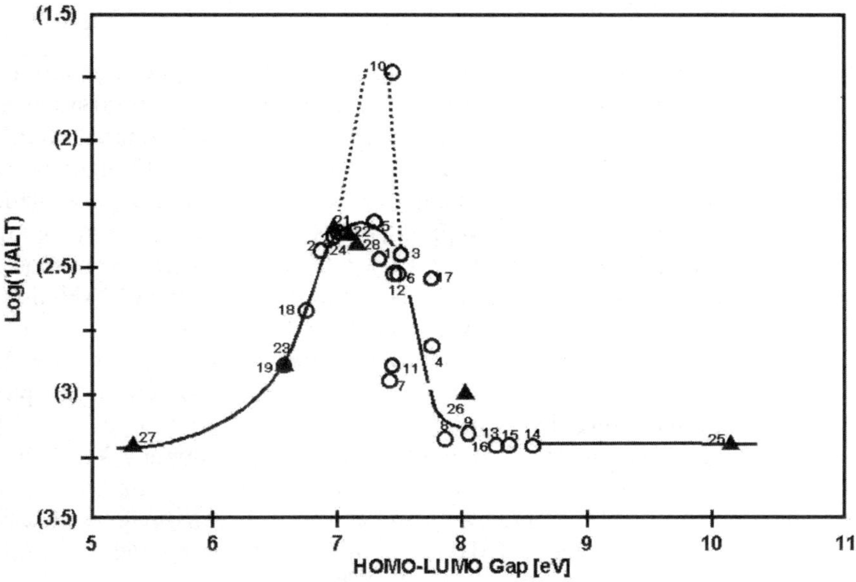

Figure 5. Variation of toxicity (log(1/ALT)) with HOMO–LUMO gap. The solid triangles denote the predicted toxicity values for some as yet untested PAHs. [Data from [108], with permission of the publisher]

Newsted and Giesy [103], and yields a predictive tool that can be used in the absence of empirical data.

Veith et al. [109,110] extended the work of Mekenyan et al. [107,108] by demonstrating that HOMO–LUMO gap energies are also excellent predictors of phototoxicity of various α-terthienyls. Even though these toxicity data did not allow for potency estimates to be corrected for tissue concentrations (as were the data used in the previously described QSAR studies), the compounds studied were accurately predicted to be toxic or non-toxic. This is notable because, although of relatively high phototoxicity potency, these α-terthienyls are chemically significantly different (e.g., they contain linked cyclopentane with sulfur substitutions rather than fused cyclohexane) than the PAHs used in Mekenyan et al.'s [107,108] initial work. Veith et al. [109,110], in a separate analysis, calculated HOMO–LUMO values for pyrene and anthracene having methyl, tert- and n-butyl, ethylene, propylene, nitro, hydroxy, and chloro substitutions. The systematic selection of these substitutions allowed the authors to make broad conclusions regarding the nature and extent of the shift in HOMO–LUMO values associated with each. In general, alkyl and hydroxyl substitution did not significantly shift the HOMO–LUMO values, whereas nitro, alkene, and chloro substitutions did. The authors point out that other factors, such as bioaccumulation and environmental half-life, will also change with substitution, and will complicate predictions of phototoxicity for these compounds in natural waters. A specific test of these model results has yet to be reported.

7.6 Photomodified toxicity

The majority of phototoxicity research in aquatic systems has been undertaken with the assumption that the primary mechanism of concern in natural systems is photosensitized photodynamic toxicity, rather than photomodified toxicity [16]. However, the work of Huang et al. [23] and Ren et al. [111] suggests that photomodification of PAHs may increase their toxicity to some aquatic plants. The approach in these studies was to photomodify PAHs in solution by treating them with UVR and then to test the toxicity of the photomodified solutions by adding them to plant growth media (Hutner's medium). Typically, PAH dissolved in 0.1% solutions of DMSO in water were irradiated at 25 to 40 μmol m^{-2} UV-B for time periods ranging from 6 h to 96 h, and then diluted to concentrations ranging from 0 to 2 mg l^{-1} in the growth medium. Based on a typical terrestrial solar spectrum, these UV-B exposure values correspond to an approximate range of 950 to 1500 μW cm^{-2}; a range approximately 3 to 5 times greater than typical in terrestrial radiation. These estimates are from three separate sources: data from mid-day summer spectral scans made in coastal California [88], similar spectral data available for Daytona Beach (source: [106]), and solar spectra generated using the SBDART [113] model discussed in the UV dose section. Acute growth effects levels (50%) for photomodified PAHs ranged from 500 to 2000 μg l^{-1} for anthracene, benzo[*a*]pyrene, fluoranthene, naphthalene, phenanthrene, and pyrene. Thresholds for effect ranged from 50 to 500 μg l^{-1}. Huang et al. produced similar results for five PAHs irradiated in natural radiation. In this study, IC_{50} estimates for PAHs irradiated for 7 or 20 days ranged from 0.2 to 2.8 mg l^{-1}, and were consistently lower for the longer irradiation durations. These natural solar radiation exposures demonstrate that reciprocity between exposure intensity (relatively high in Huang et al.'s [23] former work) and duration of irradiation (longer in this study) must be considered when evaluating effect levels.

The investigation of toxicity of photoproducts was extended by Marwood et al. [114] who exposed Lake Erie phytoplankton to anthracene and one of its primary photoproducts, 1,2-dihydroxyanthraquinone. When exposures were conducted in solar radiation, 200 μg anthracene l^{-1} caused a 50% inhibition in photosynthesis. At concentrations of 2000 μg l^{-1}, the photoproduct 1,2-dihydroxyanthraquinone reduced photosynthesis slightly when exposures were conducted in the dark, and by 50% when exposures where conducted in solar radiation. While this research was designed primarily to evaluate the usefulness of specific chlorophyll fluorescence parameters as indicators of toxic effects in phytoplankton, the results suggest that anthracene photoproducts are toxic to natural phytoplankton assemblages only at concentrations orders of magnitude greater than those occurring in surface waters. Although sensitized toxicity of anthracene was apparent in these tests, it too occurred at very high concentrations.

The potential for photoinduced toxicity (combined photosensitization and photomodification processes) in plants has been QSAR modeled by Krylov et al.

[115]. Their model, which is far too complex to present in this chapter, includes parameters for rates of photomodification of 16 PAHs, their relative photomodified toxicity, rates of uptake into leaf tissues, solar flux, triplet-state formation of intact PAHs, rates of production of modified in-situ biomacromolecules via both sensitization reactions and direct interaction of PAH photoproducts. Unlike the QSAR models discussed previously [115], this model incorporated kinetic parameters for mechanism-specific rates of deactivation of a model photosynthetic molecule (G) via type I and II photoreactions of intact PAH, as well as by PAH photoproducts. This component of the model was developed to incorporate the growth endpoints reported by Huang et al. [23,29] and Ren et al. [111]. Krylov et al. [115] confirmed the consistency of this modeling approach by completing toxicity tests of 16 intact and photomodified PAHs on *Lemna gibba*. The results indicate that, for the relatively high-concentration exposures required for toxicity of PAHs in duckweed, toxicity is best predicted by an additive model that combines both photomodified and photosensitized mechanisms of action. Additional research into photomodified PAH toxicity in *Lemna gibba*, closely related to the studies described above, includes development of a QSAR model that incorporates shape parameters [116] and identification of specific anthracene photoproducts and their toxicity [117].

Additional evidence for photomodified toxicity in plants is provided by Wiegman et al.'s azaarene phototoxicity work with the diatom species, *Phaeodactylum tricornutum* [25,93,118]. Wiegman et al. irradiated azaarenes with environmentally-realistic intensities of UVR. The EC_{50} values (for reduced photosynthesis) for quinoline, isoquinoline, acridine, and phenanthridine were reduced when exposure solutions were irradiated prior to the introduction of diatoms. The reduction of EC_{50} concentrations ranged from a factor of three to a factor of 300. Effect concentrations for these azaarenes ranged from approximately 230 $\mu g\, l^{-1}$ to 1 $mg\, l^{-1}$.

Although the concentrations of PAH in water or growth media were relatively high in most of these studies, irradiance doses were not. Even where irradiance levels were several times higher than natural radiation, the duration of irradiation was relatively short compared to natural settings where PAH contaminated sediments are exposed continuously to solar UVR during daylight hours. These studies were also completed using simple, single-chemical exposures, a condition rarely, if ever, encountered in contaminated systems. It is reasonable to assume that the complex mixtures of PAH (and other compounds) present in most contaminated sediments consist of some compounds that have the potential for photomodified toxicity. The presence of natural and anthropogenic organic material in contaminated sediments also constitutes a chromophore-rich environment where a variety of photosensitized reactions could produce toxic photomodified products at rates and concentrations similar to those used in the studies just discussed. These studies elucidate the hazard represented by photomodified toxicity, and although they do not clearly demonstrate risk, they do indicate that the potential for photomodification in contaminated locations warrants further research.

7.7 UVR exposure

The first law of photodynamics [94,95] (only absorbed wavelengths have the potential to activate photochemical processes) suggests that PAH phototoxicity will not occur in the absence of UVR, specifically wavelengths from 280 nm to about 400 nm. Of this wavelength range, the UV-A portion (315 to 400 nm) is of greatest concern because shorter wavelengths (UV-B; 280 to 315 nm), while biologically very harmful, make up only about 8% of the total UVR present, and are filtered from the water column much more effectively than longer UV wavelengths ([119] and see Chapter 3). In addition, PAHs generally absorb radiation more affectivity in the longer, UV-A wavelength range. Phototoxic potency is ultimately a function of dose – the intensity of UVR integrated over duration of exposure. In most systems, the law of reciprocity suggests that equal damage will be caused by equivalent photon doses, regardless of the rate at which they enter the system (within some reasonable bounds).

In natural aquatic systems, UVR is attenuated in the water column at a rate that generally corresponds to the concentration of DOC. Log-transformed, UV intensity values are linearly related to depth, and the slope of a best-fit line (examples are shown in Figure 6A) represents the rate at which radiation is attenuated [120–123]. Concentration and makeup of DOC can vary significantly among water bodies and over time, and the effect on subsurface UV-A can be dramatic [124–126]. For example, the absorption coefficient for a near-shore area in Lake Superior (near Duluth, MN, USA) was estimated to be -0.355 m^{-1}, whereas in a St. Louis Harbor (Duluth, MN, USA) PAH-contaminated site, the absorption slope was estimated to be -0.001 m^{-1} (unpublished data, Figure 6A). The depth at which 50% of the above-surfaces UV-A intensity would

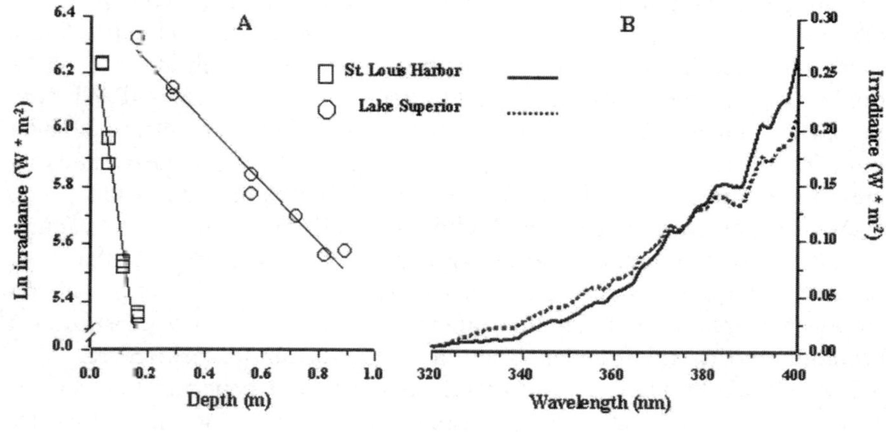

Figure 6. Absorption plots for UV-A radiation (plot A) and spectra at 50% UV penetration depth (plot B) for PAH-contaminated site in St. Louis Bay, Duluth, MN and a near-shore location in Lake Superior. The 50% penetration depths shown are 10 cm and 80 cm for the St. Louis Harbor and Lake Superior sites, respectively.

be absorbed was estimated to be 0.8 m for Lake Superior, and 0.1 m for the St. Louis Harbor location, an 8-fold difference. Given equivalent tissue PAH concentrations, and assuming that 50% of surface irradiance is sufficient to photoactivate PAH toxicity, organisms would be at risk in nearly 1 m of the Lake Superior water column, versus 10 cm in the St. Louis location water column.

These estimates of broad-spectrum UV penetration do not reflect potential differences in the spectrum of radiation that would reach PAH-exposed organisms. The differences in UV spectrum at the 50% UV penetration depth for the St. Louis Harbor and Lake Superior sites just discussed are shown in Figure 6B. These differences arise because DOC attenuates shorter UV wavelengths more efficiently than longer wavelengths and also because the makeup of DOC in different locations can vary dramatically, based on its sources in the landscape. The importance of such spectral variability has been demonstrated by Diamond et al. [90] in exposures of brine shrimp (*Artemia salina*) nauplii to three PAHs in combination with different UV spectra. In these assays, the overlap of UV-A radiation with absorbance spectra of pyrene, fluoranthene, and anthracene was manipulated using various filters. Where the radiation spectra overlapped significant potions of PAH absorbance spectra, toxicity did not differ. Where radiation and absorption spectra differed in the extent of their overlap, toxicity differed significantly. Most importantly, the variation in the spectra of UV-A used was consistent with variation possible in natural aquatic systems. The interaction of UV-A spectra and PAH absorbance spectra is summarized the equation for the phototoxicity weighting function (PWF):

$$\mathrm{PWF} = \int_{\lambda=320}^{\lambda=400} \varepsilon_\lambda I_\lambda \, \mathrm{d}\lambda$$

where: PWF = photoactivated toxicity weighting function, ε_λ = wavelength-specific molar absorptivity, and I_λ = wavelength-specific irradiance. This component of UV dose was incorporated into the Morgan and Worshawsky [102] and Newsted and Giesy [103] models discussed previously.

While this approach is a logical first approximation of actual PAH-photoactivation potential, its accuracy is limited somewhat by our knowledge of mechanisms of action involved. If, for example, mortality during PAH/UV-A exposure derives from accumulated external tissue damage, then this approach to dosimetry is likely to be acceptably accurate. However, if other mechanisms such as disruption of DNA or other macromolecule function are involved, then the spectrum of light reaching these target sites is likely to vary considerably depending on specific overlying tissue types, species, and lifestages. In this case, comparisons of toxicity between sites would be consistent if the same species was considered, but would be questionable among different species. As well as altering UV spectra, these biological components are also likely to alter the photochemistry of the toxic mechanisms occurring (e.g., by quenching PAH excited states or free radical processors).

To some extent, solar flux can be predicted in aquatic systems. However, exposure to UVR entering the water column will be greatly influenced by the life history and behavior of different species. Attenuation of UVR in the water column, as well as physical shading, creates a highly heterogeneous UV environment in which exposure is largely determined by the moment to moment location of potentially-exposed organisms. Species that reside in sediments, vegetation, other highly shaded microhabitats, or deep water during daylight will receive little UV exposure. For most motile organisms (e.g., larval fish and plankton), daily accumulated UV exposure will be a complex summation of high and low exposure periods. Except in cases where behavior is well understood and quantifiable, the risk of phototoxicity can be characterized best by setting bounding conditions for possible exposure, or by describing UVR dose in limited areas in the aquatic habitat rather than as specific estimates of expected affects.

Diamond et al. [127] have estimated UVR doses in wetlands using this approach. Typical UVR doses were estimated by first generating maximal solar radiation doses for each day using a radiative transfer model, SBDART [113]. The model produced values for the full spectrum of solar radiation, from 280 to 700 nm, for cloudless conditions. These maximal values were then modified based on cloud cover effect estimates from 30 yr of historical solar radiation data (National Renewable Energy Laboratory, Department of Energy; http://rredc.nrel gov/solar/). The values derived in this procedure were estimated daily terrestrial, spectral (2 nm increments from 280 to 700 nm) solar radiation doses. Water column doses were then derived from absorption coefficients and spectral attenuation data described by Peterson et al. [128]. Although the focus of this effort was to characterize risk of UV-B radiation effects in amphibians, the procedure is directly applicable to phototoxicity, and the resulting UV-A radiation and spectral doses could be directly incorporated into calculation of possible effects.

Because of its importance to phototoxicity, the interaction of UVR with photosensitizer has received considerable focus, and has been quantitatively incorporated into all of the QSAR and PAH absorbance/UV spectra work described above, as well as discussions of the ecological risk of PAH phototoxicity [84,86,87,89,90,104,107–110,129]. In these QSAR studies, adherence to the law of reciprocity was assumed, rather than tested specifically. Ankley et al. [130] (see Figure 7) explicitly tested the consistency of the Law of Reciprocity at predicting PAH phototoxicity by conducting assays using *Lumbriculus variegatus* exposed to several combinations of fluoranthene concentration (0, 3.7, 7.5, 15, 30, 60, 120 μg l^{-1}) and UVR intensity (16.6, 33.5, and 75.2 μW cm^{-2}). Making simplifying assumptions, that tissue concentrations are constant over the exposure duration (a 96 h uptake period preceded initiation of UV exposure) and that damage repair is negligible, Ankley et al. [130] predicted that toxicity would be described by the equation:

$$l = \frac{k_3}{D_L} R_0 I t_D$$

where l = lethality, k_3 = rate of damage accrual, D_L = the critical level of damage, R_0 = initial tissue residue, I = radiation intensity, t_D = time to death.

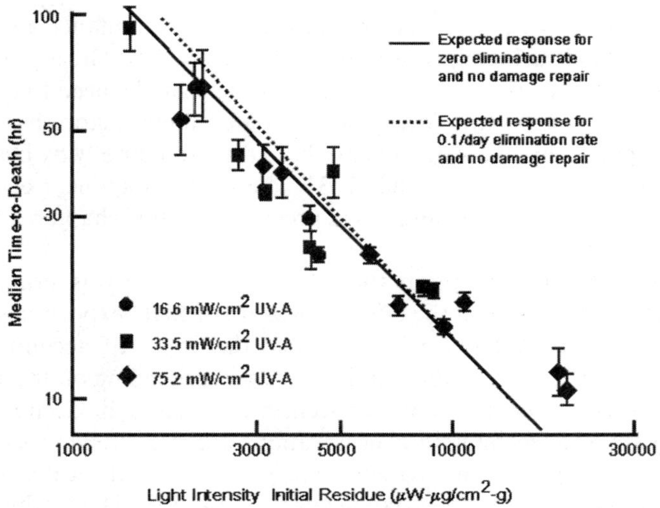

Figure 7. Time-dependent mortality of *Lumbriculus variegates* (expressed as LT50 values) versus the product of light intensity and initial tissue concentration of fluoranthene. [Data from [130], with permission of the publisher]

The results of their exposures were consistent with this prediction, as shown in Figure 7, with slight deviation at the highest UV exposure levels. This deviation is expected, as these low-PAH–high-UV exposures would be more strongly affected by the delay in the onset of mortality commonly observed in phototoxicity assays. Ankley et al. [131] extended and confirmed these results in subsequent studies by repeating exposures with pyrene, anthracene, and fluorene. As in the previous study, these PAHs, except for fluorene, which was not phototoxic, adhered to predictions based on the Law of Reciprocity. These results were in accord with earlier work of Oris and Giesy [84] in which bluegill sunfish were exposed to various concentrations of anthracene and three levels of UV-B radiation. The fit of the mortality data to a model assuming reciprocity was less definitive than in the work of Ankley et al. [130,131], possibly because of differences between the two compounds, the relatively limited data set of Oris and Giesy [84], and differences among tested populations of sunfish. These relationships were also demonstrated by Erickson et al.'s [132] mixture work, in which binary combinations of anthracene, fluoranthene, and pyrene were tested for interactive phototoxicity. The toxicity of these compounds was found to be additive (as opposed to antagonistic or synergistic), but, more importantly for this discussion, all single and mixture exposure responses adhered to the law of reciprocity.

7.8 Risk assessment for PAH phototoxicity

Ecological toxicologists characterize the probability of harmful effects occurring in natural settings by conducting risk assessments. These assessments, briefly, incorporate potency, lethality, or "hazard" of specific contaminants, and probabilities of exposure of various species in natural systems. This already complex process is confounded for phototoxicity risk assessment by the need to quantify both PAH and UVR exposure. Many of the confounding factors have been alluded to throughout this chapter, and have been discussed briefly by Diamond and Mount [89], and by Ankley et al. [133]. The UVR component of phototoxicity risk assessment has been addressed previously in this chapter, and also by Diamond et al. [127]

If the well-supported assumption that PAH phototoxicity is primarily a photosensitization process, then the critical measure of PAH exposure is tissue concentration. Because PAHs are hydrophobic they tend to accumulate in tissues to concentrations 100 to 100 000 times higher (depending on the specific PAHs) than their environmental water concentrations. These bioaccumulation factors (BAFs) are affected by lipid concentrations and metabolic processes in organisms, by fugacity process among sediment-bound PAHs, the water column, and organic suspended and dissolved material [134,135]. The pathway for accumulation of most PAHs, and other lipophyllic, potentially photoactivated compounds, is from sediment (where they tend to accumulate because of high organic content) to water, and then to aquatic organisms. Additional uptake may occur via sediment ingestion, and via the food chain.

The difficulty of estimating PAH tissue concentrations is complicated by the fact that most PAH contamination occurs as mixtures of hundreds of PAH compounds. Each of these compounds has a unique K_{ow} (organic–water partitioning coefficient), which is a reliable indicator of its tendency to remain in sediments or dissolved in water. Each compound also has a unique phototoxic potency, resistance to environmental degradation or modification, and metabolism by exposed organisms. At high concentrations, complex mixtures can also affect the solubility of their constituents, adding to the uncertainty of fugacity estimations (e.g., [136]). The phototoxicity potential of the complex mixtures typical of contaminated sites has been demonstrated in several ways, including field collection and subsequent UV treatment of PAH-exposed organisms, controlled, in situ UV exposure, and bioassays using field-collected sediments. However, the applicability of these results to broader risk assessment is limited by the unique mixture of PAHs present at these sites, and by the myriad differences in environmental factors, including penetration of UVR, temperature, carbon concentrations, etc.

The occurrence of phototoxicity in PAH-contaminated sites is nearly impossible to observe. Most highly contaminated sites are biologically depauperate, making direct observation of the toxic processes for species that would normally reside there untenable. Hence, although the potential for phototoxicity, its mechanisms and key components, has been thoroughly demonstrated, its importance in potentially affected aquatic systems has yet to be fully characterized or quantified. Thus, there is a definite need to continue studying the phenomenon due to the following factors:

(1) the large number of sites contaminated with high concentrations of PAHs,

(2) the slow degradation of most PAHs by natural processes,
(3) the continued release of PAHs via terrestrial runoff and aerial deposition, and
(4) the great potential for increased UVR exposure in aquatic systems due to environmental changes associated with global climate change.

References

1. M.L. Mitchell, D. Karentz (1993). The induction and repair of DNA photodamage in the environment. In: A.R. Young, L.O. Bjorn, J. Moan, W. Nultsch (Eds), *Environmental UV Photobiology* (pp. 345–377). Plenum Press, New York.
2. W.H. Jeffrey, D.L. Mitchell (1997). Mechanisms of UV-induced DNA damage and response in marine microorganisms. *Photochem. Photobiol.*, **65**, 260–263.
3. R.B. Setlow (1974). The wavelengths of sunlight effective in causing skin cancer: a theoretical analysis. *Proc. Natl. Acad. Sci. U.S.A.*, **71**, 3363–3366.
4. U. Strahle, S. Jesuthasan (1993). Ultraviolet irradiation impairs epiboly in zebrafish embryos: evidence for a microtubule-dependent mechanism of epiboly. *Development*, **119**, 909–919.
5. R.P. Sinha, H.D. Kumar, A. Kumar, D.-P. Häder (1995). Effects of UV-B irradiation on growth, survival, pigmentation and nitrogen metabolism enzymes in cyanobacteria. *Acta Protozool.*, **34**, 187–192.
6. P. Morliere, A. Moysan, I. Tirache (1995). Action spectrum for UV-induced lipid peroxidation in cultured human skin fibroblasts. *Free Radical Biol. Med.*, **19**, 365–371.
7. A.A. van Zeeland, A.T. Natarajan, E.A.M. Verdegaal-Immerzeel, A.R. Filon (1980). Photoreactivation of UV-induced cell killing, chromosome aberrations, sister chromatid exchanges, mutations and pyrimidine dimers in Xenopus laevis fibroblasts. *Mol. Gen. Gen.*, **180**, 495–500.
8. W.E. Daniell, H.L. Stockbridge, R.F. Labbe, J.S. Woods, K.E. Anderson, D.M. Bissel, J.D. Bloomer, R.D. Ellefson, M.R. Moore, C.A. Pierach, W.E. Schreiber, A. Tefferi, G.M. Franklin (1997). Environmental chemical exposures and disturbances of heme synthesis. *Environ. Health Perspect. Suppl.*, **105**, 37–53.
9. D.-P. Häder, H.D. Kumar, R.C. Smith, R.C. Worrest (1998). Effects on aquatic ecosystems. *J. Photochem. Photobiol. B: Biol.*, **46**, 53–68.
10. J.J. Cullen, P.J. Neale (1997). Effect of UV on short-term photosynthesis of natural phytoplankton. *Photochem. Photobiol.*, **65**, 264–266.
11. F.R. de Gruijl, J.C. Van der Leun (1994). Influence of ozone depletion on the incidence of skin cancer: quantitative prediction. In: A.R. Young, L.O. Bjorn, J. Moan, W. Nultsch (Eds), *Environmental UV Photobiology* (pp. 89–109). Plenum Press, New York.
12. S. Zigman (1993). Ocular damage by environmental radiant energy and its prevention. In: A.R. Young, L.O. Bjorn, J. Moan, W. Nultsch (Eds), *Environmental UV Photobiology* (pp. 149–181). Plenum Press, New York.
13. O. Holm-Hansen, D. Lubin, E.W. Helbling (1993). Ultraviolet radiation and its effect on organisms in aquatic environments. In: A.R. Young, L.O Bjorn, J. Moan, W. Nultsch (Eds), *Environmental UV Photobiology* (pp. 379–418). Plenum Press, New York.
14. D.M. Leech, C.E. Williamson (2001). In situ exposure to ultraviolet radiation alters

the depth distribution of *Daphnia. Limnol. Oceanogr.*, **46**, 416–420.
15. B.E. Brown, R.P. Dunne, T.P. Scoffin, M.D.A. Le Tissier (1994). Solar damage in intertidal corals. *Mar. Ecol. Prog. Ser.*, **105**, 219–230.
16. D.P. Arfsten, D.J. Schaeffer, D.C. Mulveny (1996). The effects of near ultraviolet radiation on the toxic effects of polycyclic aromatic hydrocarbons in animals and plants: a review. *Ecotoxicol. Environ. Saf.*, **33**, 1–24.
17. R.A. Larson, M.R. Berenbaum (1988). Environmental phototoxicity: solar ultraviolet radiation affects the toxicity of natural and man-made chemicals. *Environ. Sci. Technol.*, **22**, 354–360.
18. L. Santamaria, G. Prino (1964). The photodynamic substances and their mechanisms of action. *Res. Prog. Org.-Biol. Med. Chem.*, **1**, 259–336.
19. M.R. Berenbaum (1987). Charge of the light brigade: phototoxicity as defense against insects. In: J.R. Heitz, K.R. Downum (Eds), *Light-activated Pesticides* (pp. 206–216). American Chemical Society, Washington, DC.
20. E.A.J. Bleeker, S. Wiegman, P. de Voogt, M. Kraak, H.A. Leslie, E. de Haas, W. Admiraal (2002). Toxicity of azaarenes. *Rev. Environ. Contam. Toxicol.*, **173**, 39–83.
21. B. David, P. Boule (1993). Phototransformation of hydrophobic pollutants in aqueous medium I – PAHs adsorbed on silica. *Chemosphere*, **9**, 1716–1630.
22. R. Dabestani, I.N. Ivanov (1999). A compilation of physical, spectroscopic and photophysical properties of polycyclic aromatic hydrocarbons. *Photochem. Photobiol.*, **70**, 10–34.
23. X.D. Huang, D.G. Dixon, B.M. Greenberg (1993). Impacts of ultraviolet radiation and photomodification of the toxicity of polycyclic aromatic hydrocarbons to the higher plant Lemna gibba L. (duckweed). *Environ. Toxicol. Chem.*, **12**, 1067–1077.
24. K. Nikolaou, P. Masclet, G. Mouvier (1984). Sources and chemical reactivity of polynuclear aromatic hydrocarbons in the atmosphere – a critical review. *Sci. Total Environ.*, **32**, 103–132.
25. S. Wiegman, P.L.A. van Vlaardingen, W.J.G.M. Peijnenburg, S.A.M. van Beusekom, M.H.S. Kraak, W. Admiraal (1999). Photokinetics of azaarenes and toxicity of phototransformation products to the marine diatom *Phaeodactylum tricornutum. Environ. Sci. Technol.*, **33**, pp. 4256–4262.
26. J.R. Heitz, K.R. Downum (Eds) (1987). *Light-activated Pesticides.* American Chemical Society, Washington, DC.
27. L. Santamaria, G. Prino (1964). The photodynamic substances and their mechanism of action. *Res. Prog. Org.-Biol. Med. Chem.*, **1**, 259–336.
28. B.K. Sinha, C.F. Chignell (1983). Binding of anthracene to cellular macromolecules in the presence of light. *Photochem. Photobiol.*, **37**, 33–37.
29. X.-D. Huang, G.D. Dixon, B.M. Greenberg (1995). Increased polycyclic aromatic hydrocarbon toxicity following their photomodification in natural sunlight: impacts on the duckweed *Lemna gibba* L. G-3. *Ecotoxicol. Environ. Saf.*, **32**, 194–200.
30. L. Ren, L.F. Zeiler, D.G. Dixon, B.M. Greenberg (1996). Photoinduced effects of polycyclic aromatic hydrocarbons on *Brassica napus* (canola) during germination and early seedling development. *Ecotoxicol. Environ. Saf.*, **33**, 73–80.
31. M.A. Pathak (1986). Phytophotodermatitis. *Clin. Dermatol.*, **4**, 102–121.
32. M.A. Pathak, T.B. Fitzpatrick (1992). The evolution of photochemotherapy with psoralens and UVA (PUVA): 2000 BC to 1992 AD. *J. Photochem. Photobiol.*, **14** , 3–22.
33. I.R. Fahmy, H. Abu-Shady (1947). *Ammi majus* Linn: pharmacological study and isolation of a crystalline constituent, ammoidin. *Q. J. Pharm. Pharmacol.*, **20**, 281–291.
34. H.N. Nigg, H.E. Nordby, R.C. Beier, A. Dillman, C. Macias, R.C. Hansen (1993).

Phototoxic coumarins in limes. *Food Chem. Toxicol.*, **31**, 331–335.
35. G. Kavli, G. Volden (1984). Phytophotodermatitis. *Photodermatology*, **1**, 65–75.
36. B. Ljunggren (1990). Severe phototoxic burn following celery ingestion. *Arch. Dermatol.*, **126**, 1334–1336.
37. M.A. Pathak, F. Daniels, T.B. Fitzpatrick (1962). The presently known distribution of furocoumarins (psoralens) in plants. *J. Invest. Dermatol.*, **39**, 225–239.
38. G.W. Ivie (1978). Toxicological significance of plant furocoumarins. In: R. Keeler, K. Van Kampen, L. James (Eds), *Effects of Poisonous Plants on Livestock* (pp. 475–485). Academic Press, New York.
39. J.R. Heitz (1987). Development of photoactivated compounds as pesticides. In: J.R. Heitz, K.R. Downum (Eds), *Light-activated Pesticides* (pp. 1–21). American Chemical Society, Washington, DC.
40. D.J. Birmingham, M.M Key, G.E. Tublich (1961). Phototoxic bullae among celery harvesters. *Arch. Dermatol.*, **83**, 73–85.
41. C.R. Lovell (1993). *Plants and the Skin.* Blackwell Scientific, Oxford.
42. L.D. Scheel, V.B. Perone, R.L. Larkin, R.E. Kupel (1963). The isolation and characterization of two phototoxic furanocoumarins (psoralens) from diseased celery. *Biochemistry*, **2**, 1127–1131.
43. S.K. Hann, Y.-K. Park, S. Im, S.W. Byun (1991). Angelica-induced photodermatitis. *Photodermatol. Photoimmunol. Photomed.*, **8**, 84–85.
44. G.D. Pimprikar, M.J. Coign (1987). Multiple mechanisms of dye-induced toxicity in insects. In: J.R. Heitz, K.R. Downum (Eds), *Light-activated pesticides* (pp. 134–155). American Chemical Society, Washington, DC.
45. J. Kagan, W.J. Bennett, E.D. Kagan, J.L. Maas, S.A. Sweeney, I.A. Kagan, E. Seigneurie, V. Bindokas (1987). a-Terthienyl as a photoactive insecticide: toxic effects on nontarget organisms. In: J.R. Heitz, K.R. Downum (Eds), *Light-activated pesticides* (pp. 176–191). American Chemical Society, Washington, DC.
46. A. Jodlbauer, H. Tappeiner (1905). Die Beteiligung des Sauerstoffs bei der Wirkung fluorezierender Stoffe. *Dtsch. Arch. Klin. Med.*, **82**, 520–546.
47. J.J. Mottram, I. Doniach (1937). Photodynamic action of carcinogenic agents. *Nature*, **140**, 933–934.
48. J.J. Mottram, I. Doniach (1938). The photodynamic action of carcinogenic agents. *Lancet*, **1**, 1156–1159.
49. I. Doniach (1939). A comparison of the photodynamic activity of some carcinogenic and non-carcinogenic compounds. *Br. J. Exp. Pathol.*, **20**, 227–235.
50. G. Calcutt (1954). The photosensitizing action of chemical carcinogens. *Br. J. Cancer*, **8**, 177–180.
51. S.S. Epstein, M. Burroughs (1962). Some factors influencing the photodynamic response of Paramecium caudatum to 3,4-benzopyrene. *Nature*, **193**, 337–338.
52. S.S. Epstein (1963). The photodynamic activity of polycyclic hydrocarbons carcinogens. *Acta Unio Int. Cancrum*, **19**, 599–601.
53. S.S. Epstein, M. Burroughs, M. Small (1963). The photodynamic effect of the carcinogen 3,4,-benzopyrene on *Paramecium caudatum. Cancer Res.*, **23**, 35–44.
54. S.S. Epstein, M. Small, J. Koplan, N. Mantel, S.H. Hunter (1963). A photodynamic bioassay of benzo[a]pyrene with *Paramecium caudatum. J. Natl. Cancer Inst.*, **31**, 163–168.
55. S.S. Epstein, M. Small, J. Koplan, N. Mantel, H.L. Falk, E. Sawicki (1963). Photodynamic bioassay of polycyclic air pollutants. *Arch. Environ. Health*, **7**, 531–537.
56. S.S. Epstein, M. Small, H.L. Faulk, N. Mantel (1964). On the association between photodynamic and carcinogenic activities in polycyclic compounds. *Cancer Res.*, **24**,

855–861.

57. D.D. Morgan, D. Warshawsky, T. Atkinson (1977). The relationship between carcinogenic activities of polycyclic aromatic hydrocarbons and their singlet, triplet, and singlet–triplet splitting energies and phosphorescence lifetimes. *Photochem. Photobiol.*, **25**, 31–38.
58. J.W. Bowling, G.J. Leversee, P.F. Landrum, J.P. Giesy (1983). Acute mortality of anthracene-contaminated sunfish exposed to sunlight. *Aquat. Toxicol.*, **3**, 79–90.
59. J. Kagan, P.A. Kagan, H.E. Buhse, Jr. (1984). Light-dependent toxicity of α-terthienyl and anthracene toward late embryonic stages of Rana pipiens. *J. Chem. Ecol.*, **10**, 1115–1122.
60. J. Kagan, E.D. Kagan, E. Seigneurie (1986). Alpha-terthienyl, a powerful fish poison with light-dependent activity. *Chemosphere*, **15**, 49–57.
61. J. Kagan, K. Tadema-Weilandt, G. Chan, S.D. Dhawan, I.P. Jawarsky, K.A. Sudershan (1984). Oxygen requirement for near-UV mediated cytotoxicity of phenylheptatriyne to *Escherichia coli. Photochem. Photobiol.*, **39**, 465–467.
62. R. Davenport, L.R. Johnson, D.J. Schaeffer, H. Balbach (1994). Light-enhanced toxicity of TNT and some related compounds to *Daphnia magna* and *Lytechinus variagatus* embryos. *Ecotoxicol. Environ. Saf.*, **27**, 14–22.
63. L.R. Johnson, R. Davenport, H. Balbach, D.J. Schaeffer (1994). Near-ultraviolet light-enhancement of Microtox assays of trinitrotoluene and aminodinitrotoluenes. *Ecotoxicol. Environ. Saf.*, **27**, 23–33.
64. L.R. Johnson, R. Davenport, H. Balbach, D.J. Schaeffer (1994). Comparative toxicity of trinitrotoluene and aminodinitrotoluenes to *Daphnia magna, Dugesia dorotocephala,* and sheep erythrocytes. *Ecotoxicol. Environ. Saf.*, **27**, 34–49.
65. A. Zaga, E.A. Little, C.F. Rabeni, M.R. Ellersieck (1998). Photoenhanced toxicity of a carbamate insecticide to early life stage anuran amphibians. *Environ. Toxicol. Chem.*, **17**, 2543–2553.
66. A. S. Wernersson, S. Dave (1997). Phototoxicity identification by solid phase extraction and photoinduced toxicty to *Daphnia magna. Arch. Environ. Contamin. Toxicol.*, **32**, 268–273.
67. L. Cleveland, E.E. Little, R.D. Calfee, M.G. Barron (2000). Photoenhanced toxicity of weathered oil to *Mysidopsis bahia. Environ. Toxicol. Chem.*, **49**, 63–76.
68. E.E. Little, L. Cleveland, R. Calfee, M.G. Barron (2000). Assessment of the photoenhanced toxicity of weathered oil to the tidewater silverside. *Environ. Toxicol. Chem.*, **19**, 926–932.
69. M.C. Pelletier, R.M. Burgess, K.T. Ho, A. Kuhn, R.A. McKinney, S.A. Ryba (1997). Phototoxicity of individual polycyclic aromatic hydrocarbons and petroleum to marine invertebrate larvae and juveniles. *Environ. Toxicol. Chem.*, **16**, 2190–2199.
70. R. Davenport, A. Spacie (1991). Acute phototoxicity of harbor and tributary sediments from lower Lake Michigan. *J. Great Lakes Res.*, **17**, 51–56.
71. G.T. Ankley, S.A. Collyard, P.D. Monson, P.A. Kosian (1994). Influence of ultraviolet light on the toxicity of sediments contaminated with polycyclic aromatic hydrocarbons. *Environ. Toxicol. Chem.*, **13**, 1791–1796.
72. P.D. Monson, G.T. Ankley, P.A. Kosian (1995). Phototoxic response of *Lumbriculus variegatus* to sediment contaminated by polycyclic aromatic hydrocarbons, short communication. *Environ. Toxicol. Chem.*, **14**, 891–894.
73. D.S. Ireland, G.A. Burton, Jr., G.G. Hess (1996). In situ toxicity evaluations of turbidity and photoinduction of polycyclic aromatic hydrocarbons. *Environ. Toxicol. Chem.*, **15**, 574–581.
74. T. Ito (1978). Oxygen requirement for near-UV mediated cytotoxicity of phenyl-

heptatriyne to *Escherichia coli. Photochem. Photobiol.*, **28**, 493–508.
75. J.T. McCloskey, J.T. Oris (1991). Effect of water temperature and dissolved oxygen concentration on the photo-induced toxicity of anthracene to juvenile bluegill sunfish (*Lepomis macrochirus*). *Aquat. Toxicol.*, **21**, 145–156.
76. J. Kagan, A. Stokes, H.-H. Gong, R.W. Tuveson (1987). Light-dependent cytotoxicity of fluoranthene: oxygen-dependent membrane damage. *Chemosphere*, **16**, 2417–2422.
77. D.R. Kearns, A.U. Khan (1969). Sensitized photooxygenation reactions and the role of singlet oxygen. *Photochem. Photobiol.*, **10**, 193–210.
78. B.W. Gensemer, D.G. Dixon, B.W. Greenberg (1998). Amelioration of the photo-induced toxicity of polycyclic aromatic hydrocarbons by a commercial humic acid. *Ecotoxicol. Environ. Saf.*, **39**, 57–64.
79. A. Nikkilae, S. Penttinen, J.V. Kukkonen (1999). UV-B-induced acute toxicity of pyrene to the waterflea *Daphnia magna* in natural freshwaters. *Ecotoxicol. Environ. Saf.*, **44**, 271–279.
80. W. Bennett, J.L. Maas, S.A. Sweeney, J Kagan (1986). Phototoxicity in aquatic organisms: the protecting effect of beta-carotene. *Chemosphere*, **6**, 781–786.
81. A.T. Hall, J.T. Oris (1991). Anthracene reduces reproductive potential and is maternally transferred during long-term exposure in fathead minnows. *Aquat. Toxicol.*, **19**, 249–264.
82. J.E. Weinstein, K.D. Polk. Phototoxicity of anthracene and pyrene to glochidia of the freshwater mussel, *Utterbackia imbecillis. Environ. Toxicol. Chem.*, in press.
83. J.E. Weinstein (2001). Characterization of the acute toxicity if photoactivated fluoranthene to glochidia of the freshwater mussel, Utterbackia imbecillis. *Environ. Toxicol. Chem.*, **20**, 412–419.
84. J.T. Oris, J.P. Giesy (1985). The photoenhanced toxicity of anthracene to juvenile sunfish (Lepomis spp.). *Aquat. Toxicol.*, **6**, 133–146.
85. J.T. Oris, J.P. Giesy, Jr. (1986). Photoinduced toxicity of anthracene to juvenile bluegill sunfish (*Lepomis machrochirus* Rafinesque): photoperiod effects and predictive hazard evaluation. *Environ. Toxicol. Chem.*, **5**, 761–768.
86. J.T. Oris, J.P. Giesy, P.M. Allred, D.F. Grant, P.F. Landrum (1984). Photoinduced toxicity of anthracene in aquatic organisms: an environmental perspective. In: T.N. Vezirogiu (Ed.), *The Biosphere: Problems and Solution* (pp. 639–658). Elsevier Science, Amsterdam.
87. P.F. Landrum, J.P. Giesy, J.T. Oris, P.M. Allred (1985). The photoinduced toxicity of polycyclic aromatic hydrocarbons to aquatic organisms. In: J.H. Vandermeulen, S. Hrudey (Eds), *Oil in Freshwater: Chemistry, Biology, and Technology* (pp. 304–318). Pergamon, New York.
88. M.G. Barron, E.E. Little, R. Calfee, S. Diamond (2000). Quantifying solar spectral irradiance in aquatic habitats for the assessment of photoenhanced toxicity. *Environ. Toxicol. Chem.*, **19**, 920–925.
89. S.A. Diamond, D.R. Mount (1998). Evaluating the role of photo-activated PAH toxicity in risk assessment. *SETAC News, Learned Discourses*, **18**, 17–18.
90. S.A. Diamond, D.R. Mount, L.P. Burkhard, G.T. Ankley, E.A. Makynen, E.N. Leonard (2000). Effect of irradiance spectra on the photoinduced toxicity of three polycyclic aromatic hydrocarbons. *Environ. Toxicol. Chem.*, **19**, 1389–1396.
91. L.L. Holst, J.P. Giesy (1989). Chronic effects of photoenhanced toxicity of anthracene on Daphnia magna reproduction. *Environ. Toxicol. Chem.*, **8**, 933–942.
92. S.A. Diamond, J.T. Oris, S.I. Guttman (1995). Adaptation to fluoranthene exposure in a laboratory population of fathead minnows. *Environ. Toxicol. Chem.*, **14**, 1393–1400.
93. S. Wiegman (2002). *Photoenhanced Toxicity of Azaarenes to Marine Phytoplankton.*

Ph.D. Thesis (pp. 73–90). University of Amsterdam, The Netherlands.
94. A. Gilbert, J. Baggott (1991). *Essentials of Molecular Photochemistry*. Blackwell Science, Paris.
95. C.H.J. Wells (1972). *Introduction to Molecular Photochemistry*. Chapman and Hall, London (pp. 1–13).
96. K. Gollnick (1968). Type II photooxygenation reactions in solution. *Adv. Photochem.*, **6**, 1–122.
97. C.S. Foote (1991). Definition of Type I and Type II photosensitized oxidation. *Photochem. Photobiol.*, **54**, 659.
98. C.S. Foote (1987). Type I and Type II mechanisms of photodynamic action. In: J.R. Heitz, K.R. Downum (Eds), *Light-activated Pesticides* (pp. 22–38). American Chemical Society, Washington, DC.
99. D.P. Valenzeno (1987). Photomodification and singlet oxygen generation in membranes. In: J.R. Heitz, K.R. Downum (Eds), *Light-activated Pesticides* (pp. 39–57). American Chemical Society, Washington, DC.
100. M.L. Lee, D.L. Vassilaros, D.W. Later (1982). Capillary column gas chromatography of environmental polycyclic aromatic compounds. *Int. J. Environ. Anal. Chem.*, **11**, 251–262.
101. L. Santamaria, G.G. Giordano (1969). Effects of long-wave ultraviolet radiation on polycyclic hydrocarbon carcinogenesis. In: M.D. Urbach (Ed.), *The Biological Effects of Ultraviolet Radiation (with Emphasis on the Skin)* (pp. 569–580). Pergamon Press, New York.
102. D.D. Morgan, D. Warshawsky (1977). The photodynamic immobilization of *Artemia salina* nauplii by polycyclic aromatic hydrocarbons and its relationship to carcinogenic activity. *Photochem. Photobiol.*, **25**, 39–46.
103. J.L. Newsted, J.P. Giesy (1987). Predictive models for photoinduced acute toxicity of polycyclic aromatic hydrocarbons to *Daphnia magna*, Strauss (Cladocera, Crustacea). *Environ. Toxicol. Chem.*, **6**, 445–461.
104. J. Choi, J.T Oris (2000). Evidence of oxidative stress in bluegill sunfish (*Lepomis macrochirus*) liver microsomes simultaneously exposed to solar ultraviolet radiation and anthracene. *Environ. Toxicol. Chem.*, **19**, pp. 1795–1799.
105. D.W. Nebert, D.D. Petersen, A.J. Fornace (1990). Cellular responses to oxidative stress: The [AH] gene battery as a paradigm. *Environ. Health Perspect.*, **88**, 99, 13–25.
106. R. Koch (1983). Molecular connectivity index for assessing ecotoxicological behavior of organic compounds. *Toxicol. Appl. Environ. Chem.*, **6**, 87–96.
107. O.G. Mekenyan, G.T. Ankley, G.D. Veith, D.J. Call (1994). QSARs for photoinduced toxicity: I. Acute lethality of polycyclic aromatic hydrocarbons to *Daphnia magn.*, *Chemosphere*, **28**, 567–582.
108. O.G. Mekenyan, G.T. Ankley, G.D. Veith, D.J. Call (1994). QSAR estimates of excited states and photoinduced acute toxicity of polycyclic aromatic hydrocarbons. *SAR QSAR Environ. Res.*, **2**, 237–247.
109. G.D. Veith, O.G. Mekenyan, G.T. Ankley, D.J. Call (1995). QSAR evaluation of a-terthienyl phototoxicity. *Environ. Sci. Technol.*, **29**, 1267–1272.
110. G.D. Veith, O.G. Mekenyan, G.T. Ankley, D.J. Call (1995). A QSAR analysis of substituent effects on the photoinduced acute toxicity of PAHs. *Chemosphere*, **30**, 2129–2142.
111. L. Ren, X.-D. Huang, B.J. McConkey, D.G. Dixon, B.M. Greenberg (1994). Photoinduced toxicity of three polycyclic aromatic hydrocarbons (fluoranthene, pyrene, naphthalene) to the duckweed, *Lemna gibba* L. G-3, *Ecotoxicol. Environ. Saf.*, **28**,

160–171.

112. National Renewable Energy Laboratory (1992). *User's Manual: National Solar Radiation Data Base* (1961–1990). National Climatic Data Center, Asheville, NC.
113. P. Ricchiazzi, S. Yang, C. Gautier, D. Sowle (1998). SBDART: A research and teaching software tool for plane-parellel radiative transfer in the earth's atmosphere. *Bull. Am. Meteorol. Soc.*, **79**, 2101–2114.
114. C.A. Marwood, R.E.H. Smith, K.R. Solomon, M.N. Charlton, B.M. Greenberg (1999). Intact and photodified polycyclic aromatic hydrocarbons inhibit photosynthesis in natural assemblages of Lake Erie phytoplankton exposed to solar radiation. *Ecotoxicol. Environ. Saf.*, **44**, 322–327.
115. S.N. Krylov, X.-D Huang, L.F. Zeiler, G.D. Dixon, B.M. Greenburg (1997). Mechanistic quantitative structure–activity relationship model for the photoinduced toxicity of polycyclic aromatic hydrocarbons: I. Physical model based on chemical kinetics in a two-compartment system. *Environ. Toxicol. Chem.*, **16**, 2283–2295.
116. P.G. Mezey, Z. Zimpel, P. Warburton, D.P. Walker, D.G. Irvine, X.-D. Huang, D.G. Dixon, B.M. Greenberg (1998). Use of quantitative shape–activity relationships to model the photoinduced toxicity of polycyclic aromatic hydrocarbons: electron density shape features accurately predict toxicity. *Ecotoxicol. Environ. Saf.*, **7**, 1207–1215.
117. A. Mallakin, B.J. McConkey, G. Miao, B. McKibben, V. Snieckus, D.G. Dixon, B.M. Greenberg (1999). Impacts of structural photomodification on the toxicity of environmental contaminants: anthracene photooxidation products. *Ecotoxicol. Environ. Saf.*, **43**, 204–212.
118. S. Wiegman, P.L.A. van Vlaardingen, E.A.J. Bleeker, P. de Voogt, M.H.S. Kraak (2001). Photoenhnaced toxicity of azaarene isomers to the marine flagellate *Dunaliella tertiolecta*. *Environ. Toxicol. Chem.*, **20**, 1544–1550.
119. D. Lean, Attenuation of solar radiation in humic waters (1998). In: D.O. Hessen, L.J. Tranvik (Eds), *Aquatic Humic Substances: Ecology and Biogeochemistry* (pp.109–124). Springer-Verlag, Berlin.
120. M.T. Arts, R.D. Robarts, F. Kasai, M.J. Waiser, V.P. Tumber, A.J. Plante, H. Rai, H.J. De Lange (2000). The attenuation of ultraviolet radiation in high dissolved organic carbon waters of wetlands and lakes in the northern Great Plains. *Limnol. Oceanogr.*, **45**, 292–299.
121. C.E. Williamson, R.S. Stemberger, D.P. Morris, T.M. Frost, S.G. Paulsen (1996). Ultraviolet radiation in North American lakes: attenuation estimates from DOC measurements and implications for plankton communities. *Limnol. Oceanogr.*, **41**, 1024–1034.
122. I. Laurion, M. Ventura, J. Catalan, R. Psenner, R. Sommaruga (2000). Attenuation of ultraviolet radiation in mountain lakes: factors controlling the among- and within-lake variability. *Limnol. Oceanogr.*, **45**, 1274–1288.
123. I. Laurion, W.F. Vincent, D.R.S. Lean (1997). Underwater ultraviolet radiation: development of spectral models for northern high latitude lakes. *Photochem. Photobiol.*, **65**, 107–114.
124. T.A. Clair, B.G. Sayer (1997). Environmental variability in the reactivity of freshwater dissolved organic carbon to UV-B. *Biogeochemistry*, **36**, 89–97.
125. D.P. Morris, H. Zagarese, C.E. Williamson, E.G. Balseiro, B.R. Hargreaves, B. Modenutti, R. Moeller, C. Queimalinos (1995). The attenuation of solar UV radiation in lakes and the role of dissolved organic carbon. *Limnol. Oceanogr.*, **40**, 1381–1391.
126. D.P. Morris, B.R. Hargreaves (1997). The role of photochemical degradation of

dissolved organic carbon in regulating the UV transparency of three lakes on the Pocono Plateau. *Limnol. Oceanogr.*, **42**, 239–249.

127. S.A. Diamond, G.S. Peterson, J.E. Tietge, G.T. Ankley. Assessment of the risk of solar ultraviolet radiation effects on amphibians. III. Prediction of impacts in selected northern mid-western wetlands. *Environ. Sci. Technol.*, submitted.
128. G.S. Peterson, L.B. Johnson, R.P. Axler, S.A. Diamond. In situ characterization of solar ultraviolet radiation in amphibian habitats, *Environ. Sci. Technol.*, submitted.
129. G.T. Ankley, S.A. Diamond, J.E. Tietge, G.W. Holcombe, K.M. Jensen, D.L. DeFoe, R. Peterson. Assessment of the risk of solar ultraviolet radiation to amphibians. I. Dose-dependant induction of hindlimb malformations in the northern leopard frog (*Rana pipiens*). *Environ. Sci. Technol.*, submitted.
130. G.T. Ankley, R.J. Erickson, G.L. Phipps, V.R. Mattson, P.A. Kosian, B.R. Sheedy, J.S. Cox (1995). Effects of light intensity on the phototoxicity of fluoranthene to a benthic macroinvertebrate. *Environ. Sci. Technol.*, **29**, 2828–2833.
131. G.T. Ankley, R.J. Erickson, B.R. Sheedy, P.A. Kosian, V.R. Mattson, J.S. Cox (1997). Evaluation of models for predicting the phototoxic potency of polycyclic aromatic hydrocarbons. *Aquat. Toxicol.*, **37**, 37–50.
132. R.J. Erickson, G.T. Ankley, D.L. DeFoe, P.A. Kosian, E.A. Makynen (1999). Additive toxicity of binary mixtures of phototoxic polycyclic aromatic hydrocarbons to the oligochaete *Lumbriculus variegatus. Toxicol. Appl. Pharmacol.*, **154**, 97–105.
133. G.T. Ankley, L.P. Burkhard, P.M. Cook, S.A. Diamond, R.J. Erickson, D.R. Mount. Assessing risks from photoactivated toxicity of polycyclic aromatic hydrocarbons to aquatic organisms. In: *PAHs An Ecological Perspective*, P. Douben (Ed.), John Wiley & Sons, Hoboken, NJ, in press.
134. T.D. Gauthier, E.C. Shane, W.F. Guerlin, W.R. Seitz, C.L. Grant (1986). Fluorescence quenching method for determining equilibrium constants for polycyclic aromatic hydrocarbons binding to dissolved humic materials. *Environ. Sci. Technol.*, **20**, 1162–1166.
135. Y.-P. Chin, P.M. Gschwend (1992). Partitioning of polycyclic aromatic hydrocarbons to marine porewater organic colloids. *Environ. Sci. Technol.*, **26**, 1621–1626.
136. W.F. Lane, R.C. Loehr (1992). Estimating the equilibrium aqueous concentrations of polynuclear aromatic hydrocarbons in complex mixtures. *Environ. Sci. Technol.*, **26**, 983–990.

Chapter 8

Reactive oxygen species in aquatic ecosystems

David J. Kieber, Barrie M. Peake and Norman M. Scully

Table of contents

Abstract

The absorption of solar energy by dissolved organic matter (DOM) in natural waters results in a variety of photochemical transformations involving oxygen as a major reactant. These photochemical transformations generate a suite of reactive oxygen species (ROS) including the superoxide anion, the hydroxyl radical, singlet oxygen, alkoxy and peroxy radicals, the carbonate radical, and hydrogen peroxide. ROS cause numerous biogeochemical changes in aquatic ecosystems affecting the cycling of DOM, biological processes, and trace metal speciation.

In this chapter, we present a synthesis of research conducted in the study of ROS in marine and fresh waters, including a detailed discussion of recent evidence regarding the formation and removal of these compounds in the photic zone. Salient findings of this review are: (1) direct photolysis of DOM and reactions of DOM with trace metals and dissolved oxygen are one of the main sources of ROS in aquatic environments; (2) solar action spectra for ROS indicate that the highest production rates are between 290–400 nm; (3) removal pathways for most ROS are poorly known, but are likely to occur through reactions with DOM. Hydrogen peroxide is relatively unreactive towards DOM; however, it can affect DOM indirectly through photo-Fenton reactions; and (4) the impact of externally-generated ROS on aquatic organisms should be a function of their permeability through the cell membrane and inversely related to their reactivity in the water outside the cell.

8.1 Introduction

DOM absorbs nearly all of the ultraviolet radiation (UVR, 280–400 nm) in natural waters thereby controlling the attenuation of the UVR in the water column ([1–4] and Chapters 3, 6). From this perspective, DOM plays a fundamental role in controlling the ecology in the photic zone [5]. Absorption of UVR, in turn, provides the energy to break down and remineralize the DOM. It has been known for some time that when natural water samples are exposed to solar radiation their optical transparency increases due to the loss of chromophores in the organic matter [6,7]. In the case of organic-rich natural waters, it is easy to observe an increase in optical clarity and loss of the yellow–brown color of the water when it is exposed to solar radiation in the presence of dissolved oxygen. However, very little or no loss in color is observed when oxygen is removed prior to irradiation [8], which points to the important role that oxygen plays in the photooxidation of organic matter in the photic zone.

The photolysis of organic and inorganic constituents in natural waters is also an important mechanism for the production of free radicals [9]. Zafiriou and Dister [10] determined that the total production rate of radicals varied from 6 to 24×10^{-8} M h^{-1} during the summer in the Atlantic Ocean along the East Coast of the United States, employing a solar simulator with a spectral output and photon flux approximately equal to the noon time solar irradiance on a clear summer day in the Gulf of Maine. These rates translate to a very large, mean radical flux of approximately 0.7 mmol m^{-2} day^{-1} in surface waters during the summer.

The loss of yellow–brown color (and UV absorbance) in the water and free radical production are undoubtedly due to interactions of organic matter with solar UVR and visible radiation, which lead to a series of photochemical transformations involving oxygen as a major reactant (Figure 1). The importance of oxygen in the photooxidation of organic matter in natural waters is clearly evident in oxygen consumption studies, all of which show a substantial loss of dissolved oxygen when filter-sterilized natural water samples are exposed to solar radiation [11–16]. Oxygen plays a pivotal role as the initial scavenger of radicals that are produced during the irradiation of natural waters, forming an "oxygen wall" [9] and generating alkoxy and peroxy radicals ($RO^•$ and $ROO^•$, respectively) that eventually decay to stable oxygenated species. Some of the energy absorbed by DOM is dissipated through energy transfer reactions that yield singlet oxygen ($^1O_2$), and electron transfer reactions with DOM that yield the superoxide radical (O_2^-), a fraction of which disproportionates to form H_2O_2 [17]. These oxygen-containing compounds, together with the highly reactive hydroxyl radical ($^•OH$), are collectively referred to as ROS.

Many of the UVR effects that have been observed in aquatic organisms and ecosystems occur either directly or indirectly from the production and subsequent reactions of ROS. It is therefore not surprising that many ROS have been intensively studied to understand their impact on chemical and biological processes in natural waters. For example, reactions of ROS can increase the nor-

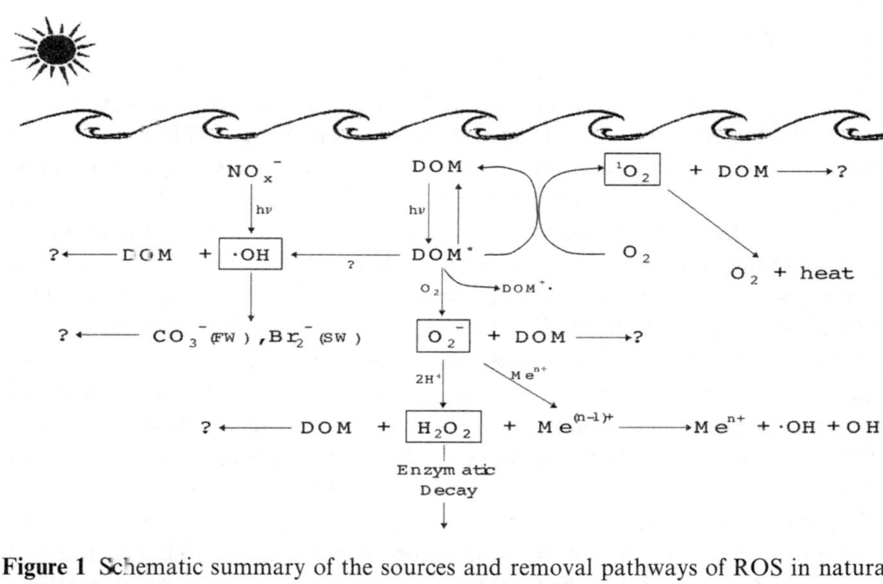

Figure 1 Schematic summary of the sources and removal pathways of ROS in natural waters including singlet oxygen ($^1O_2$), superoxide (O_2^-), hydrogen peroxide (H_2O_2) and the hydroxyl radical ($^•OH$). The main ROS are indicated by squares. Notation: FW, freshwater; SW, seawater; Me^{n+} or $Me^{(n-1)+}$, metal in the $n+$ or (n-1)+ oxidation state; NO_x^-, the nitrate or nitrite anion; and ?, unknown pathway.

mally slow rate of oxidation of some organic compounds in natural waters [18–24], change the redox state and speciation of trace metals [25–32], and cause oxidative stress to aquatic organisms [33,34]. Additionally, since ROS are involved in the photooxidation of DOM (Figure 1), they can influence the cycling of important biogeochemical elements in natural waters, as reviewed in Chapter 5 and elsewhere [6,7,35,36].

While the reaction of excited state organic matter with oxygen is the main source of many ROS in the photic zone, processes responsible for the removal of some ROS are less clear, especially when they involve DOM (Figure 1). In this chapter, we will discuss some of the processes that result in the formation and loss of ROS in marine and fresh water ecosystems, and the potential effects of externally generated ROS (i.e., photoproduced outside the cell in the surrounding water) on aquatic organisms. Intracellular production of ROS and the resultant oxidative stress that they impose on aquatic organisms is beyond the scope of this chapter and will not be discussed (see recent reviews by Josephy [37] and Vincent and Neale [34]). Analytical methods to detect these species will also not be discussed since they have been critically reviewed elsewhere [4,9,38].

8.1.1 Reaction kinetics

The concentration of a ROS measured after a given time t, $[\mathrm{ROS}]_t$, is equal to the sum of the initial concentration (i.e., at $t=0$) and the concentration photoproduced over a given period of time less the concentration that has reacted (decayed) during that time:

$$[\mathrm{ROS}]_t = [\mathrm{ROS}]_{t=0} + [\mathrm{ROS}]_{t,\mathrm{production}} - [\mathrm{ROS}]_{t,\mathrm{decay}} \tag{1}$$

In the laboratory, experimental variables such as photon flux are controlled ensuring that ROS precursors are not appreciably depleted. In this case, $[\mathrm{ROS}]_t$ rapidly attains a constant value called the steady state concentration corresponding to constant rates of formation and decay of the ROS. Invariably, this situation does not hold in natural aqueous environments involving solar irradiation because irradiance levels and biological processes (which may be a source of precursors or a sink of the ROS) will vary spatially and temporally. These variations will cause concentrations of ROS to undergo diurnal changes, as observed for H_2O_2 (vide infra). If the factors that control the rates of production and loss are known for a ROS, then spatial and temporal variations in their levels can be modeled [39].

More often than not, only the net rate of production (R) is measured, which is simply the capacity of a water sample to generate a specific ROS. This net rate (or accumulation rate) is the concentration of reactive oxygen species ([ROS]) produced during a given time (Δt) or more appropriately, photon exposure:

$$R = \Delta\,[\mathrm{ROS}]/\Delta t \tag{2}$$

The rate of a photochemical reaction is the product of the probability that an incident photon is absorbed and the probability that the absorbed photon will bring about a reaction. These probabilities are measured by the wavelength-dependent *absorbance coefficient* (a_λ) (see Chapters 5 and 6) of the DOM and the *quantum yield* (Φ_λ), respectively. The quantum yield is the efficiency of a photochemical process, and is equal to the number of moles of species formed or photolyzed divided by the number of moles of photons (Einsteins) absorbed by the chromophore. However, the complex molecular composition of DOM in aquatic environments means that there are likely to be multiple electronic transition energies and multiple precursors involved in the formation of individual ROS in natural waters, and these precursors are generally not known. Therefore, the quantum yield for the photochemical formation of ROS in natural waters is defined in terms of the DOM absorbance and is referred to as an *apparent quantum yield*, which is invariably wavelength-dependent. To express this wavelength dependence, the product (ε_λ) of the apparent quantum yield ($\Phi_{a,\lambda}$) and the absorbance coefficient (a_λ) is plotted as a function of wavelength to yield a *chemical action spectrum* [40]. The formulation of an action spectrum is an important component of photochemical models, but it can be problematic due to uncertainties in DOM absorbance measurements and the assumption that wavelength-dependent apparent quantum yields are constant with photon exposure (photon exposure is the irradiance integrated over time of exposure) when in fact they can increase or decrease [40].

Apart from the inherent efficiency of the reactions leading to the light-induced formation of a ROS as summarized by the relevant apparent quantum yield and action spectrum, the observed rate of production will depend on other factors that affect the photon exposure including water column composition and depth (Chapter 3), time of day (i.e., solar zenith angle), season, latitude (Chapter 2), and physical transport processes (Chapter 4). For more details regarding the fundamental equations used to define the rates of primary and secondary photochemical reactions and their application to aquatic systems, the reader is referred to recent reviews on this topic [41,42].

8.2 Formation and removal of ROS

8.2.1 Hydroxyl radical

The hydroxyl radical ($^{\bullet}OH$) is perhaps the most important ROS detected in natural waters. It plays a central role in transformations of organic matter in the troposphere [43], but its biogeochemical role in natural waters is poorly understood. The main source of the $^{\bullet}OH$ radical in most natural waters is from the photolysis of DOM, nitrate and nitrite [44,45], with production rates in the low to high 10^{-9} M h^{-1} range [44,46,47]. Iron and H_2O_2 can also be an important source of the $^{\bullet}OH$ radical through photo-Fenton chemistry, although this will be largely limited to iron-rich, high H_2O_2 environments such as the Suwannee or Orinoco Rivers [48,49]. The production of the $^{\bullet}OH$ radical from the photolysis of DOM is quite surprising, as there are very few known sources for the $^{\bullet}OH$ radical reported in the basic chemical literature that involve specific organic compounds. Vaughan and Blough [45] have shown that the production of the $^{\bullet}OH$ radical in a Suwannee River fulvic acid isolate occurs directly from the photolysis of DOM through an oxygen-independent pathway. They also observed that the formation of the $^{\bullet}OH$ radical occurred in the same wavelength range as the absorption band of benzoquinone. These results, along with evidence that quinones photoproduce $^{\bullet}OH$ (for review see [50]) and are a component of DOM [51,52], suggest that quinones are a potential source of the $^{\bullet}OH$ radical in natural waters.

Once formed, the $^{\bullet}OH$ radical is extremely unstable and it reacts rather indiscriminately with many organic or inorganic species at rates that are at or near the diffusion limit, either through an addition or H-atom abstraction pathway. As a result of its extreme reactivity, day time concentrations of the $^{\bullet}OH$ radical are very low in surface waters, with estimates ranging from 10^{-19} to 10^{-17} M [44,53]. While the $^{\bullet}OH$ radical is generally very reactive, as indicated by the extensive number of rate constants at or near the diffusion limit (10^{8}–10^{10} l mol^{-1} s^{-1}) [54], there are some notable exceptions of species that react relatively slowly with the $^{\bullet}OH$ radical (e.g., borate, carbon dioxide, phosphate) [54]. Likewise, while rate constants are large in many cases, there are differences in reactivity that lead to the selective loss of the $^{\bullet}OH$ radical in natural waters through reaction with only a few reactants. For example, in seawater, the $^{\bullet}OH$ radical is primarily removed through its reaction with the bromide ion, while in fresh waters with high alkalinity the bicarbonate and carbonate ions are the principal reactants and DOM predominates in low alkalinity waters [41]. These reactions result in the formation of less reactive dibromide and carbonate radicals (Figure 1) whose fates in natural waters are still unknown.

While only a few reactants are expected to control the loss of the $^{\bullet}OH$ radical in the photic zone, this does not preclude the possibility that the $^{\bullet}OH$ radical can affect the cycling of other minor species, especially those that strongly bind to DOM. Currently, the role of the $^{\bullet}OH$ radical in the transformations and cycling of DOM is poorly understood. If an organic species is present at trace levels (ca. $\leq 1 \times 10^{-6}$ M) in natural waters, then kinetic calculations suggest that the $^{\bullet}OH$ radical will not effectively remove that compound from the dissolved phase unless it is long-lived and not otherwise reactive (i.e., it is not biologically or chemically reactive). Estimates of half-lives for most trace organic compounds based on their reaction with the $^{\bullet}OH$ radical are quite long in aquatic environments located at mid latitudes (approximately 20–210 days) [46]. Shorter half-lives (ca. 7–60 days) are predicted in nitrate-rich systems, which contain higher steady state concentrations of the $^{\bullet}OH$ radical compared to nitrate-depleted waters [19]. However, the reactivity of the $^{\bullet}OH$ radical towards an organic substrate may be much higher if that compound has a heterogeneous distribution in solution due to its binding to DOM, especially if that binding is at or near the site involved in the photochemical production of the $^{\bullet}OH$ radical.

The importance of binding was demonstrated for the photolysis of the hydrophobic chlorocarbon pesticide mirex, which is predominantly bound to DOM in natural waters [55,56]. When humic acid solutions of mirex were irradiated, the mirex was photo-reduced by a humic-generated hydrated electron, which would not have occurred if mirex did not bind strongly to the DOM through hydrophobic interactions. Evidence also indicates that electrostatic interactions can affect the binding of a species to DOM, thereby affecting its photochemical reactivity. Blough [57] demonstrated that the decrease in the EPR (Electron Paramagnetic Resonance) signal during the photolysis of a series of water-soluble nitroxides in humic acid solutions was fastest for the cationic nitroxide probe (i.e., higher rate of signal loss) compared to the neutral or anionic nitroxide probes. This trend in reactivity paralleled the degree of interaction with DOM, which increased from the anionic nitroxide to the cationic nitroxide. These results indicate that it should be possible to predict the reactivity of a species with the DOM-generated $^{\bullet}OH$ radical, or other ROS, based on property–reactivity analysis employing physicochemical parameters such as the octanol water partition coefficient (K_{ow}) [58].

8.2.2 Singlet oxygen

Dissolved oxygen is present in natural waters at relatively high concentrations, generally ranging between 2.0 and 3.0×10^{-4} M. It exists primarily in the ground state triplet (denoted by $^3O_2$), with the two highest energy π electrons occupying separate molecular orbitals and having parallel spins. However, in the presence of UVR and appropriate photosensitizers, ground state oxygen is easily converted into its lowest excited singlet state ($^1O_2$) through energy transfer reactions. These reactions are often quite facile because the lowest energy level of $^1O_2$ (specifically the $^1\Delta_g$ species) is only 94 kJ mol^{-1} above the triplet ground state. One of several detailed descriptions of the physical and chemical properties of $^1O_2$ is given in Kearns [59].

There is considerable interest in the role of $^1O_2$ in the oxidation of DOM in natural waters because $^1O_2$ is reactive towards a wide range of electron-rich organic compound classes such as alkenes, sulfides and phenols (Table 1). Due to its potential importance in natural waters, structure–activity models have been developed to predict reaction rate constants for the $^1O_2$ oxidation of a series of environmentally-relevant, substituted phenols [22]. The reader is also referred to Wilkinson et al. [60] for an extensive compilation of quenching and reaction rate constants for reactions involving $^1O_2$, albeit most of the rate data are reported for solutions involving organic solvents. Singlet oxygen is also known to exert an oxidative stress in cellular systems causing toxicological effects such as lipid peroxidation and DNA damage [61].

In natural waters, $^1O_2$ is a ubiquitous ROS in the photic zone, with midday surface concentrations ranging from 10^{-15}–10^{-12} M, depending primarily on the concentration of DOM [62–64]. The primary source of $^1O_2$ in natural waters is through energy transfer reactions involving excited state triplet DOM [63], with production rates ranging from 10^{-9} to 10^{-7} M s^{-1} [62,63]. In humic isolates and natural water samples, quantum yields for $^1O_2$ production (ca. 0.005–0.03) decrease with increasing wavelength in the UVR [8,65] a trend that has also been observed for many other photoproduced species in natural waters (for reviews

Table 1. Rate constants for reaction of $^1O_2$ with selected organic compounds in aqueous solution at or near room temperature.[a] Combined quenching and reaction rate constants are denoted by*.

Compound	pH	k_r ($l\ mol^{-1}s^{-1}$)	Reference
Olefins (crocin, bixin)	7.8	$\sim 10^8$	[169]
1,4-Naphthalenedipropionate ion	NA	1.4×10^6	[170]
Hydroquinone	NA	2.5×10^7	[171]
2-Hydroxybenzoate	7.5	2.5×10^5	[172]
Phenol	8.0	1.0×10^6	[173]
4-Chlorophenol	NA	6.0×10^6	[22]
2,5-Dimethylfuran	7.0	8.2×10^8	[174]
Indole	7.2	7.0×10^7*	[175]
Alanine	8.4	2.0×10^6	[176]
Arginine	7.1	$< 1.0 \times 10^6$	[177]
Cytochrome B	7.4	1.4×10^8*	[178]
Superoxide dimutase	7.1	2.7×10^9	[179]
Histidine	7.0	9.0×10^7	[180]
Tryptophan	7.4	6.6×10^7	[181]
Diethyl sulfide	7.0	2.0×10^7	[174]
Methionine	7.0	2.1×10^7	[180]
Ascorbate	6.8	8.3×10^6*	[182]
DNA	7.0	5.1×10^5*	[183]

[a]NA: Solution pH not reported.

see [41,42]). Singlet oxygen rapidly decays back to ground state triplet oxygen almost entirely through physical quenching by water. This process effectively removes nearly all of the $^1O_2$ that is formed and limits its lifetime to approximately 4 μs in water [66]. Ground state DOM can reduce the lifetime of $^1O_2$ even further via quenching, but only at extremely high DOM concentrations [64].

Because water is so effective in removing $^1O_2$, singlet oxygen is not expected to affect the concentrations of most organic or inorganic compounds in natural waters [67], even though reaction rate constants for these reactions can be large (Table 1). Only a few compounds have been shown to react with $^1O_2$ at appreciable rates [68]. Singlet oxygen should not be important in the removal of most organic compounds from natural waters because they are present at trace levels ($\leq 1 \times 10^{-6}$ M), and will not effectively compete for $^1O_2$ relative to its rate of physical quenching. For example, the reaction of $^1O_2$ with dimethyl sulfide (DMS) is extremely slow in coastal and oligotrophic seawater (ca. 10^{-16} M s^{-1}) yielding a turnover time of approximately 10 years, based on ambient concentrations of reactants (10^{-9} M DMS, 10^{-14} M $^1O_2$) and a rate constant of 2.0×10^7 l mol^{-1} s^{-1} for diethyl sulfide (Table 1). This turnover time is orders of magnitude longer than observed DMS turnover times (e.g. 1–5 days in Equatorial Pacific surface waters [69]).

8.2.3 Superoxide radical

The superoxide radical (O_2^-)is one of the main ROS formed in sunlit natural waters, and it undoubtedly plays a key role in trace metal cycling and DOM transformations in the water column. Early studies by Petasne and Zika [17] identified possible mechanisms for the production of O_2^- in natural waters and revealed the need for research on the biogeochemical effects of O_2^- in marine and fresh waters. The primary source of O_2^- in natural waters is through the photolysis of DOM and the subsequent reduction of ground triplet state oxygen. Blough and Zepp [4], in their review of ROS in natural waters, outlined two mechanisms that may be responsible for O_2^- production from DOM photolysis. One mechanism involves the direct transfer of an electron from an excited triplet state chromophore such as DOM ($^3DOM^*$) to ground triplet state oxygen [4]:

$$^3DOM^* + {}^3O_2 \rightarrow DOM^{+\cdot} + O_2^- \quad (3)$$

The other proposed mechanism involves the reduction of ground triplet state oxygen by an aqueous solvated electron generated by excited triplet state DOM [70–72]:

$$^3DOM^* \rightarrow DOM^{+\cdot} + e_{aq}^- \quad (4)$$

$$e_{aq}^- + O_2 \rightarrow O_2^- \quad (5)$$

Recently, Thomas-Smith and Blough [73] determined that quantum yields for the production of the aqueous solvated electron in irradiated coastal DOM samples were too low to account for the production of O_2^- (as determined from the yield of H_2O_2), suggesting that reaction (3) is the main source of O_2^- in natural waters.

Typical production rates of O_2^- in seawater range from 2×10^{-12} M s^{-1} in the open ocean to 2×10^{-10} M s^{-1} in coastal water [74]. Once O_2^- is formed, approximately 50–80% disproportionates to H_2O_2 with a second-order rate constant of 2.2×10^4 l mol^{-1} s^{-1} measured in oligotrophic seawater at pH 8.3 [75]. The remaining 20–50% of O_2^- that does not disproportionate to H_2O_2 may be removed through reactions with trace metals or DOM. Superoxide interacts with DOM and a number of environmentally important trace metals in both marine and fresh waters [28,30,31,76–78]. The presence of O_2^-, for example, can result in a significant accumulation of reduced iron (Fe(II)) in surface waters [28], which in turn may initiate further chemical transformations (vide infra). The reduction of iron by O_2^- has biogeochemical implications through increasing the bioavailability of iron, particularly in some marine environments where iron limits phytoplankton growth [79,80].

It has recently been shown that organic Cu-complexes increase the decay rate of O_2^- in natural waters. Voelker et al. [30] found that organically complexed copper significantly lowered steady state O_2^- concentrations in marine waters. Goldstone and Voelker [78] also demonstrated that DOM contains a non-metallic, non-enzymatic fraction that can catalyze superoxide dismutation. When copper–DOM reactions are considered, estimated steady state concentrations of O_2^- in coastal waters are 100- to 1000-fold lower than predicted concentrations, which only consider its decay through bimolecular dismutation [78]. Thus, the photochemical redox cycling of DOM via O_2^- reactions may represent an essential step in the alteration of the optical and biochemical properties of DOM [78].

DOM is also likely to react with O_2^- at appreciable rates to form oxygenated species. However, this will depend on the pH, which regulates the relative amount of the perhydroxyl radical, $HO_2^{\bullet}$ and its conjugate base (O_2^-) [81]. Generally, the perhydroxyl radical is much more reactive than O_2^-. For example, amino acids are approximately two orders of magnitude more reactive towards $HO_2^{\bullet}$ than O_2^- [82]. Since the pK_a for the $HO_2^{\bullet}/O_2^-$ equilibrium is 4.8 in pure water and 4.6 in seawater [75], the predominant species in most natural waters will be the less reactive superoxide anion. Nevertheless, O_2^- is moderately reactive towards some inorganic and organic species (Table 2), but further studies are needed to determine possible reaction mechanisms and to identify specific organic molecules responsible for superoxide–DOM interactions in natural waters.

Table 2. Rate constants for the reaction of different organic compounds with O_2^- in aqueous solution at or near room temperature.[a]

Compound	pH	k (l mol^{-1} s^{-1})	Reference
Cysteine	7.0	$>5 \times 10^4$	[184]
L-Glutamic acid	8.7	<0.39	[185]
Glycine	8.8	<0.42	[185]
Glutathione	7.8	6.7×10^5	[186]
2-Oxoglutarate ion	10.1	<0.30	[187]
Carbonate radical ion	NA	4×10^8	[159]
Nitrite ion	NA	5×10^6	[188]
1,4-Benzoquinone	~7	9×10^8	[189]
3,4-Dihydroxybenzaldehyde	7.0	1.40×10^7	[190]
2,3-Dimethyl-1,4-benzoquinone	7.0	4.5×10^8	[191]
Ethylene	NA	2×10^5	[192]
Ethylenediaminetetraaccetate ion	9.9	<0.01	[187]
Formate ion	10.1	<0.01	[187]
Linoleate ion	11–12	0.01–0.1	[81]
L-Malate ion	10.1	<0.11	[187]
D,L-Methionine	8.3	<0.33	[185]
2-Naphthylamine	NA	1.3×10^7	[194]
Peroxidase (horseradish)	5.5	$\sim 2.5 \times 10^8$	[195]
Superoxide dismutase	6.5–9.0	$0.01–5.4 \times 10^9$	[82,193]

[a]NA: Solution pH not reported.

8.2.4 *Hydrogen peroxide*

Hydrogen peroxide is produced in all natural waters and it is one of the major products formed from the photolysis of DOM. Since Van Baalen and Marler [83] first detected H_2O_2 in the Gulf of Mexico, it has been intensively studied by numerous investigators because of its high concentrations relative to other ROS and due to its potential chemical and biological reactivity. Hydrogen peroxide can oxidize DOM through transformations involving the photo-Fenton reaction [84] and it can affect the redox chemistry of trace metals such as iron, copper and manganese [85–87], making H_2O_2 an important chemical reactant in the aquatic environment. Hydrogen peroxide is also a known cellular oxidant [88,89] and organisms devote considerable metabolic energy to remove this ROS [90].

Numerous studies have been conducted to quantify the temporal and spatial variations in H_2O_2 concentrations in both fresh water and seawater, and to assess the factors that affect its production and loss in the photic zone [83,91–99]. Variations in production and removal rates for H_2O_2 in the photic zone yield daytime H_2O_2 concentrations in the 10^{-8} M range for a variety of natural waters (Table 3). Surface water concentrations in lakes and rivers vary from approximately 5.0×10^{-8} M in clear oligotrophic systems to nearly 5.0×10^{-7} M in humic-rich lakes [92,94,100–102]. The relatively low concentrations of DOM in most marine systems give rise to H_2O_2 concentrations that rarely exceed 2.0×10^{-7} M [83,91,96,97,103,104]. Temporal changes in production and decay rates lead to diurnal variations in the net (observed) H_2O_2 concentration in the photic zone. In a typical diurnal cycle, the concentration of H_2O_2 increases in surface waters after sunrise until it reaches a maximum in the early afternoon between 1200 to 1400 h (local time). As photochemical production rates decrease during the remainder of the day and cease in the evening, the net H_2O_2 concentration decreases due to its biological delay reaching a minimum in the early morning before sunrise [91,92,96,97,100,103,105–108]. These diurnal changes in H_2O_2 concentration closely match changes in solar radiation intensity throughout the day and are especially apparent when the H_2O_2 concentration is calculated on an areal basis (mg m^{-2}). Both gradual and rapid changes in the solar irradiance (e.g., intermittent cloud cover) caused similar changes in the areal concentrations of H_2O_2 during diurnal studies conducted in a variety of Canadian shield lakes [100,108].

The half-life for the biological consumption of H_2O_2 in freshwater ecosystems ranges from a few hours for eutrophic and dystrophic systems to greater than 24 hours for oligotrophic waters [94,100,109]. In marine systems, the half-life for biological consumption of H_2O_2 can be much longer, ranging from 10 hours in coastal waters to 15 days in Antarctic seawater [99,110]. When the half-life for the biological consumption of H_2O_2 is relatively long (e.g., >24 h), then vertical mixing will heavily influence its depth distribution of H_2O_2 in natural waters [39,96,103]. Depending on the intensity of vertical mixing, the residence time (i.e., the time scale for complete turnover) of H_2O_2 in the surface mixed layer can vary from minutes in large turbulent lakes to several hours in small, thermally stratified humic-rich lakes [101,102].

On a global scale, H_2O_2 concentrations generally decrease with increasing latitude in oligotrophic waters (ca. $1.0–2.0 \times 10^{-7}$ M in subtropical regions to 3.0×10^{-8} M in polar regions) [99,111]. This trend can largely be explained in terms of latitudinal gradients in temperature and UVR, both of which decrease with increasing latitude. Temperature inversely affects rates, in part, because apparent quantum yields for the photoproduction of H_2O_2 decrease nearly two-fold per 10°C decrease in temperature [99,112]. This temperature depend-

Table 3. Daytime surface concentrations of H_2O_2 in natural waters.

Location	H_2O_2 ($\times 10^{-9}$ M)	Reference
Polar Ocean		
Antarctic Peninsula	12–21	[99,196]
Paradise Harbor	8–25	[99,196]
Weddell Sea	25–35	[99,111]
Open Ocean		
Bermuda	80–160	[126]
Equatorial Pacific	25–80	D.W. King (Pers. commun.)
Sargasso Sea	93–173	[96,105,133]
Coastal Ocean		
Gulf of Mexico	90–240	[91,123]
Mediterranean Sea	90–130	[103]
Wadden Sea, intertidal zone	1500–4500	[197]
Caribbean Sea	140–470	[93]
Peru upwelling zone	8–50	[233]
Seto Inland Sea	10–400	[107]
Port Aransas, Open Gulf	97–161	[83]
Biscayne Bay	150–275	[110]
Florida Current	55–65	[133]
Laurentian Great Lakes		
Lake Erie	66–220	[101,108,198]
Lake Ontario	38–122	[101,102,108,198]
Oligotrophic Lakes		
Canadian Shield	152–210	[92,101,102,108]
Dystrophic Lakes		
Canadian Shield	121–444	[101,102,108]
Sub-Arctic Canadian Shield	104–620	[100]
Rivers		
Patuxent	12–350	[97]
Orinoco	20–640	[93]
Shark	32–139	[133]
St. Lawrence	69–136	[102]
Geothermal waters		
Yellowstone National Park	20–490	[199]

ent decrease in production is partly compensated by a concomitant decrease in biological H_2O_2 decay rates at low temperatures (vide infra).

8.2.4.1 *Sources of H_2O_2*

The main source of hydrogen peroxide in natural waters is through its photochemical production [113]. However, as will be discussed below, other processes also affect concentrations of H_2O_2 in the water column. These include dry and wet atmospheric deposition, and biological release.

Photochemical production. The photochemical production of H_2O_2 in natural waters occurs through photoreactions initiated by absorption of ultraviolet and visible radiation by DOM [4,94]. Excited state DOM reduces dissolved oxygen to form O_2^-, which then disproportionates to H_2O_2 (Figure 1) [17,93,94,114]. Apparent quantum yields for the photochemical formation of H_2O_2 are remarkably similar in diverse marine [41,93,104,112] and fresh [41,98,115] waters. With few exceptions, apparent quantum yields decrease exponentially from approximately 10^{-3} at 290 nm to 10^{-4} at 400 nm [112]. Based on apparent quantum yield measurements in marine waters, Yocis et al. [104] and Miller [112] showed that the production of H_2O_2 was primarily due to absorption of UV-B (280–320 nm) and UV-A (320–400 nm) radiation by DOM. However, a small fraction of the total production (ca. 10–20%) occurred at wavelengths greater than 400 nm in coaastal and oligotrophic seawater [112]. Since H_2O_2 photoproduction is partly UV-B dependent, increases in UV-B radiation through stratospheric ozone thinning should affect H_2O_2 production rates. In Antarctic waters, increased UV-B resulting from ozone depletion increased predicted H_2O_2 production rates in surface waters from 20–50% [104]. Similarly, results of a modeling study by Scully et al. [116] showed that, under enhanced UV-B conditions, the *relative* increase in H_2O_2 production rates was greater in low DOM, optically clear waters compared to high DOM lake waters.

In addition to apparent quantum yield measurements, studies have been conducted to determine midday photochemical production rates of H_2O_2 in fresh water [92,94,95,98,115] and marine environments [93,103,104,117]. These studies indicate that H_2O_2 production rates are dependent on the solar irradiance, temperature (vide supra) and DOM concentrations. Cooper et al. [115] determined that H_2O_2 production rates varied significantly ($0–74 \times 10^{-7}$ M h^{-1}) when a variety of ground water samples were exposed to 6 h of solar radiation. Ground water production rates increased non-linearly with increasing dissolved organic carbon (DOC) content from 0.22 to 17.8 mg C l^{-1}. A similar trend was observed by Scully et al. [98] who found a significant, non-linear correlation between production rates and DOC content in a series of Canadian Shield lakes. Their data were fitted to a power function, and it was shown that when production rates were normalized to the concentration of DOC, they were not constant, as would be expected from a simple linear fit, but rather they increased with increasing DOC content. This result is not surprising since the chromophoric fraction of DOC in lake waters has been demonstrated to increase with increasing DOC concentration [3].

Atmospheric input. In the atmosphere, the gas phase is an important reservoir of H_2O_2 [43,118], with mixing ratios in the marine boundary layer ranging from 0.1 to 5 ppb [119,120]. At these levels, natural waters are undersaturated with respect to H_2O_2. Therefore H_2O_2 is expected to undergo a net diffusion from the atmosphere into the surface layer of the water body. For example, the diffusion of ppb levels of H_2O_2 into oligotrophic seawater results in a H_2O_2 flux at the ocean's surface (1 m^2 surface area by 1 m deep) of $0.09–4.0 \times 10^{-9}$ M h^{-1}, based on a transfer velocity of 0.6 cm s^{-1} [121]. This air-to-sea flux is significant at the sea surface, corresponding to approximately 1–40% of summertime photochemical production rates. However, this flux will be insignificant deeper in the water column due to the dilution of air-derived H_2O_2 in seawater. The air-to-sea flux of H_2O_2 does not take into account turbulent mixing and wave action, and is most likely an underestimation of the true flux. Additionally, fluxes will vary spatially and temporally due to changes in gas phase H_2O_2 concentrations, which are controlled by physical and photochemical processes in the troposphere.

Wet deposition can also be a significant source of H_2O_2 in surface waters [96,122–126]. The concentration of H_2O_2 in rain is typically 1.0×10^{-5} M [122], which is one to three orders of magnitude higher than surface concentrations in fresh water or seawater. Therefore, rain events can rapidly increase H_2O_2 concentrations in the water column. During a rainstorm in the Gulf of Mexico, Cooper et al. [123] reported that the in situ concentration of H_2O_2 increased from 8.5 to 18.5×10^{-8} M at a depth of 1 m over a period of 127 min. In another rain event in the Gulf of Mexico, the concentration at 1 m increased from 8.6 to 20.3×10^{-8} M in 25 min. Marine rain events can also increase H_2O_2 concentrations in the mixed layer down to 50 m or more [123]. Miller and Kester [96] and Kieber et al. [126] noted a 50–200% increase in surface H_2O_2 concentrations due to rain inputs into marine waters near Bermuda in the Sargasso Sea.

Other wet deposition sources include snow, melting glaciers, and run-off from snowmelt and sea ice. Snow samples collected from a remote location at Palmer Station, Antarctica had an average concentration of 5.6×10^{-6} M H_2O_2 (G.W. Miller and D.J. Kieber, unpublished results). Snowmelt run-off increased concentrations of H_2O_2 in surface seawater by more than a factor of two. In contrast, ice melt from sea ice was only slightly higher than surface seawater concentrations (G.W. Miller and D.J. Kieber, unpublished results).

Biological sources. While most if not all microorganisms actively decompose H_2O_2 (vide infra), it is somewhat surprising that some cyanobacteria and eukaryotic phytoplankton produce this compound in natural waters [105,127–131]. One important pathway for the algal mediated-formation of H_2O_2 occurs under nitrogen limiting conditions when algae acquire nitrogen by cell surface enzymatic deamination of dissolved L-amino acids (or amines) to form the ammonium cation, which is subsequently taken up by the cell. By products of this reaction, including H_2O_2 and organic species such as α-keto acids, are released into seawater and not taken up by the algae [130,131]. Not all algae produce H_2O_2 under these conditions, indicating that this process is not universal [128]. The biological production of H_2O_2 has also been noted in cultures of the icthyotoxic flagellate *Heterosigma akashiwo* [132]. Twiner and Trick observed that this toxic phytoplankton produced substantial amounts of H_2O_2 (up to 7.6 pmol min^{-1} $cell^{-1}$). The rate of peroxide production was stimulated by increasing temperature and was regulated by the availability of iron but was independent of light. Results of their study suggest that extracellular production of H_2O_2 occurs through metabolic pathways not directly linked to photosynthesis. These findings are very intriguing given that microorganisms generally decompose H_2O_2 to alleviate its toxicity.

Dark production rates for H_2O_2 were measured in the Sargasso Sea by Palenik et al. [105]. They collected samples from depths between 0 and 130 m and incubated them for 8 h in the dark. Seawater samples from the thermocline (40 to 60 m) yielded the highest H_2O_2 production rates within the first two hours (1.1×10^{-8} M h^{-1}) and averaged 1.0 to 3.0×10^{-9} M h^{-1} for the entire incubation study. Presumably algae were the source of the H_2O_2, since hydrogen peroxide production was not observed when seawater samples were first filtered through a 1 μm filter. However, the depth corresponding to the maximum biological production rate for H_2O_2 (ca. 40–60 m) did not coincide with the chlorophyll maximum (90 m). It is interesting to note that the range of production rates reported by Palenik et al. [105] is comparable to the range of particle-mediated, dark production rates determined by Moffett and Zafiriou ($0.8–2.4 \times 10^{-9}$ M h^{-1}) [106] in coastal seawater (Vineyard Sound, MA). Although microorganisms may be a potentially important source of H_2O_2 deeper in the water column (e.g., below the pycnocline) because photochemical production rates are comparatively small, these studies indicate that the biological production of H_2O_2 should be a minor source of this compound in surface waters compared to its photochemical production, which ranges from 9×10^{-9} M h^{-1} in the Mediterranean Sea [103] to greater than 134×10^{-9} M h^{-1} in high DOM coastal water from the Orinoco estuary [93].

Sources involving trace metal reactions. Hydrogen peroxide is formed from the oxidation of inorganic Cu(I) complexes through their reaction with O_2^- [77]:

$$Cu(I) + O_2^- + 2H^+ \rightarrow Cu(II) + H_2O_2 \qquad (6)$$

The observed rate constant for this overall reaction is approximately 2×10^9 l mol^{-1} s^{-1}, depending on the type of Cu(I) complex considered. In all cases, the rate of the reverse reaction is negligible [77]. Midday concentrations of Cu(I) in surface seawater range from 1.0 to 1.3×10^{-10} M in coastal and oligotrophic waters, which accounts for no more than 10 – 15% of total Cu concentrations in these waters [133]. Since Cu(II) is the predominant Cu species present in the water column, a competing reaction is the reduction of inorganic Cu(II) complexes by O_2^- ($k = 0.7 \times 10^9$ l mol^{-1} s^{-1}) yielding molecular oxygen:

$$Cu(II) + O_2^- \rightarrow Cu(I) + O_2 \qquad (7)$$

The rate for reaction (7) is approximately 30% faster than reaction (6) at a Cu(I)/Cu(II) ratio of 0.25.

Zafiriou et al. [77] observed that reactions (6) and (7) are more important in the removal of O_2^- in seawater than its bimolecular disproportionation to H_2O_2. Their conclusion is similar to that of Petasne and Zika [17] who found that approximately 20–40% of O_2^- did not disproportionate to H_2O_2 in seawater but rather decayed through other unknown reactions. Because kinetic data for representative organic complexes were not available, Zafiriou et al. [77] did not consider possible contributions from the reaction of organic copper complexes with O_2^-, even though these complexes are likely to control copper speciation in marine waters [25,85]. In a follow-up study, Voelker et al. [30] observed that naturally occurring organic copper(II)-complexes degraded O_2^- at a rate comparable to the inorganic copper(II) complexes previously studied by Zafiriou et al. [77].

8.2.4.2 H_2O_2 removal pathways

Biological removal. The major removal pathway for H_2O_2 in natural waters is through algal and bacterial consumption, presumably mediated by enzymatic processes as a detoxification mechanism. Petasne and Zika [110] observed that the biologically mediated loss of H_2O_2 was due to prokaryotic microorganisms and, to a lesser extent, eukaryotic microorganisms. The bacterium *Vibrio pelgius*, with estimated natural levels at 10^8 cells l^{-1}, removed H_2O_2 in the dark with a second order rate constant of 8.81×10^{-11} l cell^{-1} h^{-1}. This translates to a H_2O_2 loss of 2×10^{-9} M h^{-1} due to *Vibrio pelgius* at ambient H_2O_2 concentrations measured in Biscayne Bay, FL. This is a significant fraction of the total biological consumption ($5-16 \times 10^{-9}$ M h^{-1}) that was observed due to a natural assemblage of microorganisms in unfiltered Biscayne Bay seawater [110]. If samples were sterilized by autoclaving, no H_2O_2 loss was observed in the dark. In addition, H_2O_2 loss was not observed in 0.2 μm filtered seawater, but was observed after the addition of bacteria. The use of 0.2 μm filters to sterilize marine, lake or hydrothermal water samples almost completely stopped the loss of H_2O_2 in these samples [95,109]. These results are consistent with a biological rather than an abiological removal mechanism for H_2O_2. The residual loss of H_2O_2 that was observed in some filtered water samples may be due to H_2O_2 reactions with DOM or trace metals [95].

Zepp et al. [129] studied H_2O_2 cycling in nine different algal cultures, including cultures of cyanobacteria and green algae, and determined that the mean second order rate constant for the dark loss of H_2O_2 was 4×10^{-3} m^3 (mg chl *a*)$^{-1}$ h^{-1}. Using this rate constant, and typical concentrations of chlorophyll *a* of 1 and 10 mg m^{-3} (G.L. Boyer, personal communication) and H_2O_2 levels of 3 and 15×10^{-8} M in Antarctic [104] and coastal seawater [110], respectively, the calculated loss rate of H_2O_2 is 1.2×10^{-10} M h^{-1} in Antarctic seawater and 6×10^{-9} M h^{-1} in coastal seawater. Calculated loss rates are in good agreement with the average loss rate of 5×10^{-9} M h^{-1} determined at four coastal stations in the Caribbean Sea and Orinoco River outflow [93], and the loss rate of 7×10^{-11} M h^{-1} determined at an oligotrophic Antarctic station [99]. These loss rates were all determined during non-bloom, low chlorophyll conditions. However, it is important to note that algal blooms will likely yield much faster removal rates for H_2O_2 in the water column. Likewise, since these biologically-mediated decay studies were conducted in the dark, any additional effect of light on the biological removal of H_2O_2 is unknown.

Trace metal reactions. Hydrogen peroxide is an important reactive redox intermediate in natural waters for reactions involving biologically important trace metals such as iron, copper and manganese [85,87,134,135]. Perhaps the most environmentally and biologically significant aspect of H_2O_2 is its capacity to react with trace metals to form the highly reactive hydroxyl radical. As noted previously in this chapter, the hydroxyl radical can rapidly oxidize organic matter, transform anthropogenic organic pollutants into either toxic or inert compounds [21,136], and damage cell membranes in aquatic microorganisms. The toxicological response of bacterioplankton to H_2O_2 in humic-rich lakes is probably due to the production and subsequent reactions of the $^{\bullet}OH$ radical [137,138]. The high concentration of iron (>200 $\mu g\ l^{-1}$) and low pH (<5.5) that are typical of some humic-rich lakes and rivers are ideal conditions for efficient $^{\bullet}OH$ radical formation. The $^{\bullet}OH$ radical is formed by the oxidation of reduced iron via the photo-Fenton reaction (so-called because the reactants Fe(II) and H_2O_2 are formed from photochemical processes):

$$Fe(II) + H_2O_2 \rightarrow Fe(III) + OH^- + {}^{\bullet}OH \quad (8)$$

As may be expected, this reaction is affected by pH, ionic strength and temperature [86]. An analogous Fenton-type reaction is also observed for Cu(I) and H_2O_2 (for review see [135]). The pseudo-first order rate constant for this reaction in seawater is $5 \times 10^{-3}\ s^{-1}$ at 25°C and 1.0×10^{-7} M H_2O_2 [85]. This decomposition pathway is insignificant in oligotrophic seawater at subnanomolar concentrations of photochemically produced Fe(II) [139]. For example, at typical concentrations of 1.0×10^{-11} M Fe(II) and 1.0×10^{-7} M H_2O_2, the rate of loss of H_2O_2 is only 1.8×10^{-10} M h^{-1}, which is over two orders of magnitude slower than H_2O_2 production rates in these waters. Molecular oxygen will also react with Fe(II) according to reaction (9) at rates comparable to those of reaction (8):

$$Fe(II) + O_2 + \rightarrow Fe(III) + O_2^- \quad (9)$$

At a dissolved oxygen concentration of 2.1×10^{-4} M, the reported pseudo-first order rate constants for this reaction range from $5.8 \times 10^{-4}\ s^{-1}$ to $2.2 \times 10^{-3}\ s^{-1}$ [134,140–142], with an average value of $1.2 \times 10^{-3}\ s^{-1}$.

At low pH (ca. 3–5), H_2O_2 can be consumed through iron cycles associated with the oxidation of fulvic acids involving the photo-Fenton reaction (reaction 8) [84]. As previously discussed, reactions (8) and (9) may be important in humic-rich natural waters at low pH such as in black-water rivers of the southeastern United States (e.g., Suwanee River, GA), but their importance in marine waters at higher pH (typically ca. pH 8.2) is unlikely except for some organic rich coastal environments (vide supra).

Hydrogen peroxide, produced from the photolysis of organic matter, has been shown to reduce insoluble *synthetic* manganese oxides to soluble Mn(II) in natural waters, with a second order rate constant of 40 l $mol^{-1}\ s^{-1}$ [87]. However, *natural* oxides were not significantly reduced by H_2O_2. Observed rates of photochemical formation of Mn(II) in the natural oxide samples were only one-sixth of those observed with synthetic oxides. These results coupled with DOM photolysis experiments and catalase addition studies indicate that H_2O_2 reduction is only a minor source of dissolved manganese in seawater, accounting for only 10–20% of the total Mn(II) signal. The main chromophores and reductant(s) responsible for Mn photoreduction are not known, but are speculated to arise from the bacterial manganese oxides themselves [87].

Primary photolysis. Even though biologically-mediated decay is generally considered to be the main pathway for removal of H_2O_2 in seawater, direct photochemical loss also occurs:

$$H_2O_2 + h\nu \rightarrow 2^{\bullet}OH \tag{10}$$

The primary photolysis of H_2O_2 in aqueous solution yields the $^{\bullet}OH$ radical [143]. However, in natural waters, H_2O_2 has very little absorption at wavelengths greater than 290 nm and, therefore, only a very small cross section with actinic solar radiation. As a result, the photochemical decay rate of H_2O_2 is slow in natural waters (vide infra) and $^{\bullet}OH$ radical yields are small compared to other photochemical sources of $^{\bullet}OH$ (e.g., DOM, nitrate, nitrite) [47]. In the field, H_2O_2 photolysis rates are less than 5% of photochemical production rates in seawater from temperate latitudes [49], and approximately 50% of midday production rates in polar regions, based on in situ drifter experiments [104]. These contrasting results are likely due to the much lower photochemical production rates of H_2O_2 in polar waters (ca. factor of five lower) compared to lower latitudes, primarily due to differences in temperature. These results are not expected to be due to differences in photolysis rates, since the rate of primary photolysis of H_2O_2 should not be affected by temperature and should therefore be similar in these waters (and only affected by differences in the scalar irradiance).

8.2.4.3 Reactions of H_2O_2 with DOM

Based on standard state reduction potentials, H_2O_2 should be a strong oxidant in both acidic and basic solutions [144]:

$$H_2O_2 + 2H^+ + 2e^- \rightarrow 2H_2O \qquad E_H{}^{o} = 1.77\ V \tag{11}$$

$$HO_2{}^- + H_2O + 2e^- \rightarrow 3OH^- \qquad E_H{}^{o} = 0.87\ V \tag{12}$$

However, experimental results indicate that H_2O_2 is not very reactive in aqueous solution in the absence of catalysts such as oxidases, Fe(II) or light [21]. Uncatalyzed redox reactions involving H_2O_2, especially those with organic compounds, proceed very slowly in aqueous solution even though they are thermodynamically favored in some cases. Presumably these reactions are slow due to a high kinetic barrier. For example, the standard state free energy for the oxidation of dimethyl sulfide to dimethyl sulfoxide by H_2O_2 is -223 kJ mol^{-1}, indicating that this oxidation is thermodynamically quite favorable under most reaction conditions. However, this oxidation proceeds slowly with a small rate constant, 0.14 l mol^{-1} s^{-1} or less (Table 4). To further illustrate this lack of reactivity, we compiled a selection of the relatively few rate constants that have been reported for the uncatalyzed oxidation of organic species by H_2O_2 in aqueous solution (Table 4). Indeed, although H_2O_2 reacts with a wide variety of organic compounds, rate constants for these reactions are quite small (ranging from 3.0×10^{-6} to 10^2 l mol^{-1} s^{-1}) relative to comparable reactions involving other ROS (vide supra). A notable exception is the methyl radical [145], but even this rate constant (2.7×10^4 l mol^{-1} s^{-1}) is small relative to reactions of the methyl radical with other species. The poor reactivity of H_2O_2 in uncatalyzed

Table 4. Rate constants for the uncatalyzed reaction of different organic compounds with H_2O_2 in aqueous solution at or near room temperature, except where noted.[b]

Compound	pH	k (l mol^{-1} s^{-1})	Reference
Acetylacetone	9.3 (35 °C)	0.203	[200]
L-Ascorbic acid	NA[a]	6.33×10^{-5}	[201]
	0.9 – 1.6	6.28×10^{-8}	[202]
1,2 Dihydroxyanthraquinone (anizarin)	10.0	2.78×10^{-4}	[203]
Benzene sulfinate	6.4 (40 °C)	0.0213	[204]
t-Butoxy chloride	7.0	53	[205]
Carbon disulfide	9.0–11.0	10 – 100	[206]
	9.0	107	[207]
Chloramine	6.9	0.40	[208]
N-Chloromethanesulfonamide	5.2 (0 °C)	3.98	[208]
N-Chlorosuccinimide	7.7	11.0	[205]
8,8-Diapo-ψ,ψ-carotenoic acid (crocetin)	10.0	3.92×10^{-4}	[203]
N-Dichloromethanesulfonamide	5.0 (0 °C)	25.7	[208]
N,N'-Diethylthiourea	1.0	0.563	[209]
N,N'-Dimethylaniline	7.0	3.2×10^{-5}	[210]
N,N'-Dimethylbenzylamine	7.0	3.0×10^{-6}	[210]
Dimethyl sulfide	7.0	1.6×10^{-2}	[211]
	6.0–7.0	0.014–0.036	[212]
	2.0–9.0	0.018	[213]
Cysteine	7.4 (37 °C)	2.9	[214]
Dimethyl sulfoxide	2.0	2.7×10^{-6}	[213]
	13.0	8.4×10^{-5}	[213]
N,N'-Dimethylthiourea	1.0	1.55	[209]
Dithiothreital	7.4 (37 °C)	0.36	[214]
Ethylidene acetone	NA	0.077	[215]
Glutathione	7.0	0.9	[216]
Isopropoxymethylphosphoryl fluoride	7.0	4.33×10^{-4}[b]	[217]
2-Mercaptoethanol	10.0	10.2	[218]
o-Mercaptophenylacetic acid	3.9	1.84×10^{-7}[c]	[219]
Mesityl oxide	NA	0.0137	[215]
Methyl radical	NA	2.7×10^{4}	[145]
Methyl viologen (paraquat)	2.0–12.0	2.0	[220]
S-Nitrosocysteine	7.4	3.9×10^{-3}	[221]
Oxymyoglobin	5.5–10.2	20.8	[222]
Paraoxon	9.5	0.188	[223]
Phenolphthalein	9.0	1.96×10^{-3}	[224]
Pyridine	7.0	1.2×10^{-5}	[210]
Thiourea	0.5	0.563	[209]
Thioxane	2.7	2.68×10^{-3}	[225]
Tryptophan	8.4 (50 °C)	5.03×10^{-4}	[226]

[a]NA: Solution pH not reported.
[b]Pseudo first order (s^{-1}).
[c]Pseudo zero order (mol l^{-1} s^{-1}).

oxidation reactions has also been indirectly observed in natural waters. Specifically, there is no evidence for any significant H_2O_2 reactions in filtered oligotrophic waters, with low DOM and metal concentrations [92]. No loss of H_2O_2 was also observed in 0.2 μm filtered Antarctic seawater that was incubated in the dark for 30 days [99]. Therefore, although H_2O_2 is thought of as a reactive oxidant, in fact it is relatively unreactive in many natural waters, except for low pH and high DOM/metal environments. Hydrogen peroxide is, however, reactive in biological systems because organisms need to remove this source of oxidative stress, as noted in Section 8.4. In cellular systems, H_2O_2 undergoes reactions including enzymatic decay, Fenton chemistry or sulfur oxidations [90] either at the cell surface or within the cell; and it is these processes, especially enzymatic decay, that represent the main removal pathway for H_2O_2 in the water column (vide supra).

8.3 Other ROS

The photooxidation of DOM in natural waters results in the formation of a variety of other ROS in addition to those already discussed. Total production rates of alkoxy ($RO^•$) and peroxy radicals ($ROO^•$) have been determined in natural waters using probe molecules [18,146], and individual species have been inferred from the detection of carbon-centered radical precursors including the methyl, propionyl and acetyl radicals [147,148]. Carbon-centered radical precursors have also been detected in the high molecular weight fraction in DOM [148]. From these studies, it is evident that total production rates are relatively low (10^{-12} M s^{-1} range), accounting for $<1\%$ of total radical production rates determined in seawater [10]. These low rates are due to the much lower apparent quantum yields for the production of these species (ca. $\leq 10^{-5}$) compared to those of the main ROS detected in natural waters (ca. $\leq 10^{-3}$). Apart from production rate studies, no investigations have been undertaken to determine the fate of these classes of ROS in natural waters. It is expected that alkoxy and peroxy radicals will decay to stable, diamagnetic species, but the main pathways are not known. One potential decay pathway is through H-atom abstraction to form alcohols or organic peroxides. This pathway is most likely not a significant source of organic peroxides in natural waters, since they (e.g., methyl and ethyl peroxide) were not detected in the low nM range in irradiated coastal seawater [112]. Other potential products including organic acids and aldehydes have been detected (for recent reviews see [42,149]), but they have not been linked to alkoxy and peroxy precursors. Likewise, reactions of these ROS with DOM or other inorganic radicals have been postulated but not yet verified experimentally.

The carbonate radical has also been shown to be an important radical intermediate in natural waters. In freshwater environments, it is formed from the reaction of the $^•OH$ radical with the carbonate and bicarbonate ions:

$$^{\bullet}OH + HCO_3^- \rightarrow H_2O + CO_3^- \tag{13}$$

$$^{\bullet}OH + CO_3^{2-} \rightarrow OH^- + CO_3^- \tag{14}$$

and in seawater by reaction of the Br_2^- radical with the carbonate anion:

$$Br_2^- + CO_3^{2-} \rightarrow 2Br^- + CO_3^- \tag{15}$$

The carbonate radical will also form its conjugate acid, the bicarbonate radical:

$$HCO_3^{\bullet} \rightleftharpoons CO_3^{-} + H^{+} \quad (16)$$

The bicarbonate radical, $HCO_3^{\bullet}$, is more acidic than HCO_3^- (pK_a = 10.3 vs. 10^6, respectively), but the pK_a for equilibrium (16) is poorly constrained with reported values between 7.0–9.6 [150–152]. In many fresh water systems the predominant species will be the bicarbonate radical, while both the bicarbonate and carbonate radicals will be important in seawater. As may be expected, the reactivity of the carbonate radical with many organic and inorganic species will be pH dependent [153].

Huang and Mabury [154] used a radical trapping approach to determine steady state concentrations of the carbonate radical in a variety of fresh water environments. In their study, carbonate radical concentrations ranged from 5×10^{-15} to 10^{-13} M, and were a function of DOM and nitrate concentrations (which are the main sources of the $^{\bullet}OH$ radical in these waters), and alkalinity. The wavelength dependence and quantum yields for carbonate radical production are not known, but it is expected that they will mirror the action spectra for $^{\bullet}OH$ radical production, and therefore will largely be confined to wavelengths in the UV-B extending out to approximately 330–340 nm [47].

The main fate of the carbonate radical is likely through reaction with DOM [155]. This would not be surprising, since the carbonate radical behaves as a strong oxidant with a wide range of organic compounds including sulfides, phenols, aromatic anilines, amino acids and proteins. It primarily reacts through a one-electron oxidation pathway, although addition or H-atom abstraction pathways have also been reported [156,157]. The one-electron oxidation pathway yields oxidized DOM and the carbonate anion:

$$CO_3^{-} + DOM \rightarrow CO_3^{2-} + DOM^{+} \quad (17)$$

Rate constants for reaction of the carbonate radical with a variety of organic compounds in aqueous solution range from less than 10^3 to 10^9 l mol^{-1} s^{-1} (Table 5). A comprehensive tabulation of rate constants is given in Neta et al. [158]. Many of these reactions are temperature dependent characterized by an Arrhenius response [156]. Competing with DOM, the carbonate radical is fairly reactive towards inorganic species in aqueous solution (Table 5) including O_2^- and H_2O_2 [159]:

$$H_2O_2 + CO_3^{-} \rightarrow \text{products} \qquad k = 8 \times 10^5 \text{ l mol}^{-1} \text{ s}^{-1} \quad (18)$$

$$O_2^{-} + CO_3^{-} \rightarrow CO_3^{2-} + O_2 \qquad k = 4 \times 10^8 \text{ l mol}^{-1} \text{ s}^{-1} \quad (19)$$

At an O_2^- and CO_3^- concentration of 10^{-9} M and 10^{-14} M, respectively, the carbonate radical loss rate is 4×10^{-15} M s^{-1}. This removal rate is comparable to some of the faster organic reactions noted above. However, when all of the organic reactants are considered together, it is likely that they will control the fate of the carbonate radical in the photic zone. The importance of these pathways will also depend on other factors such as the concentration of DOM, which indirectly controls the production and loss of H_2O_2 and O_2^-.

Table 5. Rate constants for the reaction of different organic compounds with the carbonate radical in aqueous solution at or near room temperature.

Compound	pH	k ($l\ mol^{-1}\ s^{-1}$)	Reference
Formate ion	6.4	1.1×10^5	[227]
Methanol	13.0	6.02×10^3	[157]
Acetate ion	12.1–12.7	6.0×10^2	[228]
Propanol	13.0	4.65×10^4	[157]
Acetone	12.1–12.7	1.6×10^2	[228]
Ascorbate ion	11.4	1.2×10^9	[156]
Glucose	7.0	7×10^4	[229]
Cyclohexanol	13.0	9.37×10^4	[157]
Tetrahydrofuran	13.0	1.00×10^5	[157]
Indole	7.0	4.1×10^8	[230]
Benzene	7.0	3×10^3	[229]
Hydroquinone	11.4	1.9×10^9	[156]
Phenolate ion	11.4	1.3×10^9	[156]
Phenol	7.0	2.2×10^7	[229]
Aniline	9.0	4.6×10^8	[234]
Urea	7.0	$<1 \times 10^3$	[229]
Uracil	7.0	$<1 \times 10^4$	[229]
Butylamine	11.5	4.0×10^5	[231]
Glutathione	7.0	5.3×10^6	[229]
Thioanisole	11.0	2.3×10^7	[155]
Dithiothreitol	10.5	4.1×10^8	[232]
Glycine	7.0	$<1 \times 10^3$	[229]
Alanine	7.0	$<1 \times 10^3$	[229]
Arginine	7.0	9×10^4	[229]
Methionine	7.0	3.6×10^7	[229]
Tryptophan	7.0	7.0×10^8	[230]
Glycylglycyltryptophan	7.0	7.0×10^8	[230]
Lysozyme	7.0	5.5×10^8	[229]
Trypsin	7.0	6.8×10^8	[229]
Carbonate radical	7.0–9.0	2.0×10^7	[227]
Nitrite ion	11.4	4.8×10^5	[156]

8.4 Impact of ROS on aquatic organisms

Intracellular production of ROS such as O_2^-, H_2O_2, $^1O_2$ and the $^{\bullet}OH$ radical present some of the most pressing challenges to biological systems [37,90]. ROS react with biomolecules, such as proteins, lipids or DNA, modifying or destroying their intended functionality in a process known as oxidative stress. Organisms invest considerable cellular resources to prevent or repair oxidative damage and this is especially true for photosynthetic organisms, which generate ROS as part of the photosynthetic process [2]. While it is well documented that ROS produced inside the cell can adversely affect an organism, very little is known about the effect of externally generated ROS (i.e., species produced in the water surrounding the cell) on the growth and health of aquatic organisms.

In general, the impact of ROS on aquatic organisms should be inversely related to the reactivity of the ROS. For example, the direct effect of externally-generated $^{\bullet}OH$ radicals on aquatic microorganisms is expected to be minimal because it is extremely reactive, which results in an extremely short lifetime that does not allow for significant transport to the cell surface [160]. However, production of the $^{\bullet}OH$ radical outside the cell may exert an indirect effect on cellular systems through the production of longer-lived free radical species such as the carbonate radical (which can potentially affect algal uptake of inorganic carbon) and the dibromide radical ion (Figure 1). These longer-lived free radical species are present at higher steady state concentrations in natural waters compared to the $^{\bullet}OH$ radical (ca. 10^{-14} vs. 10^{-18} M) [4], and they are less reactive than the $^{\bullet}OH$ radical as indicated by their relatively long half-lives. As an example, the half-life of the carbonate radical is typically longer than 1.7 h in natural waters [154], which is more than enough time to allow for significant transport to the cell surface. Even though these ROS are less reactive than the $^{\bullet}OH$ radical they are nevertheless still quite reactive (vide supra) and have the potential to react at the surface of the cell, as will be discussed below.

Chemical reactivity is not the only factor to consider in assessing the impact of ROS on microorganisms. For a ROS to affect an organism it should be cell-permeable such that it reacts with and deactivates critical cellular functions. Charged radical species, including the dibromide anion, are unlikely to significantly affect cellular functions because they are not expected to diffuse through the cell membrane. Instead, these charged radicals will possibly react with extracellular surface proteins or carbohydrates, which may represent a chronic stress to microorganisms through inactivation of extracellular surface enzymes or transport proteins. Proteins with a high sulfhydryl content may be particularly susceptible, as the sulfur moieties should be highly reactive with a number of ROS such as the carbonate radical. Inactivation of transport proteins by ROS may partly explain the observed inhibition of amino acid uptake or inorganic nitrogen uptake in marine algae exposed to high UVR [161,162]. It is more likely, however, that these charged radicals will react with the much more abundant carbohydrates, which make up the cell wall or that are exuded by algae and bacteria [163]. Reaction with surface carbohydrates should result in minimal damage to the cell, and perhaps this explains why extracellular carbohydrates are produced in response to oxidative stress.

Hydrogen peroxide is not charged, unlike most of the other ROS, and therefore it will freely permeate through the cell membrane. However, it is not known whether ambient levels of H_2O_2 exert an acute or chronic stress on aquatic organisms. Studies conducted thus far point to an acute toxicological response of plankton to relatively high concentrations of H_2O_2. Barrion and Feuillade [164] found H_2O_2 to be an effective algicide in the treatment of the cyanobacteria *Oscillatoria rubescens* in lake water, but only at 10^{-3} M concentrations which are much greater than ambient concentrations. In contrast, relatively little inhibition in bacterial growth rates was observed when bacterioplankton were exposed to 10^{-9} to 10^{-6} M concentrations of H_2O_2 in humic-rich lake waters [137]. In a similar study, Mopper et al. [138] found that much higher concentrations of

H_2O_2 ($2–3 \times 10^{-6}$ M) were necessary to inhibit bacterial production in humic-rich lake water. Indirect evidence for the negative biological impact of H_2O_2 has been provided by Hessen and VanDonk [165]. They demonstrated that, when water samples with high humic content were irradiated with an artificial UVR source and subsequently inoculated with a green algal culture, growth rates were inhibited in UVR-exposed samples compared to the dark controls. The authors attributed the reduction in growth rate to the photochemical production of long-lived products such as H_2O_2 in UVR-exposed samples.

The accumulation of H_2O_2 at the surface of lakes during periods of near surface water heating may adversely affect aquatic organisms. Recent studies revealed adverse effects of solar radiation on the near surface phytoplankton communities in stratified humic lakes [166,167]. Although the negative impact may be attributed to the direct effects of UVR, the high concentrations of H_2O_2 at the surface caused by both stratification and high UVR absorbance may have also adversely influenced the phytoplankton. The evidence provided so far is inconclusive and points to a need for further research to study the chronic effects of ambient levels of H_2O_2 on plankton communities in lakes and marine systems. It is expected that the oxidative stress imposed by H_2O_2 is not due to its direct reaction with cellular constituents, but rather is the result of the production of the highly toxic $^{\bullet}OH$ radical inside the cell through the Fenton reaction [168].

ROS may also affect microorganisms through the production of toxic trace metal species. For example, the photolysis of organic Cu-complexes and interactions with O_2^- may increase the Cu bioavailability and hence Cu toxicity to phytoplankton. This interaction with transition metals is likely to be one of the main processes through which photochemically produced O_2^-, or other charged ROS, can have an adverse affect on aquatic biota; but further studies are needed to ascertain the ecological impact of these types of reactions in natural waters.

8.5 Conclusions

ROS are ubiquitous in sunlit surface waters, and they are expected to play a pivotal role in the photooxidation of DOM. However, most of what we know regarding the fate of ROS is from chemical intuition or conjecture, based on related laboratory reactions reported in the literature. For example, the overall importance of ROS in intra-humic transformations is virtually unknown, but we expect that these transformations will be important in natural waters, based on photochemical studies with hydrophobic compounds that strongly bind to DOM [55,56].

Since ROS are formed from the absorption of UVR by DOM and its subsequent photochemical decay, any changes in the atmosphere such as tropospheric warming or stratospheric ozone depletion should affect steady state concentrations of ROS in the water column. Initial studies with H_2O_2 suggest that the formation of an ozone hole will increase production rates by 20–50%. Changes in atmospheric ozone levels are also expected to affect production rates of other

ROS, especially those species whose action spectra are largely confined to the UV-B (e.g., the •OH radical [47]), but few systematic studies have been conducted to evaluate these changes.

Since many aspects of reactive oxygen chemistry are still poorly understood, it is suggested that future research should: (1) identify any significant interactions between ROS and DOM, particularly in the high molecular weight fraction; (2) determine the fate of the less reactive radicals, the dibromide radical, the carbonate radical, and alkoxy and peroxy radicals; (3) assess the impact of ROS on the health and growth of aquatic organisms as well as on ecosystem dynamics; and (4) develop models to predict the likely effect of climate change on the production and loss of ROS in the photic zone.

Acknowledgements

The authors gratefully acknowledge the National Science Foundation (OCE-0096413 and OPP-9610173, DJK) for their support of this work, Dr C. Osborn for discussions pertaining to this manuscript and G.R. Westby, Jr. for assisting with manuscript preparation.

References

1. A. Bricaud, A. Morel, L. Prieur (1981). Absorption of dissolved organic matter of the sea (yellow substance) in the UV and visible domains. *Limnol. Oceanogr.*, **26**, 43–53.
2. J.T.O. Kirk (1994). *Light and Photosynthesis in Aquatic Ecosystems.* Cambridge University Press, Cambridge.
3. N.M. Scully, D.R.S. Lean (1994). The attenuation of ultraviolet radiation in temperate lakes. *Arch. Hydrobiol. Beih Ergebn. Limnol.*, **43**, 135–144.
4. N.V. Blough, R.G. Zepp (1995). Reactive oxygen species in natural waters. In: C.S. Foote, J.S. Valentine, A. Greenberg, J.F. Liebman (Eds), *Active Oxygen in Chemistry* (pp.280–333). Blackie Academic, New York.
5. W.P. Bissett, O. Schofield, S. Glenn, J.J. Cullen, W.L. Miller, A.J. Pleuddeman, C.D. Mobley (2001). Resolving the impacts and feedback of ocean optics on upper ocean ecology. *Oceanogr.*, **14**, 30–53.
6. K. Mopper, D.J. Kieber (2000). Marine photochemistry and its impact on carbon cycling. In: S.J. de Mora, S. Demers, M. Vernet (Eds), *The Effects of UV Radiation in the Marine Environment* (pp.101–129). Cambridge University Press, New York.
7. K. Mopper, D.J. Kieber (2001). Photochemistry and the cycling of carbon, sulfur, nitrogen and phosphorous. In: D.A. Hansell, C.A. Carlson (Eds), *Biogeochemistry of Marine Organic Matter* (pp. 455–507), Academic Press, New York.
8. R.G. Zepp, N.L. Wolfe, G.L. Baughman, R.C. Hollis (1977). Singlet oxygen in natural waters. *Nature*, **267**, 421–423.
9. O.C. Zafiriou, N.V. Blough, E. Micinski, B. Dister, D.J. Kieber, J.W. Moffett (1990). Molecular probe systems for reactive transients in natural waters. *Mar. Chem.*, **30**, 45–70.
10. O.C. Zafiriou, B. Dister (1991). Photochemical free radical production rates: Gulf of Maine and Woods Hole-Miami transect. *J. Geophys. Res.*, **96**, 4939–4945.

11. C.J. Miles, P.L. Brezonik (1981). Oxygen consumption in humic-colored waters by a photochemical ferrous-ferric catalytic cycle. *Environ. Sci. Technol.*, **15**, 1089–1095.
12. M.J. Lindell, H. Rai (1994). Photochemical oxygen consumption in humic waters. *Arch. Hydrobiol. Beih Ergebn. Limnol.*, **43**, 145–155.
13. R.M.W. Amon, R. Benner (1996). Photochemical and microbial consumption of dissolved organic carbon and dissolved oxygen in the Amazon River system. *Geochim. Cosmochim. Acta*, **60**, 1783–1792.
14. B. Reitner, G.J. Herndl, A. Herzig (1997). Role of ultraviolet-B radiation on photochemical and microbial oxygen consumption in a humic-rich shallow lake. *Limnol. Oceanogr.*, **42**, 950–960.
15. H. Gao, R.G. Zepp (1998). Factors influencing photoreactions of dissolved organic matter in a coastal river of the southeastern United States. *Environ. Sci. Technol.*, **32**, 2940–2946.
16. S.S. Andrews, S. Caron, O.C. Zafiriou (2000). Photochemical oxygen consumption in marine waters: A major sink for colored dissolved organic matter. *Limnol. Oceanogr.*, **45**, 267–277.
17. R.G. Petasne, R.G. Zika (1987). Fate of superoxide in coastal seawater. *Nature*, **325**, 516–518.
18. T. Mill, D.G. Hendry, H. Richardson (1980). Free-radical oxidants in natural waters. *Science*, **207**, 886–887.
19. R.G. Zepp, J. Hoigné, H. Bader (1987). Nitrate-induced photooxidation of trace organic chemicals in water. *Environ. Sci. Technol.*, **21**, 443–450.
20. T. Mill (1989). Structure-activity relationships for photooxidation processes in the environment. *Environ. Toxicol. Chem.*, **8**, 31–43.
21. J. Hoigné (1990). Formulation and calibration of environmental reaction kinetics: Oxidation by aqueous photooxidants as an example. In: W. Stumm (Ed.), *Aquatic Chemical Kinetics* (pp.43–70). John Wiley & Sons, New York.
22. P.G. Tratnyek, J. Hoigne (1991). Oxidation of substituted phenols in the environment: A QSAR analysis of rate constants for reaction with singlet oxygen. *Environ. Sci. Technol.*, **25**, 1596–1604.
23. S. Canonica, J. Hoigné (1995). Enhanced oxidation of methoxy phenols at micromolar concentration photosensitised by dissolved natural organic material. *Chemosphere*, **30**, 2365–2374.
24. T. Mill (1999). Predicting photoreactions in surface waters. *Chemosphere*, **38**, 1379–1390.
25. J.W. Moffett, R.G. Zika (1983). Oxidation kinetics of copper(I) in seawater: Implications for its existence in the marine environment. *Mar. Chem.*, 239–251.
26. F.J. Millero, V.K. Sharma, B. Karn (1991). The rate of reduction of copper(II) with hydrogen peroxide in seawater. *Mar. Chem.*, **36**, 71–83.
27. W.L. Miller, D.W. King, J. Lin, D.R. Kester (1995). Photochemical redox cycling of iron in coastal seawater. *Mar. Chem.*, **50**, 63–77.
28. B.M. Voelker, D.L. Sedlak (1995). Iron reduction by photoproduced superoxide in seawater. *Mar. Chem.*, **50**, 93–102.
29. B.M. Voelker, F.M.M. Morel, B. Sulzberger (1997). Iron redox cycling in surface waters: Effects of humic substances and light. *Environ. Sci. Technol.*, **31**, 1004–1011.
30. B.M. Voelker, D.L. Sedlak, O.C. Zafiriou (2000). Chemistry of superoxide radical in seawater: Reactions with organic Cu complexes. *Environ. Sci. Technol.*, **34**, 1036–1042.
31. K.E. Abu-Saba, D.L. Sedlak, A.R. Flegal (2000). Indirect reduction of hexavalent chromium by copper in the presence of superoxide. *Mar. Chem.*, **69**, 33–41.

32. L. Emmenegger, R.R. Schonenberger, L. Sigg, B. Sulzberger (2001). Light-induced redox cycling of iron in circumneutral lakes. *Limnol. Oceanogr.*, **46**, 49–61.
33. S.P. Long, S. Humphries, P.G. Falkowski (1994). Photoinhibition of photosynthesis in nature. *Annu. Rev. Plant Physiol. Mol. Biol.*, **45**, 633–662.
34. W.F. Vincent, P.J. Neale (2000). Mechanisms of UV damage to aquatic organisms. In: S.J. de Mora, S. Demers, M. Vernet (Eds), *The Effects of UV Radiation in the Marine Environment* (pp 149–176). Cambridge University Press, New York.
35. R.G. Zepp, T.V. Callaghan, D.J. Erickson (1995). Effects of increased solar ultraviolet radiation on biogeochemical cycles. *Ambio*, **24**, 181–187.
36. R.G. Zepp, T.V. Callaghan, D.J. Erickson (1998). Effects of enhanced solar ultraviolet radiation on biogeochemical cycles. *J. Photochem. Photobiol.*, **46**, 69–82.
37. P.D. Josephy (1997). *Molecular Toxicology*. Oxford University Press, Oxford.
38. R.G. Zepp (1988). Environmental photoprocesses involving natural organic matter. In: F.H. Frimmel, R.F. Christman (Eds), *Humic Substances and Their Role in the Environment* (pp. 193–214). John Wiley & Sons, New York.
39. R.J. Sikorski, R.G. Zika (1993). Modeling mixed-layer photochemistry of H_2O_2: Physical and chemical modeling of distribution. *J. Geophys. Res.*, **98**, 2329–2340.
40. P.J. Neale, D.J. Kieber (2000). Assessing biological and chemical effects of UV in the marine environment: Spectral weighting functions. In: R.E. Hester, R.M. Harrison (Eds), *Issues in Environmental Science and Technology. No. 14* (pp.61–83). Royal Society of Chemistry, Cambridge.
41. N.V. Blough (1997). Photochemistry in the sea-surface microlayer. In: P.S. Liss, R.A. Duce (Eds), *The Sea Surface and Global Change* (pp.383–424). Cambridge University Press, London.
42. R.F. Whitehead, S. de Mora (2000). Marine photochemistry and UV radiation. In: R.E. Hester, R.M. Harrison (Eds), *Issues in Environmental Science and Technology. No. 14* (pp.37–60). Royal Society of Chemistry, Cambridge.
43. B.J. Finlnayson-Pitts, J.N. Pitts (2000). *Chemistry of the Upper and Lower Atmosphere: Theory, Experiments, and Application*. Academic Press, London.
44. K. Mopper, X. Zhou (1990). Hydroxyl radical photoproduction in the sea and its potential impact on marine processes. *Science*, **250**, 661–664.
45. P.P. Vaughan, N.V. Blough (1998). Photochemical formation of hydroxyl radical by constituents of natural waters. *Environ. Sci. Technol.*, **32**, 2947–2953.
46. W.R. Haag, J. Hoigné (1985). Photo-sensitized oxidation in natural water via OH radicals. *Chemosphere*, **14**, 1659–1671.
47. J.G. Qian, K. Mopper, D.J. Kieber (2001). Photochemical production of the hydroxyl radical in Antarctic waters. *Deep-Sea Res.*, **48**, 741–759.
48. R.G. Zepp, B.C. Faust, J. Hoigné (1992). Hydroxyl radical formation in aqueous reactions (pH 3–8) or iron(II) with hydrogen peroxide: The photo-Fenton reaction. *Environ. Sci. Technol.*, **26**, 313–319.
49. J.W. Moffett, O.C. Zafiriou (1993). The photochemical decomposition of hydrogen peroxide in surface waters of the Eastern Caribbean and Orinoco River. *J. Geophys. Res.*, **98**, 2307–2313.
50. A. Pochon, P.P. Vaughan, D. Gan, P. Vath, N.V. Blough, D.E. Falvey (2002). Photochemical oxidation of water by 2-methyl-1,4-benzoquinone: Evidence for a hydroxylating intermediate from laser flash photolysis and competitive trapping studies. *J. Phys. Chem.*, **106**, 2889–2894.
51. K.A. Thorn, J.B. Arterburn, M.A. Mikita (1992). ^{15}N and ^{13}C NMR investigation of hydroxylamine-derivatized humic substances. *Environ. Sci. Technol.*, **26**, 107–116.
52. D.T. Scott, D.M. Mcknight, E.L. Blunt-Harris, S.E. Kolesar, D.R. Lovely (1998).

Quinone moieties act as electron acceptors in the reduction of humic substances by humic-reducing microorganisms. *Environ. Sci. Technol.*, **32**, 2984–2989.

53. O.C. Zafiriou, M.B. True (1979). Nitrate photolysis in seawater by sunlight. *Mar. Chem.*, **8**, 33–42.
54. G.V. Buxton, C.L. Greenstock, W.P. Helman, A.B. Ross (1988). Critical review of rate constants for reactions of hydrated electrons, hydrogen atoms and hydroxyl radicals in aqueous solution. *J. Phys. Chem. Ref. Data*, **17**, 513–886.
55. S.E. Burns, J.P. Hassett, M.V. Rossi (1996). Binding effects on humic-mediated photoreaction: Intrahumic dechlorination of mirex in water. *Environ. Sci. Technol.*, **30**, 2934–2941.
56. S.E. Burns, J.P. Hassett, M.V. Rossi (1997). Mechanistic implications of the intrahumic dechlorination of mirex. *Environ. Sci. Technol.*, **31**, 1365–1371.
57. N.V. Blough (1988). Electron paramagnetic resonance measurements of photochemical radical production in humic substances. 1. Effects of O_2 and charge on radical scavenging by nitroxides. *Environ. Sci. Technol.*, **22**, 77–82.
58. M.E. Lindsey, M.A. Tarr (2000). Inhibition of hydroxyl radical reaction with aromatics by dissolved natural organic matter. *Environ. Sci. Technol.*, **34**, 444–449.
59. D.R. Kearns (1971). Physical and chemical properties of singlet molecular oxygen. *Chem. Rev.*, **71**, 395–427.
60. F. Wilkinson, W.P. Helman, A.B. Ross (1995). Rate constants for the decay and reactions of the lowest electronically excited singlet state of molecular oxygen in solution. An expanded and revised compilation. *J. Phys. Chem. Ref. Data*, **24**, 663–1021.
61. H.H. Wasserman, R.W. Murray (1979). *Singlet Oxygen*, Academic Press, New York.
62. R.G. Zepp, P. Schlotzhauer, M.R. Sink (1985). Photosensitized transformations involving electronic energy transfer in natural waters: Role of humic substances. *Environ. Sci. Technol.*, **19**, 74–81.
63. W.R. Haag, J. Hoigné (1986). Singlet oxygen in surface waters. 3. Photochemical formation and steady-state concentrations in various types of waters. *Environ. Sci. Technol.*, **20**, 341–348.
64. S.L. Holder Sandvik, P. Bilski, J.D. Pakulski, C.F. Chignell, R.B. Coffin (2000). Photogeneration of singlet oxygen and free radicals in dissolved organic matter isolated from the Mississippi and Atchafalaya River plumes. *Mar. Chem.*, **69**, 139–152.
65. W.R. Haag, J. Hoigné, E. Gassman, A.M. Braun (1984). Singlet oxygen in surface waters – Part 2: Quantum yields of its production by some natural humic materials as a function of wavelength. *Chemosphere*, **13**, 641–650.
66. M.A.J. Rodgers, P.T. Snowden (1982). Lifetime of $O_2(^1\Delta_g)$ in liquid water as determined by time-resolved infrared luminescence measurements. *J. Am. Chem. Soc.*, **104**, 5541–5543.
67. R.A. Larson, K.A. Marley (1999). Singlet oxygen in the environment. In: P. Boule (Ed.), *The Handbook of Environmental Chemistry. Reactions and Processes. Vol.* 2. *Part L. Environmental Chemistry* (pp.123–137). Springer-Verlag, New York.
68. R.M. Baxter, J.H. Carey (1982). Reactions of singlet oxygen in humic waters. *Freshwat. Biol.*, **12**, 285–292.
69. D.J. Kieber, J. Jiao, R.P. Kiene, T.S. Bates (1996). Impact of dimethylsulfide photochemistry on methyl sulfur cycling in the equatorial Pacific Ocean. *J. Geophys. Res.*, **101**, 3715–3722.
70. A.M. Fischer, J.S. Winterle, T. Mill (1987). Primary photochemical processes in photolysis mediated by humic substances. In: R.G. Zika, W.J. Cooper (Eds), *Photo-*

chemistry of Environmental Aquatic Systems (pp.141–156). American Chemical Society, Washington.

71. J.F. Power, D.K. Sharma, C.H. Langford, R. Bonneau, J. Joussot-Dubien (1987). Laser flash photolytic studies of a well-characterized soil humic substance. In: R.G. Zika, W.J. Cooper (Eds), *Photochemistry of Environmental Aquatic Systems* (pp. 157–173). American Chemical Society, Washington.
72. R.G. Zepp, A.M. Braun, J. Hoigné, J.A. Leenheer (1987). Photoproduction of hydrated electrons from natural organic solutes in aquatic environments. *Environ. Sci. Technol.*, **21**, 485–490.
73. T.E. Thomas-Smith, N.V. Blough (2001). Photoproduction of hydrated electron from constituents of natural waters. *Environ. Sci. Technol.*, **35**, 2721–2726.
74. E. Micinski, L.A. Ball, O.C. Zafiriou (1993). Photochemical oxygen activation: Superoxide radical detection and production rates in the Eastern Caribbean. *J. Geophys. Res.*, **98**, 2299–2306.
75. O.C. Zafiriou (1990). Chemistry of superoxide ion-radical (O_2^-) in seawater. I. pK_{asw} (HOO) and uncatalyzed dismutation kinetics studied by pulse radiolysis. *Mar. Chem.*, **30**, 31–43.
76. D.W. King, H.A. Lounsbury, F.J. Millero (1995). Rates and mechanism of Fe(II) oxidation at nanomolar total iron concentrations. *Environ. Sci. Technol.*, **29**, 818–824.
77. O.C. Zafiriou, B.M. Voelker, D.L. Sedlak (1998). Chemistry of the superoxide radical (O_2^-) in seawater: Reactions with inorganic copper complexes. *J. Phys. Chem. A*, **102**, 5693–5700.
78. J.V. Goldstone, B.M. Voelker (2000). Chemistry of superoxide radical in seawater: CDOM associated sink of superoxide in coastal waters. *Environ. Sci. Technol.*, **34**, 1043–1048.
79. J.H. Martin, S.E. Fitzwater (1988). Iron deficiency limits phytoplankton growth in the northeast Pacific subarctic. *Nature*, **331**, 341–343.
80. J.H. Martin, S.E. Fitzwater, R.M. Gordon (1990). Iron deficiency limits phytoplankton growth in Antarctic waters. *Global Biogeochem. Cycles*, **4**, 5–12.
81. B.H.J. Bielski, R.L. Arudi, M.W. Sutherland (1983). A study of the reactivity of HO_2/O_2^- with unsaturated fatty acids. *J. Biol. Chem.*, **258**, 4759–4761.
82. B.H.J. Bielski, D.E. Cabelli, R.L. Arudi, A.B. Ross (1985). Reactivity of HO_2/O_2^- radicals in aqueous solution. *J. Phys. Chem. Ref. Data*, **14**, 1041–1100.
83. C. Van Baalen, J.E. Marler (1966). Occurrence of hydrogen peroxide in seawater. *Nature*, **211**, 951.
84. B.M. Voelker, B. Sulzberger (1996). Effects of fulvic acid on Fe(II) oxidation by hydrogen peroxide. *Environ. Sci. Technol.*, **30**, 1106–1114.
85. J.W. Moffett, R.G. Zika (1987). Reaction kinetics of hydrogen peroxide with copper and iron in seawater. *Environ. Sci. Technol.*, **21**, 804–810.
86. F.J. Millero, S. Sotolongo (1989). The oxidation of Fe(II) with H_2O_2 in seawater. *Geochim. Cosmochim. Acta*, **53**, 1867–1873.
87. W.G. Sunda, S.A. Huntsman (1994). Photoreduction of manganese oxides in seawater. *Mar. Chem.*, **46**, 133–152.
88. A.A. Frimer, A. Forman, D.C. Borg (1983). H_2O_2 diffusion through liposomes. *Israel J. Chem.*, **23**, 442–445.
89. J.L. Pierre, M. Fontecave (1999). Iron and activated oxygen species in biology: The basic chemistry. *Biometals*, **12**, 195–199.
90. H. Seis (1985). *Oxidative Stress*, Academic Press, Orlando.
91. R.G. Zika, J.W. Moffett, R.G. Petasne, W.J. Cooper, E.S. Saltzman (1985). Spatial and temporal variations of hydrogen peroxide in Gulf of Mexico waters. *Geochim. Cosmochim. Acta*, **49**, 1173–1184.

92. W.J. Cooper, D.R.S. Lean (1989). Hydrogen peroxide concentration in a northern lake: Photochemical formation and diel variability. *Environ. Sci. Technol.*, **23**, 1425–1428.
93. C.A. Moore, C.T. Farmer, R.G. Zika (1993). Influence of the Orinoco River on hydrogen peroxide distribution and production in the Eastern Caribbean. *J. Geophys. Res.*, **98**, 2289–2298.
94. W.J. Cooper, C. Shao, D.R.S. Lean, A.S. Gordon, F.E. Scully (1994). Factors affecting the distribution of H_2O_2 in surface waters. In: L.A. Baker (Ed.), *Environmental Chemistry of Lakes and Reservoirs* (pp.391–422). American Chemical Society, Washington.
95. D.R.S. Lean, W.J. Cooper, F.R. Pick (1994). Hydrogen peroxide formation and decay in lake waters. In: G.R. Helz, R.G. Zepp, D.G. Crosby (Eds), *Aquatic and Surface Photochemistry* (pp.207–214). CRC Press, Inc, Boca Raton.
96. W.L. Miller, D.R. Kester (1994). Peroxide variations in the Sargasso Sea. *Mar. Chem.*, **48**, 17–29.
97. R.J. Kieber, G.R. Helz (1995). Temporal and seasonal variations of hydrogen peroxide levels in estuarine waters. *Est.. Coastal Shelf Sci.*, **40**, 495–503.
98. N.M. Scully, D.J. McQueen, D.R.S. Lean, W.J. Cooper (1996). Hydrogen peroxide formation: The interaction of ultraviolet radiation and dissolved organic carbon in lakewaters along a 43°–75° gradient. *Limnol. Oceanogr.*, **41**, 540–548.
99. B.H. Yocis (1995). *Dynamics of the Hydrogen Peroxide Cycle in Antarctic Waters* (M.S. Thesis). State University of New York, College of Environmental Science and Forestry.
100. N.M. Scully, W.F. Vincent (1996). Hydrogen peroxide: A natural tracer of stratification and mixing processes in subarctic lakes. *Arch. Hydrobiol. Beih Ergebn. Limnol.*, **139**, 1–15.
101. N.M. Scully, W.F. Vincent, D.R.S. Lean, S. MacIntyre (1998). Hydrogen peroxide as a natural tracer of mixing in surface layers. *Aquat. Sci.*, **60**, 169–186.
102. N.M. Scully, W.F. Vincent, D.R.S. Lean (2000). Exposure to ultraviolet radiation in aquatic ecosystems: Estimates of mixing rates in Lake Ontario and the St. Lawrence River. *Can. J. Fish. Aquat. Sci.*, **57**, 43–51.
103. K.S. Johnson, S.W. Willason, D.A. Wiesenburg, S.E. Lohrenz, R.A. Arnone (1989). Hydrogen peroxide in the western Mediterranean Sea: A tracer for vertical advection. *Deep-Sea Res.*, **36**, 241–254.
104. B.H. Yocis, D.J. Kieber, K. Mopper (2000). Photochemical production of hydrogen peroxide in Antarctic Waters. *Deep-Sea Res.*, **47**, 1077–1099.
105. B. Palenik, F.M.M. Morel (1988). Dark production of H_2O_2 in the Sargasso Sea. *Limnol. Oceanogr.*, **33**, 1606–1611.
106. J.W. Moffett, O.C. Zafiriou (1990). An investigation of hydrogen peroxide chemistry in surface waters of Vineyard Sound with $H_2{}^{18}O_2$ and $^{18}O_2$. *Limnol. Oceanogr.*, **35**, 1221–1229.
107. K. Fujiwara, T. Ushiroda, K. Takeda, Y. Kumamato, H. Tsubota (1993). Diurnal and seasonal distribution of hydrogen peroxide in seawater of the Seto Inland Sea. *Geochem. J.*, **27**, 103–115.
108. N.M. Scully, D.R.S. Lean, D.J. McQueen, W.J. Cooper (1995). Photochemical formation of hydrogen peroxide in lakes: Effects of dissolved organic carbon and ultraviolet radiation. *Can. J. Fish. Aquat. Sci.*, **52**, 2675–2681.
109. C.L. Wilson, N.W. Hinman, R.P. Sheridan (2000). Hydrogen peroxide formation and decay in iron-rich geothermal waters: The relative roles of biotic and abiotic mechanisms. *Photochem. Photobiol.*, **71**, 691–699.
110. R.G. Petasne, R.G. Zika (1997). Hydrogen peroxide lifetimes in south Florida

coastal and offshore waters. *Mar. Chem.*, **56**, 215–225.

111. R. Weller, O. Schrems (1993). H_2O_2 in the marine troposphere and seawater of the Atlantic Ocean (48°N–63°S). *Geophys. Res. Lett.*, **20**, 125–128.
112. G.W. Miller (2000). *Wavelength and Temperature Dependent Quantum Yields for Photochemical Formation of Hydrogen Peroxide in Seawater* (M.S. Thesis). State University of New York, College of Environmental Science and Forestry.
113. E.V. Shtamm, A.P. Purmal, Y.I. Skurlatov (1991). The role of hydrogen peroxide in natural aquatic media. *Russ. Chem. Rev.*, **60**, 1228–1248.
114. R.G. Zika (1981). Marine organic photochemistry. In: E.K. Dursma, R. Dawson (Eds), *Marine Organic Chemistry* (pp.299–325). Elsevier, Amsterdam.
115. W.J. Cooper, R.G. Zika, R.G. Petasne, M.C. Plane (1988). Photochemical formation of H_2O_2 in natural waters exposed to sunlight. *Environ. Sci. Technol.*, **22**, 1156–1160.
116. N.M. Scully, W.F. Vincent, D.R.S. Lean, W.J. Cooper (1997). Implications of ozone depletion for surface-water photochemistry: Sensitivity of clear lakes. *Aquat. Sci.*, **59**, 260–274.
117. D.M. Karl, J. Resing (1993). Palmer LTER: Hydrogen peroxide in the Palmer LTER region: IV. Photochemical interactions with dissolved organic matter. *Antarct. J. U.S.*, **28**, 231–234.
118. B.G. Heikes, A.L. Lazrus, G.L. Kok, S.M. Kunen, B.W. Gandrud, S.N. Gitlen, P.D. Sperry (1982). Evidence for aqueous phase hydrogen peroxide synthesis in the trophosphere. *J. Geophys. Res.*, **87**, 3045–3051.
119. B. Heikes, M. Lee, D. Jacob, R. Talbot, J. Bradshaw, H. Singh, D. Blake, B. Anderson, H. Fuelberg, A.M. Thompson (1996). Ozone, hydroperoxides, oxides of nitrogen, and hydrocarbon budgets in the marine boundary layer over the South Atlantic. *J. Geophys. Res.*, **101**, 24221–24234.
120. R. Weller, O. Schrems (1996). Photooxidants in the marine Arctic troposphere in the summer. *J. Geophys. Res.*, **101**, 9136–9147.
121. A.M. Thompson, O.C. Zafiriou (1983). Air–sea fluxes of transient atmospheric species. *J. Geophys. Res.*, **88**, 6696–6708.
122. R.G. Zika, E.S. Saltzman, W.L. Chameides, D.D. Davis (1982). H_2O_2 levels in rainwater collected in South Florida and the Bahama Islands. *J. Geophys. Res.*, **87**, 5015–5017.
123. W.J. Cooper, E.S. Saltzman, R.G. Zika (1987). The contribution of rainwater to variability of surface ocean hydrogen peroxide. *J. Geophys. Res.*, **92**, 2970–2980.
124. Y. Zuo, J. Hoigné (1992). Formation of hydrogen peroxide and depletion of oxalic acid in atmospheric water by photolysis of iron(III)-oxalato complexes. *Environ. Sci. Technol.*, **26**, 1014–1022.
125. J.D. Willey, R.J. Kieber, R.D. Lancaster (1996). Coastal rainwater hydrogen peroxide: Concentration and deposition. *J. Atmos. Chem.*, **25**, 149–165.
126. R.J. Kieber, W.J. Cooper, J.D. Willey, G.B. Avery (2001). Hydrogen peroxide at the Bermuda Atlantic time series station. Part 1: Temporal variability of atmospheric hydrogen peroxide and its influence on seawater concentrations. *J. Atmos. Chem.*, **39**, 1–13.
127. C. Van Baalen (1965). Quantitative surface plating of coccoid blue-green algae. *J. Phycol.*, **1**, 19–22.
128. B. Palenik, O.C. Zafiriou, F.M.M. Morel (1987). Hydrogen peroxide production by a marine phytoplankter. *Limnol. Oceanogr.*, **32**, 1365–1369.
129. R.G. Zepp, Y.I. Skurlatov, J.T. Pierce (1987). Algal-induced decay and formation of hydrogen peroxide in water: Its possible role in the oxidation of anilines by algae. In: R.G. Zika, W.J. Cooper (Eds), *Photochemistry of Environmental Aquatic Systems*

(pp.215–224). American Chemical Society, Washington.
130. B. Palenik, D.J. Kieber, F.M.M. Morel (1992). Dissolved organic nitrogen use by phytoplankton: The roll of cell-surface enzymes. *Biol. Oceanogr.*, **6**, 347–354.
131. S. Pantoja, C. Lee (1994). Cell-surface oxidation of amino acids in seawater. *Limnol. Oceanogr.*, **39**, 1718–1726.
132. M.J. Twiner, C.G. Trick (2000). Possible physiological mechanisms for production of hydrogen peroxide by the icthyotoxic flagellate *Heterosigma akashiwo*. *J. Plankton Res.*, **22**, 1961–1975.
133. J.W. Moffett, R.G. Zika (1988). Measurement of copper(I) in surface waters of the subtropical Atlantic and Gulf of Mexico. *Geochim. Cosmochim. Acta*, **52**, 1849–1857.
134. T.D. Waite, F.M.M. Morel (1984). Coulometric study of the redox dynamics of iron in seawater. *Anal. Chem.*, **56**, 787–792.
135. F.J. Millero (2001). *Physical Chemistry of Natural Waters*. John Wiley & Sons, Inc., New York.
136. J. Hoigné, B.C. Faust, W.R. Haag, F.E. Scully Jr., R.G. Zepp (1989). Aquatic humic substances as sources and sinks of photochemically produced transient reactants. In: I.H. Suffet, P. MacCarthy (Eds), *Aquatic Humic Substances: Influence on Fate and Treatment of Pollutants* (pp.363–381). American Chemical Society, Washington.
137. M.A. Xenopoulos, D.F. Bird (1997). Effects of acute exposure to hydrogen peroxide on the production of phytoplankton and bacterioplankton in a mesohumic lake. *Photochem. Photobiol.*, **66**, 471–478.
138. K. Mopper, A.M. Anesio, W. Granéli, G.R. Aiken (2000). Photoreactivity of humic substances and its effect on microbial availability of carbon in lake water. In: *Effects of Solar Radiation on the Abiotic and Bacterially Mediated Carbon Flux in Aquatic Ecosystems*. (A.M. Anesio, Ph.D. thesis, Chapter 5). Lund University, Lund, Sweden.
139. W.L. Miller (1990). *An Investigation of Peroxide, Iron and Bioavailibility in Irradiated Marine Waters* (Ph.D. Dissertation). Graduate School of Oceanography, University of Rhode Island.
140. C.A. Murray, G. Gill (1978). The geochemistry of iron in Puget Sound. *Geochim. Cosmochim. Acta*, **42**, 9–19.
141. F.J. Millero, S. Sotolongo, M. Izaguirre (1987). The oxidation kinetics of iron(II) in seawater, *Geochim. Cosmochim. Acta*, **51**, 793–801.
142. D.R. Kester, R.H. Byrne, Y.-J. Liang (1975). Redox reactions and solution complexes of iron in marine systems. In: T.M. Church (Ed.), *Chemistry in the Coastal Environment* (pp. 56–79). American Chemical Society, Washington.
143. J.P. Hunt, H. Taube (1952). The photochemical decomposition of hydrogen peroxide. Quantum yields, tracer and fractionation effects. *Nature*, **74**, 5999–6002.
144. G. Milazzo, S. Caroli (1978). *Tables of Standard Electrode Potentials*. Wiley-InterScience, New York.
145. P. Ulanski, G. Merenyi, J. Lind, R. Wagner, C. von Sonntag (1999). The reaction of methyl radicals with hydrogen peroxide. *J. Chem. Soc., Perkin Trans. 2*, 673–676.
146. B.C. Faust, J. Hoigné (1987). Sensitized photooxidation of phenols by fulvic acid and in natural waters. *Environ. Sci. Technol.*, **21**, 957–964.
147. D.J. Kieber, N.V. Blough (1990). Determination of carbon-centered radicals in aqueous solution by liquid chromatography with fluorescence detection. *Anal. Chem.*, **62**, 2275–2283.
148. D.J. Kieber, N.V. Blough (1992). Photoinitiated radical production in aquatic humic substances. In: *Preprint of Extended Abstracts, Division of Environmental Chemistry* (pp.84–87). American Chemical Society National Meeting, San Francisco.
149. M.A. Moran, R.G. Zepp (1997). Role of photoreactions in the formation of biologi-

cally labile compounds from dissolved organic matter. *Limnol. Oceanogr.*, **42**, 1307–1316.
150. T.E. Eriksen, J. Lind, G. Merenyi (1985). On the acid–base equilibrium of the carbonate radical. *Radiat. Phys. Chem.*, **26**, 197–199.
151. S.-N. Chen, M.Z. Hoffman (1972). Acid-base properties of the CO_3^- radical in aqueous solution. *J. Chem. Soc., Chem. Commun.*, 991–992.
152. S.-N. Chen, M.Z. Hoffman, G.H. Parsons, Jr. (1975). Reactivity of the carbonate radical toward aromatic compounds in aqueous solution. *J. Phys. Chem.*, **79**, 1911–1912.
153. S.-N. Chen, M.Z. Hoffman (1975). The effect of pH on the reactivity of the carbonate radical in aqueous solution. *Radiat. Res.*, **62**, 18–27.
154. J. Huang, S.A. Mabury (2000). Steady-state concentrations of carbonate radicals in field waters. *Environ. Toxicol. Chem.*, **19**, 2181–2188.
155. J. Huang, S.A. Mabury (2000). The role of carbonate radical in limiting the persistence of sulfur-containing chemicals in sunlit natural waters. *Chemosphere*, **41**, 1775–1782.
156. R.E. Huie, L.C.T. Shoute, P. Neta (1991). Temperature dependence of the rate constants for reactions of the carbonate radical with organic and inorganic reductants. *Int. J. Chem. Kinet.*, **23**, 541–552.
157. C.L. Clifton, R.E. Huie (1993). Rate constants for some hydrogen abstraction reactions of the carbonate radical. *Int. J. Chem. Kinet.*, **25**, 199–203.
158. P. Neta, R.E. Huie, A.B. Ross (1988). Rate constants for reactions of inorganic radical in aqueous solution. *J. Phys. Chem. Ref. Data*, **17**, 1027–1284.
159. D. Behar, G. Czapski, I. Duchovny (1970). Carbonate radical in flash photolysis and pulse radiolysis of aqueous carbonate solutions. *J. Phys. Chem.*, **74**, 2206–2210.
160. T. Mill, W. Haag, D. Karentz (1990). Estimated effects of indirect photolysis on marine organisms. In: N.V. Blough, R.G. Zepp (Eds), *Effects of Solar Ultraviolet Radiation on Biogeochemical Dynamics in Aquatic Environments. Woods Hole Oceanogr. Inst. Tech. Rept. WHOI* 90-09 (pp.89–93).
161. C.A. Bailey, R.A. Neihof, P.S. Tabor (1983). Inhibitory effect of solar radiation on amino acid uptake in Chesapeake Bay bacteria. *Appl. Environ. Microbiol.*, **46**, 44–49.
162. G. Döhler (1996). Effect of UV irradiance on utilization of inorganic nitrogen by the Antarctic diatom *Odontella weissflogii* (Janisch) grunow. *Bot. Acta*, **109**, 35–42.
163. K. Mopper, J. Zhou, K. Sri Ramana, U. Passow, H.G. Dam, D.T. Drapeau (1995). The role of surface-active carbohydrates in the flocculation of a diatom bloom in a mesocosm. *Deep-Sea Res.*, **42**, 47–73.
164. G. Barroin, M. Feuillade (1986). Hydrogen peroxide as a potential algicide for *Oscillatoria rubescens* D.C. *Water Res.*, **20**, 619–623.
165. D.O. Hessen, E. Van Donk (1994). Effects of UV radiation of humic water on primary and secondary production. *Water, Air and Soil Pollution*, **75**, 325–338.
166. I. Kaczmarska, T.A. Clair, J.M. Ehrman, S.L. MacDonald, D.R.S. Lean, K.E. Day (2000). The effect of ultraviolet B on phytoplankton populations in clear and brown temperate Canadian lakes. *Limnol. Oceanogr.*, **45**, 651–663.
167. M.A. Xenopoulos, Y.T. Prairie, D.F. Bird (2000). Influence of ultraviolet-B radiation, stratospheric ozone variability, and thermal stratification on the phytoplankton biomass dynamics in a mesohumic lake. *Can. J. Fish. Aquat. Sci.*, **57**, 600–609.
168. G.M. Rosen, B.E. Britigan, H.J. Halpern, S. Pou (1999). *Free Radicals: Biology and Detection by Spin Trapping*. Oxford University Press, Oxford.
169. G. Speranza, P. Manitto, D. Monti (1990). Interaction between singlet oxygen and biologically active compounds in aqueous solution. III. Physical and chemical

singlet oxygen quenching rate constants of 6,6′-diapocarotenoids. *J. Photochem. Photobiol. B*, **8**, 51–61.
170. B. Cazin, J.M. Aubry, J. Rigaudy (1986). Is water the best or the worst solvent for [2+4] cycloadditions of singlet oxygen to aromatic compounds? *J. Chem. Soc., Chem. Commun.*, 952–953.
171. D.O. Martire, S.E. Braslavsky, N.A. Garcia (1991). Sensitized photooxidation of dihydroxybenzenes and chlorinated derivatives. A kinetic study. *J. Photochem. Photobiol., A*, **61**, 113–124.
172. B. Kalyanaraman, S. Ramanujam, R.J. Singh, J. Joseph, J.B. Feix (1993). Formation of 2,5-dihydroxybenzoic acid during the reaction between singlet oxygen ($^1O_2$) and salicylic acid: analysis by ESR oximetry and HPLC with electrochemical detection. *J. Am. Chem. Soc.*, **115**, 4007–4012.
173. D.O. Martire, C. Evans, S.G. Bertolotti, S.E. Braslavsky, N.A. Garcia (1993). Singlet molecular oxygen [$O_2(^1\Delta_g)$] production and quenching by hydroxybiphenyls. *Chemosphere*, **26**, 1691–1701.
174. F.E. Scully, Jr., J. Hoigne (1987). Rate constants for reactions for singlet oxygen with phenols and other compounds in water. *Chemosphere*, **16**, 681–694.
175. A. Yoshimura, T. Ohno (1988). Lumiflavin-sensitized photooxygenation of indole. *Photochem. Photobiol.*, **48**, 561–565.
176. I.B.C. Matheson, J. Lee (1979). Chemical reaction rates of amino acids with singlet oxygen. *Photochem. Photobiol.*, **29**, 875–878.
177. I. Kraljic, V.A. Sharpatyi (1978). Determination of singlet oxygen rate constants in aqueous solutions. *Photochem. Photobiol.*, **28**, 583–586.
178. C. Giulivi, M. Sarcansky, E. Rosenfeld, A. Boveris (1990). The photodynamic effect of rose bengal on proteins of the mitochondrial inner membrane. *Photochem. Photobiol.*, **52**, 745–751.
179. N. Suzuki, I. Mizumoto, Y. Toya, T. Nomoto, S. Mashiko, H. Inaba (1990). Steady-state near-infrared detection of singlet molecular oxygen: A Stern–Volmer quenching experiment with luminol, superoxide dismutase, and *Cypridina luciferin* analogs. *Agric. Biol. Chem.*, **54**, 2783–2787.
180. S. Miskoski, N.A. Garcia (1993). Influence of the peptide bond on the singlet molecular oxygen-mediated ($O_2[^1\Delta_g]$) photooxidation of histidine and methionine dipeptides. A kinetic study. *Photochem. Photobiol.*, **57**, 447–452.
181. E. Reddi, M.A.J. Rodgers, J.D. Spikes, G. Jori (1984). The effect of medium polarity on the hematoporphyrin-sensitized photooxidation of L-tryptophan. *Photochem. Photobiol.*, **40**, 415–421.
182. P.T. Chou, A.U. Khan (1983). L-ascorbic acid quenching of singlet delta molecular oxygen in aqueous media: generalized antioxidant property of vitamin C. *Biochem. Biophys. Res. Commun.*, **115**, 932–937.
183. P.C. Lee, M.A.J. Rodgers (1987). Laser flash photokinetic studies of rose bengal sensitized photodynamic interactions of nucleotides and DNA. *Photochem. Photobiol.*, **45**, 79–86.
184. J.P. Barton, J.E. Packer (1970). The radiolysis of oxygenated cystine solutions at neutral pH. The role of RSSR and O_2^-. *Int. J. Radiat. Phys. Chem.*, **2**, 159–166.
185. B.H.J. Bielski, G.G. Shiue (1979). Reaction rates of superoxide radicals with essential amino acids. In: *Oxygen Free Radicals and Tissue Damage* (pp. 43–56). Excerpta Medica, New York.
186. K. Asada, S. Kanematsu (1976). Reactivity of thiols with superoxide radicals. *Agric. Biol. Chem.*, **40**, 1891–1892.
187. B.H.J. Bielski, H.W. Richter (1977). A study of the superoxide radical chemistry by

stopped-flow radiolysis and radiation induced oxygen consumption. *J. Am. Chem. Soc.*, **99**, 3019–3023.

188. R. Julian (1975). Influence des ions NO_2^- dans les radiolysis γ de solutions aqueuses neutres de tetranitromethane et formiate. *Int. J. Radiat. Phys. Chem.*, **7**, 741–748.
189. C.L. Greenstock, G.W. Ruddock (1976). Determination of superoxide (O_2^-) radical anion reaction rates using pulse radiolysis. *Int. J. Radiat. Phys. Chem.*, **8**, 367–369.
190. W. Bors, M. Saran, C. Michel (1979). Pulse-radiolytic investigations of catechols and catecholamines. 4. 3,4-dihydroxybenzaldehyde and 3,4-dihydroxyacetophenone. *J. Phys. Chem.*, **83**, 3084–3088.
191. K.B. Patel, R.L. Willson (1973). Semiquinone free radicals and oxygen. Pulse radiolysis study of one electron transfer equilibria. *J. Chem. Soc., Faraday Trans. 1*, **69**, 814–825.
192. R.A. Basson, T.A. du Plessis (1967). The influence of ethylene on the radiolytic oxidation of ferrous sulfate. *J. Chem. Soc.*, **5**, 778–781.
193. F. Lavelle, M.E. McAdam, E.M. Fielden, E. Martin, P.B. Roberts, K. Puget, A.M. Michelson (1977). A pulse-radiolysis study of the catalytic mechanism of the iron-containing superoxide dismutase from *Photobacterium leiognathi*. *J. Biochem.*, **161**, 3–11.
194. C.L. Greenstock, G.W. Ruddock (1978). Radiation activation of carcinogens and the role of ·OH and O_2^-. *Photochem. Photobiol.*, **28**, 877–880.
195. Y. Sawada, I. Yamazaki (1973). One electron transfer reactions in biochemical systems. VIII. Kinetic study of superoxide dismutase. *Biochim. Biophys. Acta*, **327**, 257–265.
196. J. Resing, G. Tien, R. Letelier, D.M. Karl, D. Jones (1993). Palmer LTER: Hydrogen peroxide in the Palmer LTER region: II. Water column distribution. *Antarct. J. U.S.*, **28**, 227–229.
197. D. Abele-Oeschger, H. Tug, R. Rottgers (1997). Dynamics of UV-driven hydrogen peroxide formation on an intertidal sandflat. *Limnol. Oceanogr.*, **42**, 1406–1415.
198. W.J. Cooper, D.R.S. Lean, J. Carey (1989). Spatial and temporal patterns of hydrogen peroxide in lake waters. *Can. J. Fish. Aquat. Sci.*, **46**, 1227–1231.
199. C.L. Wilson, N.W. Hinman, W.J. Cooper, C.F. Brown (2000). Hydrogen peroxide cycling in surface geothermal waters of Yellowstone National Park. *Environ. Sci. Technol.*, **34**, 2655–2662.
200. C.P. Patnaik, S.N. Mohapatro, A.K. Panigrahi, R.S. Panda (1987). Peroxy acid oxidations. 1. A kinetic and mechanistic study of the oxidation of acetylacetone by peroxomonophosphoric acid and hydrogen peroxide in alkaline medium. *Bull. Chem. Soc. Jpn.*, **60**, 3391–3395.
201. E. Ross, A.T. Chang (1958). Hydrogen peroxide-induced oxidation of ascorbic acid in passion fruit. *J. Agr. Food Chem.*, **6**, 610–615.
202. P.O. Kvernberg, B. Pedersen (1994). Oxidation of L-ascorbic acid with hydrogen peroxide in aqueous solution. *Acta Chem. Scand.*, **48**, 646–651.
203. K.M. Thompson, W.P. Griffith, M. Spiro (1993). Mechanism of bleaching by peroxides. Part 2. Kinetics of bleaching of alizarin and crocetin by hydrogen peroxide at high pH. *J. Chem. Soc., Faraday Trans.*, **89**, 4035–4043.
204. B.J. Lindberg (1966). Sulfinic acids. IV. Oxidation of aromatic sodium sulfinates with hydrogen peroxide. *Acta Chem. Scand.*, **20**, 1843–1862.
205. J.K. Hurst, P.A.G. Carr, F.E. Hovis, R.J. Richardson (1981). Hydrogen peroxide oxidation by chlorine compounds. Reaction dynamics and singlet oxygen formation. *Inorg. Chem.*, **20**, 2435–2438.
206. S. Elliot (1990). Effect of hydrogen peroxide on the alkaline hydrolysis of carbon

disulfide. *Environ. Sci. Technol.*, **24**, 264–267.

207. Y.G. Adewuyi, G.R. Carmichael (1987). Kinetics of hydrolysis and oxidation of carbon disulfide by hydrogen peroxide in alkaline medium and application to carbonyl sulfide. *Environ. Sci. Technol.*, **21**, 170–177.
208. D.F. Evans, M.W. Upton (1985). Studies on singlet oxygen in aqueous solution. Part 1. Formation of singlet oxygen from hydrogen peroxide with two-electron oxidants. *J. Chem. Soc., Dalton Trans.*, 1141–1145.
209. M. Hoffmann, J.O. Edwards (1977). Kinetics and mechanisms of the oxidation of thiourea and N,N′-dialkylthioureas by hydrogen peroxide. *Inorg. Chem.*, **16**, 3333–3338.
210. G. Laus (2001). Kinetics of acetonitrile-assisted oxidation of tertiary amines by hydrogen peroxide. *J. Chem. Soc. Perkin Trans. 2*, 864–868.
211. Y.G. Adewuyi, G.R. Carmichael (1986). Kinetics of hydrolysis and oxidation of dimethyl sulfide by hydrogen peroxide in alkaline medium. *Eniron. Sci. Technol.*, **20**, 1017–1022.
212. D. Shooter, P. Brimblecombe (1989). Dimethylsulfide oxidation in the ocean. *Deep-Sea Res.*, **36**, 577–585.
213. P. Amels, H. Elias, K.-J. Wannowius (1997). Kinetics and mechanism of the oxidation of dimethyl sulfide by hydroperoxides in aqueous medium. Study on the potential contribution of liquid-phase oxidation of dimethyl sulfide in the atmosphere. *J. Chem. Soc., Faraday Trans.*, **93**, 2537–2544.
214. C.C. Winterbourn, D. Metodiewa (1999). Reactivity of biologically important thiol compounds with superoxide and hydrogen peroxide. *Free Radical Biol. Med.*, **27**, 322–328.
215. C.A. Bunton, G.J. Minkoff (1949). Oxidation of α,β-unsaturated ketones with alkaline hydrogen peroxide. *J. Chem. Soc.*, 665–670.
216. Z. Abedinzadeh, M. Gardes-Albert, C. Ferradini (1989). Kinetic study of the oxidation mechanism of glutathione by hydrogen peroxide in neutral aqueous medium. *Can. J. Chem.*, **67**, 1247–1255.
217. G. Aksnes (1960). Reaction between isopropoxymethylphosphoryl fluoride and hydrogen peroxide. *Acta Chem. Scand.*, **14**, 2075–2080.
218. P.-S. K. Leung, M.R. Hoffmann (1985). Kinetics and mechanism of the oxidation of 2-mercaptoethanol by hydrogen peroxide in aqueous solution. *J. Phys. Chem.*, **89**, 5267–5271.
219. I. Pascal, D.S. Tarbell (1957). The kinetics of oxidation of a mercaptan to the corresponding disulfide by aqueous hydrogen peroxide. *J. Am. Chem. Soc.*, **79**, 6015–6020.
220. G. Levey, A.L. Rieger, J.O. Edwards (1981). Rates and mechanism for oxidation of paraquat and diquat radical cations by several peroxides. *J. Org. Chem.*, **46**, 1255–1260.
221. P.J. Coupe, D.L.H. Williams (1999). Formation of peroxynitrite from S-nitrosothiols and hydrogen peroxide. *J. Chem. Soc., Perkin Trans. 2*, 1057–1058.
222. K. Yusa, S. Keiji (1987). Oxidation of oxymyoglobin to metmyoglobin with hydrogen peroxide: Involvement of ferryl intermediate. *Biochemistry*, **26**, 6684–6688.
223. J. Epstein, M.M. Demek, D.H. Rosenblatt (1956). Reaction of paraoxon with hydrogen peroxide in dilute aqueous solution. *J. Org. Chem.*, **21**, 796–797.
224. K.M. Thompson, W.P. Griffith, M. Spiro (1993). Mechanism of bleaching by peroxides. Part 1. Kinetics of bleaching of phenolphthalein by hydrogen peroxide at high pH. *J. Chem. Soc., Faraday Trans.*, **89**, 1203–1209.
225. M.A.P. Dankleff, R. Curci, J.O. Edwards, H.-Y. Pyun (1968). The influence of solvent

on the oxidation of thioxane by hydrogen peroxide and by t-butylhydroperoxide. *J. Am. Chem. Soc.*, **90**, 3209–3218.

226. D. DeWeck, H.K. Nielsen, P.A. Finot (1987). Oxidation rate of free and protein-bound tryptophan by hydrogen peroxide and the bioavailability of the oxidation products. *J. Sci. Food Agric.*, **41**, 179–185.
227. S.-N. Chen, V.W. Cope, M.Z. Hoffman (1973). Behavior of CO_3^- radicals generated in the flash photolysis of carbonatoamine complexes of cobalt(III) in aqueous solution. *J. Phys. Chem.*, **77**, 1111–1116.
228. V.A. Kuz'min (1972). Reactions of the CO_3^- and SiO_3^- radical anions. *High Energy Chem.*, **6**, 338–339.
229. S.-N. Chen, M. Z. Hoffman (1973). Rate constants for the reaction of the carbonate radical with compounds of biochemical interest in neutral aqueous solution. *Radiat. Res.*, **56**, 40–47.
230. S.-N. Chen, M.Z. Hoffman (1974). Reactivity of the carbonate radical in aqueous solution. Tryptophan and its derivatives. *J. Phys. Chem.*, **78**, 2099–2102.
231. T.P. Elango, V. Ramakrishnan, S. Vancheesan, J.C. Kuriacose (1985). Reactions of the carbonate radical with aliphatic amines. *Tetrahedron*, **41**, 3837–3843.
232. J.L. Redpath (1973). Pulse radiolysis of dithiothreitol. *Radiat. Res.*, **54**, 364–374.
233. R.G. Zika, E.S. Saltzman, W.J. Cooper (1985). Hydrogen peroxide concentrations in the Peru upwelling area. *Mar. Chem.*, **17**, 265–275.
234. J. Huang, S.A. Mabury (2000). A new method for measuring carbonate radical activity towards pesticides. *Env. Toxicol. Chem.*, **19**, 1501–1507.

Individual and Sub-individual Effects and Responses

- UVR-induced DNA damage in aquatic organisms
- Photoprotective physiological and biochemical responses of aquatic organisms to UVR
- Photosynthesis in the aquatic environment as affected by UVR
- UVR and pelagic metazoans
- UVR-induced injuries in freshwater vertebrates
- Behavioral responses – UVR avoidance and vision

Chapter 9

UVR-induced DNA damage in aquatic organisms

Anita G.J. Buma, Peter Boelen and Wade H. Jeffrey

Table of contents

Abstract

Today much information exists on the probability of DNA damage induction in aquatic organisms as a result of natural UVR exposure. It is now generally believed that UVR-induced DNA damage is a common feature in aquatic ecosystems based on studies dominated by measures of CPD induction. The present chapter addresses some key issues related to in situ DNA damage induction and repair in a variety of aquatic organisms. The penetration of DNA effective UVR will be addressed for various water types. Furthermore, factors that may control damage induction and repair will be considered. These include mixing, ozone depletion and long term acclimation. Species specific differences, cell size effects and interactions with other environmental (stress) variables, such as nutrient limitation are also considered. Consequences of DNA damage accumulation for organisms, community structuring and trophic interactions will be discussed. Finally, some suggestions are given with respect to future research directions.

9.1 Introduction

The deleterious effects of ultraviolet radiation (UVR: 280 to 400 nm) on organisms inhabiting aquatic environments are determined by a combination of factors. The primary factor is the penetration of biologically effective radiation in the water column. Intensity and spectral composition of UVR at depth depend on physical (surface reflectance, sea state etc.) and chemical features such as dissolved organic material, pigment concentration and particle characteristics (Chapter 3). Secondly, the UVR exposure time of pelagic organisms depends on mixing phenomena, such as wind-induced vertical mixing (Chapter 4). Finally, the resultant UVR-stress will be determined by the vulnerability of the organisms to ultraviolet radiation and their capacity to ameliorate or prevent damage, for instance by repair processes, avoidance or the synthesis of UV-absorbing compounds (see Chapter 10).

Today it is generally believed that natural UVR is a strong environmental factor affecting both productivity and community structure in marine and fresh water ecosystems. In open marine waters, both UV-B (280 to 315 nm) and UV-A (315 to 400 nm) reduce phytoplankton primary production (see also Chapter 11) and bacterial production [1,2]. UVR has been demonstrated to influence the structure of marine and fresh water phytoplankton communities [3,4].

UV-B may damage essential molecules such as proteins [5,6], pigments [7,8] or DNA [9–12]. This damage can potentially affect important cellular processes such as nutrient uptake [13,14], orientation and motility [15], photosynthesis [16–18] or DNA transcription and replication [19,20]. As a result, obstructed metabolic activity can cause decreased growth rates [21,22], reproduction [23] or mortality. The molecular target sites primarily affected in situ by UVR are thought to be DNA, the photosynthetic apparatus (for phototrophic organisms), or both. For example, in Antarctic ice algae, photosynthesis was shown to be affected via reduced PSII efficiency [24,25] or changes in the RUBISCO pool [26]. A reduction in the performance of both steps in the process of photosynthesis will decrease the ability of a cell to photosynthesize, thereby hindering the carboxylation process. UV-B is strongly absorbed by DNA causing structural changes in these molecules that can interfere with vital cellular processes, such as DNA transcription and replication, resulting in mutations or cell death [19,27]. Furthermore, exposure of cells and tissues to UVR may generate indirect, oxidative DNA damage, mainly via the production of oxyradicals (see Chapter 8), leading to lesions such as strand breaks and DNA protein crosslinks. Up until a decade ago, nothing was known about the possibility of in situ DNA damage induction and accumulation in aquatic organisms. Karentz et al. [9] were the first to describe induction and repair of various DNA lesions following UV-B exposure. They showed large species specific differences in lesion induction in a variety of Antarctic marine diatoms (Figure 1), which was thought to be related with cell size. In those experiments UV-B was supplied using artificial UV-B (lamps) sources. In 1996, Jeffrey et al. [11] described, for the first time, the occurrence of UV-B-induced DNA damage in field samples of marine bacterioplankton from the Gulf of Mexico. Since then, various studies have appeared concentrating on UV-B induced DNA lesions, mainly in the marine environment, but also in fresh water systems.

Although much is known now about the effects of solar radiation on the DNA of aquatic organisms, there are also a few major limitations in the presently available information. First, almost all studies have described the presence of cyclobutane pyrimidine dimers (CPDs) only. The main reason for this is that this type of lesion is formed at the highest frequency. Furthermore, CPDs are formed exclusively as a result of UV-B. CPDs are therefore suitable indicators for UV-B

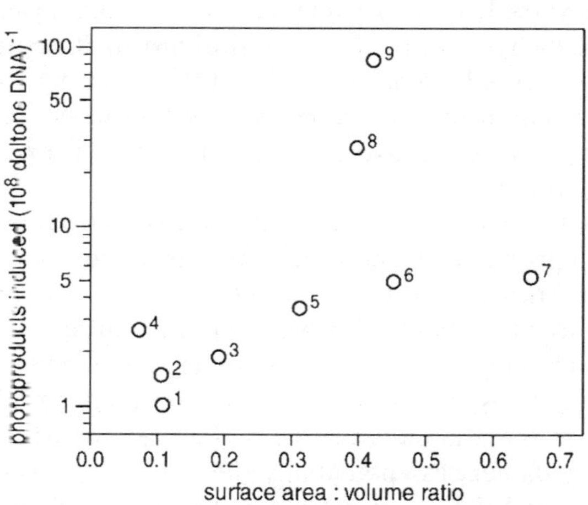

Figure 1. Relationship between surface area : volume ratios and total UV-B-induced photoproducts in 9 marine Antarctic diatoms. 1: *Coscinodiscus oculis-iridis*; 2: *Corethron cryophilum*; 3: *Odontella weissflogii*; 4: *Thalassiosira australis*; 5: *Chaetoceros convolutus*; 6: *Thalassiosira subtilis*; 7: *Chaetoceros socialis*; 8: *Chaetoceros neglectus*; 9: *Nitzschia kerguelensis*. [Redrawn after Karentz et al. 9.]

stress and are also highly relevant for investigating effects of ozone depletion. Other lesions such as the pyrimidine (6-4) pyrimidone photoproduct (6-4 PP) and its Dewar isomer, however, may be formed by UV-B, although at lower rates. Little is known about the occurrence of these lesions, as they may occur in aquatic organisms under ambient solar radiation. In addition, other possible lesions induced by both UV-B and UV-A have not yet been investigated in aquatic organisms. In particular, UV-A can induce indirect, oxidative DNA damage resulting in single strand breaks (SBB), DNA–protein crosslinks, and 8-hydroxy-guanosine (8-oxoG). Clearly, the recent progress in UV effect research in aquatic systems is greatly restricted by the type of lesion under study. Secondly, the information that has become available over the past decade has been directed mainly to pelagic marine microorganisms, such as bacterio- and phytoplankton. Relatively little is known about possible lesion formation under natural irradiance conditions in higher organisms or in fresh water systems.

The aim of this chapter is to give an overview of the current knowledge on UV-B-induced effects on the DNA of aquatic organisms. It does not aim to fully summarize the DNA damage literature, but merely to put relevant data into an ecological perspective. For this reason, most of the emphasis will be given to available field data, which, to date, has been focused almost exclusively on CPDs induced by UV-B.

9.2 UV-B induced DNA photoproducts and repair pathways

All organisms, including humans, which are regularly or occasionally exposed to natural solar radiation may be subjected to DNA damage. In fact, structural changes in DNA are thought to be among the most important deleterious biochemical consequences of UVR [9,10]. UV-B can cause dimerization of DNA bases, leading to the formation of CPDs and 6-4 PPs (Figure 2). These photoproducts block DNA transcription and replication such that only a single distortion of DNA may be sufficient to stop DNA replication. CPDs may hinder cell cycle progress and replication because they obstruct de novo synthesis of cellular components and substances required for growth and cell maintenance. As a consequence, population growth is reduced.

CPDs, especially thymine dimers (TT), are the most abundant lesions formed by natural UV-B [28]. However, 6-4 PPs may have a stronger deleterious effect on DNA transcription than CPDs. The non photoenzymatic repair of CPD and 6-4PP is susceptible to error. Therefore, repair may lead to point mutations in the genome. On top of this, longer wavelengths may result in the photoisomerization of the 6-4 PP to the Dewar pyrimidinone [28]. The action spectrum of this isomer parallels the 6-4 PP absorption spectrum [29], peaking around 310 nm, but also extending into the UV-A region. There is no information regarding the effect of the Dewar isomer on aquatic organisms. Yet it can be assumed that this form of DNA damage is as disruptive as the 6-4 PP itself. At greater depths for instance, 6-4 PP may no longer be formed, but conversion into Dewar isomers by longer wavelengths in the UV-A region may still occur. Similar to other lesions

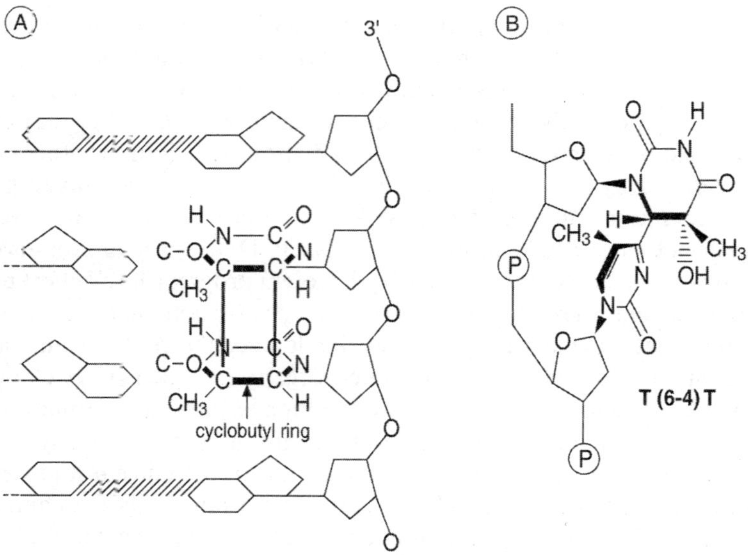

Figure 2. Structure of two photoproducts, formed by UV-B exposure to DNA. (A) Cyclobutane pyrimidine dimer (thymine dimer). (B) Pyrimidine (6-4) pyrimidone photoproduct (6-4 PP). [Redrawn from Friedberg et al. 38.]

induced by both UV-B and UV-A, such as DNA strand breaks and DNA–protein crosslinks, the Dewar isomer may thus be responsible for at least part of the biological UVR effects observed in aquatic environments [30].

Action spectra and biological weighting functions (BWFs) describe the contribution of a certain wavelength to a biological or chemical response. They may thereby help to predict how changes in the spectral composition of solar radiation alter biological or chemical responses. Action spectra are determined through experimental exposures to monochromatic radiation at different wavelengths, assuming that each wavelength contributes independently to the overall effect. In contrast, BWFs are determined through polychromatic exposures [31,32], for instance under natural irradiance conditions. Because responses to polychromatic exposures are assessed, BWFs are often considered relevant for overall responses to UV, since they incorporate damage as well as repair processes simultaneously. The action spectra of CPD and 6-4 PP formation are similar and can be described using the DNA damage action spectrum of Setlow [33] or pyrimidine dimer induction in alfalfa seedlings [34]. Generally, shorter wavelengths cause more damage than longer wavelengths. Both lesions show strongly increasing induction rates with decreasing wavelength in the UV-B region, especially below 302 nm. At higher wavelengths, however, the induction of 6-4 PPs is found to be much lower than that for CPDs.

Cells and tissues are able to handle DNA damage when conditions are favorable for repair. The main pathways to repair dimerized pyrimidines are nucleotide excision repair and photoreactivation ([35] and Chapter 10). Because nucleotide excision repair is light-independent and can therefore take place in the dark, this form of repair is commonly called "dark repair". Photoreactivation is a repair pathway where an enzyme (photolyase) reverses damaged sites in the presence of wavelengths between 350 and 450 nm. Photolyases are categorized on the basis of the chromophore involved in energy transfer. Photolyases in the folate class show maximum activity at 384 nm (UV-A region). Other known photolyases fall into the deazaflavin class with maximum activity at 440 nm (blue light), or in a "non-second-chromophore" class with maximum activity at both 370 and 450 nm. Figure 3 presents action spectra of two photolyases associated with microorganisms.

The rate of photorepair is thus directly related to the intensity of UV-A and/or visible light. Therefore, both the rate of DNA damage induction as well as that of photorepair in plankton is dependent on the prevailing light regime in the water column. CPD induction is caused primarily by UV-B, whereas photorepair of DNA damage depends on the intensity of UV-A and/or visible light. Since the attenuation of UV-B in the water column is much higher than that of UV-A or PAR ([36], Chapter 3), the induction to repair ratio is highly depth-related. For sessile organisms, deeper waters will accommodate more favorable UVR conditions as compared with shallower waters, where damage induction rates may exceed damage removal rates. In contrast, plankton organisms are moved in the upper water layer as a result of vertical mixing and exposure of the cells to the different wavelength bands will strongly fluctuate (Chapter 4). If a simple linear dose–response relationship for the induction as well as the repair of DNA damage were applicable (i.e., if the photochemical law of reciprocity would hold), mean UV-B, UV-A and visible light doses within the upper mixed surface layer (UML) would be sufficient to predict effects on plankton organisms. However, there are indications that this is not always the case, especially for repair processes [37]. Repair rates likely increase with increasing initial levels of CPDs [38]. Moreover, not only the rate of photorepair, but also the induction of the repair system is light dependent [39]. This means that under fluctuating irra-

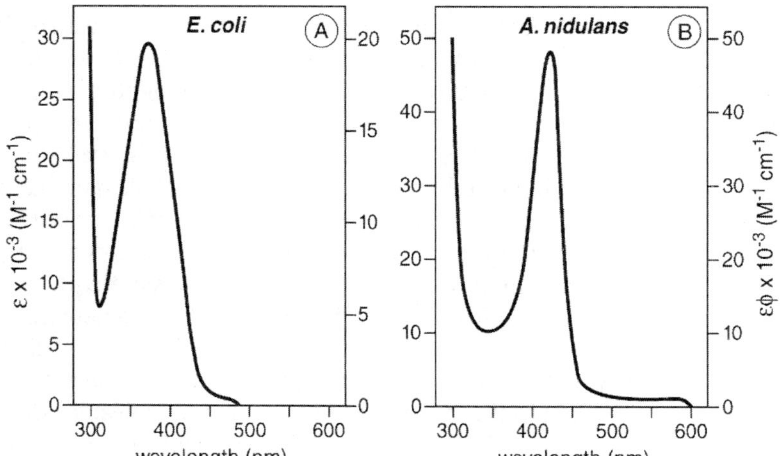

Figure 3. Absorption spectra of two photolyases (A) *Escherichia coli*; (B) *Arabidopsis* sp. [Redrawn from Friedberg et al. 38.]

diance conditions, as experienced by plankton organisms in the UML, conditions for repair may not always be optimal.

9.3 Penetration of DNA effective UVR in marine waters

In the water column, the downwelling irradiance (E_d) of the solar light field diminishes with depth due to absorption and scattering processes. In optically homogeneous waters, irradiance decreases exponentially and can be described using the vertical attenuation coefficient (K_d):

$$E_{d\lambda}(z) = E_{d\lambda}(0)e^{-K_{d\lambda}z}$$

where $E_{d\lambda}(z)$ is the level of downward irradiance at λ at z m depth, $E_{d\lambda}(0)$ the level of downwelling irradiance at λ just below the surface and $K_{d\lambda}$ the vertical attenuation coefficient of λ in m^{-1}. The value of K_d is not the same for each wavelength and depends on chemical, biological and physical parameters, such as the concentration of dissolved organic matter (DOM), pigment concentrations and the amount and type of particles present in the water column [40]. For example, the concentration of DOM in the water column strongly influences the K_d in the ultraviolet region whereas phytoplankton cells also absorb blue and red light.

It is essential to collect information on the irradiance conditions in the water column when UVR effects on aquatic organisms are considered. Since UVR effects are strongly wavelength dependent and the spectral distribution changes rapidly with depth, broadband radiometers have limited value. Accurate measurements of irradiance can only be achieved applying scanning or narrow-band spectroradiometry. To assess possible biological effects of UV-B at a certain depth, measured irradiance spectra are multiplied by an appropriate action spectrum or BWF to determine the amount of biologically effective irradiance (BEI) at that depth. If required, a biological effective dose (BED) can be calculated by integrating several measurements of BEI over time. The choice for a relevant action spectrum may be dictated by the demand for comparing weighted irradiances from various regions or systems. In addition, it will depend on the biological effect under study. For example, a scientist studying DNA damage in plankton will most likely prefer to use a DNA damage action spectrum or weighting function for calculating biologically effective doses.

Where spectroradiometry is not available, profiling radiometers may provide some information on attenuation of DNA effective wavelengths: it has been observed that the 305 nm channel of the PUV500 profiling radiometer (Biosperical Instruments Inc.) gives a good assessment of DNA effective UV-B (Jeffrey unpublished). A more direct way to determine biologically effective doses is to use a dosimeter which contains small amounts of target molecules or simple organisms. Several types of UVR dosimeters have been developed for field applications These dosimeters are based on biological material [41–44], chemical reactions [45,46] and biochemical material such as DNA solutions [11,47–50]. There are several advantages of dosimeter measurements compared to spectral radiometer measurements [47,48]. Application of dosimeters allows direct determinations of daily doses at different depths by performing underwater incubations during the whole light period. Also, a dosimeter automatically integrates the UV component of solar radiation meanwhile weighting them according to their biological effectiveness. Finally, a dosimeter is inexpensive, small, robust and portable. On the other hand, although individual dosimeters are low cost, the facilities required to process the exposed dosimeters can be expensive, and the process may be time consuming for a large number of samples [51].

Figure 4 and Table 1 present some examples of attenuation coefficients of DNA effective UV-B in various water types using DNA dosimetry. In open ocean waters, DNA effective UV-B may penetrate to significant depths, giving 1% levels (of surface irradiance) down to 25 m. For example, Central Atlantic Ocean and Gulf of Mexico waters were shown to be very UVR transparent (Table 1), although UV penetration in the Gulf of Mexico was highly variable depending on location (proximity to shore). Measured $K_{bd\text{-}eff}$ values for open ocean waters correspond with calculated K_{DNA} values reported by Smith and Baker [36] for clear ocean waters with low productivity. Waters off Curaçao, Netherlands Antilles, were also shown to be UV-B transparent, with the only exception being the eutrophic Anna Bay (Table 1). Consequently, shallow coral reef communities may be affected by UV-B, as supported by Lyons et al. [52], who measured induction of DNA damage in coral mucus and in eukaryotic and prokaryotic fractions above a coral reef. In the Gulf of Aqaba, Red Sea, no DNA

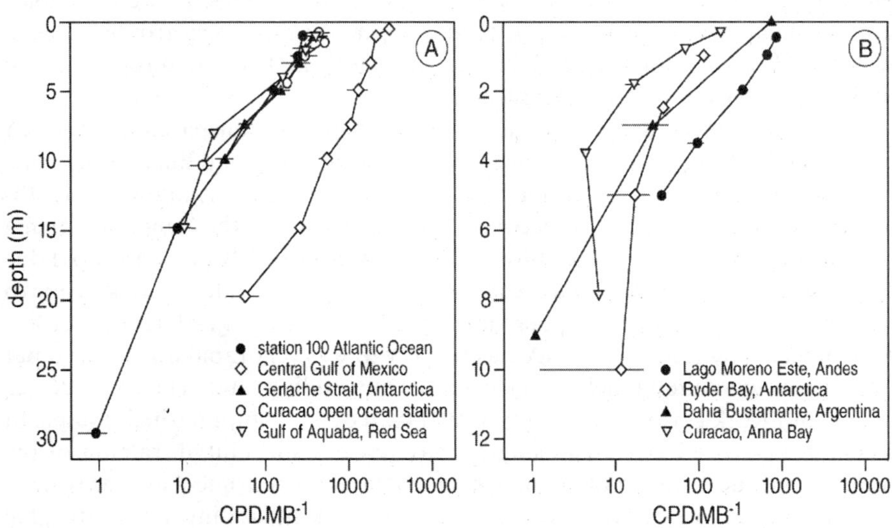

Figure 4. Attenuation of DNA effective UV-B in various waters types, measured with DNA dosimeters. (A) open ocean waters, (B) coastal waters. Open ocean (Curacao [50], Gulf of Mexico (Jeffrey, unpublished results) inshore/coastal (Antarctica [89], Jeffrey, unpublished results), Anna Bay [50], Bahía Bustamante [53]), fresh water (Andes, Helbling, unpublished results), Red Sea [95]. *y*-axes: water depth (m); *x*-axes: CPD concentrations in dosimeter DNA (CPD.MB^{-1}).

Table 1. Vertical attenuation coefficients ($K_{bd\text{-}eff}$) for DNA effective UVBR in various marine and fresh water systems, measured with DNA dosimeters. Attenuation coefficients were determined from linear regressions of natural logarithmic biologically effective irradiation against depth using the log–linear part of the curve

Location	Date	$K_{bd\text{-}eff}$	sd	*n*	Reference
Gerlache Strait, Antarctica	Oct 1995	0.13	0.014	3	Jeffrey et al., in preparation
Gerlache Strait, Antarctica	Oct 1996	0.28	0.030	6	Jeffrey et al., in preparation
Weddell–Scotia Confluence	Oct 1998	0.14	–	1	Mitchell et al., submitted
Central Gulf of Mexico	Jun 1995	0.19	–	1	Jeffrey, unpublished
Central Atlantic Ocean	Aug.1996	0.19	0.01	6	Boelen et al., 1999 [50]
Curaçao open ocean station	Nov. 1996	0.36	0.04	3	Boelen et al., 1999 [50]
Curaçao Buoy 1	April 1998	0.28	0.04	6	Boelen et al., 2001 [94]
Gulf of Aqaba, Red Sea	Sept.1998	0.46	0.04	3	Boelen et al., 2002 [95]
Coastal Gulf of Mexico (StA)	Sept 1994	0.54	–	1	Aas et al., 1996 [2]
Coastal Gulf of Mexico (StC)	Jun 1995	0.87	–	1	Aas et al., 1996 [2]
Coastal Gulf of Mexico	Sept 1994	0.51	0.040	2	Jeffrey et al., 1996 [11]
Curaçao Anna Bay	Nov. 1996	1.53	0.10	2	Boelen et al., 1999 [50]
Ryder Bay, Antarctica	Jan–Mar 1998	0.66	0.13	5	Buma et al., 2001 [89]
Kongsfjorden, Spitsbergen	June 2001	0.91	0.30	7	van de Poll, unpublished
Lago Moreno Este, Andes	Jan. 1999	0.74	–	1	Helbling unpublished

damage could be detected in dosimeter DNA below 15 m. Nevertheless, UV-B penetrated to significant depths in the Gulf of Aqaba, giving 1% depths of biological effective UV-B around 10 m. Attenuation coefficients were comparable to those measured by Regan et al. [47] in the clear oligotrophic coastal waters off Lee Stocking Island, Bahamas but slightly higher than those measured in the clear coastal waters off Curaçao [50].

Eutrophic coastal areas typically have much lower transparency for DNA effective UV-B (Table 1). In an enclosed bay in the Antarctic the presence of a phytoplankton bloom caused high $K_{bd\text{-}eff}$ values giving rise to shallow 1% depths for DNA effective radiation, between 5.4 and 9.6 m. In the temperate Bahia Bustamante, Argentina, attenuation of DNA effective UV-B was very rapid due to the presence of phytoplankton cells. $K_{bd\text{-}eff}$ could not be calculated here due to low resolution of data in the upper meters [53]. In an Arctic fjord (Kongsfjorden, Spitsbergen) attenuation of DNA effective UV-B was very rapid due to the input of sediment-containing melt water from the surrouding land. The mean $K_{bd\text{-}eff}$ over a four week period was as high as 0.91 (van de Poll, unpublished results). In summary, DNA effective wavelengths may reach a substantial fraction of the primary producers in oligotrophic systems such as open tropical marine waters, and in coral reef areas. Moreover, aquatic organisms inhabiting more eutrophic systems, characterized by higher productivity, may also experience high DNA effective irradiance. Pelagic organisms circulating in the upper meters of the water column or sessile organisms living in littoral or sub-littoral zones may be affected.

9.4 Measurement of DNA damage in marine organisms

Several methods have been described to quantify UV-B-induced DNA photoproducts. Assays can be divided into chromatographical, biochemical and immunochemical assays. In chromatographical assays, DNA is degraded by acid hydrolysis to the bases, after which the undamaged bases and photoproducts are separated by paper chromatography [54,55], thin-layer chromatography [56] or high-performance liquid chromatography [57]. In biochemical assays [58,59], photoproduct specific endonucleases are used to cleave DNA at photoproduct sites. The amount of photoproducts is then calculated from the molecular weight distribution of the single-stranded DNA fragments. Immunochemical assays use antibodies directed to a DNA photoproduct. Antibody binding can be quantified using radiolabeled DNA [9,60,61] or a secondary antibody which may be coupled to a fluorescent label [10,62], a gold particle [63] or an enzyme. This enzyme converts a substrate into a fluorescent [64] or colored product [65,66], or catalyzes a reaction resulting in the emission of light [67].

Karentz et al. [9] were the first to describe the occurrence of CPDs in (Antartic) marine phytoplankton cultures. In their study, DNA was isolated from the cells before photoproducts were quantified. Later, Buma et al. [10] developed a method to monitor UV-B-induced DNA damage in individual phytoplankton cells using immunofluorescent labeling in combination with flow cytometry. This method allows for the study of DNA damage in individual cells in relation to other flow cytometric parameters (e.g., cell size, pigments or DNA content). The application of this method for field studies, however, has some major disadvantages. First of all, results may be very difficult to interpret for complex field samples because the immunofluorescence of damaged cells may be masked by background fluorescence of other (larger) species. Moreover, small cells such as bacterio- or picophytoplankton cells may not contain enough DNA to give reliable fluorescence signals. The commonly used methods to quantify CPDs in aquatic organisms are, therefore, those where DNA is extracted from biological material collected by filtration. Often samples are fractionated during filtration to roughly distinguish between functional groups, (i.e., viruses, bacteria or phytoplankton). DNA extracts are then labeled with a CPD specific antibody, after which detection of antibody binding is done either using radiolabeling or using a secondary antibody coupled to an enzyme that converts a reaction substance resulting in luminescence [50].

9.5 Patterns of DNA damage accumulation and repair in aquatic organisms

9.5.1 Evidence from laboratory studies

Much information on DNA damage induction in aquatic organisms is based on incubation experiments under artificial UVR. Unfortunately, the ecological relevance of such studies is restricted, mainly due to the differences in spectra between artifical sources (lamps) and solar radiation. As a result, the effects of UV-B may be overestimated even when weighted UV-B irradiances are realistic. On the other hand, laboratory experiments may provide valuable information on species specific differences in UV vulnerability (see also section 9.6), the potential for repair, or effects of various environmental factors on UV vulnerability. Joux et al. [68] clearly showed species specific DNA damage accumulation in marine bacteria, mainly as a result of differences in photoreactivation potential. Earlier work showed the potential for CPD accumulation and photoenzymatic repair capacity in cyanobacteria and diatoms [9,55,69–71]. Pakker et al. [72] demonstrated 6-4 PP and CPD accumulation in the marine red macroalga *Palmaria palmata* and showed the importance of photoreactivation for successfully overcoming UV-B stress. Another study focussed on marine red algae [73] related UV-B vulnerability with patterns of CPD accumulation and repair. They found that littoral red algae were less vulnerable to CPD accumulation as compared with sub-littoral species and related UV-B vulnerability primarily with photorepair capacity.

Without directly measuring DNA damage induction in marine organisms, many studies have focussed on the potential for photoreactivation in marine organisms. Han and Kain [74] demonstrated photoreactivation by blue light in young sporophytes of several brown seaweeds. They stressed that visible light resulted in recovery from UV damage that would otherwise have caused high mortality in these organisms. Huovinen et al. [75] studied the effects of UVR on early developmental stages of the giant kelp *Macrocystis pyrifira* by exposing gametophytes to UV-B before and after germination. They found that nuclear division and translocation in zoospores after germination were most sensitive for UVR. Furthermore, recovery took place in the light, possibly as a result of photoreactivation of damaged DNA in the zoospores. Huovinen et al. [75] concluded that short exposures to UV-B may perturb or delay the development and recruitment of the giant kelp gametophytes by inhibiting their nuclear division and translocation. Coohill and Deering [76] found a clear relationship between photoreactivating wavelengths and survival of zoospores of the water mold *Blastocladiella emersonii*. Moreover, they also found that temperature largely affected photoreactivation rates, with higher temperatures causing higher survival in the spores.

CPD formation and the potential for photoreactivation have been described for fish cell lines, living fish and fish embryos [77–80]. Kouwenberg et al. [81] argued that UV induced mortality in marine zooplankton and fish could be attributed to DNA damage. Biological weighting functions for egg mortality in Atlantic cod showed a high resemblance with the wavelength dependency for DNA damage induction in DNA solutions, as published by Setlow [33]. They also demonstrated decreased egg mortality rates in the presence of photoreactivating wavelengths. In the marine planktonic copepod *Acartia omorii*, eggs were more susceptible than older life stages [82]. UV-B induced damage was alleviated under simultaneous irradiance with enhanced PAR, indicating that photoenzymatic repair plays a major role in determining UV stress in these organisms. Kouwenberg et al. [81] concluded that younger stages may be primarily affected by incident UVR in the field. Zagarese et al. [37,83] demonstrated species specific efficiency of photoenzymatic repair for various zooplankton species. Hays et al. [84] measured amphibian egg mortality as a function of photolyase activity. They found photolyase to vary 100-fold among eggs of 10 amphibian species. Moreover, hatching success was closely related with photolyase activity [84].

It is clear from these laboratory studies that photoreactivation may be a key process in determining the final UV-B response in aquatic organisms. Most organisms that have been investigated have the potential for photorepair, i.e. they exhibit photoenzymatic activity in the presence of "photoreactivation light". However, this potential for photoreactivation may not be fully exploited by aquatic organisms in their natural environment. First of all, the spectral conditions under water are highly dynamic and may not always support the photoenzymatic process. At greater depths, for instance, UV-A or PAR may remain below the thresholds for repair. Also, chemical and energetic requirements for repair may not always be sufficient. Therefore, in situ accumulation and repair patterns are needed to further clarify the effectivity of photorepair and its consequences for DNA damage accumulation in aquatic organisms.

9.5.2 Depth related CPD patterns, diel cycles and mixing effects

In pelagic organisms, UV-B exposure is to a large extent determined by mixing phenomena. As a result of (wind induced) vertical mixing, organisms are moved up and down in the upper water column. Thus, the exposure of the organisms to the different wavelengths will strongly fluctuate and will be extremely difficult to determine (see Chapter 4). With respect to DNA damage, the induction to repair ratio will also fluctuate during mixing because both processes are strongly, but in different ways, wavelength dependent. The consequences of vertical mixing for DNA damage accumulation and repair can be assessed using models such as proposed by Huot et al. [85]. They combined irradiance, mixing, and biological models to predict net DNA damage levels in the mixed layer. Model predictions were generally consistent with measurements of DNA damage levels in the field. The model showed that the depth of the mixed layer strongly influences the amount and distribution of DNA damage in the mixed layer but that the average amount of DNA damage in the whole euphotic zone was only little affected by mixing. Moreover, model calculations revealed that photoreactivation plays a significant role in preventing rapid CPD build-up in marine bacterioplankton.

Recently, many observations on in situ UV-B related DNA damage accumulation in marine organisms have been published. The smallest organisms inhabiting aquatic ecosystems are viruses. Viral lysis can be as important as flagellate grazing in causing the mortality of marine bacterial communities. UV-induced destruction of viral infectivity may, therefore, reduce virally mediated mortality of bacterio- and phytoplankton. Viruses have been shown to accumulate DNA damage in marine waters, thereby reducing viral infectivity. Both CPDs as well as 6-4 PP showed a clear depth pattern in the Gulf of Mexico [12] (6-4 PP profile shown in Figure 5). Similar patterns of damage were observed in bacterioplankton in the Southern Ocean (Jeffrey et al., unpublished results). In their study Weinbauer et al. [12] described the induction of both CPDs and 6-4 PPs in viral isolates, incubated at several water depths, but also in natural viral communities sampled at the end of the day (concentrated from specific depths). 6-4 PPs and CPDs showed a significantly similar distribution with depth. The percentage of the 6-4 PP relative to total measurable DNA damage averaged 3.1% in natural viral communities in the Gulf of Mexico. The generally higher damage levels in fixed isolates as compared with natural in situ samples indicated that mixing rates minimized photodamage accumulation in viral DNA. This conclusion was further supported by the CPD depth profiles of bacterioplankton from the Southern Ocean [86] and the Northern Gulf of Mexico [11,87] (Figure 6). During calm seas, damage was highest in the surface, decreased with depth and could be detected down to 10 m in the Gulf of Mexico. On moderately mixed days, however, no net accumulation of damage was observed, not even at the surface. In the Gerlache Strait, Antractica, damage was observed to accumulate below 20 m on a calm day. When seas approached 1 m, however, DNA damage at the end of the day was less than it was at sunrise [86]. It was postulated that damaged cells may be moved to deeper waters by mixing, where rates of photoreactivation may outweigh rates of UV-B induced damage, resulting in net decreases of DNA photoproducts.

Although to a lesser extent (see also section 9.6), microalgae also accumulate

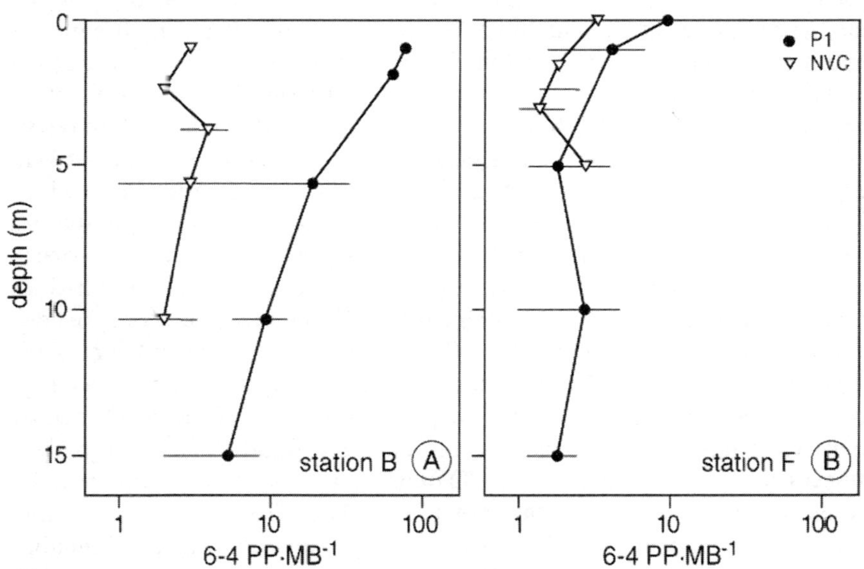

Figure 5. Depth profiles of 6-4 PP in phage DNA, at two stations (station B and F) in the Gulf of Mexico. Filled circles: PWH3a-P1 phage isolate, incubated at various water depths for an entire daily period. Open triangles: natural viral communities concentrated from specific depths at the end of the solar day. [Redrawn from Weinbauer et al. 12.]

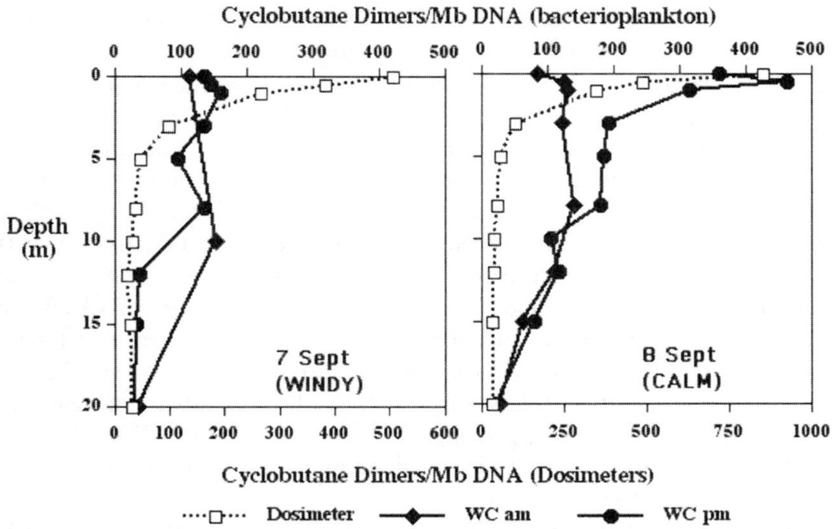

Figure 6. CPD depth profiles in bacterial assemblages and dosimeters in the Northern Gulf of Mexico, measured at 6.30h and 19.00h. Left panel: September 7, 1994. Right panel: September 8, 1994. [Redrawn from Jeffrey et al. 11.]

CPDs in the water column or in sea ice. Prezelin et al. [88] demonstrated the induction of CPDs in frazil ice algae from the Antarctic after exposure to near-surface radiation for 4 h during the morning. No damage was induced under UV-A+PAR (Photosynthetically Active Radiation) or PAR alone. A recent study carried out in the temperate Bahia Bustamante (Chubut, Argentina) area demonstrated both photosynthetic inhibition and DNA damage formation in pro- and eukaryotic pelagic organisms due to UVR [53]. In that study, plankton samples were incubated at various depths in the water column, after which the accumulated damage was measured at the end of the afternoon. CPDs accumulated rapidly in the microbial community maintained at the surface, but at depth (3 m and 6 m) a decrease was observed, indicating that damage had been removed by repair processes. Clear depth dependent relationships in CPD abundance were found during early summer in Antarctic marine bacteria and phytoplankton assemblages, the latter mainly consisting of diatoms [89]. During early summer, shelf ice melting caused stabilization of the upper water layers in these coastal waters. Here, bacteria were found to contain much higher levels of damage as compared with the eukaryote fraction (>10 μm). Nevertheless, phytoplankton exhibited CPD abundance during the whole summer in this Antarctic bay. Towards the end of the summer lower solar angles, decreasing water temperatures and higher mean wind speeds caused the water column to become deeply mixed. At that time CPDs were still present in all size fractions, but damage was distributed evenly throughout the water column.

Sessile organisms do not have the opportunity to be mixed below damaging wavelengths and are therefore likely to possess adequate strategies to reduce or prevent DNA damage. Lyons et al. [52] measured induction of DNA damage in microbial communities associated with coral mucus and observed increased CPD abundance with depth, an apparently anomalous result. DNA damage in coral mucus samples was also consistently lower than in organisms sampled in the water column at similar depths. These results suggested that sessile organisms inhabiting shallower waters rely more on photoprotective compounds such as MAAs or utilize more efficient repair systems. Unfortunately, laboratory experiments were not reported that might distinguish between these possibilities.

The depth profiles that are described above were either taken close to the UV-B peak around noon, or at the end of the afternoon, when CPD abundance was assumed to be maximal, if in agreement with the cumulative DNA effective UV-B dose. Any change in the physiology of the organism under investigation during daytime, such as the induction of repair systems, might prevent the CPD accumulation pattern from following the DNA effective dose. Therefore, diel patterns of DNA damage could give information on maximal CPD levels as they might occur during the day but also provide information on in situ repair or induction of repair processes.

Jeffrey et al. [11] demonstrated accumulation of DNA damage during the day in bacterioplankton assemblages in the Gulf of Mexico. They showed that CPD levels decreased immediately after sunset. During this period synthesis of new DNA (measured as thymidine incorporation) was not high to enough to explain the decrease in CPDs suggesting that removal of CPDs was mainly caused by excision (dark) repair and that DNA damage was not “diluted” by growth [11]. In addition, measurements of *recA* gene expression in these samples [90] and bacterioplankton from the Gerlache Strait, Antarctica [91,92], was shown to follow a clear daily pattern with maximal expression after sunset. These data collectively indicated that excision repair is essential for daily recovery from solar exposure in marine bacterioplankton.

Although there is significant evidence for the potential for photoreactivation in marine bacterioplankton [91,93], further supported by model calculations [85], direct evidence of its effectiveness in situ is limited. Boelen et al. [94] incubated CPD containing picoplankton samples at 10 m depth, where biologically effective UV-B levels were only 6% of surface levels, but UV-A and PAR (involved in photorepair) were still high (see also Table 1). A significant decrease in CPD levels was not detected, however, during the light period. Although it could not be completely ruled out that UV-A and PAR levels at 10 m depth were not sufficient to support photorepair, photoreactivation did not seem the prevailing pathway for CPD removal in these organisms. In addition, experiments carried out in the Gulf of Aqaba, Red Sea, also indicated absent or negligible photorepair, but mainly repair during the night (Figure 7). Bacterioplankton incubated in bags at the water surface as well as in situ samples showed clear daily patterns with maxima at the end of the afternoon, although the in situ samples had lower damage levels throughout [95] (Figure 7). DNA damage was also found to accumulate during the day in temperate marine bacterial and phytoplankton assemblages from Bahia Bustamante, Argentina. CPDs accumulated rapidly when samples were exposed to full solar radiation, even on cloudy days [96]. The $>10\ \mu m$ fraction (phytoplankton, mainly diatoms) also accumu-

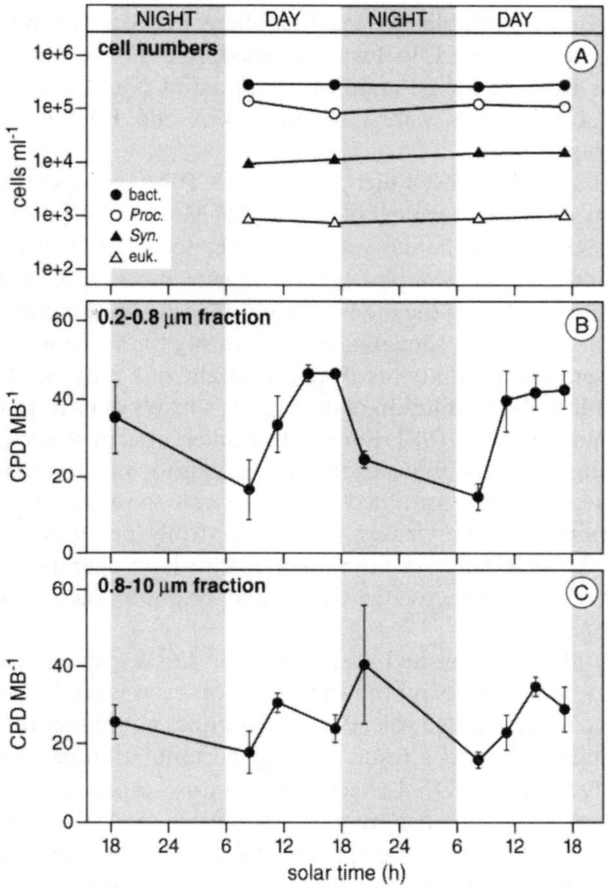

Figure 7. Diel cycles of CPD abundance in two bacterioplankton size fractions in the Gulf of Aqaba. (A) cell numbers of the major plankton groups; (B) in situ sampling at water surface 0.2–0.8 μm; (C) in situ sampling at water surface 0.8–10 μm. [Redrawn from Boelen et al. 95.]

lated damage, although at much lower rates as compared with the picoplankton fraction. Moreover, this work also showed that photosynthetic inhibition and CPD accumulation followed different daily patterns. This was also found previously for a tropical fresh water assemblage [97] (Lake Titicaca, Bolivia), when incubated at the surface for an entire light cycle. Here, CPD accumulated throughout the day, whereas photosynthetic inhibition by UV-B was virtually constant. In incubation experiments conducted at Bahia Bustamante, Argentina, no repair was found when marine microbial assemblages were incubated in photoreactivating light. Damage accumulated during morning hours was not removed during afternoon hours when samples were exposed to UV-A + PAR or PAR alone. The absence of significant photorepair, therefore, was argued to contribute to the observed rapid accumulation of CPDs during the day [96]. These observations were in accordance with similar experiments conducted in the Antarctic (Buma unpublished results) where photorepair was minor or absent in experiments where UV-B was excluded. In contrast, repair was found in situ in Bahia Bustamante in samples incubated at 3 and 6 m in the water column, where UV-B levels were low, but UV-A and PAR were obviously favoring photorepair [53].

Weinbauer et al. [12] showed diel patterns of CPDs and 6-4 PPs in natural virus communities in surface waters of the Gulf of Mexico. CPD concentrations generally increased during the day and highest concentrations were found between 15:00h and 18:00 h. Samples taken the next morning showed that the damage was removed during the night. The 6-4 PP showed a comparable trend to that found for CPDs, also showing removal during the subsequent night. The authors, however, argued that this decrease might not only be due to host-mediated repair but also to dilution of damage as a result of virus replication. In another study, Wilhelm et al. [98] showed clear differential dose responses under photoreactivating and non-photoreactivating conditions in viral infectivity. It was demonstrated that host-mediated repair was able to restore infectivity for a significant proportion of the viruses, thereby allowing the viruses to complete their lytic cycle. Most recently, in situ measurements have demonstrated that up to 52% of solar radiation-inactivated viruses may be photoreactivated in coastal marine environments.

In summary, all the published diel cycles for DNA damage induction in marine bacterio- and phytoplankton indicate that, even when photorepair occurs, it will play a rather limited role. Photorepair does not hinder rapid build up of damage during the day. As a result, damage accumulation patterns are often found to roughly follow the DNA effective UV-B dose, especially after noon.

Eggs and larvae of pelagic fish may be susceptible for UV-B induced DNA damage because they are small, transparent and occur in the upper layers of the ocean. UV-B induced CPD formation and the capacity for repair were studied in newly spawned eggs and yolk-sack larvae of northern anchovy, *Engraulis mordax*, exposed to natural radiation [99]. Eggs and larvae died when exposed to full solar irradiance. At lower levels, i.e. more natural conditions for the water column, there was a clear diel cycle of dimer concentration. This pattern closely followed solar intensity (Figure 8) and not the DNA effective dose, as found for bacteria and phytoplankton (see above). This diel cycle was thought to be due to the instantaneous interaction of damage and true photorepair, whereas dark (excision) repair was shown to be of minor importance. Photoreactivation could be stopped when samples were transferred to the dark. Unhatched embryos, spawned in the dark, also exhibited a strong photorepair response, indicating that photolyase expression in these organisms is not dependent on the previous UVR regime. Vetter et al. [99] concluded that CPD concentration at the time of sampling is a good indicator of dose rate and not of the cumulative dose and that anchovy have a highly efficient photoenzymatic repair system. In agreement with this, efficient photorepair capability in anchovy had been described before [100] showing increasing larval survival under photoreactivating conditions.

It is clear that more information is needed on the conditions affecting photoreactivation in aquatic organisms. First of all, field measurements give contradic-

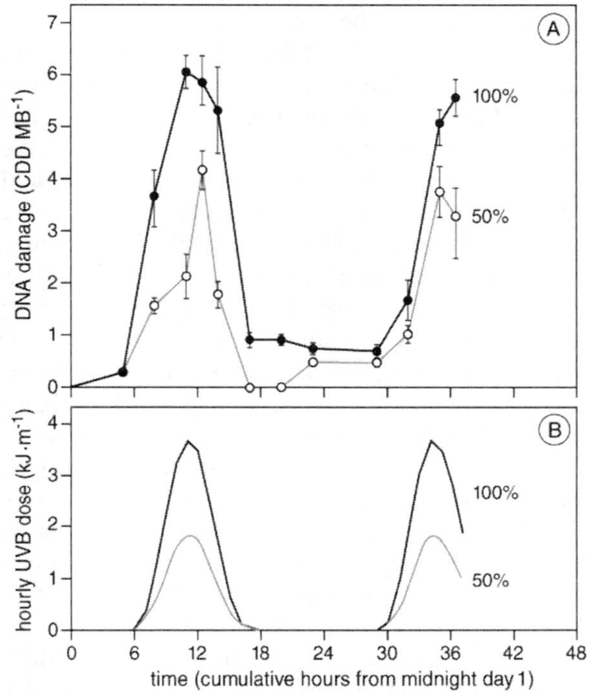

Figure 8. CPD concentrations in unhatched embryos of northern anchovy: (A) Newly spawned eggs, exposed to full solar irradiance (bold solid line) or with 50% reduction in UV-B (lighter solid line). (B) UV-B dose rates, measured as the mean of four scans (Optronics OM 752 spectroradiometer) per hour, integrated between 280 and 320 nm. [Redrawn from Vetter et al. 99.]

tory results with respect to the importance and effectiveness of photoreactivation. At the same time, highly efficient photoreactivation potential has been reported for a large variety of aquatic organisms, when exposed to artificial (light) conditions. Therefore, even when the potential for photoreactivation is present, other factors like nutrient and energy supply or differences in artificial lamp and solar spectra may interfere with the induction of repair systems. In addition, UV-B exposure might cause viability loss in aquatic organisms in situ, thereby shutting off CPD removal. This possibility of DNA damage induction beyond the capability of repair will be addressed in the next section.

9.5.3 Residual DNA damage and viability

Almost all field studies that have been carried out so far have demonstrated the presence of so-called residual DNA damage before sunrise. The occurrence of residual DNA damage indicates that damage induced during previous UV-B exposure events was not completely removed by the various repair pathways before a new UV-B exposure cycle (next day).

Residual DNA damage was found in early morning samples by Jeffrey et al.

[11] in bacterioplankton from the Gulf of Mexico and in Antarctic marine bacteria and phytoplankton [89]. The CPD levels found at the end of summer in Antarctic waters seemed to support the presence of an unrepairable fraction: significant residual CPD levels were detected despite low incident UV-B levels and a deeply mixed water column which further decreased the mean UV-B experienced by the cells. Incubation experiments also showed that CPDs induced during morning hours were not removed by photoreactivating light during the afternoon (UV-B excluded, UV-A and PAR admitted, Buma, unpublished results) or by any other repair process (i.e., dark repair). Similar results were found in the temperate marine plankton assemblages from Bahia Bustamante, Argentina [96]. In fact, residual DNA damage has been reported for every location where CPD abundance was studied. In the Gulf of Aqaba for instance (Figure 7), picoplankton size fractions retained residual DNA damage (between 14 to 43 CPD MB^{-1}) at the end of the night despite the fact that the number of CPDs decreased during the dark. This residual damage suggested that dark repair processes were not able to remove all CPDs. For a typical *Synechococcus* or bacterial cell containing circa 2.1×10^{-15} g DNA per cell [101], this would imply that between 50 and 160 CPDs per cell were still present, blocking DNA replication and cell division. Quaite et al. [34] found that at low initial dimer frequencies (less than circa 30 CPD MB^{-1}) alfalfa seedlings did not use excision repair. This level (30 CPD MB^{-1}) closely matches the residual levels that were detected in the morning samples in the Gulf of Aqaba. Another suggestion to explain the residual damage levels is that DNA damage is not uniformly distributed over all cells, but that most of the damage accumulates in a few heavily damaged cells, which are or become incapable of repair. These cells would then lose viability and eventually disappear by lysis [102].

Laboratory studies have shown that UV-B exposure causes loss of viability in marine diatoms [103–105]. Indeed, the lack of repair as reported in a number of field experiments suggest the presence of non-viable, CPD containing cells in situ, sometimes further supported by the low assimilation rates for photosynthesis, e.g., the Bahia Bustamante area, where residual CPD levels were extremely high [53]. In general, even within a population subjected to identical UV-B doses, DNA damage is not evenly distributed over cells. Buma et al. [10] showed a clear non-uniform distribution of CPD specific fluorescence in a population of diatom cells, judging from the high standard deviation of the mean for CPD specific fluorescence, using immunochemical CPD labeling in combination with flow cytometry (Figure 9). When extrapolating this to the field, most of the damage would accumulate in a limited number of cells. Because cell division cannot be completed until all the damage is repaired, part of the population might thus replicate in a normal way, while damaged cells eventually die and disappear by lysis. If viability loss plays a role, residual CPD levels would not only be determined by viability loss rates, but also by the residence time of non-viable cells in the water column. For example in the Argentinean Sea, residual CPD levels were very high, whereas CPD induction rates were not extreme. These high initial levels, therefore, might have been caused by the accumulation of non-viable cells in the water column over a prolonged period.

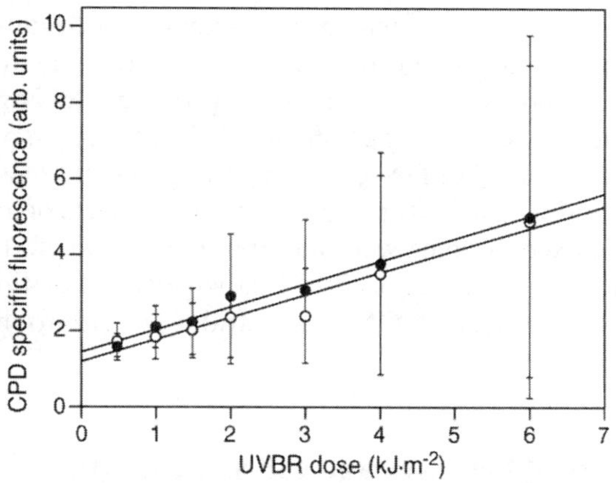

Figure 9. Dose response curve for CPD specific fluorescence against the UV-B dose in *Cyclotella* sp. cells, measured with flow cytometry. Filled circles: G1 cells, open circles: G2 cells. Error bars, standard deviations of the mean for G1 and G2 cells (at least 4000 cells analysed per condition/cell cycle stage). [Redrawn form Buma et al. 10.]

Within this context, it would be interesting to consider potential interactions between nutrient/substrate limitation and UV-B stress. Nutrient additions have been reported to decrease the sensitivity of bacterioplankton production to UVR [91]. As has been demonstrated [106] a large fraction of the bacterioplankton in marine waters is metabolically inactive as a result of substrate limitation. Phytoplankton may also experience nutrient (N, P, Fe) limitation in the open ocean or in a post-bloom situation. Suboptimal metabolic activity in these cells would hamper DNA repair, and would thereby contribute to rapid accumulation of DNA damage in these cells. In turn, damage accumulation would then decrease viability in this fraction of the community. The low repair rates and residual DNA damage levels (morning samples) in combination with low growth estimates in many bacterioplankton field studies seem to support this hypothesis.

9.5.4 Effects of varying ozone concentrations

So far, very few studies have addressed the effects of ozone depletion on DNA damage accumulation in marine organisms. Malloy et al. [107] followed DNA damage in Antarctic ichtyoplankton with ambient UV flux during austral spring. They showed that natural levels of UV-B during ozone depletion caused measurable damage to multicellular organisms occupying higher trophic levels in the Antarctic ecosystem. In particular, icefish eggs were shown to be vulnerable to CPD accumulation. Furthermore, they showed that CPD concentrations in icefish eggs closely followed the daily UV-B flux, suggesting that damage was repaired readily within a day, in accordance with the findings of Vetter et al. [99] for anchovy eggs.

More recently, Maedor et al. [108] followed daily levels of DNA damage in planktonic microorganisms incubated under ambient solar conditions, including ozone fluctuations, at Palmer Station, Antarctica. Although the patterns are complex, there does appear to be a relationship between changes in incident solar radiation caused by ozone depletion and DNA damage in plankton. Huot et al. [85] developed a model of DNA damage induction and repair in bacterioplankton in mixed and non-mixed environments. Also, effects of ozone depletion in mixed and non-mixed systems were incorporated. It was found that ozone thickness caused the largest effect on DNA damage accumulation, when compared with the effects of mixing, DOM concentration or chlorophyll concentration [85].

9.6 Species specific differences and cell size aspects

Large differences in vulnerability for DNA damage induction have been reported for aquatic organisms. Joux et al. [68] showed a high variability in UV-B responses in marine bacterial isolates as determined by survival. In contrast, CPD accumulation in the absence of repair was similar for four of five of the isolates tested. Photoreactivation kinetics were shown to be more likely candidates determining UV vulnerability. All species exhibited photoreactivation, especially under UV-A. It was concluded that UV-B may affect the microbial community structure in marine surface waters. This was supported recently in microcosm experiments carried out the Gulf of Mexico and the Southern Ocean [109].

Phytoplankton groups and species also differ in their vulnerability for UVR-induced damage. Diatoms, green algae and cyanobacteria are thought to be most resistant to UVR, followed by prymnesiophytes and other flagellates [110]. Long-term in situ experiments showed shifts in species composition in favor of more UVR resistant organisms (i.e. diatoms) in marine Antarctic phytoplankton assemblages [4 111] and a fresh water phytoplankton community [3]. Karentz et al. [9] described differences in DNA damage levels and cell survival related to differences in cell size for a variety of diatom species. It was suggested that UV-B-induced DNA damage occurs more frequently in small cells than in larger cells, due to a lower DNA screening efficiency by cell components or UVR screening compounds such as MAAs. Sommaruga and Buma [112] demonstrated large species-specific differences in DNA damage accumulation in aquatic phagotrophic protists, with representatives of the *Kinetoplastida* (bodonids) being the most vulnerable. This high vulnerability was thought to be related with the high AT content of these organisms [112]. Wiencke et al. [113] investigated CPD induction in zoospores of brown seaweeds, showing species-specific variability in CPD induction in these zoospores when exposed to artificial light. It was also assumed that DNA damage in zoospores might be higher when compared with induction rates in sporophytes. The occurrence of DNA damage in the zoospores of these seaweeds could affect the survival and selective adaptation in benthic macroalgae. Alterations in the genomic information at this early developmental stage could have important consequences for the later development of sporophytes and gametophytes [113]. Van de Poll et al. [73] demonstrated large species specific differences in UV-B vulnerability in a range of marine red macrophytes when grown under identical artificial UV-B conditions. Littoral species were highly UV-B resistant and DNA damage accumulation was negligible. In contrast, some sublittoral species showed high CPD accumulation rates due to low repair capacity. They concluded that DNA repair pathways play a major role in determining the UV sensitivity of red macrophytes. In addition, structural differences in UV tolerance between the tested species appeared to reflect their natural habitat in the water column (littoral, sub-littoral) [73].

It is generally believed that small organisms are more susceptible to DNA damage induction than larger cells. Garcia-Pichel [114] calculated that cells <2 μm cannot efficiently use UVR-absorbing compounds as sunscreens. In apparent support of this, Joux et al. [68] showed that CPD accumulation in four of five marine bacterial isolates was similar to damage accumulation in a DNA solution. Smaller organisms, therefore, would have less ability to protect themselves from UV-B induced damage. Karentz et al. [9] found a positive trend between the surface-to-volume ratio and photoproduct induction in cultures of Antarctic phytoplankton showing that smaller cells accumulated DNA damage faster than larger cells (Figure 1). If this is generally applicable, open ocean plankton would be highly vulnerable to DNA damage induction. Marine bacteria and tiny phototrophic plankton, such as prochlorophytes and *Synechococcus* spp., form the majority of the organisms in (oligotrophic) tropical marine waters, comprising up to 99% of total particulate DNA [115,116]. As found in a comparative UV-B vulnerability study, however, small open ocean phytoplankton were not per definition more vulnerable to CPD accumulation than larger phytoplankton cells from other areas (see section 9.7). On the other hand, many studies have shown that size fractions, dominated by bacteria, showed higher CPD numbers as compared with larger size fractions from the same water sample [11,94] (see also section 9.7). For example, in Antarctic marine assemblages, CPD accumulation occurred at a much higher rate in the bacterial fraction as compared with the phytoplankton fraction >10 μm (Figure 10). The same was found for a pelagic community in Bahia Bustamante, Argentina [96]. On the other hand, no significant differences between smaller and larger size fractions were found for other locations. Although Lyons et al. [52] generally detected less DNA damage in the >0.8 μm size fraction as compared to the small size fraction, on occasion, they reported higher DNA damage levels in the >0.8 μm size fraction collected in the water column from a reef area. Also, Jeffrey et al. occasionally measured more damage in the >0.8 μm fraction in the Gulf of Mexico ([11,87] and unpublished results). More recently, Maedor et al. [108] found lower levels of DNA damage in bacterioplankton than larger plankton in incubations conducted at Palmer Station. Laurion and Vincent [117] demonstrated in subarctic lakes that cyanobacteria-dominated picophytoplankton were more resistant to UV-B as could be expected from relationships based on cell size. However, in this study photosynthesis measurements were used for vulnerability assessments, which may not be comparable with vulnerability for DNA damage induction. Helbling et al. [53] argued that in general small cells

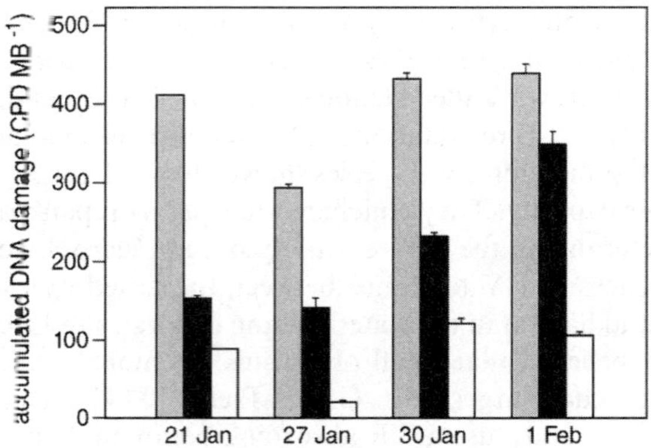

Figure 10. Plankton incubation experiments at Rothera Station (Antarctica), 1998. January 21, 30 and February 1st: clear days; January 27th: some clouds. Gray bars: incident, biologically effective UV-B (CPD Mb-1) measured with a DNA dosimeter; black bars: accumulated damage in the small size fraction (0.2 to 2 μm) (CPD Mb-1); white bars: accumulated damage in the large size fractions (> 10 μm). Samples were incubated from 9.00 until ~19.00. Error bars represent standard deviations of the mean of at least two measurements.

are more vulnerable for DNA damage induction whereas they are more resistant to damage to the photosynthetic apparatus.

9.7 Latitudinal and seasonal variability

Aquatic organisms can utilize a number of defense systems to overcome UVR stress (DNA repair, UVR screening, antioxidant enzymes and compounds, see Chapter 10). It can therefore be hypothesized that long term exposure to high incident UVR doses promotes UVR resistance in cells, species, populations or communities. For this reason it can also be hypothesized that organisms inhabiting low latitude regions have developed more efficient UVR defenses than organisms from higher latitudes [17,110,118–120]. Still, recent studies have clearly demonstrated that tropical aquatic organisms also suffer from in situ UVR stress [2,11,52,87,94,102,121]. It remains to be tested whether tropical organisms are more resistant to UVR as compared with organisms inhabiting higher latitudes. Few latitudinal or other large scale comparisons of UV-B vulnerability in aquatic organisms have been carried out [17,122]. One study revealed large differences in UVR vulnerability between regions and seasons [122], based on absolute biological weighting functions established for a large number of phytoplankton assemblages. Literature data allow for minimal comparison of UV effects due to the fact that different physiological parameters were considered (^{14}C incorporation, variable fluorescence Fv/Fm, growth rate reduction, DNA damage). The use of various types and brands of UVR meters (spectroradiometers, broad band meters, dosimeters) further complicates comparisons of field studies. Finally, even very small differences in methodological approach, for instance the application of different antibodies, may prevent direct comparisons. Therefore, proper comparisons can only be done by applying both identical methods and instruments.

In an effort to reveal latitudinal patterns of vulnerability for UV-B-mediated DNA damage accumulation, we compared CPD induction patterns for a variety of marine plankton assemblages (Table 2). For additional comparison, three high altitude lakes (Argentinean Andes) were included. Identical experiments were performed at all locations. Plankton assemblages were exposed to natural solar radiation for a daily period, after which the accumulated CPDs were measured. Simultaneously, daily DNA (biological) effective doses (BED) were measured using DNA dosimeters. Vulnerability for CPD induction in the various size fractions was assessed by calculating the Mean Damage Ratio (MDR), by normalizing CPD abundance to the level of incident biologically effective irradiance, according to

$$\mathrm{MDR} = \frac{\text{CPDs (accumulated in plankton)}}{\text{CPDs (accumulated in DNA dosimeter)}}$$

In order to distinguish between heterotrophic bacteria and the main phytoplankton groups, size fractionation was done differently for the different regions; for the (sub)tropical sites bacteria and phytoplankton were separated in fractions 0.2 to 0.8 μm and 0.8 to 10 μm (Table 2). For the other sites the large, diatom dominated fractions >10 μm were considered. This approach for comparing UV-B vulnerabilities in a variety of regions has some obvious drawbacks. First of all, samples for the experiments were kept at the surface, receiving full solar radiation for the whole experimental period. This may have introduced (overexposure) artifacts since cells may have otherwise been subjected to vertical mixing, thereby receiving a different exposure regime in situ. Furthermore, differences in species dominance at the various locations forced us to compare different floristic groups as well as geographical regions. While being aware of these important drawbacks, the advantage of the present approach was that accurate measurements of DNA damage as well as DNA effective UV-B allowed for realistic MDR calculations. Also, in spite of the floristic differences, natural representative assemblages were tested for the various geographical regions.

MDR values were significantly ($p < 0.05$) higher in the bacteria than in the larger size fractions judging from the consistently higher MDR values (Table 3). Only at the Caribbean and the Andes sites was no significant difference between small and large cells observed. Although contradicted by the Caribbean site, the bacterial fractions showed some correlation with latitude and altitude: the lowest MDR values were found for the highest mean BED (Andes) and the highest MDR for the lowest mean BED (Antarctic site). On the other hand, the Caribbean site showed relatively high MDR values, despite the fact that BEDs were high. As shown here, the observed high UV-B vulnerability does not implicate that bacteria are incapable of adjustment to high incident UVR levels. In particular, the low MDR value for the high altitude lakes hints at physiological acclimation or community change in favor of more UV-B resistant species.

The phytoplankton fractions also showed a large variability in MDR values,

Table 2. General information on the MDR experiments. MDR is defined as CPDs accumulated in plankton divided by CPDs accumulated in dosimeter DNA. Data are taken from Boelen et al. [94,95], Buma et al. [89,96], Helbling et al., unpublished results (Andes)

Location (number of experiments)	Lat./Long.	Experimental period	Habitat mean water *T*	Fractions tested	Main plankton groups
Caribbean (10)	12°N 69°W	April 1998	marine/oligotr. 27°C	small: 0.2–0.8 μm large: 0.8–10 μm	small: Prochlorophytes, bacteria large: *Synechococcus* spp.
Red Sea (2)	29°N 35°E	September 1998	marine/oligotr. 28°C	small: 0.2–0.8 μm large: 0.8–10 μm	small: Prochlorophytes, bacteria large: *Synechococcus* spp.
Andes (3)	41°S 72°W	January 1999	high altitude lakes; 19°C	small: 0.2–2 μm large: >10 μm	small: bacteria large: diatoms, colonial cyanophytes, dinoflagfellates
Argentinean Sea (4)	45°S 66°W	January 1999	marine/eutrophic 19°C	small: 0.2–2 μm large: >10 μm	small: picophytoplankton, bacteria large: diatoms
Antarctic (4)	67°S 68°W	January 1998	marine/eutrophic 2°C	small: 0.2–2 μm large: >10 μm	small: bacteria large: diatoms

Table 3. Mean damage ratios (MDR) for the various regions and size fractions

	Small		Large	
Location	**Fraction (μm)**	**MDR $\pm$ sd**	**Fraction (μm)**	**MDR $\pm$ sd**
Caribbean	0.2–0.8	0.32 ± 0.07	0.8–10	0.33 ± 0.06
Red Sea	0.2–0.8	0.15 ± 0.03	0.8–10	0.13 ± 0.01
Andes	0.2–2	0.04 ± 0.02	>10	0.03 ± 0.02
Argentinean Sea	0.2–2	0.41 ± 0.22	>10	0.28 ± 0.21
Antarctic	0.2–2	0.55 ± 0.18	>10	0.20 ± 0.09

with the Caribbean, *Synechococcus* spp. dominated fraction showing the highest MDR value. Strikingly, a significant ($p < 0.05$) difference was observed between MDR values of the Caribbean and the Red Sea, the latter also dominated by *Synechococcus* spp. The temperate and Antarctic, diatom dominated assemblages gave intermediate MDR values of 0.28 and 0.20, respectively. These values were not significantly different ($p < 0.05$). Again, phytoplankton from the high altitude lakes showed very low CPD accumulation, giving the lowest MDR value of 0.03 recorded for all experiments, suggesting acclimation to the prevailing high incident UV-B doses (Table 3). No clear latitudinal pattern was observed for the marine autotrophic assemblages. In addition, no clear relation with cell size was observed for phytoplankton. The diatoms from the temperate and the Antarctic site showed intermediate MDR values as compared with the two (sub)tropical sites. Cell size related vulnerability for CPD induction, therefore, was not demonstrated for the phytoplankton fractions.

There are several factors that could likely contribute to the vulnerability for in situ CPD induction as measured using our approach. One important factor that could affect UV-B vulnerability is the nutrient status of the cells. Theoretically, nutrient starvation could strongly hamper UV-B defense mechanisms because the synthesis of repair enzymes and screening substances requires energy and nutrients. Both picophytoplankton-dominated systems (Caribbean, Red Sea) were oligotrophic [123,124]. Therefore, at least at both picophytoplankton-dominated sites, differences in UVR vulnerability could not be explained by nutrient availability. Finally, the long and short term UVR history of the cells may have determined the level of UVR defenses prior to the incubation experiments. For example, deep vertical mixing (Caribbean) may have failed to induce UVR defense systems due to the low mean UVR exposure levels. Wind induced vertical mixing was strong in the Gulf of Aqaba too, but the sampling season (September) may have allowed the community to adjust to high UVR levels over the prolonged summer period. It is clear from the high altitude lake experiments that communities acclimate to the prevailing UVR regimes, judging from the low MDR levels recorded here. We conclude that factors other than cell size and latitude seem to determine CPD vulnerability in phytoplankton. Large parts of the world's oceans are characterized by high incident UVR levels in combination with low UVR attenuation. At the same time, nutrient (nitrate, iron) starvation commonly occurs in oceanic surface waters. The interaction between various growth limiting and stress factors (nutrient limitation, UVR stress) should therefore deserve further attention in future marine photobiology studies.

9.8 Concluding remarks

It is now evident that UV-B induced CPD accumulation is a general phenomenon in aquatic organisms. Viruses and bacteria are especially vulnerable to UV-B induced DNA damage. The small size of viruses and heterotrophic bacteria obviously favors CPD accumulation. The penetration of UV-B inside these cells is high, due to the lack of pigments, nuclear membranes or efficient UV screening by compounds such as MAAs, or, for viruses, the lack of photoreactivation potential. CPD accumulation in aquatic organisms may affect metabolic activity, growth, production, reproduction and viability. For example, Visser et al. [121] incubated seawater (<0.8 μm), consisting mainly of bacteria, in full solar radiation and demonstrated CPD induction over time, roughly following the UV-B dose. Simultaneously, reduction in bacterial production by UV-B was found to be correlated with the occurrence of DNA damage. Similar results were observed by Kase [125] within a single sample area but not between sample areas suggesting a lack of a universal relationship. For phytoplankton and other aquatic phototrophic organisms, such as macroalgae, CPD induction seems to mainly affect growth rate and not photosynthetic performance. Van de Poll et al. [73] showed a clear relationship between CPD accumulation and growth rate reduction for a range of red macrophytes. In contrast, effects of UV-B on variable fluorescence showed significantly different patterns. For phytoplankton, several studies (simulated in situ) have shown a different daily pattern for photoinhibition and CPD accumulation, indicating that these two processes are uncoupled. UV-mediated changes in the photosynthetic process, therefore, may not be indicative of DNA damage induction, or *vice versa*.

The importance of both excision repair [11,91,94] and photorepair [12,85,93,98] has been emphasized for a variety of aquatic microorganisms. In contrast, other studies have demonstrated minimal photorepair [94,96]. Therefore, more studies on DNA repair processes are needed, but in particular the factors that influence photoreactivation capacity in situ. Several data sets, available for microbes, suggest that CPDs accumulate during the day in a roughly dose-dependent manner. This indicates that repair systems may be present but not very effective, especially in the afternoon. In contrast, in higher organisms [99,107] repair seems more efficient because daily patterns follow a more dose-rate dependent pattern showing significant decreases during afternoon hours. Very little is known about wavelength dependency of photoreactivation in aquatic organisms. Therefore biological weighting functions for photoreactivation in aquatic organisms and whole assemblages are urgently needed. Furthermore, very little is known about possible thresholds for photoreactivating irradiance, or, more precisely, the dose (rate) dependency of photoreactivation. Therefore, dose–response relationships (UV-A/PAR versus repair rate) should be established for aquatic organisms.

Another factor that deserves further attention is the effect of UVR on viability. It seems from a limited number of field studies that UV-B causes viability loss in situ in a range of organisms (bacteria, phytoplankton, fish eggs). In fact the rather low repair rates that are established in incubation experiments could partly be explained by the presence of an unknown portion of non-viable organisms that have been killed during previous UV-B exposure events. In that respect photoreactivation rates might be underestimated under in situ conditions because only a fraction of the CPDs is contained in viable cells may be capable of photoenzymatic or other forms of repair. However, significant reduction in CPDs during dark periods, which could only be explained by active nucleotide excision repair [11], argue that viability may not be a dominant factor.

Finally, information on the induction and removal of lesions other than CPDs is urgently needed. Very little is known on the induction of the 6-4 PP in aquatic organisms, let alone the Dewar photoproduct. Furthermore, other lesions that may be induced by both UV-B and UV-A may be highly relevant. It is well known that UV-A may cause more than half of the total UVR effect, reducing primary and secondary production, both in marine and fresh water systems. It remains to be investigated whether or not the UV-A effect can be attributed to damage to photosystems alone (for phototrophs), or that other forms of oxidative damage are also important. The various pathways for ameliorating oxidative stress, such as the presence or induction of antioxidants, may be highly relevant as well.

DNA damage accumulates in every trophic level exposed to solar radiation. Because DNA damage accumulates in a range of trophic levels in marine and fresh water organisms very little can be concluded with respect to trophic consequences of DNA damage induction in aquatic organisms. Increased UV-B in the upper layer of the ocean might result in higher loss rates of viral infectivity. This may decrease the virus-induced mortality of bacteria and phytoplankton. On the other hand, bacteria and phytoplankton may be affected directly, so that primary and secondary production rates are depressed, although it has been found that UVR can change bacterioplankton community structure [109], assumably in favor of more UVR resistant species. Reduction of bacterial activity in the open oligotrophic ocean might have a strong impact on the microbial loop, because here production and remineralization are closely linked. Furthermore, as suggested by Davidson et al. [126] for an Antarctic marine system, changes in trophic interactions in marine organisms might strongly influence the biological pump. A reduction in bacterial activity might favor sedimentation of senescent diatoms, which would otherwise be broken down in the upper water layer. In this way, the biological pump would be stimulated. Moreover, as shown in several studies where shifts occurred in favor of diatoms, possibly as a result of a lower vulnerability for CPD induction, transportation of organic material could be enhanced as a result of (increased) UV-B. It is clear that the considerations given above are extremely speculative and without predictive meaning. In fact, many negative or positive feed back mechanisms can be envisioned, with all the potential feed-back loops that exist in aquatic food webs. Two approaches, therefore, seem to be logic options to proceed. The first one is a holistic approach where long-term effects of UVR on trophic structuring are considered, possibly in combination with molecular approaches to reveal phenomena of acclimation. The second one is ecosystem modeling, where a large number of parameters (including DNA damage accumulation and photoreactivation kinetics, production, growth) for various trophic levels and their interactive responses are incorporated.

Acknowledgements

This work was funded by the Dutch Council for Scientific Research (MEERVOUD grant to A.G.J. Buma, NAAP grant to P. Boelen) and National Science Foundation Office grants OPP9727319 and OCE 9812036 to W.H. Jeffrey.

References

1. G.J. Herndl, G. Muller-Niklas, J.Frick (1993). Major role of ultraviolet-B in controlling bacterioplankton growth in the surface layer of the ocean. *Nature*, **36**, 717–719.
2. P. Aas, M.M. Lyons, R. Pledger, D.L. Mitchell, W.H. Jeffrey (1996). Inhibition of bacterial activities by solar radiation in nearshore waters and the Gulf of Mexico. *Aquat. Microb. Ecol.*, **11**, 229–238.
3. M.L. Bothwell, D. Sherbot, A.C. Roberge, R.J. Daley (1993). Influence of natural ultraviolet radiation on lotic periphytic diatom community growth, biomass accrual and species composition: short-term versus long-term effects. *J. Phycol.*, **29**, 24–35.
4. V.E. Villafañe, E.W. Helbling, O. Holm-Hansen, B.E. Chalker (1995). Acclimatization of Antarctic natural phytoplankton assemblages when exposed to solar ultraviolet radiation, *J. Plankton Res.*, **17**, 2295–2306.
5. K. Lao, A.N. Glazer (1996). Ultraviolet-B photodestruction of a light-harvesting complex. *Proc. Natl. Acad. Sci. U.S.A.*, **93**, 5258–5263.
6. K. Garde, K. Gustavson (1999). The impact of UV-B radiation on alkaline phosphatase activity in phosphorus-depleted marine ecosystems. *J. Exp. Mar. Biol. Ecol.*, **238**, 93–105.
7. R.C. Worrest, H. van Dyke, B.E. Thomson (1978). Impact of enhanced simulated solar ultraviolet radiation upon a marine community. *Photochem. Photobiol.*, **27**, 471–478.
8. S.Z. El-Sayed, F.C. Stephens, R.R. Bidigare, M.E. Ondrusek (1990). Potential effects of solar ultraviolet radiation on Antarctic phytoplankton, UV effects on biological processes. In: N.V. Blough, R.G. Zepp, (Eds), *Effects of Solar Ultraviolet Radiation on Biogeochemical Dynamics in Aquatic Environments* (Technical Report WHOI-90-09, pp. 141–142). Woods Hole Oceanographic Institution.
9. D. Karentz, J.E. Cleaver, D.L. Mitchell (1991). Cell survival characteristics and molecular responses of Antarctic phytoplankton to ultraviolet-B radiation. *J. Phycol.*, **27**, 326–341.
10. A.G.J. Buma, E.J. van Hannen, M.J.W. Veldhuis, L. Roza, W.W.C. Gieskes (1995). Monitoring UV-B induced DNA damage in individual diatom cells by immunofluorescent thymine dimer detection. *J. Phycol.*, **31**, 314–321.
11. W.H. Jeffrey, R.J. Pledger, P. Aas, S. Hager, R.B. Coffin, R. Von Haven, D.L. Mitchell

(1996). Diel and depth profiles of DNA photodamage in bacterioplankton exposed to ambient solar ultraviolet radiation. *Mar. Ecol. Prog. Ser.*, **137**, 283–291.
12. M.G. Weinbauer, S.W. Wilhelm, C.A. Suttle, R.J. Pledger, D.L. Mitchell (1999). Sunlight-induced DNA damage and resistance in natural viral communities. *Aquat. Microb. Ecol.*, **17**, 111–120.
13. M.J. Behrenfeld, D.R.S. Lean, H.I. Lee (1995). Ultraviolet-B radiation effects on inorganic nitrogen uptake by natural assemblages of oceanic plankton. *J. Phycol.*, **31**, 25–36.
14. J.I. Goes, N. Handa, S. Taguchi, T. Hama (1995). Changes in the patterns of biosynthesis and amino acids in marine phytoplankters exposed to ultraviolet-B radiation: nitrogen limitation implicated. *Photochem. Photobiol.*, **62**, 703–710.
15. D.P. Häder (1993). Risks of enhanced solar ultraviolet radiation for aquatic ecosystems. *Prog. Phycol. Res.*, **9**, 1–32.
16. J.J. Cullen, M.P. Lesser (1991). Inhibition of photosynthesis by ultraviolet radiation as a function of dose and dosage rate results for a marine diatom. *Mar. Biol.*, **111**, 183–190.
17. E.W. Helbling, V. Villafañe, M. Ferrario, O. Holm-Hansen (1992). Impact of natural ultraviolet radiation on rates of photosynthesis and on specific marine phytoplankton species. *Mar. Ecol. Prog. Ser.*, **80**, 89–100.
18. M.J. Behrenfeld, J.T. Hardy, H. Gucinski, A. Hanneman, H.I. Lee, A. Wones (1993). Effects of ultraviolet-B radiation on primary producion along latitudinal transects in the South Pacific Ocean. *Mar. Env. Res.*, **35**, 349–363.
19. R.B. Setlow, P.A. Swenson, W.L. Carrier (1963). Thymine dimers and inhibition of DNA synthesis by ultraviolet radiation of cultured fish cells. *Science*, **142**, 1464–1465.
20. S.M. Protic, K.H. Kraemer (1985). One pyrimidine dimer inactivates expression of a transfected gene in *Xeroderma pigmentosum* cells. *Proc. Natl. Acad. Sci. U.S.A.*, **82**, 6622–6626.
21. M.J. Behrenfeld, J.T. Hardy, H.I. Lee (1992). Chronic effects of ultraviolet-B radiation on growth and cell volume of *Phaeodactylum tricornutum* (Bacillariophyceae). *J. Phycol.*, **28**, 757–760.
22. A.G.J. Buma, T. van Oyen, W. van de Poll, M.J.W. Veldhuis, W.W.C. Gieskes (2000). The high sensitivity of *Emiliania huxleyi* (Prymnesiophyceae) to ultraviolet-B radiation. *J. Phycol.*, **36**, 296–303.
23. D.P. Häder, H.D. Kumar, R.C. Smith, R.C. Worrest (1998). Effects on aquatic ecosystems. *J. Photochem. Photobiol. B: Biol.*, **46**, 53–68.
24. B.M.A. Kroon, O. Schofield, B.B. Prézelin (1994). Icecolors '93: UV-B radiation specifically decreases photosystem II (PsII) quantum yield in a field community of Antarctic ice algae exposed to natural daylight. *EOS*, **75**, 200.
25. O. Schofield, B.M.A. Kroon, B.B. Prézelin (1995). Impact of ultraviolet-B radiation on photosystem II activity and its relationship to the inhibition of carbon fixation rates for Antarctic ice algae communities. *J. Phycol.*, **31**, 703–715.
26. M.P. Lesser (1996). Acclimation of phytoplankton to UV-B radiation: oxidative stress and photoinhibition of photosynthesis are not prevented by UV-absorbing compounds in the dinoflagellate *Prorocentrum micans*. *Mar. Ecol. Prog. Ser.*, **132**, 287–297.
27. P.A. Swenson, R.B. Setlow (1966). Effects of ultraviolet radiation on macromolecular synthesis in *Escherichia coli*. *J. Mol. Biol.*, **15**, 201–219.
28. D.L. Mitchell (1995). Ultraviolet radiation damage to DNA. In: R.A. Meyers, (Ed.), *Molecular biology and Biotechnology: A Comprehensive Desk Reference* (pp. 939–943). VCH Publishers, NY.

29. D.L. Mitchell, E.S. Rosenstein (1987). The use of specific radioimmunoassays to determine action spectra for the photolysis of (6-4) photoproducts. *Photochem. Photobiol.*, **45**, 781–786.
30. D.L. Mitchell, R.S. Nairn (1989). The biology of the (6-4) photoproduct, *Photochem. Photobiol.*, **49**, 805–819.
31. J.J. Cullen, P.J. Neale, M.P. Lesser (1992). Biological weighting function for the inhibition of phytoplankton synthesis by ultraviolet radiation, *Science*, **258**, 646–650.
32. J.J. Cullen, P.J. Neale (1997). Biological weighting functions for describing the effects of ultraviolet radiation on aquatic systems. In: D.P. Häder (Ed), *The Effects of Ozone Depletion on Aquatic Ecosystems* (pp. 97–118). R.G. Landes Company.
33. R.B. Setlow (1974). The wavelengths in sunlight effective in producing skin cancer: a theoretical analysis. *Proc. Nat. Acad. Sci. U.S.A.*, **71**, 3363–3366.
34. F.E. Quaite, B.M. Sutherland, J.C. Sutherland (1992). Action spectrum for DNA damage in alfalfa lowers predicted impact of ozone depletion. *Nature*, **358**, 576–578.
35. A. Sancar, G.B. Sancar (1988). DNA repair enzymes. *Annu. Rev. Microbiol.*, **57**, 29–67.
36. R.C. Smith, K.S. Baker (1979). Penetration of UV-B and biologically effective dose-rates in natural waters. *Photochem. Photobiol.*, **29**, 311–323.
37. H.E. Zagarese, B. Tartarotti, W. Cravero, P. Gonzalez (1998). UV damage in shallow lakes: the implications of water mixing. *J. Plankton Res.*, **20**, 1423–1433.
38. E.C. Friedberg, G.C. Walker, W. Siede (1995). *DNA Repair and Mutagenesis*. ASM Press, Washington D.C.
39. H. Mitani, N. Uchida, A. Shima (1996). Induction of cyclobutane pyrimidine dimer photolyase in cultured fish cells by UVA and blue light. *Photochem. Photobiol.*, **64**, 943–948.
40. N.G. Jerlov (1976). *Marine Optics*, Elsevier Scientific Publishing Company, Amsterdam.
41. D. Karentz, L.H. Lutze (1990). Evaluation of biologically harmful ultraviolet radiation in Antarctica with a dosimeter designed for aquatic environments. *Limnol. Oceanogr.*, **35**, 245–257.
42. G. Ronto, S. Gaspar, A. Berces (1992). Phages T7 in biological UV dose measurement. *J. Photochem. Photobiol. B: Biol.*, **12**, 285–294.
43. N. Munakata, F. Morohoshi, K. Hieda, K. Suzuki, Y. Furusawa, H. Shimura, T. Ito (1996). Experimental correspondence between spore dosimetry and spectral photometry of solar ultraviolet radiation. *Photochem. Photobiol.*, **63**, 74–78.
44. G. Horneck, P. Rettberg, E. Rabbow, W. Strauch, G. Seckmeyer, R. Facius, G. Reitz, K. Strauch, J.U. Schott (1996). Biological dosimetry of solar radiation for different simulated ozone column thicknesses. *J. Photochem. Photobiol. B: Biol.*, **32**, 189–196.
45. E.M. Fleischmann (1989). The measurement and penetration of ultraviolet radiation into tropical marine water. *Limnol. Oceanogr.*, **34**, 1623–1629.
46. I. Horkay, N. Wikonkál, J. Patkó, G. Bazsa, M.T. Beck, A. Ferenczi, Z. Nagy, M. Rácz, T. Szalay (1995). SUNTEST: A chemical UVB radiation dosimeter. *J. Photochem. Photobiol. B: Biol.*, **31**, 79–82.
47. J.D. Regan, W.L. Carrier, H. Gucinski, B.L. Olla, H. Yoshida, R.K. Fujimura, R.I. Wicklund (1992). DNA as a solar dosimeter in the ocean. *Photochem. Photobiol.*, **56**, 35–42.
48. J.D. Regan, H. Yoshida (1995). DNA UVB dosimeters. *J. Photochem. Photobiol. B: Biol.*, **31**, 57–61.
49. H. Yoshida, J.D. Regan (1997). UVB DNA dosimeters analyzed by polymerase chain reactions. *Photochem. Photobiol.*, 66, 82–88.
50. P. Boelen, I. Obernosterer, A.A. Vink, A.G.J. Buma (1999). Attenuation of biologically

effective UV radiation in tropical Atlantic waters measured with a biochemical DNA dosimeter. *Photochem. Photobiol.*, **69**, 34–40.

51. A.R. Webb (1998). UV-B radiation measurements: data and methods for ecosystem research. In: C.V. Nolan, D.P. Häder, (Eds), *Role of Solar UV-B Radiation on Ecosystems* (EUR 18321, pp. 7–16). Office for Official Publications of the European Communities.
52. M.M. Lyons, P. Aas, J.D. Pakulski, L. Van Waasbergen, R.V. Miller, D.L. Mitchell, W.H. Jeffrey (1998). DNA damage induced by ultraviolet radiation in coral-reef microbial communities. *Mar. Biol.*, **130**, 537–543.
53. E.W. Helbling, A.G.J. Buma, M.K. de Boer, V.E. Villafañe (2001). In situ impact of solar ultraviolet radiation on photosythesis and DNA in temperate marine phytoplankton. *Mar. Ecol. Progr. Ser.*, **211**, 43–49.
54. R.B. Setlow, W.L. Carrier (1966). Pyrimidine dimers in ultraviolet-irradiated DNA. *J. Mol. Biol.*, **17**, 237–254.
55. P.A. O'Brien, J.A. Houghton (1982). Photoreactivation and excision repair of UV induced pyrimidine dimers in the unicellular cyanobacterium *Gloeocapsa Alpicola* (Synechosystis PCC 6308). *Photochem. Photobiol.*, **35**, 359–364.
56. K.H. Cook, E.C. Friedberg (1976). Measurement of thymine dimers in DNA by thin-layer chromatography. *Anal. Biochem.*, **73**, 411–418.
57. H.J. Niggli, P.A. Cerutti (1983). Cyclobutane-type pyrimidine photodimer formation and excision in human skin fibroblasts after irradiation with 313-nm ultraviolet light. *Biochemistry*, **22**, 1390–1395.
58. U. Kasten, D. Beyersmann, J. Dahmdaphi, A. Hartwig (1995). Sensitive nonradioactive detection of UV-induced cyclobutane pyrimidine dimers in intact mammalian cells. *Mutation Res.*, **336**, 143–152.
59. J.J. Chen, C.Z. Jiang, A.B. Britt (1996). Little or no repair of cyclobutyl pyrimidine dimers is observed in the organellar genomes of the young *Arabidopsis* seedling. *Plant Physiol.*, **111**, 19–25.
60. E. Seaman, H. van Vunakis, L. Levine (1972). Serologic estimation of thymine dimers in the deoxyribonucleic acid of bacterial and mammalian cells following irradiation with ultraviolet light and postirradiation repair. *J. Biol. Chem.*, **247**, 5709–5715.
61. H. Klocker, B. Auer, H.J. Burtcher, J. Hofmann (1982). A sensitive radioimmuno assay for thymine dimers. *Mol. Gen. Genet.*, **186**, 475–477.
62. K.C. Noz, L. Roza, W. Bergman, F. Darroudi, A.A. Schothorst (1994). UV induction of cyclobutane thymine dimers in the DNA of cultured melanocytes from foreskin, common melanocytic nevi and dysplastic nevi. *Photochem. Photobiol.*, **59**, 534–540.
63. M.U. Udono, M. Hori, M. Hirayama, H. Yoshida, T. Suematsu, T. Mori, T. Matsunaga, T. Mizuno, O. Nikaido (1993). Visualization of ultraviolet light-induced thymine dimers in DNA by immunoelectron microscopy. *Photochem. Photobiol.*, **57**, 752–754.
64. L. Roza, K.J.M. van der Wulp, S.J. McFarlane, P.H.M. Lohman, R.A. Baan (1988). Detection of cyclobutane thymine dimers in DNA of human cells with monoclonal antibodies raised against a thymine dimer-containing tetranucleotide. *Photochem. Photobiol.*, **48**, 627–633.
65. A.A. Wani, R.E. Gibson-D'Ambrosio, S.M. D'Ambrosio (1984). Antibodies to UV irradiated DNA: The monitoring of DNA damage by ELISA and indirect immunofluorescence. *Photochem. Photobiol.*, **40**, 465–471.
66. A.A. Wani, S.M. D'Ambrosio, N.K. Alvi (1987). Quantification of pyrimidine dimers by immunoslot blot following sublethal UV-irradiation of human cells. *Photochem. Photobiol.*, **46**, 477–482.

67. A.A. Vink, H.J.B.A. Bergen, O. Nikaido, R.A. Baan, L. Roza (1994). Removal of UV-induced DNA lesions in mouse epidermis soon after irradiation. *J. Photochem. Photobiol. B: Biol.*, **24**, 25–31.
68. F. Joux, W.H. Jeffrey, P. Lebaron, D.L. Mitchell (1999). Marine bacterial isolates display diverse responses to UV-B radiation. *Appl. Environ. Microbiol.*, **65**, 3820–3827.
69. M.K. Blakefield, D.O. Harris (1994). Delay of cell differentiation in *Anabaena aequalis* caused by UV-B radiation and the role of photoreactivation and excision repair. *Photochem. Photobiol.*, **59**, 204–208.
70. A.G.J. Buma, E.J. van Hannen, M.J.W. Veldhuis, W.W.C. Gieskes (1996). UV-B induces DNA damage and DNA synthesis delay in the marine diatom *Cyclotella* sp. *Sci. Mar.*, **60**, 101–106.
71. W.W.C. Gieskes, A.G.J. Buma (1997). UV damage to plant life in a photobiologically dynamic environment: the case of marine phytoplankton. *Plant Ecol.*, **128**, 16–25.
72. H. Pakker, R. Martins, P. Boelen, A.G.J. Buma, O. Nikaido, A.M. Breeman (2000). Effects of temperature on the photoreactivation of ultraviolet-B induced DNA damage in *Palmaria palmata* (Rhodophyta). *J. Phycol.*, **36**, 334–341.
73. W. H. van de Poll, A. Eggert, A.G.J. Buma, A. Breeman (2001). Effects of UV-B induced DNA damage and photoinhibition on growth of six temperate marine red macrophytes: distinct habitat related differences. *J. Phycol.*, **37**, 30–37.
74. T. Han, J.M. Kain (1993). Blue light photoreactivation in ultraviolet-irradiated young sporophytes of *Alaria esculenta* and *Laminaria saccharina* (Phaeophyta). *J. Phycol.*, **29**, 79–81.
75. P.S. Huovinen, A.O.J. Oikari, M.R. Soimasuo, G.N. Cherr (2000). Impact of UV radiation on the early development of the giant kelp (*Macrocystis pyrifira*) gametophytes. *Photochem. Photobiol.*, **72**, 308–313.
76. T.P. Coohill, R.A. Deering (1969). Ultraviolet light inactivation and photoreactivation of *Bastocladiella emersonii. Radiation Res.*, **39**, 374–385.
77. P.M. Achey, A.D. Woodhead, R.B. Setlow (1979). Photoreactivation of pyrimidine dimers in DNA from thyroid cells of the teleost *Poicilia formosa. Photochem. Photobiol.*, **229**, 305–310.
78. L. Applegate, R. Ley (1988). Ultraviolet-radiation-induced lethality and repair of pyrimidine dimers in fish embryo. *Mutat. Res.*, **198**, 85–92.
79. F.E. Ahmed, R.B. Setlow (1993). Ultraviolet radiation-induced DNA damage and its photorepair in the skin of the platyfish *Xiphophorus. Cancer Res.*, **53**, 2249–2255.
80. N. Uchida, H. Mitani, A. Shima (1995). Multiple effects of fluorescent light on repair of ultraviolet-B-induced DNA lesions in cultured goldfish cells. *Photochem. Photobiol.*, **61**, 79–83.
81. J.H.M. Kouwenberg, H.I. Browman, J.J. Cullen, R.F. Davis, J.F. St-Pierre, J.A. Runge (1999). Biological weighting of ultraviolet (280–400) induced mortality in marine zooplankton and fish. *Mar. Biol.*, **134**, 269–284.
82. D.G. Lacuna, S.I. Uye (2001). Influence of mid-ultraviolet (UVB) radiation on the physiology of the marine planktonic copepod *Acartia omorii* and the potential role of photoreactivation. *J. Plankton. Res.*, **23**, 143–155.
83. H.E. Zagarese, M. Feldman, C.E. Williamson (1997). UV-B induced damage and photoreactivation in three species of *Boeckella* (*Copepoda, Calanoida*). *J. Plankton Res.*, **19**, 357–367.
84. J.B. Hays, A.R. Blaustein, J.M. Kiesecker, P.D. Hoffman, I. Pandelova, D. Coyle, T. Richardson (1996). Developmental responses of amphibians to solar and artificial UVB sources: a comparative study. *Photochem. Photobiol.*, **64**, 449–456.

85. Y. Huot, W.H. Jeffrey, R.F. Davis, J.J. Cullen (2000). Damage to DNA in bacterioplankton: A model of damage by ultraviolet radiation and its repair as influenced by vertical mixing. *Photochem. Photobiol.*, **72**, 62–74.
86. W.H. Jeffrey, R.V. Miller, D.L. Mitchell (1997). Detection of ultraviolet radiation induced DNA damage in microbial communities of the Gerlache Strait. *Antarct. J. US*, **32**, 85–87.
87. W.H. Jeffrey, P. Aas, M.M. Lyons, R.B. Coffin, R.J. Pledger, D.L. Mitchell (1996). Ambient solar radiation-induced photodamage in marine bacterioplankton. *Photochem. Photobiol.*, **64**, 419–427.
88. B.B. Prezelin, M.A. Moline, H.A. Matlick (1998). Icecolors '93: spectral UV radiation effects on Antarctic frazil ice algae. *Ant. Res. Ser. Vol.*, **73**, 45–83.
89. A.G.J. Buma, M.K. de Boer, P. Boelen (2001). Depth distributions of DNA damage in Antarctic marine phyto- and bacterioplankton exposed to summertime ultraviolet radiation. *J. Phycol.*, **37**, 200–208.
90. R.V. Miller, W. Jeffrey, D. Mitchell, M. Elasri (1999). Bacterial responses to solar ultraviolet light. *ASM News*, **65**, 535–541.
91. W.H. Jeffrey, J.P. Kase, S.W. Wilhelm (2000). UV radiation effects on heterotrophic bacterioplankton and viruses in marine ecosystems. In: S. de Mora, S. Demers, M. Vernet (Eds), *The Effects of UV Radiation in the Marine Environment* pp. 206–236. Cambridge Univ. Press.
92. M.G. Booth, L. Hutchinson, M. Brumsted, P. Aas, R.B. Coffin, R.C. Downer, Jr., C.A. Kelley, M.M. Lyons, J.D. Pakulski, S.L. Holder Sandvik, W.H. Jeffrey, R.V. Miller (2001). Quantification of RecA as an indicator of repair potential in marine bacterioplankton communities of Antarctica. *Aquat. Microb. Ecol.*, **24**, 51–59.
93. E. Kaiser, G.J. Herndl (1997). Rapid recovery of marine bacterioplankton activity after inhibition by UV radiation in coastal waters. *Appl. Environ. Microbiol.*, **63**, 4026–4031.
94. P. Boelen, M.J.W. Veldhuis, A.G.J. Buma (2001). Accumulation and repair of UVBR mediated DNA damage in marine tropical picoplankton subjected to mixed and simulated non-mixed conditions. *Aquat. Microb. Ecol.*, **24**, 265–274.
95. P. Boelen, A.F. Post, M.J.W. Veldhuis, A.G.J.Buma (2002). Diel patterns of UVBR induced DNA damage in picoplankton size fractions from the Gulf of Aqaba, Red Sea, *Microb. Ecol.*, in press.
96. A.G.J. Buma, E.W. Helbling, M.K. de Boer, V.E. Villafañe (2001). Patterns of DNA damage and photoinhibition in temperate South-Atlantic picophytoplankton exposed to solar ultraviolet radiation. *J Photochem. Photobiol. B: Biol.*, 9–18.
97. E.W. Helbling, V.E. Villafañe, A.G.J. Buma, M. Andrade, F. Zaratti (2001). DNA damage and photosynthetic inhibition induced by solar UVR in tropical phytoplankton (Lake Titicaca, Bolivia). *Eur. J. Phycol.*, **36**, 157–166.
98. S.W. Wilhelm, M.G. Weinbauer, C.A. Suttle, R.J. Pledger, D.L. Mitchell (1998). Measurements of DNA damage and photoreactivation imply that most viruses in marine surface waters are infective. *Aquat. Microb. Ecol.*, **14**, 215–222.
99. R.D. Vetter, A. Kurtzman, T. Mori (1999). Diel cycles of DNA damage and repair in eggs and larvae of northern anchovy *Engraulis mordax*, exposed to solar ultraviolet radiation. *Photochem. Photobiol.*, **69**, 27–33.
100. S.E. Kaupp, J.R. Hunter (1981). Photorepair in larval anchovy, *Engraulis mordax*. *Photochem. Photobiol.*, **33**, 253–256.
101. R.L. Cuhel, J.B. Waterbury (1984). Biochemical composition and short term nutrient incorporation patterns in a unicellular marine cyanobacterium, *Synechococcus* (WH7803). *Limnol. Oceanogr.*, **29**, 70–374.

102. P. Boelen, M.K. de Boer, G.W. Kraay, M.J.W. Veldhuis, A.G.J. Buma (2000). UVBR induced DNA damage in natural marine picoplankton assemblages in the tropical Atlantic Ocean. *Mar. Ecol. Progr. Ser.*, **193**, 1–9.
103. D. Karentz (1994). Ultraviolet tolerance mechanisms in Antarctic marine organisms. In: C.S. Weiler, P.A. Penhale (Eds), *Ultraviolet Radiation in Antarctica: Measurements and Biological Effects* (pp. 93–110). Antarctic Res. Ser. 62, American Geophysical Union, Washington DC.
104. A.T.Davidson, D. Bramich, H. J. Marchant, A. McMinn (1994). Effects of UV-B irradiation on growth and survival of Antarctic marine diatoms. *Mar. Biol.*, **119**, 507–515.
105. A.G.J. Buma, H.J. Zemmelink, K. Sjollema, W.W.C. Gieskes (1996). UVB radiation modifies protein and photosynthetic pigment content, volume and ultrastructure of marine diatoms. *Mar. Ecol. Progr. Ser.*, **142**, 47–54.
106. J.W. Choi, B.F. Sherr, E.B. Sherr (1999). Dead or alive? A large fraction of ETS-inactive marine bacterioplankton cells, as assessed by reduction of CTC, can become ETS-active with incubation and substrate addition. *Aquat. Microb. Ecol.*, **18**, 105–115.
107. K.D. Malloy, M.A. Holman, D.L. Mitchell, H.W. Detrich (1997). Solar UV-B induced DNA damage and photoenzymatic DNA repair in Antarctic zooplankton. *Proc. Natl. Acad. Sci. U.S.A.*, **94**, 1258–1263.
108. J. Maedor, W.H. Jeffrey, J.P. Kase, J.D. Pakulski, S. Chiarello, D.L. Mitchell (2002). Seasonal fluctuations of DNA photodamage in marine plankton assemblages at Palmer Station, Antarctica. *Photochem. Photobiol.* **75**, 266–271.
109. S. Ahrens (1999). *An Analysis of the Effects of Ultraviolet Radiation on Marine Microbial Community Structure* (M.SC. Thesis). University of West Florida, Pensacola, FL.
110. M. Vernet (2000). Effects of UV radiation on the physiology and ecology of marine phytoplankton. In: S. de Mora, S. Demers, M. Vernet (Eds), *The Effects of UV Radiation in the Marine Environment* (pp. 237–279). Cambridge University Press.
111. A.T. Davidson, H.J. Marchant, W.K. de la Mare (1996). Natural UVB exposure changes the species composition of Antarctic phytoplankton in mixed-culture. *Aquat. Microb. Ecol.*, **10**, 299–305.
112. R. Sommaruga, A.G.J. Buma (2000). UV-induced cell damage is species-specific among aquatic phagotrophic protists. *J. Eukaryot. Microbiol.*, **47**, 450–455.
113. C. Wiencke, I. Gomez, H. Pakker, A. Flores-Moya, M. Altamirano, D. Hanelt, K. Bischof, F.L. Figueroa (2000). Impact of UV-radiation on viability, photosynthetic characteristics and DNA of brown algal zoospores: implications for depth zonation. *Mar. Ecol. Prog. Ser.*, **197**, 217–229.
114. F. Garcia-Pichel (1994). A model for internal self-shading in planktonic organisms and its implications for the usefulness of ultraviolet sunscreens. *Limnol. Oceanogr.*, **39**, 1704–1717.
115. J.H. Paul, W.H. Jeffrey, M. DeFlaun (1985). Particulate DNA in subtropical oceanic and estuarine planktonic environments. *Mar. Biol.*, **90**, 95–101.
116. J.G. Stockner (1988). Phototrophic picoplankton: an overview from marine and freshwater ecosystems. *Limnol. Oceanogr.*, **33**, 765–775.
117. I. Laurion, W.F. Vincent (1998). Cell size versus taxonomic composition as determinants of UV-sensitivity in natural phytoplankton communities. *Limnol. Oceanogr.*, **43**, 1774–1779.
118. K. Shibata (1969). Pigments and a UV-absorbing substance in corals and a blue-green alga living in the Great Barrier Reef. *Plant Cell Physiol.*, **10**, 325–335.

119. W.F. Wood (1987). Effect of solar ultraviolet radiation on the kelp *Ecklonia radiata*. *Mar. Biol.*, **96**, 143–150.
120. J.H. Drollet, P. Glaziou, P.M.V. Martin (1993). A study of mucus from the solitary coral *Fungia fungites* (Scleractinia: Fungiidae) in relation to photobiological UV adaptation. *Mar. Biol.*, **115**, 263–266.
121. P.M. Visser, E. Snelder, A.J. Kop, P. Boelen, A.G.J. Buma, F.C. van Duyl (1999). Effects of UV radiation on DNA photodamage and production in bacterioplankton in the coastal Carribean Sea. *Aquat. Microb. Ecol.*, **20**, 49–58.
122. P.J. Neale (2000). Spectral weighting functions for quantifying effects of UV radiation in marine ecosystems. In: S. de Mora, S. Demers, M. Vernet (Eds), *The Effects of UV Radiation in the Marine Environment* (pp. 72–101). Cambridge University Press.
123. D. Lindell, A.F. Post (1995). Ultraphytoplankton succession is triggered by deep winter mixing in the gulf of Aqaba (Eilat), Red Sea. *Limnol. Oceanogr.*, **40**, 1130–1141.
124. G.J. Gast, S. Wiegman, E. Wieringa, F.C. van Duyl, R.P.M. Bak (1998). Bacteria in coral reef water types: removal of cells, stimulation of growth and mineralization. *Mar. Ecol. Prog. Ser.*, **167**, 37–45.
125. J.P. Kase (1998). *DNA Damage and Inhibition of Bacterioplankton Production due to Solar Ultraviolet Radiation in Marine Surface Waters* (M.Sc. Thesis). University of West Florida, Pensacola, FL.
126. A.T. Davidson, van der Heijden (2000). Exposure of natural Antarctic marine plankton assemblages to ambient UV radiation promotes bacterioplankton growth and the microbial loop. *Aquat. Microb. Ecol.*, **21**, 257–264.

Chapter 10

Photoprotective physiological and biochemical responses of aquatic organisms

Anastazia T. Banaszak

Table of contents

Abstract

Tolerance, the ability to endure an external condition such as ultraviolet radiation (UVR), is dependent on the balance between the processes of damage and repair. Aquatic organisms have a variety of physiological and biochemical mechanisms available for reducing the damage incurred by exposure to UVR, including screening, quenching and repair. Screening mechanisms include both physical and chemical barriers and involve many types of UVR-absorbing compounds such as mycosporine-like amino acids (MAAs) and scytonemin, among others. When screening does not eliminate penetration of UVR into the cell, there is a variety of quenching and repair processes available to overcome damage to sensitive cellular components. Quenching is provided by various antioxidants and by carotenoids. Damaged proteins are usually replaced by de novo synthesis (turnover) whereas damaged DNA is generally repaired. Several different mechanisms exist for repairing DNA including photoreactivation, nucleotide excision repair, dimer bypass and recombinational repair. Most research on physiological and biochemical mechanisms of photoprotection has focused on screening mechanisms and in particular on the many kinds of UVR-absorbing compounds present in marine and freshwater species. There has been very little research on DNA and protein repair in aquatic organisms with most of our current knowledge of these processes being derived from bacterial systems.

10.1 Introduction

There is a wide range of tolerance by marine and freshwater organisms to UVR. For example, different strains of free-living and symbiotic dinoflagellates [1,2] demonstrate markedly variable tolerances to exposure to UVR, as do corals [3]. Resistance to exposure to UVR will depend not only on the extent of damage incurred by these damaging wavelengths, but also on the efficiency and availability of various screening mechanisms both physical and chemical, on compounds able to quench photochemically produced toxins and on mechanisms to repair damage to sensitive cellular components such as proteins and DNA [4,5]. Assessment of the tolerance of different species will not only depend on their ability to overcome damage but also on what parameters are used to assess sensitivity; normally photosynthetic parameters are monitored for autotrophic organisms and some kind of growth or survival parameter for heterotrophic organisms. Even within the same species, sensitivity to UVR can vary depending on factors such as light history and nutrient status [6]. Because autotrophic organisms must be exposed to adequate photosynthetically active radiation (PAR) so as to perform photosynthesis, they are also exposed to UVR, therefore, photoprotective mechanisms must be wavelength specific to protect against UVR while allowing for transmission of PAR.

10.2 Screening mechanisms

Screening can eliminate or at least reduce exposure to UVR by absorbing or reflecting damaging wavelengths prior to reaching UVR-sensitive cellular components. Screening may consist of the production of physical barriers such as morphological or structural features that prevent damaging wavelengths from passing or by the production of chemical compounds that absorb UVR. Usually, screening mechanisms, both physical and chemical, serve more than one purpose and thereby decrease the energy required to filter out damaging UV wavelengths.

10.2.1 Physical screening

Morphological and structural features that constitute physical barriers serve other purposes in addition to protection from damaging UVR. For example, structural features such as shells and spines also provide protection from predators. Other structural features include the production of mucous, sporopollenin and multiple-layered cell walls.

10.2.1.1 Mucus

The presence of mucus may act in physically screening cells from damage by UVR. The prymnesiophyte *Phaeocystis pouchetii* has a colonial stage in which the cells are embedded in a mucus matrix that has a high concentration of UVR-absorbing compounds that are excreted by individual cells but function in protecting the entire colony [7]. Mucus such as is excreted by *Fungia* species is believed to provide protection against sediments, desiccation and might also decrease the damaging effects of UVR by the presence of MAAs in the mucus [8,9], although the UVR screening property of MAAs in the mucus is poor [10].

10.2.1.2 Sporopollenin

Sporopollenin is a biopolymer of variable composition found in some algal cell walls, plant pollen and spores, and may function as an antimicrobial agent and provide a rigid cell wall support in large-celled species [11]. The absorption spectrum of sporopollenin has no peaks but rather increases in optical density with decreasing wavelength in the UV region. In a study of 16 species of microalgae, it was shown that 8 species, which were highly tolerant to UVR, had substantial amounts of sporopollenin that occurs in the algal cell walls and absorbs UVR, whereas the other 8 species, which were highly susceptible to UVR, contained little or no sporopollenin [11]. In addition, it was found that the sporopollenin provides constant background protection, whereas MAAs, which are also present, are inducible but exhibit a lag time in their synthesis.

10.2.1.3 Multiple cell walls

When exposed to artificial UVR for four weeks in culture, the temperate, symbiotic dinoflagellate *Symbiodinium californium* developed multiple-layered cell walls and this phenomenon disappeared after the cells were returned to culture conditions in the absence of UVR [12]. These additional cell walls were suggested to protect the UVR-sensitive cellular components from damage by UVR wavelengths. Such multiple-layered cell wall production was not observed *in hospite*, therefore the host anemone, *Anthopleura elegantissima*, which contains high concentrations of MAAs may provide sufficient protection from UVR under natural conditions.

10.2.2 Chemical screening

There are a number of different types of UVR-absorbing compounds, most of which have functions apart from photoprotection against UVR damage. Mycosporine-like amino acids are the most commonly encountered UVR-absorbing compounds in aquatic organisms; however, other compounds that absorb in the UV-A and UV-B region include scytonemin, 3-hydroxykynurenine, melanin, various secondary metabolites and fluorescent pigments.

10.2.2.1 Mycosporines

Mycosporines are a group of compounds, first identified in the mycelia of various species of fungi as the compound P310, and shown to be absent in colonies grown in darkness and closely associated with photosporogenesis [13,14]. P310 is a low molecular weight, water soluble compound with a strong absorption at 310 nm resulting from the linking of a cyclohexenone ring with the nitrogen substituent of an amino acid or amino alcohol and named mycosporine [15]. UVR has been shown to be important in inducing sporulation in a number of fungal groups [14] and mycosporine was hypothesized to act as a photoprotectant in fungal spores, which during dispersal by atmospheric transport are directly exposed to solar radiation [16]. Various mycosporines have been identified depending on the attached substituent with the only amino acids involved being serine, glutamine and glutamic acid or their corresponding amino alcohols, serinol, glutaminol and glutamicol, respectively, or the amino acid, alanine [17,18]. The cyclohexenone unit is derived from the shikimic acid pathway [19], which is the same pathway involved in the synthesis of higher plant photoprotectants such as flavonoids [20].

10.2.2.2 Mycosporine-like amino acids (MAAs)

Mycosporine-like amino acids (MAAs) are imino carbonyl derivatives of mycosporines. Originally termed 'S-320,' due to the maximum wavelength of absorption at approximately 320 nm in extracts from 5 species of *Acropora*, one species of *Pocillopora* and a species of cyanobacterium [21], these compounds were later identified as MAAs in the staghorn coral *Acropora formosa* [22]. The compounds are made up of a cyclohexenone or, more commonly, a cyclohexenimine ring with an amino acid side group. For further information on the chemistry of MAAs and of mycosporines, consult reference [18], on MAAs in coral reef organisms consult reference [23] and on MAAs in dinoflagellates consult reference [24]. In groups other than fungi, the only known mycosporines are mycosporine-glycine and mycosporine-taurine. All other identified compounds are based on the aminocyclohexenimine ring system and are collectively termed mycosporine-like amino acids [18,25]. In this chapter, to prevent confusion between mycosporines derived from fungi and those from aquatic organisms, all mycosporine or mycosporine-like compounds present in aquatic species will be referred to as MAAs.

Absorption spectra of MAAs follow a normal distribution and absorb over a width of approximately 20 nm and most marine and freshwater organisms contain a suite of MAAs thus extending the photoprotective potential across a broader spectrum. The maximum wavelength of absorption is referred to as λ_{max} and of the 19 known MAA compounds, λ_{max} ranges from 309 to 360 nm (Table 1). An additional compound, Gadusol, which absorbs maximally at 296 nm and is related to MAAs, has been found in fish eggs and the brine shrimp *Artemia* [36,42]. Differences in absorption by MAAs are determined by the substitution of different amino acids and alcohols or other amino acid functionalities to mycosporine-glycine [18]. MAAs are extracted using methanol as the sole or principal solvent and the extracts separated using reversed-phase, high-performance liquid chromatography (HPLC) with a UV detector. Quantification of MAAs uses a set of purified standards. Standards are not commercially available therefore some studies have resorted to using λ_{max} and published retention times. As more compounds are being identified, particularly those with similar retention times and λ_{max} to the more commonly encountered MAAs, there is an increasing risk of incorrectly identifying and quantifying MAAs without standards. Due to the wide variety of MAAs that can be present in an extract, spectrophotometric scans cannot be used to identify the MAAs present but are often used as a preliminary test to determine if MAAs are present or absent and in the absence of HPLC can be used to plot changes in relative absorbance.

Phyletic distribution of MAAs: MAAs have been described from a wide variety of habitats ranging from Antarctica [39] and the Arctic to temperate [43] and tropical oceans [22] as well as from tropical [44] to alpine [45] and high Arctic lakes including freshwater and terrestrial ecosystems [46,47]. MAAs occur in a wide variety of aquatic organisms, spanning phytoplankton, all of the major algal divisions, almost all invertebrate phyla, and vertebrates (Table 2). To date these compounds have been identified in most but not all phyla, but whether some of these truly do not contain MAAs or potentially that very few studies have been attempted on lesser-known phyla needs to be verified. Coelenterates contain the greatest number of MAAs (Table 2) partially because so many studies have been conducted on this phylum and because they are the group from which the majority of the MAAs have originally been identified (Table 1). Not all species within each division or phylum synthesize or accumulate MAAs. As an example, a study of 26 isolates of symbiotic dinoflagellates in culture, all exposed to the same condition of UVR and PAR, showed that 15 of the isolates synthesize MAAs and that there was no correlation between the depth of original collection and the capacity for synthesis of MAAs [62].

Table 1. Identified mycosporine-like amino acids, giving their full name, abbreviation used in Table 2, the absorption maximum (λ_{max}), the coefficient of extinction (ε for solvent used consult the appropriate reference; ND = not determined), the species from which the compound was first isolated and the reference for that identification

Full name of mycosporine-like amino acid	Abbreviation	λ_{max} (nm)	ε (dm^3 $mole^{-1}$ cm^{-1})	Species	Reference
Mycosporine-taurine	MT	309	ND	*Anthopleura eleganstissima*	26
Mycosporine-glycine	MG	310	28100	*Palythoa tuberculosa*	27
Palythine	PI	320	36200	*Palythoa tuberculosa*	28
				Chondrus yendoi	29
Palythine-serine	PS	320	10500	*Pocillopora eydouxi*	30
Palythine-threonine-sulfate	PTS	321	ND	*Stylophora pistillata*	31
Palythine-serine-sulfate	PSS	321	ND	*Stylophora pistillata*	31
N-Methylmycosporine-serine or Mycosporine-methylamine-serine	MS	325	16600	*Pocillopora eydouxi*	30
Asterina-330	AS	330	43800	*Asterina pectinifera*	32
Mycosporine-glutamic acid-glycine	MGG	330	43900	*Dysidea herbacea*	33
N-Methylmycosporine-threonine or Mycosporine-methylamine-threonine	MMT	330	33000	*Stylophora pistillata*	34
Mycosporine-2-glycine	M2G	331	ND	*Anthopleura eleganstissima*	26
Palythinol	PL	332	43500	*Palythoa tuberculosa*	35
Mycosporine-glycine-aspartic acid	MGA	332 – 334	ND	*Artemia* sp.	36
Porphyra-334 (mytilin B)	PO	334	42300	*Porphyra tenera*	37
Shinorine (mytilin A)	SH	334	ND	*Chondrus yendoi*	38
Mycosporine-glycine-valine	MV	335	ND	Not specified	39
Palythenic acid	PA	337	29000	*Halocynthia roretzi*	40
Usujirene	US	357	ND	*Palmaria palmata*	41
Palythene	PE	360	50000	*Palythoa tuberculosa*	35

Table 2. The phylogenetic distribution of UVR-absorbing compounds in marine species. + indicates that the compound has been positively identified by HPLC in at least one species in the phylum or division indicated, – indicates that an analysis by HPLC was performed but no evidence of the compound was found and a blank space indicates that no reference was found in the literature for positive identification by HPLC; ? indicates a tentative identification; abbreviations of compounds are given in Table 1, with the addition of GA (Gadusol) and the maximum absorbance, in nm, is given in brackets; data were compiled primarily from original identifications and studies involving surveys or more than one species; studies using artificial feeding, phytoplankton assemblages, partial characterization or not using HPLC analysis were not used; in addition to the references used in Table 1, the following references were consulted: [42,43,45,48–61]

	GA (286)	MT (309)	MG (310)	PI (320)	PS (320)	PTS (321)	PSS (321)	MS (325)	AS (330)	MGG (330)	MMT (330)	M2G (331)	PL (332)	MGA (332–334)	PO (334)	SH (334)	MV (335)	PA (337)	US (357)	PE (360)	SC (386)
Cyanobacteria			+	+					+				+		+	+				+?	+
Dinophyceae			+	+				+	+				+		+	+	+	+	+	+	
Bacillariophyceae			+	+					–				–		+	+	–			+	
Flagellates			+													+	+				
Protozoa			–	–					–				–		–	–	–			–	
Chlorophyta			+	+					+				+		+	+	–			–	
Phaeophyta			+	+					+				+		+	+	–			–	
Rhodophyta			+	+					+				+		+	+	–		+	+	
Anthophyta																					
Porifera			+	+					+	+			+		+	+	+	+	+?	+	
Ctenophora			–	–					–				–		–	–	–			–	
Coelenterata		+	+	+	+	+	+	+	+		+	+	+		+	+	+			+	
Platyhelminthes			+	+					–				–		+	+	+			–	
Nemertinea			+	+					–				–		+	+	+			–	
Rotifera			+	+												+			+?	+	
Chaetognatha			–	–					–				–		–	–	–			–	
Annelida			+	+					+				+		+	+	+			+	
Crustacea	+		+	+					+				+	+	+	+	+	+	+?	+	
Mollusca			+	+					+				+		+	+	+	+		+	
Bryozoa			+	+					–				–		+	+	+			+	
Echinodermata			+	+					+			+	+		+	+	+			–	
Protochordata			+	+					+				+		+	+		+		+	
Chordata			+	+					+				+		+	+	+			+	
Chordata (eyes)				+					+				+							+	
Chordata (eggs)	+																				

Biosynthesis of MAAs: The biosynthesis of mycosporines and MAAs is via the first steps of the shikimic acid pathway [63], as evidenced by the reduction or cessation of MAA accumulation in the presence of glyphosate, an inhibitor of this pathway [64]. The synthesis of MAAs can be very rapid. In the dinoflagellates, *Alexandrium excavatum* and *Prorocentrum micans* synthesis occurs within hours of transfer from low (20 μmol quanta $m^{-2}.s^{-1}$) to high (250 μmol quanta $m^{-2}s^{-1}$) light [50], which is consistent with the rapid changes in light during vertical migration [65]. In most other species, synthesis or accumulation is slower and can be on the order of weeks as was found for the dinoflagellate *Symbiodinium microadriaticum* [53] and the diatom *Thalassiosira weissflogii* [66]. In the red alga, *Porphyra umbilicalis*, 72 hours is not sufficient to induce MAA synthesis even when exposed to UVR [67]. Some species synthesize and release MAAs into the surrounding medium as shown in symbiotic dinoflagellates [53,62] and in the bloom-forming dinoflagellate *Lingulodinium* (= *Gonyaulax*) *polyedra* [68]. It has been suggested that the release of MAAs into the water column by free-living dinoflagellates may contribute to the attenuation of UVR during bloom events [68].

The shikimate pathway is only known in bacteria, fungi, algae and plants and there is no evidence of this pathway in invertebrates or vertebrates. Animals must therefore obtain MAAs through their diet [39,52,53,69] or by translocation from symbionts [53]. In the association between the dinoflagellate *Symbiodinium microadriaticum* and the jellyfish *Cassiopeia xamachana*, three MAAs are synthesized by the symbionts and the same three are exported to the host [53]. The coral *Stylophora pistillata* contains 10 different MAAs including shinorine [64], whereas the dinoflagellate symbionts in culture only synthesize shinorine [62]. Whether the symbionts *in hospite* produce all 10 compounds or whether the host, or possibly bacteria are metabolically converting shinorine into other compounds is yet to be resolved. The same five MAAs, mycosporine-glycine, shinorine, Porphyra-334, palythenic acid and palythine, although in different proportions, were found in the pteropod predator *Clione antarctica*, and in its exclusive prey, the herbivorous pteropod *Limacina helicina* [69]. The phytoplankton assemblage on which *L. helicina* feeds contains only shinorine and Porphyra-334. Although the increase in the number of MAAs may have been due to accumulation of the MAAs before the study began, there is also the possibility that there is biochemical or possibly bacterial conversion of MAAs in *L. helicina* [69]. In the association between the anemone *Anthopleura elegantissima* and the dinoflagellate *Symbiodinium californium*, MAAs were detected in host tissues but not in freshly isolated algae, the symbionts in culture nor in the culture medium, suggesting that the MAAs are derived from the diet of the anemone [53]. The same pattern may be found in the giant clam, *Tridacna crocea* [70]. Studies involving controlled diets of known MAA composition have shown that there is dietary accumulation of MAAs, in the green sea urchin, *Strongylocentrotus droebachiensis* [71], the medaka fish, *Oryzias latipes* [72], the sea hare, *Aplysia dactylomela* [73] and in the Antarctic krill, *Euphausia superba* [74]. In *S. droebachiensis*, using two controlled diets of the MAA-rich alga, *Mastocarpus stellatus* versus the MAA-deficient alga, *Laminaria saccharina*, the sea urchins fed

M. stellatus have a much higher MAA concentration in the ovaries than sea urchins fed *L. saccharina* [71]. The higher MAA content in the diet allows for transfer of photoprotectants to eggs when released into the water column and exposed to UVR [71]. Embryos from adults of *S. droebachiensis* fed *L. saccharina* had a slightly longer UVR-induced delay in the first division of the embryo than did embryos from adults fed *M. stellatus* or a combination of both algae [75]. MAAs in the embryos of *S. droebachiensis* adults fed an MAA-rich diet provided photoprotection against abnormalities induced by UV-B to at least the four-armed pluteus stage [76].

The survivorship of larvae from the coral, *Agaricia agaricites* originating from 3 m depth, which contained a 3-fold higher concentration of MAAs, was greater than for larvae originating from 24 m depth [77]. In contrast, the origin of larvae from the coral *Pocillopora damicornis* was not a factor in survival in an experiment comparing shallow (0.5 m) and deep (2–3 m) larvae exposed to UVR compared with UVR-shielded larvae. Rather, the effect of UVR appeared to be on settlement [78]. However, the depth ranges in the two studies were vastly different and the total MAA concentrations were much higher in larvae of *A. agaricites* than *P. damicornis*.

Photoprotective Function of MAAs: The photoprotective function of MAAs has mostly been inferred indirectly from their UVR absorption properties [21,22] and high molar extinction coefficients (Table 1) as well as a series of results showing that there is a direct relationship between exposure to UVR and MAA concentration. These include: (a) a number of observations in sessile species that shallow growing individuals have higher concentrations of "S-320" or MAAs than deeper growing conspecifics such as the corals *Porites lobata* [79], *Acropora* spp. [80], *Montipora verrucosa* [81], *Acropora microphthalma* [82], various Caribbean and Hawaiian species [55], *Porites astreoides* [83] and *Montastraea faveolata* [84]; (b) the presence of peaks in UV absorbance in surface bloom-forming species such as *Gonyaulax tamarensis* var *excavata* [85], *Noctiluca miliaris* [86], *Prorocentrum micans* and *Gonyaulax polyedra* [87], *Phaeocystis pouchetii* [7], *Heterocapsa triquetra* [88] and *Akashiwo sanguinea* (= *Gymnodinium sanguineum*) [56]; (c) the decrease in concentration of these compounds when UVR is filtered out and compared with UVR treated samples such as in the pennate diatoms *Psuedonitzschia* sp. and *Fragilariopsis cylindrus* [89], the red algae *Eucheuma striatum* [90] and *Chondrus crispus* [91], the jellyfish *Cassiopeia xamachana* [53], the anemone *Phyllodiscus semoni* [51], the corals *Pocillopora damicornis* [2], *Montipora verrucosa* [92] and *Porites compressa* [93], and the octocoral *Clavularia* sp. [51]; (d) the increase in concentration of these compounds when sessile species are transplanted to shallower depths such as *Montipora verrucosa* [81] and *Porites astreoides* [94]; (e) the observation that exposed portions of benthic species have higher concentrations of MAAs than shaded portions. For example, the tops of individual colonies of *Montastraea annularis* contain higher concentrations of MAAs than the sides of the colonies whereas the bottoms of the colonies have the lowest concentrations of MAAs [95] and that holothurian epidermal tissue contains higher concentrations of MAAs than do internal organs [52]; (f) the observation that there is seasonal variation in MAA concentration peaking in summer as shown for the red algae *Palmaria decipiens* [96] and *Bangia atropurpurea* [97]; and (g) the greater concentration of MAAs in algae from low latitudes compared to high latitudes thus showing a relationship between natural solar UVR doses and the concentration of MAAs [43].

Direct evidence for a photoprotective role has been shown in Porphyra-334 using measurements of quantum yield of fluorescence, intersystem crossing and photolysis, which showed that this compound does not generate radicals that would cause cellular damage [98]. MAAs have been shown to act as direct, spectrally-specific photoprotectants in the surface-blooming, red tide dinoflagellate *Akashiwo sanguinea* (= *Gymnodinium sanguineum*) [56]. Cultures of *A. sanguinea* grown in high light have a markedly lower sensitivity to UVR, as estimated by biological weighting functions (BWFs) and accumulate MAAs in higher concentrations when compared with low light grown cultures. The wavelength range of lowest sensitivity (325 to 355 nm) corresponds to the region of maximal absorbance by the MAAs. No significant differences in UV biological weight were found in *Prorocentrum micans* suggesting that MAAs provide incomplete protection in this species [99,100].

Several species have been shown to not modify "S-320" or MAA concentration in response to UVR, such as the zoanthid *Zoanthus sociatus*, in response to increased levels of UVR [101], the octocoral *Clavularia* sp. over a depth gradient [51], the coral *Montastraea annularis* on transplantation from 24 m to 12 m over 21 days [102] and the temperate anemone *Anthopleura elegantissima* in UVR-exposed versus UVR-shielded experiments [53]. Data such as these have been used to suggest that MAAs are not directly photoprotective but are rather a byproduct of other chemical reactions and that photoprotection is a secondary function. An alternative explanation is that, at least for all of the above studies that involve coelenterates, the MAAs are derived from diet and therefore rather than MAA concentration being dependent on the intensity of UVR is dependent on the concentration of MAAs in the food source.

Functions other than photoprotection that have been attributed to MAAs include antioxidant activity [103], regulation of reproduction [104] and as osmolytes [105]. The role of MAAs as osmolytes has been tested and refuted due to the small contribution that MAAs make in comparison to other osmolytes in reducing osmotic stress [106].

10.2.2.3 Scytonemin

Scytonemin occurs mostly in the extracellular, mucilaginous sheath surrounding cyanobacterial cells and occurs in every major taxonomic group of cyanobacteria and is considered to be a photoprotective compound [107]. This pigment has a molecular weight of 544-546 Da, is yellow–brown, lipid soluble and is a dimeric structure of indolic and phenolic subunits whose synthetic pathway is poorly understood. The λ_{max} of absorption is approximately 370 nm in vivo and 384 nm in acetone [107] and there is also a strong UVC-absorbing component peaking at approximately 250 nm that extends into the UV-B [108]. The synthesis of scytonemin is strongly induced on exposure to UV-A-blue and only weakly by UV-B [109]. In full-sun exposed habitats, the concentrations of scytonemin are high and in culture the rate of synthesis of scytonemin is increased by exposure to UV-A or to high photon flux [107–109].

10.2.2.4 3-Hydroxykynurenine

3-Hydroxykynurenine is a water soluble, low molecular weight, tryptophan derivative that occurs in the lens pigments of several species of marine and freshwater fish [110,111] and the cuttlefish, *Sepia officinalis* [57]. 3-Hydroxykynurenine most closely resembles screening compounds found in the lenses of primates. This compound absorbs in the UV-A region with a peak absorbance of 370 nm and may function in protecting the lens from UVR, increasing visual acuity by reducing glare, scatter and chromatic aberration (misfocusing of short wavelengths) and maximizing contrast as well as aiding in prey detection or possibly functioning as a stabilizing lens protein [111].

10.2.2.5 Melanin

Melanin absorbs at all UVR and PAR wavelengths and thus is beneficial as a sunscreen in non-photosynthetic organisms such as Arctic and Alpine cladocerans [60]. Melanin was shown to protect platyfish-swordtail hybrids of the genus *Xiphophorus* by lowering the number of dimers caused by exposure to UVR of 290, 302 or 313 nm [112] and thus acting as a UVR photoprotectant in the skin of fish. Melanin is also believed to have more than one function, including acting as a free radical scavenger and energy transducer.

10.2.2.6 Secondary metabolites

Some secondary metabolites, such as flavonoids, phlorotannins and tridentatol, appear to perform multiple ecologically important roles such as resistance to predators and pathogens, sequestration of heavy metals as well as absorption of UVR wavelengths.

Flavonoids: Flavonoids function in the protection of UVR-sensitive cellular components by specifically absorbing from 280 to 340 nm but allowing transmission of PAR to the chloroplasts so as to not diminish photosynthetic yield. Flavonoids are commonly found in the epidermis of leaves, acting to protect the underlying photosynthetic units [113] and are widely distributed in angiosperms including seagrasses and aquatic mosses. The phenylpropanoid pathway is stimulated by exposure to UVR, which results in an accumulation of flavonoids mainly in the upper epidermal cell layer in plants due to an increased transcription of a series of enzymes [114]. Phenolic compounds perform diverse roles in angiosperms including resistance to predators and to pathogens as well as recruitment of pollinators to flowers and the attraction of seed dispersal agents.

Phlorotannins: Phlorotannins are secondary metabolites analogous to the shikimate-derived condensed polyphenolics found in plants and brown algae and absorb strongly from 280 to 320 nm [115]. Exposure to UV-B was found to increase the concentration of phlorotannins in the brown alga *Ascophyllum nodosum* [116]. That phlorotannins act in photoprotection may be the reason why MAAs are virtually absent in brown algae in comparison to red and green algae [39,55,117]. Protecting cells against UVR damage is not the only function of phlorotannins as they are also known to be involved in defense against herbivores, pathogens and heavy metals.

Tridentatols: Tridentatols A to D have been described in the hydroid *Tridentata marginata*, which is associated with the pelagic *Sargassum* community [118]. Floating at the surface of the ocean results in exposure to high levels of UVR. The four tridentatol compounds have absorption maxima ranging from 313 to 342 nm, and have been hypothesized to function in photoprotection as a result of this strong absorption in the UV-A and UV-B regions. Tridentatol A has also been identified as a deterrent of predators, thereby serving a dual role [118].

10.2.2.7 Fluorescent pigments (FPs)

FPs, found in coral tissues, are host-derived pigment proteins, related to a single family of green fluorescent proteins (GFPs) that fluoresce on exposure to UVR and PAR and includes the pocilloporins [119]. FPs were found to dissipate UVR via absorption at 330 nm by fluorescence as green light (when excited at 380 nm) thus converting the damaging UVR to PAR [119, 120]. More recent work indicates that under high light conditions, pocilloporins function in protection of the photosynthetic machinery against UVR and PAR [119]. A large number of shallow-dwelling corals contain FPs, thereby reducing sensitivity to photo inhibition and bleaching in reef corals by thermal dissipation of excess photons through fluorescence and light scattering [121]. Under low light conditions, these pigments are hypothesized to enhance light capturing ability such as in the deep dwelling coral species, *Leptoseris fragilis* [122].

10.3 Quenching mechanisms

Exposure to UVR is known to result in a variety of negative effects; however, the interaction of UVR with photosensitizing molecules, some organic molecules and oxygen results in the production of toxic photoproducts including reactive oxygen species (ROS) both intracellularly and in the external environment. Toxic photoproducts have the ability to cause more damage than the UVR exposure itself. Toxic photoproducts are neutralized by various agents including antioxidants such as ascorbate, quenchers such as carotenoids and various scavenging enzymes, the levels of which are up-regulated by the presence of UVR [123–125]. Exposure to UVR can result in an increase in the production of photoreactive species on the one hand and, on the other hand, result in the induction of enzymes to neutralize these species.

10.3.1 Antioxidants

Antioxidants, quenching molecules or radical scavenging enzymes such as superoxide dismutase (SOD), ascorbate peroxidase and catalase react with and neutralize the effects of the highly toxic ROS or hydroxyl radicals (HO·) produced by the photooxidation of cell structures by UVR (see Chapter 8). Studies have shown that there is a positive relationship between the concentration of these antioxidant enzymes with PAR and UVR. In the temperate sea anemone *Anthopleura elegantissima* there was higher activity of SOD in UVR-exposed than in UVR-shielded individuals [126]. Symbiotic algae from the tropical anemone *Aiptasia pallida* exhibited higher antioxidant concentrations in UVR-exposed versus UVR-shielded anemones [123] and the effects are reversible by reciprocal transplantation of individuals taken from high light and low light conditions [124]. In the symbiotic dinoflagellate *Symbiodinium microadriaticum* it was found that SOD concentration is directly related to oxygen tension when compared in cultured cells grown under hypoxic, normoxic and hyperoxic conditions [127]. The levels of these scavenging enzymes are increased by temperature, PAR and UVR in symbiotic algae of the zoanthid *Palythoa caribaeorum* [125] and their concentration decreases with increasing depth in the coral *Acropora microphthalma* [82].

Flavonoids possess free-radical scavenging activity [128] and the MAA mycosporine-glycine also exhibits moderate antioxidant activity [103]. These compounds may be additional sources of protection against photooxidative stress due to the high concentrations of oxygen in photosynthetic symbioses. Imino-type MAAs such as palythine and shinorine had no detectable antioxidant activity [103].

10.3.2 Carotenoids and the xanthophyll cycle

Carotenoids, perhaps best known as photosynthetic accessory pigments, are also radical-trapping antioxidants, which scavenge oxygen radicals, neutralize the singlet excited state of oxygen ($^1O_2$), quench the triplet state of chlorophyll *a*, which occurs under excess light exposure, and inhibit lipid peroxidation. Increased concentrations of carotenoids were found in natural populations of the surface-blooming cyanobacterium *Microcystis aeruginosa* [129] and after exposure to artificial UV-B in cultures of the marine phytoplankter *Tetraselmis* sp. [128] and attributed to the photoprotective function of these pigments. In eukaryotes, some carotenoids are also part of the xanthophyll cycle that quenches excess photochemical energy by dissipating it as heat and thereby limiting photoinhibition [131]. The advantage of this cycle is that it has a very rapid response time to high photon flux density. In seagrasses, the xanthophyll cycle consists of a series of light-dependent reactions involving three oxygenated derivatives of carotenoids: violaxanthin, antheraxanthin and zeaxanthin [132]. During high light exposure, violaxanthin is converted into antheraxanthin and subsequently into zeaxanthin, which then accumulates, until there is a sufficient decrease in light exposure for reconversion into violaxanthin [133]. While important in protection against high photon flux density in seagrasses, it was found that the conversion into zeaxanthin was not sufficient to protect against UVR in *Halodule ovalis* [132]. Ascorbate, apart from its function as an antioxidant, also functions to reduce violaxanthin to antheraxanthin and zeaxanthin, leading to thermal dissipation of excess energy [133].

In many species of microalgae, such as dinoflagellates, diatoms and prymnesiophytes, the xanthophyll cycle pigments found in seagrasses are absent and, instead, the reaction involved is the reversible conversion of diadinoxanthin (Dn) into diatoxanthin (Dt) by de-epoxidation. High concentrations of Dn were found in natural populations of the red-tide dinoflagellate *Prorocentrum micans* and a possible role in photoprotection under high illumination was proposed [87] although interconversion between Dn and Dt was not examined. Under exposure to high photon flux densities, the free-living dinoflagellate *Alexandrium excavatum* accumulated Dt, which was reverted back into Dn under dark conditions with no new pigment synthesis involved [134]. Following exposure to artificial UVR and PAR over a 40 day period of cultures of the common coastal diatom *Thalassiosira weissflogii*, Dt was the only pigment to show a distinct increase [66]. This study also showed that the induction of the xanthophyll cycle coincides with increased photoinhibition (as inferred from decreased photochemical capacity) and decreased growth rates whereas when photoinhibition decreased, so did the concentration of Dt suggesting a photoprotective role by the xanthophyll cycle. Recovery from photoinhibition did not require the prior accumulation of MAAs but rather induction of the xanthophyll cycle and once the cells had recovered from photoinhibition, MAAs accumulated and the concentration of xanthophyll cycle pigments decreased [66]. The xanthophyll cycle also functions in dinoflagellates symbiotic with corals [135] with the interconversion of Dn and Dt showing a strong diurnal pattern [136]. In some microalgae, the xanthophyll cycle involves the conversion between violaxanthin and zeaxanthin [136]. The xanthophyll cycle pigments barely absorb in the UVR and therefore their role in preventing damage by UVR may not be directly as photoprotectants but indirectly as quenchers of toxic photoproducts.

Transparency is a common adaptation to lake and oceanic environments. As a camouflage, transparency is complicated by the presence of UVR because the presence of UVR-absorbing compounds decreases UV transparency and may reveal some organisms to predators and prey with UV vision [137]. However, in general, pigmented species are less sensitive to solar radiation than unpigmented species [60]. Planktonic crustaceans exhibit marked colouring [138]. For example, the calanoid copepod *Acanthodiaptomus denticornis* has both translucent and red-coloured morphs, whose coloring is due to carotenoid pigments derived from their algal food source. The red colored morphs were shown to have lower mortality than translucent morphs when exposed to UVR [139].

10.4 Repair mechanisms

Not all exposure to UVR can be avoided and therefore all organisms require some capacity to repair damage caused by these wavelengths. Little is known about the mechanisms that recognize and repair damage induced by UVR in marine and freshwater systems and most of what is known about repair of UVR-induced DNA damage comes from research on bacteria, which has been applied to aquatic organisms. Repair of DNA and of proteins may be induced during exposure to UVR and may continue well after the exposure to damaging wavelengths has ceased.

10.4.1 Protein repair mechanisms

Repair proteins are induced by exposure to UVR in amoebae and cyanobacteria and de novo protein biosynthesis has been shown in a variety of systems. Turnover of damaged proteins is an important mechanism in overcoming damage by UVR. Reactivation of damaged photosystems after prolonged exposure to UV-A was important in the recovery of the diatom *Melosira* sp. and of the green alga *Chlorella ellipsoidea* [140]. Photosystem II (PS II) is a protein–pigment complex that is structured to catalyze the transfer of electrons from water to plastoquinone thus releasing oxygen. At the core of PS II is a dimer of two related proteins, called D1 and D2, which bind the pigments and cofactors involved in electron transfer. The D1 protein is susceptible to PAR-induced photoinhibition and PAR driven turnover of the D1 protein is rapid through synthesis via the PS II repair cycle. Both the D1 and the D2 proteins are susceptible to degradation by UV-B especially at 300 nm and there is UVR-induced turnover of these proteins to prevent the accumulation of UVR-damaged PS II. Inhibition of protein synthesis by the antibiotic streptomycin in the diatom *Thalassiosira psuedonana* results in a greater susceptibility to photoinhibition of carbon uptake by UV-B [141]. The repair cycle can also be damaged thus limiting the turnover of functional D1 resulting in photoinhibition of photosynthesis due to a decrease in the capacity of PS II to transfer electrons [114,142]. In the aquatic cyanobacterium, *Synechococcus* sp., the genes psbAII and psbAIII are induced within 15 min of moderate exposure to UV-B. This induction results in an exchange of two distinct D1 proteins, D1:1, encoded by the constitutively-expressed psbAI, and D1:2, encoded by psbAII and psbAIII. Under the same UVR conditions, a mutant only able to express psbAI experienced a 40% drop in photosynthesis, whereas a mutant able to constitutively express psbAII and psbAIII was able to resist exposure to UV-B [142].

10.4.2 DNA repair and DNA damage tolerance mechanisms

Although the peak absorbance of DNA is at 260 nm, the absorption spectrum of DNA follows a normal distribution that extends well into the UV-B; therefore, exposure to any of these wavelengths can result in damage to DNA. UVR-induced damage can result in direct mutagenesis as well as lethal effects due to the structural changes in DNA molecules, which can interfere in the synthesis of DNA and in the transcription by RNA resulting in errors in translation of the genetic code. The most common structural change in DNA is the formation of crosslinks between bases on the same strand of DNA, the most common photoproduct of which is the formation of dimers between adjacent pyrimidine bases called cyclobutane pyrimidine dimers (CPDs). CPDs make up 50 to 80% of

UVR-induced DNA photoproducts with most of the remaining portion being made up of pyrimidine (6-4) pyrimidone dimers or 6-4 photoproducts [143]. CPDs do not cause mutagenesis as the dimers cannot base pair with other nucleotides; however, they do cause a block in DNA replication. Thus, dimer damage to DNA incapacitates its normal function of directing cellular metabolism including the replication of DNA, the transcription of genes and the synthesis of protein. There are two DNA repair pathways: photoreactivation, nucleotide excision repair and two DNA damage tolerance mechanisms: dimer bypass and recombinational repair (Figure 1). Tolerance pathways do not repair the damage but rather reduce the effect of the photoproducts on the genetic system.

10.4.2.1 Photoreactivation (DNA photorepair or photoenzymatic repair)

Photoreactivation is a single enzyme repair system that utilizes UV-A and blue light energy (385 to 450 nm) to monomerize the dimers formed between the pyrimidine bases [144]. The enzyme, photolyase, recognizes and binds to the CPDs in situ and splits the dimer using light energy thus restoring the bases to their native form (Figure 1). Photolyases can have either a folate- or a flavin-type chromophore, with absorption maxima between 350 and 450 nm. Excitation energy from UV-A or blue light is transferred from the chromophore to the active site that contains a flavin adenine dinucleotide (FAD), which subsequently transfers an electron to the dimer resulting in monomerization [144]. This pathway is present in viruses, mycoplasmas, and many eukaryotic organisms including protozoa, algae, higher plants, reptiles, amphibians, fish and marsupials and it is the major repair pathway in bacteria. In the marine environment, this pathway has been shown to be present as a repair mechanism for UVR-damaged DNA in microalgae [145,146], a red alga [147], marine and freshwater crustaceans [148–151], Antarctic bacteria [152], 12 species of Antarctic diatoms [153] and Antarctic zooplankton, including fish larvae [154] and a seagrass species, *Halodule wrightii* [155]. Organisms that depend on the repair capacity of photoreactivation also depend on the quality and quantity of light in their environment after the DNA damage has occurred.

10.4.2.2 Light-independent nucleotide excision repair (dark repair)

Another pathway available for repair of DNA damage is nucleotide excision repair or dark repair, which requires an increased production of a series of DNA replication enzymes (endonuclease, DNA polymerase, exonuclease and ligase). The enzymes recognize DNA damage possibly due to distortion of the helix, incise the DNA strand at the lesion, resynthesize the correct sequence using the information from the complementary undamaged strand of DNA and DNA polymerase, excise the lesion and close the DNA strand with ligase (Figure 1). This pathway, which does not require light energy, occurs in all types of prokaryotes and eukaryotes and is a major mechanism in mammalian cells. In contrast to photoreactivation, which repairs by directly reversing DNA damage, nucleotide excision repair reverses DNA damage by replacing the lesion with

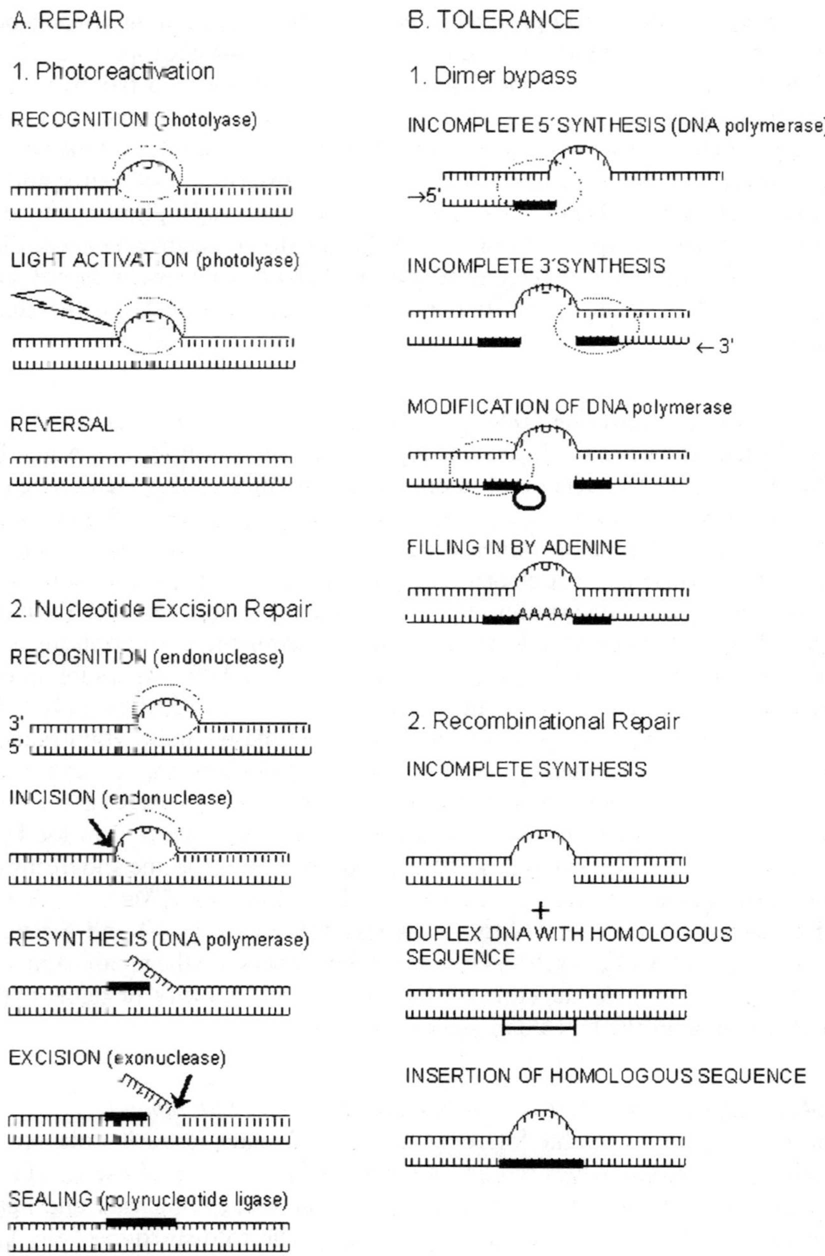

Figure 1. Schematic presentation of (A) DNA repair mechanisms: 1. Photoreactivation also known as photoenzymatic repair, and 2. Nucleotide excision repair where the lesion damaged by exposure to UV-B is reversed (photoreactivation) or expelled (nucleotide excision repair); (B) DNA damage tolerance mechanisms: 1. Dimer bypass and 2. Recombinational repair where replication proceeds around the lesion and the gap is filled in by adenine (dimer bypass) or a homologous sequence is inserted (recombinational repair).

new nucleotides. This pathway is virtually absent in Antarctic marine diatoms [153] and found in only low levels in Antarctic marine bacteria [152] but is present in *Chlamydomonas* [146].

10.4.2.3 Dimer bypass (translesion synthesis)
This pathway has been reported in the bacterium *Escherichia coli*, where gene products bind to DNA polymerase, thus altering it and resulting in insertion of adenine residues directly across from the damaged portion of the strand (Figure 1). Although, incorrect bases are inserted into the complementary strand and the damage site remains unrepaired, the advantage of dimer bypass is that it allows for DNA replication to continue.

10.4.2.4 Recombinational repair (post replication repair)
In some cases, the damaged portion of DNA is bypassed during the replication phase but otherwise replication continues. The recombinational or post replication pathway then inserts a homologous complementary DNA strand into the site opposite of the dimer damage (Figure 1). The dimer damage is left unrepaired and replication continued such that the complementary strand is error-free. This type of repair is known to occur in bacteria.

10.5 Conclusions and future directions

Efficient mechanisms to overcome the deleterious effects of UVR are even more important during variable UVR conditions such as are experienced due to depletion of the ozone layer. There are many studies on physiological and biochemical responses by marine and freshwater organisms to UVR; however, there are many more questions begging to be answered. Although a lot is known about the distribution of UVR-absorbing compounds in aquatic organisms, there are many details left unstudied, such as how the damaging wavelengths of UVR are dissipated by the different types of UVR-absorbing compounds, where these compounds are synthesized and located and the mechanisms of bacterial or biochemical interconversion of these compounds. The study of DNA damage and repair or tolerance mechanisms is virtually an open field with only a few pioneering studies. Sensitivity to UVR is species dependent and is based on the differing abilities of organisms to attenuate UVR prior to absorption by UVR-sensitive cellular components and of the ability to repair or tolerate damage to proteins and DNA in particular. More emphasis needs to be placed in this particular area to modify techniques applied to bacterial systems for use in aquatic systems. Once established, these techniques can then be used to determine the importance of repair and tolerance mechanisms in aquatic organisms relative to the importance of screening and avoidance mechanisms.

Acknowledgements

Financial support from the Consejo Nacional de Ciencias y Tecnología, México and departmental funding from the Instituto de Ciencias del Mar y Limnología, Universidad Nacional Autónoma de México are gratefully acknowledged as are comments from an anonymous reviewer. M.G. Barba-Santos aided in the preparation of Figure 1.

References

1. C. Sebastian, R. Scheuerlein, D.-P. Häder (1994). Effects of solar and artificial ultraviolet radiation on pigment composition and photosynthesis in three *Prorocentrum* species. *J. Exp. Mar. Biol. Ecol.*, **182**, 251–263.
2. P.L. Jokiel, R.H. York Jr. (1982). Solar ultraviolet photobiology of the reef coral *Pocillopora damicornis* and symbiotic zooxanthellae. *Bull. Mar. Sci.*, **32**, 301–315.
3. O. Siebeck (1988). Experimental investigation of UV tolerance in hermatypic corals (Scleractinia). *Mar. Ecol. Prog. Ser.*, **43**, 95–103.
4. S. Roy (2000). Strategies for the minimization of UV-induced damage. In: S. de Mora, S. Demers, M. Vernet (Eds), *The Effects of UV Radiation in the Marine Environment* (pp. 177–205) Cambridge University Press, Cambridge.
5. W.F. Vincent, S. Roy (1993). Solar ultraviolet-B radiation and aquatic primary production damage, protection, and recovery. *Environ. Rev.*, **1**, 1–12.
6. E. Litchman, P.J. Neale, A.T. Banaszak (2002). Increased sensitivity to ultraviolet radiation in nitrogen-limited dinoflagellates: Photoprotection and repair. *Limnol. Oceanogr.*, **47**, 86–94.
7. H.J. Marchant, A.T. Davidson, G.J. Kelly (1991). UV-B protecting compounds in the marine alga *Phaeocystis pouchetii* from Antarctica. *Mar. Biol.*, **109**, 391–395.
8. J.H. Drollet, P. Glaziou, P.M.V. Martin (1993). A study of mucus from the solitary coral *Fungia fungites* (Scleractinia : Fungiidae) in relation to photobiological UV adaptation. *Mar. Biol.*, **115**, 263–266.
9. J.H. Drollet, T. Teai, M. Faucon, P.M.V. Martin (1997). Field study of compensatory changes in UV-absorbing compounds in the mucus of the solitary coral *Fungia repanda* (Scleractinia : Fungiidae) in relation to solar UV radiation, sea-water temperature, and other coincident physico-chemical parameters. *Mar. Freshwater Res.*, **48**, 329–333
10. T. Teai, J.H. Drollet, J.-P. Bianchini, A. Cambon, P.M.V. Martin (1998). Occurrence of ultraviolet radiation-absorbing mycosporine-like amino acids in coral mucus and whole corals of French Polynesia. *Mar. Freshwater Res.*, **49**, 127–132.
11. F. Xiong, J. Komenda, J. Kopecký, L. Nedbal (1997). Strategies of ultraviolet-B protection in microscopic algae. *Physiol. Plant.*, **100**, 378–388.
12. A.T. Banaszak, R.K. Trench (1995). Effects of ultraviolet (UV) radiation on marine microalgal-invertebrate symbioses. I. Response of the algal symbionts in culture and in hospite. *J. Exp. Mar. Biol. Ecol.*, **194**, 213–232.
13. C.M. Leach (1965). Ultraviolet-absorbing substances associated with light-induced sporulation in fungi. *Can. J. Bot.*, **43**, 185–200.
14. E.J. Trione, C.M. Leach, J.T. Mutch (1966). Sporogenic substances isolated from fungi. *Nature*, **212**, 163–164.
15. J. Favre-Bonvin, N. Arpin, C. Brevard (1976). Structure de la mycosporine (P 310). *Can. J. Chem.*, **54**, 1105–1113.
16. H. Young, V.J. Patterson (1982). A UV protective compound from *Glomerella cingulata* – a mycosporine. *Photochemistry*, **21**, 1075–1077.

17. J. Bernillon, M.-L. Bouillant, J.-L. Pittet, J. Favre-Bonvin, N. Arpin (1984). Mycosporine glutamine and related mycosporines in the fungus *Pyronema omphalodes*. *Phytochemistry*, **23**, 1083–1087.
18. W.M. Bandaranayake (1998). Mycosporines: Are they nature's sunscreens? *Nat. Prod. Rep.*, **15**, 159–172.
19. J. Favre-Bonvin, J. Bernillon, N. Salin, N. Arpin (1987). Biosynthesis of mycosporines: mycosporine glutaminol in *Trichothecium roseum*. *Phytochemistry*, **26**, 2509–2514.
20. M. Tevini, A.H. Teramura (1989). UV-B effects on terrestrial plants. *Photochem. Photobiol.*, **50**, 479–487.
21. K. Shibata (1969). Pigments and a UV-absorbing substance in corals and a blue–green alga living in the Great Barrier Reef. *Plant Cell Physiol.*, **10**, 325–335.
22. W.C. Dunlap, B.E. Chalker (1986). Identification and quantitation of near-UV absorbing compounds (S-320) in a hermatypic scleractinian. *Coral Reefs*, **5**, 155–159.
23. W.C. Dunlap, J.M. Shick (1998). Ultraviolet radiation-absorbing mycosporine-like amino acids in coral reef organisms: A biochemical and environmental perspective. *J. Phycol.*, **34**, 418–430.
24. A.T. Banaszak, R.K. Trench (2001). Ultraviolet sunscreens in dinoflagellates. *Protist*, **152**, 93–101.
25. N. Arpin, M.L. Bouillant (1981). Light and mycosporines. In: G. Turian, H. R. Hohl (Eds), *The Fungal Spore, Morphogenetic Controls* (pp. 435–454). Academic Press, London.
26. W.R. Stochaj, W.C. Dunlap, J.M. Shick (1994). Two new UV-absorbing mycosporine-like amino acids from the sea anemone *Anthopleura elegantissima* and the effects of zooxanthellae and spectral irradiance on chemical composition and content. *Mar. Biol.*, **118**, 149–156.
27. S. Ito, Y. Hirata (1977). Isolation and structure of a mycosporine from the zoanthid *Palythoa tuberculosa*. *Tetrahedron Lett.*, **28**, 2429–2430.
28. S. Takano, D. Uemura, Y. Hirata (1978). Isolation and structure of a new amino acid, palythine, from the zoanthid *Palythoa tuberculosa*. *Tetrahedron Lett.*, **26**, 2299–2300.
29. I. Tsujino, K. Yabe, I. Sekikawa, N. Hamanaka (1978). Isolation and structure of a mycosporine from the red alga *Chondrus yendoi*. *Tetrahedron Lett.*, **16**, 1401–1402.
30. T.T. Teai, P. Raharivelomanana, J.-P. Bianchini, R. Faure, P.M.V. Martin, A. Cambon (1997). Structure de deux nouvelles iminomycosporines isolées de *Pocillopora eydouxi*. *Tetrahedron Lett.*, **38**, 5799–5800.
31. J. Wu Won, B.E. Chalker, J.A. Rideout (1997). Two new UV-absorbing compounds from *Stylophora pistillata*: Sulfate esters of mycosporine-like amino acids. *Tetrahedron Lett.*, **38**, 2525–2526.
32. H. Nakamura, J. Kobayashi, Y. Hirata (1981). Isolation and structure of a 330 nm UV-absorbing substance, Asterina-330 from the starfish *Asterina pectinifera*. *Chem. Lett.*, 1413–1414.
33. W.M. Bandaranayake, J.E. Bemis, D.J. Bourne (1996). Ultraviolet absorbing pigments from the marine sponge *Dysidea herbacea*: Isolation and structure of a new mycosporine. *Comp. Biochem. Physiol.*, **115C**, 281–286.
34. J. Wu Won, J.A. Rideout, B.E. Chalker (1996). Isolation and structure of a novel mycosporine-like amino acid from the reef-building corals *Pocillopora damicornis* and *Stylophora pistillata*. *Tetrahedron Lett.*, **36**, 5255–5256.
35. S. Takano, D. Uemura, Y. Hirata (1978). Isolation and structure of two new amino acids, palythinol and palythene, from the zoanthid *Palythoa tuberculosa*. *Tetrahedron Lett.*, **49**, 4909–4912.
36. P.T. Grant, C. Middleton, P.A. Plack, R.H. Thomson (1985). The isolation of four

aminocyclohexenimines (mycosporines) and a structurally related derivative of cyclohexane-1,3-dione (Gadusol) from the brine shrimp, *Artemia*. *Comp. Biochem. Physiol.*, **80B**, 755–759.
37. S. Takano, A. Nakanishi, D. Uemura, Y. Hirata (1979). Isolation and structure of a 334. nm UV-absorbing substance, Porphyra-334. from the red alga *Porphyra tenera* (Kjellman). *Chem. Lett.*, 419–420.
38. I. Tsujino, K. Yabe, I. Sekikawa (1980). Isolation and structure of a new amino acid, shinorine, from the red alga, *Chondrus yendoi* Yamada et Mikami. *Bot. Mar.*, **23**, 65–68.
39. D. Karentz, F.S. McEuen, M.C. Land, W.C. Dunlap (1991). Survey of mycosporine-like amino acid compounds in Antarctic marine organisms: Potential protection from ultraviolet exposure. *Mar Biol.*, **108**, 157–166.
40. J. Kobayashi, H. Nakamura, Y. Hirata (1981). Isolation and structure of a UV-absorbing substance 337 from the ascidian *Halocynthia roretzi*. *Tetrahedron Lett.*, **22**, 3001–3002.
41. I. Sekikawa, C. Kubota, T. Hiraoki, I. Tsujino (1986). Isolation and structure of a 357 nm UV-absorbing substance, usujirene, from the red alga *Palmaria palmata* (L.) O. Kuntze. *Jpn. J. Phycol.*, **34**, 185–188.
42. P.T. Grant, P.A. Plack, R.H. Thomson (1980). Gadusol, a metabolite from fish eggs. *Tetrahedron Lett.*, **21**, 4043–4044.
43. U. Karsten, T. Sawall, D. Hanelt, K. Bischof, F.L. Figueroa, A. Flores-Moya, C. Wiencke (1998). An inventory of UV-absorbing mycosporine-like amino acids in macroalgae from polar to warm-temperate regions. *Bot. Mar.*, **41**, 443–453.
44. R.A. Kinzie III, A.T. Banaszak, M.P. Lesser (1998). Effects of ultraviolet radiation on primary productivity in a high altitude tropical lake. *Hydrobiologia*, **385**, 23–32.
45. R. Sommaruga, F. Garcia-Pichel (1999). UV-absorbing mycosporine-like compounds in planktonic and benthic organisms from a high-mountain lake. *Arch. Hydrobiol.*, **144**, 255–269.
46. A. Quesada, W.F. Vincent, D.R.S. Lean (1999). Community and pigment structure of Arctic cyanobacterial assemblages: The occurrence and distribution of UV-absorbing compounds. *FEMS Microb. Ecol.*, **28**, 315–323.
47. F. Xiong, J. Kopecký, L. Nedbal (1999). The occurrence of UV-B absorbing mycosporine-like amino acids in freshwater and terrestrial microalgae (Chlorophyta). *Aquat. Bot.*, **63**, 37–49.
48. H. Nakamura, J. Kobayashi, Y. Hirata (1982). Separation of mycosporine-like amino acids in marine organisms using reversed-phase high-performance liquid chromatography. *J. Chromatogr.*, **250**, 113–118.
49. W.C. Dunlap, D. McB. Williams, B.E. Chalker, A.T. Banaszak (1989). Biochemical photoadaptation in vision: U.V.-absorbing pigments in fish eye tissues. *Comp. Biochem. Physiol.*, **93B**, 601–607.
50. J.I. Carreto, M.O. Carignan, G. Daleo, S.G. De Marco (1990). Occurrence of mycosporine-like amino acids in the red-tide dinoflagellate *Alexandrium excavatum*: UV-photoprotective compounds?. *J. Plankton Res.*, **12**, 909–921.
51. J.M. Shick, M.P. Lesser, W.R. Stochaj (1991). Ultraviolet radiation and photo-oxidative stress in zooxanthellate Anthozoa: The sea anemone *Phyllodiscus semoni* and the octocoral *Clavularia* sp. *Symbiosis*, **10**, 145–173.
52. J.M. Shick, W.C. Dunlap, B.E. Chalker, A.T. Banaszak, T.K. Rosenzweig (1992). Survey of ultraviolet radiation-absorbing mycosporine-like amino acids in organs of coral reef holothuroids. *Mar. Ecol. Prog. Ser.*, **90**, 139–148.
53. A.T. Banaszak, R.K. Trench (1995). Effects of ultraviolet (UV) radiation on marine

microalgal-invertebrate symbioses. II. The synthesis of mycosporine-like amino acids in response to exposure to UV in *Anthopleura elegantissima* and *Cassiopeia xamachana*. *J. Exp. Mar. Biol. Ecol.*, **194**, 233–250.
54. T. Teai, J.H. Drollet, J.-P. Bianchini, A. Cambon, P.M.V. Martin (1997). Widespread occurrence of mycosporine-like amino acid compounds in scleractinians from French Polynesia. *Coral Reefs*, **16**, 169–176.
55. A.T. Banaszak, M.P. Lesser, I.B. Kuffner, M. Ondrusek (1998). Relationship between ultraviolet (UV) radiation and mycosporine-like amino acids (MAAs) in marine organisms. *Bull. Mar. Sci.*, **63**, 617–628.
56. P.J. Neale, A.T. Banaszak, C.R. Jarriel (1998). Ultraviolet sunscreens in *Gymnodinium sanguineum* (Dinophyceae): Mycosporine-like amino acids protect against inhibition of photosynthesis. *J. Phycol.*, **34**, 928–938.
57. N. Shashar, F.I. Hárosi, A.T. Banaszak, R.T. Hanlon (1998). UV radiation blocking compounds in the eye of the cuttlefish *Sepia officinalis*. *Biol. Bull.*, **192**, 187–188.
58. A. Gröniger, R.P. Sinha, M. Klisch, D.-P. Häder (2000). Photoprotective compounds in cyanobacteria, phytoplankton and macroalgae – a database. *J. Photochem. Photobiol. B: Biol.*, **58**, 115–122.
59. T.A. Moisan, B.G. Mitchell (2001). UV absorption by mycosporine-like amino acids in *Phaeocystis antarctica* Karsten induced by photosynthetically available radiation. *Mar. Biol.*, **138**, 217–227.
60. B. Tartarotti, I. Laurion, R. Sommaruga (2001). Large variability in the concentration of mycosporine-like amino acids among zooplankton from lakes located across an altitude gradient. *Limnol. Oceanogr.*, **46**, 1546–1552.
61. J.I. Carreto, M.O. Carignan, N.G. Montoya (2001). Comparative studies on mycosporine-like amino acids, paralytic shellfish toxins and pigment profiles of the toxic dinoflagellates *Alexandrium tamarense*, *A. catanella* and *A. minutum*. *Mar. Ecol. Prog. Ser.*, **223**, 49–60.
62. A.T. Banaszak, T. LaJeunesse, R.K. Trench (2000). The synthesis of mycosporine-like amino acids (MAAs) by cultured, symbiotic dinoflagellates. *J. Exp. Mar. Biol. Ecol.*, **249**, 219–233.
63. R. Bentley (1990). The shikimate pathway – A metabolic tree with many branches. *Crit. Rev. Biochem. Mol. Biol.*, **25**, 307–384.
64. J.M. Shick, S. Romaine-Lioud, C. Ferrier-Pagès, J.-P. Gattuso (1999). Ultraviolet-B radiation stimulates shikimate pathway-dependent accumulation of mycosporine-like amino acids in the coral *Stylophora pistillata* despite decreases in its population of symbiotic dinoflagellates. *Limnol. Oceanogr.*, **44**, 1667–1682.
65. J.I. Carreto, S.G. De Marco, V.A. Lutz (1989). UV-absorbing pigments in the dinoflagellates *Alexandrium excavatum* and *Prorocentrum micans*. Effects of light intensity. In: T. Okaichi, D.M. Anderson, T. Nemoto (Eds), *Red Tides: Biology, Environmental Science, and Toxicology* (pp. 333–336). Elsevier Science, New York.
66. L. Zudaire, S. Roy (2001). Photoprotection and long-term acclimation to UV radiation in the marine diatom *Thalassiosira weissflogii*. *J. Photochem. Photobiol. B: Biol.*, **62**, 26–34.
67. A. Gröniger, C. Hallier, D.P. Häder (1999). Influence of UV radiation and visible light on *Porphyra umbilicalis*: Photoinhibition and MAA concentration. *J. Appl. Phycol.*, **11**, 437–445.
68. M. Vernet, K. Whitehead (1996). Release of ultraviolet-absorbing compounds by the red-tide dinoflagellate *Lingulodinium polyedra*. *Mar. Biol.*, **127**, 35–44.
69. K. Whitehead, D. Karentz, J.I. Hedges (2001). Mycosporine-like amino acids (MAAs) in phytoplankton, a herbivorous pteropod (*Limacina helicina*), and its pteropod

predator (*Clione antarctica*) in McMurdo Bay, Antarctica. *Mar. Biol.*, **139**, 1013–1019.

70. M. Ishikura, C. Kato, T. Maruyama (1997). UV-absorbing substances in zooxanthellate and azooxanthellate clams. *Mar. Biol.*, **128**, 649–655.
71. A.K. Carroll, J.M. Shick (1996). Dietary accumulation of UV-absorbing mycosporine-like amino acids (MAAs) by the green sea urchin (*Strongylocentrotus droebachiensis*). *Mar. Biol.*, **124**, 561–569.
72. D.S. Mason, F. Schafer, J.M. Shick, W.C. Dunlap (1998). Ultraviolet radiation-absorbing mycosporine-like amino acids (MAAs) are acquired from their diet by medaka fish (*Oryzias latipes*) but not by SKH-1 hairless mice. *Comp. Biochem. Physiol.*, **120A**, 587–598.
73. T.H. Carefoot, D. Karentz, S.C. Pennings, C.L. Young (2000). Distribution of mycosporine-like amino acids in the sea hare *Aplysia dactylomela*: Effect of diet on amounts and types sequestered over time in tissues and spawn. *Comp. Biochem. Physiol.*, **126C**, 91–104.
74. S.J. Newman, W.C. Dunlap, S. Nicol, D. Ritz (2000). Antarctic krill (*Euphausia superba*) acquire a UV-absorbing mycosporine-like amino acid from dietary algae. *J. Exp. Mar. Biol. Ecol.*, **255**, 93–110.
75. N.L. Adams, J.M. Shick (1996). Mycosporine-like amino acids provide protection against ultraviolet radiation in eggs of the sea urchin *Strongylocentrotus droebachiensis*. *Photochem. Photobiol.*, **64**, 149–158.
76. N.L. Adams, J.M. Shick (2001). Mycosporine-like amino acids prevent UVB-induced abnormalities during early development of the green sea urchin *Strongylocentrotus droebachiensis*. *Mar. Biol.*, **138**, 267–280.
77. D.F. Gleason, G.M. Wellington (1995). Variation in UVB sensitivity of planula larvae of the coral *Agaricia agaricites* along a depth gradient. *Mar. Biol.*, **123**, 693–703.
78. I.B. Kuffner (2001). Effects of ultraviolet (UV) radiation on larval settlement of the reef coral *Pocillopora damicornis*. *Mar. Ecol. Prog. Ser.*, **217**, 251–261.
79. J.E. Maragos (1972). *A Study of the Ecology of Hawaiian Reef Corals* (Ph.D. Thesis, 290 pp). University of Hawaii, Honolulu.
80. W.C. Dunlap, B.E. Chalker, J.K. Oliver (1986). Bathymetric adaptations of reef-building corals at Davies Reef, Great Barrier Reef, Australia. III. UV-B absorbing compounds. *J. Exp. Mar. Biol. Ecol.*, **104**, 239–248.
81. G. Scelfo (1985). Relationship between solar radiation and pigmentation of the coral *Montipora verrucosa* and its zooxanthellae. In: P.L. Jokiel, R.H. Richmond, R.A. Rogers (Eds), *Coral Reef Population Biology* (pp. 440–451). Sea Grant Coop. Rep. Hawaii University, Hawaii.
82. J.M. Shick, M.P. Lesser, W.C. Dunlap, W.R. Stochaj, B.E. Chalker, J. Wu Won (1995). Depth-dependent responses to solar ultraviolet radiation and oxidative stress in the zooxanthellate coral *Acropora microphthalma*. *Mar. Biol.*, **122**, 41–51.
83. J.E. Corredor, A.W. Bruckner, F.Z. Muszynski, R.A. Armstrong, R. García, J.M. Morell (2000). UV-absorbing compounds in three species of Caribbean zooxanthellate corals: Depth distribution and spectral response. *Bull. Mar. Sci.*, **67**, 821–830.
84. M.P. Lesser (2000). Depth-dependent photoacclimatization to solar ultraviolet radiation in the Caribbean coral *Montastraea faveolata*. *Mar. Ecol. Prog. Ser.*, **192**, 137–151.
85. C.S. Yentsch, C.M. Yentsch (1982). The attenuation of light by marine phytoplankton with specific reference to the absorption of near-UV radiation. In: J. Calkins (Ed.), Plenum Press, *The Role of Solar Ultraviolet Radiation in Marine Ecosystems* (pp. 691–700). New York.

86. W.M. Balch, F.T. Haxo (1984). Spectral properties of *Noctiluca miliaris* Suriray, a heterotrophic dinoflagellate. *J. Plankton Res.*, **6**, 515–525.
87. M. Vernet, A. Neori, F.T. Haxo (1989). Spectral properties and photosynthetic action in red-tide populations of *Prorocentrum micans* and *Gonyaulax polyedra*. *Mar. Biol.*, **103**, 365–371.
88. S.-Å. Wängberg, A. Persson, B. Karlson (1997). Effects of UV-B radiation on synthesis of mycosporine-like amino acid and growth in *Heterocapsa triquetra* (Dinophyceae). *J. Photochem. Photobiol. B: Biol.*, **37**, 141–146.
89. E.W. Helbling, B.E. Chalker, W.C. Dunlap, O. Holm-Hansen, V.E. Villafañe (1996). Photoacclimation of antarctic marine diatoms to solar ultraviolet radiation. *J. Exp. Mar. Biol. Ecol.*, **204**, 85–101.
90. W.F. Wood (1989). Photoadaptive responses of the tropical red alga *Eucheuma striatum* Schmitz (Gigartinales) to ultra-violet radiation. *Aquat. Bot.*, **33**, 41–51.
91. U. Karsten, L.A. Franklin, K. Lüning, C. Wiencke (1998). Natural ultraviolet radiation and photosynthetically active radiation induce formation of mycosporine-like amino acids in the marine macroalga *Chondrus crispus* (Rhodophyta). *Planta*, **205**, 257–262.
92. R.A. Kinzie III (1993). Effects of ambient levels of solar ultraviolet radiation on zooxanthellae and photosynthesis of the reef coral *Montipora verrucosa*. *Mar. Biol.*, **116**, 319–327.
93. I.B. Kuffner (2001). Effects of ultraviolet radiation and water motion on the reef coral *Porites compressa* Dana: A flume experiment. *Mar. Biol.*, **138**, 467–476.
94. D.F. Gleason (1993). Differential effects of ultraviolet radiation on green and brown morphs of the Caribbean coral *Porites astreoides*. *Limnol. Oceanogr.*, **38**, 1452–1463.
95. F.Z. Muszynski, A. Bruckner, R.A. Armstrong, J.M. Morell, J.E. Corredor (1998). Within-colony variations of UV absorption in a reef building coral. *Bull. Mar. Sci.*, **63**, 589–594.
96. A. Post, A.W.D. Larkum (1993). UV-absorbing pigments, photosynthesis and UV exposure in Antarctica: Comparison of terrestrial and marine algae. *Aquat. Bot.*, **45**, 231–243.
97. U. Karsten, J.A. West (2000). Living in the intertidal zone – seasonal effects on heterosides and sun-screen compounds in the red alga *Bangia atropurpurea* (Bangiales). *J. Exp. Mar. Biol. Ecol.*, **254**, 221–234.
98. F.R. Conde, M.S. Churio, C.M. Previtali (2000). The photoprotector mechanisms of mycosporine-like amino acids. Excited-state properties and photostability of porphyra-334. in aqueous solution. *J. Photochem. Photobiol. B: Biol.*, **56**, 139–144.
99. M.P. Lesser (1996). Acclimation of phytoplankton to UV-B radiation: Oxidative stress and photoinhibition of photosynthesis are not prevented by UV-absorbing compounds in the dinoflagellate *Prorocentrum micans*. *Mar. Ecol. Prog. Ser.*, **132**, 287–297.
100. M.P. Lesser (1996). Erratum. *Mar. Ecol. Prog. Ser.*, **141**, 312.
101. G. Scelfo (1985). The effects of visible and ultraviolet solar radiation on a UV-absorbing compound and chlorophyll *a* in a Hawaiian zoanthid. *5th International Coral Reef Congress* (pp. 107–112). Tahiti.
102. D.F. Gleason, G.M. Wellington (1993). Ultraviolet radiation and coral bleaching, *Nature*, **365**, 836–838.
103. W.C. Dunlap, Y. Yamamoto (1995). Small-molecule antioxidants in marine organisms: Antioxidant activity of mycosporine-glycine. *Comp. Biochem. Physiol.*, **112B**, 105–114.
104. W.M. Bandaranayake, A. Des Rocher (1999). Role of secondary metabolites and

pigments in the epidermal tissues, ripe ovaries, viscera, gut contents and diet of the sea cucumber *Holothuria atra*. *Mar. Biol.*, **133**, 163–169.
105. A. Oren (1997). Mycosporine-like amino acids as osmotic solutes in a community of halophilic cyanobacteria. *Geomicrobiol. J.*, **14**, 231–240.
106. A. Portwich, F. Garcia-Pichel (1999). Ultraviolet and osmotic stresses induce and regulate the synthesis of mycosporines in the cyanobacterium *Chlorogloeopsis* PCC 6912. *Arch. Microbiol.*, **172**, 187–192.
107. F. Garcia-Pichel, R.W. Castenholz (1991). Characterization and biological implications of scytonemin, a cyanobacterial sheath pigment. *J. Phycol.*, **27**, 395–409.
108. J.G. Dillon, R.W. Castenholz (1999). Scytonemin, A cyanobacterial sheath pigment, protects against UVC radiation: Implications for early photosynthetic life. *J. Phycol.*, **35**, 673–681.
109. F. Garcia-Pichel, N.D. Sherry, R.W. Castenholz (1992). Evidence for an ultraviolet sunscreen role of the extracellular pigment scytonemin in the terrestrial cyanobacterium *Chlorogloeopsis* sp. *Photochem. Photobiol.*, **56**, 17–23.
110. R.J.W. Truscott, J.A. Carver, A. Thorpe, R.H. Douglas (1992). Identification of 3-Hydroxykynurenine as the lens pigment in the Gourami *Trichogaster trichopterus*. *Exp. Eye Res.*, **54**, 1015–1017.
111. A. Thorpe, R.H. Douglas, R.J.W. Truscott (1993). Spectral transmission and short-wave absorbing pigments in the fish lens – I. Phylogenetic distribution and identity. *Vision Res.*, **33**, 289–300.
112. F.E. Ahmed, R.B. Setlow (1993). Ultraviolet radiation-induced DNA damage and its photorepair in the skin of the platyfish *Xiphophorus*. *Cancer Res.*, **53**, 2249–2255.
113. R.P. Trocine, J.D. Rice, G.N. Wells (1982). Photosynthetic response of seagrasses to UVA radiation and the influence of visible light intensity. *Plant Physiol.*, **69**, 341–344.
114. M.A.K. Jansen, V. Gaba, B.M. Greenberg (1998). Higher plants and UV-B radiation: Balancing damage, repair, and acclimation. *Trends Plant Sci.*, **3**. 131–135.
115. N.M. Targett, T.M. Arnold (1998). Predicting the effects of brown algal phlorotannins on marine herbivores in tropical and temperate oceans. *J. Phycol.*, **34**, 195–205.
116. H. Pavia, G. Cervin, A. Lindgren, P. Åberg (1997). Effects of UV-B radiation and simulated herbivory on phlorotannins in the brown alga *Ascophyllum nodosum*. *Mar. Ecol. Prog. Ser.*, **157**, 139–146.
117. K. Hoyer, U. Karsten, T. Sawall, C. Wiencke (2001). Photoprotective substances in Antarctic macroalgae and their variation with respect to depth distribution, different tissues and developmental stages. *Mar. Ecol. Prog. Ser.*, **211**, 117–129.
118. J.J. Stachowicz, N. Lindquist (1997). Chemical defense among hydroids on pelagic *Sargassum*: Predator deterrence and absorption of solar UV radiation by secondary metabolites. *Mar. Ecol. Prog. Ser.*, **155**, 115–126.
119. S.G. Dove, O. Hoegh-Guldberg, S. Ranganathan (2001). Major colour patterns of reef-building corals are due to a family of GFP-like proteins. *Coral Reefs*, **19**, 197–204.
120. S. Kawaguti (1969). Effect of the green fluorescent pigment on the productivity of the reef corals. *Micronesica*, **5**, 313.
121 A. Salih, A. Larkum, G. Cox, M. Kühl, O. Hoegh-Guldberg (2000). Fluorescent pigments in corals are photoprotective. *Nature*, **408**, 850–853.
122. D. Schlichter, H.W. Fricke (1991). Mechanisms of amplification of photosynthetically active radiation in the symbiotic deep-water coral *Leptoseris fragilis*. *Hydrobiologia*, **216**, 389–394.
123. M.P. Lesser, J.M. Shick (1989). Effects of irradiance and ultraviolet radiation on

photoadaptation in the zooxanthellae of *Aiptasia pallida*: Primary production, photoinhibition, and enzymic defenses against oxygen toxicity. *Mar. Biol.*, **102**, 243–255.

124. M.P. Lesser, J.M. Shick (1989). Photoadaption and defenses against oxygen toxicity in zooxanthellae from natural populations of symbiotic cnidarians. *J. Exp. Mar. Biol. Ecol.*, **134**, 129–141.
125. M.P. Lesser, W.R. Stochaj, D.W. Tapley, J.M. Shick (1990). Bleaching in coral reef anthozoans: Effects of irradiance, ultraviolet radiation, and temperature on the activities of protective enzymes against active oxygen. *Coral Reefs*, **8**, 225–232.
126. J.A. Dykens, J.M. Shick (1982). Oxygen production by endosymbiotic algae controls superoxide dismutase activity in their animal host. *Nature*, **297**, 579–580.
127. J.L. Matta, R.K. Trench (1991). The enzymatic response of the symbiotic dinoflagellate *Symbiodinium microadriaticum* (Freudenthal) to growth *in vitro* under varied oxygen tensions. *Symbiosis*, **11**, 31–45.
128. C.A. Rice-Evans, N.J. Miller, G. Paganga (1997). Antioxidant properties of phenolic compounds. *Trends Plant Sci.*, **2**, 152–159.
129. H.W. Paerl, J. Tucker, P. T. Bland (1983). Carotenoid enhancement and its role in maintaining blue–green algal (*Microcystis aeruginosa*) surface blooms. *Limnol. Oceanogr.*, **28**, 847–857.
130. J.I. Goes, N. Handa, S. Taguchi, T. Hama (1994). Effect of UV-B radiation on the fatty acid composition of the marine phytoplankter *Tetraselmis* sp.: relationship to cellular pigments. *Mar. Ecol. Prog. Ser.*, **114**, 259–274.
131. W.F. Vincent, A. Quesada (1994). Ultraviolet radiation effects on cyanobacteria: Implications for Antarctic microbial ecosystems. In: C. S. Weiler, P. A. Penhale (Eds), *Ultraviolet Radiation in Antarctica: Measurements and Biological Effects* (pp. 111–124). American Geophysical Union, Washington, D.C.
132. S.P. Dawson, W.C. Dennison (1996). Effects of ultraviolet and photosynthetically active radiation on five seagrass species. *Mar. Biol.*, **125**, 629–638.
133. B. Demmig-Adams, W.W. Adams III (1992). Photoprotection and other responses of plants to high light stress, *Annu. Rev. Plant Physiol. Plant Mol. Biol.*, **43**, 599–626.
134. S. Demers, S. Roy, R. Gagnon, C. Vignault (1991). Rapid light-induced changes in cell fluorescence and in xanthophyll-cycle pigments of *Alexandrium excavatum* (Dinophyceae) and *Thalassiosira pseudonana* (Bacillariophyceae): A photo-protection mechanism. *Mar. Ecol. Prog. Ser.*, **76**, 185–193.
135. I. Ambarsari, B.E. Brown, R.G. Barlow, G. Britton, D. Cummings (1997). Fluctuations in algal chlorophyll and carotenoid pigments during solar bleaching in the coral *Goniastrea aspera* at Phuket, Thailand. *Mar. Ecol. Prog. Ser.*, **159**, 303–307.
136. B.E. Brown, I. Ambarsari, M.E. Warner, W.K. Fitt, R.P. Dunne, S.W. Gibb, D.G. Cummings (1999). Diurnal changes in photochemical efficiency and xanthophyll concentrations in shallow water reef corals: Evidence for photoinhibition and photoprotection. *Coral Reefs*, **18**, 99–105.
137. S. Johnsen, E.A. Widder (2001). Ultraviolet absorption in transparent zooplankton and its implications for depth distributon and visual predation. *Mar. Biol.*, **138**, 717–730.
138. O. Siebeck (1978). Ultraviolet tolerance of planktonic crustaceans. *Verh. Internat. Verein. Limnol.*, **20**, 2469–2473.
139. J. Ringelberg, A.L. Keyser, B.J.G. Flik (1984). The mortality effect of ultraviolet radiation in a translucent and in a red morph of *Acanthodiaptomus denticornis* (Crustacea, Copepoda) and its possible ecological relevance. *Hydrobiologia*, **112**, 217–222.

140. D.-S. Kim , Y. Watanabe (1994). Inhibition of growth and photosynthesis of freshwater phytoplankton by ultraviolet A (UVA) radiation and subsequent recovery from stress. *J. Plankton Res.*, **16**, 1645–1654.
141. M.P. Lesser, J.J. Cullen, P.J. Neale (1994). Carbon uptake in a marine diatom during acute exposure to ultraviolet B radiation: Relative importance of damage and repair. *J. Phycol.*, **30**, 183–192.
142. D. Campbell, M.-J. Eriksson, G. Öquist, P. Gustafsson, A.K. Clarke (1998). The cyanobacterium *Synechococcus* resists UV-B by exchanging photosystem II reaction-center D1 proteins. *Proc. Natl. Acad. Sci. U.S.A.*, **95**, 364–369.
143. D.L. Mitchell, R.S. Nairn (1989). The biology of the (6-4) photoproduct. *Photochem. Photobiol.*, **49**, 805–819.
144. A. Sancar, G.B. Sancar (1988). DNA repair enzymes. *Annu. Rev. Biochem.*, **57**, 29–67.
145. P. Halldal, O. Taube (1972). Ultraviolet action and photoreactivation in algae. In: A.C. Giese (Ed.), *Photophysiology* (pp. 163–188). Academic Press, New York.
146. G.D. Small, C.S. Greimann (1977). Repair of pyrimidine dimers in ultraviolet-irradiated *Chlamydomonas*. *Photochem. Photobiol.*, **25**, 183–187.
147. H. Pakker, C.A.C. Beekman, A.M. Breeman (2000). Efficient photoreactivation of UVBR-induced DNA damage in the sublittoral macroalga *Rhodymenia psuedopalmata* (Rhodophyta). *Eur. J. Phycol.*, **35**, 109–114.
148. D.M. Damkaer, D.B. Dey (1983). UV damage and photoreactivation potentials of larval shrimp, *Pandalus platyceros*, and adult euphausiids, *Thysanoessa raschii*. *Oecologia*, **60**, 169–175.
149. T. Naganuma, T. Inoue, S. Uye (1997). Photoreactivation of UV-induced damage to embryos of a planktonic copepod. *J. Plankton Res.*, **19**, 783–787.
150. H.E. Zagarese, M. Feldman, C.E. Williamson (1997). UV-B-induced damage and photoreactivation in three species of *Boeckella* (Copepoda, Calanoida). *J. Plankton Res.*, **19**, 357–367.
151. H.E. Zagarese, C.E. Williamson, T.L. Vail, O. Olsen, C. Queimalinos (1997). Long-term exposure of *Boeckella gibbosa* (Copepoda, Calanoida) to in situ levels of solar UVB radiation. *Freshwat. Biol.*, **37**, 99–106.
152. D. Karentz (1994). Ultraviolet tolerance mechanisms in Antarctic marine organisms. In: C.S. Weiler, P.A. Penhale (Eds), *Ultraviolet Radiation in Antarctica: Measurements and Biological Effects* (pp. 93–110). American Geophysical Union, Washington, D.C.
153. D. Karentz, J.E. Cleaver, D.L. Mitchell (1991). Cell survival characteristics and molecular responses of Antarctic phytoplankton to ultraviolet-B radiation. *J. Phycol.*, **27**, 326–341.
154. K.D. Malloy, M.A. Holman, D. Mitchell, H.W. Detrich III (1997). Solar UVB-induced DNA damage and photoenzymatic DNA repair in Antarctic zooplankton. *Proc. Natl. Acad. Sci. U.S.A.*, **94**, 1258–1263.
155. R.P. Trocine, J.D. Rice, G.N. Wells (1981). Inhibition of seagrass photosynthesis by ultraviolet-B radiation. *Plant Physiol.*, **68**, 74–81.

Chapter 11

Photosynthesis in the aquatic environment as affected by UVR

Virginia E. Villafañe, Kristina Sundbäck, Félix L. Figueroa and E. Walter Helbling

Table of contents

Abstract

Since the discovery of the Antarctic ozone "hole", many studies have been conducted to determine the effects of enhanced UV-B (280–315 nm) on photosynthetic rates of autotrophic organisms. It is accepted now that even natural levels of UVR (280–400 nm) are stressful for some autotrophic organisms. In this chapter we will summarize what we know about the effects of UVR on the photosynthesis of aquatic organisms. Here we consider three major groups – phytoplankton, microphytobenthos (MPB), macroalgae/marine angiosperms – which differ in many ways, especially in regard to their habitats. While phytoplankton live in the water column, MPB and macroalgae occupy the benthic environment. This creates substantial differences with respect to the amount and quality of radiation that they receive. Thus, although there is a common and general response to UVR of these autotrophic organisms – i.e., inhibition of photosynthesis – there are differences among the groups studied. These are mainly due to differences in the radiation conditions to which cells are exposed, as well as to the specific sensitivity/acclimation of the organisms under study. To evaluate the overall response of aquatic primary producers to UVR, it is crucial to consider the temporal scale of experimentation, to allow enough time for repair mechanisms and acclimation to UVR. Thus, short-term experiments frequently give an insight about the worst-case scenario for UVR effects on photosynthesis. We also review in this chapter the effects of UVR upon some related physiological processes (e.g., nutrient incorporation/assimilation, pigment synthesis/bleaching) and morphology (e.g., cell size) that may in turn affect the photosynthetic performance. Finally, to determine the impact of natural and increased levels of UVR upon aquatic ecosystems, we consider the interactive effects of other variables (pH, carbon dioxide concentrations, temperature, etc.) with UVR. Consequences for aquatic autotrophic organisms of increased UV-B levels due to ozone depletion events are still uncertain, but changes in biogeochemical cycles, community structure, and trophic web dynamics can be expected.

11.1 Introduction

The photosynthetic process in aquatic ecosystems is responsible for fixing approximately 40% of our planet's yearly amount of carbon available for the production of new living matter, with about 48.5 Pg C yr^{-1} fixed in the aquatic ecosystems [1–3]. Carbon fixation in the aquatic environment, mediated by the utilization of solar radiation, takes place in both the water column and the benthos. While water-column autotrophic organisms (mainly phytoplankton) are responsible for most of the share in carbon fixation, benthic organisms (i.e., macrophytes and microalgal communities) are involved in about 10% of the total production [4,5]. Although this latter amount is globally less than that due to phytoplankton, marine macrophytes also provide food (directly or through detritus) to a wide variety of invertebrates and fish in the coastal ecosystems [6]. Benthic microalgal communities, on both hard and soft substrata, also serve a crucial ecological function in shallow freshwater and marine habitats. They constitute the local basis of the food webs in shallow areas, which are recognized as having high secondary production (e.g., of fish and their prey). In these areas, the microphytobenthic (MPB) communities may account for 50% or more of the total primary production, equalling or exceeding the productivity of the water column [7].

Even though solar radiation is attenuated in the water column (see Chapter 3) it penetrates to a depth that will vary, among other things, according to the location (e.g., oceanic vs. coastal), latitude and concentration of particulate and dissolved matter. The euphotic zone in the water column (i.e., 1% of surface PAR, 400–700 nm) can vary from few centimetres in estuarine waters or lakes with a heavy load of DOM [8,9] to more than 100 m in the open ocean [10]. Hence ultraviolet radiation (UVR, 280–400 nm) can penetrate accordingly to comparable depths [11] (see also Chapter 3). In coastal waters, biologically effective ultraviolet B radiation (UV-B, 280–315 nm) reaches only to 1 m depth, as in the Baltic Sea [12], whereas in the Mediterranean it can penetrate as deep as 20 m [13]. This variability is also observed in other environments. For example, in a study carried out in freshwater Japanese ponds and lakes, Hodoki and Watanabe [14] determined that the 1% of surface UV-B varied from 0.3 to 2 m, depending mostly on the concentration of chlorophyll-*a* (chl-*a*) and particulate organic carbon present in the water body. The photic zone in the benthic environment extends to ca. 3 mm into the sediment. Fiber optic microsensor measurements have shown that UVR can penetrate down to at least 1.25 mm of this zone, and through scattering it can even exceed the incoming UVR by up to 50% [15,16]. In addition, as the water column in estuaries and embayments is often shallow, and regularly absent in intertidal areas, UVR can reach high levels at the sediment surface. Thus, and in view of this background, UVR should be considered a very important environmental factor that can affect different metabolic and physiological processes in autotrophic organisms living in the water column and in the benthos.

In this chapter we will discuss the role of UVR in affecting the photosynthetic process in phytoplankton, MPB, and macroalgae. This is especially important as the effects of UVR on the photosynthesis of these organisms may have a considerable impact on higher trophic levels of the aquatic ecosystem (Chapter 12), as well as in climate change (Chapter 17) and biogeochemical cycles (Chapter 5).

11. Methodology to assess UVR effects on photosynthesis

11.2.1 Exposure of samples

In order to assess UVR effects on photosynthesis, three approaches for exposing algae to UVR are used. These include (1) natural solar radiation, modified by various filters that selectively screen off certain wavebands of radiation; (2) natural solar radiation which is supplemented with artificial UVR from lamps, and (3) fully artificial radiation, implying laboratory experiments. UVR experiments at their best require both that the target organisms are exposed to as realistic a light field as possible, and that high-quality measurements of radiation are obtained. The most realistic results are probably gained from experiments performed under natural solar radiation; artificial radiation sources, however, have also been shown to be very useful for studying mechanistic aspects of UVR responses.

11.2.1.1 In situ incubations

In this type of incubations, the samples are exposed to solar radiation in their natural habitat and at their natural in situ depth. To assess the effects of ambient UVR, this approach often involves three types of radiation fields, achieved by filters, i.e. PAR + UV-B + UV-A, PAR + UV-A, and only PAR (see section 11.2.3). Although in situ incubations will result in the most realistic responses, they certainly have the constraint of being conditioned by weather conditions. Therefore, comparatively few in situ studies on the effects of UVR on algal photosynthesis have been conducted, particularly in rough-weather areas, such as the Arctic [17] and Antarctica [18,19].

Phytoplankton can be exposed to an in situ field of radiation by using UV-transparent (see section 11.2.3) bottles hanging from a line or tubes placed in trays (Figure 1) which are incubated at different depths in the water column [18,20]. One disadvantage with this approach is that phytoplankton cells are kept at a fixed depth for the entire incubation period (e.g., few hours), thus receiving a constant proportion of the surface incident radiation. In the water column, however, cells are moving within the upper mixed layer (UML) and thus exposed to a variable field of irradiance [21] (see Chapter 4). So far, few studies have addressed the importance of mixing rates on the phytoplankton photosynthesis [21–23], and with the exception of the experiments performed by Marra [24] on the effects of PAR, we are not aware of such studies done under in situ conditions.

Fixed screens with different filtering capacities have been frequently used to study the in situ effect of ambient UVR on shallow-water benthic microalgae in streams [25], lakes [26] and marine habitats [27,28]. In a four-month in situ experiment on UVR effects on MPB communities of a microtidal bay, Wulff et al. [29] used 80 × 80 cm screens placed in wooden frames that were pressed into the sediment. This type of field set-up, however, requires frequent cleaning and careful monitoring of the radiation field below the screens.

In the case of marine macrophytes (macroalgae and marine angiosperms), most of the in situ experimentation has been conducted in the intertidal zone, where access to growing plants is relatively easy [30–33]. Subtidal populations have received less attention due to the complications of working in situ at different depths, especially in high latitude zones [33,34]. Several authors [30–32,35] have investigated effects of UVR on macroalgal photosynthesis by incubating algae in their natural environment and monitoring daily variation in photosynthesis and irradiance under different radiation treatments using a similar set up as those described for MPB experiments (Figure 1). More recently,

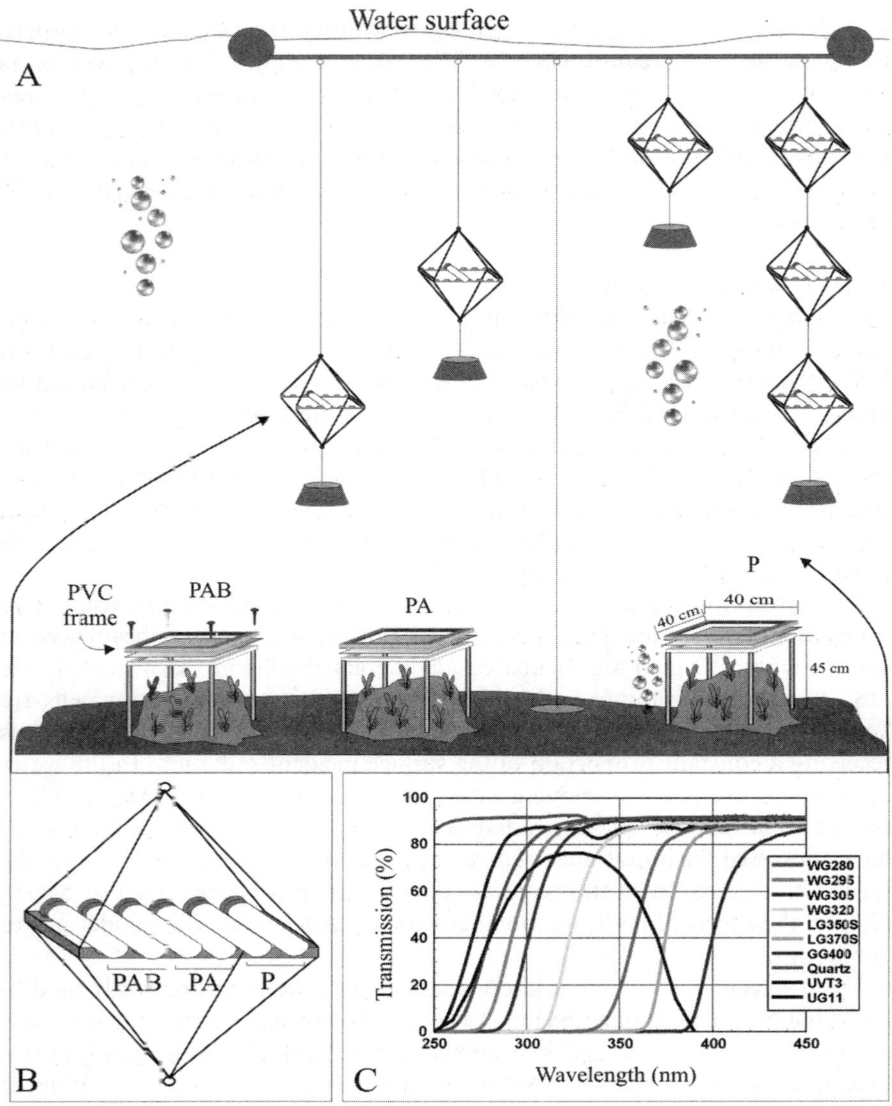

Figure 1. Schematic representation of in situ incubation for phytoplankton and benthic algae. (A) General disposition of trays with tubes and filters for cutting off different portions of the solar spectrum; in the bottom a set up for benthic algae incubation is presented. (B) Close up of one tray containing duplicate quartz tubes for three different radiation treatments: PAB, unfiltered solar radiation; PA, PAR + UV-A and P, only PAR. (C) Transmission characteristics of various materials and filters used in photobiological experimentation.

however, efforts have been devoted to analyze in situ photosynthetic activity of subtidal algae. This experimental design has consisted of determining the effective quantum yield by using an underwater fluorometer [36–39]. An alternative approach has been to incubate marine plants for several days at their natural growth site, and after that period the algae were collected and the quantum yield measured using a non-submersible fluorometer [35,40–42].

11.2.1.2 Simulated in situ incubations

Considering the practical difficulties of in situ incubations, outdoor incubations in temperature-controlled containers (e.g., on deck of research vessels, or in flow-through systems on land sites) have been used as an alternative approach. This incubation method is suitable for both short-term (hours) and long-term (days–weeks) experiments carried out with microalgae [43–47], as well as with macroalgae [32,48–54]. This set up is often used for determining a worst–case scenario, as samples are exposed to surface (i.e., maximum) incident irradiance. Therefore, neutral density filters are often used to approximately simulate the attenuation of solar radiation in the water column. These filters, however, do not mimic the differential spectral attenuation that actually occurs in the water column (Chapter 3), and samples are generally exposed to higher UV-B/UV-A/PAR ratios than they would normally experience. It is particularly important to approach realistic ratios between UV-B, UV-A and PAR, as DNA repair mechanisms depend on those ratios [55,56] (see Chapter 9).

In contrast to phytoplankton, simulated in situ incubations imply fairly realistic light conditions for MPB in the intertidal or littoral zone. This is particularly true when incubating intact sediment cores, as the sediment will provide natural refuges for benthic microalgae, such as motile diatoms and cyanobacteria [46,47,57]. Similar approaches have also been used for hard substrata, often involving colonization of artificial substrata [58].

11.2.1.3 Supplemented UV-B or UVR

As with several UV experiments carried out with terrestrial organisms [59,60], experimental treatments on aquatic organisms have included the enhancement of ambient UV-B. In some cases, these treatments simulate ozone depletion events. Such experiments, in which natural solar radiation is enhanced by artificial UVR, have been done with phytoplankton [61–63] and microphytobenthos [64]. A few studies have included simultaneously exclusion and enhancement of UV-B [47,65]. A shortcoming, however, in the majority of experiments using elevated levels of UV-B, has been the use of fixed levels of UV-B for few hours per day. Moreover, the levels of enhancement have varied greatly, from moderate (~20%) to ca. 100% above ambient, often resulting in unnatural ratios between PAR and UVR, thus making comparisons between experiments difficult. As mentioned before, it is crucial for ecologically relevant studies that the spectral composition of the radiation is realistic [66]. One way to achieve this is to provide additional UV-B so that it mirrors the natural dose curve, as has been used in terrestrial studies [67]. This is possible with a system in which the intensity given is controlled by a computer system linked to a UV-B sensor that continuously measures ambient UV-B levels. This type of set-up allows the simulation of low levels of enhanced UV-B (5–20%) as observed during ozone depletion events, and has been used to study the UV-B response of both MPB [65,68] and phytoplankton [63].

11.2.1.4 Artificial radiation

Various artificial radiation sources have been used to assess UVR effects on aquatic autotrophic organisms. An assorted number of them are commercially available, such as fluorescent and halogen lamps. So far, most studies carried out with artificial radiation sources have been done with the main objective to determine the impact of UVR at fixed irradiances [69], or in combination with neutral density screens and cut-off filters to obtain biological weighting functions (BWFs) [70,71]. In order to determine the sensitivity of intertidal and subtidal algae, Dring et al. [72,73] used a solar simulator, in which different levels of ozone reduction can be arranged, and Röttgers [74] had used a similar system to address the response of phytoplankton cultures to changes in UVR. However, it has been found very difficult to mimic the solar radiation spectrum in these types of experiments; in fact very few of the light sources can give reliable results in photobiological research [75]. Moreover, one should be extremely cautious when extrapolating the results obtained in this way to the natural environment.

11.2.2 *Materials and filters*

A combination of different materials and filters are normally used to separate different wavebands of the incident irradiance spectrum. In most of the experiments conducted either in the field or in the laboratory, it is customary to use tubes or vessels made of a material transparent to UVR, such as Quartz, Plexiglas, or Teflon. There are many types of filters that are broadly used in photobiological research, ranging from "film type" filters, such as Ultraphan, Folex, Mylar-D, and acetate, to "glass type" filters such as Schott, Hoya and Oriel. Representative spectra of the transmission characteristics of commonly used filters and materials are shown in Figure 1. In general, the materials are long pass filters, and thus they screen off the energy of the lower wavelengths. However, there are filters that allow the energy of just a portion of the spectrum to pass, as is the case of the UG11 filters (see Figure 1).

11.2.3 *Variables measured and experimental approaches*

Various experimental approaches have been used to evaluate the impact of UVR on different cell processes (Figure 2). The evolution of oxygen [76,77] and incorporation of radiocarbon [20,78] have been widely used not only to determine the productivity of a water body, but also to assess the impact of UVR [18,22,30,43,49,79–84]. In addition, oxygen microsensors [85] have been shown to be practical tools for high-resolution measurements of UVR effects in sediments and microfilms [46,64], particularly in combination with optical microsensors measuring UVR [16].

In recent years, pulse amplitude modulated (PAM) chlorophyll fluorescence associated with the photosystem II (PS II) has become a useful tool for evaluation of photosynthesis [86–90]. In fact, chlorophyll fluorescence can be an

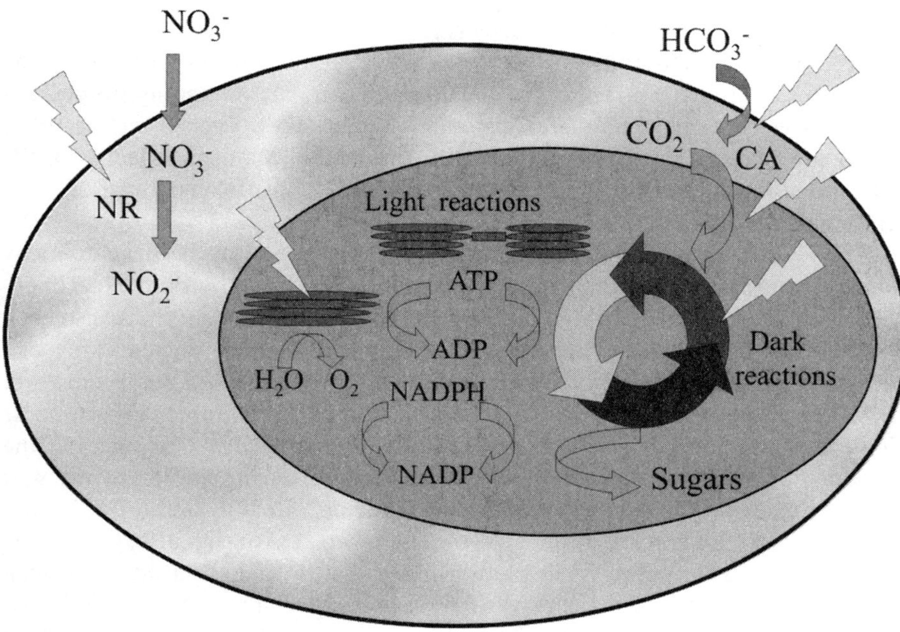

Figure 2. Diagram of a eukaryotic algal cell indicating different processes that could be influenced by UVR and that directly or indirectly affect the photosynthetic process.

indicator of different functional levels in photosynthesis, such as photon capture by light-harvesting pigments, primary light reactions, thylakoid electron transport reactions, dark-enzymatic stroma reactions and slow regulatory feedback processes [91]. The relationship between oxygen evolution and chlorophyll fluorescence in different organisms has also been demonstrated [92,93]. Photosynthetic activity has been estimated as chlorophyll fluorescence in macroalgae growing in a variety of water bodies, as in the Arctic [40,94], Antarctic [33,95], North Sea [72], Chinese Sea [96], Mediterranean Sea [31,50,51,97], tropical [98] or Patagonian [99,100]. Taking into account the differences in photosynthetic organization between macroalgae and higher plants, an optimization of the PAM instrumentation has been needed to measure accurately the low chlorophyll fluorescence emission of macroalgae [88,101]. Furthermore, the presence of phycobilisomes in the light-harvesting system of red algae results in generally lower fluorescence values than that measured in green- and brown-algae [101]. Due to the increased sensitivity of the PAM fluorescence instrumentation in recent years, this technique has been also used to study UV-B effects on MPB [65], as well as to address UVR effects on phytoplankton [74,102,103].

Studies on the effects of UVR upon phytoplankton have been conducted using both natural communities and monospecific cultures [17,18,21,22,43,44,81–84,103–113]. The exposure of samples has included in situ [19,44,81,111] and simulated in situ incubations [21,43,44,82], as well as the use of artificial radiation [83,104]. Short-term studies have been generally performed in periods of less than one day, implying that no acclimation has been allowed, and, hence, some of the observed effects represent the worst-case scenario. Still, the majority of UVR studies on phytoplankton photosynthesis have been done using this approach, and they provide a base of comparison among species and different ecosystems. Long-term experiments (i.e., days, weeks), on the other hand, are a preferable choice when making predictions about the effects of UVR on an ecological scale; however, relatively few studies have been performed using this approach [83,107,110,114–118].

The response of benthic microalgae to UVR has mainly been assessed by studying natural or semi-natural communities in situ, or in outdoor experimental flumes (see references below), although laboratory experiments have also been made [119,120]. Basically, two types of studies have been conducted: (1) experiments where communities have been allowed to colonize on hard substrata [58,121] and, (2) experiments where intact natural communities in sediments have been studied [29,47]. These two approaches differ in the aspect that the former allows UVR to exert a selective pressure during early growth and succession, which is not the case when studying already established, dense communities with no or little net growth. These two approaches also differ in the choice of target variables. While photosynthetic rate (^{14}C incorporation, oxygen microprofiles) and photochemistry (PAM) have been monitored for MPB in sediments, accrual of biomass (as chl-*a*, or algal cells) has been the most commonly measured variable, particularly in long-term experiments on periphyton on hard substrata (see references in section 11.4).

Finally, the impact of UV-B radiation on marine macrophytes has been mostly conducted on individual species and not on the whole community. The criteria to select species for experimentation/analyses have varied: (a) they are key species due to their contribution to primary production, or because they create a habitat for other marine plants and invertebrates, as the seagrass *Posidonia oceanica* in the Mediterranean Sea [42,122], *Laminaria* beds in the North Sea [123], or *Macrocystis* on the Pacific coast of California [4], (b) they represent a high share of macroalgal biomass in the ecosystem, as *Ulva* in eutrophic coastal waters [124] and, (c) they are commercially important as *Porphyra* sp., *Gelidium sequipedale*, *Macrocystis pyrifera* or *Chondrus crispus* [30,38,125,126].

11.3 Effects of UVR on phytoplankton photosynthesis

In the following paragraphs we summarize the status of our knowledge about both short-term and long-term effects upon phytoplankton photosynthesis. However, several reviews dealing with the impact of UVR on phytoplanktonic organisms have been published [45,127–130], so we encourage the reader to refer to them for more specific details that are not addressed here.

11.3.1 Short-term effects

One of the best-known effects of solar radiation upon phytoplanktonic organisms is photoinhibition, which refers to the reduction of photosynthetic rates at relatively high irradiances [131]. Many studies have used production–irradiance (P–I) curves to determine phytoplankton photoinhibition due to high PAR levels [132,133]; in addition, research has been carried out to determine the additional effects of UVR, not only in tropical [111] but also in temperate [81] and polar regions [19,133]. Interestingly, it has been shown that the relative effect of UVR (i.e., as compared to the PAR control) is sometimes higher at lower irradiances. On bright days, when high PAR levels already inhibit the photosystem, UVR produces a relatively lesser effect. This observation, however, depends on many variables, such as the light history of the cells and species composition. In addition, when phytoplankton cells are exposed to increased levels of solar radiation they may show a threshold for inhibition, which is followed by a steep increase in photosynthetic inhibition at mid-irradiances, levelling off at higher irradiance values [43,83]. However, in some cases no discernible threshold was determined [17,108].

In general, when in situ incubations are done, UVR causes a sharp decrease in photosynthetic rates (as compared with the PAR-only treatment) especially in surface waters (Figure 3). Even though UV-B radiation is more effective per unit energy (see Chapter 2), and hence potentially more damaging than those at longer wavelengths, many studies conducted in different locations have shown that UV-A is responsible for most of the photosynthetic inhibition, just because their natural levels are much higher [19,79,111]. Photosynthesis inhibition decreases with depth, depending, among other things, on water transparency, presence of microorganisms, as well as on incident radiation. The depth distribution of photosynthesis inhibition is highly variable and hence, surface values are not good indicators of the total inhibition in the water column as it has been demonstrated in a comparison between freshwater and seawater environments from mid-latitudes and sub-Antarctic areas [134]. Furthermore, when evaluating the integrated photosynthesis inhibition, it is more important to consider the extent of the euphotic zone that is inhibited (e.g., optical depth), rather than the physical depth at which the inhibition is observed.

Inhibition of photosynthesis due to UVR is highly variable, depending on the irradiance/doses received by the cells, their specific sensitivity and acclimation potential, as well as the interaction with other variables that can mask the observed effects (mixing, temperature, pH, etc.). The daily integrated loss of carbon fixation in the euphotic zone in Antarctic waters was calculated to be about 4.9%, under normal ozone column concentrations [19]. At the time of ozone depletion events, which are responsible for a relative increase in incident solar UV-B (see Chapter 2), there was a greater photosynthetic inhibition – reducing daily aquatic primary production by an additional ~4–12% [18,19]. However, taking into consideration the magnitude and timing of ozone depletion events, the yearly loss of carbon fixation in the Southern Ocean due to these processes was estimated to be <0.15% [21]. In addition, some studies [21,22] have demonstrated that the effects of mixing, i.e., fluctuating radiation regimes (Chapter 4), are more important in affecting photosynthesis than the variations in ozone levels. Studies conducted with temperate phytoplankton

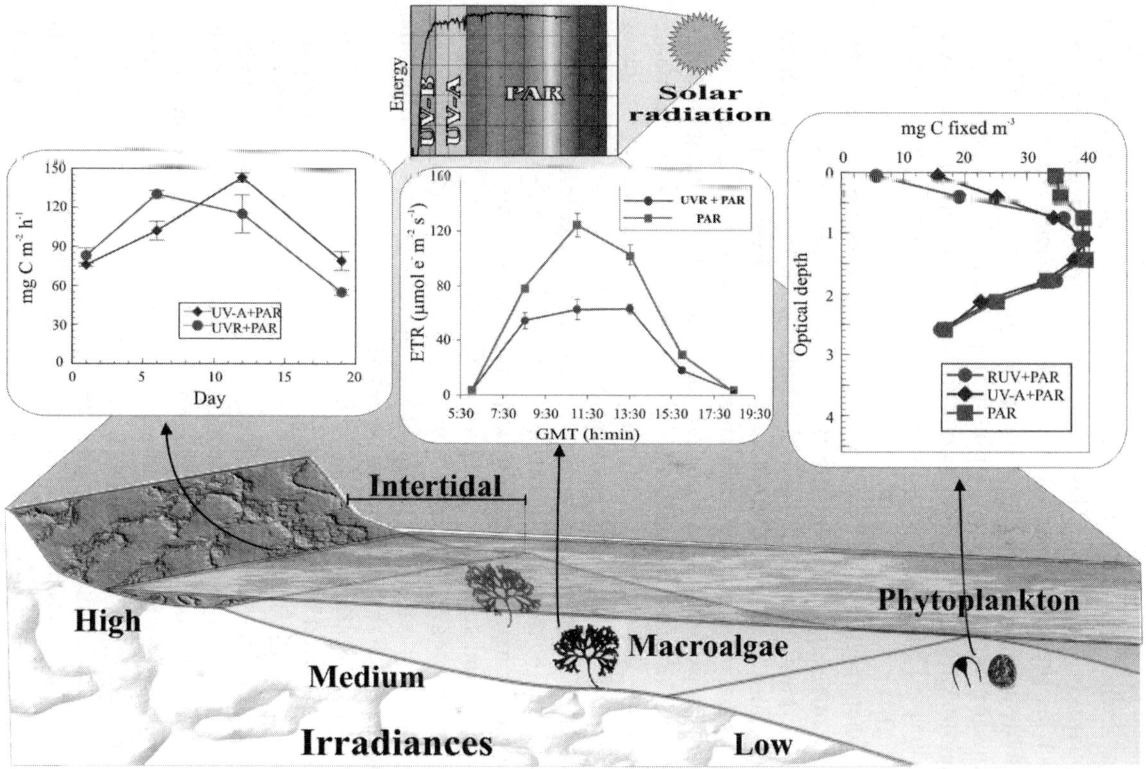

Figure 3. Schematic illustration of potential effects of solar radiation on phytoplankton, MPB and macroalgae photosynthesis. For benthic algae, three zones are defined in term of the irradiance received: high, medium and low. While the high irradiance zone is clearly defined as the intertidal zone, the boundaries between medium and low radiation are uncertain and will vary among different water bodies, species considered, etc. The graphs are examples of UVR-induced photoinihibition in the three groups of algae considered here.

[135], simulating mixing conditions for Patagonian waters, showed an UVR-induced reduction of photosynthetic rates when the UML extended to a relatively small portion of the euphotic zone ($Z_{UML}/Z_{Eph} < 0.5$). When mixing was deep ($Z_{UML}/Z_{Eph} > 0.8$), and mean PAR levels were low, phytoplankton were able to use UV-A radiation for carbon fixation. The use of solar UV-A at low PAR irradiances has also been observed in Californian waters, with a 10–20% increase in photosynthesis due to this effect [136].

In terms of photosynthesis, studies have demonstrated that tropical phytoplanktonic species are more resistant to UVR than those from polar environments [43,83,111], probably due to their evolutionary light history with naturally high radiation levels. In addition, tropical organisms had a higher irradiance threshold for photosynthesis inhibition [83] than polar species [43,137], thus providing an additional evidence of their resistance to high UVR levels. Solar radiation increases with altitude [138] and thus photosynthesis in lakes located at high altitudes might exhibit enhanced inhibition. The inhibition of photosynthesis, however, depends not only on the irradiance received at the lake surface, but also on the differences on water temperature, attenuation coefficients and phytoplankton composition among other variables [139]. Biological Weighting Functions (BWFs) [71,106] had also implied the higher resistance of tropical organisms to UVR [83] as compared to those from polar environments [109,137]. Figure 4 shows a comparison of different BWFs calculated for different geographic locations – Arctic, Antarctica, tropical lakes and temperate latitudes.

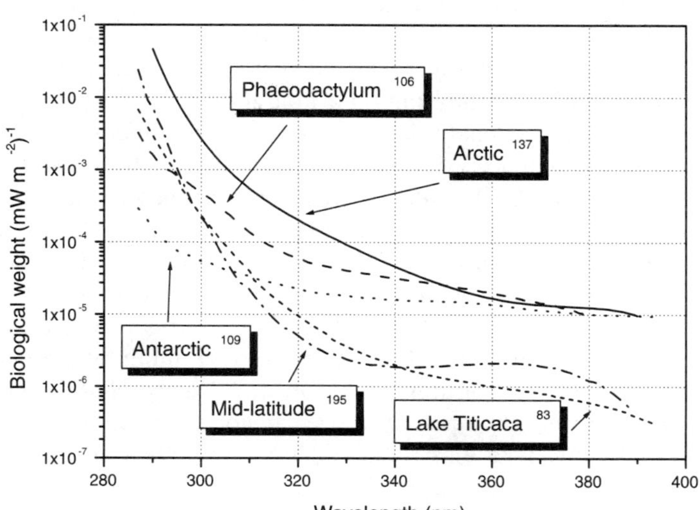

Figure 4. Representative biological weighting functions of a laboratory culture and phytoplankton assemblages from different environments. The numbers in the labels indicate the reference from where the data were obtained.

11.3.2 Long-term effects

During long-term experiments, species have the potential to acclimate to new radiation conditions, and processes such as DNA repair and synthesis of photoprotective compounds may occur [107,118,140,141] (see also Chapter 10). One of the best ways to test UVR effects on aquatic autotrophic organisms on a long-term basis is by using a "model ecosystem" or mesocosms [129], in which a parcel of the aquatic body is isolated and allowed to progress under similar conditions as in the natural environment. The main restriction of these experiments is that it is not possible to completely simulate natural conditions – e.g., water movements are restricted and larger organisms are normally excluded. Hence, one should be cautious when interpreting results obtained in these experiments, as other factors (e.g., immigration) are important components when addressing UVR effects from an ecological point of view (see also Chapter 4). Experiments carried out in polar areas [137] showed that, at the beginning of experimentation, both Arctic and Antarctic phytoplankton cells were significantly inhibited by UVR. This inhibition, however, did not increase as the experiment progressed, and growth rates (based either on chl-*a* content or carbon incorporation) were not significantly different between the UVR + PAR and the PAR treatments [137]. Kim and Watanabe [79] found that even though short-term exposure to UVR provoked a significant decrease of chl-*a* and photosynthetic rates in two freshwater phytoplankton species, *Melosira* sp. and *Chlorella ellipsoidea*, under prolonged UV-A exposure; however, the algae acclimatized by reactivation of the photosystem and enhanced cellular chlorophyll synthesis. Results from long-term exposure of freshwater phytoplankton are also very variable with no effects determined in an Alpine location [142], low impact of UVR in a community from a Canadian lake [143] and significant changes in phytoplankton composition in a lake from the Andes region [144].

Interactive effects of UVR with other ecological variables are important when addressing photosynthetic inhibition on a long-term basis. In particular, temperature seems to play a crucial role. For example, the temperate dinoflagellate *Prorocentrum micans* had a maximum decrease in photosynthetic rates after 21 days of exposure to solar UVR [145], whereas Antarctic phytoplankton had this maximum inhibition after 9 days [115]. In addition, research has been conducted to address the interactive effects of UVR and nutrient limitation. There was variability in the responses, with studies that revealed that nutrient-limited cultures were more sensitive to UV-B than those nutrient-replete [105,146]; however, Behrenfeld et al. [147] did not find growth inhibition produced by UV-B in nitrogen-limited cultures. Bergeron and Vincent [148] determined growth rates in different phytoplankton size categories present in a P-enriched system in a Subarctic lake and found different responses according to the wavebands to which cells were exposed.

11.4 Effects of UVR on microphytobenthos photosynthesis

Although benthic microalgal communities include the same major algal taxa as the pelagic communities, there is a crucial difference. The density of autotrophic, as well as heterotrophic microorganisms is several orders of magnitude higher in benthic communities, resulting in microbial mats or biofilms. These are characterized by steep physical, chemical and biological gradients, leading to a close spatial and temporal coupling of turnover processes within the mat system [149]. Thus, it can be expected that the responses to UVR of these communities are rather complex.

11.4.1 Short-term effects

Decreased photosynthetic rates (measured as ^{14}C uptake, oxygen production, or chlorophyll fluorescence) appear to be the most frequently observed short-term effect for MBP, particularly at enhanced UV-B levels [15,47,57,64,65,68,120]. These results are, however, ecologically relevant only when realistic, moderate increases of UV-levels are used. Wulff et al. [68] found a 50% decrease in ^{14}C-uptake of MPB on sand when UV-B was increased by 15% above ambient (23% when biologically weighted according to Cullen et al. [106]), though only under nutrient-depleted conditions. Using oxygen microsensors, Bebout and Garcia-Pichel [15] found a dramatic (50–90%) decrease in gross photosynthesis of the surface layers of a cyanobacterial mat (*Microcoleus chthonoplastes*) under moderate UV-B irradiances (0.35–0.79 W m^{-2}). This decrease was also related to an active downward migration in response to UV-B. Non-invasive fluorescence measurements on natural diatom biofilms (dominated by *Gyrosigma balticum*) exposed to supplemented UV-B (7 and 15% above ambient) resulted in a sequence of responses, starting by significantly increased effective quantum yield $\Phi_{PS\ II}$ (probably reflecting downward migration), followed by a reduction in maximum quantum yield of PS II (F_v/F_m) and minimal fluorescence (F_o) [65].

Observed short-term responses of MPB photosynthesis to ambient UV levels are less clear-cut, when compared with supplemented UV-B, and appear to vary with substrate type and community density. For a muddy sediment, no significant effects of ambient UV-B on either carbon uptake or oxygen microprofiles of a diatom mat (dominated by large motile species) were found [46]. In a sandy sediment, on the other hand, both carbon uptake and allocation of a community dominated by small (mainly non-motile) diatoms and cyanobacteria decreased significantly under ambient UV-B, although only at the end of a 3-week experiment [47]. Finally, no effects of ambient UV-B on photosynthetic rates were observed in freshwater stream periphyton [150].

11.4.2 Long-term effects

Growth, measured as the accrual of biomass or chl-*a*, is the variable most often studied for long-term UV-B effects on MPB. The clearest negative effects of UV-B on MPB growth have been observed for periphyton colonizing artificial substrata [27,28,58,121,151]. However, in the majority of these experiments, significant negative effects on growth (30–100% decrease) were only found during the first few weeks. After this UV-inhibition phase, statistically significant negative effects disappeared, or were even reversed [25,26,58,121,152] (note that some of these studies excluded both UV-A and UV-B). In one case, the explanation for the reversed effect was the higher sensitivity of grazers than their prey [58,153, see Chapter 15]. When experimenting with already established periphyton communities, however, no detrimental effects were observed [150,154].

The effect of UV-B on the growth of natural, established MPB communities inhabiting marine sediments shows a different general pattern than the above-mentioned colonization experiments. In sediments, significant effects appear to be fewer, they are more frequently found for rate variables (photosynthesis, C-allocation) than for state variables (biomass, pigment and species composition), and they occur later during the experiment (after 1–2 weeks) (Figure 3) [29,47,64]. The delay of effects may partly be due to increasing nutrient limitation in the course of the experiments caused by the experimental set-up, particularly when working with sandy sediments, which are generally poorer in nutrients than fine sediments (see [47] and section 11.8.1). However, an intriguing question is why the observed effects on rate variables were not reflected more clearly in the state variables? Besides biological reasons, there could be methodological reasons. For example, if a deeper sediment layer than the actual layer affected by UV-B is sampled, there might be a "dilution" of effects (see further details in Wulff et al. [29]).

Peletier et al. [119] concluded, from laboratory experiments with diatom species isolated from intertidal sediments, that ambient (or even future increased) UV-B is unlikely to affect sediment-inhabiting MPB. Although it may appear that experiments on intact sediment MPB do at least partly support this conclusion [46], they do not fully rule out the role of (even ambient) UV-B as a controlling factor [47,65,68]. Odmark et al. [47] found that, while the removal of UV-B created a response, moderate enhancement of UV-B had less obvious effects. This suggests that ambient UV-B can indeed be a factor exerting a selective pressure on MPB particularly in sandy sediments, whereas an increased UV-B exposure due to ozone depletion would not severely affect the type of MPB community studied. However, given that UV-B, at present level, is a selective force in MPB communities of sandy sediments, there is no a priori reason to assume that the communities should respond to a less degree to an increase in UV-B levels. Moreover, early successional growth phases of sediment communities are indeed, like periphytic communities, susceptible to moderately enhanced (15%) UV-B levels [68]. Epipelic communities on sediments of oligotrophic lakes have also shown a significant response to ambient UV-B [26].

11.4.3 Are UV-B effects on microphytobenthos habitat-specific?

The benthic habitats in the above-cited experiments differ in several aspects, such as the type of substratum, water movement and nutrient status. Also the community properties, such as the level of productivity and composition of the food webs (e.g., importance of grazers) vary greatly. Are UVR responses then habitat-specific? The answer appears to be yes, although similarities in responses also exist. When MPB on both hard and soft substrata (and phytoplankton) were studied simultaneously in alpine oligotrophic lakes in Canada [26,121], it was found that attached periphyton was affected by ambient UV-B, while epipelon of the sediment and phytoplankton remained unaffected. On the other hand, Vinebrook and Leavitt [154] found that ambient UVR had no effect at all on epilithon, while a significant stimulating effect was found for epipelon. This contradicting result was explained by the fact that established epilithon, and not early successional stages, were studied in the latter experiment. An indication of the importance of the substratum type was also found for sediment MPB in a microtidal brackish-water area in Sweden. More variables were affected by UV-B in the sandy than in the muddy sediment [46,47,64]. This suggests that muddy sediments may function as a refuge for MPB (shallow photic zone, dominance of motile diatoms), while in the sandy sediment UVR penetrates deeper and the MPB is dominated by small-sized attached species.

Among community properties, grazing pressure is obviously an important controlling factor for the susceptibility of the MPB to UV-B. Heavily grazed periphyton communities become thin and are thus more sensitive to UV-B [120]. On the other hand, if the grazers are more sensitive to UV-B than the algae, the algal growth will benefit from UV-B [153, Chapter 15].

On a larger geographical scale, climate, latitude/elevation, and general nutrient status of the ecosystem may explain differences in the UV-B responses of MPB. As discussed before, early colonization stages of MPB are more susceptible to UV-B than already established, thick communities. Thus, UV-B can be expected to be a more important controlling factor in for example a cold climate where colonization events are more frequent due to ice and scouring [121]. Similarly, freshwater periphyton in mid-latitudes [150,155] may be more resistant to current UV-B stress than periphyton communities of higher latitudes [26,58]. The combined effects of climate warming and increased input of dissolved organic matter (DOM) have been suggested to moderate the effects of UVR-increase in alpine oligotrophic lakes [26]. However, climate change in combination with acidification may also increase the exposure of organisms to UV-B, particularly in clear, shallow lakes and streams [156].

11.5 Effects of UVR on marine macrophyte photosynthesis

Studies on comparative primary productivity of marine macrophytes under different scenarios of UV climate are rather scarce; moreover, there is a diffuse picture of their photoadaptive strategies. Taking into account the distinct origin and the morpho-functional divergences of macroalgal species, a common adaptive strategy is unlikely. Thus, a number of responses can be determined among species. However, a general pattern is observed: under natural radiation levels they show daily photoinhibition – a decrease in the photosynthetic rates/yield [48,131], at least at high zenith angles [48–51]. In most cases, high PAR irradiances at noon cause a decrease in photosynthetic rates [88,90], but UVR also contributes largely to this process [30,72,77,157,158].

In intertidal algae in particular, the highest photoinhibition values (mainly due to PAR) are found when low tide coincides with local noon [88,97]. Even algae harvested from rock pools, where they are normally exposed to extreme solar irradiances, show signs of photoinhibition after prolonged periods of exposure [30,90]. Under these conditions, increases in temperature and partial desiccation of algal thallus also contribute to the observed photoinhibition [38,87]. Deep-water algae and those adapted to shaded environments are inhibited even faster when exposed to direct solar radiation [90].

Recovery of photosynthesis – measured as an increase in fluorescence quantum yield – starts when irradiance begins to decrease, but remains still at saturating levels. Recovery is species-specific and occurs faster in sun-adapted algae than in algae growing at deep or shaded locations and then transferred to the surface. In the eulittoral red algae *Porphyra leucosticta* [30], *Asparagopsis armata* and *Felmanophycus rayssae* [97] from southern Spain, recovery of photosynthesis occurs immediately after a decrease of only 10–20% of solar radiation. However, the brown alga *Padina boryana* recovers with a 30% irradiance decline, whereas in *Sargassum polycystum* a reduction of 70% in the incident radiation is required [96]. In their review on red macroalgae, Figueroa and Gómez [38] reported photoinhibition of ~30–80% at noon, but most of the species showed full recovery in the afternoon. In contrast, only partial recovery was observed in red algae from the North Sea [72] or from Patagonian waters (Helbling et al., unpublished). The recovery of macroalgae after UVR exposure (as compared to the PAR control) is highly variable, with little recovery found in *Macrocystis pyrifera* [159] and in *Gelidium sesquipedale* [31], and high recovery with beneficial effects of UV-B in the brown alga *Dictyota dichotoma* [35] and in the marine angiosperm *Posidonia oceanica* [42]. These specific responses provide important information, as the recovery kinetics gives insights into the photoadaptive strategies of macroalgae and their light-stress tolerance capacity. Thus, those algae capable of dynamic (reversible) photoinhibition under high solar radiation levels and with a rapid recovery capacity will have competitive advantages as compared to those without any efficient photoprotection mechanism.

The ability for dynamic photoinhibition during exposure to high radiation, as well as the general degree of photosynthetic adaptation of individual species to different light regimes influences the upper depth distribution of algal zonation [86,88,94,98]. In fact, several taxa and life history stages of inter- and subtidal polar algae show a strong correlation between their depth distribution and their capacity to cope with high radiation stress [33,34,77,94]. Thus, species growing in the upper subtidal zone show in general more tolerance to high solar radiation levels, especially to UVR, than algae from deeper waters

[40,72,73,77,157,160].

Relatively few studies have been conducted on a long-term basis to determine the effects of UVR on photosynthesis and growth of macroalgae [54,151]. In experiments carried out with *Ulva* sp., UV-B caused a decrease of both growth rates and photosynthesis during the first week of exposure to solar radiation, but UV-A stimulated growth as compared to the PAR treatment [54]. However, after two weeks of exposure, no differences were observed between treatments, a fact that hints to the action of acclimation mechanisms, which protect algae against UV stress (see below and Chapter 10).

11.6 Carbon and nitrogen allocation

There is evidence that UVR, especially UV-B, affects carbon allocation in aquatic autotrophic organisms. This has important consequences for food web dynamics, as these changes will affect growth and, consequently, the availability of food for other trophic levels, such as bacteria and heterotrophic microorganisms (see Chapters 5 and 15). Changes in lipid, protein, polysaccharide, and fatty acid levels due to UVR have been determined in some phytoplanktonic and MPB organisms [47,161–167]. These studies have especially highlighted the variations in responses, according to the specific sensitivity of the organisms. For example, Buma et al. [168], working with three marine diatoms, found a significant increase in cell protein content when cells were exposed to low UV-B doses, whereas the opposite occurred at higher doses. Veen et al. [169], working with a chlorophyte, demonstrated an increase in cell protein levels when cells were exposed to UVR. Skerratt et al. [167] exposed the diatom *Odontella* to UV-B radiation and found a reduction in lipid content whereas in *Chaetoceros* an increase was found. Goes et al. [162,163], working with diverse phytoplanktonic species, found changes in the rates and sizes of storage and structural carbohydrates and polyunsaturated fatty acids when exposed to UV-B. Moreover, Döhler [170,171] found UVR-induced changes in pool sizes of diverse amino acids of several Antarctic and temperate marine phytoplankton species with UV-A causing, in general, an increase in their levels, whereas UV-B produced the opposite effect. These results agree with those obtained by Goes et al. [172], who also showed that UV-B caused changes in amino acid concentrations within the cell. Finally, studies performed to determine UVR effects on the ATP content of Antarctic phytoplankton showed a reduction in this component when cells were exposed to UVR [173], but Döhler and Biermann [161], working with a marine diatom, did not find any effect.

Studies performed with MPB communities in sediments have also demonstrated changes in carbon allocation as a result of UVR exposure. The most frequently observed change was a larger relative allocation to proteins at UV-B exposure [29,64,68]. This can be interpreted as a larger proportion of fixed carbon spent on growth when carbon fixation decreases, as microalgal cells tend to retain synthesis of proteins rather than storage products under adverse environmental conditions [174]. Other UV-B effects on carbon allocation were related to polar lipids, which were lower under enhanced UV-B [47]. In macroalgae it was also found that UV-B radiation affects carbon and nitrogen allocation, although very few studies have been done in this regard. For example, Altamirano et al. [54] found that more than 78% of the seasonal changes in the internal content of carbon and nitrogen in the green macroalga *Ulva rigida* were explained by seasonal changes of UV-B.

11.7 Mechanisms to reduce the effects of UVR on photosynthesis

Adaptation to UVR assumes the existence of mechanisms that protect the organism or reduce the deleterious effects. According to Roy [140] four basic mechanisms allow an organism to cope with a stressful situation, i.e., UVR exposure, (1) avoidance, (2) reducing the stress by a physiological behavioral mechanism – e.g., through the synthesis of UV-absorbing compounds, (3) repairing the damage produced and, (4) acclimating to stress when allowed enough time. More details on these mechanisms can be found in other chapters of this book.

Avoidance mechanisms seem to be a common strategy against exposure to high levels of UVR. For microalgae living in soft substrata, such as motile cyanobacteria and diatoms [15,56,65], this involves downward vertical migration. Bebout and Garcia-Pichel [15] showed that, by migrating down to 300 μm depth, cyanobacteria could reduce their UV exposure to 10% of that at the sediment surface. For benthic diatoms, the observed downward movement of *Gyrosigma balticum* at high light levels was first suggested to be related to PAR rather than UV-B [46,64]. However, a subsequent fluorescence study indicated that the migration could in fact be a direct response to UV-B [65]. Avoidance can also be achieved by means of circadian rhythms that allow an organism to swim down at noon to depths where radiation intensities are low, as occur in some dinoflagellates [175]. However, it should be considered that UVR can alter the motility and phototaxis of some autotrophic organisms, as found in several microalgal species [5]. Moreover, in other organisms, loss of flagella has also been reported [176]. Thus, in some sensitive organisms, avoidance mechanisms can be severely altered by UVR exposure.

Another strategy to minimize the effects of UVR is through the presence of UV-screening compounds. The most studied compounds are those collectively named mycosporine-like amino acids (MAAs), which are found in many marine and freshwater autotrophic and heterotrophic organisms [177,178] (see also Chapter 10). Evidence of their protective role upon physiological mechanisms remains still unclear, and in some cases it seems that they just provide partial protection, as in some cyanobacteria [179]. In other cases, though, MAAs have been proved to be an effective protection mechanism [107,180] so that photosynthesis in phytoplanktonic cells with higher amounts of MAAs was less inhibited. In benthic diatoms, however, the production of such protective substances does not appear to be a major strategy. Although MAAs have been detected in MPB of shallow-water subtidal sediments, the concentrations are low and show no significant increase at UV exposure (ambient or increased) [29,46,47,64], which agrees with the findings of Peletier et al. [119]. Jeffrey et al. [181] tested 152 algal species and found that diatoms generally had low concentrations of UV-screening compounds as compared with other algal groups. Moreover, Helbling et al. [107] found that pennate diatoms (which usually dominate benthic diatom communities) contained less MAAs than centric diatoms. Other compounds may also have a protective role, functioning as UV-screening agents (see Chapter 10). For example, scytonemin is a UV-absorbing extracellular substance found in the sheath of cyanobacterial filaments [182]. In addition, high concentrations of carotenoids as a result of UVR exposure have been observed in diatom mats [65], and some cyanobacteria as well as chlorophytes [162], suggesting an UV-protecting function of these pigments.

MAAs have also been reported in green, red and brown macroalgae from tropical, temperate and polar regions [55,125,158,177,183,184]. The concentration of MAAs in macroalgae has been found to be related to depth zonation and UV exposure [177]. Their accumulation seems to be higher under high than under low daily irradiance values (i.e. different latitudes), and moreover, generally higher in intertidal than in subtidal algae [184–186]. In addition, this accumulation seems to be a wavelength-dependent process [125,185,187,188], and an UV-B-mediated increment of these compounds has been shown in a variety of algae [155,189]. In *Chondrus crispus*, both UV-A and UV-B stimulated a strong accumulation of shinorine, whereas the content of palythinol and palythine was mainly stimulated by PAR, indicating a MAA-specific induction triggered by these wavelengths [125]. In *Palmaria palmata*, on the other hand, and when exposed only to PAR, a 6-fold increase in the porphyra-334 concentration was observed; the treatment receiving PAR + UV-A gave similar results plus an accumulation of shinorine; under full solar radiation, accumulation of porphyra-334, shinorine and palythine was observed [185]. In addition, in *Chondrus crispus*, pre-exposure to blue light followed by growth under natural UV-A led to a 7-fold increase in the synthesis of shinorine as compared with growth without the blue light pre-treatment [188]. So, it has been hypothesized that there are two photoreceptors for MAAs synthesis in *C. crispus*, one for blue light and one for UV-A, which act synergistically [188]. In macroalgae, other types of potentially protective compounds are also found, such as phlorotanins in brown algae [190] and coumarins in the green alga *Dasycladus* [191,192].

In addition, and while UVR-mediated DNA damage occurs in aquatic autotrophic organisms [168,193–196], repair mechanisms of the DNA molecule (see Chapter 9) are also present [193]. However, the presence of one or other mechanism (i.e., photoreactivation, nucleotide excision repair or recombination repair) is clearly dependant on the species under study and the radiation conditions at which the cells are exposed (see Chapter 9).

Finally, acclimation mechanisms to cope with high UVR intensities are important in several aquatic organisms. These usually occur on a long-term basis, when organisms have been exposed for enough time to the stress factor (UVR). One of these acclimation mechanisms is the previously mentioned synthesis of MAAs, as found in some natural populations and cultures of phytoplankton

[107,110,141]. However, the synthesis of UV-absorbing compounds is not a general response, and several species do not show an increase of MAAs content even after several weeks of exposure to UVR [197,198]. Acclimation can also occur through a change in the community composition [110], so that those species more adapted to a particular light regime will dominate. For example, a natural Antarctic phytoplankton population dominated by flagellates (80% in terms of carbon biomass) changed to a diatom dominated population when receiving UVR + PAR, whereas in those samples receiving only PAR, small flagellates still dominated. However, diverse responses are observed at different sites. For example, Mousseau et al. [199] in their study conducted with an estuarine community also observed changes in diversity when samples were exposed to different radiation treatments. A shift from a diatom-dominated community to small flagellates occurred more rapidly in the treatment receiving enhanced UV-B as compared to those receiving natural UV-B levels. Clearly, responses are strongly species-specific and depend on radiation levels and quality to which organisms are exposed.

11.8 Other photosynthesis-related effects

There are a number of UVR effects that are closely related to the photosynthetic performance of aquatic primary producers. These effects are due to the couplings between radiation – especially UVR – and a number of morphological and biochemical factors within the cell [200]. Thus, for example, radiation-induced changes in nutrient uptake, synthesis and allocation of metabolic compounds, motility/orientation, and cell morphology will result in variations in photosynthetic rates. In the following paragraphs we will outline some of these effects – nutrient incorporation and enzyme activities related to carbon and nitrogen metabolism, accumulation or damage on pigments. Specific effects, such as DNA damage, which may induce a reduction in growth rates [195], and hence affect overall primary productivity, are addressed in Chapter 9.

11.8.1 Nutrient incorporation/assimilation and enzyme activities

Growth of aquatic autotrophic organisms is dependent not only on carbon assimilation, but also on the incorporation and assimilation of nitrogen, phosphate, sulfur and several micronutrients [201]. In general, it is considered that UVR – especially UV-B – is an inhibitor of uptake processes (especially nitrogenous), whereas UV-A stimulates or exerts no significant effects on the uptake of these ions [170,171]. In particular, studies carried out with phytoplanktonic organisms have demonstrated that nitrate and ammonium uptakes are affected by UVR [202–204]. Furthermore, Döhler [170,171,203,205], working with several Antarctic and North Sea phytoplanktonic species, has pointed out the diversity of responses among the organisms tested. Thus, samples dominated by the prymnesiophyte *Pheaocystis pouchetii* were very sensitive to UV-B doses (in terms of ^{15}N-ammonium uptake), and so were those containing *Ceratium* sp., *Coscinodiscus* sp. and *Noctiluca* sp. ^{15}N-nitrate uptake was not or only slightly affected by UV-B irradiances [203]. On the other hand, in experiments conducted with North Sea natural phytoplankton populations, Döhler and Hagmeier [206] found that UV-A radiation stimulated ^{15}N-ammonium uptake. Fewer studies have addressed the effects of UVR on P-uptake of phytoplanktonic cells. Hessen et al. [176], working with the chlorophyte *Chlamydomonas reinhardtii* found a stimulation under low UV-B doses (< 3.6 kJ m^{-2} at 312 nm), but higher inhibition when UV-B doses were higher. In addition, studies on UV-B effects carried out in both sandy and muddy sediments have suggested that the nutrient availability may be an important factor for the susceptibility of MPB communities to UV-B exposure [47]. Wulff et al. [68] designed an experiment to test this hypothesis, and showed that the availability of nutrients indeed can act to mitigate the effects of UV-B on a microbenthic community on a sandy substratum.

Some studies have also addressed the UVR effects on nutrient incorporation in marine macroalgae [52,90]. In particular, these studies have focused on UVR effects upon carbonic anhydrase (CA) and nitrate reductase (NR) activities [32,192,207–210]. These are important enzymes involved in the incorporation of carbon and nitrogen within the cell [211], thus any stress factor that affects them will ultimately influence photosynthesis. Studies carried out with algae collected from southern Spain [32,192,208,210] found daily variations (i.e., circadian rhythms) in NR and CA activities. In *Dasycladus vermicularis* it was found that these variations were antagonistic during the onset of solar radiation, although these changes only partially matched those of photosynthesis [192], suggesting that these processes are affected differentially by UVR. In long-term studies, it has been shown that UV-A radiation stimulated NR activity, and UV-B decreased both nitrate uptake and NR activities [209,212]. On the other hand, UV-B radiation seems to stimulate CA activity in eulittoral algae but not in subtidal [13,209]. In addition, experiments were conducted to determine the effects of UVR on the activity of Calvin cycle enzymes, such as ribulose-1,5-biphosphate carboxylase/oxygenase (RUBISCO) and glyceraldehyde-3-phosphate dehydrogenase (G3PDH), and in Arctic macroalgae it was found that the photosynthetic activity decreased due to the negative effects of UVR upon these enzymes [213].

11.8.2 Pigments

Several researchers have reported the decrease of photosynthetic pigments due to exposure to UVR [214–216]. This reduction can be due to a combination of factors, such as the inhibition of de novo synthesis and the natural turnover of pigments, or directly to photobleaching [216]. Bleaching can occur not only because of UVR, but also due to exposure to high PAR intensities [216]; it is species-specific and also depends on the spectral characteristics of the radiation treatments imposed to the cells. Helbling et al. [214], working with several marine phyto-plankton cultures, found that *Nannochloris oculata* (Eustigmatophyceae) had a decrease in chl-*a* content of 30, 60 and 80% under PAR only, PAR + UV-A, and PAR + UVR, respectively, after being exposed for 4.5 h to solar radiation. The prymnesiophyte *Isochrysis galbana*, on the other hand, did not experience significant changes in chl-*a* content (for the same radiation treatments) even after 7 h of exposure. Other experiments have also demonstrated the differential sensitivity to UVR of various pigments [217], with the phycobiliproteins being especially sensitive to these wavelengths [218].

Absolute amounts of photosynthetic pigments, commonly used as an estimator of growth in autotrophic organisms, seem to be also affected by UVR. During a long-term experiment, Helbling et al. [43] simulated ozone depletion events by moving Antarctic phytoplankton towards the Equator, so that the samples were exposed not only to increased levels of UVR, but also to natural changes in the relative proportions of UV-B and UV-A. They found a decrease in the growth rate of Antarctic phytoplankton exposed to UVR as compared to that exposed to the PAR-only treatment. However, growth rates were not significantly different when the samples were incubated under UVR levels similar to those found at their sampling site in the Antarctic. Data on long-term experiments conducted in both polar areas [137] showed that even though photosynthesis was initially affected by UVR (day 1), growth rates, evaluated either as carbon fixation or chl-*a* content, did not show any significant differences. In general, studies have demonstrated that different growth responses due to UVR exposure occur not only among taxa [198,219], but also within the same genus. For example, in the chlorophyte *Dunaliella salina* growth rate was not affected by UVR, whereas in *D. tertiolecta* it was significantly reduced after 3 days of exposure [198].

The differential sensitivity of pigments to UVR has also been studied for MPB organisms. Phycobilins of cyanobacterial mats appear to be more sensitive than chlorophylls and carotenoids, the latter often increasing at UV-B exposure [56,57,217]. However, in experiments conducted on intact sediment communities dominated by diatoms, no changes in pigment composition (expressed as ratios to chl-*a*) were observed [29,46,47,64,68], with one exception: higher carotenoid content was observed at enhanced UV-B levels in a *Gyrosigma* mat, probably reflecting a UV-B-protecting strategy [65].

In marine macroalgae, various responses were also found when addressing the effects of UVR upon various pigments. Exposure of *Porphyra umbilicalis* to artificial UVR levels decreased chl-*a* and phycocyanin concentrations by 65 and 67%, respectively, whereas carotenoids and phycoerythrin decreased by as much as 75 and 82%, respectively [220]. Furthermore, and under ambient levels, UVR not only decreased the concentration of chl-*a* and biliproteins in the red alga *Porphyra leucosticta*, but the pattern of daily variation was also affected [30]. The damage of photosynthetic pigments by UVR in *P. leucosticta* was suggested to be the cause of a decrease in photosynthetic rates. However, in *Macrocystis pyrifera*, it was found that the main light-harvesting complex of this alga, the fucoxanthin–chlorophyll protein, was the specific site for UV damage [158]. Finally, in *Ulva rigida* [221] and *Dasycladus vermicularis* [191] the content of chlorophyll and carotenoids was significantly higher in the presence of UV-B than that in the control (PAR only), suggesting the presence of an efficient protective-pigment mechanism.

11.8.3 Cell morphology and size

When evaluating the photosynthetic responses to UVR of diverse organisms, some studies have revealed the importance of cell size [43,81,82,112,177]. In phytoplanktonic organisms, it was found that, although there is certainly variability in responses, small cells – with a relatively high surface to volume ratio – are more resistant to photosynthesis inhibition but more vulnerable to DNA damage [81,82,195]. On the other hand, and provided that microplanktonic cells (20–200 μm) do not have high concentrations of UV-absorbing compounds, they are more vulnerable to UVR (in terms of photosynthesis). This has been demonstrated in a comparative study carried out in the Andean lakes [82], where it was found that larger phytoplanktonic cells had a higher kinetics of inhibition and hence were more affected by UVR than smaller cells. For MPB organisms, on the other hand, there are contradictory findings as to whether UV-B-related changes in species composition are related to cell size or are due to taxon-specific sensitivity [82,119,222]. As seen for planktonic algae, increasing size may occur both on an individual species level, as cell division is hampered [177,197], and on community level, as species with larger cell-size could be favoured [116,222]. For MPB there is some indication for the latter, but not for the former. Bothwell et al. [58] found that large, stalked diatom species increased their dominance at UV-B exposure during periphyton succession, and Vinebrook and Leavitt [121] found that the growth of the small-sized diatom *Achanthes minutissima* was suppressed under UV-B exposure.

Besides size, the morphology also seems to influence the response of algae to solar radiation. This has been shown particularly for macroalgae. A comparison between the red algae *Porphyra leucosticta* and *Rissoella verruculosa*, which have comparable zonation patterns at intertidal sites, shows the different photoprotective strategies of these algae [30,32]. This is probably related to different absorption properties because of the thallus thickness and pigment composition. *P. leucosticta* has a thin thallus consisting of one cell layer in which light transmits rapidly and homogenously towards the harvesting complexes, whereas in *R. verruculosa*, which has a more complex structure, some scattering of photons through the multilayered thallus (self-shading) may take place. This was evident when the algae were exposed to full solar radiation, and in *Porphyra* UVR accounted for about 30% of the total photoinhibition, whereas no effects were observed in *Rissoella*. In addition, some studies hint about the importance different life stages, which are closely related with size. Although studies have focused on the macrothallus or adult stages, it is expected that UVR stress would be more evident in the microscopic life stages (single- and few-celled), mainly due to their structural simplicity. These studies, in addition, bring about important consequences for algal zonation. For example, depth distribution patterns of large kelps have been frequently thought to reflect the light requirements of establishment stages (spores, embryos, etc.). Paradoxically, most studies performed to address the relationship between the physiological performance under different light environments have been done with the large sporophytes, whereas spore adaptation has been largely overlooked [223]. The question of whether early developmental stages of macroalgae, particularly spores, are more susceptible to UVR than larger life history phases has been less addressed [73]. If this is so, it is reasonable to think that the physiological adaptation of spore stages (such as the ability to acclimate to different light climates) will have consequences for the whole population dynamics [224].

11.9 UVR effects on aquatic photosynthesis: conclusions and future research

UV can reduce photosynthetic rates of both micro- and macroalgae by *direct effects* on the photosynthetic apparatus through (1) pigment photobleaching in the photosynthetic antenna, (2) reduction of proteins in photosystem II, (3) decrease of enzyme activity in the Calvin cycle, and (4) inhibition of carbonic anhydrase activity, as well as via *indirect effects*, such as DNA damage. Few studies have analyzed the effects of UVR on photosynthetic responses other than carbon assimilation, for example nitrogen or phosphorus assimilation. Thus, analyzing the effects of UVR on such integrated metabolic processes should become an experimental effort of high priority.

A fact that complicates the study of UVR effects on photosynthetic organisms is that ozone depletion is occurring parallel to other global environmental changes, such as the increase in CO_2 and temperature, as well as the increasing eutrophication and acidification of natural waters. In order to predict the effects of increased UV-B radiation on aquatic ecosystems it is necessary to take into account the changes in other environmental factors. The increase in UV-B levels does affect algal physiology and ecology, including biogeochemical cycles in the coastal zone and enhanced radiation may have a significant global-scale climatic impact [225] (see also Chapters 5 and 17). Changes in productivity or diversity of aquatic primary producers due to elevated UV-B levels are likely to bring about alterations on several trophic levels of coastal marine food webs. Therefore, changes in community structure and ecosystem function can be expected.

Even *ambient* UVR levels can have a significant effect on benthic and water-column algal communities. For example, a general feature of MPB response is that ambient UV-B levels can exert a selective pressure on *early successional* stages on both hard and soft substrata. However, systems appear to vary greatly in susceptibility, depending on climate, the availability of refuges and nutrients, as well as the level of productivity and structure of the food webs. The local ecological implication of the initial selective pressure during early colonization will thus depend on the general importance of colonization events in relation to UV-B exposure. In the case of macroalgae, UV-B may also function as a selective pressure at the time of early colonization and recruitment. Consequently, macroalgal zonation can be determined by the different resistance against UV-B radiation of spore germination and growth of young plants [223].

On a longer time scale, algal resistance, shielding properties of the habitat, and trophic cascades may counteract UV-B effects on the primary producer level. For example, increasingly more studies on already established benthic microalgal and phytoplankton communities suggest a UV-B- effect minor to that first expected from short-term experiments. Thus, it is still very difficult to draw predictive conclusions about a general, long-term effect of UV-B on aquatic primary producers. This appears to apply particularly to systems where primary production depends mainly on benthic microalgae. Despite the fact that these communities consist of small organisms, with a rapid turnover, we perhaps still need to address system-level responses to UV-B on even longer time scales (years), such as in experiments conducted on terrestrial communities [60]. However, even results from these terrestrial field experiments show that observed effects are not unambiguous, particularly as effects can be largely modified by other environmental factors (temperature and nutrient availability), which may even reverse the initial UVR effects.

We are now at the stage where we can conclude that UVR affects all types of aquatic primary producers, although the long-term response at the community level may be highly variable and modified by both environmental and biological factors. Future experimental approaches must include the interaction between different environmental factors in the scenario of ozone depletion. Only then can we expand our knowledge on the effects of increased UV-B, not only at organism level, but also at the community and ecosystem level, which is crucial for understanding the consequences for aquatic biodiversity and productivity.

Acknowledgements

This work was supported by Fundación Antorchas (Projects No. 13887-83 and No. 13955-3), ANPCyT-PROALAR (Project No. 104), FEDER (Project 1FD97-0824) and Fundación Playa Unión. We thank R. Gonçalves for his help with computer drawings. This is Contribution No. 34 of Estación de Fotobiología Playa Unión.

References

1. C.B. Field, M.J. Behrenfeld, J.T. Randerson, P. Falkowski (1998). Primary production of the biosphere: Integrating terrestrial and oceanic components. *Science*, **281**, 237–240.
2. M.J. Behrenfeld, P.G. Falkowski (1997). Photosynthetic rates derived from satellite-based chlorophyll concentration. *Limnol. Oceanogr.*, **42**, 1–20.
3. P.G. Falkowski (1994). The role of phytoplankton photosynthesis in global biogeochemical cycles. *Photosynth. Res.*, **39**, 235–258.
4. K. Mann, A.R.O. Chapman (1975). Primary production of marine macrophytes. In: J.P. Cooper (Ed.), *Photosynthesis and Productivity in Different Environments* (pp. 207–223). Cambridge University Press, Oxford.

5. D.-P. Häder, R.C. Worrest, H.D. Kumar, R.C. Smith (1995). Effects of increase solar ultraviolet radiation on aquatic ecosystems. *Ambio*, **24**, 174–180.
6. D.O. Duggins, C.A. Simenstand, J.A. Estes (1989). Magnification of secondary production by kelp detritus in coastal marine ecosystem. *Science*, **245**, 555–561.
7. G.J.C. Underwood, J. Kromkamp (1999). Primary production by phytoplankton and microphytobenthos in estuaries. *Adv. Ecol. Res.*, **29**, 93–153.
8. P.J. Neale (2001). Effects of ultraviolet radiation on estuarine phytoplankton production: Impact of variations in exposure and sensitivity to inhibition. *J. Photochem. Photobiol. B: Biol.*, **62**, 1–8.
9. D.P. Morris, H.E. Zagarese, C.E. Williamson, E.G. Balseiro, B.R. Hargreaves, B. Modenutti, R. Moeller, C. Queimaliños (1995). The attenuation of solar UV radiation in lakes and the role of dissolved organic carbon. *Limnol. Oceanogr.*, **40**, 1381–1391.
10. J.T.O. Kirk (1994). *Light and Photosynthesis in Aquatic Ecosystems.* Cambridge University Press, United Kingdom.
11. R.C. Smith, K.S. Baker (1979). Penetration of UV-B and biologically effective doserates in natural waters. *Photochem. Photobiol.*, **29**, 311–323.
12. H. Piazena, D.-P. Häder (1997). Penetration of solar UV and PAR into different waters of the Baltic Sea and remote sensing of phytoplankton. In: D.-P Häder (Ed.), *The Effect of Ozone Depletion on Aquatic Ecosystems* (pp.45–96). R.G. Landes Company, Academic Press, Austin.
13. F.L. Figueroa (1998). Underwater light and primary production. In: C.V. Nolan, D.-P. Häder (Eds), *Role of Solar UV-B Radiation on Ecosystems* (Ecosyst. Res. Rep., **30**, pp. 121–133). Ecosystem Research, Luxemburg.
14. Y. Hodoki, Y. Watanabe (1998). Attenuation of solar ultraviolet radiation in eutrophic freshwater lakes and ponds. *Jpn. J. Limnol.*, **59**, 27–37.
15. B.M. Bebout, F. Garcia-Pichel (1995). UV B-induced vertical migrations of cyanobacteria in a microbial mat. *Appl. Environ. Microbiol.*, **61**, 4215–4222.
16. F. Garcia-Pichel, B.M. Bebout (1996). Penetration of ultraviolet radiation into shallow water sediments: High exposure for photosynthetic communities. *Mar. Ecol. Prog. Ser.*, **131**, 257–262.
17. E.W. Helbling, H.C. Eilertsen, V.E. Villafañe, O. Holm-Hansen (1996). Effects of UV radiation on post-bloom phytoplankton populations in Kvalsund, North Norway. *J. Photochem. Photobiol. B: Biol.*, **33**, 255–259.
18. R.C. Smith B.B. Prézelin, K.S. Baker, R.R. Bidigare, N.P. Boucher, T.L. Coley, D. Karentz, S. MacIntyre, H.A. Matlick, D. Menzies, M. Ondrusek, Z. Man, K.J. Waters (1992). Ozone depletion: Ultraviolet radiation and phytoplankton biology in Antarctic waters, *Science*, **255**, 952–959.
19. O. Holm-Hansen, E.W. Helbling, D. Lubin (1993). Ultraviolet radiation in Antarctica: Inhibition of primary production. *Photochem. Photobiol.*, **58**, 567–570.
20. O. Holm-Hansen, E.W. Helbling (1995). Técnicas para la medición de la productividad primaria en el fitoplancton. In: K.Alveal, M.E. Ferrario, E.C. Oliveira, E. Sar (Eds), *Manual de Métodos Ficológicos* (pp. 329–350). Universidad de Concepción, Concepción
21. E.W. Helbling, V.E. Villafañe, O. Holm-Hansen (1994). Effects of ultraviolet radiation on Antarctic marine phytoplankton photosynthesis with particular attention to the influence of mixing. In: C.S. Weiler, P.A. Penhale (Eds). *Ultraviolet Radiation in Antarctica: Measurements and Biological Effects* (pp. 207–227). American Geophysical Union, Washington DC.
22. P.J. Neale, R.F. Davis, J.J. Cullen (1998). Interactive effects of ozone depletion and vertical mixing on photosynthesis of Antarctic phytoplankton. *Nature*, **392**, 585–589.

23. J. Köhler, M. Schmitt, H. Krumbeck, M. Kapfer, E. Litchman, P.J. Neale (2001). Effects of UV on carbon assimilation of phytoplankton in a mixed water column. *Aquat. Sci.*, 294–309.
24. J. Marra (1978). Phytoplankton photosynthetic response to vertical movement in a mixed layer. *Mar. Biol.*, **46**, 203–208.
25. P.M. Kiffney, W.H. Clements, T.A. Cady (1997). Influence of ultraviolet radiation on the colonization dynamics of a Rocky Mountain stream benthic community, *J. North Am. Benthol. Soc.*, **16**, 520–530.
26. R.D. Vinebrook, P.R Leavitt (1999). Differential responses of littoral communites to ultraviolet radiation in an alpine lake, *Ecology*, **80**, 223–237.
27. S. Reizopoulou, P. Santas, D. Danielidis, D.-P. Häder, R. Santas (2000). UV effects on invertebrate and diatom assemblages of Greece. *J. Photochem. Photobiol. B: Biol.*, **56**, 172–180.
28. S.N. Francoeur, R.L. Lowe (1998). Effects of ambient ultraviolet radiation on littoral periphyton: Biomass accrual and species-specific responses. *J. Freshwat. Ecol.*, **13**, 29–37.
29. A. Wulff, C. Nilsson, K. Sundbäck, S.-Å. Wängberg, S. Odmark (1999). UV radiation effects on microbenthos – a four month field experiment. *Aquat. Microb. Ecol.*, **19**, 269–278.
30. F.L. Figueroa, S. Salles, J. Aguilera, C. Jiménez, J. Mercado, B. Viñegla, A. Flores-Moya, M. Altamirano (1997). Effects of solar radiation on photoinhibition and pigmentation in the red alga *Porphyra leucosticta* Thur. in Le Jol. *Mar. Ecol. Prog. Ser.*, **151**, 81–90.
31. I. Gómez, F.L. Figueroa (1998). Effects of solar UV stress on chlorophyll fluorescence kinetics of intertidal macroalgae from southern Spain: a case study in *Gelidium species*. *J. Appl. Phycol.*, **10**, 285–294.
32. A. Flores-Moya, I. Gómez, B. Viñegla, M. Altamirano, E. Pérez-Rodríguez, C. Maestre, R.M. Caballero, F.L. Figueroa (1998). Effects of solar radiation on the endemic Mediterranean red alga *Rissoella verruculosa*: photosynthetic performance, pigment content and the activities of enzymes related to nutrient uptake. *New Phytol.*, **139**, 673–683.
33. K. Bischof, D. Hanelt, H. Tüg, U. Karsten, P. Brouwer, C. Wiencke (1998). Acclimation of brown algal photosynthesis to ultraviolet radiation in Arctic coastal waters (Spitsbergen, Norway). *Polar Biol.*, **20**, 388–395.
34. K. Bischof, D. Hanelt, C. Wiencke (1998). UV radiation can affect depth-zonation of Antarctic macroalgae. *Mar. Biol.*, **131**, 597–605.
35. A. Flores-Moya, D. Hanelt, F.L. Figueroa, M. Altamirano, B. Viñegla, S. Salles (1999). Involvement of solar UV-B radiation in recovery of photoinhibited photosynthesis in the brown alga *Dictyota dichotoma* (Hudson) Lamouroux. *J. Photochem. Photochem. B: Biol.*, **49**, 129–135.
36. S. Beer, B. Vilenkin A. Weil , M Vete, L. Susel, A. Eshel (1998). Measuring photosynthetic rates in seagrasses by pulse amplitude modulated (PAM) fluorometry. *Mar. Ecol. Prog. Ser.*, **174**, 293–300.
37. P.J. Ralph, R. Gademann, W.C. Dennison (1998). In situ seagrass photosynthesis measured using a submersible, pulse-amplitude modulated fluorometer. *Mar. Biol.*, **132**, 367–373.
38. F.L. Figueroa, I. Gómez (2001). Photosynthetic acclimation to solar UV radiation of marine red algae from the warm-temperate coast of southern Spain: a review. *J. Appl. Phycol.*, **13**, 235–248.
39. S. Beer, M. Björk (2000). Measuring rates of photosynthesis of two tropical seagrasses

by pulse amplitude modulated (PAM) fluorometry. *Aquat. Bot.*, **66**, 69–76.

40. D. Hanelt, C. Wiencke, W. Nultsch (1997). Influence of UV radiation on the photosynthesis of Arctic macroalgae in the field. *J. Photochem. Photobiol. B: Biol.*, **30**, 179–187.
41. D.-P. Häder, M. Lebert, F.L. Figueroa, C. Jiménez, B. Viñegla, E. Pérez-Rodríguez (1998). Photoinhibition in Mediterranean macroalgae by solar radiation measured on by PAM fluorescence. *Aquat. Bot.*, **61**, 225–236.
42. F.L. Figueroa, C. Jiménez, B. Viñegla, E. Pérez-Rodríguez, J. Aguilera, A. Flores-Moya, M. Altamirano, M. Lebert, D.-P. Häder (2002). Effects of UV solar radiation on photosynthesis of the marine angiosperm *Posidonia oceanica* (L.) Delile from Southern Spain. *Mar. Ecol. Progr. Ser.*, **230**, 59–70.
43. E.W. Helbling, V.E Villafañe, M Ferrario, O. Holm-Hansen (1992). Impact of natural solar radiation on rates of photosynthesis and on specific marine phytoplankton species. *Mar. Ecol. Prog. Ser.*, **80**, 89–100.
44. M. Vernet, E.A. Brody, O. Holm-Hansen, B.G. Mitchell (1994). The response of Antarctic phytoplankton to ultraviolet radiation: absorption, photosynthesis, and taxonomic composition. In: C.S. Weiler, P.A. Penhale (Eds), *Ultraviolet Radiation in Antarctica: Measurements and biological effects* (pp. 43–158). American Geophysical Union, Washington DC.
45. O. Holm-Hansen, D.Lubin (1994). Solar ultraviolet radiation: effects on rates of CO_2 fixation in marine phytoplankton. In: N.E. Tolbert, J. Preiss (Eds), *Regulation of atmospheric CO_2 and O_2 by photosynthetic carbon metabolism* (pp. 55–74). Oxford University Press, Oxford.
46. K. Sundbäck, C. Nilsson, S. Odmark, A. Wulff (1996). Does present UV-B radiation influence marine diatom-dominated microbial mats? A case study. *Aquat. Microb. Ecol.*, **11**, 151–159.
47. S. Odmark, A. Wulff, S.-Å. Wängberg, C. Nilsson, K. Sundbäck (1998). Effects of UV-B radiation in a microbenthic community of a marine shallow-water sandy sediment. *Mar. Biol.*, **132**, 335–345.
48. D.-P. Häder, M. Lebert, J. Mercado, J. Aguilera, S. Salles, A. Flores-Moya, C. Jiménez, F.L. Figueroa (1996). Photosynthetic oxygen production and PAM fluorescence in the brown alga *Padina pavonica* measured in the field under solar radiation. *Mar. Biol.*, **127**, 61–66.
49. D.-P. Häder, H. Herrmann, R. Santas (1996). Effects of solar radiation and solar radiation deprived of UV-B and total UV on photosynthetic oxygen production and pulse amplitude modulated fluorescence in the brown alga *Padina pavonia. FEMS Microbiol. Ecol.*, **19**, 53–61.
50. D.-P. Häder, M. Lebert, A. Flores-Moya, C. Jiménez, J. Mercado, S. Salles, J. Aguilera, F.L. Figueroa (1997). Effects of solar radiation on the photosynthetic activity in the red alga *Corallina elongata* Ellis et Soland, *J. Photochem. Photobiol. B: Biol.*, **37**, 197–202.
51. D.-P. Häder, H. Herrmann, J. Schäfer, R. Santas (1997). Photosynthetic fluorescence induction and oxygen production in two Mediterranean *Cladophora* species measured on site. *Aquat. Bot.*, **56**, 253–264.
52. L.A. Franklin, R.M. Forster (1997). The changing irradiance environment: Consequences for marine macrophyte physiology, productivity and ecology. *Eur. J. Phycol.*, **32**, 207–232.
53. C.W. Grobe, T.M. Murphy (1998). Solar ultraviolet-B radiation effects on growth and pigment composition of the intertidal alga *Ulva expansa* (Setch.) S. and G. (Chlorophyta), *J. Exp. Mar. Biol. Ecol.*, **225**, 39–51.

54. M. Altamirano, A. Flores-Moya, F. Conde, F.L. Figueroa (2000). Growth seasonality, photosynthetic pigments, and carbon and nitrogen content in relation to environmental factors: a field study of *Ulva olivascens* (Ulvales, Chlorophyta). *Phycologia*, **39**, 50–58.
55. D. Karentz (1994). Ultraviolet tolerance mechanisms in Antarctic marine organisms. In: C.S. Weiler, P.A. Penhale (Eds), *Ultraviolet Radiation in Antarctica: Measurements and Biological Effects* (pp. 93–110). American Geophysical Union, Washington DC.
56. A. Quesada, W.F. Vincent (1997). Strategies of adaptation by Antarctic cyanobacteria to ultraviolet radiation. *Eur. J. Phycol.*, **32**, 335–342.
57. F. Garcia-Pichel, W. Castenholz (1994). On the significance of solar ultraviolet radiation for the ecology of microbial mats. In: L. Stal, P. Caumette (Eds), *Microbial Mats* (NATO ASI Series, pp. 77–84). Vol. G 35, Springer-Verlag, Berlin.
58. M.L. Bothwell, D. Sherbot, A.C. Roberge, R.J. Daley (1993). Influence of natural ultraviolet radiation on lotic periphytic diatom community growth, biomass accrual, and species composition: short-term versus long-term effects. *J. Phycol.*, **29**, 24–35.
59. M.M. Caldwell, A.H. Teramura, M. Tevini, J.F. Bornman, L.O. Björn, G. Kulandaivelu (1995). Effects of increased solar ultraviolet radiation on terrestrial plants. *Ambio*, **24**, 166–173.
60. L.O. Björn, T.V. Callaghan, C. Gehrke, D. Gwynn-Jones, J.A. Lee, U. Johanson, M. Sonesson, N.D. Buck (1999). Effects of ozone depletion and increased ultraviolet-B radiation on northern vegetation. *Polar Res.*, **18**, 331–337.
61. S.Z. El-Sayed, F.C. Stephens, R.R. Bidigare, M.E. Ondrusek (1990). Potential effects of solar ultraviolet radiation on antarctic phytoplankton. In: N.V. Blough, R.G. Zepp (Eds), *Effects of Solar Ultraviolet Radiation on Biogeochemical Dynamics in Aquatic Environments* (pp. 141–142). WHOI 90-09, Massachussets.
62. B. Mostajir, T. Sime-Ngando, S. Demers, C. Belzile, S. Roy, M. Gosselin, J.P. Chanut, S. de Mora, J. Fauchot, F. Vidussi, M. Levasseur (1999). Ecological implications of changes in cell size and photosynthetic capacity of marine Prymnesiophyceae induced by ultraviolet-B radiation. *Mar. Ecol. Progr. Ser.*, **187**, 89–100.
63. S.-Å. Wängberg, A. Wulff, C. Nilsson, U. Stagell (2001). Impact of UV-B radiation on microalgae and bacteria: a mesocosm study with computer modulated UV-radiation addition. *Aquat. Microb. Ecol.*, **25**, 75–86.
64. K. Sundbäck, S. Odmark, A. Wulff, C. Nilsson (1997). Effects of UV-B radiation on a marine benthic diatom mat. *Mar. Biol.*, **128**, 171–179.
65. G.J.C. Underwood, C. Nilsson, K. Sundbäck, A. Wulff (1999). Short-term effects of UV-B radiation on chlorophyll fluorescence, biomass, pigments, and carbohydrate fractions in a benthic diatom mat. *J. Phycol.*, **35**, 656–666.
66. R. Santas, A. Koussoulaki, D.-P. Häder (1997). In assessing biological UV-B effects, natural fluctuations of solar radiation should be taken into account. *Plant Ecol.*, **128**, 93–97.
67. A.A. McLeod (1997). Outdoor supplementation systems for studies of the effects of increased UV-B radiation. *Plant Ecol.*, **128**, 78–92.
68. A. Wulff, S. Å. Wängberg, K. Sundbäck, C. Nilsson, G. Underwood (2000). Effects of UV-B radiation on a marine microphytobenthic community growing on a sand-substratum under different nutrient conditions. *Limnol. Oceanogr.*, **45**, 1144–1152.
69. F.L. Figueroa, J.M. Blanco, F. Jiménez-Gómez, J. Rodríguez (1997). Effects of ultraviolet radiation on carbon fixation in Antarctic nanophytoflagellates. *Photochem. Photobiol.*, **66**, 185–189.
70. J.J. Cullen, P.J. Neale (1994). Ultraviolet radiation, ozone depletion, and marine photosynthesis. *Photosynth. Res.*, **39**, 303–320.

71. P.J Neale, D.J. Kieber (2000). Assessing biological and chemical effects of UV in the marine environment: Spectral weighting functions. In: R.E. Hester, R.M. Harrison (Eds), *Causes and Environmental Implications of Increased UV-B Radiation* (pp. 61–83). The Royal Society of Chemistry, Cambridge.
72. M.J. Dring, A. Wagner, J. Boeskov, K. Lüning (1996). Sensitivity of intertidal and subtidal red algae to UVA and UVB radiation, as monitored by chlorophyll fluorescence measurements: influence of collection depth and season, and length of irradiation. *Eur. J. Phycol.*, **31**, 293–302.
73. M.J. Dring, V. Makarov, E. Schoschina, M. Lorenz, K. Lüning (1996). Influence of ultraviolet-radiation on chlorophyll fluorescence and growth in different life-history stages of three species of *Laminaria* (Phaeophyta). *Mar. Biol.*, **126**, 183–191.
74. R. Röttgers (1999). *A physiological study on photoacclimation of marine planktonic algae to different ultraviolet-B irradiances.* (Ph. D. Thesis 95 pp.). Universität Bremen, Germany.
75. D.B. Brown, A.E. Peritz, D.L. Mitchell, S. Chiarello, J. Uitto, F.P. Gasparro (2000). Common fluorescent sunlamps are an inappropriate substitute for sunlight. *Photochem. Photobiol.*, **72**, 340–344.
76. R.H. Aalderink, R. Jovin (1997). Estimation of the photosynthesis/irradiance (P/I) curve parameters from light and dark bottle experiments. *J. Plankton Res.*, **19**, 1713–1742.
77. D. Hanelt, B. Melchersmann, C. Wiencke, W. Nultsch (1997). Effects of high light stress on photosynthesis of polar macroalgae in relation to depth distribution. *Mar. Ecol. Progr. Ser.*, **149**, 255–266.
78. E. Steemann Nielsen (1952). The use of radio-active carbon (C14) for measuring organic production in the sea. *J. Cons. Perm. Int. Explor. Mer*, **18**, 117–140.
79. D.-S. Kim, Y. Watanabe (1994). Inhibition of growth and photosynthesis of freshwater phytoplankton by ultraviolet A (UVA) radiation and subsequent recovery from stress. *J. Plankton Res.*, **16**, 1645–1654.
80. J. Beardall, T. Berman, S. Markager, R. Martinez, V. Montecino (1997). The effects of ultraviolet radiation on respiration and photosynthesis in two species of microalgae. *Can. J. Fish. Aquat. Sci.*, **54**, 687–696.
81. E.W. Helbling, A.G.J. Buma, M.K. de Boer, V.E. Villafañe (2001). In situ impact of solar ultraviolet radiation on photosynthesis and DNA in temperate marine phytoplankton. *Mar. Ecol. Prog. Ser.*, **211**, 43–49.
82. E.W. Helbling, V.E. Villafañe, E.S. Barbieri (2001). Sensitivity of winter phytoplankton communities from Andean lakes to artificial ultraviolet-B radiation. *Rev. Chil. Hist. Nat.*, **74**, 273–282.
83. E.W. Helbling, V.E. Villafañe, A.G.J. Buma, M. Andrade, F. Zaratti (2001). DNA damage and photosynthetic inhibition induced by solar UVR in tropical phytoplankton (Lake Titicaca, Bolivia). *Eur. J. Phycol.*, **36**, 157–166.
84. P.J. Neale, J.J. Cullen, R.F. Davis (1998). Inhibition of marine photosynthesis by ultraviolet radiation: Variable sensitivity of phytoplankton in the Weddell–Scotia Confluence during the austral spring. *Limnol. Oceanogr.*, **43**, 433–448.
85. N.P. Revsbech (1989). An oxygen microelectrode with a guard cathode. *Limnol. Oceanogr.*, **34**, 474–478.
86. W.J. Henley, G. Levavasseur, L.A. Franklin, S.T. Lindley, J. Ramus, C.B. Osmond (1991). Diurnal responses of photosynthesis and fluorescence in *Ulva rotundata* acclimated to sun and shade in outdoor culture. *Mar. Ecol. Prog. Ser.*, **75**, 19–28.
87. D. Hanelt (1992). Photoinhibition of photosynthesis in marine macrophytes of the South China Sea. *Mar. Ecol. Progr. Ser.*, **82**, 199–206.

88. D. Hanelt (1996). Photoinhibition of photosynthesis in marine macroalgae. *Sci. Mar.*, **60** (Suppl. 1), 243–248.
89. U. Schreiber, T. Endo, H. Mi, K. Asada (1995). Quenching analysis of chlorophyll fluorescence by the saturation pulse method: Particular aspects relating to the study of eukaryotic algae and cyanobacteria. *Plant Cell Physiol.*, **36**, 873–882.
90. D.-P. Häder, F.L. Figueroa (1997). Photoecophysiology of marine macroalgae. *Photochem. Photobiol.*, **66**, 1–14.
91. U. Schreiber, U. Schliwa, W. Bilger (1986). Continuous recording of photochemical and non-photochemical chlorophyll fluorescence quenching with a new type of modulation fluorometer. *Photosynth. Res.*, **10**, 51–62.
92. B. Genty, J. Briantais, N.R. Baker (1989). The relationship between the quantum yield of photosynthetic electron transport and quenching of chlorophyll fluorescence. *Biochim. Biophys. Acta*, **990**, 87–92.
93. I.A. Flameling, J. Kromkamp (1998). Light dependence of quantum yields for PSII charge separation and oxygen evolution in eukaryotic algae. *Limnol. Oceanogr.*, **43**, 284–297.
94. D. Hanelt (1998). Capability of dynamic photoinhibition in Arctic macroalgae is related to their depth distribution. *Mar. Biol.*, **131**, 361–369.
95. D. Hanelt, M.J. Jaramillo, W. Nultsch, S. Senger, R. Westermeir (1994). Photoinhibition as a regulative mechanism of photosynthesis in marine algae of Antarctica. *Ser. Cient. INACH*, **44**, 67–77.
96. D. Hanelt, J. Li, W. Nultsch (1994). Tidal dependence of photoinhibition of photosynthesis in marine macrophytes of the South China sea. *Bot. Acta*, **107**, 66–72.
97. C. Jiménez, F.L. Figueroa, S. Salles, J. Aguilera, J. Mercado, B. Viñegla, A. Flores-Moya, M. Lebert, D.-P. Häder (1998). Effects of solar radiation on photosynthesis and photoinhibition in red macrophytes from an intertidal system of southern Spain. *Bot. Mar.*, **41**, 329–338.
98. L.A. Franklin, G.G.R. Seaton, C.E Lovelock, A.W.D. Larkum (1996). Photoinhibition of photosynthesis on a coral reef. *Plant Cell Environm.*, **19**, 825–836.
99. D.-P. Häder, M. Lebert, E.W. Helbling (2000). Photosynthetic performance of the chlorophyte *Ulva rigida* measured in Patagonia on site. *Rec. Res. Develop. Photochem. Photobiol.*, **4**, 259–269.
100. D.-P. Häder, M. Lebert, E.W. Helbling (2001). Effects of solar radiation on the Patagonian macroalgae *Enteromorpha linza* (L.) J. Agardh – Chlorophyceae. *J. Photochem. Photobiol. B: Biol.*, **62**, 43–54.
101. C. Büchel, C. Wilhelm (1993). In vivo analysis of slow chlorophyll fluorescence induction kinetics in algae: progress, problems and perspectives. *Photochem. Photobiol.*, **58**, 137–148.
102. O. Schofield, B.M.A. Kroon, B.B. Prézelin (1995). Impact of ultraviolet-B radiation on photosystem II activity and its relationship to the inhibition of carbon fixation rates for Antarctic ice algae communities. *J. Phycol.*, **31**, 703–715.
103. C.A. Marwood, R.E.H. Smith, J.A. Furgal, M.N. Charlton, K.R. Solomon, B.M. Greenberg (2000). Photoinhibition of natural phytoplankton assemblages in Lake Erie exposed to solar ultraviolet radiation. *Can. J. Fish. Aquat. Sci.*, **57**, 371–379.
104. B. Bühlmann, P. Bossard, U. Uehlinger (1987). The influence of longwave ultraviolet radiation (u.v.-A) on the photosynthetic activity (14C-assimilation) of phytoplankton. *J. Plankton Res.*, **9**, 935–943.
105. J.J. Cullen, M.P. Lesser (1991). Inhibition of photosynthesis by ultraviolet radiation as a function of dose and dosage rate: results for a marine diatom. *Mar. Biol.*, **111**, 183–190.

106. J.J. Cullen, P.J. Neale, M.P. Lesser (1992). Biological weighting function for the inhibition of phytoplankton photosynthesis by ultraviolet radiation. *Science*, **258**, 646–650.
107. E.W. Helbling, B.E. Chalker, W.C. Dunlap, O. Holm-Hansen, V.E. Villafañe (1996). Photoacclimation of antarctic marine diatoms to solar ultraviolet radiation. *J. Exp. Mar. Biol. Ecol.*, **204**, 85–101.
108. M.J. Behrenfeld, J. Hardy, H. Gucinski, A. Hanneman, H. Lee, A. Wones (1993). Effects of ultraviolet-B radiation on primary production along latitudinal transects in the South Pacific Ocean. *Mar. Environm. Res.*, **35**, 349–363.
109. P.J. Neale, M.P. Lesser, J.J. Cullen (1994). Effects of ultraviolet radiation on the photosynthesis of phytoplankton in the vicinity of McMurdo station, Antarctica. In: C.S. Weiler, P.A. Penhale (Eds), *Ultraviolet Radiation in Antarctica: Measurements and Biological Effects* (pp. 125–142). American Geophysical Union, Washington DC.
110. V.E. Villafañe, E.W. Helbling, O. Holm-Hansen, B.E. Chalker (1995). Acclimatization of Antarctic natural phytoplankton assemblages when exposed to solar ultraviolet radiation. *J. Plankton Res.*, **17**, 2295–2306.
111. V.E. Villafañe, M. Andrade, V. Lairana, F. Zaratti, E.W. Helbling (1999). Inhibition of phytoplankton photosynthesis by solar ultraviolet radiation: Studies in Lake Titicaca, Bolivia. *Freshwater Biol.*, **42**, 215–224.
112. I. Laurion, W.F. Vincent (1998). Cell size versus taxonomic composition as determinants of UV-sensitivity in natural phytoplankton communities. *Limnol. Oceanogr.*, **43**, 1774–1779.
113. A.T. Banaszack, P.J. Neale (2001). Ultraviolet radiation sensitivity of photosynthesis in phytoplankton from an estuarine environment. *Limnol. Oceanogr.*, **46**, 592–603.
114. R.C. Worrest (1982). Review of literature concerning the impact of UV-B radiation upon marine organisms. In: J. Calkins (Ed.), *The Role of Solar Ultraviolet Radiation in Marine Ecosystems* (pp. 429–457). Plenum Press, New York.
115. M.P. Lesser, P.J. Neale, J.J. Cullen (1996). Acclimation of Antarctic phytoplankton to ultraviolet radiation: Ultraviolet-absorbing compounds and carbon fixation. *Mol. Mar. Biol. Biotechnol.*, **5**, 314–325.
116. S.-Å. Wängberg, J.-S. Selmer, K. Gustavson (1996). Effects of UV-B radiation on biomass and composition in marine phytoplankton communities. *Sci. Mar.*, **60** (Suppl. 1), 81–88.
117. C. Hazzard, M.P. Lesser, R.A. Kinzie III (1997). Effects of ultraviolet radiation on photosynthesis in the subtropical marine diatom, *Chaetoceros gracilis* (Bacillariophyceae). *J. Phycol.*, **33**, 960–968.
118. O. Holm-Hansen (1997). Short- and long-term effects of UVA and UVB on marine phytoplankton productivity. *Photochem. Photobiol.*, **65**, 266–268.
119. H. Peletier, W.W.C. Gieskes, A.G.J. Buma (1996). Ultraviolet-B radiation resistance of benthic diatoms isolated from tidal flats in the Dutch Wadden Sea. *Mar. Ecol. Prog. Ser*, **135**, 163–168.
120. A.E. McNamara, W.R Hill (2000). UV-B irradiance gradient affects photosynthesis and pigments but not food quality of periphyton. *Freshwater Biol.*, **43**, 649–662.
121. R.D. Vinebrook, P.R. Leavitt (1996). Effects of ultraviolet radiation on periphyton in an alpine lake. *Limnol. Oceanogr.*, **41**, 1035–1040.
122. G. Pergent, J.M. Romero, C. Pergent-Martini, M.A. Mateo, C.F. Bouderesque (1994). Primary production, stocks and fluxes in the Mediterranean seagrass *Posidonia oceanica*. *Mar. Ecol. Progr. Ser.*, **106**, 139–146.
123. K. Lüning (1990). *Seaweeds: Their Environment, Biogeography, and Ecophysiology.*

Wiley, New York.
124. W. Schramm, P.H. Nienhuis (1996). Marine benthic vegetation. Recent changes and effect of eutrophication. *Ecological Studies* (Vol. 123). Springer-Verlag, Berlin.
125. U. Karsten, L.A. Franklin, K. Lüning, C. Wiencke (1998). Natural ultraviolet and photosynthetic active radiation induce formation of mycosporine-like amino acids in the marine macroalga *Chondrus crispus* (Rhodophyta). *Planta*, **205**, 257–262.
126. I. Gómez, F.L. Figueroa, I. Sousa-Pinto, B. Viñegla, E. Pérez-Rodríguez, C. Maestre, S. Coelho, A. Felga, R. Pereira (2001). Effects of UV radiation and temperature on photosynthesis as measured by PAM fluorescence in the red alga *Gelidium pulchellum* (Turner) Kützing. *Bot. Mar.*, **44**, 9–16.
127. W.F. Vincent, S. Roy (1993). Solar ultraviolet-B radiation and aquatic primary production: damage, protection, and recovery. *Environ. Rev.*, **1**, 1–12.
128. M. Vernet, R.C. Smith (1997). Effects of ultraviolet radiation on the pelagic Antarctic ecosystem. In: D.-P. Häder (Ed.), *The effects of ozone depletion on aquatic ecosystems* (pp.247–265). R.G. Landes Company, Academic Press, Austin.
129. S.-Å. Wängberg, J.-S. Selmer (1997). Studies of effects of UV-B radiation on aquatic model ecosystems. In: D.-P. Häder (Ed.), *The effects of ozone depletion on aquatic ecosystems* (pp. 199–213). R.G. Landes Company, Academic Press, Austin.
130. M. Vernet (2000). Effects of UV radiation on the physiology and ecology of marine phytoplankton. In: S. de Mora, S. Demers, M. Vernet (Eds), *The effects of UV radiation in the marine environment* (pp. 237–278). Cambridge University Press, Cambridge.
131. C.B. Osmond (1994). What is photoinhibition? Some insights from comparisons of shade and sun plants. In: N.R Baker, J.R Bowyer (Eds), *Photoinhibition of Photosynthesis, from Molecular Mechanisms to the Field* (pp. 1–24). Bios Scientific Publ., Oxford.
132. P.J. Neale, P.J. Richerson (1987). Photoinhibition and the diurnal variation of phytoplankton photosynthesis – I. Development of a photosynthesis-irradiance model from studies of in situ responses. *J. Plankton Res.*, **9**, 167–193.
133. E.W. Helbling, V.E. Villafañe, O. Holm-Hansen (1995). Variability of phytoplankton distribution and primary production around Elephant Island, Antarctica, during 1990–1993. *Polar Biol.*, **15**, 233–246.
134. V.E. Villafañe, E.W. Helbling, H.E. Zagarese (2001). Solar ultraviolet radiation and its impact on aquatic ecosystems of Patagonia, South America. *Ambio*, **30**, 112–117.
135. E.S. Barbieri, V.E. Villafañe, E.W. Helbling (2002). Experimental assessment of UVR effects upon marine phytoplankton photosynthesis under variable radiation regimes. *Limnol. Oceanogr.*, **47**, 1648–1655.
136. B.B. Prézelin, J. Kerfoot, J. Gorga, T. McGovern, J. Robiart (2000). In situ spectral response of phytoplankton communities to regional and seasonal variability in solar UV radiation off coastal California: SUPACC 1998. In: *Associated Workshop to the SPARC* 2000 *Impacts of UV Radiation on Aquatic and Terrestrial Ecosystems.* Mar del Plata, Argentina, abstract, p. 29.
137. E.W. Helbling, V.E. Villafañe (2001). UVR effects on phytoplankton primary production: A comparison between Arctic and Antarctic marine ecosystems. In D.O Hessen (Ed.), *UV-Radiation and Arctic Ecosystems* (pp. 203–226). Springer-Verlag.
138. M. Blumthaler, W. Rehwald (1992). Solar UV-A and UV-B fluxes at two alpine stations at different altitudes. *Theor. Appl. Climatol.*, **46**, 39–44.
139. P.J. Neale, E. Litchman, C. Sobrino, C. Callieri, G. Morabito, V. Montecino, Y. Huot, P. Bossard, C. Lehmann, D. Steiner (2001). Quantifying the response of phytoplankton photosynthesis to ultraviolet radiation: Biological weighting func-

tions versus in situ measurements in two Swiss lakes. *Aquat. Sci*, **63**, 265–285.
140. S. Roy (2000). Strategies for the minimisation of UV-induced damage. In: S. de Mora, S. Demers, M. Vernet (Eds), *The effects of UV radiation in the marine environment* pp. 177–205. Cambridge University Press, Cambridge.
141. L. Zudaire, S. Roy (2001). Photoprotection and long-term acclimation to UV radiation in the marine diatom *Thalassiosira weissflogii*. *J. Photochem. Photobiol. B. Biol.*, **62**, 26–34.
142. S. Halac, M. Felip, L. Camarero, S. Sommaruga-Wögrath, R. Psenner, J. Catalan, R. Sommaruga (1997). An in situ enclosure experiment to test the solar UVB impact on plankton in a high-altitude mountain lake. I. Lack of effect on phytoplankton species composition and growth. *J. Plankton Res.*, **19**, 1671–1686.
143. I. Laurion, D. R. S. Lean, W. F. Vincent (1998). UVB effects on a plankton community: Results from a large-scale enclosure assay. *Aquat. Microb. Ecol.*, **16**, 189–198.
144. S. Cabrera, M. López, B. Tartarotti (1997). Phytoplankton and zooplankton response to ultraviolet radiation in a high-altitude Andean lake: short- versus long-term effects. *J. Plankton Res.*, **19**, 1565–1582.
145. M.P. Lesser (1996). Acclimation of phytoplankton to UV-B radiation: oxidative stress and photoinhibition of photosynthesis are not prevented by UV-absorbing compounds in the dinoflagellate *Prorocentrum micans*. *Mar. Ecol. Progr. Ser.*, **132**, 287–297.
146. M.P. Lesser, J.J. Cullen, P.J. Neale (1994). Carbon uptake in a marine diatom during acute exposure to ultraviolet B radiation: Relative importance of damage and repair. *J. Phycol.*, **30**, 183–192.
147. M.J. Behrenfeld, H. Lee II, L.F. Small (1994). Interactions between nutritional status and long-term responses to ultraviolet-B radiation stress in a marine diatom. *Mar. Biol.*, **118**, 523–530.
148. M. Bergeron, W.F. Vincent (1997). Microbial food web responses to phosphorus supply and solar UV radiation in a subarctic lake. *Aquat. Microb. Ecol.*, **12**, 239–249.
149. H.W. Paerl, J.L. Pinckney (1996). A mini-review of microbial consortia: Their roles in aquatic production and biogeochemical cycling. *Microbiol. Ecol.*, **31**, 225–247.
150. W. Hill, S.M. Dimick, A.E. McManara, C.H. Branson (1997). No effects of ambient UV radiation detected in periphyton and grazers. *Limnol. Oceanogr.*, **42**, 769–774.
151. R. Santas, A. Korda, C. Lianou, P. Santas (1998). Community responses to UV radiation. I. Enhanced UV-B effects on biomass and community structure of filamentous algal assemblages growing in a coral reef mesocosm. *Mar. Biol.*, **131**, 153–162.
152. R. Santas, P. Santas, C. Lianou, A. Korda (1998). Community responses to UV radiation. II. Effects of solar UV-B on field-grown diatom assemblages of the Carribean. *Mar. Biol.*, **131**, 163–171.
153. M.L. Bothwell, D.M.J. Sherbot, C.M. Pollock (1994). Ecosystem response to solar ultraviolet-B radiation: Influence of trophic-level interactions. *Science*, **265**, 97–100.
154. R.D. Vinebrook, P.R. Leavitt (1998). Direct and interactive effects of allochthonous dissolved organic matter, inorganic nutrients, and ultraviolet radiation on an alpine littoral food web. *Limnol. Oceanogr.*, **43**, 1065–1081.
155. D.M. DeNicola, K.D. Hoagland (1996). Effects of solar spectral irradiance (visible to UV) on a prairie stream epilithic community. *J. North Am. Benth. Soc.*, **15**, 155–169.
156. D.W, Schindler, P.J. Curtis, B.R. Parker, M.P. Stainton (1996). Consequences of climate warming and lake acidification for UV-B penetration in North American boreal lakes. *Nature*, **379**, 705–708.

157. A.W.D. Larkum, W.F. Wood (1993). The effect of UV-B radiation on photosynthesis and respiration of phytoplankton, benthic macroalgae and seagrasses. *Photosynth. Res.*, **36**, 17–23.
158. W.F. Wood (1989). Photoadaptative responses of the tropical red alga *Euchema striatum* Schmitz (Gigartinales) to ultraviolet radiation. *Aquat. Bot.*, **33**, 41–51.
159. S.K. Clendennen, R.C. Zimmerman, D.A. Powers, R.S. Alberte (1996). Photosynthetic responses of the giant kelp *Macrocystis pyrifera* (Phaeophyceae) to ultraviolet radiation. *J. Phycol.*, **32**, 614–620.
160. E.A. Drew (1974). Light inhibition of photosynthesis in macroalgae. *Br. Phycol. J.*, **9**, 217–218.
161. G. Döhler, I. Biermann (1994). Impact of UV-B radiation on the lipid and fatty acid composition of synchronized *Ditylum brightwelli* (West) Grunow. *Zs. Naturforsch. C Biosci.*, **49**, 607–614.
162. J.I. Goes, N. Handa, S. Taguchi, T. Hama (1994). Effect of UV-B radiation on the fatty-acid composition of the marine-phytoplankton *Tetraselmis* sp – relationship to cellular pigments. *Mar. Ecol. Prog. Ser.*, **114**, 259–274.
163. J.I. Goes, N. Handa, K. Suzuki, S. Taguchi, T. Hama (1997). Photosynthetic production of organic compounds in Antarctic marine phytoplankton in presence of UV-B radiation: Implications for the Antarctic marine food web. *Proc. Natl. Inst. Polar Res. Symp. Polar Biol.* (Tokyo), 18–23.
164. N.-H. Norsker, G. Støttrup (1994). The importance of dietary HUFAs for fecundity and HUFA content in the harpacticoid, *Tisbe holothuriae* Humes. *Aquaculture*, **125**, 155–166.
165. K.S. Wang, T.J. Chai (1994). Reduction in omega-3 fatty acids by UV-B irradiation in microalgae. *J. Appl. Phycol.*, **6**, 415–421.
166. M.T. Arts, H. Rai (1997). Effects of enhanced ultraviolet-B radiation on the production of lipid, polysaccharide, and protein in three freshwater algal species. *Freshwater. Biol.*, **38**, 597–610.
167. J.H. Skerratt, A.D. Davidson, P.D. Nichols, T.A. McMinn (1998). Effect of UV-B on lipid content of three Antarctic marine phytoplankton. *Phytochemistry*, **49**, 999–1007.
168. A.G.J. Buma, H.J. Zemmelink, K. Sjollema, W.W.C. Gieskes (1996). UV-B radiation modifies protein and photosynthetic pigment content, volume and ultrastucture of marine diatoms. *Mar. Ecol. Prog. Ser.*, **142**, 47–54.
169. A. Veen, M. Reuvers, P. Roncak (1997). Effects of acute and chronic UV-B exposure on a green alga: A continuous culture study using a computer-controlled dynamic light regime. In: J. Rozema, W.W.C. Gieskes, S.C. van de Geijn, C. Nolan, H. de Boois (Eds), *UV-B and Biosphere* (pp. 28–40). Kluwer Academic Publishers, Belgium.
170. G. Döhler (1997). Effect of UVB radiation on utilization of inorganic nitrogen by Antarctic microalgae. *Photochem. Photobiol.*, **66**, 831–836.
171. G. Döhler (1997). Assimilation of inorganic nitrogen by Antarctic and temperate marine phytoplankton species under UV stress. *Russ. J. Plant Physiol.*, **44**, 518–525.
172. J.I. Goes, N. Handa, S. Taguchi, T. Hama (1995). Changes in the patterns of biosynthesis and composition of amino acids in a marine phytoplankter exposed to ultraviolet-B radiation: Nitrogen limitation implicated. *Photochem. Photobiol.*, **62**, 703–710.
173. J.H. Vosjan, G. Döhler, G. Nieuwland (1990). Effects of UV-B irradiance on the ATP content of microorganisms of the Weddell Sea Antarctica. *Neth. J. Sea Res.*, **25**, 391–394.

174. E. Marañon, E. Fernández, R. Anadón (1995). Patterns of macromolecular synthesis by natural phytoplankton assemblages under changing upwelling regimes: In situ observations and microcosm experiments. *J. Exp. Mar. Biol. Ecol.*, **188**, 1–28.
175. M.M. Tilzer (1973). Diurnal periodicity in the phytoplankton assemblage of a high mountain lake. *Limnol. Oceanogr.*, **18**, 15–30.
176. D.O. Hessen, E. Van Donk, T. Andersen (1995). Growth responses, P-uptake and loss of flagella in *Chlamydomonas reinhardtii* exposed to UV-B. *J. Plankton Res.*, **17**, 17–27.
177. D. Karentz, F.S. McEuen, M.C. Land, W.C. Dunlap (1991). Survey of mycosporine-like amino acid compounds in Antarctic marine organisms: Potential protection from ultraviolet exposure. *Mar. Biol.*, **108**, 157–166.
178. W.C. Dunlap, J.M. Shick (1998). Ultraviolet radiation-absorbing mycosporine-like amino acids in coral reef organisms: a biochemical and environmental perspective. *J. Phycol.*, **34**, 418–430.
179. F. Garcia-Pichel, C.E. Wingard, R.W. Castenholz (1993). Evidence regarding the UV-sunscreen role of a mycosporine-like compound in the cyanobacterium *Gloeocapsa* sp. *Appl. Environ. Microb.*, **59**, 170–176.
180. P.J. Neale, A.T. Banaszak, C.R. Jarriel (1998). Ultraviolet sunscreens in *Gymnodinium sanguineum* (Dinophyceae): Mycosporine-like amino acids protect against inhibition of photosynthesis. *J. Phycol.*, **34**, 928–938.
181. S.W. Jeffrey, H.S. MacTavish, W.C. Dunlap, M. Vesk, K. Groenewoud (1999). Occurrence of UVA- and UVB-absorbing compounds in 152 species (206 strains) of marine microalgae. *Mar. Ecol. Prog. Ser.*, **189**, 35–51.
182. F. Garcia-Pichel, N.D. Sherry, R.W. Castenholz (1992). Evidence for an ultraviolet sunscreen role of the extracellular pigment scytonemin in the terrestrial cyanobacterium *Chlorogloeopsis* sp. *Photochem. Photobiol.*, **56**, 17–23.
183. Y. Tsujino, K. Yabe, Y. Sekikawa (1980). Isolation and structure of a new amino acid, shinorine, from the red alga *Chondrus yendoi* Yamada et Mikami. *Bot. Mar.*, **23**, 65–67.
184. U. Karsten, T. Sawall, D. Hanelt, K. Bishop, A. Flores-Moya, F.L. Figueroa, C. Wiencke (1998). Contents of UV absorbing mycosporine-like aminoacids in macroalgae from polar to warm-temperate regions. *Bot. Mar.*, **41**, 443–453.
185. U. Karsten, C. Wiencke (1999). Factors controlling the formation of UV-absorbing mycosporine-like amino acids in the marine red alga *Palmaria palmata* from Spitsbergen (Norway). *J. Plant Physiol.*, **155**, 407–415.
186. K. Hoyer, U. Karsten, T. Sawall, C. Wiencke (2001). Photoprotective substances in Antarctic macroalgae and their variation with respect to depth distribution, different tissues and development stages. *Mar. Ecol. Progr. Ser.*, **211**, 117–129.
187. L.A. Franklin, I. Yakovleva, U. Karsten, K. Lüning (1999). Synthesis of mycosporine-like amino acids in *Chondrus crispus* (Florideophyceae) and the consequences for sensitivity to ultraviolet B radiation. *J. Phycol.*, **35**, 682–693.
188. L.A. Franklin, G. Kräbs, R. Kuhlenkamp (2001). Blue light and UV-A radiation control the synthesis of mycosporine-like amino acids in *Chondrus crispus* (Florideophyceae). *J. Phycol.*, **37**, 257–270.
189. X. Molina, V. Montecino (1996). Acclimation to UV irradiance in *Gracilaria chilensis* Bird, MacLachlan et Oliveira (Gigartinales, Rhodophyta). *Hydrobiologia*, **326/327**, 415–420.
190. H. Pavia, G. Cervin, A. Linfgren, P. Aberg (1997). Effects of UV-B radiation and simulated herbivory on phlorotanins in the brown algae *Ascophyllum nodosum*. *Mar. Ecol. Progr. Ser.*, **157**, 139–146.

191. E. Pérez-Rodríguez, I. Gómez, U. Karsten, F.L. Figueroa (1998). Effects of UV radiation on photosynthesis and excretion of UV-absorbing compounds of *Dasycldus vermicularis* (Dasycladales, Chlorophyta) from southern Spain. *Phycologia*, **37**, 379–387.
192. I. Gómez, E. Pérez-Rodríguez, B. Viñegla, F.L. Figueroa, U. Karsten (1998). Effects of solar radiation on photosynthesis, UV-absorbing compounds and enzyme activities of the green alga *Dasycladus vermicularis* from southern Spain. *J. Photochem. Photobiol. B: Biol.*, **47**, 46–57.
193. D.L. Mitchell, D. Karentz (1993). The induction and repair of DNA photodamage in the environment. In: A.R. Young, L.O. Björn, J. Moan, W. Nultsch, (Eds), *Environmental UV Photobiology* (pp. 345–377). Plenum Press, New York.
194. A.G.J. Buma, A.H. Engelen, W.W.C. Gieskes (1997). Wavelength dependent induction of thymine dimers and growth rate reduction in the marine diatom *Cyclotella* sp. exposed to ultraviolet radiation. *Mar. Ecol. Progr. Ser.*, **153**, 91–97.
195. A.G.J. Buma, E.W. Helbling, M.K. de Boer, V.E. Villafañe (2001). DNA damage patterns in temperate South-Atlantic picophytoplankton assemblages exposed to solar ultraviolet radiation. In: E.W. Helbling, C.L. Ballaré and V.E. Villafañe (eds), *Impacts of Ultraviolet Radiation on Aquatic and Terrestrial Ecosystems*, Special Issue of *J. Photochem. Photobiol. B: Biol.*, **62**, 9–18.
196. W.H. van de Poll, A. Eggert, A.G.J. Buma, A.M. Breeman (2001). Effects of UV-B induced DNA damage and photoinhibition on growth of temperate marine red macrophytes: Habitat related differences in ultraviolet-B tolerance. *J. Phycol.*, **37**, 30–37.
197. M.J. Behrenfeld, J.T. Hardy, H. Lee III (1992). Chronic effects of ultraviolet-B radiation on growth and cell volume of *Phaeodactylum tricornutum* (Bacillariophyceae), *J. Phycol.*, **28**, 757–760.
198. V.E Villafañe, J.I. Albarracín, E.W. Helbling (2000). Long-term effects of UVR on growth rates and synthesis of UV-absorbing compounds in marine phytoplanktonic species, Associated Workshop to the SPARC 2000 *Impacts of UV Radiation on Aquatic and Terrestrial Ecosystems* (pp. 32–33). Mar del Plata, Argentina, abstract.
199. L. Mousseau, M. Gosselin, M. Levasseur, S. Demers, J. Fauchot, S. Roy, P. Z. Villegas, B. Mostajir (2000). Effects of ultraviolet-B radiation on simultaneous carbon and nitrogen transport rates by estuarine phytoplankton during a week-long mesocosm study. *Mar. Ecol. Progr. Ser.*, **199**, 69–81.
200. D.O. Hessen, H.J. De Lange, E. Van Donk (1997). UV-induced changes in phytoplankton cells and its effects on grazers. *Freshwat. Biol.*, **38**, 513–524.
201. C.S. Lobban, P.J. Harrison (1997). *Seaweed Ecology and Physiology*. Cambridge University Press, Cambridge.
202. G. Döhler (1987). Effect of irradiation on nitrogen metabolism in marine diatoms and phytoplankton. *Oceanis*, **13**, 487–493.
203. G. Döhler (1992). Impact of UV-B radiation on uptake of 15N-ammonia and 15N-nitrate by phytoplankton of the Wadden Sea. *Mar. Biol.*, **112**, 485–489.
204. G. Döhler, T. Buchmann (1995). Effects of UV-A and UV-B irradiance on pigments and 15N-ammonium assimilation of the haptophycean *Pavlova*. *J. Plant Physiol.*, **146**, 29–34.
205. G. Döhler (1995). Effects of UV-A and UV-B radiation on pigments and assimilation of 15N-ammonium and 15N-nitrate by natural phytoplankton of the Wadden Sea. *Beitr. Biol. Pflan.*, **69**, 1–16.
206. G. Döhler, E. Hagmeier (1997). UV effects on pigments and assimilation of 15N-ammonium and 15N- nitrate by natural marine phytoplankton of the North Sea.

Bot. Acta, **110**, 481–488.
207. G. Döhler, E. Hagmeier, C. David (1995). Effects of solar and artificial UV irradiation on pigment and assimilation of 15N ammonium and 15N nitrate by macroalgae. *J. Photochem. Photobiol. B: Biol.*, **30**, 179–187.
208. B. Viñegla, F.L. Figueroa, C. Maestre (2000). Variations in the total carbonic anhydrase and in situ nitrate reductase activities under solar radiation on a daily basis: consequences on C:N interactions. In: F.M. Cánovas, F.J. Florencio (Eds), *Avances del metabolismo del nitrógeno: Bioquímica, Fisiología y Biología Molecular* (pp.319–326). Servicio de Publicaciones de la Universidad de Málaga.
209. B. Viñegla, C. Maestre, F.L. Figueroa (2000). Response to UV radiation level on enzymatic activities (total carbonic anhydrase and in situ nitrate reductase) in marine macroalgae and seagrasses. In: F.M. Cánovas, F.J. Florencio (Eds), *Avances del metabolismo del nitrógeno: Bioquímica, Fisiología y Biología Molecular* (pp. 341–342). Servicio de Publicaciones de la Universidad de Málaga.
210. F.L. Figueroa, B. Viñegla (2001). Effects of solar UV radiation on photosynthesis and enzyme activities (carbonic anhydrase and nitrate reductase) in marine macroalgae from southern Spain. *Rev. Chil. Hist. Nat.*, **74**, 237–249.
211. D.H. Turpin (1991). Effect of inorganic N availability on algal photosynthesis and carbon metabolism. *J. Phycol.*, **27**, 14–20.
212. B. Viñegla (2000). *Efecto de la Radiación Ultravioleta Sobre Actividades Enzimáticas Relacionadas con el Metabolismo Del carbono y Nitrógeno en Macroalgas y Fanerógamas marinas.* (Ph.D. Thesis, pp. 404). Universidad de Málaga, Spain.
213. K. Bischof, D. Hanelt, C. Wiencke (2000). Effects of ultraviolet radiation on photosynthesis and related enzyme reactions of marine macroalgae. *Planta*, **211**, 555–562.
214. E.W. Helbling, S. Avaria, J. Letelier, V. Montecino, B. Ramirez, M. Ramos, W. Rojas, V.E. Villafañe (1993). Respuesta del fitoplancton marino a la radiación ultravioleta en latitudes medias (33° S), *Rev. Biol. Mar.*, **28**, 219–237.
215. S. Gerber, D.-P. Häder (1995). Effects of enhanced solar irradiation on chlorophyll fluorescence and photosynthetic oxygen production of five species of phytoplankton. *FEMS Microbiol. Ecol.*, **16**, 33–42.
216. H. Maske, M. Latasa (1997). Solar ultraviolet radiation dependent decrease of particle light absorption and pigments in lake phytoplankton. *Can. J. Fish. Aquat. Sci.*, **54**, 697–704.
217. A. Quesada, J.-L. Mouget, W.F. Vincent (1995). Growth of Antarctic cyanobacteria under ultraviolet radiation: UVA counteracts UV-B inhibition. *J. Phycol.*, **31**, 242–248.
218. R.P. Sinha, M. Lebert, A. Kumar, H.D. Kumar, D.-P. Häder (1995). Spectroscopic and biochemical analyses of UV effects on phycobiliproteins of *Anabaena* sp. and *Nostoc carmium*. *Bot. Acta*, **108**, 87–92.
219. P.L. Jokiel, R.H. York (1984). Importance of ultraviolet radiation in photoinhibition of microalgal growth. *Limnol. Oceanogr.*, **29**, 192–199.
220. J. Aguilera, C. Jiménez, F.L. Figueroa, M. Lebert, D.-P. Häder (1999). Effect of ultraviolet radiation on thallus absorption and photosynthetic pigments in the red alga *Porphyra umbilicalis*. *J. Photochem. Photobiol. B: Biol.*, **48**, 75–82.
221. M. Altamirano, A. Flores-Moya, F.L. Figueroa (1999). Long-term effect of natural sunlight under various ultraviolet radiation conditions on growth and photosynthesis of intertidal *Ulva rigida* (Chlorophyceae) cultivated in situ, *Bot. Mar.*, **43**, 119–126.
222. F. Garcia-Pichel (1994). A model for internal self-shading in planktonic organisms and its implications for the usefulness of ultraviolet sunscreens. *Limnol. Oceanogr.*,

39, 1704–1717.

223. P.S. Huovinen, A.O.J. Oikari, M.R. Soimasuo, G.N. Cherr (2000). Impact of UV radiation on the early development of giant kelp (*Macrocystis pyrifera*) gamethophytes. *Photochem. Photobiol.*, **72**, 308–313.
224. D.C. Reed (1990). The effects of variable settlement and early competition on patterns of kelps recruitments. *Ecology*, **71**, 776–787.
225. J.R. Kelly (1986). How might enhanced levels of solar UVB radiation affect marine ecosystems?. In: J.G. Titus (Ed.), *Effects of Changes in Stratospheric Ozone and Global Climate Change*, United Nations Environment Programme and US Environmental Protection Agency.

Chapter 12

UVR and pelagic metazoans

Dag O. Hessen

Table of contents

Abstract

Pelagic metazoans cover organisms from rotifers to whales that inhabit a variety of localities from alpine pond to oceans. Knowledge on specific responses on UVR is restricted to very few taxa or species, yet the fact that all organisms share a basic set of biochemical and cellular properties and have a limited range of protective mechanisms and responses, allows for some generalizations across taxa. The main discussion in this chapter is devoted to the best examined groups: zooplankton and fish. From the early development of life on this planet without oxygen (and ozone), UVR has probably been a key player in the early evolution of aquatic life. As an agent of mutations, UVR may operate directly at the gene level, but also by shaping phenotypic traits and behavior by adaptation. The fact that a range of energetically costly protective mechanisms has been evolved is a strong argument for the ecological and evolutionary relevance of UVR. This is further supported by in situ studies that confirm a variety of direct and indirect UVR effects under natural conditions. Such studies do, however, also confirm a wide range of responses (or lack of such) between and within organisms, suggesting that a key effect of changes UVR climate could be community shifts. For some organisms that appear rather insensitive to UVR, increased UVR could in fact be beneficial, yet the multitude of effects on food webs within ecosystems and the entangled bank of life are hard to predict.

Based on the present literature is may be concluded that UVR may play a key role for pelagic metazoans, and that it often plays in concert with other ambient factors that may be mutually strengthening.

12.1 Introduction

What do pelagic metazoans have in common, except for sharing an aquatic habitat? Not too much, apparently. Their habitats range from the tropics to polar regions, from deep sea to alpine ponds, and from highly transparent oceans and oligotrophic lakes to brownish, humic lakes or eutrophic localities with very high UV attenuation. They cover numerous species and taxa over several phyla from jellyfish to fish and mammalians, from the minute rotifers or nauplii to the largest creatures on the planet. In addition, there is also a wide variety of benthic or sessile animals with larval stages that for brief periods belong to the pelagic community. Within the pelagic community there is a vast number of spatial niches, life cycle strategies, morphology and color. There are also generation times ranging from few days to many years, implying widely different growth rates, longevity and metabolic turnover. This multitude of evolutionary strategies has several bearings on the role of UVR, and could lead to the conclusion that the search for general phenomena in marine and freshwater ecosystems, as well as among the highly diverse pelagic community is overly optimistic. Hopefully this chapter will provide evidence that this is not the case, that UV is a common challenge to all pelagic animals and that this challenge can be met by a finite number of strategies.

First of all; the role of salinity per se is probably minor, except that UVR mediated membrane damage could cause an osmotic distortion that would be most pronounced at low salinity. Also ecosystem attributes such as longitude or latitude, depth, size or transparency of habitats are all a matter of gradients. These are important gradients, but does not call for distinct categories of aquatic habitats. Thus when treating major taxa, there is made no straight division between ecosystems, although the special challenges in some habitats have been stressed. There are some pronounced differences between major taxa, however, that call for special attention. Although basic cellular processes, and thus the UVR mediated damage, are uniform and most organisms share a set of cellular means of UV-protection, there are still some inherent differences attributed to level of organism organization and life cycle strategies.

It is also important to recall that there may be pronounced variability in UV-responses also within taxa or even species. The individuals of many species undergo ontogenetic shifts that may create bottleneck periods with regard to UV-susceptibility (e.g., hatching, larval stages, periods of rapid growth, moulting) that yet are poorly understood. There may also be strong variations in abiotic or biotic factors such as temperature, oxygen, ionic composition or nutritional status, burden of parasites or predators. In fact these intra-species properties may be more prominent than inter-species differences. In general, small organisms could be more susceptible than larger ones since all body components, including reproductive organs, can be reached by UVR. Also, rapid cell division or growth, implying a very active gene expression, could be a major determinant of susceptibility [1]. Thus is would be expected that, within species, eggs and embryos would be the most vulnerable stages. On the other hand organisms with a short life span and short generation time may be less vulnerable to accumulated damage with age, hence for such organisms it would pay off to invest in protective means for eggs and offspring on the expense of somatic tissue.

There is a scattered, and in general limited, knowledge on the effect of UVR for most aquatic organisms, and for a number of major categories there is no empirical support at all for judgement of UV-effects. Hence a systematic screening of all groups will be impossible, and this chapter will rather focus on general aspects of UVR in major aggregated and functionally related groups and draw on existing knowledge within these groups to generalize predictions. Although care should be taken when extrapolating effects from one taxon to another, some general knowledge may nevertheless be derived since the biological effects of UV at the cellular level are fairly similar across taxa. This means that some of these general effects need not to be reiterated for all systematic groups. The major questions are how the animals are able to cope with this stress, i.e., what kind of protective means and life cycle strategies are evolved – and at which costs? What are the ecological and evolutionary consequences of UVR? How could this affect communities and food webs?

This book covers an extensive list of UV-mediated damage to a wide range of aquatic organisms. Some chapters are devoted to specific types of protection or damage, with emphasis on particular taxa. Thus although reiterating some of these aspects, the main part of this chapter is devoted to truly planktonic metazoans and aspects of their ways of coping with UVR that is not extensively covered in other chapters.

The first and most important question is whether there are reasons to believe that UVR indeed is an ecologically or evolutionary relevant factor for pelagic animals? If so, does this mean that UVR poses a significant stress on these organisms? To really be concerned with present or future UVR, these basic questions demand a *yes*. For the sake of argument, this chapter will first address the variety of defense mechanisms that are evolved to cope with UVR. The very presence of such mechanisms in pelagic animals is in itself a strong argument for UVR as a major player in an ecological and evolutionary context. The various kinds of UV-mediated cellular damage will be the focus of other chapters; hence the scope here will be the ecological and evolutionary consequences. Later in this chapter we will proceed to case studies that evaluate the role of UVR for particular organisms. One particular challenge is that while UVR may pose both stress and constraints on organism performance, these sub-lethal effects may be hard to trace in nature. Pelagic organisms are evolved under some UV-stress and hence should be able to tolerate this, but at some costs that may be quite subtle. Nevertheless, there are several observations of in situ effects of UVR, and these are of course particularly important for the final judgement of the ecological role of UVR.

12.2 Direct damage and means of protection (lines of defense)

The general, direct effects of UVR at the cellular level are rather uniform within the animal kingdom. These include first of all DNA-damage, membrane damage and a range of other cellular injuries that may be caused by intracellular photoproducts. They also include immunosuppression, yet the responses here may be more different across phyla, especially between invertebrates and vertebrates. Finally skin lesion, cancers and eye-damage (cataract) may be common responses in vertebrates. These effects may sum up in reduced fitness of various kinds, ranging from death to a slight decrease in life span, growth or rate of reproduction. These various types of damage will not be reiterated here. They may, however, be used as examples of reported responses when direct effects on different organisms are treated more specifically towards the end of this chapter.

The fact that animals in most regards share the same nucleic acids and other macromolecules, the same set of enzymes and proteins in general and in fact much of the same cellular machinery also generates fairly general action spectra. Although action spectra with sufficient resolution have been provided for a very limited number of aquatic organisms [2–4], they do all have the common property of rapidly increasing effects towards the lower end of the spectrum, and very limited effects in the PAR-area. In fact for DNA damage the relative biological response to wavelengths beyond 310 nm is negligible [5]. Of specific interest for aquatic organisms are thus the optical properties of the water column per se related to the depth distribution of the animals. If there is a strong attenuation in the short-wave area (as would be the case for localities rich in

DOC), this will be the main determinant of the biologically effective dose, i.e., the sum of the action spectrum and spectral irradiance. Especially for freshwaters, but also to some extent for coastal areas, DOC is by far the most important determinant of UVR attenuation [6,7] and changes in DOC could be far more important for pelagic animals than predicted changes in stratospheric ozone.

Animals may cope with the supposed challenges of UVR in different ways, and, with regard to the direct UVR effects, several lines of defense may be identified that play a different role in different organisms. There may either be seasonal or spatial means of UVR avoidance by timing of reproductive season, depth distribution or diurnal migration. A second line of defense is UV screening compounds (e.g. pigments, mycosporine-like amino acids). Third, there are several enzymes and other macromolecules serving as radical scavengers that in various ways cope with ambient or intracellular harmful photo-products. A last major defense would be the various means of enzymatic photo-repair that is a common property of all organisms.

12.2.1 Seasonal and spatial responses

Commonly, such strategies will be a trade-off between positive and negative effects of solar radiation. Since photosynthetically active radiation (PAR) is the key determinant of primary production, the productivity at the base of the food web will be closely related wavelengths in the range of 400–600 nm. The pelagic grazers that harvest this phytoplankton yield will thus invariably expose themselves also to the shorter wavelengths. Planktonic crustaceans do, however, commonly have a pronounced diurnal vertical migration that may range from a few to hundreds of metres. The downward migration during daytime is mostly attributed the risk of predation from visual predators (mostly fish), but in fact short-wave light may serve not only as the proximate cause of diurnal migration, but also as the ultimate cause as demonstrated by migratory behavior also in the absence of predators.

First, zooplankton may adapt or alter their *seasonal or spatial distribution* to reduce the UV-stress. Seasonal life cycle adaptations to avoid periods of peak solar intensity may very well be a strategy for UV-exposed and sensitive organisms, yet this is not explicitly demonstrated. Diurnal vertical migration is, however, commonly accredited to direct UVR [8,9]. A critical question is whether organisms have a sufficient spectral resolution to separate UVR from, for example, blue light. This may be important to respond to an increased UVR under constant PAR. Such behavioral responses clearly are important evolutionary traits for swimming animals, and could affect both productivity and trophic interactions; the topic will be fully covered in another chapter, and will thus only briefly be touched upon here to illustrate some ecological implications.

A high spectral resolution and sensitivity is probably common in most pelagic organisms, but again special interest has been devoted to the freshwater cladoceran *Daphnia*. For *Daphnia*, sensitivity to UVR was suggested in the early works of Koehler [10] and Merker [11] as cited in Smith and Macagno (1990)

[12]. The thorough study of Smith and Macagno [12] revealed four peaks in the spectral sensitivity in *D. magna*, at 348, 434, 525 and 608 nm. Peak sensitivity in the ultraviolet was found at 340 nm for *Daphnia*, and UVR gave a strong negative phototaxis as opposed to visible light [13]. Flamarique and Browman [14] demonstrated a wavelength dependent orientation in polarized light among *Daphnia*, which was also different between different species. Experiments have shown that *Daphnia* can detect UVR and respond with downward migration [8,9,13,15], and that pigmented clones migrate less than non-pigmented clones [9,15]. Evidence so far do indicate that aquatic invertebrate can detect UV-A but, probably, not UV-B.

In situ observations support the potential direct effect of detrimental radiation on vertical distribution of zooplankton in Arctic localities. Hebert and Emery [16] reported different patterns of spatial distribution among melanic and hyaline clones of *D. pulex* and *D. middendorffiana* in North-American Arctic. Melanic clones ranged freely through the water column, while unpigmented clones were restricted to the pond bottom under high irradiances. In support of this it was observed that in high Arctic localities at Svalbard the diurnal migration pattern depends on levels of pigmentation and weather conditions. *Daphnia* in these localities are commonly heavily melanized (see below), but some deeper localities may house transparent (hyaline) clones. For a strictly transparent clone it was observed that the entire population was concentrated at, or close to, the sediment surface during days with clear sky (making sampling by net hauls sometimes nearly impossible), while animals were evenly distributed in the water columns on cloudy days [17]. For heavily pigmented animals in a shallow adjacent pond, no sign of downward migration was recorded even under bright sun. A pronounced diurnal migration of *Daphnia*, *Bosmina* and copepods in alpine localities devoid of vertebrate predators was accredited to UVR avoidance [18].

One important aspect in this regard is the trade-off between pigmentation, migration, UV-damage and predation. These aspects have some ecological and evolutionary implications, since increased pigmentation allows animals to stay in upper (warm) layers where food is abundant even under high UVR. On the other hand the pigmentation has its costs both in terms of energy expenditure and visibility that may enhance the risk of predation. These strategies will be more closely examined when discussing the role of pigments (below).

The general ability to respond to short-wave radiation by downward migration is also seen in fish [19]. When placed in the quartz cylinders with three replicate treatments of visible, visible plus UV-A, and visible plus UV-A and B, cod larvae distributed themselves evenly throughout the vertical extent of the cylinder (15 cm) under the visible and visible +UV-A treatments. In the treatments exposed to visible +UV-A and B larvae were consistently found at the bottom of the cylinders, particularly at peak solar intensity. This was observable even on the first few days of the treatments when mortality or morbidity was not a factor. Such UVR avoidance also seems, for fish, to rely on UV-A receptors that have been found in a variety of fish species [20–22], yet as with crustacean invertebrates fish also seem unable to detect UV-B.

12.2.2 *Photoprotective pigmentation*

A second line of defense is the photoprotective pigments, of which mycosporine-like amino acids (MAAs), carotenoids and melanins are the most prominent in aquatic animals (Chapter 10). The conspicuous presence of pigmentation in light-exposed animals clearly aims at a higher UVR tolerance. In fact the very presence of these UV-protective properties strongly suggests that UV is a potential stress factors, since all protective means have their cost. This is clearly seen as a general decrease in pigmentation with increasing depths, reaching the extreme in deep-water or cave-dwelling animals that may be almost completely devoid of skin pigmentation. Zooplankton, being susceptible to visual predators, face a dual challenge. Strong pigmentation implies high visibility and thus a high risk of predation. Lack of pigmentation, on the other hand, could render the animals more susceptible to UVR damage. There should thus be a selection towards invisible UV-screens.)

12.2.2.1 *MAAs*

A heterogeneous category of UV-screening compounds are collectively labeled MAAs. These are widespread in shallow water organisms [23,24] absorb radiation primary in the 310–360 nm range and seem to be primarily associated with UV-stress. Probably most of the MAAs in heterotrophs are derived via food from autotrophs [25,26]. MAAs are also found in freshwater zooplankton, yet there may be conspicuous differences among taxa. Ethanol extracts from alpine copepods contained compounds absorbing in the range for MAA (peak absorbance at 330nm), while no such were found in Arctic or alpine populations of *Daphnia* (Figure 1, Borgeraas and Hessen, unpublished data). This is in support of Sommaruga and Garcia-Pichel [27] who reported frequently high concentrations of MAAs in alpine calanoid copepods, moderate levels in sympatric rotifers, while virtually no MAAs were detected in cladocera. This points to some important taxonomic differences in pigmentation strategies, at least between copepods and cladocera, corresponding to what is seen for carotenoids. Helbling et al. [28] also demonstrated the key role of MAAs for copepods exposed to extreme UVR such as *Boeckella titicacae*, which, as the name hints, is an inhabitant of lake Titicaca (3810 m a.s.l). They also demonstrated a downregulation of the MAA synthesis under reduced UVR exposure, suggesting that this is a dynamic trait. The more general properties of MAAs are covered in Chapter 10.

12.2.2.2 *Melanin*

Melanins are complex macromolecules that serve as highly efficient screens for short wave radiation across the animal kingdom. They are widespread in all groups of metazoans, and have at least two key ecological functions: camouflage and UV-protection. The potential role of melanin in UV-protection is perhaps best illustrated by freshwater zooplankton. Some cladocerans have developed a highly conspicuous carapace melanization that appears to be a unique adaptation to UVR. Highly UV-exposed populations of various species and clones of the *Daphnia pulex* complex and *Daphnia longispina* may frequently have a dark

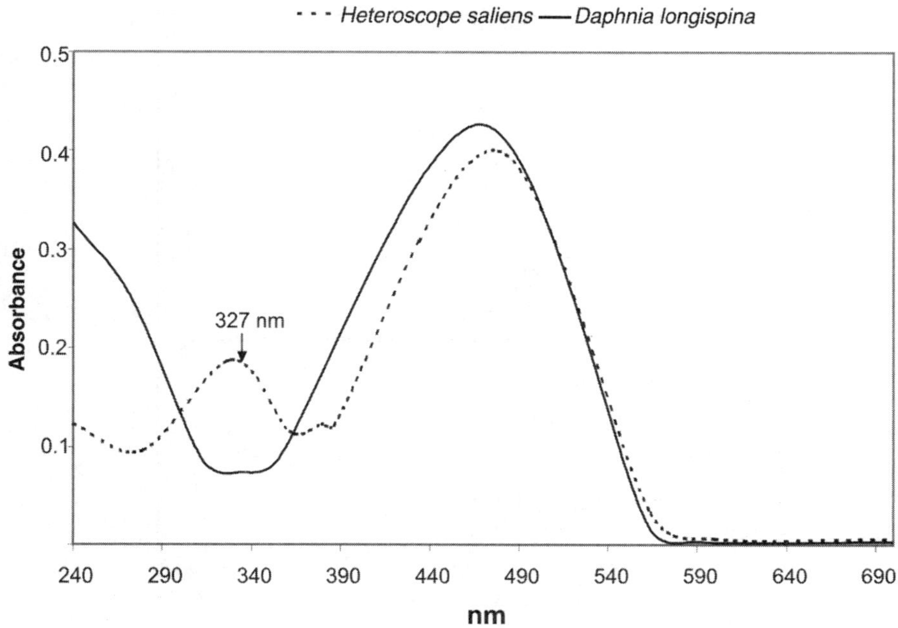

Figure 1. Example of absorption spectra of aqueous ethanol extracts of *D. longispina* and the copepod *Heterocope saliens* showing the peak of the carotenoid absorption maximum at 476 nm present in both species and of UV-absorbing compounds at 327 nm in the copepod.

appearance that most often is caused by a carapace melanization [16,29]. Within a region these clones often occur in the clearest ponds and are replaced by non-melanic clones when vegetation cover increases or water transparency decreases due to increased concentrations of DOC.

That carapace melanization in zooplankton is, apparently, a unique property of Arctic and alpine cladocera is somewhat puzzling since the ability is shared among different clones and taxa. In fact almost all arthropoda may synthesize melanin for eyes and specific structures. All *Daphnia* species may synthesize melanin for protection of their resting eggs (ephippia), yet among the crustaceans the ability of carapace melanization seem restricted to Arctic and alpine cladocera and a few other crustacean taxa. Aquatic insects may commonly possess melanization, as do the vertebrates.

The fact that melanized animals are far more tolerant to UVR than their non-melanized relatives, [9,16,29,30] and Figure 2, strongly supports the view of a primarily UV-protective function. This is further supported when comparing UVR transmission through the carapace of hyaline and melanic animals. Melanized carapaces offer a highly efficient UV-absorption, especially with the shorter wavelengths, relative to transparent ones (Figure 3). Another argument for the role of melanins in UV-protection is the fact that melanin synthesis involves great costs both in terms of energy demands and increased risk of predation (see below). The fact that this kind of extensive melanization occurs almost exclusively in alpine and arctic localities could be accredited to the common UV-

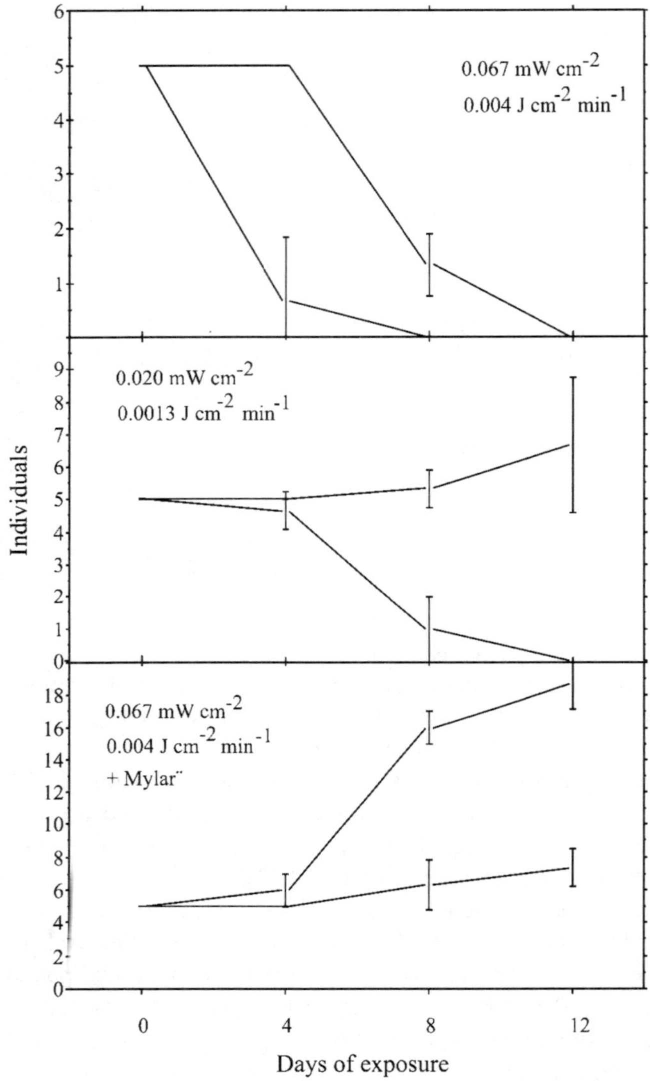

Figure 2. Population development of melanized (filled symbols) and hyaline *Daphnia* under high light intensity (upper panel) medium light intensity (middle panel) or high light intensity plus Mylar sheet (lower panel). Five juvenile individuals were added in triplicate for each treatment, and population development followed for 12 days, covering one reproductive cycle. Light was provided with a 15 W Wilber-Lourmat lamp with peak intensity at 312 nm, and blue–white light (100 μE m^{-2} was provided for photorepair. [From Hessen 29.]

transparency and shallowness in these habitats, rendering the animals vulnerable to high levels of UVR. For Arctic areas there will also be a continuous diurnal UV-exposure during summer. Also, these habitats commonly have low (or no) predation pressure from fish.

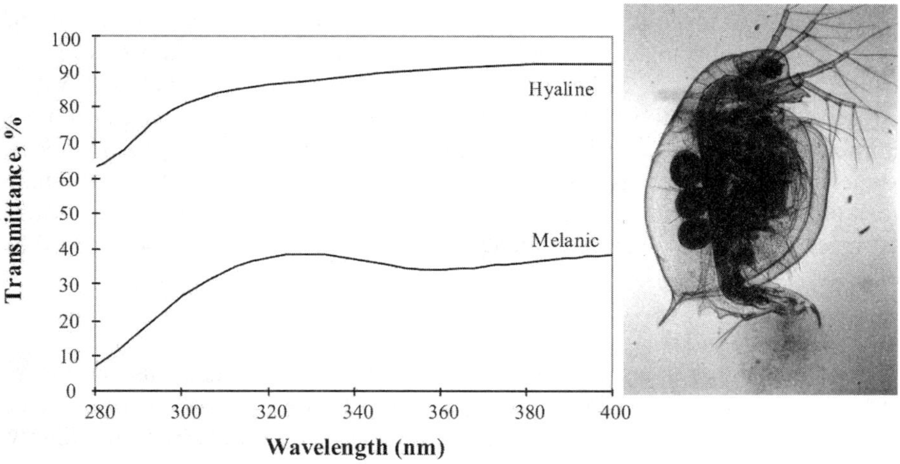

Figure 3. Spectral transmittance of irradiation of *Daphnia* carapaces with different levels of melanization. Hyaline: dorsal part of hyaline carapace. Melanic: dorsal part of pigmented animal (Hessen et al. [114]). A hyaline individual of *Daphnia* is shown in the insert. [Photo: D. Hessen.]

The role of melanin in other pelagic metazoans has been less well studied. Melanic zooplankton are generally rare both in freshwater and marine areas since this would dramatically increase the risk of predation, yet there are some surface dwelling species that may have a conspicuous dark coloration such as some species of Arctic winged snails (opistobranchs) with high levels of melanin (Hessen unpublished).

The role of melanins in fish may be more obscure. A key function is control of color pattern in the skin that may serve both as sexual signals and camouflage. In teleost fish, a melanin-concentration hormone (MCH) regulates adaptive color change, and the MCH activity may be related to photoperiod [31]. To what extent UVR plays a role in this regulation is not settled. There are probably several components in the outer layers of fish skin that protect it against UV, and a direct relation between the amount of these compounds and UVR-induced erythema in fish has been revealed [32,33]. On the other hand, the direct role of melanin in UV-protection may be less important, although the role of melanins in this regard may be highly species-specific. One study comparing albino and pigmented fish (medeka, *Oryzias latipes*) revealed no difference in UV-B induced mortality, yet both morphs had similar amounts of colorless UVR absorbing compounds in their skin [33]. The presence of MAAs has been verified in this species [25]. This suggests that, at least for this species, melanin does not play a major role in UVR protection. Although the major photoprotective role of melanins is simple sunscreening, melanin precursors may also serve as anti-oxidants, and in fact the activity of key anti-oxidants like super-oxide dismutase has been linked to the presence of melanin in fish skin [34].

In amphibians, the dark coloration in eggs and tadpoles is chiefly accredited to the presence of melanin, and the presence of melanin together with a suite of other UV-B absorbing compounds renders at least the tadpoles of the common frog, *Rana temporaria*, rather insensitive to UVR [35]. Nevertheless, UVR is suggested as one (among several) key factors explaining the dramatic decline of amphibians worldwide [36]. Although melanin without doubt plays a key role in some mammals (such as humans), the corresponding role of skin and hair melanins for aquatic mammals is less obvious. Presumably these animals would be more at risk of either eye damage such as cataract or indirect (food web) effects.

12.2.2.3 Carotenoids

Carotenoids may serve a dual role in photoprotection in organisms, serving as either anti-oxidant or radical scavenger, and offering protection from direct photon flux by quenching. The conspicuous red coloration of alpine and highly light-exposed plankton organisms was recognized in early work such as that of Merker [11] and Brehm [37]. This coloration is caused by high levels of tissue carotenoids, and the role of carotenoids in photoprotection in clear low-land localities has been also convincingly demonstrated [38,39]; yet carotenoids, apparently, are aimed more towards the UV-A and blue. Carotenoids are present in all groups of crustaceans; however, at highly variable levels and carotenoid composition. The major groups of carotenoids identified in calanoid copepods are astaxantin, cryptoxanthin, echinenone and hydroxyechinone-like fractions, all probably derived through food from algal β-carotene precursors [39]. Partali et al. [40] recorded a total of 11 different carotenoids in *Daphnia magna*, some in trace amounts only. They also demonstrated how the carotene profiles in *Daphnia* could vary with food source.

While the role of carotenoid photoprotection seems well justified in copepods, it is more obscure in the cladocera [16,41]. Sub-Arctic alpine copepods (*Heterocope*) were found to have ten times more carotenoids than sympatric populations of cladocerans, and even low-land transparent copepods have higher carotenoid levels than highly light-exposed *Daphnia* [41]. Carotenoids are also widespread in fish, notably anadromous salmonids, yet the role of carotenoids in photoprotection in these species is not settled.

12.2.2.4 Other UVR screening compounds

The above-mentioned major UV screening molecules or pigments are not an exhaustive list of potentially UV-blocking or UV-absorbing compounds. Most organic and inorganic structures in exuvia, skin or epidermis may serve such a role although not particularly evolved for this purpose. There might be, however, several unidentified substances aimed specifically at UV protection. For instance the egg shell or chorion of cod eggs do apparently offer some UV protection, since fewer cyclobutane pyrimidine dimers (a kind of DNA damage caused only by UV-B, Chapter 9) are found in eggs compared with recently hatched embryos [42]. This is apparently not a common feature of all species, however, since no corresponding difference between eggs and hatched embryos was detected in Northern anchovy [42,43].

12.2.2.5 UV, pigments, predation and evolutionary trade-offs

Pelagic metazoans commonly face multiple challenges with regard to photoprotective strategies. Since the adaptive role of pigments is to allow for a surface dwelling life, this is indeed in conflict with the risk of predation, which normally is highest under high light conditions (surface layers). The synthesis of pigments is an energy demanding process. This is particularly well demonstrated for zooplankton with a melanized exuvia, where the melanin needs to be re-synthesized for each moult, and where growth rates are lower in pigmented animals as compared with their hyaline conspecifics [29]. Thus the melanized clones are assumed to be competitively inferior under moderate or low UV-stress. Moreover, these dark animals are more visible to visual predators, and when co-occurring with hyaline clones, as may occur in some alpine lakes, there is a strong selective preference among fish for the melanized individuals [44].

Similarly, Hansson [45] found that the level of carotenoid pigmentation in copepods was up to ten times higher in lakes without predatory fishes than where fishes are present. Moreover, animals from the same population exposed to either UVR or predators displayed a 10% difference in pigmentation after only four days, suggesting that pigmentation is an inducible trait. The latter is a particularly intriguing observation, since it demonstrates that adaptations to either UV or predation in terms of pigmentation do not necessarily require evolutionary time scales. For melanized *Daphnia*, most clones may rapidly shut down their melanin synthesis in the absence of UV [19]. The extent to which the presence of fish or fish kairomones may induce shifts in pigmentation strategies (as commonly observed for morphological features) is not known. Non-melanized *Daphnia* are apparently not capable of switching into melanized under prevailing UVR, however [19], suggesting that this in a non-inducible property that probably depends on a slight genetic modification.

Also, zooplankton without a conspicuous pigmentation may show a depth distribution and adaptation to different UV regimes at different depths. Comparison of a large number of epipelagic (15 species) and mesopelagic (19 species) organisms revealed pronounced differences in their tissue transparency to different wavelengths, yet all species were considered transparent [46]. In general, the tissues from epipelagic species had lower UV transparency as compared with mesopelagic species, demonstrating that more subtle regulation of intracellular UV-exposure may occur without increasing the risk of predation.

12.2.3 Anti-oxidants

A third defense mechanism are intracellular processes such as repair of DNA damage and the production of quenching agents and anti-oxidant enzymes that neutralize reactive oxygen species (ROS) produced by UV [47]. Examples include carotenoids (CAR), involved in quenching of activated photosensitizers and singlet oxygen, superoxide dismutase (SOD), which eliminates the superoxide radical, catalase (CAT), which detoxifies hydrogen peroxide to oxygen and water, and glutathione transferase (GST), which neutralizes peroxidized macromolecules and detoxifies breakdown products after lipid peroxidations [48]. Antioxidants are linked with resistance against UVR in plants, microorganisms and mammalian cells and skin tissue [47,49,50], but little is known about antioxidant protection in pelagic metazoans.

A survey of bulk carotenoids, as well as the enzymes SOD, CAT and GST, in populations of various *Daphnia* species, ranging from coastal rock-pools in Southern Norway to high Arctic populations [51] did not reveal any strong pattern in enzyme activities that could be attributed to species affinities, habitat, or tentative UV-exposure of pigmentation (Figure 4). Since a far higher UV-susceptibility was revealed among hyaline clones relative to melanic clones [19], an a priori assumption would be that the lack of carapace melanization should be compensated for by other means of photoprotection like increased activity of essential anti-oxidants, but in general this was not the case. Studies of natural alpine and Arctic populations of *Daphnia* did, however, reveal diurnal patterns in levels of anti-oxidants that could be related to solar exposure (Hessen and Borgeraas unpublished). Tissue concentrations of GST and SOD, less so for CAT, displayed a pattern with low levels in the afternoon, following after UV-exposure, and a gradual build-up during the night. Correspondingly a number of anti-oxidants have also been screened in fish skin [34], yet their bearing on UVR susceptibility is not settled.

The above tested anti-oxidants do not provide an exhaustive list of potential scavengers of oxidants in aquatic animals, however, and there may be other UV-protective enzymes that play a key role.

12.2.4 Recovery and DNA repair

A last major defense would be the various means of enzymatic photorepair that is probably a common property of all organisms. These general effects on DNA, proteins and membranes and the corresponding cellular repair mechanisms will not be reiterated here. It has long been known that longer wavelengths, notably in the UV-A and blue, can counteract UVR damage by repair of DNA. This effect is, among others, one reason why laboratory experiments tend to overestimate UVR damage if sufficient spectral quality or quantity is accounted for. There are a number of studies that point to the role of photorepair or photoreactivation in marine animals from copepods [52–54], shrimps and eupausides [55,56] to fish [56–58]. Also, for eggs of cod, a pronounced photorecovery was found [2]. There are comparatively fewer studies for freshwaters but, for *Daphnia*, Siebeck and Böhm [59] not only demonstrated a pronounced species-specific difference in UV-susceptibility, they also provided data on spectral effects and the effects of recovery radiation. While spectral sensitivity for *Daphnia* followed a general CIE action spectrum, peak recovery radiation was in the blue region (420–440 nm). They found that recovery radiation (light repair) strongly increased when recovery radiation was provided after UV-exposure. This positive effect was further increased when recovery radiation was provided also under UV-exposure. The effect of recovery radiation rapidly decreased with

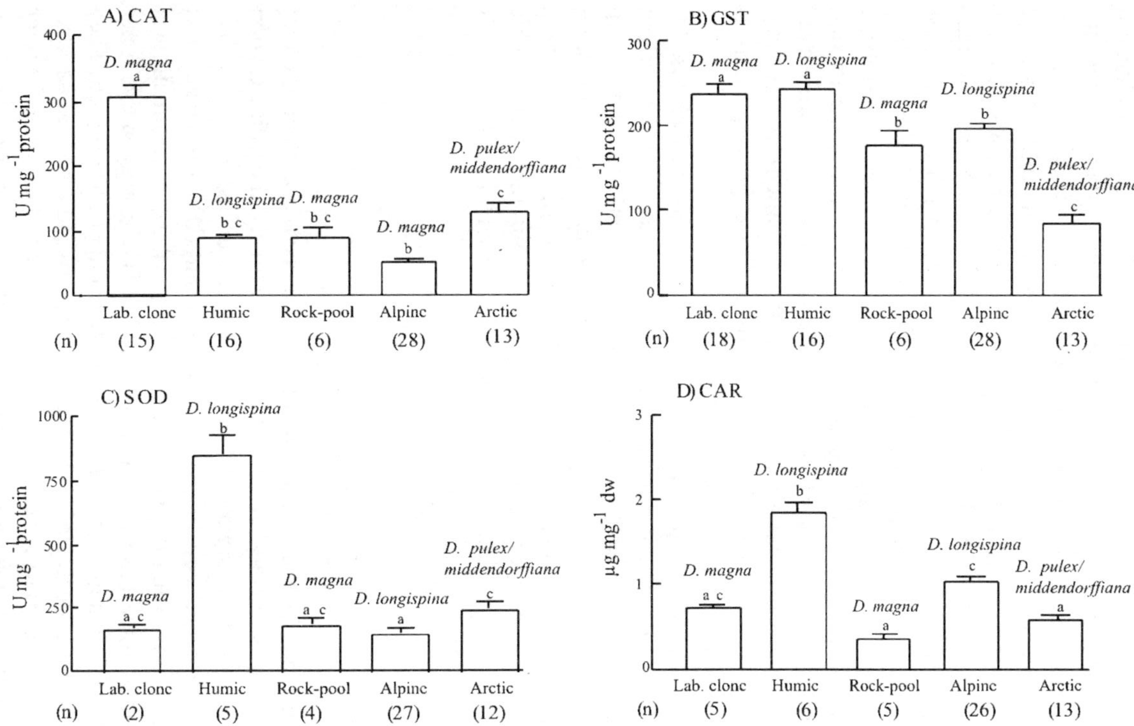

Figure 4. Specific activities (U mg^{-1} protein) of (A) catalase, (B) glutathione transferase, (C) superoxide dismutase and (D) bulk carotenoids in *Daphnia* from five different areas of Norway. The lab clone had not received UVR for several generations, Kj. Putten is a humic lake with high levels of DOC and low UVR; Hvaler represents shallow and highly UVR exposed rock-pool populations, Finse represents oligotrophic alpine ponds, and Svalbard represents high Arctic ponds at 79 °N. [From Borgeraas and Hessen, 51, with permission of Kluwer Publishers.]

increasing post-exposure darkness. For freshwater copepods, a pronounced species-specific variability in photorecovery was found [60].

12.3 Indirect effects

There are several ways by which aquatic organisms may be *indirectly* influenced by UVR. UVR plays a major role in surface water chemistry that in numerous ways may affect the aquatic biota. Breakdown or oxidation of complex macromolecules, notably humus molecules, may both induce availability of organic substratum for microbial heterotrophs [61, Chapter 8]. This may also liberate mineral nutrients such as N and P, and in fact the effects on biogeochemical cycling of key elements like C, N and P in surface waters is one of the less recognized yet most important properties of UVR in many shallow waters. Further, UVR may also react with organic macromolecules and cause the release a number of potentially limiting, but also potentially harmful, metal species that may affect aquatic biota. Finally UVR promotes the formation of a wide array of photoproducts in water (Chapter 8 and below). This set of important mechanisms, all of which may have profound effects on surface dwelling organisms, is one aspect that clearly is unique to aquatic organisms.

A range of food web effects may also be considered as indirect effects. Although direct UVR damage on particular organisms may be minor, there could nevertheless be pronounced effects if lower trophic levels are affected (bottom-up effects). Alternatively if competitors or predators suffer comparatively more from UVR, this could yield an *apparent* positive effect of UVR on other species or taxa [62]. While, especially, the effects of UVR on aquatic humus and heterotrophic bacteria as well as food web effects will be considered elsewhere, we shall here just briefly touch upon these indirect mechanisms.

12.3.1 ROS in water

Aquatic organisms may be exposed to ambient UV-induced ROS formed either in tissue or in surface waters by photon reaction with dissolved organic carbon [63]. The role of dissolved organic carbon (DOC) in surface waters is thus twofold; first of all DOC constitutes a highly efficient UV absorbent, and UV-attenuation in most aquatic environments is primarily a direct function of DOC [6]. Secondly, however, the trapping of high levels of energy in the upper few centimetres produces a number of biologically harmful photoproducts. They include excited triplet state DOM, solvated electrons, organic radical cations, superoxide, singlet oxygen, hydroxyl radicals and peroxy radicals (see Chapter 8). Since these compounds may only last nanoseconds, molecular probes have been used to quantify the rate of production (see review by Zafiriou et al. [64]). H_2O_2 is produced photochemically when UVR strikes DOM but is more long-lived since its decay is principally biological in most systems. In lakes, concentrations are much higher than in marine systems and may reach, during midday, in excess of 1000 nM, whence DOC concentrations may be a main determinant [63]. The highly toxic carbon monoxide is also generated in both fresh and marine waters when DOM is exposed to solar radiation.

The role of such indirect effects in ambient waters for the aquatic biota is not settled. Experiments with high UV-doses indicated negative effects on phytoplankton but no effect on zooplankton (*Daphnia*) in water with high levels of DOC [65]. UV exposure experiments along a gradient of DOC (humus) with spectral distribution and doses close to natural outdoor maximum surface irradiation clearly suggested an overall net positive effect of DOC, and no detectable negative indirect effects either on phyto- or zooplankton [66]. This could suggest that UV mediated effects of radicals and strong oxidants in the water are minor compared with direct, intracellular damage. Borgeraas and Hessen [67] did, however, observe an increased catalase activity in *Daphnia* with increased DOC and increased UV, suggesting an enhanced need for peroxide detoxification. It should be recalled that most of the photo-products are very short-lived, and it is difficult to separate intracellular from extracellular effects during UV exposure even experimentally, since exposure to pre-irradiated water will not capture the effects of short-lived photoproducts. Knowledge of the biological role of this multitude of photoproducts is still scarce, but it is definitely a potential stressor that is not shared with terrestrial biota.

12.3.2 Food web effects

Obviously any organism, even those rather insensitive for UVR, may be positively or negatively affected to the extent that other members of the aquatic community are affected. Even harmful effects of UVR on a species may turn into net positive effects provided that the major competitor or predator is more adversely affected – and vice versa. UVR tolerant grazers of predators may suffer if preferred food organisms are vulnerable to UVR [62]. The complexity of most food webs does not allow for a clear judgement of such effects, yet there are some general observations indicating that such effects could be profound.

For zooplankton–phytoplankton interactions several different mechanisms could operate; first the phytoplankton productivity could be reduced, and there could be community shifts owing to different UV-susceptibility among different species or taxa [68–70] or cells could be morphologically or biochemically altered, causing reduced nutrient quality for the grazers [71–73]. This first set of mechanisms is related to direct effects on the autotrophs, and will not extensively be covered here. It should be emphasized, however, that there are a number of rather subtle indirect mechanisms that could affect various trophic levels via the food web. While both phytoplankton community structure [69,70] as well as cell membranes [71] may alter both ingestion and assimilation of algae, an even more important mechanism could be caused by the fact that, even at low dose rates, the total lipid content of some microalgae may be reduced [74] and this effect includes the polyunsaturated fatty acids (PUFAs) [71,75,76]. For zooplankton and fish larvae, the only source of these key fatty acids for membrane functioning, growth and development is dietary. Since they cannot synthesize them de novo, they must be obtained through prey organisms [e.g., 77,78].

Finally, the grazers themselves may be affected, causing reduced grazing pressure on phytoplankton. Such effects have rarely been studied, yet there is evidence that UVR may reduce grazing activity in marine [79] and freshwater [80] nanoflagellates. Laboratory experiments demonstrated a pronounced decline in grazing rates shortly after UVR exposure in *Daphnia*, and a gradual post-treatment recovery after several hours (Figure 5).

12.4 Ambient parameters and their interaction with UV

There are a number of biotic and abiotic factors that may be of particular relevance for the pelagic environment with regard to UV-effects. Some of these have already been touched upon, such as concentrations of DOC, which will not only be the main determinant of UV-attenuation but also may promote the formation of harmful photo-products. Oxygen, temperature and ionic strength (in freshwaters) may also be determinants of UV-effects. System productivity may be important in various ways. High primary production (and thus increased cell numbers of phytoplankton) will aim at reducing UV-exposure. Food access could also affect the physiological status of the animals and thus their UV-susceptibility. There is little general information on such physiological responses, however. Rather than screening the long list of potential effects, we shall focus on a few parameters of particular relevance.

12.4.1 Oxygen and temperature

Oxygen and temperature in several ways affect UVR susceptibility. High p_{O_2} could aim at increasing the oxidative stress, and low temperature could slow down kinetic reactions such as anti-oxidant expression in poikilotherm animals. Shallowness and wind exposure promote oxygen saturation in surface waters. High primary production may cause pronounced vertical gradients in O_2-concentrations, and diurnal oscillations. Since oxidative damage such as lipid peroxidation is intimately linked with O_2-concentrations both inside and outside cells [47], and UVR is a major mediator of such oxidative damage, this parameter may be of vital importance for aquatic organisms. Temperature may also show strong vertical gradients, and naturally also shifts along altitudinal and longitudinal gradients. An a priori assumption would be that low temperatures in particular could increase the UV-susceptibility in heterotrophs as it does for autotrophs [81,82]. These two key parameters may also interact with each other.

Studies on phototrophic organisms show that UV-effects may be balanced between photochemical damage and biosynthetic repair. This balance will shift increasingly towards damage with decreasing temperature [83]. There are very few data on corresponding temperature induced trade-off in eukaryotic heterotrophs. Borgeraas and Hessen [67] tested the UV-susceptibility in *Daphnia*

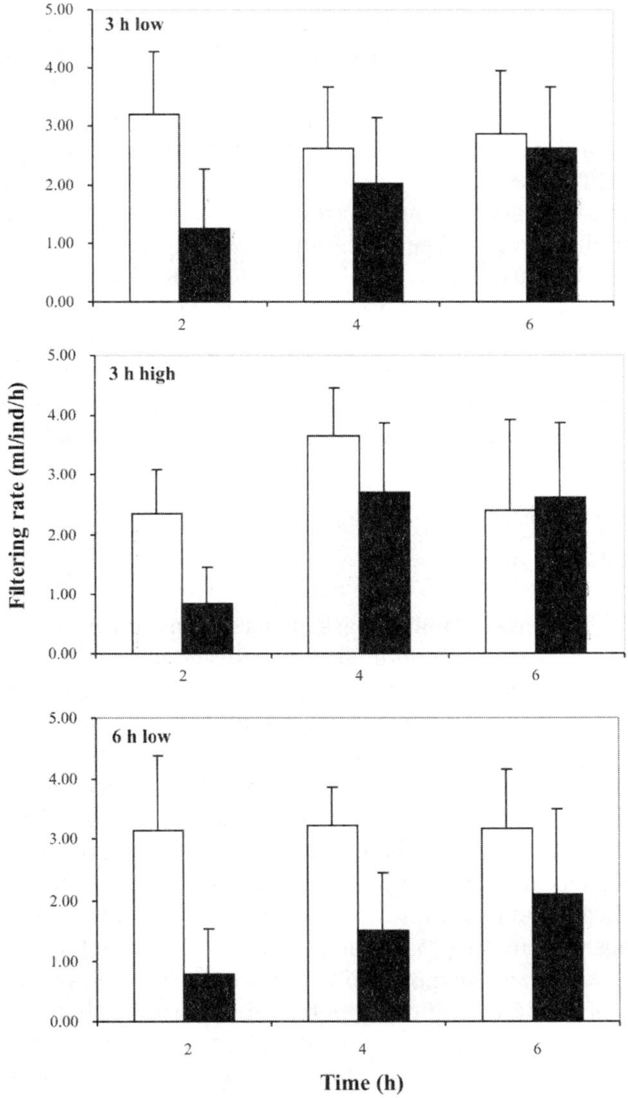

Figure 5. Post-treatment recovery in grazing rates of *Daphnia* after different exposure to UVR. Filled bars represent UVR exposed individuals, open bars are unexposed controls. Light was provided by a 100 W xenon lamp with spectral properties close to daylight after passing through a cellulose acetate filter. At "high light" the dose-rate over 300–315 nm and 300–400 nm band was 1.1 and 35.9 W m^{-2} respectively, while 282 W m^{-2} for PAR (400–700 nm). 'Low light" represents dose-rates reduced by 50%. All data are means ($\pm$SD) of three replicates, each consisting of 10 animals. Grazing rates were measured as the decrease in algal cell number over time by use of a Scatron flow cytometer. [Hessen and Brudevoll, unpublished.]

magna along gradients of oxygen saturation and temperatures. Increased oxygen concentrations over the range 5.6–14 mg $O_2\,l^{-1}$ did not cause increased mortality under UVR, however, and a lowered temperature (range 6–18 °C) did in fact decrease UV-susceptibility. Neither oxygen nor temperature caused any significant effects on anti-oxidant expression. Low temperatures may slow down UV-induced mortality in several ways. Although repair and detoxification mechanisms may be impaired at low temperatures, so may also activation processes such as ROS metabolism and lipid peroxidation. Low temperatures may also change the physiological status of the animal. *D. magna* is not a cold adapted species, however, and the tested clone had been raised to high temperatures (18 °C for years). Thus these tests may not be relevant for cold adapted species, but for *Daphnia* they are in support of those of Abele et al. [84], who reported that exposure to elevated temperatures and hydrogen peroxide elicits oxidative stress and anti-oxidant response in the Antarctic intertidal limpet *Nacella concinna*.

12.4.2 UV and water chemistry

A number of different water quality parameters such as concentration (and type) of DOC, pH, salinity and sub-optimal or insufficient concentrations of specific ions or minerals could affect the biotic responses to UVR. The presence of DOC is in most cases the main determinant of UV attenuation and effects on freshwater metazoans (see Chapter 3), and yet increased DOC also yields increased production of ambient free radicals and oxidants; the net effect on the biota is in general positive. The role of inorganic parameters is less well known. Under laboratory conditions, swollen body tissue may be observed in UV treated freshwater invertebrates animals, which could indicate osmotic disorder [15]. This could indicate that ionic content or salinity per se could be one determinant to UV tolerance, but Hessen [15] found no effect of salinity in the range from 250 to 1000 $\mu S\,cm^{-1}$ in *D. pulex* exposed to artificial UVR (peak wavelength 312 nm).

Low levels of calcium is another potential co-stressor that could severely increase the UV-susceptibility of zooplankton. Calcium concentrations of <10 mg l^{-1} seem sub-optimal for Ca-demanding *Daphnia* species and UV-B tolerance was significantly reduced over a gradient from 10 to 0.5 mg Ca l^{-1} in both *D. magna* and high Arctic *D. tenebrosa* [85]. Whether this effect could be accredited to some physiological interaction between UV stress and Ca metabolism (i.e., membrane damage and distorted uptake of Ca) or merely to the additional effect of two physiologically independent stressors remains unresolved. Nevertheless, it is obvious that the highly variable ionic content and Ca concentration in freshwater localities, ranging from values around 0.1 mg Ca l^{-1} in very dilute soft-water localities to commonly >20 mg Ca l^{-1} in hardwater lakes, could also be a determinant of the effects of UVR on zooplankton communities.

12.5 Taxa specific responses/evidence for in situ effects of UVR on pelagic metazoans

Over the past two decades there has been a rapidly growing body of literature on UV-effects. A major share of this literature deals with laboratory studies or otherwise artificial exposure or artificial conditions. For obvious reasons, laboratory experiments rarely mimic ambient conditions with regard to water temperature, oxygen or food. Even more troublesome is the fact that spectral qualities, recovery radiation, dose or dose-rate may differ from natural conditions. In particular, the principle of *reciprocity* of UVR (or lack of such) is commonly ignored or not tested for. Commonly, high intensity over short periods may generate a different pattern of damage than comparatively lower doses over long periods. Also, we are commonly faced with a long list of *potential* effects, this chapter being no exception, but there are comparatively few studies that really provide evidence for UVR effects in situ. One particular challenge is that while present-day UV may pose both stress and constraints on organism performance, these sub-lethal effects are not easily captured. Life is rarely optimal for organisms in nature, and such effects of UVR will work in parallel with sub-optimal access to food, sub-optimal water quality etc. Sometimes the sum of such sub-optimal parameters becomes lethal. Pelagic organisms have evolved under some UV-stress and hence should be able to tolerate this, but at some costs that can only be quantified by assessment of fitness in the absence of UVR in general or more specifically UV-B. Nevertheless, there are several observations of in situ effects of UVR, and these are particularly important for the final judgement of the ecological role of UVR.

12.5.1 Case studies of UVR and zooplankton

As stated above, the presence of a conspicuous and often costly pigmentation by carotenoids or melanin is strongly indicative of UV as a major evolutionary force. Similarly the vertical distribution and migration of zooplankton in lakes or ponds devoided of predators also supports this view. There is also evidence that geographical distribution in cases may be governed by the UV regimes. In fact Williamson et al. [86] provided evidence that macrozooplankton community structure in a set of lakes along a deglaciation chronosequence in Glacier Bay Alaska could be attributed to the UV attenuation in these lakes. Terrestrial succession in the watersheds of these lakes results in increasing DOC content over time. Due to the primary role of DOC in controlling UV attenuation in lakes, the oldest lakes supported more UV-susceptible species. Previous studies by the same author have demonstrated that UVR effects on zooplankton in general are strongly influenced by DOC-concentrations in lakes [87].

Several other studies have demonstrated detrimental effects of UVR in situ on a variety of zooplankton taxa [30, 60,87–90]. Clearly there are not only major differences among taxa and species, the same species may also show different susceptibility to UVR, suggesting that enhanced tolerance may be induced. Siebeck and Böhm [59] found that natural UVR strongly affected *Daphnia* species, but that individuals from a clear, alpine locality were far more tolerant compared with individuals from a large lowland lake. Correspondingly Hebert and Emery [16] found surface UVR to be lethal for transparent *Daphnia*, while melanized clones or morphotypes were largely unaffected. Zagarese et al. [60] reported a remarkable difference in UVR susceptibility even within the same genus of calanoid copepods, and while this study found *Boeckella gracilipes* to be sensitive for in situ exposure, Cabrera et al. [88] found that this species was highly tolerant in an alpine lake where other species were very sensitive. The use of a solar simulator calibrated to closely mimic natural exposure to UVR yielded intriguing differences among several cladocerans, an ostracod and an amphipod. The 96 h LD_{50} estimates ranged from 4.2 to 84.0 μW cm^{-2} [91], with the ostracod *Cyprinotus incongruens* as by far the most tolerant species. Somewhat surprisingly, the epineustonic *Scapholebris kingii* was most sensitive to UVR, with no difference between two color morphs.

There is also evidence based on in situ studies that there is a different spectral response among species. For example, Williamson et al. [87] found that while the cladocerans *Daphnia* and *Diaphanosoma* responded both to UV-A and UV-B, the sympatric copepod *Diaptomus* was affected by UV-B only. In one of the few reported studies on pelagic insects, Williamson et al. [92] found that the midge larvae *Chaeoborus* (which is highly transparent and normally shows a pronounced diurnal migration) not only was very sensitive to UVR in general, but that is responded mostly to UV-A. The fact that some species are very sensitive to UV-A suggests that organelles or other macromolecules other than DNA can be the main targets.

Perhaps the most comprehensive studies on marine zooplankton and fish were undertaken in a series of experiments where in situ studies on eggs and embryos of atlantic cod (*Gadus morhua*) and eggs and adult copepods (*Calanus finmarchicus*) were combined with experimental studies [2-4,93–95]. One key conclusion from these comparative studies was that the copepod was far more susceptible to UVR than eggs or fry of cod. Both were strongly dependent on K_d, determined by levels of DOC (Figure 6). Eggs of *Calanus finmarchicus* and Atlantic cod were incubated under the sun, with and without the UV-B and/or UV-A wavebands. UV-exposed eggs exhibited a lower percent hatching compared to those protected from UVR: UVR had a strong negative impact on *C. finmarchicus* eggs. Further, the percent hatching in UV-B-exposed eggs was not significantly lower than that in eggs exposed to UV-A only, and, under natural solar radiation, UV-A appeared to be more detrimental to *C. finmarchicus* embryos than was UV-B. The strong effect of UV-A is simply an effect of higher absolute levels of UV-A compared to UV-B. A highly wavelength-dependent mortality was found for *C. finmarchicus*, with the strongest effects occurring under exposures to wavelengths below 312 nm. At the shorter wavelengths (<305 nm) UV-B-induced mortality was strongly dose-dependent, but not significantly influenced by dose-rate.

The BWFs derived for UV-B-induced mortality in *C. finmarchicus* were simi-

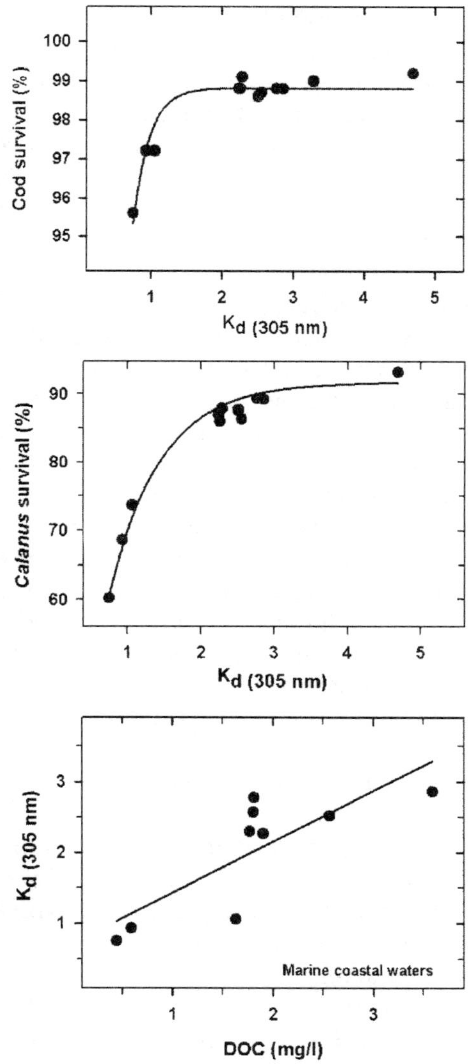

Figure 6. (Top) K_d vs. modeled survival of Atlantic cod (*Gadus morhua*) embryos exposed to UVR in a mixed water column. (Middle) K_d at 305 nm vs. modelled survival of *Calanus finmarchicus* embryos exposed to UVR in a mixed water column. (Bottom) Dissolved organic carbon (DOC) vs. diffuse attenuation coefficient (K_d) at 305 nm from field measurements in the estuary and Gulf of St. Lawrence, Canada. The straight line is the regression, the curved lines are the 95% confidence intervals. [Modified from Browman and Vetter 19, with permission from the Estuarine Research Federation.]

lar in shape to the action spectrum of naked DNA. Further, the wavelength-dependence of DNA damage was similar to that for the mortality effect. These observations suggest that UV-induced mortality in *C. finmarchicus* was a direct result of DNA damage. It was concluded that UVR could cause egg mortality as high as 32% under natural conditions [4]. A model that included the BWFs, vertical mixing of eggs, meteorological and hydrographic conditions, and ozone depletion indicated that UV-induced mortality in the *C. finmarchicus* egg population could be as high as 51% [94]. These values are certainly maximum estimates, but nevertheless points to a potential susceptibility of this key species in northern marine waters. Also, for marine zooplankton species it must be assumed that UVR susceptibility varies over a wide range. For instance zoea larvae of the American lobster (*Homarus americanus*) were virtually insensitive to natural UVR [96].

While most attention has been devoted to the effects of UVR on crustacean zooplankton (and a few observations on rotifers), there are other members of the pelagic community and evidence exists that pelagic opistobranchs may suffer direct damage under natural conditions [97].

Most of these studies have addressed direct effects and notably mortality, but there are also some reports on more subtle and sublethal effects of UVR on zooplankton such as developmental anomalies in nauplii [3]. Such effects have also been observed in crab larvae and euphausides [53] and juvenile cladocerans (Hessen unpublished) under artificial UVR. Also, general reproductive problems such as decreased number of offspring or skewed sex ratios have commonly been reported under experimental conditions (see Zagarese and Williamson [98] for review), suggesting that severe mutations or distorted embryogenesis probably is an ontogenetic bottleneck with regard to UVR.

12.5.2 Case studies of UVR and vertebrates

There are also several studies that have attempted to verify the potential effects on fish under natural conditions, covering a wide range of effects, from cellular effects such as DNA-damage to skin lesions, cataracts or spawning failure and death [3,99–103]. To these add a large number of experiments with artificial radiation. The vulnerability among fish to UVR expressed as sun-burn skin lesions has long been recognized [104]. Such effects have been observed in a variety of fishes in shallow waters [e.g., 105–107], yet some of these observations have been made on fish in captivity and may thus not represent natural conditions. Such effects may expose the fish to pathogen invasions as UVR also depresses the immune system in fish [108,109]. Lens damage (cataract) has also been observed in fish exposed to in situ UVR, but is apparently confined to very shallow waters [110].

A study on UVR susceptibility in cod eggs was undertaken in parallel with the exposure of *Calanus finmarchicus* reported above [2,4,111]. While the BWFs for cod eggs resembled that of naked DNA, and UVR effects were indeed observed under natural solar intensities, it was nevertheless concluded that cod eggs and embryos were far less susceptible to natural UVR compared with the tested copepods. While a model evaluation suggested that a more than 50% increase in damage could occur under realistic ozone depletion scenarios for *Calanus*, this was almost negligible for cod eggs or embryos. This does not imply that UVR is irrelevant for cod recruitment, but suggests that indirect mechanisms could be superimposed on the direct effects.

Cod eggs and larvae are damaged by UV-A and UV-B and possess the typical means of repairing DNA damage via photorepair [2,19]. While cod eggs and larvae clearly carry out photorepair, it was found that the capacity for repair would not be adequate for full repair before the onset of new damage on the following day. This low capacity for photorepair can lead to a greater multi-day accumulation of DNA damage than currently observed for temperate fishes [43]. There are clearly species differences in rates of DNA damage and repair. These differences may account for why some larvae appear to conform to dose reciprocity while others do not. In the northern anchovy, *Engraulis mordax*, a baseline level of CPDs remains after the first exposure but in general CPD levels do not accumulate over many days [43]. Anchovy larvae do not obey dose-reciprocity relationships [100,101]. Cod, with a more limited capacity for photo-repair, accumulate damage over multiple days and adhere more closely to dose reciprocity under the conditions tested [4].

It is important to point out that variability in cloud cover, water quality, and vertical distribution and displacement within the mixed layer (Chapter 4) are all likely to have a greater effect on the flux of UV-B radiation to which the eggs of zooplankton and fishes are exposed than will ozone layer depletion at these latitudes. Thus, although UV-B radiation can have negative impacts (direct effects) on pelagic animals, it must be viewed as only one amongst many environmental factors that produce the mortality typically observed in the planktonic early life stages of these organisms. For fish species whose early life stages are distributed throughout the mixed layer, it seems most likely that UV-B would represent only a minor source of direct mortality for the population, but sublethal UVR might for instance work in concert with pathogens due to immunosuppression.

With the exception of amphibians, that at least may spend part of their life history in surface water, there is little information on in situ effects of UVR. One could expect that, in particular, populations of polar pinnipeds could suffer both from retinal damage and immunosuppression as a result of an increased number of incidents with low ozone, but this has not yet been confirmed. Amphibians, however, have been the subject of particular interest, since their populations have suffered widespread declines and extinctions in recent decades, and UVR has been suggested as one of the potential causes [36]. It is demonstrated that UVR can cause a variety of damage in natural populations such as egg mortality, retinal damage and altered patterns of distribution [112]. As for other species, the UVR sensitivity is widely different among species. Apparently UVR may work in concert with a range of other ambient factors (drainage, toxins, pathogens) [113], and this is probably how UVR works in general.

Can strong conclusions now be drawn as to the effect of UVR on pelagic metazoans? For some species yes but for most species no. This is still a field where most work is yet to be done, and where, probably, the major effects are rather subtle and sub-lethal. Judging from the general knowledge on UVR and autotrophs, it is reasonable to believe that present day UVR strongly modifies not only productivity and community composition, but also food web interactions. We still know little about adaptations to UVR and their costs, but like the structuring effect of past competition, present day pelagic food webs correspondingly bear the footprints of the "ghost of UVR in the past".

References

1. T. Naganuma, T. Inoue, S. Uye (1997). Photoreactivation of UV-induced damage to embryos of a planktonic copepod. *J. Plankton Res.*, **19**, 783–787.
2. J.H.M. Kouwenberg, H.I. Browman, J.J. Cullen, R.F. Davis, J.-F. St-Pierre, J.A. Runge (1999). Biological weighting of ultraviolet (280–400 nm) induced mortality in marine zooplankton and fish. I. Atlantic cod (*Gadus morhua*) eggs. *Mar. Biol.*, **134**, 269–284.
3. J.H.M. Kouwenberg, H.I.Browman, J.A. Runge, J.J. Cullen, R.F. Davis, J.-F. St-Pierre (1999). Biological weighting of ultraviolet (280–400 nm) induced mortality in marine zooplankton and fish. II. *Calanus finmarchicus* eggs. *Mar. Biol.*, **134**, 285–293.
4. H.I. Browman, C.A. Rodriquez, F. Beland, J.J. Cullen, R.F. Davis, J.H.M. Kouwenberg, P.S. Kuhn, B. McArthur, J.A. Runge, J.-F. St-Pierre, R.D. Vetter (2000). Impact of ultraviolet radiation on marine crustacean zooplankton and ichthyoplankton: a synthesis of results from the estuary and Gulf of St. Lawrence, Canada. *Mar. Ecol. Prog. Ser.*, **199**, 293–311.
5. R.B. Setlow (1974). The wavelengths in sunlight effective in producing skin cancer: a theoretical analysis. *Proc. Natl. Acad. Sci. U.S.A.*, **71**, 3363–3366.
6. N.M. Scully, D.R.S. Lean (1994). The attenuation of ultraviolet radiation in temperate lakes. *Arch. Hydrobiol. Beih. Ergebn. Limnol.*, **43**, 135–144.
7. D.P. Morris, H. Zagarese, C.E. Williamson, E.G. Balseiro, B.R. Haragreaves, B. Modenutti, R. Moeller, C. Queimalinos (1995). The attenuation of solar UV radiation in lakes and the role of dissolved organic carbon. *Limnol. Oceanogr.*, **40**, 1381–1391.
8. D.M. Leech, C.E. Williamson (2001). In situ exposure to ultraviolet radiation alters the depth distribution of *Daphnia*. *Limnol. Oceanogr.*, **46**, 416–420.
9. S.C. Rhode, M. Pawlowski, R. Tollrian (2001). The impact of ultraviolet radiation on the vertical distribution of zooplankton of the genus *Daphnia*. *Nature*, **415**, 69–72.
10. O. Koehler (1924). Über das Farbensehen von *Daphnia magna* Strauss. *Z. Vergl. Physiol.*, **1**, 84–174.
11. E. Merker (1940). Der Lichtschutz in belebtem Wasser. *Int. Rev. Ges. Hydrobiol.*, **40**, 174–196.
12. K.C. Smith, E.R. Macagno (1990). UV photoreceptors in the compound eye of *Daphnia magna* (Crustacea: Branchiopoda). A fourth spectral class in single omatidia. *J. Comp. Physiol. A*, **166**, 597–606.
13. U.C. Storz, R.J. Paul (1998). Phototaxis in water fleas (*Daphnia magna*) is differently influenced by visible and UV light, *J. Comp. Physiol. A*, **183**, 709–717.
14. I.N. Flamarique, H.I. Browman (2000). Wavelength-dependant polarization orientation in *Daphnia*. *J. Comp. Physiol. A*, **186**, 1073–1087.
15. D.O. Hessen (1994). *Daphnia* responses to UV-B light. *Arch. Hydrobiol. Beih.*, **43**, 185–195.
16. P.D.N. Hebert, C.J. Emery (1990). The adaptive significance of cuticular pigmentation in *Daphnia*. *Funct. Ecol.*, **4**, 703–710.
17. D.O. Hessen, J. Borgeraas, K. Kessler, U.H. Refseth (1999). UV-B susceptibility and

photoprotection of Arctic *Daphnia morphotypes. Polar Res.*, **18**, 345–352.

18. D.O. Hessen (1993). DNA-damage and pigmentation in alpine and arctic zooplankton as bioindicators of UV-radiation. *Verh. Int. Verein. Limnol.*, **25**, 482–486.
19. H.I. Browman, R.D. Vetter (2001). Impacts of solar ultraviolet radiation on crustacean zooplankton and ichthyoplankton: Case Studies from Sub-Arctic marine Ecosystems. In: D.O. Hessen (Ed.), *UV Radiation and Arctic Ecosystems* (pp: 261–306). Springer-Verlag.
20. J.E.G. Downing, M.B.A. Djamgoz, J.K. Bowmaker (1986). Photoreceptors of a cyprinid fish, the roach: morphological and spectral charcteristics. *J. Comp. Physiol. A*, **159**, 859–868.
21. H.I. Browman, C.W. Hawyshyn (1992). Thyroxine induces a precocial loss of ultraviolet photosensitivity in rainbow trout (*Oncorhynchus mykiss*, Teleostei). *Vision Res.*, **32**, 2303–2312.
22. L. Beaudet, H.I. Browman, C.W. Hawryshyn (1993). Optic nerve response and retinal structure in rainbow trout of different sizes. *Vision Res.*, **33**, 1739–1746.
23. D. Karentz, F.S. McEuen, W.C. Dunlap (1991). Survey of mycosporine-like amino acid compounds in Antarctic marine organisms: potential protection from ultraviolet exposure. *Mar. Biol.*, **108**, 157–166.
24. J.B. McClintock, D. Karentz (1997). Mycosporine-like amino acids in 38 species of subtidal marine organisms from McMurdo Sound. Antarctica, *Antarct. Sci.*, **9**, 392–398.
25. D.S. Mason, F. Schafer, J.M. Shick, W.C. Dunlap (1998). Ultraviolet radiation-absorbing mycosporine-like amino acids (MAAs) are acquired from their diet by medaka fish (*Oryzias latipes*) but not by SKH-1 hairless mice. *Comp. Biochem. Physiol. A.*, **120**, 587–598.
26. T.C. Carefoot, D. Karentz, S.C. Pennings, C.L. Young (2000). Distribution of mycospirine-like amino acids in the sea hare *Aplysia dactylomela*: effect of diet on amounts and types sequestered over time in tissue and spawn. *Comp. Biochem. Physiol. C*, **126**, 91–104.
27. R. Sommaruga, F.Garcia-Pichel (1999). UV-absorbing mycosporine-like compounds in planktonic and benthic organisms from a high-mountain lake. *Arch. Hydrobiol.*, **144**, 255–269.
28. E.W. Helbling, F. Zaratti, L.O. Sala, E.R. Palenque, C.F. Menchi, V.E. Villafañe (2002). Mycosporine like aminoacids protect the copepod *Boeckella titicacae* (Harding) against high levels of solar UVR. *J. Plankton Res.*, **24**, 225–234.
29. D.O. Hessen (1996). Competitive trade-off strategies in Arctic *Daphnia* linked to melanism and UV-B stress. *Polar Biol.*, **16**, 573–579.
30. I D. Zellmer (1998). The effect of solar UVA and UVB on subarctic *Daphnia pulicaria* in its natural habitat. *Hydrobiologia*, **379**, 55–62.
31. V.E. Lyon, B.I. Baker (1993). The effect of photoperiod on plasma levels of melanin concentrating hormone in the trout. *J. Neoroendokrinol.*, **5**, 493–499.
32. D.L. Fabacher, E.E. Little (1998). Photoprotective substance occurs primarily in outer layers of fish skin. *Environ. Sci. Poll. Res.*, **5**, 4–6.
33. D.L. Fabacher, E.E. Little, G.K. Ostrander (1999). Tolerance of an albino fish to ultraviolet-B radiation. *Env. Sci. Poll. Res.*, **6**, 69–71.
34. T. Nakano, M. Sato, M. Takeuchi (1993). Superoxide-dismutase activity in the skin of fish. *J. Fish Biol.*, **43**, 492–496.
35. R. Hofer, C. Mokri (2000). Photoprotection in tadpoles of the common frog, *Rana temporaria. J. Photochem. Photobiol. B: Biol.*, **59**, 48–53.
36. A.R. Blaustein, J.M. Kiesecker, D.P. Chivers, D.G. Hokit, A. Marco, L.K. Belden, A.

Hatch (1998). Effects of ultraviolet radiation on amphibians: Field experiments. *Am. Zool.*, **38**, 799–812.
37. V. Brehm (1938). Die Rotfarbung von Hochgebirgsorganismen. *Biol. Rev.*, **13**, 307–318.
38. N.G. Hairston Jr. (1979). The effect of temperature on carotenoid photoprotection in the copepod *Diaptomus nevadensis. Comp. Biochem. Physiol. A*, **62**, 445–448.
39. J. Ringelberg, A.L. Keyser, B.J.G. Flik (1984). The mortality effect of ultraviolet radiation in a red morph of *Acanthodiaptomus denticornis* (Crustacea: Copepoda) and its possible ecological relevance. *Hydrobiologia*, **112**, 217–222.
40. V. Partali, Y. Olsen, P. Foss, S. Liaaen-Jensen (1985). Carotenoids in food chain studies – I. Zooplankton (*Daphnia magna*) response to a unialgal (*Scenedesmus acutus*) carotenoid diet, to spinach, and to yeast diets supplemented with individual carotenoids. *Comp Biochem. Physiol.*, **82B**, 767–772.
41. D.O. Hessen, K. Sørensen (1990). Photoprotective pigmentation in alpine zooplankton populations. *Aqua Fennica*, **20**, 165–170.
42. R.D. Vetter, A.L. Kurtzman, E.A. Lynn, K.W. Patterson, R.C. Smith (2002). Solar ultraviolet radiation causes DNA damage in pelagic fish larvae under natural conditions. *Proc Natl. Acad. Sci. U.S.A.*, in press.
43. R.D. Vetter, A.L. Kurtzman, T. Mori (1999). Diel cycles of DNA damage and repair in eggs and larvae of Northern anchovy, *Engraulis mordax*, exposed to solar ultraviolet radiation. *Photochem. Photobiol.*, **69**, 27–33.
44. H. Saegrov, A. Hobaek, J.H. L'Abee Lund (1996). Vulnerability of melanic *Daphnia* to brown trout predation. *J. Plankton Res.*, **18**, 2113–2118.
45. L.A. Hansson (2000). Induced pigmentation in zooplankton: a trade-off between threats from predation and ultraviolet radiation. *Proc. Royal Soc. Lond. Ser B*, **267**, 2327–2331.
46. S. Johnsen, E.A. Widder (2001). Ultraviolet absorption in transparent zooplankton and its implications for depth distribution and visual predation. *Mar. Biol.*, **138**, 717–730.
47. B. Halliwell, J.M.C. Gutteridge (1989). *Free Radicals in Biology and Medicine*. Clarendon Press, Oxford.
48. B. Ketterer, D.J. Meyer (1989). Glutathione transferases: a possible role in detoxication and repair in DNA and lipid hydroperoxides. *Mutat.. Res.*, **214**, 33–40.
49. C.J. Bertling, F. Lin, A.W. Girotti (1996). Role of hydrogen peroxide in the cytotoxic effects of UVA/B radiation on mammalian cells. *Photochem. Photobiol.*, **64**, 137–142.
50. R. Kerb, J. Brockmöller, T. Reum, I. Roots, I. (1997). Deficiency of glutathione S-transferases T1 and M1 as heritable factors of increased cutaneous UV sensitivity. *J. Invest. Dermatol.*, **108**, 229–232.
51. J. Borgeras, D.O. Hessen (2002). Species-specific and clone-specific expression of anti-oxidants and pigmentation in *Daphnia* spp., *Hydrobiologia*, in press.
52. D.B. Dey, D.M. Damkaer, G.A. Heron (1988). UV-B dose/dose-rate responses of seasonally abundant copepods of Puget Sound. *Oecologia*, **76**, 321–329.
53. D.M. Damkaer, D.B. Dey, G.A. Heron, E.F. Prentice (1980). Effecs of UV-B radiation on near-surface zooplankton of Puget Sound. *Oecologia*, **44**, 149–58.
54. D.M. Damkaer (1982). Possible influences of solar UV radiation in the evolution of marine zooplankton. In: J. Calkins J (Ed.), *The Role of Solar Ultraviolet Radiation in Marine Ecosystems* (pp. 701–706). Plenum Press, New York.
55. D.M. Damkaer, D.B. Dey (1983). UV damage and photoreactivation potentials of larval shrimp, *Pandalus platyceros*, and adult euphausiids, *Thysanoesa raschii. Oecologia*, **60**, 169–175.

56. K.D. Malloy, M.A. Holman, D. Mitchell, H. William Detrich (1997). Solar UV-B-induced DNA damage and photoenzymatic DNA repair in antarctic zooplankton. *Proc. Natl. Acad. Sci. U.S.A.*, **94**, 1258–1263.
57. S.E. Kaup, J.R. Hunter (1981). Photorepair in larval anchovy. *Engraulis mordax*. *Photochem Photobiol.*, **33**, 253–256.
58. J.D. Regan, W.L. Carrier, C. Samet, B.L. Olla (1982). Photoreactivation in two closely related marine fishes having different longevities. *Mech. Ageing Develop.*, **18**, 59–66.
59. O. Siebeck, U. Böhm (1994). Challenges for an appraisal of UV-B effects upon planktonic crustaceans under natural conditions with a non-migrating (*Daphnia pulex* obtusa) and a migrating cladocran (*Daphni galeata*). *Arch. Hydrobiol. Beih.*, **43**, 197–206.
60. H.E. Zagarese, M. Feldman, C.E. Williamson (1997). UV-B-induced damage and photoreactivation in three species of *Boeckella* (Copepoda, Calanoida). *J. Plankton Res.*, **19**, 357–367.
61. M.J. Lindell, W. Granéli, L.J. Tranvik (1995). Enhanced bacterial growth in response to photochemical transformation of dissolved organic matter. *Limnol. Oceanogr.*, **40**, 195–199.
62. M.L. Bothwell, D.M.J. Sherbot, C.M. Pollock (1994). Ecosystem response to solar ultraviolet-B radiation: influence of trophic-level interactions. *Science*, **265**, 97–100.
63. W.J. Cooper, C. Shao, D.R.S. Lean, A.S. Gordon, N.E. Scully (1994). Factors affecting the distribution of H_2O_2 in surface waters. In: L.A. Baker, (Ed.), *Environmental Chemistry of Lakes and Reservoirs* (pp. 391–422). American Chemical Society, Washington.
64. O.C. Zafiriou, N.V. Blough, E. Micinski, B. Dister, D. Kieber, J. Moffett (1990). Molecular probe systems for reactive transients in natural-waters. *J. Mar. Chem.*, **30**, 45–70.
65. D.O. Hessen, E. Van Donk (1994). UV-radiation of humic water; effects on primary and secondary production. *Water Air Soil Poll.*, **74**, 1–14.
66. D.O. Hessen, P.J. Færøvig (2001). The photo-protective role of humus-DOC for *Selenastrum* and *Daphnia*. *Plant Ecol.*, **154**, 225–235.
67. J. Borgeraas, D.O. Hessen (2000). UV-B induced mortality and antioxidant enzyme activities in *Daphnia magna* at different oxygen concentrations and temperatures. *J. Plankton Res.*, **22**, 1167–1183.
68. D. Karentz, J.E. Cleaver, D.L. Mitchell (1991). Cell survival characteristics and molecular responses of Antarctic phytoplankton to ultraviolet-B radiation. *J. Phycol.*, **27**, 326–341.
69. D. Karentz, M.L. Bothwell, R.B. Coffin, A. Hanson, G.J. Herndl, S.S. Kilham, M.P. Lesser, M. Lindell, R. Moeller, D.P. Morris, P.J. Neale, R.W. Sanders, C.S. Weiler, R.C. Wetzel (1994). Impact of UV-B radiation on pelagic freshwater ecosystems: Report of working group on bacteria and phytoplankton. *Arch. Hydrobiol. Beih.*, **43**, 31–69.
70. E. Van Donk, B. Faafeng, H.J. De Lange, D.O. Hessen (2001). Differential sensitivity to natural ultraviolet radiation among phytoplankton species in Arctic lakes (Svalbard, Norway). *Plant Ecol.*, **154**, 213–223.
71. D.O. Hessen, H.J. DeLange, E. Van Donk (1997). UV-induced changes in phytoplankton cells and its effects on grazers. *Freshwat. Biol.*, **38**, 513–524.
72. J.D. Scott, L. Chalker-Scott, A.E. Foreman, M. D'Angelo (1999). *Daphnia pulex* fed UVB-irradiated *Chlamydomonas reinhardtii* show decreased survival and fecundity. *Photochem. Photobiol.*, **70**, 308–313.
73. H.J. DeLange, M. Lürling (2002). Effects of UV-B irradiated algae on zooplankton

grazing. *Hydrobiologia*, in press.
74. M.T. Arts, H. Rai (1997). Effects of enhanced ultraviolet-B radiation on the production of lipid, polysaccharide and protein in three freshwater algal species. *Freshwat. Biol.*, **38**, 597–610.
75. J.I. Goes, N. Handa, S. Taguchi, T. Hama (1994). Effect of UV-B radiation on the fatty acid composition of the marine phytoplankton *Tetraselmis* sp: Relationship to cellular pigments. *Mar. Ecol. Progr. Ser.*, **114**, 259–274.
76. K.S. Wang, T. Chai (1994). Reduction in omega-3 fatty acids by UV-B irradiation in microalgae. *J. Appl. Phycol.*, **6**, 415–421.
77. J.R. Rainuzzo, K.I. Reitan, Y. Olsen (1997). The significance of lipids at early stages of marine fish: a review. *Aquaculture*, **155**, 103–115.
78. K.I. Reitan, J.R. Rainuzzo, G. Øie, Y. Olsen (1997). A review of the nutritional effects of algae in marine fish larvae. *Aquaculture*, **155**, 207–221.
79. C.A. Ochs (1997). Effects of UV radiation on grazing by two marine heterotrophic nanoflagellates on autotrophic picoplankton. *J. Plankton Res.*, **19**, 1517–1536.
80. R. Sommaruga, A. Oberleiter, R. Psenner (1996). Effect of UV radiation on the bacterivory of a heterotrohic nanoflagellate. *Appl. Enviromn. Microbiol.*, **62**, 4395–4400.
81. P.J. Neale, R.F. Davis, J.J. Cullen (1998). Interactive effects of ozone depletion and vertical mixing on photosynthesis of Antarctic phytoplankton. *Nature*, **392**, 585–589.
82. R. Rae, W.F. Vincent (1998). Effects of temperature and ultraviolet radiation on microbial food web structure: potential responses to global change. *Freshwat. Biol.*, **40**, 747–758.
83. J.C. Roos, W. Vincent (1998). Temperature dependence of UV radiation effects on Antarctic cyanobacteria. *J. Phycol.*, **34**, 118–125.
84. D. Abele, H. Großpietsch,, H.O. Pörtner (1998). Temporal fluctuations and spatial gradients of environmental PO_2, temperature, H_2O_2 and H_2S in its intertidal habitat trigger enzymatic antioxidant protection in the capitellid worm *Heteromastus filiformis*. *Mar. Ecol. Prog. Ser.*, **163**, 179–191.
85. D.O. Hessen, N.E.W. Alstad Rukke (2000). Increased UV-susceptibility in *Daphnia* at low calcium concentrations. *Limnol. Oceanogr.*, **45**, 1834–1838.
86. C.E. Williamson, O.G. Olson, S.E. Lott, N.D. Walker, D.R. Engstrom, B.R. Hargreaves (2001). Ultraviolet radiation and zooplankton community structure following deglaciation in Glacier Bay, Alaska. *Ecology*, **82**, 1748–1760.
87. C.E. Willamson, H. Zagarese, P.C. Schulze, B.R. Hargraves, J. Seva (1994). The impact of short-term exposure to UV-B radiation on zooplankton communities in north temperate lakes. *J. Plankton Res.*, **16**, 205–218.
88. S. Cabrera, M. López, B. Tartarotti (1997). Phytoplankton and zooplankton response to ultraviolet radiation in a high altitude Andean lake: short- versus long-term effects. *J. Plankton Res.*, **19**, 1565–1582.
89. R. Sommaruga, R. Psenner (1997). Ultraviolet radiation in a high mountain lake of the Austrian Alps: Air and underwater measurements. *Photochem. Photobiol.*, **65**, 957–963.
90. B. Tartarotti, S. Cabrera, R. Psenner, R. Sommaruga (1999). Survivorship of *Cyclops abyssorum* tatricus (Cyclopoida, Copepoda) and *Boeckella gracilipes* (Calanoida, Copepoda) under ambient levels of solar UVB radiation in two high-mountain lakes. *J. Plankton Res.*, **21**, 549–560.
91. R.D. Hurtubise, J.E. Havel, E.E. Little (1998). The effects of ultraviolet-B radiation on freshwater invertebrates: Experiments with a solar simulator. *Limnol. Oceanogr.*, **43**, 1082–1088.

92. C.E. Williamson, B.R. Hargreaves, P.S. Orr, P.A. Lovera (1999). Does UV radiation play a role in changes in predation and zooplankton community structure in acidified lakes?. *Limnol. Oceanogr.*, **44**, 774–783.
93. H.I. Browman, R.D. Vetter, C. Alonso Rodriguez, J.J. Cullen, R.F. Davis (2002). UV (280–400 nm) induced DNA damage in eggs and larvae of *Calanus finmarchicus* Gunnerus (Copepoda) and Atlantic cod (*Gadus morhua*). *Photochem. Photobiol.*, in press.
94. P. Kuhn, H.I. Browman, J.-F. St-Pierre, B. McArthur (1999). Penetration of ultraviolet radiation in the waters of the estuary and Gulf of St. Lawrence. *Limnol. Oceanogr.*, **44**, 710–716.
95. F. Béland, H.I. Browman, C. Alonso Rodriguez, J.-F. St-Pierre (1999). The effect of solar ultraviolet radiation (280–400 nm) on the eggs and larvae of Atlantic cod (*Gadus morhua*). *Can. J. Fish. Aquat. Sci.*, **56**, 1058–1067.
96. C.A. Rodriguez, H.I. Browman, J.-F. St-Pierre (2000). High survival of neustonic zoea I larvae of American lobster *Homarus americanus* following short-term exposure to ultraviolet radiation (280 to 400 nm). *Mar. Ecol. Progr. Ser.*, **193**, 305–309.
97. D.B. Carlini, J.D. Regan (1995). Photolyase activities of *Elysa tuca*, *Bursatella leachii*, and *Haminaea antillarum* (Mollusca: Opisthobranchia). *J. Exp. Mar. Biol. Ecol.*, **189**, 219–232.
98. H.E. Zagarese, C.E. Williamson (2000). Impact of solar UV radiation on zooplankton and fish. In: S. de Mora, S. Demers, M. Vernet (Eds), *The Effects of UV Radiation in the Marine Environment* (pp. 279–309). Cambridge University Press, Cambridge.
99. J.R. Hunter, S.E. Kaupp, J.H. Taylor (1982). Assessment of effects of UV radiation on marine fish larvae. In: J. Calkins (Ed.), *The Role of Solar Ultraviolet Radiation in Marine Ecosystems* (pp. 459–493). Plenum Press, New York.
100. J.R. Hunter, S.E. Kaupp, J.H. Taylor (1981). Effects of solar and artificial ultraviolet-B radiation on larval northen anchovy, *Engraulis mordax*. *Photochem. Photobiol.*, **34**, 477–486.
101. J.R. Hunter, J.H. Taylor, H.G. Moser (1979). Effect of ultraviolet irradiation on eggs and larvae of the northern anchovy, *Engraulis mordax*, and the Pacific mackerel, *Scomber japonicus*, during the embryonic stage. *Photochem. Photobiol.*, **29**, 325–338.
102. E.L. Bass, S.N. Sistrun (1997). Effect of UV-A radiation on development and hatching success in *Oryzias latipes*, the Japanese medaka. *Bull. Environ. Contam. Toxicol.*, **59**, 537–542.
103. C.E. Williamson, S.L. Metzgar, P.A. Lovera, R.E. Moeller (1997). Solar ultraviolet radiation and the spawning habitat of yellow perch, *Perca flavescens*. *Ecol. Appl.*, **7**, 1017–1023.
104. C. Dunbar (1959). Sunburn in fingerling rainbow trout. *Progr. Fish Cult.*, **21**, 74.
105. A.M. Bullock, R. Coutts (1985). The impact of solar ultraviolet radiation upon the skin of rainbow trout, *Salmo gairdneri* Richardson, farmed at high altitude in Bolivia. *J. Fish Dis.*, **8**, 263–272.
106. R. Bergham, A.M. Bullock, M. Karakiri (1993). Effects of solar radiation on the population dynamics of juvenile flatfish in the shallows on Wadden Sea. *J. Fish Biol.*, **42**, 329–345.
107. V. S. Blazer, D.L. Fabacher, E.E. Little, M.S. Ewing, K.M. Kocan (1997). Effects of ultraviolet-B rediation on fish: histologic comparison of a UVB-sensitive and a UVB-tolerant species. *J. Aquat. An. Health*, **9**, 132–143.
108. H.M. Salo, T.M. Aaltonen, S.E. Markkula, E.I. Jokinen (2000). Ultraviolet B irradiation modulates immune system of fish (*Rutilus rutilus*, Cyprinidae). II. Blood.

Photochem. Photobiol., **71**, 65–70.

109. E.I. Jokinen, H.M. Salo, T.M. Aaltonen, S.E. Markkula (2001). Ultraviolet B irradiation modulates immune system of fish (*Rutilus rutilus*, Cyprinidae). III. Lymphocytes. *Photochem. Photobiol.*, **73**, 505–512.
110. A.P. Cullen, C.A. Monteith, McMaster (1993). Damage to the rainbow-trout (*Oncorhyncus mycisi*) lens following an acute dose of UVB. *Curr. Eye Res.*, **12**, 97–106.
111. C.A. Rodriguez, H.I. Browman, J.A. Runge, J.-F. St-Pierre (2000). Impact of solar ultraviolet radiation on hatching of a marine copepod, *Calanus finmarchicus. Mar Ecol Progr Ser.*, **193**, 85–93.
112. K.V. Fite, A. Blaustein, L. Bengtson, P.E. Hewitt (1998). Evidence of retinal light damage in *Rana cascadae*: A declining amphibian species. *Copeia*, **4**, 906–914.
113. J.M. Kiesecker, A.R. Blaustein, L.K. Belden (2001). Complex causes of amphibian population declines. *Nature*, **410**, 681–684.
114. D.O. Hessen, J. Borgeraas, J.B. Ørbaek (2002). Responses in pigmentation and anti-oxidant expression in arctic *Daphnia* along gradients of DOC and UV exposure. *J. Plankton. Res.*, in press.

Chapter 13

UVR-induced injuries in freshwater vertebrates

Edward E. Little and David Fabacher

Table of contents

Abstract

UVR can cause injuries to freshwater organisms that can lead to mortality. Freshwater fish and amphibians are especially susceptible to the harmful effects of UVR. Sunburn in cultured fish has been the most frequently reported UV-induced injury in freshwater organisms. The wavelengths typically responsible for UV-induced injuries are in the UV-B wavelength range (280–320 nm). Sunburn-like injuries have also been observed in certain amphibian life stages and probably occur to some degree in other aquatic organisms. Sunburn in nature has not been widely reported, probably because aquatic organisms often succumb to opportunistic infections shortly after acquiring this injury, and thus are not readily observed. UV sensitivity exhibited by an organism can be linked to the concentration and distribution of photoprotective pigments contained in the integument in addition to efficiency of photorepair mechanisms. A number of climatic conditions can contribute to UV-induced injuries, including extended periods of elevated UV associated with ozone depletion, changes in cloud cover or extent of sunny conditions, and global warming that may give rise to increased water clarity because of water column stratification. Habitat characteristics, in turn, also play a role in the risk of UV-induced injuries. These characteristics include not only latitude and altitude of the organism's habitat, but also the degree of shading provided by canopy or substrate cover. In addition, the chemical composition of the aquatic habitat (particularly the composition and concentration of DOC) can influence UV penetration into the water column. Moreover, UVR in the aquatic habitat is considerably more dynamic than at the Earth's surface, and can vary by orders of magnitude depending on water clarity. The presence of certain chemical contaminants can additively or synergistically increase UV injury in organisms, even in habitats having low UVR. Reports of UV-induced injuries in aquatic organisms will probably increase with ozone depletion and global warming. Thus, further research is needed in this area, especially research on sublethal UV-induced effects including behavior, growth, and reproduction.

13.1 Introduction

There are a variety of reports in the literature concerning UV-induced injuries in aquatic organisms, particularly fish [1,2]. An excellent treatise on the vulnerability of fish and amphibians to UVR was recently published [3]. Laboratory studies with different fish species have shown that the short-wave UV-B portion of the solar spectrum is the most damaging. The most commonly reported harmful effect of solar radiation in fish has been skin damage, which is often referred to as sunburn (Figure 1). Fish skin is sensitive to UV-B because the skin generally lacks a keratinized outer layer, has dividing cells in all layers of the epidermis, and normally does not contain protective melanin-containing cells in the epidermis [1].

A variety of reports have documented the harmful effects of UVR on fish.

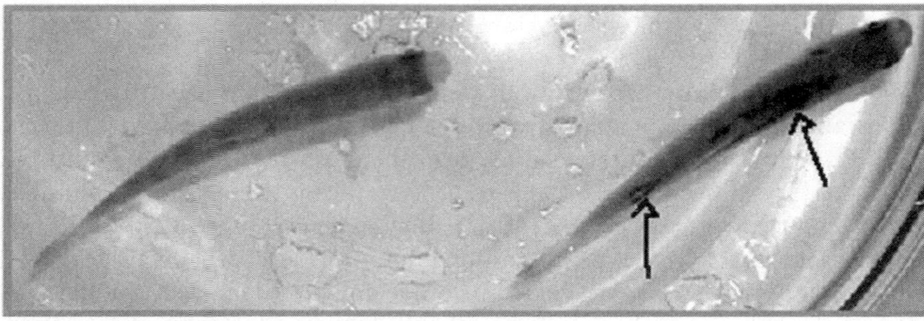

Figure 1. UV-exposed rainbow trout (*Oncorhychus mykiss*) on right showing darkened sunburn on dorsal surface (arrow), an unexposed fish is on left.

When the eggs and alevins of sockeye salmon (*Oncorhynchus nerka*) were irradiated with UVR, histologic examination of the alevins revealed changes in the epidermis and fibroelastic layers of the irradiated skin [4]. Dunbar [5] observed necrotic areas behind the head and around the base of the dorsal fin in rainbow trout (*Oncorhynchus mykiss*) fingerlings within three days when the fish were exposed to solar UV after being transferred from a hatchery to outdoor ponds. DeLong et al. [6] observed sunburn lesions and subsequent fungal infections at the site of these lesions in chinook salmon (*Oncorhynchus tshawytscha*). Bullock and Roberts [7] observed sunburn lesions in rainbow trout and Atlantic salmon (*Salmo salar*) fry exposed to solar radiation. Bullock et al. [8] observed solar radiation induced sunburn in koi carp (*Cyprinus carpio*) held outdoors in clean water with no shading cover, but the symptoms subsided in water that had algae or contained aquatic plant cover. Bullock and Coutts [9] observed solar radiation induced lesions on the dorsal surface of broodstock rainbow trout held on a high altitude fish farm in Bolivia. In another report Bullock [10] mentions other cases of probable solar radiation induced skin lesions in farmed finfish. Dermatitis was observed in two species of cultured salmonids held in unshaded raceways in British Columbia [11]. Ramos et al. [12] observed sunburn-like dorsal skin lesions in juvenile paddlefish (*Polyodon spathula*) held in outdoor raceways. The number of mucus secreting goblet cells was significantly reduced in the dorsal epidermis of two cyprinids and two salmonids exposed to both simulated and solar UV-B [13]. Goblet cells on the ventral side of the fish were unaffected. The authors speculated that the reduction in number of goblet cells could result in less mucus production with a subsequent reduction in non-specific defenses. "Sunburn cells" can be observed histologically in UV-B-injured fish skin and are characterized by dense granules of fragmented nuclear material [10,14,15]. These cells can be classified as either Type A or B and are differentiated by their morphological appearance under light microscopy [10], where the morphology of Type A cells most closely resembles that described for human skin, while the morphology of Type B cells closely resembles cells in eye and brain lesions in larval anchovy and mackerel as described by Hunter et al. [16].

Much of the information on UV-B-affected fish has come from laboratory studies. For example, Lahontan cutthroat trout (*Oncorhynchus clarki henshawi*) were much more sensitive to simulated solar UV-B than were razorback suckers (*Xyrauchen texanus*) [15]. Cutthroat trout showed darkening of the dorsal skin resulting from UV-B-induced melanosome dispersion with 48 h of exposure. Histological observations included sloughing of mucous cells, epidermal and dermal necrosis and edema, and in some cases secondary fungal infection. Cells that appeared to be Type B sunburn cells were occasionally observed in cutthroat trout. More free melanosomes were observed in the skin of UV-B exposed cutthroat trout. Conversely, the response of razorback suckers was very different. Razorback suckers showed no visible signs of sunburn after 72 h of UV-B exposure. Nevertheless, histologic examination revealed that cellular necrosis had occurred, but not nearly to the extent that it had in cutthroat trout. There was an increase in epidermal thickness as a result of large low-electron-dense cells that appeared to be club cells. There were some focal areas of increased melanocyte accumulation in the dermis. No sunburn cells were observed. Brown trout (*Salmo trutta*) were also found to be sensitive to simulated UV-B exposure [14]. Loss of cell layers occurred in exposed fish by increased surface cell sloughing or by detachment of the outer and middle cellular layers. Mucus cells disappeared from the epidermis of exposed fish. In addition, sunburn cells were also observed, although no distinction could be made as to whether they were Type A or B. Channel catfish (*Ictalurus punctatus*) fingerlings were found to be highly sensitive to simulated solar UV-B [17]. After 24 h exposure all the body surfaces were slightly darker than in control fish. Thinning of the most dorsal epidermis was accompanied by edema and occasionally both Type A and B sunburn cells were observed. Focal necrosis and sloughing of the outer epidermal layer was widespread by 48 h exposure (Figure 2). Three fish were dead and all remaining live fish appeared to have fungal infection on the dorsal skin. At 72 h exposure all remaining fish were dead. There appeared to be a time dependent decrease in the number of mucus cells in the epidermis.

UV injuries are not limited to fish. For example, aquatic tiger salamander (*Ambystoma tigrinum*) larvae were highly sensitive to UV-B, and developed lesions over the dorsal skin areas (Figure 3) within seven days of exposure to simulated UV irradiances as low as 2 μW cm^{-2} [Carey personal communication]. Cellular damage was found in the epidermis of surviving alpine newt (*Triturus alpestris*) larvae exposed to both simulated and solar UVR [18]. Using a combination of ambient solar UV and simulated UV-B, Flamarique et al. [19] observed substantial skin damage and prominent eye lens opacities in tadpoles of two frog species, *Hyla regilla* and *Rana aurora*. Darkened dorsal areas were evident among certain pelagic aquatic invertebrates prior to mortality during simulated UV exposure [20].

In addition to skin injuries, increased UV-B exposure of fish in natural waters and shallow water hatchery pools and raceways may cause eye lens damage in the form of cataracts [21]. Cataracts, evident in observations of blinded brown trout in European streams (http://s.o.w.tripod.com/dyingbrowns.htm) have also been observed in eye lenses of hatchery reared lake trout (*Salvelinus namaycush*) exposed to solar UV [22–24].

A

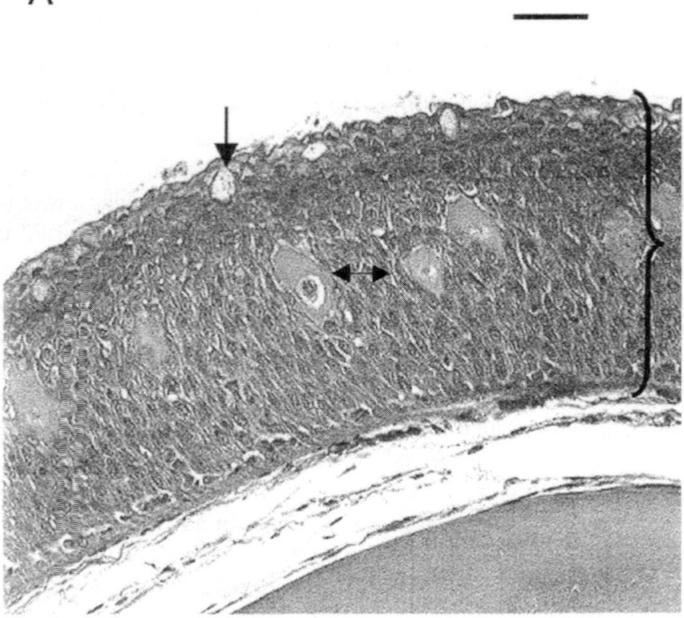

B

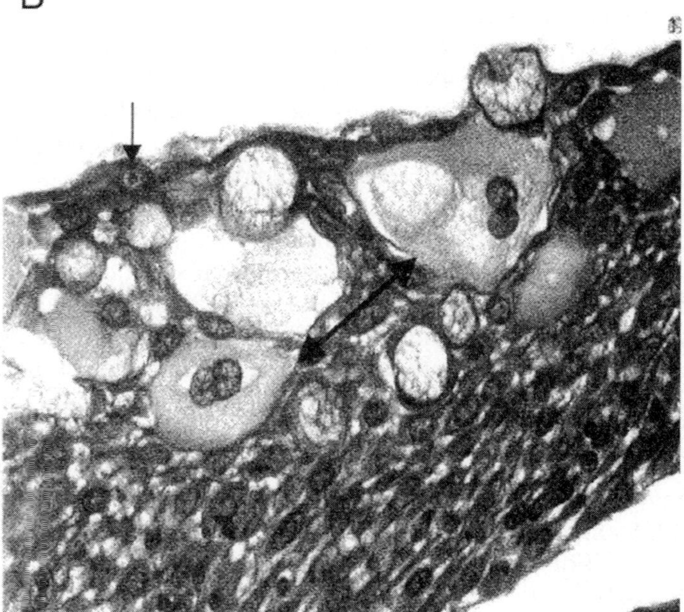

Figure 2. (A) Dorsal skin of unexposed channel catfish (*Ictalurus punctatus*). Epidermis and dermis close to the dorsal-most portion is 10 or more cells thick and contains some mucous (single arrow) and club (double arrow) cells. Scale bar equals 20 μm. (B) Dorsal-lateral skin of channel catfish exposed to UV-B radiation for 24 h. Club cells are often vacuolated and hypertrophied (double arrow) in this group. Sunburn cells (single arrow) with perinuclear halo are also observed. Scale bar equals 10 μm. [From Ewing et al. 17.]

Figure 3. Tiger salamander (*Ambystoma tigrinum*) following 48 h of UV exposure. Note extensive fungal infections (arrows) over skin surface.

13.2 Impact on natural populations

Injuries among natural fish populations have not been frequently reported, possibly because affected fish succumb to opportunistic infections shortly after acquiring this injury and are not readily observed. Nonetheless, solar UV was found to affect the survival of bluegill (*Lepomis macrochirus*) embryos in natural lakes and the selection of spawning habitat and hatching success of yellow perch (*Perca flavescens*) [25,26]. Coho salmon (*Oncorhynchus kisutch*) stocks that develop in relatively sunny rain-free areas of British Columbia appear to be declining compared to more northern coastal stocks that develop under less sunny conditions [27]. The declines are thought to result from solar UVR injuries to early life stages that develop in shallow freshwater streams. UVR has also been implicated in the decline of marine fish such as cod (*Gadus morhua*) in the North Atlantic where water column UV levels were found to be at irradiance levels harmful to developing embryos [28].

Brown trout (*Salmo trutta*) endemic to certain clear water streams in Europe have been declining in recent years (http://s.o.w.tripod.com/dyingbrowns.htm). Abnormally dark colored and blind brown trout have frequently been observed in these streams while other fish species in these streams such as rainbow trout and grayling appear unaffected. Many of these streams were channelized and are shallow with little or no canopy cover or protective in-water substrate. Conversely, unchannelized streams that appear to support healthy brown trout populations have rocky bottoms substrates as well as shade from tree canopy that can provide places for fish to avoid excessive solar radiation. These unchannellized streams also had deep pools and were often tea-colored from tannic substances. Such observations provide circumstantial evidence that solar UV may be affecting brown trout in certain streams in Europe.

13.3 Injuries can become sites of infection

Fish exposed to UV-B may experience a depressed immune system and subsequent vulnerability to infection by pathogens [29,30]. Epidermal lesions from sunburn allow invasion by pathogens, especially fungi [2,31]. During laboratory studies, fungal hyphae were frequently observed at the margins of the sunburned fish skin within one or two days of the initial sunburn [2]. These fungal infections progressed over the dorsal surface of the fish and the fish died soon after. Extensive fungal infection was also noted among tiger salamander larvae that had developed skin lesions during UV exposure [Carey personal communication]. In mammals, UV-B-induced immunosuppression may occur through isomerization of urocanic acid from the *trans* to the *cis* form [32–34]. The *cis* form of urocanic acid may modulate cell-mediated immunity by binding to receptors [35,36]. Urocanic acid occurs in mammalian skin predominantly as the *trans* isomer. Upon UV-B irradiation of the skin, urocanic acid isomerizes to the *cis* isomer along with concurrent suppression of the immune response. In mammals this is thought to be a protective mechanism by preventing uncontrolled autoimmune destruction of sun-damaged skin cells [34]. Exposure to high levels of UV-B could cause hyperstimulation of this mechanism with subsequent increased susceptibility to pathogens. In a preliminary study, a substance that appeared to be *trans*-urocanic acid was found in the skin of UV-B-exposed and unexposed rainbow trout [37]. Thus, a similar mechanism of immunosuppression may be induced by UV-B in fish. Although it remains to be established that urocanic acid functions similarly in fish epidermis as it does in mammals, the detection of this chemical raises the possibility that localized immunosuppression may also be responsible for epidermal infections observed in fish and other aquatic organisms [37].

Peptides on the skin surface have been shown to play an immunological role by binding to pathogens in amphibians [38–40]. The amino acid composition of these peptides may be vulnerable to photo-decomposition or photo-transformation by UV causing the peptide to be less effective as bactericidal or fungicidal agents. In such cases, UV could directly lead to increased vulnerability to epidermal infection. Sustained exposure to solar radiation may play an important role in the initiation of disease outbreaks in freshwater fish. Suppression of the immune system may occur in UV-B-exposed fish, making the fish more susceptible to disease organisms. Conversely, low-level infection by pathogens may increase the vulnerability of the skin of fish to ambient levels of solar UV-B. For example, Bullock [41] observed increased radiation damage in the skin of plaice (*Pleuronectes platessa*), a marine fish infested with an ectoparasite that also occurs on the skin of cultured salmonids reared in freshwater. Apparently, very low doses of UV-B were sufficient to enhance the rate of breakdown in skin structure initiated by the ectoparasite. Thus, conditions that affect the integrity of the epidermis prior to, and during, exposure to solar UVR can lead to increased susceptibility of the fish to UV-B.

13.4 Photoprotective mechanisms

Protection of freshwater aquatic organisms from UV-induced injury is dependent on a variety of factors that can function as photoprotective mechanisms. When UVR breaches photoprotective mechanisms in sufficient amount, UV-induced injury will occur. Aquatic organisms vary in their tolerance to UV exposure. There is a likely interplay between the ecological niche occupied by an organism and its UV sensitivity. Throughout an organism's life stages, its habitats and habits will likely complement the organism's tolerance for UV. Nocturnal or crepuscular activity regimes would clearly limit UV exposure, as would the organism's selection of UV-limiting habitats. Fish species have adapted to certain levels of UV and probably exhibit different ranges of tolerance to solar UVR. Species naturally adapted to high levels of solar radiation exposure would be more tolerant of high UV-B levels than species adapted to low levels of solar radiation. Indeed, razorback suckers, a fish species naturally adapted to high solar UV levels, were tolerant of simulated UV-B and did not develop sunburn after 21 days of exposure [42]. In contrast, sunburn was observed in rainbow trout and Lahontan cutthroat trout within two days, and Apache trout (*Oncorhynchus apache*) after five days exposure to simulated UV-B [2].

The sensitivity to UVR will vary with the fish's acclimation to solar radiation. Increases in water clarity and changes in water chemistry (particularly decreases in organic carbon content) can heighten UV intensity in the water column [43]. Even though fish may detect UV-B when applied in laboratory studies, they probably do not directly avoid solar UV-B [44], but indirectly avoid UV-B by seeking positions lower in the water column or by seeking shade to avoid intense visible or UV-A. Since solar visible and UV-A would not increase with ozone depletion, higher levels of solar UV-B would go undetected by fish and could cause harmful effects [45]. However, fish that exploit surface or shallow water habitats probably evolved adaptations to tolerate the high UV irradiance of such habitats [46].

There are also physiological characteristics that underlie an organism's tolerance to UV-B including the efficiency of photorepair mechanisms for damaged DNA. DNA is particularly vulnerable to UV because it induces the formation of cross-linkages, or dimers, in the pyrimidine base thymine. Such cross-linkages include cyclobutane-type dimers of thymine, cytosine, and uracil; pyrimidine adducts; photohydrates; and DNA–protein cross-links [47] can interfere with DNA replication and protein synthesis necessary for cell division in growth and replacement and can lead to the development of tumors, as well as lesions. Most organisms are capable of repairing the DNA damage induced by UV-B through excision repair, photoreactivation, and post-replication repair ([47], see also Chapter 9). Among these, photoreactivation is promoted by the DNA photolyase, an enzyme that binds to the cyclobutane dimer, becomes activated by absorbing photons from UV-A and visible light, then cleaves the dimer from the ring before unbinding [48]. Blaustein et al. [49] found that the amount of this enzyme in embryonic amphibians is directly correlated with the UV-B tolerance of that species. Although this correlation does not demonstrate a higher rate of dimer repair among tolerant organisms compared to sensitive organisms, it does suggest a cellular basis for photorepair efficiency. However, results from studies with the wood frog (*Rana sylvatica*) indicated variation in response to photolyase depending on environmental conditions and led Smith et al. [50] to conclude that estimating amphibian photorepair is a complicated process and that previous conclusions regarding the relationship between photorepair and amphibian population decline must be reevaluated. Photorepair is likely to be ubiquitous among fish given the range of species for which evidence for photorepair has been found, including goldfish [51], anchovy larvae [52] and fathead minnow [53]. Photorepair efficiency in fish varied by as much as 500% between two closely related species [54].

Excision repair involves damage recognition, incision of the DNA chain near the site of the lesion as DNA is excised and resynthesized around the damaged site, and ligation following detachment of DNA polymerase [48]. Species may be capable of both types of repair mechanisms and may vary as to which one predominates. For example, much of the DNA repair occurred during daylight hours through photorepair, and remaining repair occurred in darkness through excision repair [55].

Regardless of an organism's efficiency for photorepair such mechanisms are not completely efficient, so do not entirely ensure against potential injury. Therefore, additional protective mechanisms are adaptive. These include pigments that sequester the highly reactive oxygen free radicals or other reactive species that are generated by UV and are responsible for DNA damage and other cellular injuries ([56], see also Chapter 8). UV-shielding pigments also provide a means of protection from UV. These can be extrinsic as well as intrinsic filters. Aquatic animals utilize available environmental factors that provide photoprotection, such as DOC, which reduce environmental UV levels, and as a consequence the dose the organism receives. Aquatic organisms also incorporate UV-absorbing or reflective pigments in their integument.

13.4.1 Photoprotective substances in aquatic organisms

There are several types of photoprotective substances found in aquatic organisms (see Chapter 10). Photoprotective pigments have certain general characteristics as outlined below [57]. Briefly, most photoprotective substances share the π-electron systems that occur in conjugated bond structures such as alternating single and double bonds in linear molecules, and in aromatic and cyclic compounds containing electron resonance. Overlapping orbits of π-electrons have absorption maxima in the UV region that causes an energetic transition of π-electrons to anti-bonding π^*-electron orbits. Alteration in the structure of a conjugated molecule changes the absorbance characteristics and therefore the irradiance spectra that are attenuated. Longer wavelengths are absorbed by larger molecules and shifts in wavelengths absorbed occur as the number of conjugated bonds or number of substituents is increased. Absorption increases as more side chains and substitutions are added to the molecular structure. The non-bonding electrons of oxygen, nitrogen, and halogen atoms become a part of the resonance of the ring structure and shift the absorption maximum. Over the course of evolution, changes in the molecular structure have developed for specific absorbance characteristics. An organism can screen a broad spectrum of UV-B and UV-A wavelengths by synthesizing a range of photo-absorbing molecules.

Mycosporine-like amino acids, another type of photoprotective substance, have been found in a diversity of organisms ranging from bacteria to fish [57]. Up to 19 kinds of mycosporine-like amino acids have been identified. Certain organisms contain several of these substances that broadly screen UV [58]. The concentrations of these substances increase proportionately with the intensity of UV irradiance they are exposed to [58–60]. Mycosporine-like amino acids are probably not synthesized by fish and invertebrates, but acquired through the diet, especially from grazing on algae [61]. Gadusol, also believed to be photoprotective, is structurally related to mycosporine-like amino acids and is found in the eggs of cod and Mediteranian fish [62] and in brine shrimp [63].

Photoprotective melanins are found in a diversity of vertebrate and invertebrate organisms and are polymers formed from 5,6-dihydroxyindole, a phenolic and indolic compound [64]. Melanins are complex molecules that broadly absorb UV and visible radiation, but show no specific absorption maximum; however, their absorption increases with decreasing wavelength [65,66]. Melanin is produced in melanophores that then deposit melanin on subcellular organelles called melanosomes, which are often positioned above the nucleus [67]. Exposure to UV can cause an increase in melanin production and the number of melanosomes during long-term exposures to UV [57]. Fish typically have no melanin in the epidermis leaving this skin layer relatively unprotected from UVR, and appear to rely on a colorless compound(s) secreted in the mucus covering the body to provide epidermal photoprotection [68]. In frogs (Figure 4), however, melanin does occur in the epidermis, where photoprotection appears to be related to the amount of melanin as well as its distribution [69]. UV-B-tolerant boreal toads (*Bufo boreas*) have a distinct double layer of melanin. Related Woodhouse's toads (*Bufo woodhousii*) that live in lower altitude habitats also have a double melanin layer, but the melanin appears to be diffuse and less concentrated. Nocturnal gray tree frogs (*Hyla versicolor*) have a single layer of melanin, and are sensitive to UV-B. Tiger salamanders (*Ambystoma tigrinum*) have a diffuse and limited distribution of melanocytes and are highly sensitive to UV-B [Carey personal communication].

Carotenoid pigments occur widely among crustacean zooplankton, though the composition and quantities vary with species [55]. Carotenoids are produced by algae, bacteria and plants, and transferred to invertebrate and vertebrates through the food chain [70]. Carotenoids are thought to be the main photoabsorbing pigment in copepods. Copopods show greater concentrations of carotenoids than do cladocerans, and those living in UV intense environments have higher concentrations than organisms from low UV habitats [71]. Caro-

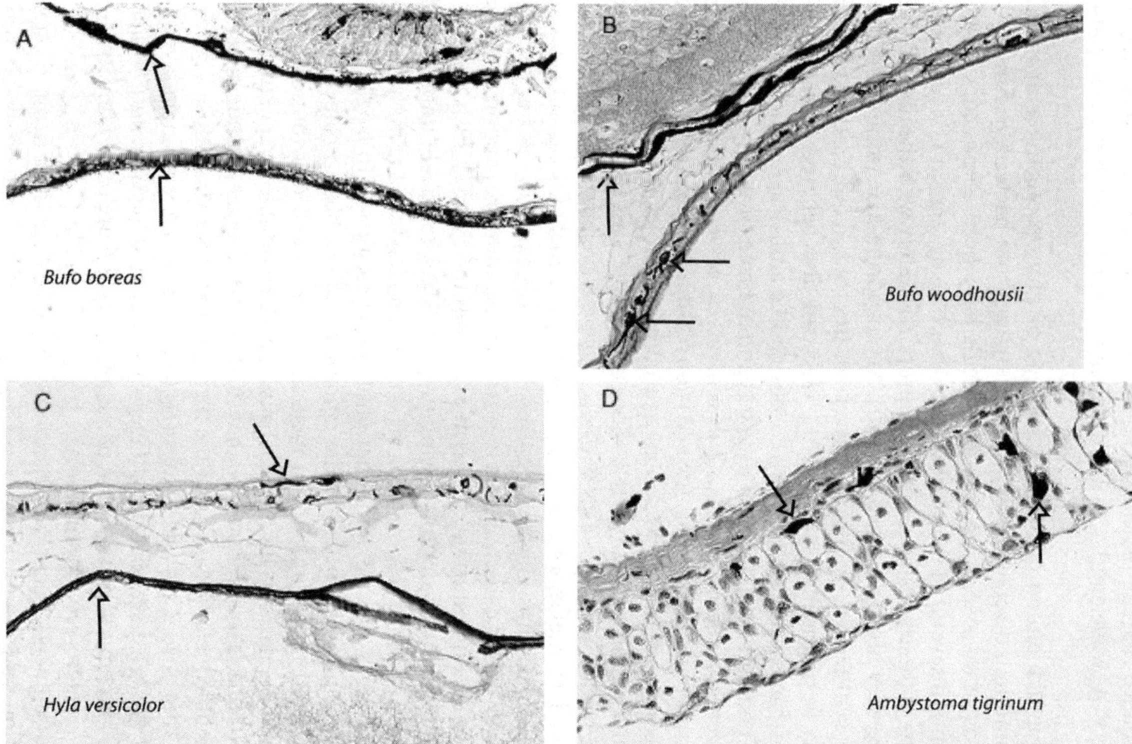

Figure 4. Distribution of melanin in the larval skin of (A) the boreal toad (*Bufo boreas*), (B) Woodhouse's toad (*Bufo woodhousii*), (C) gray tree frog (*Hyla versicolor*), (D) tiger salamander (*Ambystoma tigrinum*). Note dense layer of melanin below skin (single arrow) in the toad and frog larvae, and the melanin distribution in the epidermal layer (double arrow) which is most dense in the UV tolerant boreal toad, and least dense in the nocturnal (and UV sensitive) gray tree frog. Melanin distribution in tiger salamander larvae is diffuse throughout the dermis and epidermis (arrows) and far less dense compared to the frogs and toads. [(A,B) modified from Little et al. [69]; (D) from Carey unpublished data.]

tenoids are often distributed throughout the body of copepods, whereas cladocerans tend to concentrate carotenoids in ovarian lipids and eggs [55]. In addition to carotenoids, melanin is also present in the cuticle of cladocerans and tends to be more pronounced among species living at high altitudes [72,73]. Carotenoids have limited UV-filtering capacity, however, and in cladocerans the carotenoids are thought to play an important photoprotective role by sequestering oxygen free radicals [55,56].

Freshwater fish vary in their tolerance of UV-B [2]. Tolerance of fish exposed to simulated UV-B appeared to be unrelated to melanin pigmentation but related to the amount of an unidentified colorless photoprotective substance in the skin [42,74]. This photoprotective substance appeared to be localized in the outer dorsal skin layers (Table 1), which includes the epidermis and overlying mucus, but was also detected in the eyes and gills of UV-tolerant razorback suckers (*Xyrauchen texanus*) [68]. The following examples illustrate how the amount of this photoprotective substance in a given fish species is related to the UV tolerance of that species. Channel catfish were found to be extremely sensitive to simulated UV-B, darkening within 24 h of exposure and having no detectable photoprotective substance [17]. Razorback suckers exposed to solar simulated UV-B did not develop sunburn after 21 days of exposure, while rainbow trout sunburned within 48 h [68]. When sections of the dorsal skin were removed from unexposed fish and methanol extracts scanned in a spectrophotometer, there was more photoprotective substance in the extracts of razorback suckers than rainbow trout [42]. Thus, there was a direct relationship between the amount of photoprotective substance and the period of time in which these fish developed sunburn (Table 2). The photoprotective substance appeared to have functioned as a natural sunscreen and protected razorback suckers from the harmful effects of simulated solar UV-B. In contrast to melanin, which offers some photoprotection at any given wavelength, the photoprotective

Table 1. Absorption maximum (λ_{max}) and amount of photoprotective substance in methanol extracts of various tissues from rainbow trout (*Oncorhychus mykiss*) and razorback suckers (*Xyrauchen texanus*). [from Fabacher and Little 68]

Tissue	Species	λ_{max} (nm)[a]	Amount[b]
Outer dorsal skin layers	rainbow trout	294.0 [0.0]	0.10 [0.0]
	razorback suckers	294.5 [0.2]	0.44 [0.05][c]
Inner dorsal skin layers	rainbow trout	290.2 [0.4]	0.03 [0.01]
	razorback suckers	292.3 [0.6]	0.12 [0.01][c]
Eyes	rainbow trout	not detected	–
	razorback suckers	293.5 [0.3]	0.02 [0.003]
Gills	rainbow trout	not detected	–
	razorback suckers	293.5 [0.3]	0.02 [0.002]
Liver	rainbow trout	not detected	–
	razorback suckers	not detected	

[a]Mean λ_{max} [standard error] for six fish.
[b]Mean absorbance units/milligram wet weight (au/wt) [standard error] of tissue for six fish.
[c]Photoprotective factor of tissue differs significantly between species.

Table 2. Relative amount of sunscreen present in fish skin, average UV exposure time to induce sunburn, and vulnerability of fish species to UV injury; compiled from Little and Fabacher [2], Ewing et al. [17], Fabacher and Little [68,74]

Species	Amount of sunscreen	Days of exposure to sunburn	Vulnerability to UV injury
Razorback sucker	100	>21	low
Pigmented Medaka	59	10	medium
Albino Medaka	59	10	medium
Rainbow trout	23	2	high
Channel catfish	ND	1	very high

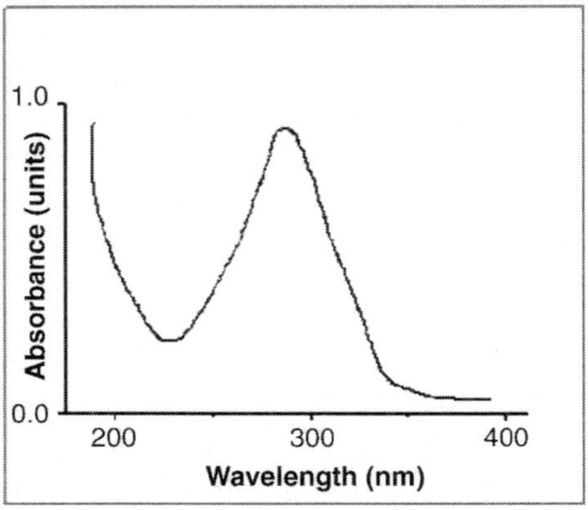

Figure 5. Spectrophotometric UV absorbance of photoprotective substance from the skin of razorback sucker (*Xyrauchen texanus*). Absorbance maximum occurs at 294 nm, the broad shoulders of the absorbance curve would provide protection from UV wavelengths less than and greater than 294 nm. [From Fabacher and Little 68.]

substance would offer a large amount of protection in the UV-B wavelength range because the absorption maximum is around 294 nm and the slope of the shoulders of this peak covers many of the UV-B wavelengths (Figure 5).

When cutthroat trout and razorback suckers were exposed to simulated UV-B, cutthroat trout (*Oncorhynchus clarki henshawi*) showed grossly visible signs to exposure (dorsal skin darkening) by 48 h [15]. Razorback suckers, however, did not show any visible signs of sunburn during the entire 72 h of exposure. Cutthroat trout had considerably less of the photoprotective substance than did razorback suckers. Histologic examination of cutthroat trout dorsal skin showed sloughing of mucous cells, necrosis and edema in both epidermis and dermis, and, in some cases, secondary fungal infection (Figure 6). Conversely, histologic observation of razorback sucker skin revealed that necrosis had occurred, but the severe sloughing and necrosis observed in cutthroat trout skin had not occurred (Figure 7). There was an increase in razorback sucker epidermal thickness, apparently resulting from hypertrophy and hyperplasia of large cells containing large central regions of low electron density. These

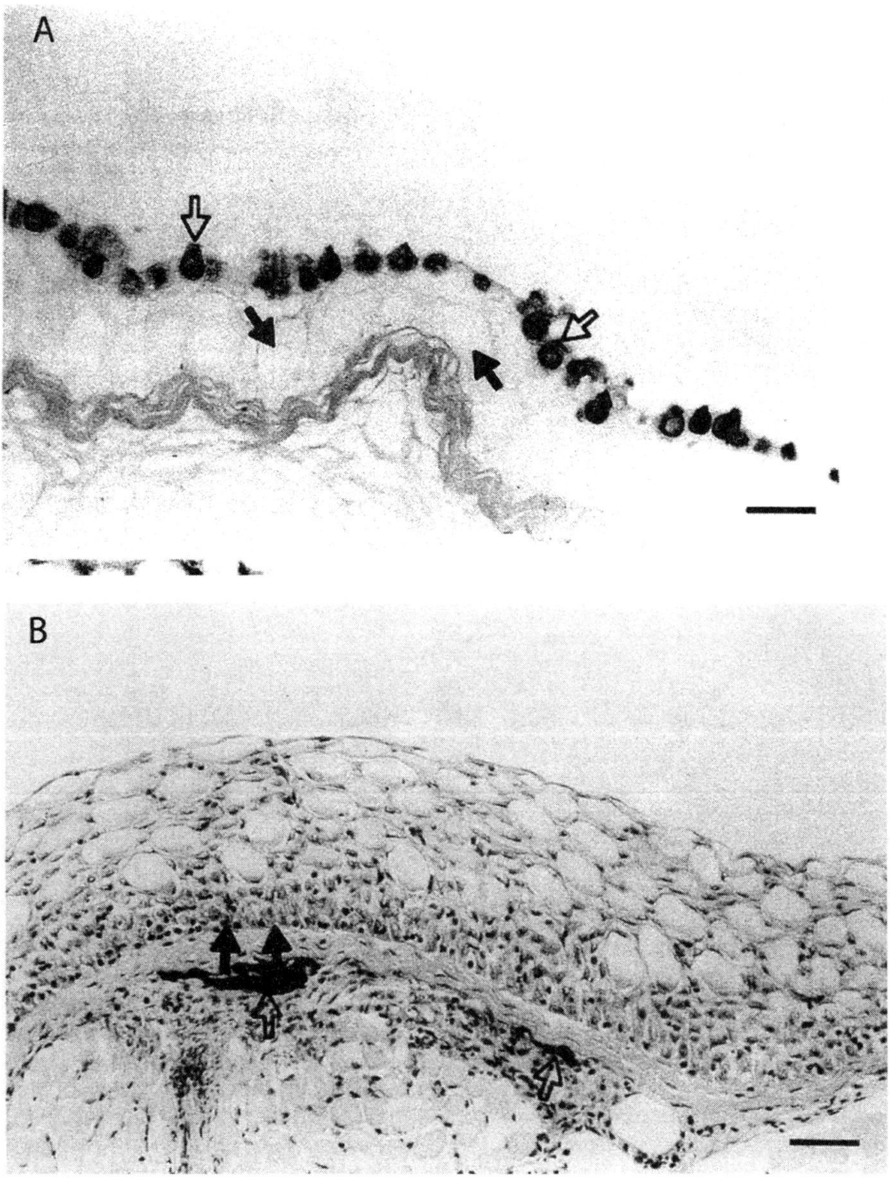

Figure 6. (A) Skin of cuttroat trout (*Oncorhynchus clarki henshawi*) not exposed to UV. Mucous cells (open arrows) are scattered throughout the epidermis. Melanocytes (filled arrows) are seen on either side of the stratum compactum. Bar equals 50 μm. (B) Skin after 72 h of UV exposure. Note: epidermis is lifted off the basement membrane, and the area between the dermis and epidermis is filled with necrotic cells, inflammatory cells, and exudates. A few sunburn cells (arrows) with glassy, pyknotic nuclei and perinuclear halos can be seen. Bar equals 19 μm. [From Blazer et al. 15.]

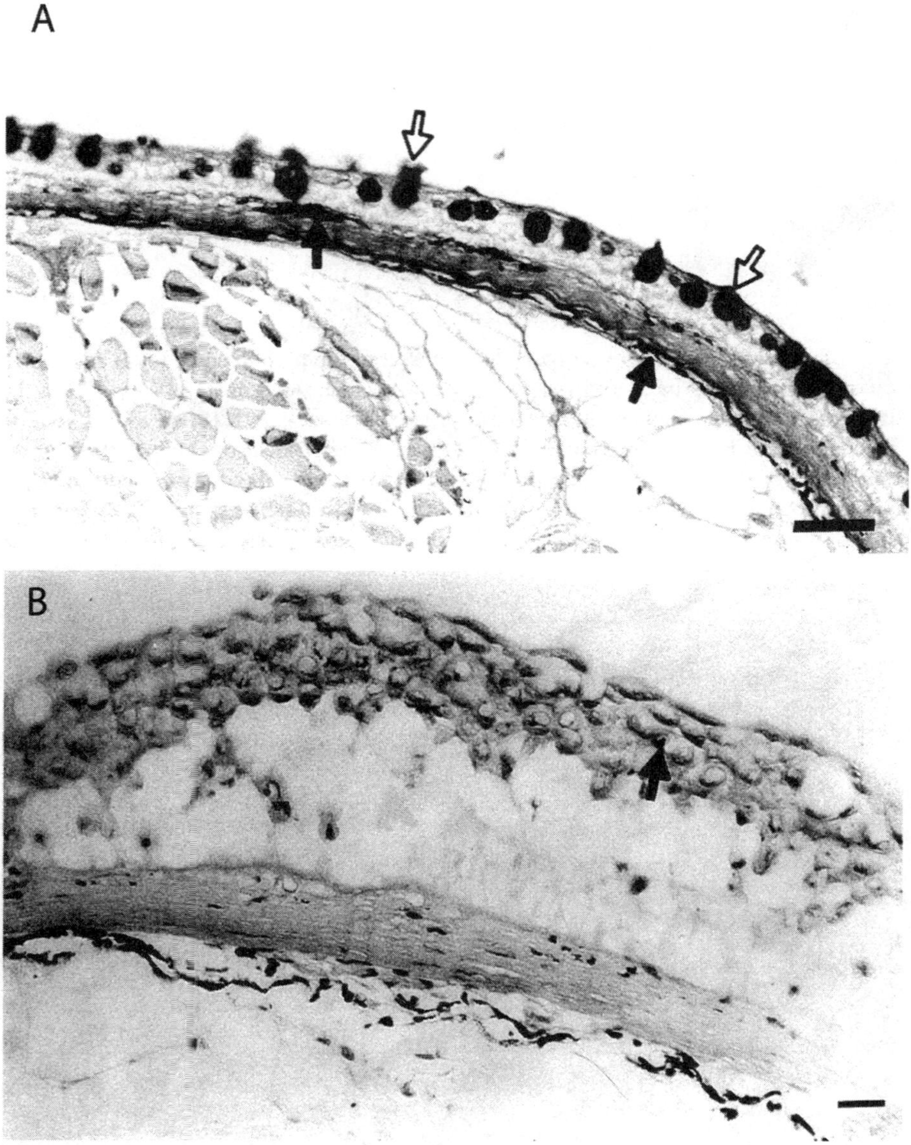

Figure 7. (A) Skin of razorback suckers (*Xyrauchen texanus*) not exposed to UV. Mucous cells (open arrows) line the very outer surface of the epidermis. A major portion of the epidermis is composed of large PAS-negative cells (filled arrows). Bar equals 50 μm. (B) Skin of razorback sucker after 72 h of UV exposure. The epidermis is thickened to hyperplasia and hypertrophy of the large PAS-negative cells. There are many leucocytes infiltrating the basal portion of the epidermis (filled arrows), melanocyte accumulations below the dermis (open arrows) and inflammation in the hypodermis. The changes observed were reversible and non-lethal. Bar equals 50 μm. [From Blazer et al. 15.]

cells appeared to be club cells and were larger near the surface of the epidermis and may contain the photoprotective substance that appeared to afford razorback suckers a great deal of protection from UV.B injury.

The photoprotective substance, or its precursor, may be of dietary origin and trophically accumulated, secreted by the cells of the epidermis, and concentrated in the epidermis and overlying mucus. However, it has not been determined whether this photoprotective substance is trophically accumulated or induced by ambient UV-B, or whether nature has selected for tolerant individuals with large amounts of this photoprotective substance. The photoprotective substance would act as a sunscreen and effectively block UV-B from causing sunburn in fish that had enough of it in the dorsal skin. If the photoprotective substance is trophically accumulated, dietary accumulation of this compound in fish skin could determine the UV-B tolerance of fish to sunburn. This could be very important in hatchery management in formulating diet supplementation to insure the success of fish stocks released in the wild.

13.5 Factors controlling UV injury

Injuries induced by exposure to UVR are dose-dependent. A number of factors can influence dose including intensity of exposure, spectral composition of the irradiance, and duration of exposure. Each of these may be influenced by climate and habitat.

13.5.1 Climate conditions

Stratospheric ozone depletion has been the focus of many investigations concerning UV impacts. Depletion of ozone concentration in the stratosphere reduces the filtering capacity of the ozone layer resulting in increased UV irradiance reaching the Earth's surface. Ozone depletion of about 5% has occurred in the past decade over the United States, but has been considerably greater in the Southern hemisphere [75,76]. In addition, brief periods (1–2 days) of elevated UV can occur over localized regions as a result of irregularities in atmospheric ozone concentrations.

Increased UV dose can occur as a result decreased cloudiness. For example, the number of sunny days in Central America has increased significantly over the past decade [77]. Such changes in weather patterns were found to significantly correlate with the loss of amphibian species. Although clear sky irradiance would not likely vary greatly from day to day, aside from changes associated with solar angle, the cumulative UV dose would increase considerably compared to doses that occur under cloudy conditions. In addition, temperature and soil moisture would likely decline under such conditions. Individually and in combination, these stressors could be harmful and contribute to population decline.

Increased surface temperatures, possibly as a result of global warming trends, could lead to the stratification of the water column. For example, nutrients released from submerged aquatic vegetation could stay in the lower portion of the water column. This could then result in fewer algae in the upper water column. This could result in greatly increased water clarity and UV irradiance of the upper water column where a majority of juvenile forms of aquatic organisms occur. Surface waters are highly productive and numerous organisms would be affected by increases in surface irradiance.

13.5.2 Habitat characteristics

Degradation and destruction of aquatic habitats often include the removal of trees, rocks, aquatic vegetation, and other structures that protect aquatic organisms from solar radiation. There would be a greater solar impact in the water column in these habitats and greater potential of increased exposure of aquatic organisms to solar UV-B. Habitat degradation and destruction that include the loss of shading structures may be more important than ever with expected increases in solar UV-B. Thus, it is important that aquatic organisms have shade that protects them from much of the solar radiation spectrum.

Aquatic organisms have adapted to certain levels of UV-B and exhibit different levels of tolerance to UV-B. Species naturally adapted to high levels of solar radiation exposure would be more tolerant to high levels of UV-B than species not naturally adapted to high levels, especially in clear, shallow water. The presence of shading structures, including vegetation in riparian and littoral zones, is important. There should be sufficient numbers of trees along a stream bank or pond edge to provide sufficient canopy shading over the water column. Floating and submerged aquatic vegetation also provides protection from solar UV-B. Dewatering of the water column by irrigation or other diversion, or channelization results in erosion of stream banks and excess deposition in side channels can result in overexposure of organisms to solar UV-B. Aquatic organisms can also be exposed to sudden and intense levels of UV-B when they inhabit water that is turbid a large part of the year and then clears up during the summer. In general, any activity that increases water clarity, such as the release of reservoir water or the presence of introduced zebra mussels (*Dreissena polymorpha*), could increase exposure of the vast majority of freshwater aquatic organisms to harmful levels of solar UV-B.

Water quality characteristics of the habitat can also affect the UV-B dose received by aquatic organisms. DOC concentration plays a major role in limiting UV-B in the water column and has been extensively investigated [78,43,79]. Investigations in Minnesota show that slight reductions in DOC resulted in dramatic increases in water column UV-B irradiance (Figure 8). DOC originates from diverse sources, especially terrestrial vegetation. The chemical composition of DOC may vary considerably among watersheds, reflecting the unique vegetation and soil chemistry of the site [80]. Therefore, it is likely that the UV filtering characteristics will vary as well. Moreover, UV irradiance in the aquatic habitat is considerably more dynamic than at the Earth's terrestrial surface, and can vary by orders of magnitude depending on DOC concentrations. Thus, conditions

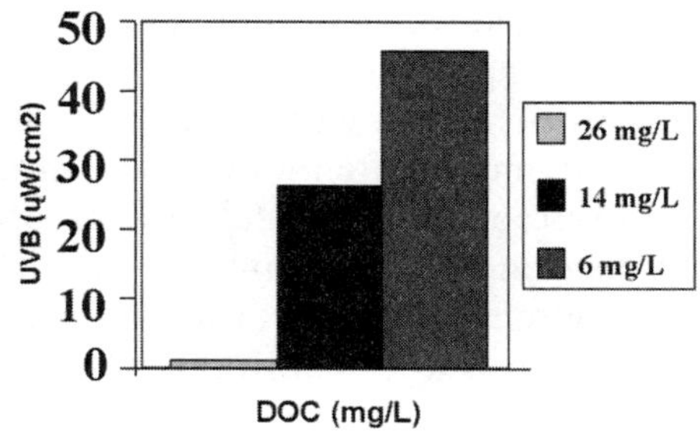

Figure 8. Average total UVB irradiance at a 10 cm water depth in amphibian habitats in Minnesota having dissolved carbon concentrations ranging from 6 to 26 mg L^{-1}.

that influence DOC will significantly influence UV dose. Acidification from acid deposition breaks down humic and fulvic acids making them less effective in screening UV [79].

In aquatic habitats UV can interact additively or synergistically with certain contaminants, increasing their toxicity and severity of injury ([81], see Chapter 7). Chemicals of anthropogenic origin that have molecular characteristics similar to photoprotective substances may be altered by absorbed UV. This interaction may generate free radicals or singlet oxygen that can alter DNA, enzymes, or lipoproteins leading to cellular injury and rapid death. For example, polycyclic aromatic hydrocarbons (PAHs) and other components of crude and refined petroleum increase in toxicity by as much as 10000-fold in the presence of UV [82]. Unexpected acute mortality occurred among bluegill sunfish treated with the PAH anthracene and exposed to solar UV in outdoor artificial streams [83]. It was concluded that solar UV significantly enhanced the toxicity of anthracene to the fish. In a subsequent study with juvenile sunfish (*Lepomis* spp.), the acute toxicity of anthracene was photoenhanced by simulated UVR [84]. The authors also observed severe necrosis and loss of epidermal cellular layers in affected fish. UV may also change the chemical structure of the substance to a more toxic form. UV breaks down ferrocyanide compounds to release free cyanide, which is toxic to fish and amphibians [Figure 9 – 85,86]. Pesticides, plastics, and pharmaceuticals may also be transformed into more toxic substances [87]. Thus, photo-

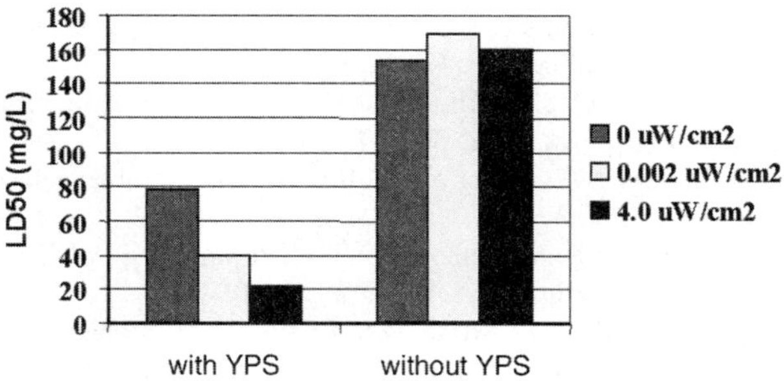

Figure 9. Average mortality of rainbow trout (*Oncorhynchus mykiss*) exposed to a fire-retardant chemical with and without ferrocyanide (YPS) alone and in the presence of UV. [From Little and Calfee 85.]

sensitization and sunburn-like lesions can occur at solar irradiance levels that would otherwise be harmless.

It is apparent that a variety of factors, acting singly or as multiple stressors, can contribute to UV-induced injury in freshwater organisms. Each species of freshwater organism will be susceptible to harmful levels of solar UVR depending on the conditions present at a certain point in time. If conditions are appropriate for UVR to penetrate sensitive cellular molecules, and if cellular repair mechanisms are unable to keep up with the rate of cellular damage, UV-induced injury is inevitable.

References

1. A.M. Bullock (1982). The pathological effects of ultraviolet radiation on the epidermis of teleost fish with reference to the solar radiation effects in higher animals. *Proc. R. Soc. Edinb.*, **81B**, 199–210.
2. E.E. Little, D. L. Fabacher (1994). Comparative sensitivity of rainbow trout and two threatened salmonids, Apache trout and Lahontan cutthroat trout, to ultraviolet-B radiation. *Archiv. Hydrobiol.*, **43**, 217–226.
3. R. Hofer (2000). Vulnerability of fish and amphibians to ultraviolet radiation. *Res. Adv. Photochem. Photobiol.*, **1**, 265–282.
4. G.M. Bell, W.S. Hoar (1950). Some effects of ultraviolet radiation on sockeye salmon eggs and alevins. *Can. J. Res.*, **28**, 35–43.
5. C.E. Dunbar (1959). Sunburn in fingerling rainbow trout, *Prog. Fish-Cult.*, **21**, 74.
6. D.C. DeLong, J.E. Halver, W.T. Yasutake (1958). A possible cause of "sunburn" in fish. *Prog. Fish-Cult.*, **20**, 111–113.
7. A.M. Bullock, R.J. Roberts (1981). Sunburn lesions in salmonid fry: a clinical and histological report. *J. Fsh Dis.*, **4**, 271–275.
8. A.M. Bullock, R.J. Roberts, P. Waddington, W.D.A. Bookless (1983). Sunburn lesions in koi carp. *Vet. Rec.*, **112**, 551.
9. A.M. Bullock, R.R. Coutts (1985). The impact of solar ultraviolet radiation upon the skin of rainbow trout, *Salmo gairdneri* Richardson, farmed at high altitude in Bolivia. *J. Fish Dis.*, **8**, 263–272.
10. A.M. Bullock (1988). Solar ultraviolet radiation: a potential environmental hazard in the cultivation of farmed finfish. In: J.F. Muir, R.J. Roberts (Eds), *Recent Advances in Aquaculture* (Vol. 3, pp. 139–224). Croom Helm, London.
11. J.R. Brocklebank, R.D. Armstrong (1994). Solar dermatitis in hatchery-reared salmonids in British Columbia. *Can. Vet. J.*, **35**, 651–652.
12. K.T. Ramos, L.T. Fries, C.S. Berkhouse, J.N. Fries (1994). Apparent sunburn of juvenile paddlefish. *Prog. Fish.-Cult.*, **56**, 214–216.
13. K. Kaweewat, R. Hofer (1997). Effect of UV-B radiation on goblet cells in the skin of different fish species. *J. Photochem. Photobiol. B: Biol.*, **41**, 222–226.
14. C. Noceda, S.G. Sierra, J.L. Martinez (1997). Histopathology of UV-B irradiated brown trout *Salmo trutta* skin. *Dis. Aquat. Org.*, **31**, 103–108.
15. V.S. Blazer, D.L. Fabacher, E.E. Little, M.S. Ewing, K.M. Kocan (1997). Effects of ultraviolet-B radiation on fish: histologic comparison of a UVB-sensitive and a UVB-tolerant species. *J. Aquat. An. Health*, **9**, 132–143.
16. J.R. Hunter, J.H. Taylor, G.H. Moser (1979). Effect of ultraviolet radiation on eggs and larvae of the northern anchovy, *Engraulis mordax*, and the Pacific mackeral,

Scomber japonicus, during the embryonic stage. *Photochem. Photobiol.*, **29**, 325–338.
17. M.S. Ewing, V.S. Blazer, D.L. Fabacher, E.E. Little, K.M. Kocan (1999). Channel catfish response to ultraviolet-B radiation. *J. Aquat. An. Health*, **11**, 192–197.
18. A.M. Nagl, R. Hofer (1997). Effects of ultraviolet radiation on early larval stages of the Alpine newt, *Triturus alpestris*, under natural and laboratory conditions. *Oecologia*, **110**, 514–519.
19. I.N. Flamarique, K. Ovaska, T.M. Davis (2000). UV-B induced damage to the skin and ocular system of amphibians. *Biol. Bull.*, **199**, 187–188.
20. R.D. Hurtubise (1996). *The Effects of Ultraviolet-B Radiation on Aquatic Invertebrates.* (M.Sc. Thesis). Southwestern Missouri State University.
21. A.P. Cullen, C.A. Monteith-McMaster (1993). Damage to the rainbow trout (*Oncorhyncus mykiss*) lens following an acute dose of UVB. *Curr. Eye Res.*, **12**, 97–106.
22. L.N. Allison (1962). Cataract among hatchery-reared lake trout. *Prog. Fish-Cult.*, **24**, 155.
23. L.N. Allison (1963). Cataract in hatchery trout. *Trans. Am. Fish. Soc.*, **92**, 34–38.
24. E.W. Steucke Jr., L.H. Allison, R.G. Piper, R. Robertson (1968). Effects of light and diet on the incidence of cataract in hatchery-reared lake trout. *Prog. Fish-Cult.*, **30**, 220–226.
25. C. Gutierrez-Rodriguez, C.E. Williamson (1999). Influence of solar radiation on early life-history stages of the bluegill. *Lepomis macrochirus. Environ. Biol. Fish.*, **55**, 307–319.
26. C.E. Williamson, S.L. Metzgar, P.A. Lovera, A.R.E. Moeller (1997). Solar ultraviolet radiation and the spawning habitat of yellow perch, *Perca flavescens. Ecol. Appl.*, **7**, 1017–1023.
27. C. Walters, B. Ward (1998). Is solar radiation responsible for declines in marine survival rates of anadromous salmonids that rear in small streams?. *Can. J. Fish. Aquat. Sci.*, **55**, 2522–2538.
28. M.P. Lesser, J.H. Farrell, C.W. Walker (2001). Oxidative stress, DNA damage and p53 expression in the larvae of Atlantic cod (*Gadus morhua*) exposed to ultraviolet (290–400 nm) radiation. *J. Exp. Biol.*, **204**, 157–164.
29. M.G. Zeeman, W.A. Brindley (1981). Effects of toxic agents upon fish immune systems: a review. In: R.P. Sharma (Ed.), *Immunologic Considerations in* Toxicology *2* (pp. 1–60). CRC Press, Boca Raton, FL.
30. J.F. Knowles (1992). The effect of chronic radiation on the humoral immune response of rainbow trout *Oncorhynchus mykiss. Int. J. Radiat. Res.*, **62**, 239–248.
31. A.D. Pickering, R.H. Richards (1980). Factors influencing the structure, function, and biota of the salmonid epidermis. *Proc. R. Soc. Edinb.*, **79B**, 93–104.
32. E.C. DeFabo, F.P. Noonan (1983). Mechanism of immune suppression by ultraviolet radiation in vivo. I. Evidence for the existence of a unique photoreceptor in skin and its role in photoimmunology. *J. Exp. Med.*, **157**, 84–98.
33. E.C. DeFabo, F.P. Noonan, J.E. Frederick (1990). Biologically effective doses of sunlight for immune suppression at various latitudes and their relationship to changes in stratospheric ozone. *Photochem. Photobiol.*, **52**, 811–817.
34. F.P. Noonan, E.C. DeFabo (1992). Immunosuppression by ultraviolet-B radiation: initiation by urocanic acid. *Immunol. Today*, **13**, 250–254.
35. M. Norval, J.W. Gilmour, J.T. Simpson (1990). The effect of histamine receptor antagonists on immunosupppression induced by the *cis*-isomer of urocanic acid. *Photodermatol. Photoimmunol. Photomed.*, **7**, 243–248.
36. E.W. Palaszynski, F.P. Noonan, E.C. DeFabo (1992). *Cis*-urocanic acid down-regulates the induction of adenosine 3,5'-cyclic monophosphate by either *trans*-urocanic

acid or histamine in human dermal fibroblasts *in vitro*. *Photochem. Photobiol.*, **55**, 165–171.

37. D.L Fabacher, E.E. Little, S.B. Jones, E.C. DeFabo, L.J. Webber (1994). Ultraviolet-B radiation and the immune response of rainbow trout. In: J.S. Stolen, T.C. Fletcher, (Eds), *Modulators of Fish Immune Responses: Models for Environmental Toxicology, Biomarkers, Immunostimulators* (Vol. 1, pp. 205–217). SOS Publications, Fair Haven, New Jersey.
38. C. Carey, N. Cohen, L. Rollins-Smith (1999). Amphibian declines: an immunological perspective. *Dev. Comp. Immunol.*, **23**, 459–472.
39. L. Jacob, M. Zasloff (1994). Potential therapeutic applications of magninins and other anti-microbial agents of animal origin. In: J. Marsh, J.A. Goods, (Eds), *Antimicrobial peptides* (Vol. 186, pp. 197–223). Ciba Foundation Symposium, John Wiley and Sons, Chichester, UK.
40. P. Nicolas, A. Mor (1995). Peptides as weapons against mirco-organisms in the chemical defense system of vertebrates. *Annu. Rev. Microbiol.*, **49**, 277–304.
41. A.M. Bullock (1985). The effect of ultraviolet-B radiation upon the skin of the plaice, *Pleuronectes platessa* L., infested with the bodonid ectoparasite *Ichthyobodo necator* (Henneguy, 1883). *J. Fish. Dis.*, **8**, 547–550.
42. D.L. Fabacher, E.E. Little (1995). Skin component may protect fishes from ultraviolet-B radiation. *Environ. Sci. Pollut. Res.*, **2**, 30–32.
43. N.H. Skully, D.R.S. Lean (1994). The attenuation of ultraviolet radiation in temperate lakes. *Arch. Hydrobiol. Beih.*, **43**, 135–144.
44. C.W. Hawryshyn (1992). Polarization vision in fish. *Am. Sci.*, **80**, 164–175.
45. C.E. Williamson (1995). What role does UV-B radiation play in fresh-water ecosystems? *Limnol. Oceanogr.*, **40**, 386–392.
46. C.E. Williamson (1996). Effects of UV radiation on freshwater ecosystems. *Int. J. Environ. Stud.*, **51**, 245–256.
47. M. Tevini (1993). Moecular biological effects of ultraviolet radiation. In: M. Tevini (Ed.), *UV-B radiation and Ozone Depletion* (pp.1–15). Lewis Publishers, Boca Raton, FL.
48. D.L. Mitchell D. Karentz (1993). The induction and repair of DNA photodamage in the environment. In: A.R. Younge, J. Moan, L. Bjorn, W. Nultsch (Eds), *Environmental UV Photobiology* (pp. 345–377). Plenum Press, New York.
49. A.R. Blaustein, P.D. Hoffman, D.G. Hokit, J.M Kiesecker, S.C. Walls, J.B. Hays (1994). UV repair and resistance to solar UV-B in amphibian eggs: A link to population declines? *Proc. Natl Acad. Sci.* U.S.A., **91**, 1791–1795.
50. M.A. Smith, C.M. Kapron, M. Berrill (2000). Induction of photolyase activity in wood frog (*Rana sylvatica*) embryos. *Photochem. Photobiol.*, **72**, 575–578.
51. A. Shima, R.B. Setlow (1984). Survival and pyrimidine dimers in cultured fish cells exposed to concurrent sun lamp ultraviolet and photoreactivating radiators. *Photochem Photobiol.*, **39**, 49–56.
52. S. Kaup, J. Hunter (1981). Photorepair in larval anchovy *Engraulis mordax*. *Photochem. Photobiol.*, **33**, 253–256.
53. L.A. Applegate, R.D. Ley (1988). Ultraviolet radiation-induced lethality and repair of pyrimidine dimers in fish embryos. *Mutat. Res.*, **198**, 85–92.
54. J.D. Regan, W L. Carrier, C. Samet, B.L. Olla (1982). Photoreactivation in two closely related marine fishes having different longevities. *Mech. Ageing Dev.*, **18**, 59–66.
55. O. Siebeck, T.L. Vail, C.E. Williamson, R. Vetter, D. Hessen, H. Zagarese, E. Little, E. Balseiro, B.Modenutti, J. Seva, A. Schumate (1994). Impact of UV-B radiation on zooplankton and fish in pelagic freshwater systems. *Arch. Hydrobiol. Beih. Ergebn.*

Limnol., **43**, 101–114.
56. N.I. Krinsky (1979). Carotinoid protection against oxidation. *Pure Appl. Chem.*, **51**, 649–660.
57. C.S. Cockell, J. Knowland (1999). Ultraviolet radiation screening compounds. *Biol. Rev.*, **74**, 311–345.
58. D. Karentz, F.S. McEuen, M.C. Land, W.C. Dunlap (1991). Survey of mycosporine-like amino acids in Antarctic marine organisms: Potential protection form ultraviolet exposure. *Mar. Biol.*, **108**, 157–166.
59. D.F. Gleason (1993). Differential effects of ultraviolet radiation on green and brown morphs of Caribbean coral *Porites astreoides*. *Limnol. Oceanogr.*, **38**, 1452–1463.
60. J.M. Shick, M.P. Lesser, W.C. Dunlap, W.R. Stochaj, B.E. Chalker, J. WuWaon (1995). Depth-dependent responses to solar ultraviolet radiation and oxidative stress in zooanthellate coral a *Acopora microphthalm*. *Mar. Biol.*, **122**, 42–51.
61. J.M. Shick, W.C. Dunlap, R.M. Larson (1994). Coral reef holothuriods (Echinodermata) accumulate UV photoprotectants from their diet. *Am. Zool.*, **33**, 62.
62. P.T. Grant, P.A. Plack (1980). Gadusol, a metabolite from fish eggs. *Tetrahedron Lett.*, **21**, 4043–4044.
63. P.T. Grant, C Middleton, P.A. Plack, R.H. Tomson (1985). The isolation of four aminocyclohexenimines (mycrosporins) and a structurally related derivation of cyclohexane-1,3-dione (gadusol) from the brine shrimp *Artemia*. *Comp. Biochem. Physiol.*, **80B**, 755–759.
64. N. Kollias, R.M. Sayre, L. Zeise, M.R. Chedekel (1991). Photoprotection by melanin. *J. Photochem. Photobiol. B: Biol.*, **9**, 135–160.
65. P.R. Crippa, V. Cristofoletti, N. Romeo (1978). A band model for melanin deduced from optical absorption and photoconductivity experiments. *Biochim. Biophys. Acta*, **538**, 164–170.
66. J.A. Menon, S. Persad, H.F. Haberman, C.J. Kurian (1983). A comparative study of the physical and chemical properties of melanins isolated from red and black hair. *J. Invest. Dermatol.*, **80**, 202–206.
67. B.A. Gilchrest, H.V Park, M.S. Eller, M. Yaai (1996). Mechanisms of ultraviolet-induced pigmentation. *Photochem. Photobiol.*, **63**, 1–10.
68. D.L. Fabacher, E.E. Little (1998). Photoprotective substance occurs primarily in outer layers of fish skin. *Environ. Sci, Pollut. Res.*, **5**, 4–6.
69. E.E. Little, R.D. Calfee, D.L. Fabacher, C. Carey, V.S. Blazer, E.M. Middleton (2002). Effects of ultraviolet-B radiation on toad early life stages. *Environ. Sci. Pollut. Res.*, www.scientificjournals.com/onlinefirst/espr.
70. G. Britton, T.W. Goodwin (1981). *Carotinoid Chemistry and Biochemistry*. Pergammon Press, Oxford.
71. D.O. Hessen, K. Sorenson (1990). Photoprotective pigments in alpine zooplankton populations. *Aqua Fenn.*, **20**, 165–170.
72. C. Luecke, W.J. O'Brien (1981). Phototoxicity and fish predation: Selective factors in color morphs of Hetterocope. *Limnol. Oceanogr.*, **26**, 454–460.
73. P.D.N Hebert, D.B. McWalter (1983). Cuticular pigmentation in Arctic *Daphnia*: adaptive diversification of asexual lineages. *Am. Nat.*, **122**, 286–291.
74. D.L. Fabacher, E.E. Little (1999). Tolerance of an albino fish to ultraviolet-B radiation. *Environ. Sci. Pollut. Res.*, **6**, 69–71.
75. J.R. Herman, P.K. Bhartia, J. Kiemke, Z. Ahmad, D. Larko (1996). UV-B increases (1979–1992) from decreases in total ozone. *Geophys. Res. Lett.*, **23**, 2117–2120.
76. J.R, Herman, N. Krotkov, E. Celarier, D. Larko, G. Lobow (1999). Distribution of UV radiation at the Earth's surface from TOMS-measured UV-backscattered radiances.

J. Geophys. Res., **104**, 12059–12076.

77. E.M. Middleton, J.R. Herman, E.A. Celarier, J.W. Wilkinson, C. Carey, R.J. Rusin (2001). Evaluating ultraviolet radiation exposure with satellite data at sites of amphibian declines in Central and South America. *Cons. Biol.*, **15**, 914–929.
78. D.R.S. Lean (1997). Influence of ultraviolet-B radiation on aquatic ecosystems. In: E. Little (Ed.), *Environmental Toxicology and Risk Assessment*: *Vol* 7. *ASTM STP* 1333 (pp 1–20). American Society for Testing and Materials, West Conshohocken, Pennsylvania.
79. D.W. Schindler, P.J. Curtis, B.R. Parker, M.P. Stainton (1996). Consequences of climate warming and lake acidification for UV-B penetration in North American boreal lakes. *Nature*, **379**, 705–708.
80. S.A. Diamond, G.S. Peterson, J.E. Tietge, G.T. Ankley (2002). Assessment of the risk of solar ultraviolet radiation to amphibians. III. Prediction of impacts in selected northern mid-western wetlands. *Environ. Sci. Technol.*, **36**, 2866–2874.
81. E.E. Little, R. Calfee, L. Cleveland, R. Skinker, A. Zaga-Parkhurst, M.G. Barron (2000). Photo-enhanced toxicity in amphibians: Synergistic interactions of solar ultraviolet radiation and aquatic communities. *J. Iowa Acad. Sci.*, **107**, 67–71.
82. J.T. Oris, J.P. Geisy (1987). The photoinduced toxicity of polycyclic aromatic hydrocarbons to larvae of the fathead minnow (*Pimephales promelas*). *Chemosphere*, **16**, 1396–1404.
83. J.W. Bowling, G.J. Leversee, P.F. Landrum, J.P. Giesy (1983). Acute mortality of anthracene-contaminated fish exposed to sunlight. *Aquat. Toxicol.*, **3**, 79–90.
84. J.T. Oris, J.P. Giesy (1985). The photoenhanced toxicity of anthracene to juvenile sunfish (*Lepomis* spp.). *Aquat. Toxicol.*, **6**, 133–146.
85. E.E. Little, R.D. Calfee (April, 2000). *The Effects of UVB Radiation on the Toxicity of Fire-fighting Chemicals to Fish and Amphibians* (Report to the U.S. Forest Service Wildland Fire Chemical Systems Program). Missoula Technology Development Center, Missoula, Montana.
86. G.E. Burdick, M. Lipschuetz (1950). Toxicity of ferro- and ferricyanide solutions to fish and determination of the cause of mortality. *Trans. Am. Fish. Soc.*, **78**, 192.
87. A. Zaga, E.E. Little, C.F. Rabeni, M.R. Ellersieck (1998). Photoenhanced toxicity of a carbamate insecticide to early life stage amphibians. *Environ. Toxicol. Chem.*, **17**, 2022–2035.

Chapter 14

Behavioral responses – UVR avoidance and vision

Dina M. Leech and Sönke Johnsen

Table of contents

Abstract

Light-mediated behaviors, such as visual foraging and migration, have been the focus of numerous studies spanning a wide variety of taxa; however, the role of UVR in these and other behaviors has generally been given less attention.

Recently, the effects of UVR on aquatic ecosystems have been examined more closely as a result of increasing UV-B radiation reaching the Earth's surface due to stratospheric ozone depletion. UVR is now known to penetrate deeply into many freshwater and marine systems, and organisms occupying all trophic levels are susceptible to damage or mortality from UVR exposure. Behavioral avoidance is one means by which organisms can reduce exposure to damaging radiation. Both laboratory and field experiments have demonstrated that many species are negatively phototactic to UV and shorter-wavelength visible light. In addition, UV photoreceptors have been reported in a variety of fish and invertebrates, suggesting that UV vision may be prominent in aquatic organisms. These UV photoreceptors are thought to be used for navigation, communication, enhanced foraging, and possibly UVR avoidance. Given the presence of negative phototactic behaviors as well as UV vision, UVR may be an important factor influencing migration and abundance patterns as well as predator–prey and intraspecific interactions.

14.1 Introduction

Behavioral responses to light have long been of interest to aquatic scientists, both freshwater and marine. Light-mediated behaviors such as mate recognition, visual foraging, and especially vertical migration are the focus of numerous studies spanning a wide diversity of taxa [1–3]. However, the role of UVR in these and other behaviors has only recently been more closely examined.

Until recent decades, UVR was not thought to be an important factor influencing aquatic ecosystems, as it was believed to rapidly attenuate through the water column. UVR is now known to penetrate deeply into many freshwater and marine systems, with dissolved organic carbon (DOC) as one of the primary factors regulating UV attenuation [4,5]. In addition, aquatic organisms occupying all trophic levels from viruses and phytoplankton to zooplankton and fish are susceptible to damage or mortality from UVR [6–9]. UVR may directly affect organisms via cellular and tissue damage, genetic mutation, or mortality; or it may indirectly affect organisms by constraining them to suboptimal habitats where temperature and food concentrations may be low and predation risk high. Tolerance to UVR differs among species [8–12] and therefore some species are more likely to respond behaviorally to damaging UVR than others. Consequently, UVR can alter species composition and trophodynamics within an ecosystem, possibly shifting communities towards more UV-tolerant species [13].

There are three means by which organisms can respond to potential UVR damage: (1) avoidance, (2) photoprotection, and (3) photorepair [14]. The extent to which organisms use each mechanism differs both within and among taxa. For example, among freshwater organisms, many species of the cladoceran *Daphnia* are capable of photorepair while copepods such as *Diaptomus oregonensis* and *Acanthodiaptomus denticornis* depend more on photoprotective compounds [15,16]. In the southern hemisphere, three species of the freshwater calanoid copepods within the genus *Boeckella* vary in their use of photoprotection versus photorepair [17]. Differences in photorepair and photoprotection are also seen among marine organisms. Photorepair in two closely related marine fish, the tautog *Tautog onitis* and the cunner *Tautogolabrus adspersus*, appears to be related to longevity, with the longer-lived tautog possessing greater photorepair capabilities than the shorter-lived cunner [18]. In Antarctica, where the ozone hole is the greatest, photoprotection by mycosporine-like amino acids is prevalent in several marine organisms from algae and invertebrates to fish [19]. Although our understanding of the photorepair and photoprotection capabilities of aquatic organisms is increasing, less is known about behavioral avoidance of UVR in nature.

While some wavelengths of UVR are damaging, others are potentially beneficial to aquatic organisms. For example, UV photoreceptors have been described in a variety of aquatic organisms from bacteria to fish [20]. The adaptive significance of these UV photoreceptors is not fully understood; however, research suggests that they may enhance navigation, communication, and foraging [20]. It is also possible that UV photoreceptors may help organisms to avoid depths at which damaging wavelengths are present.

This chapter first describes the underwater UV environment. The different types of phototactic responses, such as positive versus negative phototaxis, are then described and related to UV tolerance as well as UV vision. Finally, implications for behavioral responses to UVR are addressed, including the role of UVR in diel vertical migration (DVM) and predator–prey interactions.

14.2 The underwater UV environment

Solar radiation is both absorbed and scattered as it penetrates through water (see Chapter 3). As a result, downwelling irradiance decreases with depth, with shorter and longer wavelengths attenuating more rapidly than the wavelength of peak transmission (which is generally found from 470–550 nm). While particulates and the water itself contribute somewhat to the attenuation of UVR, absorption by DOC is one of the primary factors regulating variations in UV attenuation in aquatic ecosystems [4,5]. In systems with high DOC, UVR is attenuated rapidly while, in systems with low DOC, UVR can penetrate deeply into the water column. In 25% of lakes in several regions of North America, 1% attenuation depths (the depth to which 1% of surface irradiance penetrates) were estimated to be greater than 4 m for 320 nm and greater than 10 m for 380 nm [21]. In the clearest ocean waters, 1% attenuation depths are estimated to be 50 m for 320 nm and approximately 200 m for 380 nm (Figure 1) [22], and there is evidence that increased levels of UV-B are entering the oceans as stratospheric

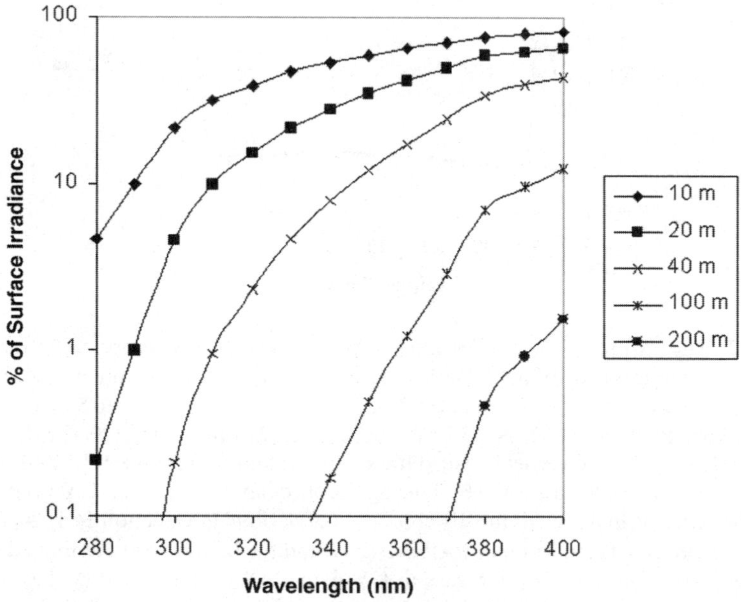

Figure 1. Percent of surface irradiance present at depth in the clearest ocean waters. Percent surface irradiance was determined using diffuse attenuation coefficients derived from Smith and Baker [22]. Measurements of irradiance were taken with a submersible spectoradiometer in the Sargasso Sea and the Central Equatorial Pacific.

ozone decreases [23]. Thus, biologically relevant UVR is present at considerable depths in many freshwater and marine ecosystems. UV-sensitive organisms may avoid depths at which damaging wavelengths are present but may seek depths at which potentially beneficial wavelengths (i.e., used for photorepair and UV vision) are present.

In addition to DOC, other factors influence the depth to which UVR penetrates, including season, latitude, sea state, time of day, cloud cover, and turbidity (Chapters 2 and 3). Light intensity and spectral composition are both affected by each of these factors, creating potential "light niches". For example, relative quantities of UVR are greater during crepuscular periods (i.e. dawn and dusk) than daylight hours (Figure 2) due to the increasing proportion of high-UV skylight in the total irradiance [24]. Many species of larval fish that possess UV vision feed primarily during crepuscular periods [24–26]. These twilight hours may provide an "optical foraging niche" for fish predators with UV vision, enhancing target-background contrast. Indeed, near the surface of the ocean, up to 40% of the photons in the horizontal and downward directed lines of sight are UV-A [27], potentially silhouetting prey (Figure 3).

14.3 Behavioral responses to UVR

Behavioral responses to radiation often vary with wavelength. Some

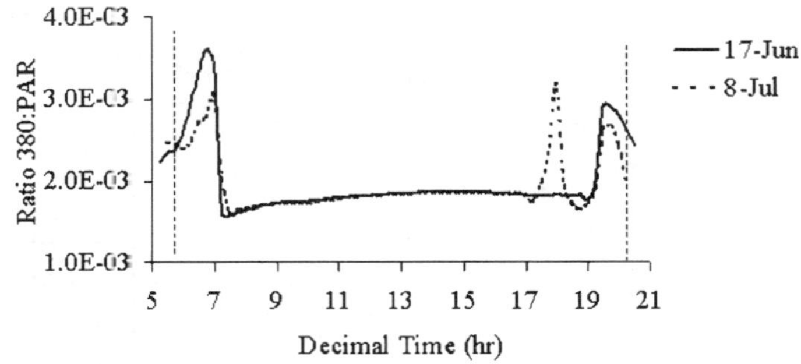

Figure 2. The ratio of UV-A (380 nm) to photosynthetically active radiation (PAR, 400–700 nm) in terms of W m^{-2}. Data were collected during the summer of 2001 with a UV radiometer (model Biospherical GUV–521) located at the Lacawac Sanctuary in the Pocono Mts., PA, USA (41.23 N, 75.21 W). Sunrise (5.27 h) and sunset (20.37 h) for 17 June 2001 are denoted by the vertical dashed lines. Sunrise and sunset on 8 July 2001 occurred at 5.36 h and 20.37 h, respectively. During crepuscular periods, the UV:PAR ratio is higher because the light field is mostly composed of skylight (see Section 14.1). As the sun's elevation increases, the amount of PAR increases and the light field in dominated by solar radiation. Note that a similar increase in UV-A-to-PAR occurs when patches of clouds pass over the sun. This is shown between 17.0–19.0 h on 8 July 2001.

Figure 3. Simultaneous images taken at (a) green (490–560 nm) and (b) ultraviolet (350–380 nm) wavelengths. Note the bright background in the UV image that silhouettes fish strongly, even against the reef. [Taken from Losey et al. 27.]

wavelengths induce positive phototaxis or movement towards a light source while other wavelengths induce negative phototaxis or movement away from a light source. For many species, exposure to UVR (280–400 nm) and shorter-wavelength visible (i.e., blue light, 400–440 nm) light induces negative phototaxis. These negative phototactic or avoidance behaviors correspond to wavelengths that are also known to be potentially damaging or lethal [11,16]. In motile organisms, both vertical and horizontal movements have been observed in avoidance of exposure to damaging radiation [28–31], while in less motile organisms covering behaviors are exhibited [32]. Many organisms, such as sea anemones, sea urchins, and sea cucumbers, cover themselves with shells, rocks, and other materials during peak periods of irradiance. Hiding within burrows and among rocks and macrophyte beds, as seen in many amphibian and larval fish species, also helps organisms reduce exposure to damaging radiation.

14.3.1 Laboratory experiments

Thus far, behavioral responses to damaging light have primarily been examined in the laboratory using artificial UV radiation sources. Experiments have been conducted on a variety of organisms from both freshwater and marine systems occupying all trophic levels. At the lower trophic levels, both phytoplankton and protozoa have been shown to exhibit negative phototaxis to UVR. For example, individual cells within mats of the filamentous marine cyanobacteria *Microcoleus chthonoplastes* were shown to migrate to greater depths in response to increased UV-B exposure [28]. The red-colored freshwater ciliate *Blepharisma japonicum* responded with backward swimming when exposed to wavelengths within the UV-B range but began swimming forward when exposed to visible light at 580 nm [33]. It is important to investigate the behavior of organisms at these lower trophic levels as their response to UVR may directly or indirectly influence responses of those at higher trophic levels.

Laboratory experiments have clearly demonstrated that the wavelength of incident radiation is an important behavioral cue for zooplankton. Certain freshwater cladocerans become more agitated and negatively phototactic in the presence of blue light but remain calm and positively phototactic to red light [29]. These "color dances" of Cladocera were hypothesized to cue zooplankton to high concentrations of algal food, which typically filters out short wavelengths greater than longer wavelengths (i.e., "red-dance" keeps individuals in place, "blue-dance" promotes wandering). However, it was also suggested that the patterns of the dances may explain patterns of diurnal vertical migration. More recent studies with monochromatic radiation have demonstrated that *Daphnia magna* are positively phototactic to visible light (421–600 nm) and negatively phototactic to UVR (260–380 nm) with maximal sensitivity at 340 nm [34]. Copepods have also shown UV avoidance behavior in the laboratory. In small experimental enclosures examining horizontal movements, the freshwater cyclopoid *Cyclops serrulatus* was found to avoid exposure to UV-B radiation (280–320 nm) [10]. This study also noted that UV behavioral responses correlated well with UV tolerance (i.e., UV-sensitive organisms avoid UV-B exposure, see Section 14.3.4). UVR avoidance behaviors were also detected in the marine echinoid larva *Dendraster excentricus* exposed to an artificial UV–visible radiation source (315–700 nm) [30].

Certain stream-dwelling organisms have been shown to be negatively phototactic to UVR in laboratory microcosm experiments. Macroinvertebrates that inhabit or feed on the tops or sides of rocks, such as larval stages of mayflies, caddisflies, and blackflies, exhibited increased drift to more shaded areas when exposed to increased UV-B radiation [31]. Drifting was 60–70% less in the UV-B shielded controls.

Interestingly, deep-sea crustaceans also respond behaviorally to UVR. Tethered individuals of the oplophorid shrimp *Systellaspis debilis* respond to changes in ambient UVR by pitching, changing swimming speed, and moving their feeding appendages [35]. Possible explanations for behavioral responses to UV in deep-sea crustaceans are discussed in Section 14.4.2.

For some organisms, short exposures to UVR inhibit movement altogether. For example, following exposure to artificial UV-B, veligers and post-veligers of the zebra mussel *Dreissena polymorpha* ceased all swimming and crawling motions. However, exposure to UV-A and visible light had no effect on behavior [36]. A similar delay in phototaxis was noted for the green algae *Volvox aureus* exposed to both artificial and solar UV-B radiation [37].

Although these studies provide valuable information regarding organismal responses to varying wavelengths of radiation (i.e., action spectra), they do not tell us how animals respond to natural levels of solar radiation. Artificial lamps generally do not exactly replicate the solar spectrum. UV-B lamps often have greater output in the UV-B range compared to the solar spectrum. In order to supplement UV-A and visible light, UV-A and cool white lamps are used in laboratory setups, and these lamps often have less output in the UV-A and visible range than solar radiation. The total intensity of these lamps in terms of energy or quanta may be similar to solar radiation, but the spectral composition varies greatly (i.e., skewed towards the shorter wavelengths). Solar simulators come the closest to replicating both the intensity and spectral output of the sun; however, these instruments are very expensive, only irradiate a small area, and are only used by a handful of laboratories.

14.3.2 Field experiments

Few field studies have examined behavioral responses of organisms to natural solar radiation. One of the difficulties in these studies is determining whether a behavior is in response to UVR or visible light. High UV systems are also high visible light systems, both of which are known to be potentially damaging [1,8,38]. In addition, many animals have a separate suite of responses to varying levels of visible light. Typically, experimental enclosures are constructed of materials that vary in UVR transmittance. Commonly used materials that transmit full solar radiation include polyethylene, quartz, and acrylic plastics such as OP-4 (CYRO Industries) and UVT (Spartech, Inc. formerly Townsend/Glasflex), all of which can be expensive. UV-blocking materials include Mylar® D and acrylic plastics such as OP-2 (CYRO Industries) and Plexiglas®. While these materials vary in their UV transparency, they have similar transparencies in the visible range. Therefore, using a combination of these materials, behaviors and/or survival can be examined in the presence of full solar radiation, in the absence of UV-B radiation, and in the absence of UV-B and UV-A radiation. It is important to note that these experiments do not provide information concerning responses to a single wavelength; instead, they examine the effect of removing particular wavebands (i.e., UV-B or UV-B and UV-A). Because they remove entire wavebands, the UV blocking materials also unavoidably change the total irradiance, which can confound results. However, the difference in total irradiance between the UV- transparent and UV-blocking materials is often less than 10%. Finally, except for quartz, which is extremely expensive and hard to fabricate, the usual UV-transparent materials tend to block a significant fraction (25–50%) of UV-B.

In the field, solar UV-B has been demonstrated to inhibit motility and oriented movement in phytoplankton such as eukaryotic flagellates, blue–green algae or cyanobacteria, and gliding green algae [39]. When motility is compromised, phytoplankton are at risk of being exposed to greater light intensities, which may result in a bleaching of pigments; or they may be exposed to reduced light intensities, which may result in a reduction in photosynthetic rates. Exposure to increased or decreased irradiances also depends on mixing processes as well as the buoyancy of the individual cells (see also Chapter 4).

Recent field studies have also reported that zooplankton exhibit UVR avoidance in nature. The first evidence of a vertical avoidance response of *Daphnia* to solar UVR was recently published [40]. In the presence of full solar radiation, *D. pulicaria* rapidly descended from the surface waters (1.5 m) of a high-UV lake. In the absence of UV-B and shorter wavelength UV-A radiation (<380 nm), the majority of *D. pulicaria* remained in the surface waters. Thus, a stronger negative phototactic response was detected in the presence of UVR than in the absence of UVR. Negative phototactic behaviors have also been observed in a population of *D. catawba* inhabiting a high-UV lake located in the Pocono Mts., PA, USA (Figure 4) [41]. Experiments conducted in this study demonstrated that, in some cases, *D. catawba* actually swim towards the surface waters in the absence of UVR in spite of the probable presence of fish kairomones (Figure 4). These field results for *Daphnia* are supported by smaller scale experiments conducted in the laboratory [42,43].

Although *Daphnia* often displayed a preference for the surface waters in the absence of UVR, the response was variable, with mean depths of *Daphnia* increasing in the absence of UVR [41]. The reason for this variability is unknown. One explanation is that irradiance differed among experiments. Although the experiments in this study were not designed to specifically test zooplankton responses to irradiance, preliminary observations suggest that as irradiance increased, *Daphnia* responded with increased negative phototaxis. Because the acrylic used to construct the UV-blocking columns did transmit some longer wavelength UV-A (50% transmittance at 384 nm), this may be an avoidance response to either longer wavelength UV-A light or visible light [40]. Other species, such as the freshwater copepod *Diaptomus nevadensis*, the marine copepod *Acartia tonsa*, the cladoceran *Daphnia magna*, and the hydromedusan *Polyorchis penicillatus* also become negatively phototactic in response to increasing irradiance, both in the UV and visible range [1,38].

The harpacticoid copepod *Tigriopus californicus*, which lives in shallow tide

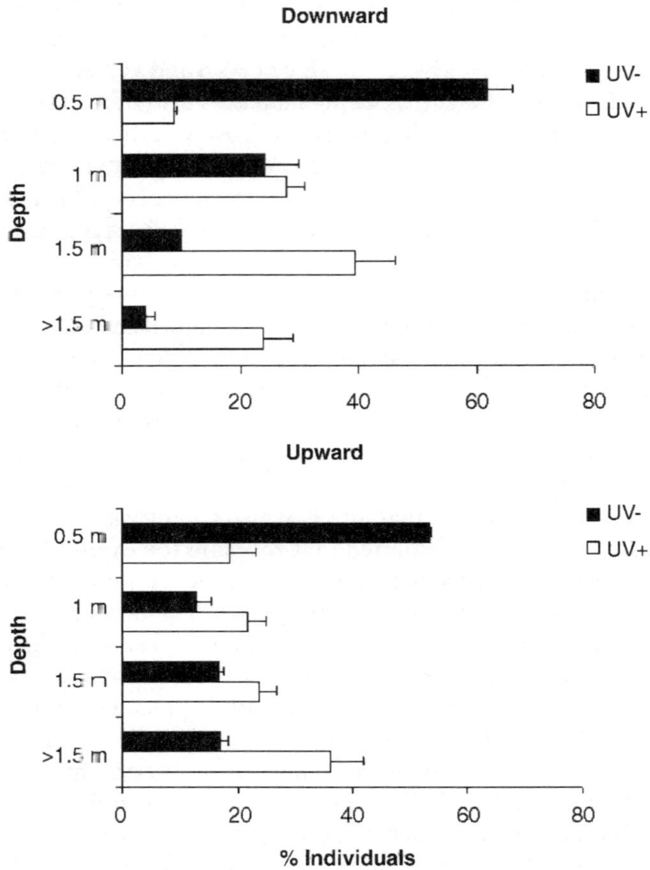

Figure 4. A comparison of the downward and upward migrations of *Daphnia catawba* in the presence and absence of UVR. There were three UV-transparent columns and three UV-opaque columns. Each column was suspended 10 cm below the surface of Lake Giles, a high-UV lake located in northeastern PA, USA. The downward experiment was conducted on 14 July 2000 and the upward on 2 August 2000. Mean solar irradiance was measured with a LICOR model LI–200SA pyranometer near solar noon (1300 h) when the experiments were conducted. Mean solar irradiance equaled 659 W m^{-2} on 14 July 2000 and 694 W m^{-2} on 2 August 2000.

pools, was found to aggregate in shaded regions of pools at midday but show no preference at dawn and dusk [44]. These same authors used lab experiments to demonstrate that *T. californicus* responds more to UV-B than to visible radiation and suggest that they may possess UV photoreceptors.

Small stream invertebrates have also been noted to respond negatively to UVR in nature. Blackfly larvae appear to exhibit a diurnal emigration, or migration out of UV-exposed stream channels, during periods of peak irradiance but return to UV-exposed regions as irradiance levels decrease [45]. In streams that were experimentally shielded from UVR exposure, however, larvae remained in the stream channels throughout the day. Larvae were allowed to move freely between the treatments and on average larval densities in the UV-shielded channels were 161–168% greater than those in the UV-exposed channels.

Differences in the spawning depths of yellow perch, *Perca flavescens*, in a high-versus a low-UV lake suggest that yellow perch also avoid UV exposure. Spawning depth was reported to be deeper in a high-UV lake (median = 3.2 m) compared to a low-UV lake (median = 0.4 m) [46]. In addition, yellow perch eggs were incubated at the surface of each lake in a modified reciprocal transplant experiment. Eggs were exposed to full solar radiation, shielded from UV-B, or kept in the dark. In the high-UV lake, all eggs perished in all the light treatments, but survival time was longer (2–4 days) for eggs in the UV-B shielded treatment. Furthermore, those collected from the high-UV lake survived longer than those collected from the low-UV lake. Most eggs (>96%) incubated in the light treatments of the low UV lake as well as the dark controls of both lakes survived to hatching. Comparable results, using a similar experimental design, were reported for the bluegill *Lepomis machrochirus* in which the median nesting depth was observed to be deeper in a high UV lake compared to a low UV lake [47].

It is more difficult to perform behavioral experiments in the open ocean. The approach has been to observe the distribution of organisms in relation to their photic environment combined with laboratory experiments examining UV tolerance and phototaxis. Both ascidians and sea urchins were shown to exhibit UV avoidance. The distribution of the solitary ascidian *Corella inflata* varied with exposure to direct solar radiation, particularly UVR exposure, with populations conspicuously absent from unshaded areas [48]. Laboratory experiments confirmed that UVR is lethal to all life history stages of *C. inflata*, with the younger stages being most vulnerable. In addition, none of the life stages possessed UV-absorbing photoprotective compounds. The sea urchins *Arbacia punctulata* and *Lytechinus variegatus* were shown to be negatively phototactic to bright solar radiation but positively phototactic to white light [49]. These data are consistent with the observation that echinoplutei migrate to deeper depths in the water column during peak periods of irradiance [30,50], but this response could also be related to other factors such as predator avoidance. The sea urchin *Strongylocentrotus droebachiensis* shades or covers itself in response to UVR exposure, particularly in response to UV-B or a combination of UV-B and UV-A [32]. Covering behavior was also shown to increase with increasing intensity of UVR exposure. In some sea urchin species, covering behavior has been observed to vary diurnally, with the greatest response during peak irradiance [49,51].

14.3.3 Relation to UV tolerance, pigmentation, and photorepair

Behavioral responses to UVR appear to be related to UV tolerance (i.e., defined as the sum of an organism's photoprotection (pigmentation) and photorepair capabilities). For example, during periods of high UV, organisms occupying the surface waters of Lake Giles, a high UV lake in the Pocono Mts., PA, USA, were found to be more UV-tolerant than those inhabiting deeper waters during the day [12]. Laboratory experiments with the ostracod *Cypris* sp. demonstrated that this species is highly tolerant to UV-B exposure and actually showed a behavioral preference or positive phototaxis towards UV-B irradiance [10]. In the same study, the protozoan *Paramecium aurelia* was also shown to be highly tolerant and positively phototactic to UV-B irradiance [10].

The action spectrum of phototaxis in copepods has been demonstrated to depend on pigmentation. Within the visible light spectrum, *Diaptomis nevadensis* swimming speeds were faster in blue light compared to red light [38]. In addition, less pigmented individuals were more responsive to changes in wavelength than pigmented individuals [52]. Similar results have been reported for melanized *Daphnia* within the UV spectrum [43].

14.4 UV vision and photoreception

UV vision has been documented in a variety of terrestrial organisms including insects, birds, amphibians, reptiles, and mammals [53–55]. It is therefore not surprising that many aquatic organisms also perceive light in the UV spectrum. Most UV photoreceptors in aquatic organisms have been described in fish species; however, UV photoreceptors have also been reported in bacteria and algae as well as some species of protozoans, annelids, cnidarians, and crustaceans (Table 1).

14.4.1 Relation to habitat and age

Many UV photoreceptors have a maximum absorbance peak in the UV-A range but UV-B photoreceptors have been documented in some species (Table 1). One explanation for the rarity of UV-B vision is that UV-B radiation is potentially more damaging to the eye. For instance, cataracts are reported to occur in several fish species inhabiting shallow waters [56]. Seeing in the UV-A may therefore be less detrimental to the eye; however, prolonged exposure to UV-A radiation may also be potentially damaging, albeit less than UV-B. In addition, since eyes are photon-, not energy-counters, seeing in the UV-A provides more light than in the UV-B. However, visible light provides more photons than UV, making UV vision a curious trait (See section 14.4.2).

Some authors have suggested that UV photoreceptors vary with habitat such that peak absorbance correlates with wavelengths present in their photic environment [57–59]. In some cases, species such as the rudd *Scardinius erythrophthalmus* [60] and the brown trout, *Salmo trutta* [61], display seasonal changes in spectral sensitivity that correspond to seasonal changes in the photic environment associated with daylength and temperature. Most of these shifts are in the longer wavelengths with shorter wavelength sensitivity remaining the same [62]. Behavior shifts are also suggested to contribute to shifts in spectral sensitivity, such as foraging at the surface during summer months and in deeper strata during winter months [60].

UV photoreceptors in some fish species not only vary with habitat but with age as well. Many fish species, such as *Lepomis gibbosus*, *Perca flavescens*, and *Salmo* sp., possess UV photoreceptors as larvae but lose them with maturity [63,64]. This loss of UV photoreception coincides with a habitat shift from the surface waters to more demersal waters in addition to a change in diet from small to larger zooplankton prey and/or fish [63–65]. In some species of salmonoids, however, UV photoreceptors disappear during earlier life history stages and reappear in adults. For example, ultraviolet cones and UV sensitivity in the sockeye salmon, *Oncorhynchus nerka*, diminished during smoltification and reappeared at the late juvenile or adult stage [66]. The author also noted that the arrangement of the UV cones in the retina of the adult sockeye salmon was similar to those of saltwater salmon, *O. tshawytsha* and *O. keta*, collected while migrating back to natal streams or spawning in streams. This suggests that UV photoreceptors may assist in navigation during migrations. Goldfish and species of cyprinids retain their UV photoreceptors as adults. These species experience little to no change in habitat or diet and therefore a change in the spectral sensitivity of their photoreceptors would not be expected. Ontogenetic changes in spectral sensitivity among aquatic species other than fish are less well known.

14.4.2 Adaptive significance

The adaptive role of UV vision is not completely understood. In some organisms, UV photoreceptors assist in navigation and orientation, associated with the e-vector of the polarized light field [67,68], while in others they have been demonstrated to enhance color discrimination [55,69]. Recognition and communication between conspecifics and mates at UV wavelengths has been speculated in species of coral reef fish [20,27]. Recently, the epithelial mucus of several marine fish species was found to contain UV-absorbing compounds, which may be seen by fish with UV vision [70]. Consequently, it is suggested that one fish may see another as "tanned" or "untanned", potentially playing an important role in visual communication.

UV photoreceptors are also thought to help in the detection of prey during visual foraging by enhancing prey contrast [20,63,64]. Planktonic prey, such as *Daphnia* and *Diaptomus*, absorb solar radiation in the near-UV [71]. Because of this, these zooplankton may appear darker than their surrounding background. In addition, planktonic prey also scatter light and may appear lighter or darker depending on the direction of illumination, shape, and refractive index differences (Figure 5) [71]. Larvae of the phantom midge *Chaoborus trivittatus* reflect blue light greater than longer wavelength red light and it is predicted that the reflectance curve will shift towards shorter wavelengths as the angle of incidence increases [72]. These differences in reflectance were hypothesized to reduce visibility to visual feeding fish and therefore reduce mortality. This would be true for fish without UV photoreceptors, but increased reflectance at shorter wavelengths may increase visibility to foragers with UV vision.

Laboratory experiments have demonstrated that larval fish do feed better in

Table 1. Survey of the distribution of UV photoreceptors among aquatic organisms. (This list is not all-inclusive.) Microspectrophotometry is abbreviated as MSP.

Organism	Method	Wavelength of maximum response or absoption (nm)	Reference
Bacteria			
Mutant, *Escherichia coli*	behavior	396–450	[89]
Purple eubacterium, *Ecotothiorhodospira halophila*	behavior	N/A	[90]
Saltwater bacterium, *Halobacterium halbium*	behavior	280, 370	[91]
Phytoplankton			
Cyanobacterium, *Chologloeopsis*	physiology, MAAs induction	310	[76]
Green alga rhizoid, *Bryopsis plumosa*	physiology, MAAs induction	260, 310	[92]
Protozoans			
Ciliates			
Chlamydodon mnemosyne	behavior	360	[93]
Blepharisma japonicum	behavior	N/A	[33]
Annelids			
Alciopid worm, *Torrea candida*	electrophysiology	400	[77]
Cnidarians			
Sea anemone, *Anthopleura xanthogrammica*	behavior	360	[91]
Molluscs			
Giant clam, *Tridacna* sp.	electrophysiology	360	[94]
Crustaceans			
Cladoceran, *Daphnia magna*	behavior	348	[75]
Harpacticoid copepod, *Tigriopus californicus*	behavior	N/A	[44]
Ectoparasitic copepod, *Lepeophtheirus salmonis*	behavior	352–400	[95]
Crayfish, *Procambarus clarkia*	MSP	440	[53]
Mantis Shrimp, *Pseudosquilla ciliate*	MSP	400	[96]
Deep sea oplophroid shrimp			
Systellaspis debilis	behavior	410	[35]
Janicella spinacauda	electrophysiology	370	[97]

Oplophorus spinosus	electrophysiology	400	[97]
Oplophorus gracilirostris	electrophysiology	400	[97]
Estuarine intertidal crabs			
Sesarma reticulatum	behavior	360	[98]
S. cinereum	behavior	360	[98]
Pinnotheres ostreum	behavior	360	[98]
Prawn, *Palaemonetes vulgaris*	electrophysiology	390	[99]
Spiny lobster, *Panulirus argus*	MSP, electrophysiology	370	[100]
Horseshoe crab, *Limulus polyphemus*	electrophysiology	360	[101]
Fishes			
Clupeomorpha			
Northern anchovy, *Engraulis mordax*	MSP	358, 365	[102]
Ostariophysi			
Minnow, *Phoxinus laevis*	operant conditioning	response down to 365	[103]
Roach, *Rutilus rutilus*	MSP	355–360	[104]
Goldfish, *Carassius auratus*	heart-rate conditioning	365	[105]
Carp, *Cyprnus carpio*	MSP, behavior	377	[106]
Danio, *Danio aequipinnatus*	electophysiology	358	[107]
Eastern Golden Shiner, *Notemigonus crysoleucas*	MSP	355	[27]
Rudd, *Scardinius erythrophthalmus*	MSP	355–360	[108]
Japanese dace, *Tribolodon hakonensis*	MSP	350–370	[109]
Salmoniformes			
Rainbow trout, *Oncorhynchus mykiss*	heart-rate conditioning	390	[110]
Brown trout, *Salmo trutta*	MSP	355	[65]
Atlantic salmon, *Salmo salar*	MSP	360	[111]
Sockeye salmon, *Oncorhynchus nerka*	MSP, electrophysiology	N/A	[66]
Acanthopterygii			
Guppy, *Poecilia latipinna*	MSP	412	[112]
Guppy, *P. reticulata*	MSP	411	[112]
Sunfish, *Lepomis* spp.	MSP	360–370	E. Loew, (personal commun.)
Yellow perch, *Perca flavescens*	MSP	385	[113]
Atlantic Silverside, *Menidia menidia*	MSP	365	[114]
Killifish, *Fundulus heteroclitus*	MSP	363	115]
California topsmelt, *Atherinops affinis*	MSP	355	[116]

Table 1. (*cont.*)

Organism	Method	Wavelength of maximum response or absoption (nm)	Reference
Grunion, *Leuresthes tenuis*	MSP	355	[116]
Kelp greenling, *Hexagrammos decagrammus*	MSP	350, 358	[102]
White spotted greenling, *H. steilen*	MSP	364	[102]
Lingcod, *Ophiodon elongates*	MSP	359	[102]
Puget Sound sculpin, *Artedius meanyi*	MSP	363, 375	[102]
Cabezon, *Scorpaenichthys marmorutus*	MSP	364	[102]
Rock prickleback, *Xiphister mucosus*	MSP	364	[102]
Dwarf wrymouth, *Lyconectes aleutensis*	MSP	355	[102]
Wolf-eel, *Anarrhichthys oceliatus*	MSP	378	[102]
Pacific sandfish, *Trichodon trichodon*	MSP	359	[102]
Atlantic halibut, *Hippoglossus hippoglossus*	in situ hybridization	N/A	[117]
Cichlid, *Metriaclima zebra*	MSP	368	[118]
Damselfish, *Dascyllus albisella*	MSP	>400	[119]
Pomacentridae,			
Dascyllus trimaculatus			
Pomacentrus coelestris			
Chromis punctipinnis	MSP	360	[120]
Reptiles			
Red-eared terrapin, *Pseudemys scripta elegans*	electrophysiology	360	[121]
Caspian terrapin, *Mauremys caspica*	electrophysiology	360	[121]

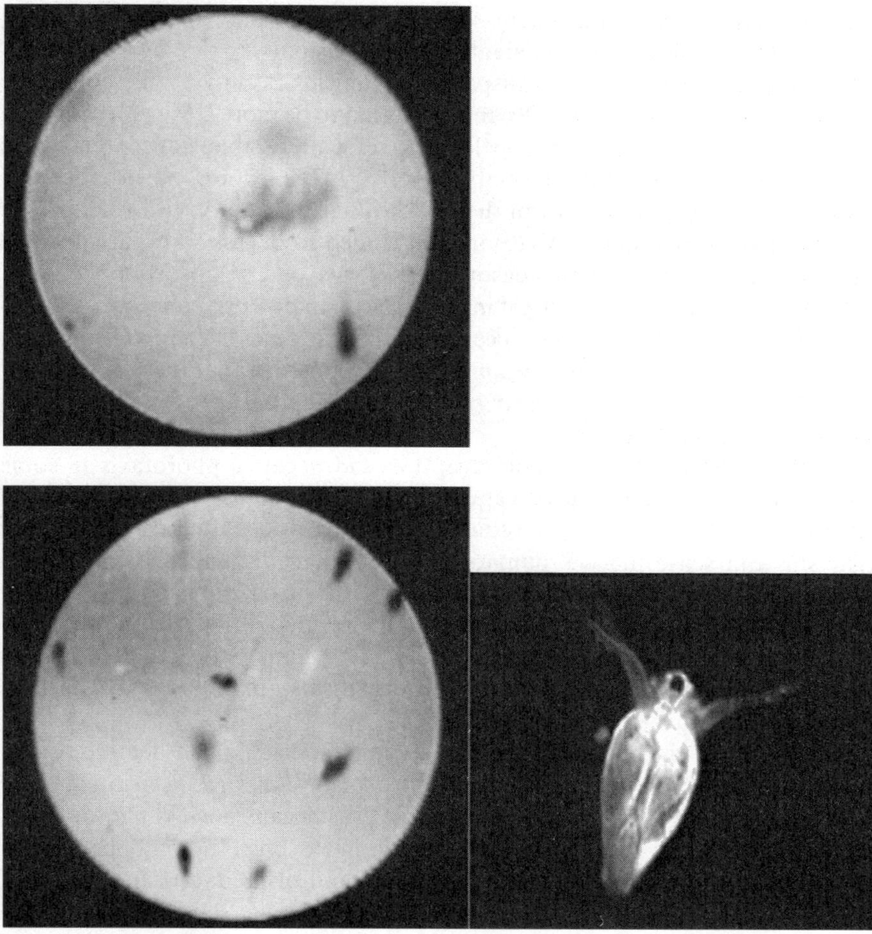

A B

Figure 5. UV images taken with a UV video camera sensitive between 320 nm and 410 nm. (A) Image taken in Oneida Lake, NY, USA showing *Daphnia* sp. in silhouette against the brighter skylight. Freshwater copepods *Diaptomus siscilis* are also shown and appear darker because they contain a dense, UV-absorbing (orange) pigment. (B) Image of *Daphnia* sp. showing UV scatter 90° from the direction of artificial UV illumination from a xenon light source. [Photos provided by E.R. Loew and W.N. McFarland.]

the presence of UV-A wavelengths [63,64] and can feed under monochromatic UV-A [63]. However, recent experiments with trout suggest that UV photoreceptors do not enhance foraging under natural levels of solar radiation [73]. In field experiments conducted in Patagonia, Argentina (41°08′S, 71°25′W) with rainbow trout, *Oncorhynchus mykiss*, the removal of UV wavelengths from solar radiation had no effect on the number of prey eaten or on prey preference. These experiments were run outdoors between 1000–1300 h local time. It is not known if a difference would have been noticed during crepusclular periods when relative UV levels are higher and planktivory is more challenging.

It has been suggested that increased absorbance in prey species in the UV range due to photoprotective pigments increases visibility to predators, especially in transparent organisms. Transparent organisms occupying the epipelagic zone in the Northwest Atlantic Ocean were found to be more UV absorbent than those occupying the deeper mesopelagic zone, while visible transparency was similar for organisms inhabiting both regions [74]. However, absorbance was greatest in the UV-B range not in the UV-A range where UV vision occurs. In addition, species with high UV-absorption tended to be less transparent in the visible range. For both these reasons, the effects of UV absorption on UV visibility were predicted to be slight in comparison to potential photoprotection.

UV photoreceptors have also been identified in several zooplankton prey, such as the cladoceran *Daphnia magna* [75]. It is possible that these UV photoreceptors may also serve a means of predator avoidance. However, this hypothesis has yet to be fully tested.

The presence of both UV photoreception and negative phototaxis in some species suggest that UV photoreceptors may help animals to avoid depths at which levels of damaging solar radiation are high. Indeed, it is not known if organisms can sense the UV damage they are incurring and respond appropriately without the aid of UV photoreceptors. In the cyanobacterium *Chlogloeopsis*, a UV-B photoreceptor is linked to the production of the photoprotective compound shinorine, a mycosporine-like amino acid [76]. Induction efficiency of shinorine was greatest when organisms are exposed to UV-B at 310 nm.

Curiously, UV vision is also noted in some mesopelagic and benthic organisms where little to no UVR is present. One explanation for UV vision at these depths is that many deep-sea fishes and some crustaceans possess photophores, light emitting organs with maximum emission in the blue, that may be used to communicate information between conspecifics and/or predators and prey. UV photoreceptors in these species have significant blue sensitivity, and the emissions of the photophores correlate well with the maximum transmission of the water as well as with the maximum sensitivity of the visual pigments [59]. These organisms are also known to be vertical migrators and it is suggested that UV photoreceptors may be used to detect varying ratios of shorter to longer wavelengths that would occur at sunrise and sunset, which could trigger the organisms to ascend and descend if enough solar radiation were available [35]. In the alciopid worm, *Torrea candida*, it is suggested that UV photoreceptors are used as a depth gauge [77].

14.5 Implications for behavioral responses to UVR

The distribution and abundance of organisms can have a profound effect on an ecosystem's structure and function. Nutrient cycling, predator–prey interactions, and community structure may all be influenced by distribution patterns. As such, numerous studies have been conducted to understand factors influencing vertical and horizontal distribution and abundance [2,78]. However, UVR has historically received less attention. Implications of behavioral responses to UVR in diel vertical migration and predator–prey interactions are discussed further below.

14.5.1 Diel vertical migration

One of the most interesting behavioral responses to solar radiation is the phenomenon of zooplankton diel vertical migration (DVM). Large zooplankton often exhibit strong migrations during the day to deeper, darker depths in the water column. Smaller zooplankton, in turn, remain in the surface waters during daylight and migrate to the deeper waters at night to avoid predation or interference by larger zooplankton [52,79,80]. Many hypotheses have been proposed to explain these patterns. Some of the earliest works on DVM demonstrated that solar radiation was a potentially important proximate as well as an ultimate factor inducing migrations [81–83]. These experiments, however, were conducted in the laboratory and no field studies were conducted to demonstrate a clear link between damaging solar radiation and zooplankton migration patterns in nature. Consequently, other factors such as temperature, food, and especially predation have typically been more widely studied and identified as the primary factors inducing DVM [78].

In spite of the importance of predators inducing migrations, predation alone does not explain the variety of DVM patterns observed in nature [84,85]. For example, vertical migrations have been detected in organisms inhabiting fishless systems [52]. Most of these systems tend to be high alpine or desert lakes in which damaging solar radiation can be intense. Several experiments have shown that ambient levels of UVR can lead to a decrease in survival as well as a decrease in growth and reproduction in both freshwater and marine organisms [6,8,14,86], and negative phototactic behavior has been demonstrated in the laboratory and field [28–38,40–51]. Given these recent findings, UVR may be more important than previously thought in influencing the vertical migration and distribution of organisms [34,84], serving as both a proximate and an ultimate cause of DVM.

Indeed, zooplankton often migrate deeper than the depths to which damaging UV-B radiation penetrates, in both freshwater and marine systems. While damaging UV-B may not be present, UV-A radiation continues to penetrate through the water column. For example, in the open ocean, the 1% level of 375 nm is four times as deep as the 1% level of 310 nm (Jerlov type II oceanic water) [87], and in freshwater lakes, UV-A penetration can be two times or greater [21]. Given that many fish species use UV-A light to forage, zooplankton may migrate to deeper depths in order to avoid visually feeding predators with UV-A photoreceptors. Many freshwater and marine species, however, continue to migrate to even deeper depths, suggesting that other factors besides UVR, such as temperature and predation, are inducing migrations.

14.5.2 Predator–prey interactions

While both UV-B and UV-A can be damaging, UV-B is generally more damaging than UV-A radiation per photon. The UV photoreceptors in many species peak in the UV-A range (see Table 1). If animals are cueing to UV-A wavelengths that penetrate more deeply into the water column, they would be protected from potentially more damaging UV-B found closer to the surface. These alterations in depth to prevent UVR exposure may in turn influence the overlap of predator and prey in both time and space. For example, UV-tolerant zooplankton may find refuge from larval fish predators, which are susceptible to UV damage [46,86], in the surface waters of high UV systems.

Many species of larval fish have retinal cones that perceive UV-A (350–370 nm) and these are thought to help larvae locate and capture their prey [63,64]. However, some prey species also have UV-A photoreceptors. Responses to UV-A wavelengths in these organisms may therefore also be a means of predator avoidance in the surface waters. In this case, predation may be the ultimate cause of DVM but UV-A light would be the proximate cue. Further investigation is needed to test these types of hypotheses.

14.6 Future directions

As levels of stratospheric ozone continue to decrease, future increases in UVR reaching the Earth's surface are predicted [88]. UVR has been shown to be damaging to many aquatic organisms from bacteria to fish [6–9], and UV avoidance behavior has been observed in several species [28–51]. Yet responses to future changes in the underwater UVR environment are largely unknown.

The presence of UV photoreceptors in such a wide variety of freshwater and marine organisms suggests that UV vision is prominent in aquatic ecosystems. Further experimentation is needed to identify potential UV photoreceptors as well as action spectra for behavioral responses to varying wavelengths of light. In addition, field experiments are needed to understand responses to natural solar radiation. As seen in the feeding experiments with rainbow trout [73], laboratory results may not always match those reported in the field. Interpreting responses of organisms to solar radiation may require an integration of scientists working in the fields of vision ecology, behavioral ecology, as well as bio-optics.

UVR is only one of many potential stressors acting on aquatic communities. Other stressors such as pH, temperature, competition, predation, and food limitation can also influence the vertical and seasonal abundance and distribution of aquatic organisms. UVR is likely to interact with these stressors through a variety of mechanisms. For example, high UVR levels in the surface waters of low DOC systems may force animals into deeper waters where habitats are suboptimal due to lower temperatures or greater risk of predation. Further investigation is needed to understand how UVR interacts with these other important abiotic and biotic stressors.

Acknowledgements

We thank Bruce Hargreaves for his assistance with solar radiation data collection and analysis. We also thank Ellis Loew for our conversations concerning UV vision in fish and predator–prey interactions. Gabriella Grad, Robert Moeller, and Angela Padeletti also provided many thoughtful comments on earlier revisions of the manuscript. George Losey, Tom Cronin, Ellis Loew, and William McFarland generously provided UV photos.

References

1. R.B. Forward (1988). Diel vertical migration: zooplankton photobiology and behavior. *Oceanogr. Mar. Biol. Ann. Rev.*, **26**, 361–393.
2. W. Lampert (1989). The adaptive significance of diel vertical migration of zooplankton. *Funct. Ecol.*, **3**, 21–27.
3. I. Huse, J.C. Holm (1993). Vertical distribution of Atlantic salmon (*Salmo salar*) as a function of illumination. *J. Fish Biol.*, **43**, 147–156.
4. J.T.O. Kirk, B.R. Hargreaves, D.P. Morris, R.B. Coffin, B. David, D. Fredickson, D. Karentz, D.R.S. Lean, M.P. Lesser, S. Madronich, J.H. Morrow, N.B. Nelson, N.M. Scully (1994). Measurements of UV-B radiation in two freshwater lakes: an instrument comparison. *Arch. Hydrobiol. Ergebn. Limnol.*, **43**, 71–99.
5. D.P. Morris, H. Zagarese, C.E. Williamson, E.G. Balseiro, B.R. Hargreaves, B. Modenutti, R.E. Moeller, C. Queimalinos (1995). The attenuation of solar UV radiation in lakes and the role of dissolved organic carbon. *Limnol. Oceanogr.*, **40**, 1381–1391.
6. J.J. Karanas, H. van Dyke, R.C. Worrest (1979). Midultraviolet (UV-B) sensitivity of *Acartia clausii* Giesbrecht (Copepoda), *Limnol. Oceanogr.*, **24**, 1104–1116.
7. J.J. Cullen, P.J. Neale, M.P. Lesser (1992). Biological weighting functions for the inhibition of phytoplankton photosynthesis by ultraviolet radiation. *Science*, **258**, 646–650.
8. O. Siebeck, T. Vail, C.E. Williamson, R. Vetter, D. Hessen, H. Zagarese, E. Little, E. Balseiro, B. Modenutti, J. Seva, A. Shumate (1994). Impact of UV-B radiation on zooplankton and fish in pelagic freshwater ecosystems. *Arch. Hydrobiol. Beih. Ergebn. Limnol.*, **43**, 101–114.
9. C.E. Williamson (1995). What role does UV-B radiation play in freshwater ecosystems?. *Limnol. Oceanogr.*, **40**, 386–392.
10. J.A. Barcelo, J. Calkins (1978). Positioning of aquatic microorganisms in response to visible light and simulated solar UV-B irradiation. *Photochem. Photobiol.*, **29**, 75–83.
11. R.C. Worrest (1982). Review of literature concerning the impact of UV-B radiation upon marine organisms. In: J. Calkins (Ed.), *The Role of Solar Ultraviolet Radiation in Marine Ecosystems* (pp. 429–457). Plenum, New York.
12. D.M. Leech, C.E. Williamson (2000). Is tolerance to UV radiation in zooplankton related to body size, taxon, or lake transparency? *Ecol. Appl.*, **10**, 1530–1540.
13. M.L. Bothwell, D.M.J. Sherbot, C.M. Pollock (1994). Ecosystem response to solar ultraviolet-B radiation: influence of trophic-level interactions. *Science*, **265**, 97–100.
14. H.E. Zagarese, C.E. Williamson (1994). Modeling the impacts of UV-B radiation on ecological interactions in freshwater and marine ecosystems. In: R.H. Biggs, M.B.E. Joyner (Eds), *Stratospheric Ozone Depletion/UV-B Radiation in the Biosphere* (pp.

315–328). Springer-Verlag, New York.
15. J. Ringelberg, A.L. Keyser, B.J.G. Flik (1984). The mortality effect of ultraviolet radiation in a red morph of *Acanthodiaptomus denticornis* (Crustacea: Copepoda) and its possible ecological relevance. *Hydrobiologia*, **112**, 217–222.
16. O. Siebeck, U. Bohm (1991). UV-B effects on aquatic animals. *Ver. Int. Verein. Theor. Angew. Limnol.*, **24**, 2773–2777.
17. H.E. Zagarese, M. Feldman, C.E. Williamson (1997). UV-B induced damage and photoreactivation in three species of *Boeckella* (Copepoda, Calanoida). *J. Plankton Res.*, **19**, 357–367.
18. J.D. Regan,, W.L. Carrier, C. Samet, B.L. Olla (1982). Photoreactivation in two closely related marine fishes having different longevities. *Mech. Aging Develop.*, **18**, 59–66.
19. D. Karentz, F.S McEuen, M.C. Land, W.C. Dunlap (1991). Survey of mycosporine-like amino acid compounds in Antarctic marine organisms: potential protection from ultraviolet exposure. *Mar. Biol.*, **108**, 157–166.
20. N. Shashar (1994). UV vision by marine animals: mainly questions. In: D. Gulko, P.L. Jokiel (Eds), *Ultraviolet Radiation and Coral Reefs* (HIMB Tech. Report #41, pp. 201–206). UNIHI-Sea Grant-CR-95-03.
21. C.E. Williamson, R.S. Stemberger, D.P. Morris, T.M. Frost, S.G. Paulsen (1996). Ultraviolet radiation in North American lakes: attenuation estimates from DOC measurements and implications for plankton communities. *Limnol. Oceanogr.*, **41**, 1024–1034.
22. R.C. Smith, K.S. Baker (1981). Optical properties of the clearest natural waters (200–800 nm). *Appl. Opt.*, **20**, 177–184.
23. J.B. Kerr, C.T. McElroy (1993). Evidence for large upward trends of ultraviolet-B radiation linked to ozone depletion. *Science*, **262**, 1032–1034.
24. I. Novales-Flamarique, A. Hendry, C.W. Hawryshyn (1992). The photic environment of a salmonid nursery lake. *Exp. Biol.*, **169**, 121–141.
25. J.F Haney (1989). Diel patterns of zooplankton behavior. *Bull. Mar. Sci.*, **43**, 583–603.
26. L.J. Shoeneck, C.E. Williamson, M.E. Stoeckel (1990). Diel periodicity and selectivity in the feeding rate of the predatory copepod *Mesocyclops edax*. *J. Plankton Res.*, **12**, 29–40.
27. G.S. Losey, T.W. Cronin, T.H. Goldsmith, D. Hyde, N.J. Marshall, W.N. McFarland (1999). The UV visual world of fishes: a review. *J. Fish Biol.*, **54**, 921–943.
28. B.M. Bebout, F. Garcia-Pichel (1995). UV-B induced vertical migrations of cyanobacteria in a microbial mat. *Appl. Environm. Microbiol.*, **61**, 4215–4222.
29. F.E. Smith, E.R. Baylor (1953). Color responses in the cladocera and their ecological significance. *Am. Nat.*, **87**, 49–55.
30. J.T. Pennington, R.B. Emlet (1986). Ontogenetic and diel vertical migration of a planktonic echinoid larva, *Dendraster excentricus* (Eschscholtz): occurrence, causes, and probable consequences. *J. Exper. Mar. Biol. Ecol.*, **104**, 69–95.
31. P.M. Kiffney, E.E. Little, W.H. Clements (1997). Influence of ultraviolet-B radiation on the drift response of stream invertebrates. *Freshwat. Biol.*, **37**, 485–492.
32. N.L. Adams (2001). UV radiation evokes negative phototaxis and covering behavior in the sea urchin *Strongylocentrotus droebachiensis*. *Mar. Ecol. Progr. Ser.*, **213**, 87–95.
33. F. Lenci, G. Checcucci, F. Ghetti, D. Gioffre, A. Sgarbossa (1997). Sensory perception and transduction of UV-B radiation by the ciliate *Blepharisma japonicum*. *Biochim. Biophys. Acta*, **1336**, 23–27.
34. U.C. Storz, R.J. Paul (1998). Phototaxis in water fleas (*Daphnia magna*) is differently

influenced by visible and UV light. *J. Comp. Physiol. A*, **183**, 709–717.
35. T.M. Frank, E.A. Widder (1994). Evidence for behavioral sensitivity to near-UV light in the deep sea crustacean *Systellaspis debilis*. *Mar. Biol.*, **118**, 279–284.
36. L. Chalker-Scott, J. Scott, J. Titus (1994). Brief exposure to ultraviolet radiation inhibits locomotion of veligers and juvenile *D. polymorpha*. In: *37th Conference of the International association for Great Lakes research and Estuarine Research Federation: Program and Abstracts* (pp. 166). IAGLR, Buffalo, NY.
37. M.K. Blakefield, J. Calkins (1992). Inhibition of phototaxis in *Volvox aureus* by natural and simulated solar ultraviolet light. *Photochem. Photobiol.*, **55**, 867–872.
38. N.G. Hairston Jr. (1976). Photoprotection by carotenoid pigments in the copepod *Diaptomus nevadensis*. *Proc. Nat. Acad. Sci. U.S.A.*, **73**, 971–974.
39. D.-P. Häder (1993). Effects of enhanced solar ultraviolet radiation on aquatic ecosystems. In: M. Tevini (Ed.), *UV-B Radiation and Ozone depletion: Effects on Humans, Animals, Plants, Microorganisms and Materials* (pp. 155–192). Lewis Publishers.
40. D.M. Leech, C.E. Williamson (2001). In situ exposure to solar UV radiation alters the depth distribution of *Daphnia*. *Limnol. Oceanogr.*, **46**, 416–420.
41. D.M. Leech, A.T. Padeletti, C.E.Williamson. Behavioral responses to UV radiation vary across zooplankton taxa. *Limnol. Oceanogr.*, submitted.
42. D.O. Hessen (1994). *Daphnia* responses to UV-light. *Arch. Hydrobiol. Beih. Ergebn. Limnol.*, **43**, 185–195.
43. S. Rhode, M. Pawlowski, R. Tollrian (2001). The impact of ultraviolet radiation on the vertical distribution of zooplankton genus *Daphnia*. *Nature*, **412**, 69–72.
44. G.G. Martin, C. Speekmann, S. Beidler (2000). Photobehavior of the harpacticoid copepod *Tigriopus californicus* and the fine structure of its nauplius eye. *Inv. Biol.*, **119**, 110–124.
45. W.F. Donahue, D.W. Schindler (1998). Diel emigration and colonization responses of blackfly larvae (Diptera: Simuliidae) to ultraviolet radiation. *Freshwat. Biol.*, **40**, 357–365.
46. C.E. Williamson, S.L. Metzgar, P.A. Lovera, R.E. Moller (1997). Solar ultraviolet radiation and the spawning habitat of yellow perch, *Perca flavescens*. *Ecol. Appl.*, **7**, 1017–1023.
47. C. Gutierrez-Rodriguez, C.E. Williamson (1999). Influence of solar radiation on early life-history stages of the bluegill sunfish, *Lepomis machrochirus*. *Environ. Biol. Fishes*, **55**, 307–319.
48. B.L. Bingham, N.B. Reyns (1999). Ultraviolet radiation and distribution of the solitary ascidian *Corella inflata*. *Biol. Bull.*, **196**, 94–104.
49. D.T. Sharp, I.E. Gray (1962). Studies on factors affecting the local distribution of two sea urchins, *Arbacia punctulata* and *Lytechinus variegatus*. *Ecology*, **43**, 309–313.
50. J.L.J. Eastwood (1972). *The Development and Photic Behavior of Lytechinus Variegatus Echinoplutei Larvae* (Ph.D. thesis). Lehigh University, Bethlehem, PA, USA.
51. N. Millot (1975). The photosensitivity of echinoids. In: F.S. Russell, M. Yonge (Eds), *Advances in marine biology* (pp. 1–52). Academic Press, New York.
52. N.G. Hairston (1980). The vertical distribution of diaptomid copepods in relation to body pigmentation. In: W.C. Kerfoot (Ed.), *Evolution and Ecology of Zooplankton Communities* (pp. 98–110). University Press, Hanover, New Hampshire.
53. M.J. Tovee (1995). Ultra-Violet photoreceptors in the animal kingdom: their distribution and function. *TREE*, **10**, 455–459.
54. T.H. Goldsmith, G.D. Bernard (1985). Visual pigments in invertebrates. *Photochem. Photobiol.*, **42**, 805–809.

55. G.H. Jacobs (1992). Ultraviolet vision in vertebrates. *Am. Zool.*, **32**, 544–554.
56. A.P. Cullen, C.A Monteith-McMaster (1993). Damage to rainbow trout (*Oncorhynchus mykiss*) lens following acute dose of UV-B. *Curr. Eye Res.*, **14**, 359–377.
57. J.N. Lythgoe (1984). Visual pigments and environmental light. *Vision Res.*, **24**, 1539–1550.
58. J.S. Levine, E.F MacNichol (1979). Visual pigments in telost fishes: Effects of habitat, microhabitat and behavior on visual system evolution. *Sensory Proc.*, **3**, 95–131.
59. J.K. Bowmaker (1990). Visual pigments of fishes. In: R.H. Douglas, M.B.A. Djamgoz (Eds), *The Visual System of Fish* (pp. 81–108). Chapman and Hall Press, London.
60. W.B.A. Muntz, A W. Wainwright (1978). Annual cycles in the light environments and visual mechanisms of fishes. In: J.E. Thorpe (Ed.), *Rhythmic Activity of Fishes* (pp. 105–129). Academic Press, London.
61. W.B.A. Muntz, G.S.V. Mouat (1984). Annual variations in the visual pigments of brown trout inhabiting lochs providing different light environments. *Vision Res.*, **24**, 1575–1580.
62. A.V. Whitemore (1988). *The Visual Pigments and Action Spectra of Five Types of Cone in the Retina of the Rudd, Scardinius erythrophthalmus*, (PhD. Thesis). University of London.
63. E.H. Loew, W.N McFarland, E.L. Mills, D. Hunter (1993). A chromatic action spectrum for planktonic predation by juvenile yellow perch, *Perca flavescens*. *Can. J. Zool.*, **71**, 384–386.
64. H.I. Browman, I. Novales-Flamarique, C. Hawryshyn (1993). Ultraviolet photoreception contributes to prey search behavior in two species of zooplanktivorus fishes. *J. Exp. Biol.*, **186**, 187–198.
65. J.K. Bowmaker, Y.W. Kuntz (1987). Ultraviolet receptors, tetrachromatic color vision and retinal mosaics in the brown trout (*Salmo trutta*): age-dependent changes. *Vision Res.*, **27**, 2101–2108.
66. I. Novales Flamarique (2000). The ontogeny of ultraviolet sensitivity, cone disappearance and regeneration in the sockeye salmon *Oncorhynchus nerka*. *J. Exp. Biol.*, **203**, 1161–1172.
67. C.W. Hawryshyn (1992). Polarization vision in fish. *Am. Sci.*, **80**, 164–175.
68. D.C. Parkyn, C.W. Hawryshyn (1993). Polarized light sensitivity in the rainbow trout (*Oncorhynchus mykiss*): characterization from multiunit responses in the optic nerve. *J. Comp. Physiol.*, **172**, 493–500.
69. D. Coughlin, C.W. Hawryshyn (1994). The contribution of ultraviolet and short-wavelength sensitive cone mechanisms to color vision in rainbow trout. *Brain Behavior. Evol.*, **43**, 219–232.
70. J.P. Zamzow, G.S. Losey. Ultraviolet radiation absorbance by coral reef fish mucus: photo-protection and visual communication. *Environ. Biol. Fish*, **63**, 41–47.
71. E.R. Loew, W.N. McFarland (1990). The underwater visual environment. In: R.H. Douglas, M.B.A. Djamgoz (Eds), *The Visual System of Fish* (pp. 1–44). Chapman and Hall Press, London.
72. L.A. Giguere, R.L. Dunbrack (1990). Thin layer interference may reduce the visibility of transparent phantom midge (*Chaoborus trivittatus*) to predators. *Can. J. Fish. Aquat. Sci.*, **47**, 1043–1046.
73. V. Rocco, J.P. Barriga, H. Zagarese, M. Lozada (2002). How much does ultraviolet radiation contribute to the feeding performance of rainbow trout, *Oncorhynchus mykiss*, juveniles under natural illumination? *Environ. Biol. Fish.*, **63**, 223–228.
74. S. Johnsen, E.A Widder (2001). Ultraviolet absorption in transparent zooplankton and its implications for depth distribution and visual predation. *Mar. Biol.*, **138**,

717–730.

75. K.C. Smith, E.R. Macagno (1990). UV photoreceptors in the compound eye of *Daphnia magna* (Crustacea, Branchiopoda): A fourth spectral class in single ommatidia. *J. Comp. Physiol. A*, **166**, 597–606.
76. A. Portwich, F. Garcia-Pichel (2000). A novel prokaryotic UVB photoreceptor in the cyanobacterium *Chlorogloeopsis* PCC6912, *Photochem. Photobiol.*, **71**, 493–498.
77. G. Wald, S. Rayport (1977). Vision in annelid worms. *Science*, **196**, 1434–1439.
78. W.E. Neill (1990). Induced vertical migration in copepods as a defense against invertebrate predation. *Nature*, **345**, 524–526.
79. M.D. Ohman (1990). The demographic benefits of diel vertical migration by zooplankton. *Ecol. Monogr.*, **60**, 257–281.
80. A.G. Huntsman (1924). Limiting factors for marine animals. I. The lethal effect of sunlight. *Contr. Can. Biol.*, **2**, 82–88.
81. A.B. Klugh (1930). The effect of the ultraviolet component of the sun's radiation upon some aquatic organisms. *Can. J. Res.*, **2**, 312–317.
82. A.R. Moore (1912). Concerning negative phototropism in *Daphnia pulex*. *J. Exp. Zool.*, **13**, 573–575.
83. W.C. Kerfoot (1985). Adaptive value of vertical migration: comments on the predation hypothesis and some alternatives. *Contr. Mar. Sci.*, **27**, 91–113.
84. J. Ringelberg (1999). The photobehavior of *Daphnia* spp. as a model to explain diel vertical migration in zooplankton. *Biol. Rev.*, **74**, 397–423.
85. C. Loose, E. von Elert, P. Dawidowicz (1993). Chemically-induced diel vertical migration in *Daphnia*: a new bioassay for kairomones exuded by fish. *Arch. Hydrobiol.*, **126**, 329–337.
86. C.E. Williamson, H.E. Zagarese, P.C. Schulze, B.R. Hargreaves, J. Seva (1994). The impact of short-term exposure to UV-B radiation on zooplankton communities in north temperate lakes. *J. Plankton Res.*, **16**, 205–218.
87. N.G. Jerlov (1976). *Optical Oceanography*, Elsevier Publishing Company, Amsterdam.
88. S. Madronich (1994). Increase in biologically damaging UV-B radiation due to stratospheric ozone reductions: a brief review. *Arch. Hydrobiol. Ergebn. Limnol.*, **43**, 17–30.
89. H. Yang, H. Inokuchi, J. Adler (1995). Phototaxis away from blue light by an *Escherichia coli* mutant accumulating protoporphyrin IX. *Proc. Nat. Acad. Sci. U.S.A.*, **92**, 7332–7336.
90. W. Sprenger, W.D. Hoff, J.P. Armitage, K. Hellingwerf (1993). The eubacterium *Ectothiorhodospira halophila* is negatively phototactic, with a wavelength dependence that fits the absorption spectrum of the photoactive yellow protein. *J. Bacteriol.*, **175**, 3096–3105.
91. R. Menzel (1979). Spectral sensitivity and color vision in invertebrates. In: H. Hutrum (Ed.), *Comparative Physiology and Evolution in Invertebrates* (pp. 503–580). Springer-Verlag, New York.
92. M. Iseki, S. Wada (1995). Action spectrum in the ultraviolet region for phototropism of Bryopsis rhizoids. *Plant Cell. Physiol.*, **36**, 1033–1040.
93. M. Selbach, H.W. Kuhlmann (1999). Structure, fluorescent properties and proposed function in phototaxis of the stigma apparatus in the ciliate *Chlamydodon mnemosyne*. *J. Exp. Biol.*, **202**, 919–928.
94. L.A. Wilkens (1984). Ultraviolet sensitivity in hyperpolarizing photoreceptors of the giant clam *Tridacna*. *Nature*, **309**, 446–448.
95. I. Novales Flamarique, H.I. Browman, M. Belanger, K. Boxaspen (2000). Ontogenetic

changes in visual sensitivity of the parasitic salmon louse *Lepeophtheirus salmonis*. *J. Exp. Biol.*, **203**, 1649–1659.

96. T.W. Cronin, N.J. Marshall (1989). A retina with at least ten spectral types of photoreceptors in a stomatopod crustacean. *Nature*, **339**, 137–140.
97. T.M. Frank, J.F. Case (1988). Visual spectral sensitivities of bioluminescent deep-sea crustaceans. *Biol. Bull.*, **175**, 261–273.
98. R. Forward, T.W. Cronin (1979). Spectral sensitivity of larvae from intertidal crustaceans. *J. Comp. Physiol.*, **133**, 311–315.
99. G. Wald, E.B. Seldin (1968). Spectral sensitivity of the common prawn *Palaemonetes vulgaris*. *J. Gen. Physiol.*, **51**, 694–700.
100. D.R. Cummins, D.M. Chen, T.H. Goldsmith (1984). Spectral sensitivity of the spiny lobster *Panulirus argus*. *Biol. Bull.*, **166**, 269–276.
101. G. Wald, J.M. Krainin (1963). The median eye of *Limulus*: an ultraviolet receptor. *Proc. Nat. Acad. Sci. U.S.A.*, **50**, 1011–1017.
102. L.L. Britt, E.R. Loew W.N. McFarland (2001). Visual pigments of PNW fish larvae. *J. Exp. Biol.*, **204** 2581–2587.
103. F. Schiemanz (1924). Über den Farbensinn der Fische. *Z. Vergl. Physiol.*, **1**, 175–200.
104. J.A. Avery, J.K. Bowmaker, M.B.A. Djamgoz, J.E.G. Downing (1983). Ultraviolet sensitive receptors in a freshwater fish. *J. Physiol., London*, **334**, 23P.
105. C.W. Hawrynshyn, R. Beauchamp (1985). Ultraviolet photosensitivity in goldfish: an independent UV retinal mechanism. *Vision Res.*, **25**, 11–20.
106. C.W. Hawryshyn F.I. Harosi (1991). Ultraviolet photoreception in carp: MSP and behaviorally determined action spectra. *Vision Res.*, **31**, 567–576.
107. A.G. Palacios, T.H. Goldsmith, G.D. Bernard (1996). Sensitivity of cones from a cyprinid fish (*Danio aequipinnatus*) to ultraviolet and visible light. *Vision Neurosci.*, **13**, 411–21.
108. A.V. Whitmore, J.K. Bowmaker (1989). Seasonal variation in cone sensitivity and short-wave absorbing visual pigments in the rudd. *J. Comp. Physiol.*, **166**, 103–115.
109. F.I. Harosi, Y. Hashimoto (1983). Ultraviolet visual pigment in a vertebrate: a tetrachromatic cone system in the dace. *Science*, **222,** 1021–1023.
110. C.W. Hawrynshyn, M.G. Arnold, D.J. Chaisson, P.C. Martin (1987). Developmental changes in ultraviolet photosensitivity in rainbow trout. *Social Neurosci. Abs.*, **13**, 1298.
111. Y.W. Kunz (1987). Tracts of putative ultraviolet receptors in the retina of the two year old brown trout and the Atlantic salmon. *Experientia*, **43**, 1202–1204.
112. J.S. Levine, E.F. MacNichol (1982). Color vision in fishes. *Sci. Am.*, **246**, 140–149.
113. E.R. Loew, C.M. Wahl (1991). A short wavelength sensitive cone mechanism in juvenile yellow perch, *Perca flavescens*. *Vision Res.*, **31**, 353–360.
114. I. Novales Flamarique, F.I. Harosi (1999). Photoreceptor pigments of the Blueback Herring (*Alosa aestevalis*, Clupeidae) and the Atlantic Silverside (*Menidia menidia*, Atherinidae). *Biol. Bull.*, **197**, 235–236.
115. I. Novales Flamarique, F.I. Harosi (2000). Photoreceptors, visual pigments, and ellipsosomes in the killifish, *Fundulus heteroclitus*: A microspectrophotometric and histological study. *Visual Neurosci.*, **17**, 403–420.
116. E.R. Loew, R.A. McAlary, W.N. McFarland (1996). Ultraviolet visual sensitivity in the larvae of two species of marine atherinid fishes. In: P.H., Lenz, D.K. Hartline, J.E. Purcell, D.L. Macmillan (Eds), *Zooplankton: Sensory Ecology and Physiology* (pp. 195–210). Gordon and Breach Publishers, Amsterdam.
117. J.V. Helvik, O. Drivenes, T. Harboe, H.-C. Seo (2001). Photoreceptor cells in Atlantic halibut retina. *J. Exp. Biol.*, **204**, 2553–2559.

118. K.L. Carlton, F.I. Harosi, T.D. Kocher (2000). Visual pigments of African cichlid fishes: evidence for ultraviolet vision from microspectrophotometry and DNA sequences. *Vision Res.*, **40**, 879–890.
119. G.S. Losey, P.A. Nelson, J.P. Zamzow (2000). Ontogeny of spectral transmission in the eye of the tropical damselfish and possible effects on UV vision. *Environ. Biol. Fishes*, **59**, 21–28.
120. W.N. McFarland, E.R. Loew (1994). Ultraviolet visual pigments in marine fishes of the family Pomacentridae. *Vision Res.*, **34**, 1393–1396.
121. J. Ammermuller, A. Itzhaki, R. Weiler, I. Perlman (1998). UV-sensitive input to horizontal cells in the turtle retina. *Eur. J. Neurosci.*, **10**, 1544–1552.

Community and Ecosystem Perspectives

- UVR and its effects on species interactions
- Past UVR environments and impacts on lakes
- UVR effects on aquatic ecosystems: a changing climate perspective

Chapter 15

UVR and its effects on species interactions

Ruben Sommaruga

Table of contents

Abstract

Species interactions are crucial to understand the control of population growth and community structure. This chapter presents a brief and critical review of what is known about the effects of UVR (280–400 nm) on species interactions in aquatic ecosystems with emphasis on competition and predation/herbivory. Information on other species interactions such as symbiosis, parasitism, and disease are also briefly reviewed. The existing information indicates that UVR acts as a selective force in pioneer communities of transparent and shallow ecosystems strongly influencing competition output between species at the base of the food web and community structure. However, whether more UV-tolerant species could replace sensitive ones in established communities of natural environments remains uncertain. Examples of positive and negative feedbacks between populations of prey and predators/grazers caused by UVR have been found, but the present information does not ascertain as to whether these mechanisms are widespread in natural ecosystems. Despite the important advance during the last years in our understanding of how ambient and enhanced levels of UV-B radiation (280–320 nm) influence species interactions and trophic relationships, there is still a major gap of knowledge, which is partially attributed to the complexity and biological variability of the species response to UVR, but also to methodological caveats. Consequently, many of the scenarios and hypotheses stated shortly after the discovery of the stratospheric ozone reduction still remain in dispute. Without further research on this topic and the use of more realistic ecological approaches, our assessment of the impact of UVR at the community and ecosystems levels will remain fragmentary and recommendations for sound policy decisions impracticable.

15.1 Introduction

In analyzing the role played by UVR on species interactions, I have included five categories based on the mechanism, namely competition, predation–herbivory, mutualism, parasitism, and disease [1]. The definition and use of these categories in the scientific literature has been flexible as asserted by the interactions between species included under the categories of mutualism, disease, and parasitism [2]. Nevertheless, competition, predation, grazing, and parasitism are among the most important processes in ecology, because they are crucial to understand control mechanisms of population growth and community structure [3].

As we have seen in previous chapters, the bulk of information about the effects of UVR on aquatic organisms has been gathered in studies where species interactions were not considered. Although aquatic ecologists have obtained information on how sensitive different organisms are, including those considered as keystone species, comparatively little effort has been addressed to study potential positive and negative feedbacks caused by UVR on species interactions. Yet, this information is necessary to understand the community and ecosystem response to present and increased levels of incident UVR. Particularly, knowledge at the community level is a pre-requisite for the application of concepts like stability and recovery [4] that has been sometimes wrongly applied in studies of UV impact based on single species.

In general, the characteristics of UVR as an environmental stressor differ from other toxic agents such as pesticides or other man-made chemical substances that were not previously found in the environment. As evidenced by the several strategies developed among different forms of life to obtain protection and to repair damage, solar UVR has been an important selective factor during evolution of life on Earth. Moreover, the existence of a temporal pattern in UV irradiances, of a natural vertical gradient of UVR in the water column, of physical refuges, as well as the dual role of UVR (i.e., having negative and positive effects) impart a different nature to the interaction between UVR and aquatic organisms. Although UVR may affect species interactions in a similar way as a toxic substance does, the ecological buffer or adaptive response of aquatic communities to the effects of UVR may well be larger than that to xenobiotics.

Solar UVR may affect species interactions by direct and indirect ways. Figure 1 depicts some examples of hypothetical changes in population size over time of two interacting species with different UV-sensitivity, where the resulting effect

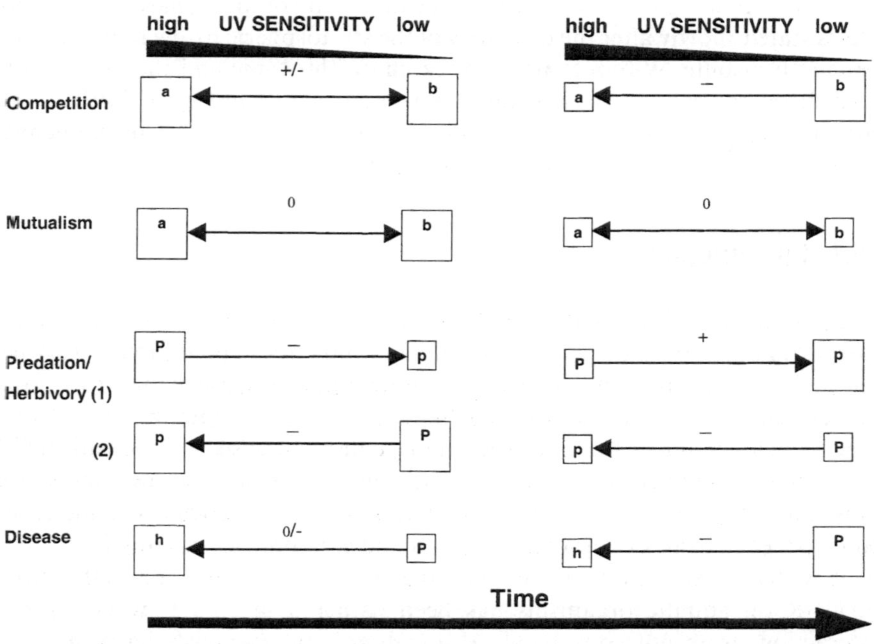

Figure 1. Hypothetical changes in size population over time as affected by differential sensitivity to UVR in five types of species interactions. Size of a population is represented by small or large squares. The 0, +, and – represent the effects of one species on the other (e.g., 0 means a neutral effect of population a on population b). P: predator or parasite, p: prey, and h: host.

has been categorized into neutral, positive, or negative [5]. Despite its simplicity and the fact that species interactions are seldom one-to-one in natural food webs, which are usually characterized by a high linkage density (i.e., the average number of interactions per species in the web), Figure 1 provides a framework to analyse the effect of UVR on species interactions and to compare these scenarios with results obtained from the scientific literature.

In this chapter, I shall critically review our knowledge of the link between UVR and the different categories of species interactions mentioned above, trying to lay emphasis on the overall net result in population size for the respective interactions, and, wherever possible, presenting general patterns, identifying major gaps, and future research directions.

15.2 UVR, competition, and changes in species composition

In a pioneering study, Jokiel [6] observed that the UV-tolerant branching sponge *Callyspongia diffusa* replaced the UV-sensitive sponge *Zygomycale parishi* in shallow (< 3 m depth) reefs of Kaneohe Bay, Hawaii. He hypothesized that metabolic costs to obtain UV tolerance could place species at a competitive disadvantage in shaded environments, but on the other hand could offer a selective advantage in competition for space in sunlit areas. In a simple and elegant experiment, Jokiel [6] tested whether UV tolerance in *C. diffusa* offers a competitive advantage against *Z. parishi* in the presence of UVR. After 7 days, *Z. parishi* grew over *C. diffusa* in both experimental setups. However, while in the UV-shielded treatment the median tissue overgrowth was 7 mm, it was only 1 mm under full solar radiation. Within 2 months, *Z. parishi* overwhelmed *C. diffusa* in the UV-shielded treatment, but the latter species remained healthy in the UV-exposed one. Although this study did not test wavelength-specific effects of solar UVR on competition, it identified probably for the first time the importance of UVR as an environmental variable potentially affecting species competition in aquatic ecosystems.

One often expected effect of increased levels of incident UV-B radiation in aquatic ecosystems is a change in species composition particularly, of primary producers [7–10]. The rationale behind this hypothesis is based on the different sensitivity to UVR found among species of planktonic and benthic algal communities (see Chapter 11). Changes in species composition within a community are hypothesized to occur by replacement of UV-sensitive species by resistant ones, which occupy similar (trophic) niches [7,9,10]. A change in species composition can take place directly if the population of a UV-sensitive species does not survive to UVR levels above its tolerance threshold or indirectly if outcompeted by more tolerant species. There is strong evidence from several studies in marine and freshwater systems indicating that UVR plays a major role in shaping the structure of communities during the early colonization and succession of many aquatic habitats/ecosystems through selection against less UV-tolerant species [11–18]. However, could enhanced levels of incident UV-B radiation per se lead to the extinction of species in an established (mature) community? Or is it more probable that enhanced (ambient) UV-B levels could cause changes in community structure by an alteration in the population size of UV-sensitive and more resistant species as a consequence of different growth rates and competition [19]? Testing the first hypothesis (i.e., extinction) in natural communities is difficult because generally there is a lack of references to what to compare present population/community structure, particularly in places like Antarctica where enhanced UV-B levels have been experienced during the austral spring for more than 20 years. Thus, most studies, except those following a paleo-approach, have tested whether UV-B radiation offers a competitive advantage to tolerant species in long-term micro/mesocosm experiments where UV-B has been excluded and/or artificially enhanced.

Table 1 presents a summary of studies done to test the hypothesis of changes in taxonomic composition in phytoplankton (for which most information is available). The data to address this hypothesis are of uneven quality and the studies differ in experimental design, sophistication level, and statistic strength, so their interpretation in some cases is problematic. Thus, for example, when grazers were present but their abundance was not controlled or their food spectrum not assessed, their effect on changes in phytoplankton species composition will be difficult to discern from those potentially caused by UV-B radiation. Moreover, an additional limitation in the methodology used in exclusion experiments is the distinction between UV-B and UV-A effects. Separation between the effects of these wavebands is generally accomplished by the use of the polyester foil Mylar D (DuPont de Nemours & Co. Inc.). This material, however, cuts off only part of the biologically effective UV-B radiation. For example, when the transmittance of Mylar D (23 μm thickness, 50% transmittance at 316 nm) is multiplied by the solar spectrum for $\sim$40 °N latitude near summer solstice, it cuts off 60% of UV-B ($<$320 nm) or only 56% of the biologically effective radiation when the biological weighting function for *Daphnia pulicaria* mortality is used [20]. Furthermore, this value will change depending on the thickness of Mylar D used, which is seldom reported in the experimental design although it strongly affects the cut-off wavelength in the UV-B range.

One of the first studies on this topic was done by Worrest [21] with estuarine phytoplankton (Yaquina Bay, Oregon) exposed in small microcosms (15 L, depth: 0.30 m) to natural solar radiation of wavelengths $>$380 nm plus enhanced UV-B radiation. The phytoplankton dominated by diatoms changed after 4 weeks (sampling was done only at the beginning and the end of the experiment) with an apparent increase in the dominance of *Chaetoceros* sp. and a decrease of *Skeletonema costatum*. Similar findings have been reported in two studies with phytoplankton from the Gullmar Fjord, Sweden, exposed in small aquaria (18 L and 40 L, depth: 0.23–0.49 m) to artificial UVR or solar radiation plus enhanced UV-B [22,23]. In contrast, in two experiments done in the west coast of Sweden (Gullmar Fjord) with large enclosures (6 m^3, depth: 3.5 m) shifts in phytoplankton species composition were not observed, even in the enhanced UV-B treatments [24]. In microcosm experiments using small containers (1 L) with phytoplankton cultures isolated from Seal Island (Antarctica), changes in taxonomic composition were observed only when exposed to high solar UVR fluxes typical

Table 1. Summary of studies using microcosms or mesocosms to investigate changes in taxonomic composition of phytoplankton caused by UVR; see text for more details on each study

Habitat	Exposure conditions	Container volume (L)	Duration (d)	Changes	Comments	Reference
Estuarine	solar radiation + enhanced UV-B	15	28	yes	natural assemblage	[21]
Estuarine	artificial UVR	18	7	yes	natural assemblage	[22]
Estuarine	solar radiation + enhanced UV-B	40	10	yes	natural assemblage	[23]
Estuarine	solar radiation + enhanced UV-B	6000	8–11	no	natural assemblage	[24]
Marine	solar radiation	1	5–16	yes	mixed cultures	[25]
Marine	solar radiation	2	15	yes	natural assemblage	[26]
Marine	solar radiation	0.5	8	yes	6 co-occurring species	[27]
Freshwater	solar radiation	1000	16	no	natural assemblage	[29]
Freshwater	solar radiation	300	30	no	natural assemblage	[13]
Freshwater	solar radiation + enhanced UV-B	20000	44	no	natural assemblage	[30]
Freshwater	artificial UVR	600	56	no	natural assemblage	[31]

for the tropics, but not under ambient UV irradiances [25]. In another study with surface phytoplankton collected in Arthur Harbor, Antarctica, and exposed to solar radiation in 2 L flasks, important changes in taxonomic composition were observed already after 4 days [26]. In the presence of UVR, the original assemblage composition dominated by flagellates (their experiment #2) shifted by day 13 to the dominance of diatoms. Davidson et al. [27] performed competition experiments in nutrient-rich media using six co-occurring phytoplankton species isolated from the Southern Ocean that were exposed in small bags (0.5 L) to natural solar radiation in an outdoor tank. Their results indicated that overall growth and production by this artificial community was not affected by UVR. However, UV-B caused changes in the growth rate of some species. Thus, for example, growth rate of four diatom species did not change significantly but that of the flagellate stage of *Phaeocystis antarctica* decreased in the presence of UV-B. On the other hand, the growth of the colonial form of *P. antarctica* was enhanced when exposed to UV-B. Changes in species composition were elicited after 2 d exposure and by day 8 the proportion of the colonial form of *P. antarctica* increased mainly at the expense of *Chaetoceros simplex*, although extinction was not observed. These results contrasted with previous studies by Karentz [9] and Karentz and Spero [28] showing that growth of the colonial form of *Phaeocystis* sp. declined in the presence of UV-B radiation. In a field study at the marginal ice zone of the Bellinghausen Sea, *Phaeocystis* populations appeared to be negatively affected by increased levels of UV-B during the "ozone hole", but this did not offer a competitive advantage to co-occurring diatoms species [28].

Results from an experiment with 1 m^3 enclosures (depth: 0.95 m) in a transparent alpine lake from the Austrian Alps indicated no significant differences in species composition after 16 days between the UV-B-shielded and -exposed treatments [29]. Although there were important changes in the proportion of co-occurring species, for example, a decrease in the chrysophyte *Chromulina* sp. and an increase in the chlorophyte *Dyctiosphaerium* sp., the change in dominant species was not caused by UV-B radiation. In another alpine system (Pipit Lake, Canada), no changes in species composition in phytoplankton assemblages, consisting mainly of picocyanobacteria, chrysophytes, cryptophytes, and dinoflagellates, were observed during a 30 days enclosure (0.3 m^3, depth: 0.7 m) experiment where UV-B was excluded [13]. Experiments with large enclosures (20 m^3, depth: 1 m) placed in the littoral zone of mesotrophic Jack's Lake, Canada, showed no evidence for collapse of specific phytoplankton populations or any large-scale taxonomic shift under ambient, UV-B-excluded, or -enhanced treatments [30]. In another long-term experiment (8 weeks) with indoor microcosms (600 L) receiving artificial UVR, no effects of UV-B radiation on species composition, abundance or biovolume of phytoplankton (and other planktonic and benthic communities) was observed [31].

Certainly, a small database as the one presented above may lead to generalizations that are not correct. However, three factors that appear to explain the contrasting results of these studies are the variations in UVR transparency of the water in which the experiments are done, the prior exposure regimes of the species in their place of origin, and the size of the experimental container used in the studies. The growth rate of phytoplankton species originating from sunlit habitats appears to be less or not at all affected when exposed to UV-B radiation [25,32, see also 33 for a review on other photosynthetic organisms]. Thus, exclusion of UV-B radiation or even its enhancement may not offer a competitive advantage to species already adapted to high solar UV-B irradiances. Beside the obvious disadvantages of the enclosure approach, like, for instance, the elimination of advection and diffusion, the size of the enclosure has a major effect on natural avoidance mechanisms, such as vertical displacement, and on the characteristics of the radiation field experienced by the organisms. Thus, in small-sized enclosures, organisms are exposed to a uniform field of UV radiation due to the short path-length that solar radiation needs to travel before reaching an algal cell [24]. This situation, however, differs largely from natural conditions, particularly in turbid waters (e.g., estuaries) where the water column is characterized by a strong gradient of UV irradiance and spectral characteristics. Together with the absence of mixing that may minimize the UV effect (see Chapter 4) and the long-term (days to weeks) exposure, it is not surprising that significant shifts in species composition caused by UV-B have generally been observed in experiments with small enclosures.

Examples of studies with natural communities of benthic microalgae or with communities of periphyton growing on artificial substrates for long periods are less common. However, in contrast to the dramatic changes observed when pioneer species colonizing new substrates are exposed to UVR, these studies suggest that neither ambient nor enhanced UV-B radiation significantly affects algal species composition [34–37].

Finally, an alternative approach to test the hypothesis of change in species composition has been to look at changes in the dominance of algal species that remain preserved in the sediments, for instance diatoms. In this approach, the main advantage is that the 'historical' reference or initial assemblage structure can be reconstructed in most cases (see Chapter 16). On the other hand, it may be difficult to isolate the effect of UVR from other environmental changes, except when recognition of present UV-sensitive species with a long sediment record is possible. Results by McMinn et al. [38] showed that changes during 20 years (~1971 to 1991) in the relative abundance of diatom taxa analysed in three sediment cores from anoxic basins in Vestfold Hills, Antarctica, were not distinguishable from long-term natural variability. However, as the authors acknowledged, the study was done in a coastal area where a thick ice-cover is present at time of phytoplankton growth and therefore it was not representative of the zone affected by the ozone reduction (see also [39] for other critics on this study). Nevertheless, this approach remains an interesting alternative to explore.

15.3 Herbivory and predation: the complex response of trophic interactions to UVR

In the previous section, the response to UVR of populations at one trophic level (basal species) was considered. The interaction of UVR, however, with more than one trophic level adds substantial complexity to the possible responses, with the potential occurrence of positive and negative feedbacks (Figure 1). Both prey and predator populations might be affected by UVR, and, if so, the net effect will depend on the relative tolerance threshold of the interacting species. Yet, as soon as we consider more than one interaction between species, responses in the food web are expected to be much more complex than depicted in Figure 1, including potential changes in population size at different trophic levels. Thus, for example, in the hypothetical aquatic food web depicted in Figure 2, a potential reduction in population size of the UV-sensitive species 5 feeding on UV-sensitive basal species 7 and UV-tolerant species 8 and eaten by species 3 may increase the competitive advantage of species 8 and at the same time reduce the population size of species 3 but increase those of 2 and 1.

Bothwell et al. [40] found the first direct evidence of complex interactions in the food web during a colonization experiment with freshwater periphyton growing in artificial flumes of 1 cm depth located outdoor in British Columbia, Canada. They observed that short-term effects of UVR (mainly UV-A) caused inhibition of diatom growth and accrual rate (chlorophyll-*a*). However, after the third week, UVR reduced the number of algal grazers (chironomid larvae mainly *Cricotopus bicinctus* and *Orthocladius* sp.), and by the fourth week, the initial negative effect on algal biomass was reversed. This food web, however, was relatively simple with mainly one herbivore species. Other studies with periphyton in natural streams have failed to confirm the positive feedback described above. For example, in a 28-day study in Otter Creek, Nebraska, ambient levels of UVR did not affect algae or herbivores colonizing tiles submersed at a depth of 8–22 cm [41]. The authors argued that lack of significant

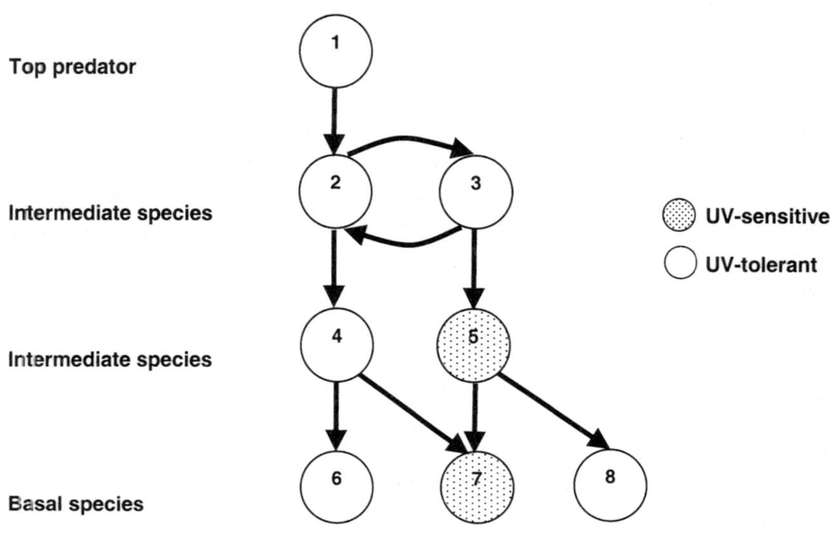

Figure 2. Hypothetical aquatic food web consisting of 8 species having 9 from 64 possible species interactions. [Modified from 2.]

differences in herbivore densities between treatments might have been due to the presence of long UV-A wavelengths in the "UV-excluded treatment", which have been suggested by Bothwell et al. [40] to elicit UV-avoidance responses of invertebrates. In three colonizing experiments done in the upper White Oak Creek, Tennessee, periphyton and herbivores, mainly the snail *Elimia clavaeformis*, were not significantly affected by ambient UVR levels [42, see also critics and discussions in 43 and 44]. In an experiment lasting 30 days in the Cache la Poudre River, Colorado, a negative effect of UVR on periphyton biomass accrual and abundance of invertebrates colonizing artificial substrates submersed at ~10 to 40 cm depth was only apparent at the end of the experiment [45]. However, the authors concluded that it was unclear whether the effect on invertebrates was caused by UVR, by interaction with other invertebrates, or higher algal biomass in the UV-excluded treatment. They speculated that ending the experiment after 30 days and interrupting successional shifts in algal species might have avoided a positive feedback of UVR on algal biomass. McNamara and Hill [36] suggested that the different responses of periphyton observed in the studies mentioned above are related to the presence of more UV-resistant communities in streams at low elevations of mid-latitudes than at higher elevation or latitudes. This seems to be counterintuitive because UV-B fluxes increase with altitude (see Chapter 2). On the other hand, several authors have argued that one possible explanation for the dissimilar results obtained are the different exposure characteristics of organisms in artificial flumes and natural streams, particularly the shallow depth and exclusion of higher-level predators that may exacerbate the effects of UVR on periphyton and insect larvae [15,41,45].

A lack of indirect effects mediated by UVR has also been observed in a colonizing experiment in an alpine lake [13]. While development of epilithon (mainly diatoms and cryptophytes) was suppressed by UV-A and UV-B radiation, zoobenthos like the sediment-dwelling *Gammarus lacustris* and chironomids with burrowing habits were not affected. There are only a few studies in coastal marine environments addressing this topic. During an experiment in the coast of Greece, biomass of colonizing benthic algae (mainly pennate diatoms) and species community structure were affected by ambient levels of UV-B, but invertebrate biomass was not [15].

Autecological studies with freshwater and marine heterotrophic nanoflagellates (HNF) have provided some evidence for a positive feedback between UVR and prey populations. Sommaruga et al. [46] reported that artificial and natural UVR (mainly UV-B but also UV-A) strongly reduced bacterivory rates of the freshwater HNF *Bodo saltans*. In laboratory experiments, they found that mortality rates (i.e., negative growth rates) of bacteria in the UV-B-excluded or dark treatments were higher than in the presence of UV-B. Furthermore, depending on predator density, even positive bacterial growth rates were observed in the presence of UV-B [46]. Similar evidence was obtained by Ochs [47,48] in laboratory experiments with the marine HNF *Paraphysomonas bandaiensis* and *P. imperforata* grazing on two strains of the picocyanobacterium *Synechococcus* sp. Prey population size was always higher in those treatments where grazing by

HNF was more affected by UVR. On the other hand, studies with natural protist assemblages exposed to ambient and enhanced UV-B levels have provided mixed results. In a study with microbial communities from two arctic systems, although ambient UV-B levels negatively affected growth of some ciliates species, community-grazing rates were not [49]. In a 16-day mesocosm experiment with a microbial food web (zooplankton excluded) from a UV-transparent alpine lake, negative effects of ambient UV-B radiation were observed on HNF growth and bacterivory rates. However, this did not result in higher bacterial abundance suggesting that bacteria were negatively affected as well [50]. In an experiment with a microbial food web (organisms >240 μm excluded) from the St. Lawrence Estuary, Canada, enhanced UV-B radiation reduced significantly the populations of large phytoplankton and ciliates after 7 days [51]. The increase in prey abundance, mainly in HNF, was interpreted as a release from predation pressure by ciliates [52]. Reductions in HNF bacterivory rates were not observed until the 7th day [53] when bacterial abundance increased [54]. In other two mesocosm studies with estuarine communities (including zooplankton), no major effects to enhanced UV-B levels were observed except for phytoplankton in one of the studies, while positive feedbacks among different components including fish larvae were not found [55,56].

Another potential effect of UVR on the predator–prey interaction is when the prey population has a higher UV-sensitivity than the predator, potentially leading to a negative feedback (Figure 1 case 2 of predation/herbivory). Investigations on this possible scenario have been based on the observation that the UV-B-irradiated green alga *Selenastrum capricornutum* was ingested by *Daphnia magna* but digested with lower efficiency than those in the control without UV-B [57]. This effect was significant only after ≥ 12 h irradiation with artificial UV-B radiation (max. at 312 nm). The authors hypothesized that changes in both mucous secretion and in thickness of the cell wall were responsible for the lower digestibility. In a later study, other species of phytoplankton (*Chlamydomonas reinhardtii*, *Scenedesmus acutus*, *S. subspicatus*, and *Cryptomonas pyrenoidifera*) cultured in the presence of a high UV-B dose were assessed for qualitative and quantitative cell changes and their effect on life-history parameters of *D. pulex* [58]. Beside reduction of algal growth rates by UV-B, important changes in the nutritional quality of the algae (e.g., total lipid concentration and fatty acids composition) were also observed. The intrinsic growth rate of *D. pulex* kept in the dark, however, was only significantly affected when feeding on the UV-irradiated *S. subspicatus*. Changes in intrinsic growth rate were mainly caused by a smaller number of offspring in the UV-B treatment. On the other hand, UV-B-irradiated *C. reinhardtii* and *C. pyrenoidifera* caused a reduction in the length of newborns in the first clutch. Changes in survival of *D. pulex* were not observed in all cases. Interestingly, in a similar experiment, but including another culture strain of *C. reinhardtii*, the survival of *D. pulex* neonates kept in the dark was strongly affected when feeding on UV-B-irradiated algae [59]. *Daphnia* feeding on UV-B-irradiated algae also showed reduced intrinsic growth rates, clutch number and size. In this experiment, the growth rate of *C. reinhardtii* was only affected at the beginning but after 7 days it was similar to the control and changes in total lipids concentration were not observed. The results of these studies, although interesting, are difficult to interpret with regard to the net response of changes in population size. Considering that species of *Daphnia* like *D. magna* are known to be UV-B-sensitive [60,61], additional parallel experiments with *D. magna* and *D. pulex* concomitantly exposed to UVR would have been necessary to evaluate this interaction. On the other hand, the contrasting results obtained using the same species but different strains [58,59] stress the large biological variability found in response to the exposure to UVR.

In the only one study with freshwater periphyton, enhanced UV-B levels reduced photosynthesis and photosynthetic pigments, but algal nutritional quality (as measured by cell N and P content) and growth of juveniles of the snail *Physella gyrina* fed with UV-B-irradiated periphyton were not altered [36].

15.4 Mutualism and UVR: symbiosis of algae-invertebrates and algae-protists

Most of our knowledge on the interaction between UVR and mutualistic associations is based on studies on algal-invertebrate symbiosis, particularly on scleractinian corals and their endosymbiotic dinoflagellates, the so-called zooxanthellae from the genus *Symbiodinium*. Recently, the scientific literature on this topic has been extensively reviewed [33,62,63]. Here, I will only briefly highlight the most important aspects regarding potential population changes in this association as affected by UVR and review the information for other symbiotic relationships.

Bleaching or discoloration in corals has increased dramatically in tropical areas over the past 20 years. This phenomenon is the result of the expulsion or loss of endosymbionts or at least their pigments. Although not necessarily lethal to the coral, widespread bleaching may cause massive death of coral reefs [64,65]. The role of solar UVR as responsible for coral bleaching remains controversial [33]. However, independently of the factor(s) that may cause bleaching, expulsion of endosymbiotic zooxanthellae will expose them directly to the potential negative effects of UVR. Several studies have shown that UVR inhibits the growth of different species of *Symbiodinium* when isolated from the host, although species-specific differences in sensitivity have also been observed [33]. Furthermore, UVR severely depresses photosynthetic rates in freshly isolated zooxanthellae from corals or other reef organisms, but this effect is small or absent *in hospite*. For example, in *Prochloron* sp., a prokaryotic microalgal symbiont of a colonial tropical ascidia, photosynthesis was strongly inhibited in isolation but not in the host [66]. The different sensitivity between isolated and *in hospite* forms appears to be related to protection given by the host through the accumulation in their tissue of sunscreens such as mycosporine-like amino acids (see Chapter 10).

Symbiosis between algae and protists is widespread, for example, among marine and freshwater planktonic ciliates and large foraminifera. In some cases, the whole cell of different groups of algae lives as endosymbionts in the host, while other species preserve only the plastids. Although the latter type of association is not a "true symbiosis", functionally, plastids represent a source of photosynthetic products for the host, similarly to true algal endosymbionts [67]. Surprisingly, effects of UVR on algal–protozoan symbiotic associations have hardly been studied despite their important role in food webs, particularly as primary producers in oligotrophic systems. Martin-Webb [68] performed UV-exclusion experiments with natural ciliate assemblages including *Mesodinium rubrum* and *Laboea strobila* collected from a shallow area (Georg Bank, 40–42 °N) on the continental shelf, NW Atlantic. The haptorid ciliate *M. rubrum* contains a cryptophyte symbiont and is an important primary producer in coastal areas [69], while the large-sized *L. strobila* is a conspicuous plastidic oligotrichid in temperate waters [70]. Results from these experiments indicated a lower UV-B sensitivity in symbiotic than other ciliates from this coastal area [68].

Large foraminifera from several families have endosymbionts represented by chlorophytes, rhodophytes, dinoflagellates, or pennate diatoms. The type of algal symbionts appears to influence the optimal depth occupied by some species of foraminifera [71]. Yet, bleaching, particularly of the reef-dwelling *Amphistegina gibbosa*, has been observed in populations of subtropical waters like the Florida Keys [72]. Indirect evidence suggests that, similar to bleaching in corals, UVR may be contributing to this phenomenon [72,73]. Thus, for example, *A. gibbosa* shows a seasonal bleaching cycle with maxima during the summer solstices, preceding maximum summer water temperature by *ca.* two months. Moreover, bleaching in this species is observed in remote areas, where pollution is unlikely to be a contributing factor [72,73]. Interestingly, affected *A. gibbosa* is highly predated by the foraminifer *Floresina amphiphaga* and also often found infested by cyanobacteria. These observations were never recorded before the detection of bleaching in this species [73].

15.5 The interaction between UVR and parasites

This type of interaction is obviously restricted to ectoparasites or to the free stadium of endoparasites. Although UVR is generally associated with negative effects, it may also play a positive role on species interactions. Thus, for example, the ectoparasite copepod *Lepeophtheirus salmonis* (salmon lice) uses photoreceptors to avoid UVR and eventually to optimise host finding (e.g., by utilizing UV contrast vision [74]).

Most of our knowledge on the interaction between UVR and parasites, however, is related to viruses, which, although considered obligatory parasites, resemble a predator–prey interaction [75]. Due to the obligatory use of the host metabolic machinery to produce new viral copies and the impossibility to repair themselves, viruses are very vulnerable to several stressors when occurring in the water column. Among other environmental factors, solar UVR affects viruses negatively by reducing their infectivity [76]. Loss of viral infectivity after exposure to solar radiation seems to be mainly caused by damage to the viral genome, although indirect damage to the capsid has also been suggested to result in inactivation [77]. Wavelengths <320 nm are generally the most effective ones to cause viral inactivation [78], although UV-A radiation [79] and wavelengths <556 nm [77] have been found to inactivate viruses as well. Like in many other planktonic groups, different viruses appear to have different tolerance towards solar radiation [77,78,80], but the reason(s) for this remains unclear. Kellogg and Paul [81] found that the degree of UV damage of six marine vibriophages was negatively correlated with the G+C content and suggested that the increase of thymine dimer targets increases their sensitivity by reducing the ability to repair the damage, a hypothesis previously proposed for bacteria by Singer and Ames [82]. The DNA damage, however, can be repaired after infection takes place using the host repair mechanisms. Thus, different repair mechanisms or efficiencies may also explain the variability observed in virus inactivation rates. The infectivity of phages can be restored inside bacteria, either through a specific host-repair-machinery (=photoreactivation) [83,84] or by a virus encoded repair system [85,86]. The light-dependent repair mechanism of bacteria seems to be crucial to restore the infectivity of natural aquatic viruses [83,84,87]. Therefore, the potential recovery of viruses makes it difficult to predict the overall effect of UVR in this interaction. Moreover, the inactivation–recovery process is further complicated by the fact that the physiological status of bacteria can be also impaired by UVR [88,89].

The physical disruption/destruction of the viral particle by high-energy photons is another mechanism that can account for loss of viral abundance [90]. The exact mechanism of this destruction, however, is not well understood and the experimental results gathered with different viruses are inconclusive [91,92].

Finally, another potential interplay between UVR and viruses occurs when they coexist with their host in a type of mutualistic relationship, where the nucleic acid of the virus is integrated in the genome of the host and is replicated with it (lysogenic state). Ultraviolet C radiation produced by germicidal lamps (max. at ~254 nm) has normally been used, among other stressors, to induce the shift from lysogenic to lytic state in a complex mechanism involving the DNA repair SOS system of the host [93]. However, natural or simulated solar UVR seems not to be very efficient in this process [94,95].

15.6 UV radiation and infection diseases

Parasitism is an important ecological interaction that may cause dramatic changes in the host population size. As discussed above, solar UVR has the potential to act directly or indirectly in this process, for example by damaging the parasite or by causing damage to the host and increasing its susceptibility to infections. About the latter type of interaction, our knowledge is restricted mainly to studies on fish and amphibians. Solar UV-B radiation is known to cause injury to the skin (sunburn), reduction of goblet cells (mucus secreting cells), and epidermal hyperplasia in fish although sensitivity is species- and developmental stage-specific [96,97]. The damaged skin tissue is usually susceptible to bacterial and parasite infections. Particularly, *Saprolegnia*, an oomycete, is a common opportunistic facultative parasite of freshwater fish [96]. Infection by *Saprolegnia* causes loss of epithelial integrity and tissue destruction due to cellular necrosis or dermal and epidermal damage [98,99]. Infections may result from direct UV-B damage to the skin or from suppression of the immune system. In the case of the parasite *Saprolegnia*, the decrease in the secretion of mucus appears to be crucial for the infection as it acts as the primary physical barrier [100]. However, UV-B radiation may have a strong immunosuppressive effect on fish, probably weakening their resistance to infectious agents in relation to impairment of the non-specific immune defense [101]. Nevertheless, secondary parasitic infections of fish by *Saprolegnia* after UV-B exposure appear to have been only documented for laboratory studies [102]. On the other hand, results from field observations and experiments have shown that increased UV-B exposure of western toads embryo, *Bufo boreas* caused by reduction in water depth at oviposition sites is related to higher infection by *S. ferax* [103,104]. For example, *S. ferax*-associated mortality (i.e. the proportion of dead to hatching embryos) was higher than 50% at water depths <20 cm depth but less than 19% in water deeper than 45 cm [104].

15.7 Summary and concluding remarks

Taken together, the results presented in section 15.2 suggest that the extinction of entire populations of basal species by enhanced UV-B levels seems improbable in established aquatic communities. In transparent and shallow aquatic ecosystems, UVR is undoubtedly a major force shaping the structure of pioneer communities. However, whether enhanced UV-B fluxes could offer a competitive advantage to tolerant species of phytoplankton in natural environments remains uncertain. A major effort is needed to understand the underlying physiological mechanisms resulting in the observed changes or lack of changes in community structure. Although there is probably no perfect experimental design to test the direct and indirect effects of UVR, ecological studies should resemble the conditions to which organisms are exposed to solar radiation. The contrasting results obtained with enclosure experiments and the highlighted methodological caveats call for extreme caution in extrapolating previous results on changes in species composition to natural environments. In connection to scenarios of shift in taxonomic composition, it has often been anticipated that a change in phytoplankton (or other community) species composition will have a major impact on higher trophic levels and cause altered patterns of trophic dynamics [7,8,105]. This Eltonian perspective of ecosystem functioning may not necessarily apply even under the worst-case scenario of population extinction. Analyses of food web studies where species have been removed, and predictions of the food-web theory, suggest that consequences for higher trophic levels will depend on both the functioning role of the species (e.g., a keystone species) and the complexity of the food web [106]. Thus, for example, the extinction of a species in a simple food web with few dominant species may have dramatic consequences for higher trophic levels (resulting potentially in other extinctions), while in a complex community the effects will be small. These predictions are further supported by the observation that linkage density in food webs increases with their size [107], and that regardless of the size of the food web there is a nearly constant ratio (~2 to 3) of prey to each predator species [108].

The existing information about indirect effects mediated by UVR on trophic interactions (section 15.3) suggest that positive feedbacks as observed in artificial flumes with benthic organisms and in laboratory studies with microorganisms are not the rule in natural systems. Certainly, more ecological studies are needed before we can consider them as important processes occurring in aquatic ecosystems. Particularly, a combination of autecological and synecological approaches could be fruitful in view of the large difference in species sensitivity observed. Assessments where entire components are considered as "black boxes" will mask the species' response. On the other hand, it can be anticipated that for planktonic groups with uncertain or difficult taxonomy this would be a difficult task. Regarding indirect effects of UVR on grazers mediated through algae, there is an urgent need to do experiments under more realistic UV exposure conditions considering the combined effect of UVR on grazers. A less explored interaction is when UVR acts together with predation as countervailing selective pressure on aquatic organisms that obtain protection through pigmentation but at the same time increase their conspicuousness to predators [109,110]. The effect of UVR on food quality, particularly on polyunsaturated fatty acids are thought to play a major role in the food web of shallow and clear waters as these compounds are essential for a balanced growth in herbivores [111]. Consequently, studies considering the effect of changes in food quality and life history traits of invertebrates as affected by UVR are a promising research area. Finally, the effects of UVR on anti-predator behavioural responses as evidenced for amphibians [112] need to be investigated on different groups of organisms including the direct effect of UVR on chemical signals (e.g. kairomones) important in predator–prey relationships.

Although there is an increasing number of studies addressing the role of UVR on the mortality of symbiotic corals, our knowledge of UV effects on other types of symbioses in marine and freshwater systems is scarce (section 15.4). It seems reasonable to hypothesize that beside the well-established advantage of endosymbiosis for survival in nutrient-poor waters, symbionts, for instance, of protozoans, may also offer protection against UV damage by providing photoprotecting compounds such as mycosporine-like amino acids. The finding that symbiotic ciliates are less UV-sensitive than other ciliate species supports such assumptions. The study of the association between phototrophic endosymbionts and ciliates living in the illuminated zone of anoxic marine sandy sediments could be particularly interesting in this regard.

Information on the effects of UVR on parasites and diseases in aquatic systems is mainly restricted to viruses and *Saprolegnia* spp. As we have seen in section 15.5, the interaction among UVR, viruses, and their hosts is extremely complex including direct and indirect effects. The use of models may help to explore the response of this system under UV stress. For example, the time needed to intercept a host depends on the product of contact rate, host population density, and the inverse probability of infection per contact [90]. Results from a random encounter model predict that viruses of common (abundant) species may have an advantage by requiring less time to contact their host and consequently by receiving a lower UV dose in exposed habitats [113]. Consequently, viruses from bacteria should be less exposed to damaging UV irradiances than those of phytoplankton. However, host specificity may reduce the effective population size that could be infected. Obviously, this is a field where more research is needed to define the net response of the interaction.

The example of *Saprolegnia*-associated mortality on amphibians represents a good example of how other synergistic processes, like climatic warming may exacerbate negative effects of ambient UVR (see Chapter 17). Whereas secondary parasitic infection by *Saprolegnia* after UV exposure seems to be more common in captive fish, this parasite is responsible for a high mortality in natural populations of amphibians. Nevertheless, it remains to be established how sensitively *Saprolegnia* species react to increased UV-B fluxes.

This brief review clearly indicates that ecologists still have much to learn about the interactions of UVR in the functioning of aquatic ecosystems and, at the same time, much to contribute to this topic. Although it is obvious that the role of UVR on species interactions is now recognized by aquatic ecologists and considered essential for assessing the ecosystem response, our gap of knowledge is still large. One consequence of this situation is that many predictions about potential effects of enhanced UV-B fluxes on aquatic ecosystems remain only speculations. This must change rapidly in the near future, considering that scientific knowledge alone does not lead to political decisions, and that policy based on a weak scientific basis is doomed [2].

Acknowledgements

I am indebted to Dr. Elena Martin-Webb for providing unpublished information on UV sensitivity of symbiotic ciliates and Dr. Pamela Hallock for information on bleaching in foraminifera. I also thank Drs. Craig Williamson, Horacio Zagarese, Robert G. Wetzel, Nils Ekelund, and Roland Psenner for valuable comments and several colleagues and students for improving the clarity of the text and figures. This endeavor was supported by the Austrian Science Foundation (FWF 14153-BIO).

References

1. P.A. Abrams (1987). On classifying interactions between populations. *Oecologia*, **73**, 272–281.
2. C.J. Krebs (2001). *Ecology: The Experimental Analysis of Distribution and Abundance*. Benjamin Cummings, Adison Wesley Longman, Inc., San Francisco.
3. J.M. Cherrett (1990). Key concepts: The results of a survey of our members' opinions.

In: J.M. Cherrett (Ed.), *Ecological Concepts: The Contribution of Ecology to an Understanding of the Natural World* (pp.1–16). Blackwell Scientific Publ., Oxford.
4. R.P. McIntosh (1998). The myth of community as organism. *Perspec. Biol. Med.*, **41**, 426–438.
5. H.T. Odum (1983). *Systems Ecology: An Introduction.* Wiley, New York.
6. P.L. Jokiel (1980). Solar ultraviolet radiation and coral reef epifauna. *Science*, **207**, 1069–1071.
7. R.C. Worrest (1983). Impact of solar ultraviolet-B radiation (290–320 nm) upon marine microalgae. *Physiol. Plant.*, **58**, 428–434.
8. D.-P. Häder, R.C. Worrest (1991). Effects of enhanced solar ultraviolet radiation on aquatic ecosystems. *Photochem. Photobiol.*, **53**, 717–725.
9. D. Karentz (1994). Ultraviolet tolerance mechanisms in Antarctic marine organisms. In: C.S. Weiler, P.A. Penhale (Eds), *Ultraviolet Radiation and Biological Research in Antarctica* (Antarctic Research Series 62, pp. 93–110). American Geophysical Union, Washington, D.C.
10. D.-P. Häder, H.D. Kumar, R.C. Smith, R.C. Worrest (1998). Effects on aquatic ecosystems. *J. Photochem. Photobiol. B: Biol.*, **46**, 53–68.
11. M.L. Bothwell, D. Sherbot, A.C. Roberge, R J. Daley (1993). Influence of natural ultraviolet radiation on lotic periphytic diatom community growth, biomass accrual, and species composition: short-term versus long-term effects. *J. Phycol.*, **29**, 24–35.
12. R. Santas, D. Danielidis, C. Lianou (1997). UV-B radiation and depth interaction during primary succession of marine diatom assemblages of Greece. *Limnol. Oceanogr.*, **42**, 986–991.
13. R.D. Vinebrooke, P.R. Leavitt (1999). Differential responses of littoral communities to ultraviolet radiation in an alpine lake. *Ecology*, **80**, 223–237.
14. S.N. Francoeur, R.L. Lowe (1998). Effects of ambient ultraviolet radiation on littoral periphyton: Biomass accrual and taxon-specific responses. *J. Freshwat. Ecol.*, **13**, 29–37.
15. S. Reizopoulou, P. Santas, D. Danielidis, D.-P. Häder, R. Santas (2000). UV effects on invertebrate and diatom assemblages of Greece. *J. Photochem. Photobiol. B: Biol.*, **56**, 172–180.
16. A. Wulff, S.-Å.Wänberg, K. Sundbäck, C. Nilsson, G.J.C. Underwood (2000). Effects of UVB radiation on a marine microphytobenthic community growing on a sand-substratum under different nutrient conditions. *Limnol. Oceanogr.*, **45**, 1144–1152.
17. B. Higley, H.J. Carrick, M.T. Brett, C. Luecke, C.R. Goldman (2001). The effect of ultraviolet radiation and nutrient additions on periphyton biomass and composition in a sub-alpine lakes (Castle Lake, USA). *Internat. Rev. Hydrobiol.*, **86**, 147–163.
18. C.E. Williamson, O.G. Olson, S.E. Lott, N.D. Walker, D.R. Engstrom, B. Hargreaves (2001). Ultraviolet radiation and zooplankton community structure following deglaciation in Glacier Bay, Alaska. *Ecology*, **82**, 1748–1760.
19. W.F. Vincent, S. Roy (1993). Solar ultraviolet-B radiation and aquatic primary production: damage, protection, and recovery. *Environ. Rev.*, **1**, 1–12.
20. C.E. Williamson, P.J. Neale, G. Grad, H.J. De Lange, B.R. Hargreaves (2001). Beneficial and detrimental effects of UV radiation on aquatic organisms: implications of variation in spectral composition. *Ecol. Appl.*, **11**, 1843–1857.
21. R.C. Worrest (1981). Impact of UV-B radiation upon estuarine microcosms. *Photochem. Photobiol.*, **33**, 861–867.
22. S.-Å.Wängberg, J.-S. Selmer, K. Gustavson (1996). Effects of UV-B radiation on biomass and composition in marine phytoplankton communities. *Sci. Mar.*, **60**, 81–88.

23. S.-Å. Wängberg, A. Wulff, C. Nilsson, U. Stagell (2001). Impact of UV-B radiation on microalgae and bacteria: a mesocosm study with computer modulated UV-B radiation addition. *Aquat. Microb. Ecol.*, **25**, 75–86.
24. S.-Å. Wängberg, K. Garde, K. Gustavson, J.-S. Selmer (1999). Effects of UVB radiation on marine phytoplankton communities. *J. Plankton Res.*, **21**, 147–166.
25. E.W. Helbling, V. Villafañe, M. Ferrario, O. Holm-Hansen (1992). Impact of natural ultraviolet radiation on rates of photosynthesis and on specific marine phytoplankton species. *Mar. Ecol. Prog. Ser.*, **80**, 89–100.
26. V.E. Villafañe, E.W. Helbling, O. Holm-Hansen, B.E. Chalker (1995). Acclimatization of Antarctic natural phytoplankton assemblages when exposed to solar ultraviolet radiation. *J. Plankton Res.*, **17**, 2295–2306.
27. A.T. Davidson, H.J. Marchant, W.K. de la Mare (1996). Natural UVB exposure changes the species composition of Antarctic phytoplankton in mixed culture. *Aquat. Microb. Ecol.*, **10**, 299–305.
28. D. Karentz, H.J. Spero (1995). Response of a natural *Phaeocystis* population to ambient fluctuations of UVB radiation caused by Antarctic ozone depletion. *J. Plankton Res.*, **17**, 1771–1789.
29. S. Halac, M. Felip, L. Camarero, S. Sommaruga-Wögrath, R. Psenner, J. Catalan, R. Sommaruga (1997). An in situ enclosure experiment to test the solar UV-B impact on microplankton in a high altitude mountain lake: I) lack of effect on phytoplankton species composition and growth. *J. Plankton Res.*, **19**, 1671–1687.
30. I. Laurion, D.R.S. Lean, W.F. Vincent (1998). UVB effects on a plankton community: results from a large-scale enclosure assay. *Aquat. Microb. Ecol.*, **16**, 189–198.
31. H.J. DeLange, A.M. Verschoor, R. Gylstra, J.G.M. Cuppen, E. Van Donk (1999). Effects of artificial ultraviolet-B radiation on experimental aquatic microcosms. *Freshwat. Biol.*, **42**, 545–560.
32. P.L. Jokiel, R.H. York (1984). Importance of ultraviolet radiation in photoinhibition of microalgal growth. *Limnol. Oceanogr.*, **29**, 192–199.
33. J.M. Shick, M.P. Lesser, P.J. Jokiel (1996). Effects of ultraviolet radiation on corals and other coral reef organisms. *Global Change Biol.*, **2**, 527–545.
34. K. Sundbäck, C. Nilsson, S. Odmark, A. Wulff (1996). Does ambient UV-B radiation influence marine diatom-dominated microbial mats? A case study. *Aquat. Microb. Ecol.*, **11**, 151–159.
35. K. Sundbäck, S. Odmark, A. Wulff, C. Nilsson, S.-Å. Wängberg (1997). Effects of enhanced UVB radiation on a marine benthic diatom mat. *Mar. Biol.*, **128**, 171–179.
36. E. McNamara, W.R. Hill (2000). UV-B irradiance gradients affects photosynthesis and pigments but not food quality of periphyton. *Freshwat. Biol.*, **43**, 649–662.
37. S. Odmark, A. Wulff, S.-Å. Wängberg, C. Nilsson, K. Sundbäck (1998). Effect of UVB radiation in a microbenthic community of a marine shallow-water sandy sediment. *Mar. Biol.*, **132**, 335–345.
38. A. McMinn, H. Heijnis, D. Hodgson (1994). Minimal effects of UVB radiation on Antarctic diatoms over the past 20 years. *Nature*, **370**, 547–549.
39. M.L. Bothwell, D. Karentz, E.J. Carpenter (1995). No UVB effect? *Nature*, **374**, 601.
40. M.L. Bothwell, D.M.J. Sherbot, C.M. Pollock (1994). Ecosystems response to solar ultraviolet-B radiation: Influence of trophic-level interactions. *Science*, **265**, 97–100.
41. D.M. DeNicola, K.D. Hoagland (1996). Effects of solar spectral irradiance (visible to UV) on a prairie stream epilithic community. *J. N. Am. Benthol. Soc.*, **15**, 155–169.
42. W.R. Hill, S.M. Dimick, A.E. McNamara, C. A. Branson (1997). No effects of ambient UV radiation detected in periphyton and grazers. *Limnol. Oceanogr.*, **42**, 769–774.
43. W.F. Donahue, J.J. Clare (1999). Effects of ultraviolet radiation on trophic interac-

tions not detected? *Limnol. Oceanogr.*, **44**, 470–473.
44. W.R. Hill, A.E. McNamara (1999). Sun angle, site selection, parameter choice, and generalizing UV effects in freshwater benthic ecosystems (Reply to the comment by Donahue and Clare). *Limnol. Oceanogr.*, **44**, 474–475.
45. P.M. Kiffney, W.H. Clements, T.A. Cady (1997). Influence of ultraviolet radiation on the colonization dynamics of a Rocky Mountain stream benthic community. *J. N. Am. Benthol. Soc.*, **16**, 520–530.
46. R. Sommaruga, A. Oberleiter, R. Psenner (1996). Effect of UV radiation on the bacterivory of a heterotrophic nanoflagellate. *Appl. Environ. Microbiol.*, **62**, 4395–4400.
47. C.A. Ochs (1997). Effects of UV radiation on grazing by two marine heterotrophic nanoflagellates on autotrophic picoplankton. *J. Plankton Res.*, **19**, 1517–1536.
48. C.A. Ochs, L.P. Eddy (1998). Effects of UV-A (320–399 nanometers) on grazing pressure of a marine heterotrophic nanoflagellate on strains of the unicellular cyanobacteria *Synechococcus* spp. *Appl. Environ. Microbiol.*, **64**, 287–293.
49. S. Wickham, M. Carsten (1998). Effects of ultraviolet-B radiation on two arctic microbial food webs. *Aquat. Microb. Ecol.*, **16**, 163–171.
50. R. Sommaruga, B. Sattler, A. Oberleiter, A. Wille, S. Sommaruga-Wögrath, R. Psenner, M. Felip, L. Camarero, S. Pina, R. Gironés, J. Catalán (1999). An in situ enclosure experiment to test the solar UV-B impact on microplankton in a high-altitude mountain lake: II. Effects on the microbial food web. *J. Plankton Res.*, **21**, 859–876.
51. B. Mostajir, T. Sime-Ngando, S. Demers, C. Belzile, S. Roy, M. Gosselin, J.-P. Chanut, S. de Mora, J. Fauchot, F. Vidussi, M. Levasseur (1999). Ecological implications of changes in cell size and photosynthetic capacity of marine Prymnesiophyceae induced by ultraviolet-B radiation. *Mar. Ecol. Prog. Ser.*, **187**, 89–100.
52. B. Mostajir, S. Demers, S. de Mora, C. Belzile, J.-P. Chanut, M. Gosselin, S. Roy, P.Z. Villegas, J. Fauchot, J. Bouchard, D. Bird, P. Monfort, M. Levasseur (1999). Experimental test of the effect of ultraviolet-B radiation in a planktonic community. *Limnol. Oceanogr.*, **44**, 586–596.
53. K. Chatila, S. Demers, B. Mostajir, J.-P. Chanut, P. Monfort (1999). Bacterivory of a natural heterotrophic protozoan community exposed to different intensities of ultraviolet-B radiation. *Aquat. Microb. Ecol.*, **20**, 59–74.
54. K. Chatila, S. Demers, B. Mostajir, M. Gosselin, J.-P. Chanut, P. Monfort, D. Bird (2001). The responses of a natural bacterioplankton community to different levels of ultraviolet-B radiation: A food web perspective. *Microb. Ecol.*, **41**, 56–68.
55. A.A. Keller, P. Hargraves, H. Jeon, G. Klein-MacPhee, E. Klos, C. Oviatt, J. Zhang (1997). Effects of ultraviolet-B enhancement on marine trophic levels in a stratified coastal system. *Mar. Biol.*, **130**, 277–287.
56. R.M. Forster, H. Schubert (2001). The effects of ultraviolet radiation on the planktonic community of a shallow, eutrophic estuary: results of mesocosms experiments. *Helgol. Mar. Res.*, **53**, 23–34.
57. E. Van Donk, D.O. Hessen (1995). Reduced digestibility of UV-B stressed and nutrient-limited algae by *Daphnia magna*. *Hydrobiologia*, **307**, 147–151.
58. H.J. De Lange, E. Van Donk (1997). Effects of UVB-irradiated algae on life history traits of *Daphnia pulex*. *Freshwat. Biol.*, **38**, 711–720.
59. J.D. Scott, L. Chalker-Scott, A.E. Foreman, M. D'Angelo (1999). *Daphnia pulex* fed UVB-irradiated *Chlamydomonas reinhardtii* show decreased survival and fecundity. *Photochem. Photobiol.*, **70**, 308–313.
60. R.D. Hurtubise, J.E. Havel, E.E. Little (1998). The effect of ultraviolet-B radiation on

freshwater invertebrates: Experiments with a solar simulator. *Limnol. Oceanogr.*, **43**, 1082–1088.
61. A. Cywinska, D. Crump, D. Lean (2000). Influence of UV radiation on four freshwater invertebrates. *Photochem. Photobiol.*, **72**, 652–659.
62. W.C. Dunlap, J.M. Shick (1998). Ultraviolet radiation-absorbing mycosporine-like amino acids in coral reef organisms: A biochemical and environmental perspectives. *J. Phycol.*, **34**, 418–430.
63. D.F. Gleason (2001). Ultraviolet radiation and coral communities. In: C.S. Cockell, A.R. Blaustein (Eds), *Ecosystems, Evolution and Ultraviolet Radiation* (pp. 118–149). Springer-Verlag, New York.
64. R.L. Hayes, P.G. Bush (1990). Microscopic observation of recovery in the reef-building scleractinian coral, *Montastrea annularis* after bleaching on a Cayman reef. *Coral Reefs*, **8**, 203–209.
65. J.H. Drollet, M. Faucon, P.M.V. Martin (1995). Elevated sea-water temperature and solar UV-B flux associated with two successive coral mass bleaching events in Tahiti. *Mar. Freshwat. Res.*, **46**, 1153–1157.
66. L. Dionisio-Sese, T. Maruyama, S. Miyachi (2001). Photosynthesis of *Prochloron* as affected by environmental factors. *Mar. Biotechnol.*, **3**, 74–79.
67. M. Laval-Peuto (1991). Endosymbiosis in the Protozoa: Session summary. In: P.C. Reid, C.M. Turley, P.H. Burkill (Eds), *Protozoa and their Role in Marine Processes* (NATO ASI Ser. G: Ecol. Sci. vol. 25, pp.143–160). Springer-Verlag, Berlin.
68. E.B. Martin-Webb (1999). *The Effect of Solar Ultraviolet Radiation on Marine Planktonic Ciliates Survival and the Penetration of Ultraviolet Irradiance on Georges Bank* (Ph.D. Thesis). University of Rhode Island.
69. T. Lindholm (1985). *Mesodinium rubrum* a unique photosynthetic ciliate. *Adv. Aquat. Microbiol.*, **3**, 1–48.
70. M. Laval-Peuto, F. Rassoulzadegan (1988). Autofluorescence of marine planktonic Oligotrichina and other ciliates. *Hydrobiologia*, **159**, 99–110.
71. J.L. Lee (1991). Brief perspective on the autoecology of marine protozoa. In: P.C. Reid, C.M. Turley, P H. Burkill (Eds), *Protozoa and their Role in Marine Processes* (NATO ASI Ser. G: Ecol. Sci. vol. 25, pp. 181–194). Springer-Verlag, Berlin.
72. P. Hallock, H.K. Talge, E.M. Cockey, R.G. Muller (1995). A new disease in reef-dwelling foraminifera: Implications for coastal sedimentation. *J. Foraminifer. Res.*, **25**, 280–286.
73. P. Hallock (2000). Symbionts-bearing foraminifera: harbingers of global change? *Micropaleontology*, **46**, 95–104.
74. K.A. Aarseth, T.A. Schram (1999). Wavelength-specific behaviour in *Lepeophtheirus salmonis* and *Calanus finmarchicus* to ultraviolet and visible light in laboratory experiments (Crustacea: Copepoda). *Mar. Ecol. Prog. Ser.*, **186**, 211–217.
75. J.M. Bohannan, R.E. Lenski (1997). Effect of resource enrichment on a chemostat community of bacteria and bacteriophage. *Ecology*, **78**, 2303–2315.
76. C.A. Suttle, F. Chen (1992). Mechanisms and rates of decay of marine viruses in seawater. *Appl. Environ. Microbiol.*, **58**, 3721–3729.
77. L.W. Sinton, R.K. Finley, P.A. Lynch (1999). Sunlight inactivation of fecal bacteriophages and bacteria in sewage-polluted seawater. *Appl. Environ. Microbiol.*, **65**, 3605–3613.
78. M.G. Weinbauer, S.W. Wilhelm, C.A. Suttle, R.J. Pledger, D.L. Mitchell (1999). Sunlight-induced DNA damage and resistance in natural viral communities. *Aquat. Microb. Ecol.*, **17**, 111–120.
79. D.R. Garza, C.A. Suttle (1998). The effect of cyanophages on the mortality of

Synechococcus spp. and selection for UV resistant viral communities. *Microb. Ecol.*, **36**, 281–292.
80. R.T. Noble, J.A. Fuhrman (1997). Virus decay and its causes in coastal waters. *Appl. Environ. Microbiol.*, **63**, 77–83.
81. C.A. Kellog, J.A. Paul (2002). Degree of ultraviolet radiation damage and repair capabilities are related to G+C content in marine vibriophages. *Aquat. Microb. Ecol.*, **27**, 13–20.
82. C.E. Singer, B.N. Ames (1970). Sunlight ultraviolet and bacterial DNA base ratios. *Science*, **170**, 822–826.
83. M.G. Weinbauer, S.W. Wilhelm, C.A. Suttle, D.R. Garza (1997). Photoreactivation compensates for UV damage and restores infectivity to natural marine virus communities. *Appl. Environ. Microbiol.*, **63**, 2200–2205.
84. S.W. Wilhelm, M.G. Weinbauer, C.A. Suttle, R.J. Pledger, D.L. Mitchell (1998). Measurements of DNA damage and photoreactivation imply that most viruses in marine surface waters are infective. *Aquat. Microb. Ecol.*, **14**, 215–222.
85. M. Furuta, J.O. Schrader, H.S. Schrader, T.A. Kokjohn, S. Nyaga, A.K. McCullough, R.S. Lloyd, D.E. Burbank, D. Landstein, L. Lane (1997). *Chlorella* virus PBCV-1 encodes a homologue of the bacteriophage T4 UV damage repair gene denV. *Appl. Environ. Microbiol.*, **63**, 1551–1556.
86. J.J. Shaffer, L.M. Jacobsen, J.O. Schrader, K.W. Lee, E.L. Martin, T.A. Kokjohn (1999). Characterization of *Pseudomonas aeruginosa* bacteriophage UNL-1, a bacterial virus with a novel UV-A-inducible DNA damage reactivation phenotype. *Appl. Environ. Microbiol.*, **65**, 2606–2613.
87. S.W. Wilhelm, M.G. Weinbauer, C.A. Suttle, W. H. Jeffrey (1998). The role of sunlight in the removal and repair of viruses in the sea. *Limnol. Oceanogr.*, **43**, 586–592.
88. R. Sommaruga, I. Obernosterer, G.J. Herndl, R. Psenner (1997). Inhibitory effect of solar radiation on thymidine and leucine incorporation by freshwater and marine bacterioplankton. *Appl. Environ. Microbiol.*, **63**, 4178–4184.
89. S.W. Wilhelm, R.E.H. Smith (2000). Bacterial carbon production in Lake Erie is influenced by viruses and solar radiation. *Can. J. Fish. Aquat. Sci.*, **57**, 317–326.
90. A.G. Murray, G.A. Jackson (1982). Viral dynamics: a model of the effects of size, shape, motion and abundance of single-celled planktonic organisms and other particles. *Mar. Ecol. Prog. Ser.*, **89**, 103–116.
91. C.A. Suttle, A.M. Chan, C. Feng, D.R. Garza (1993). Cyanophages and sunlight: a paradox. In: R. Guerrero, C. Pedrós-Alió (Eds), *Trends in Microbial Ecology* (pp.303–307). Spanish Society for Microbiology, Barcelona.
92. K.E. Wommack, R.T. Hill, T.A. Muller, R.R. Colwell (1996). Effects of sunlight on bacteriophage viability and structure. *Appl. Environ. Microbiol.*, **62**, 1336–1341.
93. G.C. Walker (1984). Mutagenesis and inducible responses to deoxyribonucleic acid damage in *Escherichia coli*. *Microb. Rev.*, **48**, 60–93.
94. R.M. Wilcox, J.A. Fuhrman (1994). Bacterial viruses in coastal seawater: lytic rather lysogenic production. *Mar. Ecol. Prog. Ser.*, **114**, 35–45.
95. H.L. Schrader, J.O. Schrader, J.J. Walker, N.B. Bruggeman, J.M. Vanderloop, J.J. Shaffer, T.A. Kokjohn (1997). Effects of host starvation on bacteriophage dynamics. In: R.Y. Morita (Ed.), *Bacteria in Oligotrophic Environments: Starvation-survival Lifestyle* (pp. 368–385). Chapman & Hall, N.Y.
96. O. Siebeck, T.L. Vail, C.E. Williamson, R. Vetter, D.O. Hessen, H. Zagarese, E.E. Little, E. Balseiro, B. Modenutti, J. Seva, A. Shumate (1994). Impact of UV-B radiation on zooplankton and fish in pelagic freshwater ecosystems. *Arch. Hydrobiol. Beih. Ergebn. Limnol.*, **43**, 101–114.

97. K. Kaweewat, R. Hofer (1997). Effect of UV-B radiation on globet cells in the skin of different fish species. *J. Photochem. Photobiol. B: Biol.*, **41**, 222–226.
98. G.A. Neish (1977). Observations on saprolegniasis of adult sockeye salmon. *Oncorhynchus nerka* (Walbaum), *J. Fish Biol.*, **10**, 513–522.
99. D.W. Bruno, B.P. Wood (1994). *Saprolegnia* and other Oomycetes. In: P.T.K. Woo, D.W. Bruno (Eds), *Fish diseases and disorders, Vol.* 3, *Viral, Bacterial and Fungal Infections* (pp. 599–659). CABI Publishing, Wallingford, Oxon, UK.
100. L.G. Willoughby (1989). Continued defense of salmonid fish against *Saprolegnia* fungus after its establishment, *J. Fish. Dis.*, **12**, 63–67.
101. H.M. Salo, T.M. Aaltonen, S.E. Markkula, E.I. Jokinen (1998). Ultraviolet B irradiation modulates the immune system of fish (*Rutilus rutilus*, Cyprinidae). I. Phagocytes. *Photochem. Photobiol.*, **67**, 433–437.
102. L. Fabacher, E.E. Little, S.B. Jones, E.C. De Fabo, L.J. Webber (1994). Ultraviolet-B radiation and the immune response of rainbow trout. In: J.S. Stolen, T.C. Fletcher (Eds), *Modulators of Fish Immune Responses: Models for Environmental toxicology, Biomarkers, Immunostimulators* (Vol. 1, pp. 205–217). SOS Publications, Fair Haven, N.J.
103. J.M. Kiesecker, A.R. Blaustein (1995). Synergism between UV-B radiation and a pathogen magnifies amphibian embryo mortality in nature. *Proc. Natl. Acad. Sci. U.S.A.*, **92**, 11049–11052.
104. J.M. Kiesecker, A.R. Blaustein, L.K. Belden (2001). Complex causes of amphibian populations declines. *Nature*, **410**, 681–684.
105. R.C. Worrest (1994). Aquatic systems (Freshwater and Marine). In: R.H. Biggs, M.E.B. Joyner (Eds), *Stratospheric Ozone Depletion/UV-B Radiation in the Biosphere* (NATO ASI Series, vol. 118, pp. 151–153). Springer-Verlag, Berlin.
106. S.L. Pimm (1991). *The Balance of Nature? Ecological Issues in the Conservation of Species and Communities.* The University of Chicago Press, Chicago.
107. K. Havens (1992). Scale and structure in natural food webs. *Science*, **257**, 1107–1109.
108. N.D. Martinez (1991). Artifacts or attributes? Effects of resolution on the Little Rock Lake food web. *Ecol. Monogr.*, **61**, 367–392.
109. S.G. Morgan, J.H. Christy (1996). Survival of marine larvae under the countervailing selective pressures of photodamage and predation. *Limnol. Oceanogr.*, **41**, 498–504.
110. L.-A. Hansson (2000). Induced pigmentation in zooplankton: a trade-off between threats from predation and ultraviolet radiation. *Proc. R. Soc. Lond. B*, **267**, 2327–2331.
111. D.O. Hessen, H.J. De Lange, E. Van Donk (1997). UV-induced changes in phytoplankton cells and its effects on grazers. *Freshwat. Biol.*, **38**, 513–524.
112. L.B. Katz, J.M. Kiesecker, D.P. Chivers, A.R. Blaustein (2000). Effects of UV-B radiation on anti-predator behavior in three species of amphibians. *Ethology*, **106**, 921–931.
113. A.G. Murray, G.A. Jackson (1993). Viral dynamics II: A model of the interaction of ultraviolet light and mixing processes on virus survival in seawater. *Mar. Ecol. Prog. Ser.*, **102**, 105–114.

Chapter 16

Past UVR environments and impacts on lakes

Peter R. Leavitt, Dominic A. Hodgson and Reinhard Pienitz

Table of contents

Abstract

Paleoecological reconstructions of past irradiance regimes provide critical insights into the scales, causes and consequences of temporal variability in UVR. Such retrospective analyses using microfossil remains (diatoms), photo-protective pigments (scytonemin) or bulk sedimentary characteristics (% organic matter) can allow reconstruction of past concentrations of UVR-absorbing dissolved organic carbon compounds (DOC), spectral irradiance regimes (UVR : PAR ratio), lake transparency (depth of UVR penetration), as well as algal and invertebrate responses to altered UVR flux. Application of fossil approaches to lakes at treeline show both that climatically-induced changes in the export of terrestrial DOC to lakes are up to 100-fold more effective than modern ozone depletion at altering biological exposure to UVR, and that naturally occurring droughts can increase UVR exposure in lakes by up to 10-fold. When applied to Arctic and Antarctic lakes, preliminary fossil analyses suggest that historical variability in UVR exposure has been high, but that lakes may have received twice as much UVR prior to ~3000–4000 years ago as they do at present. Finally, when used in combination with long-term environmental monitoring, historical reconstructions have proven valuable at identifying the importance of UVR relative to other stressors in regulating lake production. Thus, although further research is required to validate fossil interpretations, paleoecological analyses can provide critical information on the role of UVR in regulating lake structure and function.

16.1 Introduction

Most laboratory and short-term field experiments show that exposure to intense UVR has detrimental effects on cellular processes [1], individual survival [2], and population growth rates [3,4]. However, demonstration of substantial UVR impacts on entire lake ecosystems has proven elusive to date, possibly because of variable species sensitivity to UVR [2,4], habitat specificity of biotic responses [3], incomplete reciprocity of UVR doses [5], countervailing effects of individual wavelengths [5], or complex trophic interactions [6–8]. Potential underestimation of UVR impacts may have arisen also because initial investigations focused on minor increases in UVR exposure arising from ozone depletion [e.g., 9], rather than variability in terrestrial dissolved organic matter (DOM) fluxes, the principle control of UVR penetration into lakes [10,11]. However, with the demonstration that major environmental disturbances such as droughts [12,13], lake acidification [14], long-term climate change [15] and even cosmic events [16,17] greatly alter inputs, metabolism and degradation of photo-protective DOM, it seems likely that unique, ecosystem-level impacts of UVR will be identified [e.g., 18].

One approach to assessing the potential impacts of UVR on aquatic ecosystems is to quantify the magnitude and consequences of past variations in UVR. In this regard, paleoecological analyses can provide valuable insights into the history of UVR exposure and its potential impacts on aquatic ecosystems [19]. Modern lake surveys show that planktonic and benthic community composition varies predictably as a function of the key determinants of UVR exposure (lake depth, water clarity, DOM content) as well as the environmental drivers that regulate DOM fluxes and optical properties (e.g., climate, pH) [reviewed in 20]. Further, multivariate analysis of sub-fossil (1–5 yr old) assemblages of diatoms in lake sediments have been used to develop statistical models (termed *transfer functions*) of the relationships between present-day species composition and environmental conditions that allow quantitative reconstruction of past lake conditions, including DOM content and optical regime [e.g., 21–23]. Consequently, application of these models to sedimentary sequences can be used to quantify historical patterns of UVR variability over 100s to 1000s of years, potentially at annual resolution [15,23]. Because past episodes of UVR exposure appear to be greater than those arising from many modern processes [15,24,25], these analyses may also offer insights into the unique impacts of UVR on ecosystem processes [e.g., 26,27], including those occurring immediately after lake formation when DOM inputs are lowest [2,28].

The goal of this chapter is to review how paleoecological techniques can be used to reconstruct past UVR environments and impacts on lakes. In the first section of the chapter, we briefly summarize UVR impacts on key aquatic biota, as well as the main controls of UVR exposure. Next we critically review the main microfossil and biogeochemical methods used to reconstruct past UVR environments. Third, we use a series of case studies to illustrate the insights obtained from historical reconstructions of irradiance environments and ecosystem responses. Finally, we identify new research avenues that we feel may bring important insights into the role UVR in regulating the structure and function of aquatic ecosystems and that may illustrate the wide range of past irradiance environments. In this review, we use DOC to indicate dissolved organic matter measured as organic carbon concentration, chromophoric DOM (CDOM) to indicate similar measurements based on optical properties of water, and DOM for cases where carbon or optics are not directly measured, or when describing general sources and fluxes of dissolved organic compounds.

16.2 Impacts of UVR on lake biota

16.2.1 Natural controls of UVR in lakes

Biotic exposure to UVR varies as a function of solar production, atmospheric attenuation, orbital precession, seasonal factors (ice cover), meteorological conditions, lake chemistry (DOM content, pH, salinity), basin depth, the presence of physical refuges, and individual behavior. Variations in solar production of UVR are thought to be minor except on scales of planetary evolution (e.g., supernova, neutron star merger), and are usually ignored in ecological and most evolutionary studies [16,29]. While increased sunspot activity can deplete stratospheric ozone [30], instrumental records suggest such changes account for

~9% of inter-annual variation in UVR [31]. Similarly, volcanic eruptions or meteor impacts can increase or reduce UVR transmission through the atmosphere depending on the relative abundance of dust, aerosols and chemicals ejected from the Earth's crust (e.g., O_3-depleting Cl) [17]. In contrast, greater variation in stratospheric ozone and atmospheric UVR penetration has been recorded over short timescales (e.g., 26% in 3 days; [18]) due to transient changes in local ozone content, cloud cover, aerosols, particulates and pollutants [32]. However, with the exception of meso-scale depletions of ozone in polar regions [33,34] and variations in snow, ice and debris cover on frozen lakes [e.g., 35], most recent research suggests that the largest changes in exposure of aquatic biota arise from variation in landscape and in-lake processes related to DOM biogeochemistry [12,13,15,24].

Comparison of physical and chemical factors regulating UVR penetration into water show that variation in concentrations of chromophoric dissolved organic compounds is the main factor regulating biotic exposure to energetic irradiance in most lakes [10,11,36–38]. In general, changes in DOM influx provide the single largest source of variation [e.g., 13], although variation in source water content can be important in marine-influenced systems [39]. Physical attenuation by algae or particulates may be important also in either highly unproductive [40], hypereutrophic or turbid systems [10,38], or in shallow lakes where the carbon-specific attenuation of DOM is low [41]. Because of the strong negative exponential relation between UVR transmission and wavelength, variations in mass-specific attenuation of DOM [41,42] and spectral attenuation characteristics ("S"; [43]) influence exposure of biota in polar, alpine, acidified or saline lakes where terrestrial inputs of DOM are low, DOM is highly degraded, or where phytoplankton are a quantitatively important source of DOM [41,44].

Finally, virtually all surveys of modern and sub-fossil algal communities show that lake depth is a critical factor influencing algal community composition [45]. While not explicitly linked to irradiance to date, we hypothesize that spatial and temporal variation in basin morphology is a key factor regulating biotic exposure to UVR on a biogeographic and historical basis, through the provision of deepwater refugia [15] or by controlling the development of thermal stratification ("UVR traps"; [18,46]). Further, as a consequence of the supremacy of in situ processes in regulating biotic exposure to UVR, we suggest that historical changes in UVR impacts are most likely to arise from variations in catchment and lake characteristics [cf., 28], rather than from factors which influence UVR transmission to the lake surface [15,24].

16.2.2 Impacts of UVR on aquatic biota

Most short-term experimental and observational research concludes that intense UVR, particularly UV-B (280–315 nm), damages cellular structures, decreases metabolic efficiency and reduces growth or survival of individuals [1,47]. Direct inhibition of algal photosynthesis and growth appears to be greatest when phytoplankton are trapped in surface waters [18,48; Chapter 11], restricted from exploiting refugia by hard substrates [3,6,49], or are exposed to photochemical toxins [50,51]. However, algal sensitivity to UVR varies greatly among taxa [e.g., 4], partly because of taxon-specific differences in cell structure [52], photoprotective pigment content [53], efficiency of cellular repair mechanisms [54] or behavioral exploitation of refugia [55,56]. Similarly, pelagic invertebrates are differentially impacted by UVR [57–59; Chapter 12], reflecting variations in species' photoprotective pigmentation [60], body size [58], repair mechanisms [5,61], or avoidance behavior [62]. Finally, recent evidence suggests that UVR may alter the reproductive success of both macrophytes [63] and predaceous fishes [7,8], thereby altering both trophic interactions and habitat structure within lake ecosystems.

Comparison of experimental investigations suggests that while the absorbance of high-energy irradiance must necessarily initiate cellular damage [64], complex indirect effects may be the predominant mechanism by which UVR effects are expressed. Unfortunately, no consistent pattern has been noted for integrated food-web or ecosystem response to changes in UVR exposure [3,6,9,57,65], probably because most experimental studies have been conducted on physical and temporal scales that are tractable for experimentation, but which may lack ecological realism. Such scale dependent problems may include insufficient experimental duration (days–weeks), unrealistic irradiance regimes (continuous light; spectral composition; instantaneous change in flux), absence of key food-web components (predators, competitors) and simplified physical structure (refugia, sediments, plants). While these approaches clearly identify the potential for UVR impacts to alter biotic interactions, research on broader physical and temporal scales is required to evaluate the importance of UVR relative to other processes in regulating ecosystem structure [66]. In this regard, we feel that paleoecological analyses may provide significant insights into the role of high-energy irradiance in structuring aquatic communities. Below we review the main methods for reconstruction of past UVR environments and use a series of case studies to illustrate UVR impacts on ecosystem organization.

16.3 Paleoecological methods for UVR reconstruction

Lake sediments for reconstruction of past UVR environments can be collected using any standard paleoecological technique including gravity, piston, percussion or freeze-coring (reviewed in [67]), although subsequent treatment of the sample must consider the type of analysis used for environmental reconstruction [68]. In general, sediment samples should not be exposed to direct light, warm temperatures ($>4\,^{\circ}C$), extreme pH or active biological communities. While freeze-coring often minimizes degradation of biogeochemical fossils and provides the highest temporal resolution, this technique is limited to use in short-term analyses (<1 m sediment depth) and may be inappropriate for some delicate microfossils. In cases where millennium-scale UVR reconstructions are required, sediments are collected most often using Livingston, percussion or advanced piston-coring techniques [67]. Although many paleoecologists subsequently sample cores at relatively coarse intervals (> 1 cm), limits to temporal resolution are set by the rate of sediment deposition at the site, and the mechanical skill of the sampling process. Ages of individual sediment samples may be determined using a combination of naturally occurring radioisotopes, including ^{137}Cs (post-1950 AD), ^{210}Pb (post-1850 AD), ^{14}C (pre-1850 dates) or U : Th ratio (pre-25 000 years), together with optically stimulated- or thermo-luminescence analyses where quartz grains are present [67]. Additional chronological control may be obtained from annual laminae (varves) and from quantification of natural (tephra, post-glacial clays, pollen spectra) or anthropogenic (Pb stable isotopes, soot particles, contaminants) sedimentary markers of known age. Further details of core collection and processing for individual fossils are presented within Last and Smol [67].

16.3.1 Microfossil indices of past UVR environments

Over the last decade, considerable progress has been made in the development of ecological calibration studies and development of statistical inference techniques to reconstruct past environmental variables from paleoecological data [69]. This widely used *transfer function* approach depends on strong statistical correlations between organismal abundance and measured chemical or biophysical variables in modern lakes, and assumes that these relationships remain valid equally in the past. In the case of UVR reconstructions, this approach is based on the striking changes in the chemical (nutrients, DOC) and physical (mixing regimes, water transparency) properties of lakes across arctic and alpine treelines, patterns which are reflected also in the zonal distribution of freshwater diatoms (class Bacillariophyceae) [22,70–72]. Multivariate statistical models describing variance in community composition as a function of measured environmental gradients are developed using survey data before being applied to historical reconstructions at individual sites.

In the circumpolar region of the Northern Hemisphere, diatom-based inference models have been developed for the quantitative reconstruction of DOC or CDOM [22,70,72], water color [73,74], and total organic carbon [74,75]. Because DOC is a limnological variable that is highly correlated with lake catchment vegetation and soils [e.g., 76–78], reconstructions of DOC based on the siliceous fossils of lacustrine diatoms can be used as a proxy for past vegetation shifts and climate [reviewed in 79]. When combined with statistical descriptions of the relationship between DOC content and UVR penetration [e.g., 15], these models provide the starting point for more detailed reconstructions of past UVR environments and the changes in terrestrial environments that regulate DOC flux.

To develop a UVR transfer function, between 25 and 100 lakes are selected to lie along obvious gradients of terrestrial vegetation, with care given to select lakes with similar morphology (depth) and hydrologic regime (closed basins, no peatlands) but with contrasting levels of the chemical parameter of interest, in this case UVR-absorbing DOC. Description of diatom response to predominant environmental gradients begins with concomitant collection of surface sediments (upper 0.25–0.5 cm) and a wide range of chemical, physical and biological parameters. Ideally, selection of environmental variables is based on prior knowledge of the main gradients in lake characteristics, particularly those related to DOC biogeochemistry. The best inference models result when there are relatively few strong gradients of environmental change, or when multiple gradients exhibit a high degree of linear correlation (e.g., DOC, CDOM, color). Extraction of diatom fossils from sediments follows standard protocols, beginning with digestion of the organic and carbonate sediment matrix using 30% H_2O_2 or mixtures of either nitric (HNO_3) and sulfuric acid (H_2SO_4), or potassium dichromate ($K_2Cr_2O_7$) and sulfuric acid techniques [80]. Normally, acidified sub-samples ($\sim$1 cm^3 wet sediment) are heated for 2 h, at ca. 80°C, before repeated centrifugation and decanting with distilled water to neutralize the suspensions. Different concentrations of each diatom suspension are then deposited onto cover slips and left to dry before being mounted onto microscope slides using a permanent resin (Naphrax® or Hyrax®). Diatom enumeration is carried out using light microscopy (1000–1250 × magnification), with a minimum of $\sim$500 valves counted in each sample to characterize fossil assemblage composition.

Development of diatom-based models for the reconstruction of DOC and other variables involves a three-step analytical approach [22,70]. First, multivariate statistics are used to identify the main environmental factors correlated with changes in diatom community composition. Common approaches include the use of canonical (direct gradient) ordinations that assume either a unimodal (Canonical Correspondence Analysis, CCA) or linear (Redundancy Analysis, RDA) change in species abundance along environment gradients (Figure 1A). Usually both species abundance (% or concentration) and environmental variables will require transformations to normalize variance prior to analysis. Additionally, abundances are often centred and standardized prior to ordination in order to improve the ease of biological interpretation. Species–environmental relationships are often summarized using an ordination bi-plot in which axes are constrained to be linear combinations of measured environmental variables, and species or lakes are plotted in the ordination space (Figure 1B). This approach allows the relationship between species and environmental change to be modeled directly [81]. Environmental variables are added to the ordination using stepwise or forward selection and are retained only if they independently explain a significant ($p < 0.05$) amount of variance in fossil assemblage composition based on ordinations constrained to that variable alone. The significance of both ordination axes and individual variables is determined usually using Monte Carlo tests with 500 to 1000 iterations [82].

Assuming that DOC is identified as an important factor regulating variability in diatom community composition among survey lakes, the second step is to develop statistical descriptions of the mean (*optimum*) and variance (*tolerance*) in environmental conditions that regulate species abundance. In the case of UVR reconstructions, the responses of modern diatoms to a DOC gradient are

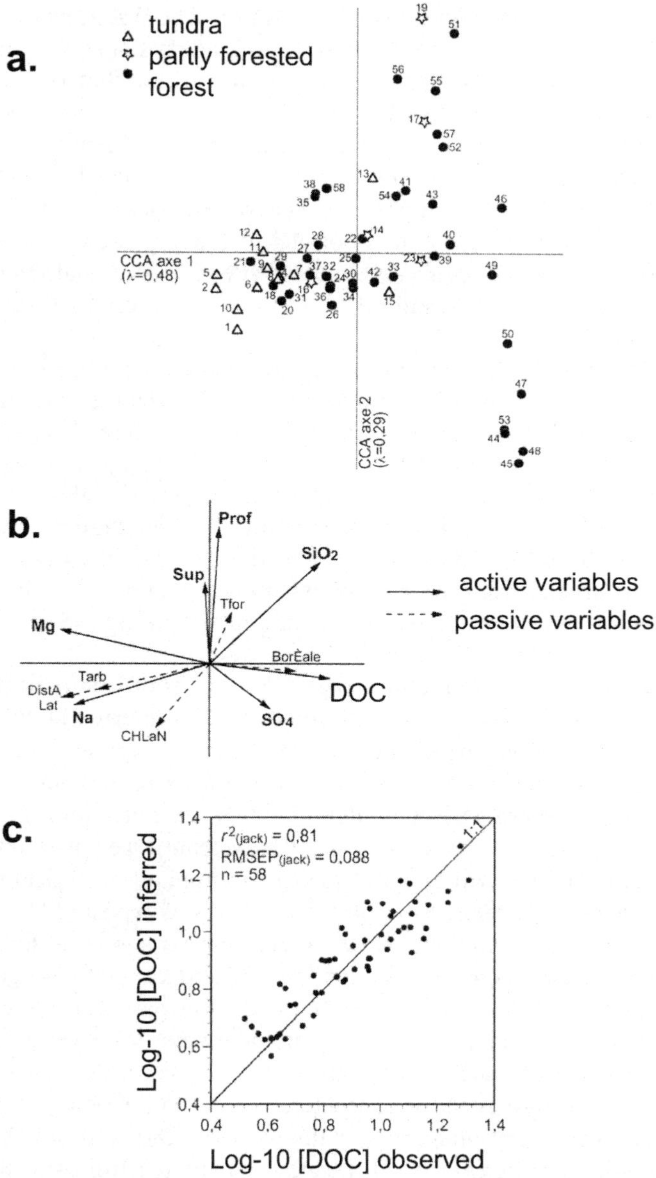

Figure 1. Canonical correspondence analysis of diatom community composition for lakes arranged along a gradient across treeline in the subartic of northern Québec, Canada [22]. (a) Ordination biplot of sub-fossil diatom assemblages from 57 lakes lying in tundra (triangle), partly-forested (star) and forested catchments (solid). (b) Relations of environmental gradients in lake chemistry to changes in diatom community composition showing that changes in DOC concentrations are positively correlated to the main direction of community variance (i.e., CCA axis 1). (c) Diatom-based inference model (*transfer function*) describing the relationship between measured DOC and that inferred using weighted-average calibration and regression approaches. See text for additional detail. [Figure modified from [15] with permission.]

modeled using a weighted-average (WA) calibration. The WA approach assumes that each species exhibits a unimodal response to gradients of DOC, with highest relative abundance under conditions that are optimal for that taxon's growth [83]. The technique is insensitive to a poor fit of the unimodal model and is suited, therefore, to paleoecological studies in which sediments contain taxa produced in a variety of habitats throughout the year, while chemical data are obtained from less intensive sampling. In general, estimates of species optima and tolerance improve with the number of sites included in the survey, the length of the environmental gradient, and evenness of site distribution along the gradient. Further details on the WA calibration approach are given by Birks [69] and Hall and Smol [84].

In step three, past DOC concentrations are reconstructed from analyses of changes in fossil assemblages by applying modeled species–environment relationships using a WA regression approach (Figure 1C). Here, estimates of past DOC are inferred by multiple regression of species DOC optima, weighted by the relative abundance of taxa, but downweighted by the variability (tolerance) in the species–DOC relationship. The performance of these diatom–DOC transfer functions can be evaluated using randomization procedures (bootstrapping), while problems arising from poor fossil assemblage analogues within modern diatom communities can be assessed using dissimilarity indices (e.g., ANALOG; Line and Birks, unpublished program).

Finally, inference of past regimes of UVR and photosynthetically active radiation (PAR) regimes requires conversion of diatom-inferred DOC estimates into reconstructions of past irradiance environments. This goal is accomplished best by using bio-optical models that are based on DOC–irradiance relationships and on the response curves of algae for DNA damage and inhibition of photosynthesis [15,25]. Here, estimates of wavelength-weighted underwater UVR exposure (T^*) are based on spectral attenuation and biological weighting to allow quantitative estimation of potential impacts of exposure to past UVR. Further, this powerful approach allows direct comparison of potential UVR impacts among causal factors including O_3 loss and climate change [15,85]. However, in general, weighted transparency estimates based on DOC must be considered a lower boundary to variability in underwater UVR exposure, due to the influence of other physical factors (see above).

In the analysis of past spectral characteristics, T^* is defined as $\int 1/K(\lambda)\ \varepsilon(\lambda)\ E_{orel}(\lambda)\ F(\lambda)\ d\lambda$, where the integral is evaluated over 280–400 nm. $K(\lambda)$ is the diffuse attenuation coefficient at wavelength λ calculated from statistical relationships with DOC in the survey lakes [85]. $\varepsilon(\lambda)$ is the biological weighting factor for DNA damage (T^*_{DNA}) [86] or for inhibition of photosynthesis by UVR (T^*_{PI}) [87], and is expressed on a relative scale ($\varepsilon = 1.0$ at 300 nm). $E_{0rel}(\lambda)$ is the normalized surface irradiance at the study location ($E_{0rel} = 1.0$ at 400 nm), while $F(\lambda)$ is the factor of enhancement in surface radiation flux for a given level of stratospheric ozone depletion (set to 1.0 for 330 Dobson Units). T^* values are calculated at 1 nm intervals and are summed from 280 to 400 nm to give a total T^* for UVR. The main advantage to the T^* approach is that it does not require estimates of absolute UVR flux in order to evaluate the relative magnitude of

UVR impacts arising from past environmental change. Finally, although not explicitly attempted to date, such spectral irradiance models may be combined with diatom-inferred estimates of past lake depth to develop quantitative estimates of photon flux and biological exposure, assuming relatively small variations in past atmospheric transmission of UVR.

As with all inference model based approaches, reconstructions of past DOC and UVR levels using fossil diatoms are subject to problems arising from poor modern analogues for past diatom assemblages, taxonomic inconsistencies, insufficient range of past environmental variation, and multiple, co-linear environmental gradients [84]. Fortunately, inter-investigator harmonization of fossil taxonomy has reduced difficulties with species identification, while comparisons among multiple cores suggest that single, central sites capture the main components of historical variation, assuming little effect of riverine inputs, sediment redistribution or groundwater springs [e.g., 88]. As discussed in Pienitz et al. [89], the absence of modern analogues appears most problematic during the early Holocene, when terrestrial sources of DOC are poorly developed and benthic *Fragilaria* are the most common fossil diatoms. Similarly, inference models can perform poorly, despite strong environmental gradients, if variation in reconstructed variables (e.g., DOC) arise from multiple regulatory mechanisms (e.g., forest and peatland C sources). Although model performance can be improved through judicious site selection [e.g., 22], problems can arise during historical reconstructions unless multiple indicators are used to evaluate the relative importance of other potential forcing factors. In the case of DOC reconstructions, the use of peatland-specific microfossils [90], siliceous phytoliths from terrestrial plants [91], aquatic pollen [25] and near-infrared characterization of carbon sources [92] may provide insights into the sources and optical properties of terrestrial DOM. Similarly, use of multiple fossil indicators, each with different environmental sensitivity (e.g., invertebrates, pigments, diatoms) can help identify bias arising from confounding environmental variables. Finally, although within-lake variance in reconstructed variables is rarely as great as that expressed among lakes in surveys, multivariate evaluation of the main direction of variance in fossil assemblages can be used to identify historical eras in which environmental gradients are weak and reconstructions may be less reliable.

16.3.2 Other microfossil metrics of past UVR exposure

Recent observations show that some zooplankton deposit photo-protective pigments in their exoskeletons and resting eggs in response to high-energy irradiance [93,94]. Because these remains often are well preserved in lake sediments [95], analysis of the relative proportion of pigmented and hyaline remains in cores may be a valuable index of the relative exposure of taxa to UVR, particularly in response to long-term variations in UVR penetration arising from development of terrestrial DOM sources (D. Hessen, University of Olso, unpublished data). To date, no study has attempted to test this hypothesis. Similarly, because UVR impacts diminish with increased zooplankton size [58], analyses of fossil assemblage size structure may reveal changes in the average exposure of the population through time. However, we advise caution in both approaches, due to the potentially confounding influences of size-selective, visually-orienting predators such as fish. As shown elsewhere, colonization of fishless habitats by vertebrate planktivores results in rapid elimination of both large-bodied and highly-coloured crustacean zooplankton, events which are clearly recorded in lake sediment archives [95]. Similarly, interpretation of fossil size structure may be complicated by the observation that many of the most sensitive organisms (rotifers, early instars) leave a poor fossil record. However, in spite of these caveats, we feel that there is significant unexploited potential for microfossil remains from invertebrates to be used as indices of population response to past UVR regimes.

16.3.3 Biogeochemical indices of past UVR exposure

Past UVR environments can be inferred from analyses of fossil photo-protective pigments produced by eukaryotic algae, phototrophic bacteria and other organisms [e.g., 24]. The basic principles underlying this approach are that many organisms produce sunscreen compounds in response to intense UVR, and that these compounds are deposited in sediments following the death of the organism. Potential fossil indicator compounds include scytonemin and its derivatives from cyanobacteria [24,96], mycosporin-like amino acids from algae and invertebrates (MAAs; [97]), and melanin from zooplankton, such as *Daphnia* [93]. However, because preservation of these pigments is rarely complete [see 98], absolute concentrations of pigment cannot be used to quantitatively estimate past UVR flux. Instead, most reconstructions have quantified the concentrations of UVR-absorbing pigments relative to those of ubiquitous compounds that measure total algal abundance (e.g., carotenoids, chlorophylls [chls]). This approach assumes that all measured pigments degrade at similar rates. Given this assumption, the index should record the past exposure of an "average" organism, and should be greatest when a high proportion of the primary producer population is exposed to damaging levels of UVR [24,26]. Naturally, exposure of other trophic levels may vary independently of algal exposure as a consequence of the adaptive strategies of individual animals (see above).

To date, most pigment-based reconstructions of irradiance have used water-insoluble compounds derived from cyanobacteria to estimate historical changes in the UVR exposure of phototrophic populations in response to climate change and human activities. Although other approaches are possible, high-performance liquid chromatography has been the main analytical method used to quantify both past UVR exposure and biotic responses. Leavitt and Hodgson [68] have provided a comprehensive review of the main methods used to isolate, identify and quantify fossil pigments. Here, we provide a brief overview of the main methods used in our laboratories to reconstruct past UVR environments.

Once removed from a lake, sediments should be frozen (< -20°C) in the dark under an inert atmosphere (N_2, Ar, CO_2) or vacuum until isolation and quantification of pigments. In order to improve the reproducibility of pigment extraction, well-mixed sediment sub-samples should be freeze-dried under a hard vacuum (<0.1 Pa) for 24–48 h. Lipid-soluble pigments are extracted from the bulk sediments by soaking powdered sediments in a mixture of degassed acetone:methanol:water (80:15:5, by volume) for 24 h in the dark and under an inert atmosphere at 0°C. Pigment concentrations are most often quantified by reversed-phase high-performance liquid chromatography (RP-HPLC), which separates complex mixtures according to the relative attraction of individual pigments for the non-polar stationary phase (both coating and support material) and the polar mobile solvent phase. So far, most UVR reconstructions have been based on polar pigments such as scytonemin or related compounds [24] that pass through the HPLC column rapidly and are among the first compounds to be detected. In most cases, efficient analysis of large sample numbers minimally requires an autosampler, high-pressure pumps (>20 kPA), high-resolution column, and in-line photo-diode array spectrophotometer (300–800 nm range). Additional components may include in-line detectors of pigment fluorescence or mass-selective spectrometric detectors [e.g., 99]. Both the analytical system detailed by Mantoura and Llewellyn [100] modified by Leavitt and Findlay [101] and that of Wright et al. [102] have proven robust in isolating UVR-absorbing pigments from sediments of 500 lakes (see below), despite some limitations in resolving power [103, but see 104]. Analytical separation is achieved using either a two- or three-stage solvent system in which pigment extracts are introduced to the chromatographic column, and solvent polarity is systematically altered to sequentially isolate compounds of progressively decreasing polarity. At a minimum, accurate quantification of pigment abundance requires separation of marker compounds from contaminants, identification of a pigment's true identity, and calibration of the HPLC system with an authentic standard of known purity. Further details on HPLC calibration and pigment quantification are provided by Leavitt and Hodgson [68].

Past UVR penetration has been measured as a ratio of UVR-absorbing pigments : algal carotenoids, an index which is linearly related to the depth of UVR penetration in whole-lake experiments [24,26]. To date, most of our reconstructions have been based on the UVR-absorbing pigment, C_a, which has a mass of 635 according to mass spectrometric determinations using negative ion-atmospheric pressure chemical ionization techniques [68]. Visualization of this compound is improved substantially by first dissolving the whole extract into an injection solution containing an aged (3 months) solution of Sudan II dye before injection into the HPLC system [26,68]. Similar reconstructions can be achieved using the sum of scytonemin and its derivatives, compounds that preserve in lake sediments for over 100 000 years (D. Hodgson, unpublished data). Abundance of UVR-absorbing pigments is expressed relative to total algal biomass in order to distinguish whether photo-protectant production arises from a unique population (e.g., surface dwelling) or represents a general response of the phototrophic community to UVR [24]. Total algal abundance can be measured as changes in the concentration (nmoles pigment g^{-1} dry sediment or nmoles pigment g^{-1} organic matter) or accumulation rate (nmoles pigment cm^{-2} yr^{-1}) of β-carotene, a chemically-stable carotenoid ubiquitous in algae, chl *a* and its pheopigment derivatives, or the sum of individual algal group indicators including alloxanthin (cryptophytes), diatoxanthin (diatoms), colonial cyanobacteria (myxoxanthophyll or echinenone) and chlorophytes (lutein).

Surveys of modern lake communities suggest that pigment-based UVR indices are elevated only when a substantial portion of the algal community is exposed to potentially-damaging levels of irradiance. Indices range from 10–800% in clear lakes, depending on the presence of refugia, lake depth, circulation patterns and the distribution of algal biomass. Leavitt et al. [24] noted that UVR-absorbing compounds were abundant, and UVR indices elevated (>50%), in mountain lakes that were both shallow (<5 m maximum) and of low DOM content (<1.5 mg DOC l^{-1}). As recorded by Sommaruga et al. [38], these lakes often lie above treeline, have a high proportion of their volume exposed to >1% ambient UVR, and exhibit variable and often low C-specific attenuation of UVR by DOM [40]. Such a lack of physical refuge from UVR presumably requires phototrophic organisms to produce photo-protective compounds in order to reduce cellular oxidation from UVR-produced singlet oxygen and free-radicals [64].

Algal exposure to UVR may also be increased by the presence of fine particulates which, while reducing total irradiance, act to increase ratios of UVR : PAR because photon scattering is less wavelength dependent than is its absorbance by DOM (e.g., *S* in [43]). Exposure also increases because sedimentation of particulates scours DOM from the water column leading to lower absolute concentrations [e.g., 38], while adsorption of DOM to particles causes a "package effect" that decreases absorbance per unit pathlength of water without altering the specific absorbance of individual DOM molecules. Thus, for photosynthetic organisms, the need to remain in light necessitates exposure to high levels of UVR and the production of photo-protective compounds. In contrast, the presence of surface blooms (e.g., eutrophic lakes) does not seem to elevate UVR indices, probably because water column circulation reduces exposure and because productive lakes often have high levels of UVR-absorbing DOM [24,26].

Reconstruction of past UVR environments from sedimentary pigment profiles has a number of significant challenges before UVR indices can be quantitatively related to past photon flux. First, quantification of UVR-absorbing compounds is difficult because of considerable analytical requirements (HPLC, MS). To date, most reconstructions rely on partly characterized pigments produced by benthic cyanobacteria [24,26,27], but little is known of the precise structure of these compounds, or of their distribution among organisms [64]. Similarly, reconstruction of past UVR environments from scytonemin requires isolation and quantification of a series of derivatives using advanced mass spectrometric techniques [68,105]. Second, comparison of decadal historical records with annually resolved pigment profiles suggests that some post-depositional transformation occurs, and that very recent deposits (<3 yr old) may not provide reliable indices of recent UVR environments [26]. Similarly, complete diagenesis of photo-protectant, chl and carotenoid pigments can occur, particularly in deep lakes (>100 m) with very low sedimentary organic matter contents (<2%) where degradation of pigments is often extensive [98]. Finally, more research is required to standardize expressions of UVR flux. Presently, we favour a relative ratio of UVR-absorbing pigments to ubiquitous, chemically-stable carotenoids in order to correct for variation in fossil abundance arising from single populations (e.g., surface dwelling) or changes in total algal sedimentation [24,26]. However, while sensitivity analyses suggest that most historical changes in a UVR index arise from variation in deposition of photo-protectant compounds, the reciprocal nature of the ratio at least allows for the possibility of artifacts arising from changes in total algal abundance (e.g., trophic interactions, pH, etc.) independent of variations in past UVR exposure. As with most paleoecological analyses, this problem can be best alleviated through the use of multiple sedimentary proxies of DOC or UVR.

16.3.4 Sedimentary organic matter as an index of past UVR penetration

Reconstruction of UVR exposure from analyses of fossil diatoms and pigments requires levels of training or analytical expertise that are likely to preclude their widespread adoption as methods to evaluate UVR impacts on lake ecosystems. Consequently, recent research has begun to investigate the use of bulk sedimentary constituents as retrospective predictors of water-column DOC concentrations and UVR [27]. In general, sediment organic content is determined as % mass loss-on-ignition (LOI) at 500°C for 1 h [106]. To date, surveys of both shallow alpine [24] and subarctic [80] lakes distributed across treeline revealed strong linear correlations (Pearson $r > 0.75$) between sediment organic matter (% LOI) and the dissolved organic matter content (mg DOC l^{-1}) of overlying waters. In these cases, surveys spanned gradients of terrestrial vegetation development, from bare rock catchments to drainages with well-developed coniferous forests, and represent a wide range in supply of terrestrial-derived DOM [21,89]. Based on this relationship, past DOC concentrations can be reconstructed from organic matter (% LOI) profiles in sediment cores from regional lakes [27], while depths of UVR penetration (as 1% surface irradiance) can be estimated using published optical models [11,14]. When this UVR depth is expressed as a proportion of area-weighted basin depth, an index of past UVR exposure and lake sensitivity can be calculated [cf., 38].

Accurate use of bulk sedimentary organic matter to reconstruct past UVR penetration requires strong correlations between dissolved and sedimentary organic matter in modern surveys, relatively consistent optical characteristics of DOM, and that these relations remain valid in the past. Positive correlations between water column and sedimentary organic matter may be reinforced by several mechanisms. First, because most water-column C resides in DOM, its sedimentation as colloids or precipitates is a major process increasing the organic matter content of bottom deposits. Second, while sedimentary organic matter is also a function of lake production and catchment erosion [107], these processes will tend to reinforce DOM–sediment relationships due to either

DOM production (algal, macrophyte sources of DOM) or enhanced removal (adsorption on fine particulates). Finally, in shallow lakes, sediments may act as a substantial source of DOM due to turbulent mixing and resuspension of DOM-rich interstitial waters. Unfortunately, high variability in the specific attenuation characteristics of DOM has been documented from a number of lake surveys [40,43], and may limit the accuracy of sediment-based UVR reconstructions in some very clear lakes (<1 mg DOC l^{-1}). As well, as with all paleoecological reconstructions, assumptions of constant regression relationships through time need to be viewed critically. However, the high agreement observed between pigment- and sediment-based UVR reconstructions (see below) suggests that this simple technique holds much promise, particularly in extreme environments where biochemical or microfossil preservation may be poor (e.g., early Holocene, saline lakes).

16.4 Fossil evidence of past UVR environments in lakes

Present concern about the interactive effects of ozone depletion, DOM-degrading acidic precipitation and global warming has led several investigators to conclude that extreme variations in the biogeochemistry of DOM and its impacts on UVR penetration may be the most significant challenge to aquatic ecosystem integrity and function [e.g., 66,108]. Because the range of historical variance in DOM flux and UVR attenuation is often greater than that arising from modern processes [15,24], paleoecological reconstructions of past UVR and its impacts on lake ecosystems may provide essential insights into the role of high energy irradiance in structuring aquatic ecosystems. Here we use a *case study* approach to demonstrate the value of such retrospective analyses, particularly in the case of long-term environmental change, early lake evolution and human impacts. As our intent is to stimulate research in this area, our examples include both research with well-documented mechanistic explanations, and provocative new studies which, although less substantiated, have the potential to greatly improve our understanding of past UVR environments and their impacts on lakes.

16.4.1 Holocene climate change

To address the potential impact of long-term climate change relative to that of ozone depletion, Pienitz and Vincent [15] combined paleolimnological analyses with bio-optical models based on present-day conditions in lakes of northern Canada. Specifically, they estimated past underwater light conditions from DOC concentrations that were inferred from fossil diatom assemblages preserved in Holocene sedimentary deposits from a lake near the treeline (Queen's Lake) in the central Northwest Territories, Canada (Figure 2). Analysis of fossil pollen records indicate that this region of the continent underwent deglaciation ca. 8000

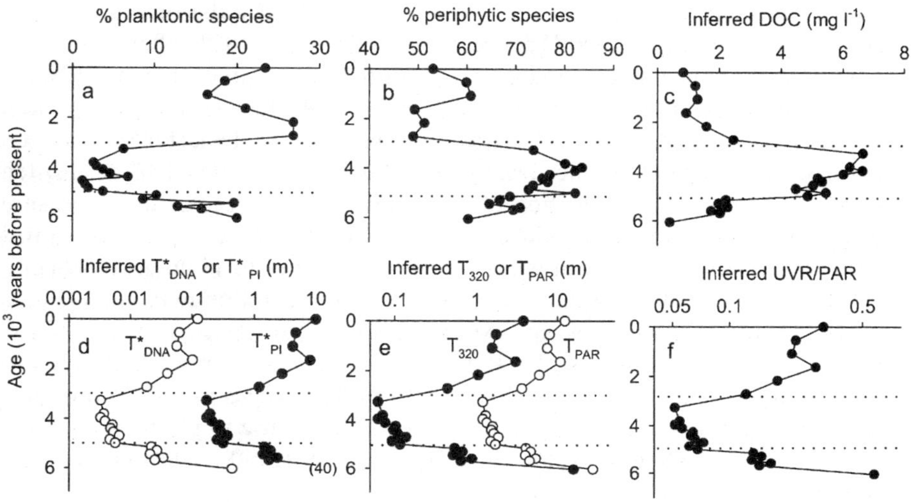

Figure 2. Changes in fossil diatom community structure and inferred optical conditions in Queen's Lake, Northwest Territories (Canada)[15]. Diatom data are expressed as a percentage of the total number of valves in each sample associated with planktonic or benthic taxa. Diatom species data were used to infer DOC concentration (mg l^{-1}), biologically weighted UV exposure (T^*_{PI} or T^*_{DNA}) and underwater spectral balance (water column transparency for 320 nm UVR [T_{320}], PAR [T_{PAR}], and the ratio between the two [UVR/PAR]) over the last 6000 years. The dotted lines delimit the period of mid-Holocene maximum forest cover. This analysis demonstrates that biotic exposure to UVR varies substantially due to changes in catchment vegetation and DOC supply, and that, during mid-Holcene climatic warm periods, UVR exposure declined two-orders of magnitude. Climatic cooling at ~3000 yr BP reduced DOC inputs by reducing soil development and DOC supply, and led to increases in UVR penetration that are up to 4000-fold more significant than those expected to arise from a moderate (30%) ozone depletion. [Figure reprinted by permission from *Nature* [15], copyright 2000, Macmillan Magazines Ltd.]

yr BP, and that terrestrial vegetation was sparse and tundra-like for the first 3000 years of lake existence [109].

Diatom community structure and inferred DOC levels showed three distinct and abrupt changes during the history of Queen's Lake. First, analyses showed that both diatom biomass and inferred DOC concentrations were low (<2 mg DOC l^{-1}) during the first ~3000 years of lake existence, with particularly few fossils recovered from sediments >6500 yr BP. This initial period was followed by a major shift in fossil species composition and inferred chemical conditions ca. 5000 yr BP, with increased ratios of periphytic : planktonic taxa to >70% total diatom assemblage. This second period also corresponded to a major increase in algal production, recorded as the sediment mass-specific concentration of diatom valves, as well as a three-fold increase in inferred DOC levels. Based on fossil pollen analyses, Pienitz and Vincent [15] argued that changes in lake chemistry and production resulted from climatic warming that stimulated treeline advance and increased forest density for about 2000 years. Finally, diatom-based reconstructions indicated that DOC concentrations declined >85% after 3000 yr BP, concomitant with climatic cooling and a southward retreat of treeline [109].

The large and rapid changes in DOC imply that Queen's Lake experienced major shifts in the underwater optical environment over the last 6000 years as a consequence of climate-induced variation in forest development [89]. Consistent with this hypothesis, application of bio-optical models derived from measurements in high-latitude waters showed that the inferred DOC shifts were equivalent to a two order-of-magnitude decrease in exposure to biologically-effective UVR between 6000 yr BP and the mid-Holocene vegetation maximum (Figure 2). In contrast, the most recent 3000 years were characterised by a >50-fold increase in levels of damaging UVR, with recent inferences agreeing closely with present-day estimates of UVR exposure. Overall, changes in DOC concentrations arising from climatic variability increased exposure to photosynthetically damaging UVR 4000-fold more than did a moderate (30%) decline in stratospheric ozone levels [15].

Climatic control of past UVR exposure has also been identified as a key factor regulating lake production and algal community composition in montane lakes at treeline [27]. For example, geochemical and palynological analysis of sediments demonstrates that Crowfoot Lake, Alberta (51°26′N,116°31′W), lay above treeline during the Younger Dryas (ca. 11 100–10 100 ^{14}C yr BP), was a subalpine lake for the next ~6000 years, then returned to its present position near timberline following regional climatic cooling ca. 4000 yr BP [110,111]. Analyses of fossil pigments confirmed that algal abundance was reduced 10- to 25-fold during periods of high UVR exposure, inferred from both fossil pigment- and bulk organic matter-based estimates of irradiance penetration (Figure 3). Through the use of coupled DOC-UVR optical models, it was shown that algal abundance was reduced whenever the depth of UVR penetration (as 1% surface irradiance) exceeded mean lake depth, and deepwater refugia were lost, especially early in the lakes history (>10 500 yr BP) and following climatic cooling at ~3500 yr BP [27]. The authors argued provocatively that lake production was suppressed by a combination of low DOM and high UVR rather than by variations in mineral nutrient flux because patterns of fossil pollen, modern lake evolution [28], and terrestrial nutrient cycling were inconsistent with regulation of lake production by N and P, the major mineral nutrients [27]. Further, as changes in UVR exposure were linked to long-term climatic variability, and because increased UVR exposure occurred despite substantial pools of terrestrial DOM, both diatom- and pigment-based analyses suggest that future global warming may increase UVR penetration, alter gross community composition, and strongly suppress the primary production of many boreal lakes.

16.4.2 Early lake evolution

Recently, several lines of evidence have combined to suggest that biotic exposure to UVR should be greatest immediately after deglaciation, prior to the develop-

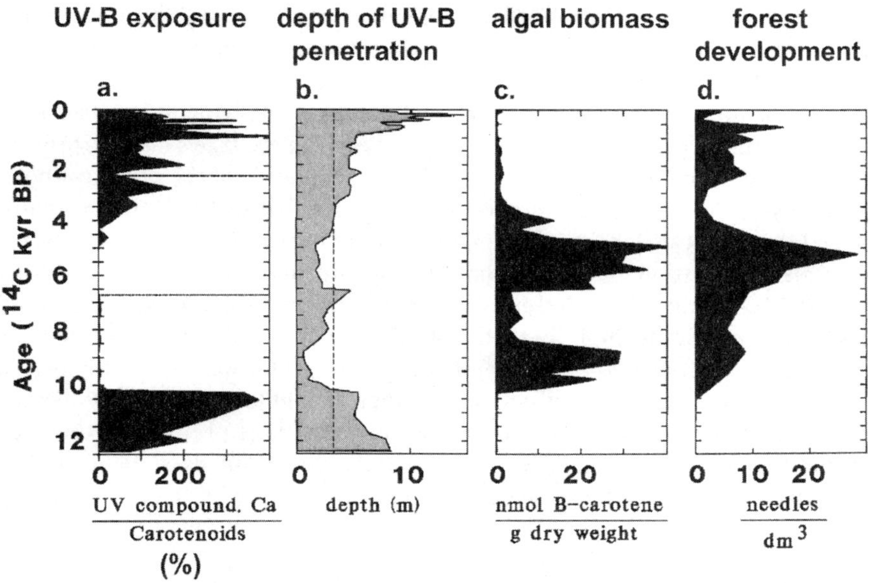

Figure 3. Change in past UVR exposure inferred from pigments (a), depth of UV-B penetration inferred from sedimentary organic matter content (b), and total algal abundance (c) in sediments from Crowfoot Lake, Alberta, Canada [27]. Forest development (d) was estimated from changes in concentration of locally-derived tree needles [from 111]. Algal exposure was measured as the ratio of UVR-absorbing pigment C_a : carotenoids (alloxanthin, diatoxanthin, lutein). Depth of UVR penetration (as 1% surface irradiance, in m) was inferred from historical changes in the organic matter content of Crowfoot Lake sediments. Total algal abundance was estimated from concentrations of the ubiquitous carotenoid *β*-carotene. This analysis demonstrates that algal abundance was low whenever UVR penetration was great during both the recent and early Holocene periods, when local forests were absent or poorly developed. Production of photo-protective pigments was greatest when the depth of UVR penetration exceeded the mean depth of Crowfoot Lake (vertical dashed line) during ~11 300–10 050 ^{14}C yr BP and ~4000 ^{14}C yr BP-present. Rates of algal decline ca. 4000 yr BP reflect reduced DOC supply at high elevations and are similar to those expected to occur as a result of DOC declines arising because of global warming at low elevations [108]. See text for details. [Figure modified from 27.]

ment of terrestrial sources of UVR-absorbing DOM. First, optical models show that DOM from terrestrial sources is the single most important factor regulating UVR penetration within a lake [10,11]. Second, whole-lake experiments and empirical studies show that variations in terrestrial DOM supply and mineralization are more significant factors regulating UVR exposure than is modern stratospheric ozone depletion [12,14]. Third, analysis of lake chronoseries from Glacier Bay, Alaska, demonstrate that DOC content is low for at least the first century following modern lake formation [28] and that this initial high UVR exposure can structure biotic communities [2,112]. Finally, the observations that diatoms are particularly sensitive to changes in UVR exposure [e.g., 49,57], and that these taxa are rare in early post-glacial sediments [89], suggest together that extreme UVR transparency is a common mechanism directing the early evolution of glacial lake ecosystems.

Quantification of past UVR environments using fossil pigments has been used to document that exposure to UVR is greatest early in the lake's history, prior to development of regional forests (Figure 4). Here, analysis of the complete post-glacial history of three lakes in sub-humid central British Columbia, Canada, showed that UVR-absorbing algal pigments were present for at least 1000 years following deglaciation, but were absent from sediments at all other times during the past 12 000 years [27]. Presently, UVR penetration is inconsequential at all sites (< 10 cm) due to high levels of DOM (> 10 mg DOC l^{-1}). In addition, reconstructions showed that algal biomass was 10-fold lower during the period of elevated UVR penetration than at any other time in lake history. At all sites, sharp reductions in UVR penetration and increased lake production occurred concomitant with the development of terrestrial carbon sources ca. 10 700 yr BP, consistent with the hypothesis that terrestrial carbon is the key factor regulating

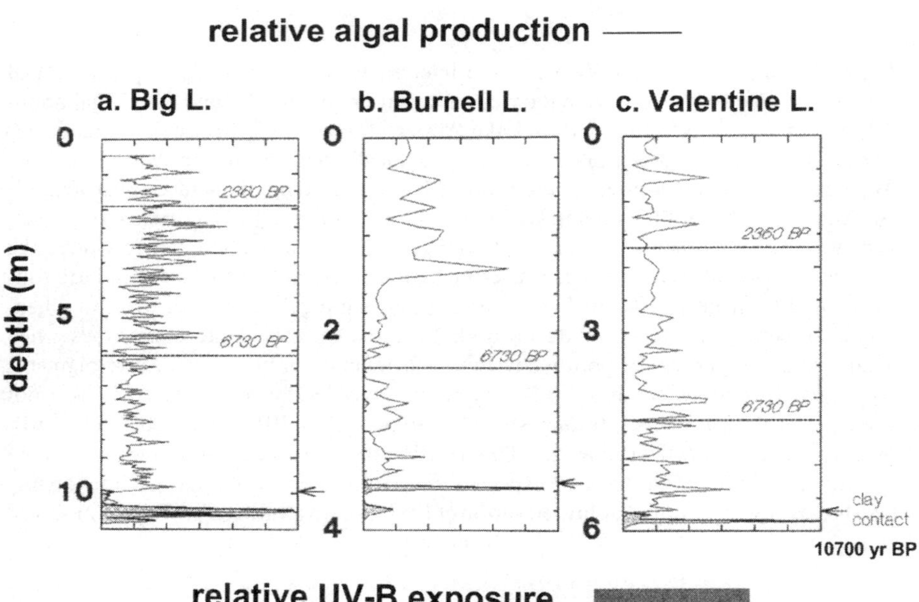

Figure 4. Changes in UVR exposure (shaded) and total algal abundance (line) in Big (a), Burnell (b), and Valentine lakes (c), British Columbia, Canada [27]. UVR exposure and total algal abundance as in Figure 3, except for scale change. Approximate chronological control of Burnell and Valentine lakes was provided by volcanic ash and basal clay–organic contact layers only, and by 14 accelerator mass spectrometric (AMS) determinations of ^{14}C activity and sediment age in Big Lake [27, not shown here]. Comparison among lakes demonstrates that algal biomass is low when UVR penetration is great, and that such periods occur immediately following deglaciation in all cases (see also Figure 3). [Figure modified from 27.]

irradiance regimes in lakes [11].

In a similar study, Saulnier-Talbot et al. [25] used a diatom-based paleo-optical approach to estimate past depths of UVR penetration in a coastal basin (Lake Kachishayoot) following its isolation from the marine waters of Hudson Bay in subarctic Québec, Canada. Prior comparisons of optical environments in coastal systems have revealed that shifts from marine to freshwater conditions are accompanied by increased DOC, changes in C-specific UVR attenuation and declines in UVR penetration [39]. Consistent with these modern analyses, abrupt increases in diatom-inferred DOC concentrations and water color coincided with the retreat of postglacial marine waters and arrival of spruce trees within the local catchment [25]. Their analyses also revealed large changes in the underwater irradiance environment over the course of the post-glacial period, from extremely high UVR exposure after the initial formation of the lake and its isolation from the sea, to an order-of-magnitude lower exposure following development of spruce forests in the catchment. Interestingly, the use of additional macrofossil markers allowed investigators to show that UVR penetration remained high even following development of alternative DOC sources such as *Sphagnum* mats. These results further support the hypothesis that development of local conifer populations represents the critical step altering spectral irradiance characteristics of northern lakes.

16.4.3 Historical changes in polar UVR flux

High latitude aquatic ecosystems may be particularly susceptible to UVR both because of extremely low concentrations of photo-protective DOM and because of natural and anthropogenic mechanisms that lead to stratospheric ozone depletion [21]. For example, in the Antarctic, strong westerly circulation each winter causes a circumpolar vortex that isolates part of the stratosphere, allowing it to cool and subsequently form thin high clouds that contain chlorine (Cl) and bromine. These elevated concentrations of active Cl, mainly derived from chlorofluorocarbons (CFCs), are known to catalyze the reaction of ozone to molecular oxygen ($2O_3 + 2ClO \rightarrow 3O_2 + 2ClO$), leading to the spring ozone hole. Thereafter, the stratosphere warms and the polar vortex breaks up, allowing the O_3-depleted stratosphere to mix with mid-latitude air and replenish polar O_3 concentrations. Depletion of stratospheric ozone is currently estimated at 4–6% per decade over northern mid-latitudes [33,113], and is expected to increase fluxes of UV-B at least until the middle of the 21st century [34].

Despite sophisticated understanding of the anthropogenic mechanisms that regulate ozone depletion, little is known of the causes or magnitude of natural variations in UVR flux in polar regions. Fortunately, several factors may make high latitude lakes particularly amenable to reconstruction of past irradiance regimes. First, many lakes have no terrestrial source of UVR-absorbing DOM within their catchments [e.g., 21,114], therefore post-glacial changes in UVR exposure would be expected to arise solely from changes in solar production, atmospheric transmission or lake depth. Second, as a consequence of their shallow nature and low thermal capacity, ponds and shallow lakes are ice-free for part of the summer thereby allowing biological responses to changes in irradiance environments. Finally, these small systems have often accumulated sediments and fossils continuously throughout the Holocene, particularly in more moderate climates of coastal regions [115,116]. In particular, diatoms and pigments are often well preserved, probably because the sites are frozen completely and biological inactive for much of the year. As a consequence of these factors, investigators have recently begun exploiting polar sedimentary deposits to better understand past variability in UVR environments.

Surveys of shallow lakes and ponds in eastern Antarctica have revealed that modern sedimentary environments preserve fossil pigments and that the abundance of photo-protective compounds increases as a function of algal exposure to UVR (Hodgson et al. unpublished). For example, the Larsemann Hills region (69°23′ S, 76°53′ E) is the second largest of four major ice-free oases on the east coast of Antarctica and contains more than 150 lakes and ponds. Minimal cloud cover, very transparent waters ($K_{d_{UV\text{-}B}}$ 0.21–0.35 m^{-1}; [117]), and a 2–3 month ice-free season allow UV-B penetration beyond maximum lake depth and select for filamentous cyanobacteria mats [118] that produce both MAAs and scytonemin to reduce UVR impacts [119]. A survey of 70 lakes and ponds has revealed that photo-protective compounds are abundant in benthic communities of shallower lakes (< 6 m depth). Although MAAs were found to be degraded in the first few centimetres of sedimentary deposits, scytonemin was well-preserved for 1000s of years, as oxidized (yellow–green), reduced (red) forms and partly degraded forms [105].

Preliminary analysis of ^{14}C-dated Holocene cores suggests both that UVR flux was greater prior to ~4000 yr BP and that historical variability in UVR exposure was expressed on a variety of temporal scales (Figure 5). In this study, scytonemin and its derivatives were summed and expressed as a ratio with algal carotenoids, an index which has been shown to be linearly related to the depth of UVR penetration when calculated using similar compounds [24,26]. In addition, scytonemin was expressed as a function of both total xanthophyll carotenoids, compounds which disperse excess heat energy during photosynthesis [120], and total chls to roughly estimate the proportion of metabolic effort expended by cells on photo-protection versus production. Although the magnitude of historical change varied among indices, in all cases photo-protective pigments were relatively more abundant in sediments older than ~4000 years, particularly compared with those deposited between ~2000 and ~700 yr BP. While the absence of modern increases in UVR exposure suggests that recent ozone depletion has not greatly altered natural variation in UVR flux, the low resolution of these core analyses may preclude the detection of very recent events. However, regardless of the chronological resolution of the core, it appears likely that biotic exposure to UVR has varied by at least 400% during the past 12 000 years, with ancient levels possibly reflecting increased solar production arising from long-term changes in planetary orbit. Further paleoecological research is being conducted to determine whether changes in exposure arise from variation in UVR flux to the lake surface, or from changes in lake characteristics (transpar-

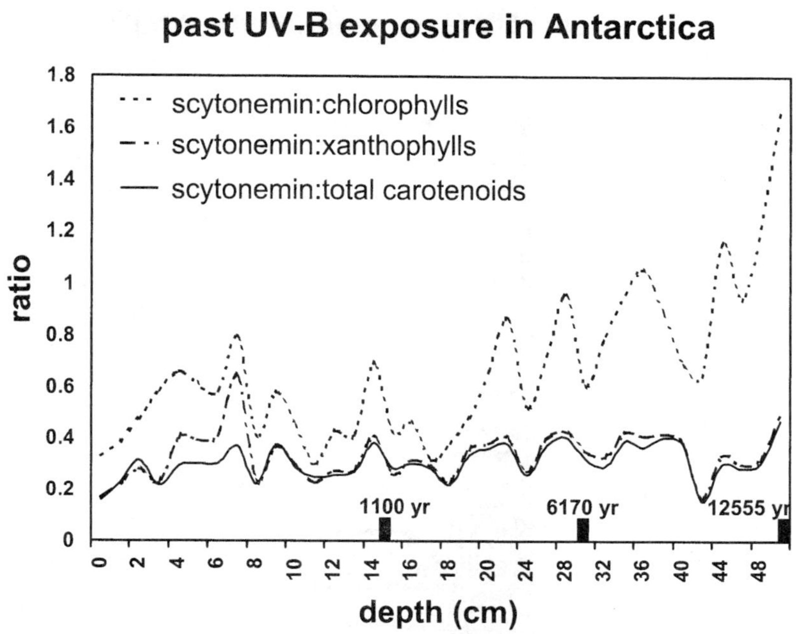

Figure 5. Historical change in past UVR exposure at Larsemann Hills, Antarctica (69°23′S, 76°53′E), during the past 12 500 years [Hodgson et al. unpublished data]. UVR exposure estimated as in Figure 3, using ratios of scytonemin : total chlorophylls (- - -), scytonemin : xanthophyll carotenoids (– - -) and scytonemin : total carotenoids (solid line). Similar indices have been shown to be linearly related to the depth of UV-B penetration (as 1% surface irradiance) based on whole-lake experiments [24,26]. All pigment and derivative concentrations were quantified using high-performance liquid chromatography and mass spectrometry [68]. Sediment age was estimated from three accelerator mass spectrometric determinations of ^{14}C activity. Analyses suggest that algal exposure to UVR was at least two-fold greater prior to ~4000 yr BP than at present, and that UVR exposure varied ~400% during the Holocene.

ency, depth, ice cover).

Unpublished analyses of fossil pigments in Arctic ponds also suggests that UVR flux declined after ~4000 yr BP (Figure 6; Leavitt et al. unpublished data). Further, comparison among fossil markers suggest that variance in UVR exposure arose from changes in atmospheric processes rather than from variation in lake properties. Our analysis is based on a 150 cm (of 223 total) sedimentary sequence recovered from Col Pond on Cape Herschel, Ellesmere Island, Canada (78°37′N, 74°42′W), a primary reference region for high Arctic research [115]. Col Pond occupies the central plateau of Cape Herschel, is the first pond to thaw, and is the least nutrient-rich of local sites. Further, Col Pond has been subject to previous biological [114] and paleoecological analyses [115] which show that diatom communities had been stable for at least 4000 years, but that community composition had altered during the past ~150 years due to human activity. Because diatom composition is extremely sensitive to both changes in lake depth [reviewed in 121] and DOC [reviewed in 79], the absence of community vari-

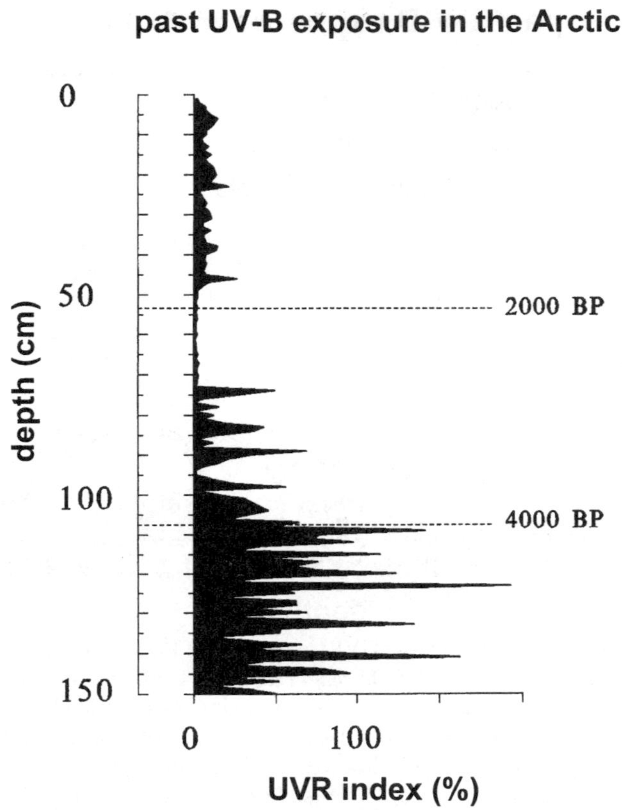

Figure 6. Historical change in past UVR exposure at Col Pond, Ellesmere Island, Canada (78°37′N, 74°42′W), during the past ~6000 years [Leavitt et al. unpublished data]. UVR exposure estimated as in Figure 3, using ratios of C_a : carotenoids. All pigment and derivative concentrations quantified using high-performance liquid chromatography [68]. Sediment age estimated by assuming constant mass accumulation rates since pond formation ca. 8500 yr BP[122]. No organic matter could be recovered for determinations of ^{14}C activity. Analyses suggest that algal exposure to UVR was at least 3-fold greater prior to ~4000 yr BP than at present. Timing of declines in UVR exposure is similar to those seen in the Antarctic (Figure 5) and coincide with climatic cooling and UVR change at other latitudes [15,27].

ation through much of the core suggests that there has been no persistent change in either lake depth or UVR transparency. Because the pond lies in a rock basin, lacks submerged vegetation, is very shallow (~1 m) and has sediments with low organic matter content (~2% LOI) we infer that UVR has likely penetrated throughout the water column since pond inception.

Sedimentary pigment analyses showed that photoprotective compounds were abundant and UVR indices were elevated in the oldest sediments, but declined ~3-fold to modern values ca. 3000–4000 yr BP (Figure 6). As sedimentary organic matter content also declined at this time, we infer that reductions in UVR did not arise from increased DOC content within the lake (see methods above). Similarly, variations in UVR index did not reflect changes in cyanobacterial abundance, as concentrations of myxoxanthophyll from those algae were constant through the period of inferred UVR decline. Unfortunately, chronological control is very poor at Col Pond, and sediment ages are estimated only approximately from published marine emergence curves for Cape Herschel [122] and by assuming that sediment accumulation rates were constant through time. With this caveat in mind, it is interesting to note that both the timing and direction of UVR change was similar at both poles, suggesting that fossil pigments may record changes in global irradiance regimes. Consistent with this hypothesis, timing of UVR declines corresponded to the onset of the modern, comparatively cool climate [123,124] and altered UVR regimes at mid-latitudes [15] and high elevations (Figure 2; [27]). Because our UVR indices are scaled by total algal production, they should be independent of changes in growing season duration arising from global cooling. We are presently conducting further research at other, less marine-influenced, sites to determine whether these historical patterns represent changes in regional irradiance due to cloud cover or whether sediments may be recording true variability in solar production or stratospheric transmission of UVR.

16.4.4 Rapid variation in UVR environments

In addition to long-term changes in UVR regimes arising from variations in global climate and carbon biogeochemistry, aquatic ecosystems are subject to extremely rapid alterations in UVR penetration due to natural and human-induced mechanisms. In particular, attention has focused on the combined impacts of ozone depletion, global warming and acidic precipitation, the so-called *triple whammy* of environmental disturbance [66,108]. Lake acidification by anthropogenic mineral acids both increases the rate of DOM removal from the water column [14], and reduces the specific attenuation of remaining DOM [42], leading to order-of-magnitude increases in UVR penetration [14]. In contrast, declines in stream flow during droughts reduce export of terrestrial and wetland DOM to lakes, while increasing water residence times, thereby causing more thorough DOM mineralization and precipitation and increased UVR penetration [13,125]. Further, mineral sulfur stored in shallow sediments can be oxidized during droughts to reform acids that deplete DOM and allow 3-fold increases in UV-B penetration [12]. Although lakes may recover from individual disturbances, concern is mounting that ecosystems may not be resilient to multiple concurrent stressors [108], perhaps resulting in rapid fundamental changes in lake organization (*state change*; [126]). Once again, we propose that sedimentary analyses can provide valuable insights into how human activities interact with other stressors to regulate the impact of irradiance on lakes.

Analyses of fossil pigments in the sediments of alpine lakes have shown that variation in UVR exposure arising from regional droughts can be greater than that attributed to human depletion of stratospheric ozone [24]. Prior paleoecological analyses have demonstrated the presence of droughts at low elevations and cool temperatures at treeline during ~1850–1900, both conditions which reduce terrestrial export of DOM to lakes [cf., 13]. As expected, deposition of photo-protective pigments was greater during drought intervals than at other times (Figure 7). When expressed relative to total algal abundance (as fossil carotenoids), UV-B exposure was found to be 5- to 10-fold greater than that in the most recent sediments, and as much as 4-fold greater than that seen in the most transparent Rocky Mountain lake. Extension of this analysis to other sites has demonstrated that this UVR event was widespread among regional lakes. Given that up to 200 000 present-day North American lakes have DOM levels typical of montane lakes ($\ll$2 mg DOC l^{-1} [14]), that DOM-depleting continental-scale droughts have been more intense during the recent past [127], and that global warming will likely intensify droughts [108], these analyses suggest that UVR impacts on lakes may be widespread in the future.

Both pigment- and diatom-based reconstructions of past UVR environments have demonstrated that anthropogenic acidification alters the fundamental irradiance regime of lakes [26,128]. For example, Dixit et al. [23] applied DOC-inference models to high-resolution cores from three central Canadian lakes receiving acidic precipitation and showed that DOC declined up to 75% and

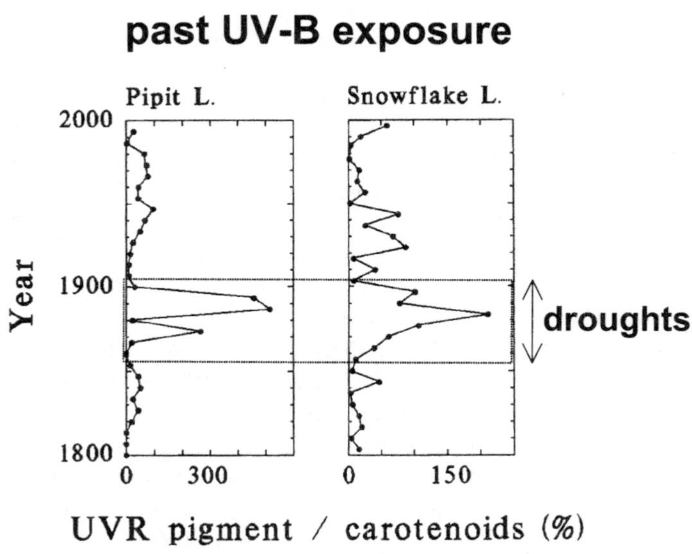

Figure 7. Historical changes in UVR exposure in alpine Snowflake and Pipit lakes, western Canada [24]. Past UVR exposure estimated as in Figure 3, using ratios of C_a : carotenoids. All pigment and derivative concentrations quantified using high-performance liquid chromatography [68]. A total of four regional lakes demonstrate that UVR exposure increased ~10-fold during periods of lowland droughts and upland cool temperatures, both factors that reduce export of terrestrial DOM to lakes [24]. In contrast, sites contain little evidence of increased UVR exposure arising from recent ozone depletion. Figure modified from [24]. [Reprinted by permission from *Nature* [24], copyright 2000, Macmillan Magazines Ltd.]

penetration of UV-B increased up to 2.5-fold as a result of acid emissions from local smelters (Figure 8). In general, fossil inferences showed excellent agreement with historical measurements of DOC concentrations, greatly improving researcher's confidence in historical reconstructions. Interestingly, this study also

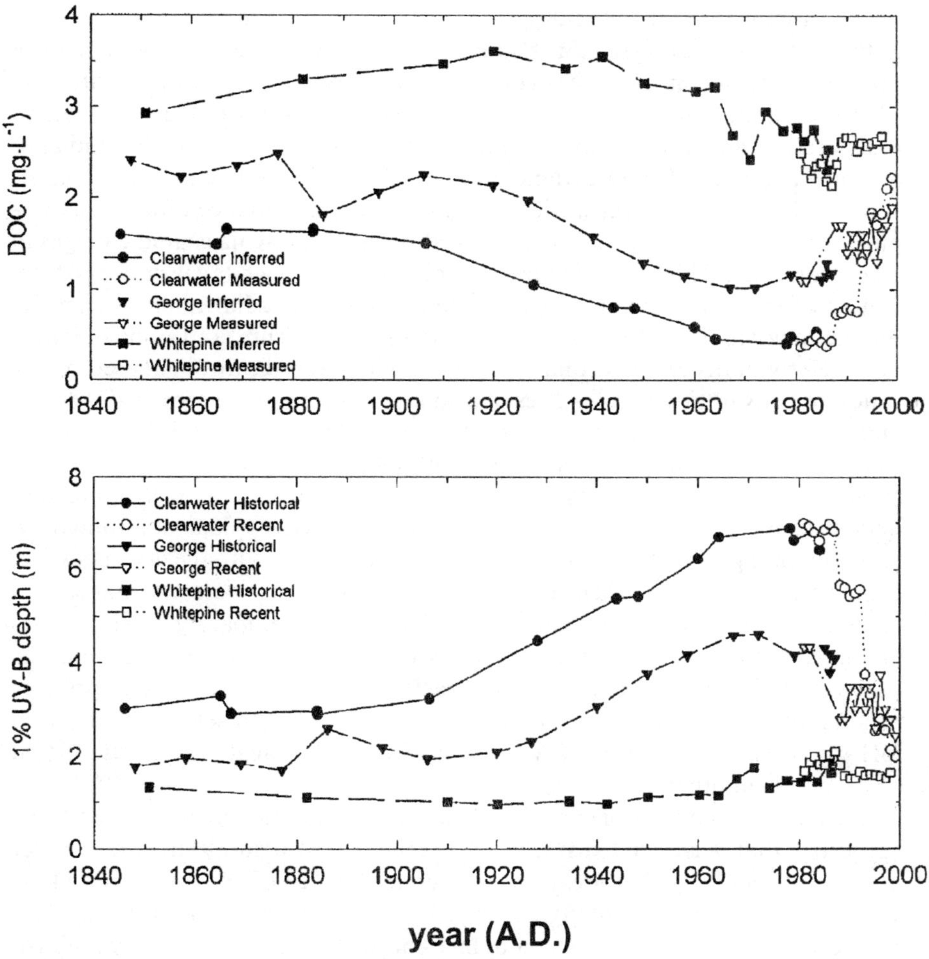

Figure 8. Historical changes in concentrations of UVR-absorbing DOC (top panel) and the depth of penetration of UV-B (as 1% surface irradiance; lower panel) for Clearwater (circles), George (triangle) and Whitepine (square) lakes, Ontario, Canada [23]. Filled symbols are inferred from analyses of fossil diatoms, open symbols are direct historical observations. Analyses demonstrate that lake acidification arising from regional smelting activities led to declines in DOC and up to 6-fold increases in the depth of UV-B penetration. In general, clear lakes were most susceptible to acidic precipitation, but were not most naturally variable in the past. Instead, forest fires are hypothesized to increase the variability in pre-mining DOC concentrations in impacted lakes. [Figure modified from [23] and reprinted with permission.]

showed that lake sensitivity to DOC loss, baseline variability (pre-emission), and degree of recovery all varied greatly among sites, with naturally-transparent lakes exhibiting the greatest sensitivity to humans, but not the highest natural variability. Instead, authors concluded that local fires may have selectively impacted certain lakes, leading to reductions in export of photo-protective DOM from watersheds [cf., 128].

Paleoecological analyses can also be used to quantify the relative impacts of multiple stressors on lake production and structure. For example, Leavitt et al. [26] used annual fossil pigment profiles, 19 year-long historical records, and multivariate statistics to measure unique algal responses to changes in irradiance and lake chemistry during a whole-lake acidification experiment. Their analysis showed that 80–83% of historical variance in the abundance of fossil chlorophylls and carotenoids could be explained statistically by measured changes in lake properties. This study further demonstrated that while effects of pH accounted for 50% of change in algal communities, irradiance (12%) and its interactions with lake chemistry (20%) were significant, substantial and independent determinants of algal community variability. Specifically, increased penetration of PAR during initial stages of acidification stimulated growth of benthic and metalimnetic algal populations, whereas 8-fold increases in UV-B penetration during severe acidification phases (pH $<$ 4.5) greatly suppressed growth of sensitive taxa (e.g., chrysophytes) and lake production. Fossil analyses further demonstrated that the deposition of photo-protective pigments similar to scytonemin (e.g., C_a) exhibited a strong linear correlation with the depth of UVR penetration ($r^2 = 0.70$) when UVR-absorbing pigments were normalized to changes in total algal abundance [26]. Finally, by comparing fossil profiles among acidified lakes, the authors proposed a conceptual model to explain the high variability in ecosystem response to acidification; UVR suppressed primary production in clear lakes ($<$3 mg DOC l^{-1}) or those with severe acidification (pH $<$ 4.5), whereas increased PAR transmission during acidification stimulated algal growth in other instances.

In contrast to impacts of drought and acidic precipitation, recent research has suggested that terrestrial disturbance associated with forest fires or timber harvest may increase DOM export to lakes and reduce UVR penetration [128] so long as soils are not totally destroyed [13,125]. Consistent with this view, paleoecological analysis of a chain of lakes in the Great Plains of central North America suggested that UVR penetration declined 5-fold in upstream lakes, but not those further downstream, concomitant with development of regional agriculture [129]. Although patterns are consistent with a 50% decline in soil C content and the ability of sequential lake basins to trap and remove organic matter through sedimentation, interpretations of changes in UVR index require analysis of mixing regime of these polymictic lakes in order to distinguish whether declines in UVR exposure represent increased water-column DOM, or reductions in mean algal exposure to UVR due to lake eutrophication.

16.5 Conclusions

To date, paleoecological approaches to reconstruction of past UVR regimes and their impacts on lakes have demonstrated that changes in biological exposure to UVR caused by climate change are at least 10-fold greater than those associated with moderate (30%) ozone depletion [e.g., 15,24]. In all cases, climatic variability acts to change export of UVR-absorbing DOM from land, although the direction of lake response depends on whether warming reduces hydrologic export [cf., 125] or results in development of new C sources (e.g., soils; [47]). These patterns suggest that all glacial lakes have experienced severe UVR stress early in their evolution, and that this may be a key factor limiting early lake productivity. Similarly, preliminary comparison of historical irradiance regimes in both polar regions suggests that past UVR exposure is intrinsically variable due to both short-term processes and, potentially, millennium-long solar or atmospheric variability (Figures 5, 6). Finally, because human alteration of global C cycling and atmospheric processes is so pervasive [108], and because ecosystem and environmental monitoring efforts continue to decline, we foresee an expanded role for such retrospective analyses to identify past environmental change and to quantify the relative importance of multiple stressors on ecosystem structure and function [26].

Several avenues of future research may benefit from historical reconstructions of past UVR. For example, little is known of how long-term changes in irradiance impact trophic interactions in natural ecosystems [e.g., 6], yet most lake sediments archive a wide range of invertebrate and plant fossils that allow quantitative reconstruction of past grazer communities, predator–prey interactions, and even estimates of fish abundance [95]. Similarly, hydrated lime is frequently added to acidified lakes to mitigate impacts of low pH, yet this approach has long been known to greatly reduce water-column DOM concentration through adsorption onto carbonate particles [130]. Reconstruction of potential UVR impacts arising from the use of lime has not yet been attempted, but may provide critical information to explain unexpected delays in lake recovery from acidification. Similarly, quantitative analyses of the rate of change of fossil assemblages in response to natural and anthropogenic forcing is required to better formulate management plans for ecosystem protection. For example, preliminary analysis of declines in lake production following mid-Holocene climate change (Figure 3) suggests that past rates of ecosystem response ($>1\%\ yr^{-1}$) are similar to those forecast for future global warming [e.g., 125] and may provide unexpectedly good analogues for its impacts [27]. Thus, when combined with ecosystem models, long-term monitoring, ecosystem experiments and short-term manipulations, paleoecology provides modern ecologists and environmental scientists with key insights into the scale, causes and consequences of temporal variability in UVR.

Acknowledgements

This research was supported by NSERC grants to P.R.L., by NSERC and FCAR grants to R.P., and by NERC and ASAC grants to D.A.H. D.A.H. thanks Wim Vyverman, Elie Verleyen, Angela Squier, Brendan Keely and Pedro Montiel for many useful discussions. P.R.L. thanks Marianne Douglas and John Smol for sediments from Col Pond and for useful discussions on Arctic pond ecology. R.P. thanks Warwick Vincent for stimulating discussions on arctic limnology.

References

1. D. Karentz, M.L. Bothwell, R.B. Coffin, A. Hanson, G.L. Herndl, S.S. Kilham, M.P. Lesser, M. Lindell, R.E. Moeller, D.P. Morris, P.J. Neale, R.W. Sanders, C.S. Weiler, R.G. Wetzel (1994). Impact of UV-B radiation on pelagic freshwater ecosystems: Report of working group on bacteria and phytoplankton. *Arch. Hydrobiol. Ergebn. Limnol.*, **43**, 31–69.
2. C.E. Williamson, O.G. Olson, S.E. Lott, N.D. Walker, D.R. Engstrom, B.R. Hargreaves (2001). Ultraviolet radiation and zooplankton community structure following deglaciation in Glacier Bay, Alaska. *Ecology*, **82**, 1748–1760.
3. R.D. Vinebrooke, P.R. Leavitt (1999). Differential responses of littoral communities to ultraviolet radiation in an alpine lake. *Ecology*, **80**, 223–237.
4. F. Xiong, L. Nedbal, A. Neori (1999). Assessment of UV-B sensitivity of photosynthetic apparatus among microalgae: Short-term laboratory screening versus long-term outdoor exposure. *J. Plant Physiol.*, **155**, 54–62.
5. C.E. Williamson, P.J. Neale, G. Grad, H.J. De Lange, B.R. Hargreaves (2001). Beneficial and detrimental effects of UV on aquatic organisms: Implications of spectral variation. *Ecol. Appl.*, **11**, 1843–1857.
6. M.L. Bothwell, D.M.J. Sherbot, C.M. Pollock (1994). Ecosystem response to solar ultraviolet-B radiation: Influence of trophic-level interactions. *Science*, **265**, 97–100.
7. C.E. Williamson, S.L. Metzgar, P.A. Lovera, R.E. Moeller (1997). Solar ultraviolet radiation and the spawning habitat of yellow perch, *Perca flavescens*. *Ecol. Appl.*, **7**, 1017–1023.
8. C. Gutierrez-Rodriguez, C.E. Williamson (1999). Influence of solar ultraviolet radiation in early life-history stages of the bluegill sunfish, *Lepomis macrochirus*. *Environ. Biol. Fish.*, **55**, 307–319.
9. R. Smith, B.B. Prézelin, K.S. Baker, R.R. Bidigare, N.P. Boucher, T. Coley, D. Karentz, S. MacIntyre, H.A. Matlick, D. Menzies, M. Ondrusek, Z. Wan, K.J. Waters (1992). Ozone depletion: Ultraviolet radiation and phytoplankton biology in Antarctic waters. *Science*, **255**, 952–959.
10. N.M. Scully, D.R.S. Lean (1994). The attenuation of ultraviolet radiation in temperate lakes. *Arch. Hydrobiol. Ergebn. Limnol.*, **43**, 135–144.
11. D.P. Morris, H. Zagarese, C.E. Williamson, E.G. Balseiro, B.R. Hargreaves, B. Modenutti, R. Moeller, C. Queimalinos (1995). The attenuation of solar UV radiation in lakes and the role of dissolved organic carbon. *Limnol. Oceanogr.*, **40**, 1381–1391.
12. N.D. Yan, W. Keller, N.M. Scully, D.R.S. Lean, P.J. Dillon (1996). Increased UV-B penetration in a lake owing to drought-induced acidification. *Nature*, **381**, 141–143.
13. D.W. Schindler, P.J. Curtis, S.E. Bailey, B.R. Parker, K.G. Beaty, M.P. Stainton (1997). Climate-induced changes in the dissolved organic carbon budget of boreal lakes. *Biogeochemistry*, **36**, 9–28.
14. D.W. Schindler, P.J. Curtis, B.R. Parker, M.P. Stainton (1996). Consequences of climatic warming and lake acidification for UV-B penetration in North American boreal lakes. *Nature*, **379**, 705–708.

15. R. Pienitz, W.F. Vincent (2000). Effect of climate change relative to ozone depletion on UV exposure in subarctic lakes. *Nature*, **404**, 484–487.
16. C.S. Cockell (1999). Crises and extinction in the fossil record – a role for ultraviolet radiation? *Paleobiology*, **25**, 212–225.
17. C.S. Cockell, A.R. Blaustein (2000). "Ultraviolet spring" and the ecological consequences of catastrophic impacts. *Ecol. Lett.*, **3**, 77–81.
18. M.A. Xenopoulos, Y.T. Prairie, D.F. Bird (2000). Influence of ultraviolet-B radiation, stratospheric ozone variability, and thermal stratification on the phytoplankton biomass dynamics in a mesohumic lake. *Can. J. Fish. Aquat. Sci.*, **57**, 600–609.
19. J.P. Smol, B.F. Cumming (2000). Tracking long-term changes in climate using algal indicators in lake sediments. *J. Phycol.*, **36**, 986–1011.
20. E.F. Stoermer, J.P. Smol (1999). *The Diatoms: Applications for the Environmental and Earth Sciences.* Cambridge University Press, Cambridge.
21. W.F. Vincent, R. Pienitz (1996). Sensitivity of high-latitude freshwater ecosystems to global change: temperature and solar ultraviolet radiation. *Geosci. Canada*, **23**, 231–236.
22. M-A. Fallu, R. Pienitz (1999). Diatomées lacustres de Jamésie-Hudsonie (Québec) et modèle de reconstitution des concentrations de carbone organique dissous. *Ecoscience*, **6**, 603–620.
23. S.S. Dixit, W. Keller, A.S. Dixit, J.P. Smol (2001). Diatom-inferred dissolved organic carbon reconstructions provide assessments of past UV-B penetration in Canadian Shield lakes. *Can. J. Fish. Aquat. Sci.*, **58**, 543–550.
24. P.R. Leavitt, R.D. Vinebrooke, D.B. Donald, J.P. Smol, D.W. Schinder (1997). Past ultraviolet radiation environments in lakes derived from fossil pigments. *Nature*, **388**, 457–459.
25. É. Saulnier-Talbot, R. Pienitz, W.F. Vincent (2003). Holocene lake succession and palaeo-optics of a subartic lake, northern Québec (Canada). *Holocene*, **13**, in press.
26. P.R. Leavitt, D.L. Findlay, R.I. Hall, J.P. Smol (1999). Algal responses to dissolved organic carbon loss and pH decline during whole-lake acidification: Evidence from paleolimnology. *Limnol. Oceanogr.*, **44**, 757–773.
27. P.R. Leavitt, B.F. Cumming, J.P. Smol, M. Reasoner, R. Pienitz, D. Hodgson (submitted). Climatic control of UV radiation impacts on lakes. *Geology*.
28. D.R. Engstrom, S.C. Fritz, J.E. Almendinger, S. Juggins (2000). Chemical and biological trends during lake evolution in recently deglaciated terrain. *Nature*, **408**, 161–166.
29. C.S. Cockell (2000). The ultraviolet history of the terrestrial planets – implications for biological evolution. *Planet. Space Sci.*, **48**, 203–214.
30. J.A.E. Stephenson, M.W.J. Scourfield (1991). Importance of energetic solar protons in ozone depletion. *Nature*, **352**, 137–139.
31. J.K. Angell, J. Korshover (1973). Quasi-biennial and long-term fluctuations in total ozone. *Month. Weather Rev.*, **101**, 426–443.
32. S. Madronich, R.L. McKenzie, M.M. Caldwell, L.O. Björn (1995). Changes in ultraviolet radiation reaching the earth's surface. *Ambio*, **24**, 143–152.
33. J.R. Herman, R. McPeters, D. Larko (1993). Ozone depletion at northern and southern latitudes derived from January 1979 to December 1991 total ozone mapping spectrometer data. *J. Geophys. Res.*, **98**, 12783–12793.
34. World Meteorological Organization (1999). *Scientific Assessment of Ozone Depletion*: 1998. WMO, Geneva.
35. C. Belizile, W.F. Vincent, J.A.E. Gibson, P. Van Hove (2001). Bio-optical characteristics of the snow, ice and water column of a perennially ice-covered lake in the High Arctic. *Can. J. Fish. Aquat. Sci.*, **58**, 2405–2418.
36. C.E. Williamson, R.S. Stemberger, D.P. Morris, T.M. Frost, S.G. Paulsen (1996).

Ultraviolet radiation in North American lakes: Attenuation estimates from DOC measurements and implications for plankton communities. *Limnol. Oceanogr.*, **41**, 1024–1034.
37. I. Laurion, W.F. Vincent, D.R.S. Lean (1997). Underwater ultraviolet radiation: Development of spectral models for northern high latitude lakes. *Photochem. Photobiol.*, **65**, 107–114.
38. R. Sommaruga, R. Psenner, E. Schafferer, K.A. Koinig, S. Sommaruga-Wögrath (1999). Dissolved organic carbon concentration and phytoplankton biomass in high-mountain lakes of the Austrian Alps: Potential effect of climatic warming on UV underwater attenuation. *Arc. Antarc. Alp. Res.*, **31**, 247–253.
39. D. Conde, L. Aubriot, R. Sommaruga (2000). Changes in UV penetration associated with marine intrusions and freshwater discharge in a shallow coastal lagoon of the southern Atlantic Ocean. *Mar. Ecol. Prog. Ser.*, **207**, 19–31.
40. I. Laurion, M. Ventura, J. Catalan, R. Psenner, R. Sommaruga (2000). Attenuation of ultraviolet radiation in mountain lakes: Factors controlling the among- and within-lake variability. *Limnol. Oceanogr.*, **45**, 1274–1288.
41. M.T. Arts, R.D. Roberts, F. Kusai, W.J. Waiser, V.P. Tumber, A.J. Plante, H. Rai, H.J. de Lange (2000). The attenuation of ultraviolet radiation in high dissolved organic carbon waters of wetlands and lakes on the northern Great Plains. *Limnol. Oceanogr.*, **45**, 292–299.
42. W.F. Donahue, D.W. Schindler, S.J. Page, M.P. Stainton (1998). Acid-induced changes in DOC quality in an experimental whole-lake manipulation. *Environ. Sci. Technol.*, **32**, 2954–2960.
43. S. Markager, W.F. Vincent (2000). Spectral light attenuation and the absorption of UV and blue light in natural waters. *Limnol. Oceanogr.*, **45**, 642–650.
44. D.M. McKnight, E.D. Andrews, R.L. Smith, R. Dufford (1994). Aquatic fluvic acids in algal-rich Antarctic ponds. *Limnol. Oceanogr.*, **39**, 1972–1979.
45. R.I. Hall, P.R. Leavitt, J.P. Smol, N. Zirnhelts (1997). Comparison of diatoms, fossil pigments and historical records as measures of lake eutrophication. *Freshwat. Biol.*, **38**, 401–417.
46. M.A. Xenopoulos, D.W. Schindler (2001). The environmental control of near-surface thermoclines in boreal lakes. *Ecosystems*, **4**, 699–707.
47. R. Sommaruga (2001). The role of solar UV radiation in the ecology of alpine lakes. *J. Photochem. Photobiol. B: Biol.*, **62**, 35–42.
48. W.F. Vincent, P.J. Neale, P.J. Richardson (1984). Photoinhibition: Algal responses to bright light during diel stratification and mixing in a tropical alpine lake. *J. Phycol.*, **20**, 201–211.
49. E.M. Watkins, D.W. Schindler, M.A. Turner, D. Finlay (2001). Effects of solar ultraviolet radiation on epilithic metabolism, and nutrient and community composition in a clear-water boreal lake. *Can. J. Fish. Aquat. Sci.*, **58**, 2059–2070.
50. R.G. Zepp, T.V. Callaghan, D.J. Erikson (1998). Effects of enhanced solar ultraviolet radiation on biogeochemical cycles. *J. Photochem. Photobiol. B: Biol.*, **46**, 69–82.
51. M.A. Xenopoulos, D.F. Bird (1997). Effect of acute exposure to hydrogen peroxide on the production of phytoplankton and bacterioplankton in a mesohumic lake. *Photochem. Photobiol.*, **65**, 471–478.
52. F.S. Xiong, J. Komenda, J. Kopecky, L. Nedbal (1997). Strategies of ultraviolet-B protection in microscopic algae. *Physiol. Plant.*, **100**, 378–388.
53. S.W. Jeffrey, H.S. MacTavish, W.C. Dunlap, M. Vesk, K. Groenewoud (1999). Occurrence of UVA- and UVR-absorbing compounds in 152 species (206 strains) of marine microalgae. *Mar. Ecol. Prog. Ser.*, **189**, 35–51.

54. W.F. Vincent, S. Roy (1993). Solar ultraviolet-B radiation and aquatic primary production: Damage, protection, and recovery. *Environ. Rev.*, **1**, 1–12.
55. W.F. Vincent, R.W. Castenholtz, M.T. Downes, C. Howard-Williams (1993). Antarctic cyanobacteria: Light, nutrients, and photosynthesis in the microbial mat environment. *J. Phycol.*, **29**, 745–755.
56. F. Garcia-Pichel, M. Mechling, R.W. Castenholtz (1994). Diel migrations of microorganisms within a benthic, hypersaline mat community. *Appl. Environ. Microbiol.*, **60**, 1500–1511.
57. R.D. Vinebrooke, P.R. Leavitt (1998). Direct and interactive effects of allochthonous dissolved organic matter, inorganic nutrients, and ultraviolet radiation on an alpine littoral food web. *Limnol. Oceanogr.*, **43**, 1065–1081.
58. D.M. Leech, C.E. Williamson (2000). Is tolerance to UV radiation in zooplankton related to body size, taxon, or lake transparency. *Ecol. Appl.*, **10**, 1530–1540.
59. D.L. Wubben, E. Vareschi, I. Hawes, G. Bodeker (2001). Differential UVB-sensitivities of five New Zealand freshwater zooplankton species. *NZ J. Mar. Freshwat. Res.*, **35**, 635–645.
60. B. Tartarotti, I. Laurion, R. Sommaruga (2001). Large variability in the concentration of mycosporine-like amino acids among zooplankton from lakes located across an altitude gradient. *Limnol. Oceanogr.*, **46**, 1546–1552.
61. G. Grad, C.E. Williamson, D.M. Karapelou (2001). Zooplankton survival and reproduction responses to damaging UV radiation: a test of reciprocity and photoenzymatic repair. *Limnol. Oceanogr.*, **46**, 584–591.
62. D.M. Leech, C.E. Williamson (2001). In situ exposure to ultraviolet radiation alters the depth distribution of *Daphnia. Limnol. Oceanogr.*, **46**, 416–420.
63. R. Rae, D. Hanelt, I. Hawes (2001). Sensitivity of freshwater macrophytes to UV radiation: Relationship to depth zonation in an oligotrophic New Zealand lake. *Mar. Freshwat. Biol.*, **52**, 1023–1032.
64. C.S. Cockell, J. Knowland (1999). Ultraviolet radiation screening compounds. *Biol. Rev.*, **74**, 311–345.
65. A. McMinn, H. Heijnis, D. Hodgson (1994). Minimal effects of UVB radiation on Antarctic diatoms over the past 20 years. *Nature*, **370**, 547–549.
66. D.W. Schindler (1998). A dim future for boreal waters and landscapes. *Bioscience*, **48**, 157–164.
67. W.M. Last, J.P. Smol (2001). *Tracking Environmental Change using Lake Sediments. Volume* 1: *Basin Analysis, Coring, and Chronological Techniques*. Kluwer Academic Press, The Netherlands.
68. P.R. Leavitt, D. Hodgson (2001). Sedimentary pigments. In: J.P. Smol, H.J.B. Birks, W.M. Last (Eds), *Tracking Environmental Change using Lake Sediments Volume* 3: *Terrestrial, Algal, and Siliceous Indicators* (pp. 295–325). Kluwer, The Netherlands.
69. H.J.B. Birks (1995). Quantitative palaeoenvironmental reconstructions. In: D. Maddy, J.S. Brew (Eds), *Statistical Modelling of Quaternary Science Data* (Technical Guide 5, pp. 161–254). Cambridge University Press, Cambridge.
70. R. Pienitz. J.P. Smol (1993). Diatom assemblages and their relationship to environmental variables in lakes from the boreal forest-tundra ecotone near Yellowknife, Northwest Territories, Canada. *Hydrobiologia*, **269/270**, 391–404.
71. J. Weckström, A. Korhola, T. Blom (1997). The relationship between surface-water temperature and diatom assemblages in 30 subarctic lakes from northern Fennoscandia: A potential tool for paleotemperature reconstructions. *Arc., Alpine Res.*, **29**, 75–92.
72. T.E. Laing, J.P. Smol (2000). Factors influencing diatom distributions in circumpolar

treeline lakes of northern Russia. *J. Phycol.*, **36**, 1035–1048.
73. T. Korsman, H.J.B. Birks (1996). Diatom-based water chemistry reconstructions from northern Sweden: A comparison of reconstruction techniques. *J. Paleolimnol.*, **15**, 65–77.
74. H. Seppä, H. Weckström (1999). Holocene vegetational and limnological changes in the Fennoscandian tree-line area as documented by pollen and diatom records from Lake Tsuolbmajavri, Finland. *Ecoscience*, **6**, 621–635.
75. P. Rosen, R.I. Hall, T. Korsman, I. Renberg (2000). Diatom transfer functions for quantifying past air temperature, pH, and total organic carbon concentration from lakes in northern Sweden. *J. Paleolimnol.*, **24**, 109–123.
76. D.R. Engstrom (1987). Influence of vegetation and hydrology on the humus budgets of Labrador lakes. *Can. J. Fish Aquat. Sci.*, **44**, 1306–1314.
77. C.S. Forsberg (1992). Will an increased greenhouse impact in Fennoscandia give rise to more humic and coloured lakes? *Hydrobiologia*, **229**, 51–58.
78. R.G. Wetzel (1992). Gradient-dominated ecosystems: Sources and regulatory functions of dissolved organic matter in freshwater ecosystems. *Hydrobiologia*, **229**, 181–198.
79. A.F. Lotter, R. Pienitz, R. Schmidt (1999). Diatoms as indicators of environmental change near arctic and alpine treeline. In: E.F. Stoermer, J.P. Smol (Eds), *The Diatoms: Applications for the Environmental and Earth Sciences* (pp. 205–226). Cambridge University Press, Cambridge.
80. R. Pienitz, J.P. Smol, H.J.B. Birks (1995). Assessment of freshwater diatoms as quantitative indicators of past climatic change in the Yukon and Northwest Territories, Canada. *J. Paleolimnol.*, **13**, 21–49.
81. P. Legendre, L. Legendre (1998). *Numerical Ecology*. Elsevier Science, Amsterdam.
82. C.J. F. ter Braak, P. Similauer (1998). *CANOCO v.* 4.0. Microcomputer Power, Ithaca, NY.
83. H.J.B. Birks (1990). Diatoms and pH reconstructions. *Phil. Trans. Roy. Soc. London B*, **327**, 263–278.
84. R.I. Hall, J.P. Smol (1999). Diatoms as indicators of lake eutrophication. In: E.F. Stoermer, J.P. Smol (Eds), *The Diatoms: Applications for the Environmental and Earth Sciences* (pp. 128–168). Cambridge University Press, Cambridge.
85. W.F. Vincent, I. Laurion, R. Pienitz (1998). Arctic and Antarctic lakes as optical indicators of global change. *Ann. Glaciol.*, **27**, 691–696.
86. R.B. Setlow (1974). The wavelengths in sunlight effective in producting skin cancer: A theoretical analysis. *Proc. Nat. Acad. Sci. U.S.A.*, **71**, 3363–3366.
87. J.J. Cullen, P.J. Neale, M.P. Lesser (1992). Biological weighting function for the inhibition of phytoplankton photosynthesis by ultraviolet radiation. *Science*, **258**, 646–650.
88. N.J. Anderson (1998). Variability of diatom-inferred phosphorus profiles in a small lake basin and its implications for histories of lake eutrophication. *J. Paleolimnol.*, **20**, 47–55.
89. R. Pienitz, J.P. Smol, G.M. MacDonald (1999). Paleolimnological reconstruction of Holocene climatic trends from two boreal treeline lakes, Northwest Territories, Canada. *Arc. Antarc. Alp. Res.*, **31**, 82–93.
90. M.S.V. Douglas, J.P. Smol (1987). Siliceous protozoan plates in lake sediments. *Hydrobiologia*, **154**, 13–23.
91. D.R. Piperno (2001). Phytoliths. In: J.P. Smol, H.J.B. Birks, W.M. Last (Eds), *Tracking Environmental Change using Lake Sediments. Volume* 3: *Terrestrial, Algal, and Siliceous Indicators* (pp. 235–251). Kluwer, The Netherlands.

92. P. Rosen, E. Dabakk, I. Renberg, M. Nilsson, R.I. Hall (2000). Near-infrared spectrometry (NIRS): a new tool for inferring past climatic changes from lake sediments. *Holocene*, 161–166.
93. P.D.N. Hebert, C.J. Emery (1990). The adaptive significance of cuticular pigmentation in *Daphnia. Funct. Ecol.*, **4**, 703–710.
94. D.O. Hessen, J. Borgeraas, K. Kessler, U.H. Refseth (1999). UV-B susceptibility and photoprotection of arctic *Daphnia* morphotypes. *Polar Res.*, **18**, 345–352.
95. E. Jeppesen, P. Leavitt, L. DeMeester, J.P. Jensen (2001). Functional ecology and palaeolimnology: Using cladoceran remains to reconstruct anthropogenic impact. *Trends Ecol. Evol.*, **16**, 191–198.
96. F. Garcia-Pichel, R.W. Castenholtz (1991). Characterization and biological implications of scytonemin, a cyanobacterial sheath pigment. *J. Phycol.*, **27**, 395–409.
97. R. Sommaruga, F. Garcia-Pichel (1999). UV-absorbing mycosporine-like compounds in planktonic and benthic organisms from a high-mountain lake. *Arch. Hydrobiol.*, **144**, 255–269.
98. K. Cuddington, P.R. Leavitt (1999). An individual-based model of pigment flux in lakes: Implications for organic biogeochemistry and paleoecology. *Can. J. Fish. Aquat. Sci.*, **56**, 1964–1977.
99. D.A. Hodgson, S.W. Wright, N. Davies (1997). Mass spectrometry and reverse phase HPLC methods for the identification of degraded fossil pigments in lake sediments and their application in palaeolimnology. *J. Paleolimnol.*, **18**, 335–350.
100. R.F.C. Mantoura, C.A. Llewellyn (1983). The rapid determination of algal chlorophyll and carotenoid pigments and their breakdown products in natural waters by reversed-phase high-performance liquid chromatography. *Anal. Chim. Acta*, **151**, 297–314.
101. P.R. Leavitt, D.L. Findlay (1994). Comparison of fossil pigments with 20 years of phytoplankton data from eutrophic Lake 227, Experimental Lakes Area, Ontario. *Can. J. Fish. Aquat. Sci.*, **51**, 2286–2299.
102. S.W. Wright, S.W. Jeffrey, R.F.C. Mantoura, C.A. Llewellyn, T. Bjørnland, D. Repeta, N.A. Welschmeyer (1991). Improved HPLC method for analysis of chlorophylls and carotenoids from marine phytoplankton. *Mar. Prog. Ecol. Ser.*, **77**, 183–196.
103. S.W. Jeffrey, S.W. Wright, M. Zapata (1999). Recent advances in HPLC pigment analysis of phytoplankton. *Mar. Freshwat. Res.*, **50**, 879–896.
104. R.L. Airs, J.E. Atkinson, B.J. Keely (2001). Development and application of high resolution liquid chromatographic method for the analysis of complex pigment distributions. *J. Chromatogr. A*, **917**, 167–177.
105. A.H. Squier (2000). *Fossil Pigments as Marker for Environmental Change in Antarctic Lakes* (MSc. Thesis). Department of Chemistry, University of York, UK.
106. W.E. Dean (1974). Determination of carbonate and organic matter in calcareous sediments and sedimentary rocks by loss on ignition: Comparison with other methods. *J. Sediment Petrol.*, **44**, 242–248.
107. D.J. Rowan, J. Kalff, J.B. Rasmussen (1992). Profundal sediment organic matter concentration and physical character do not reflect lake trophic status, but rather reflect inorganic sedimentation and exposure. *Can. J. Fish. Aquat. Sci.*, **49**, 1431–1438.
108. D.W. Schindler (2001). The cumulative effects of climate warming and other human stresses on Canadian freshwaters in the new millennium. *Can. J. Fish. Aquat. Sci.*, **58**, 18–29.
109. G.M. MacDonald, T.W.D. Edwards, K.A. Moser, R. Pienitz, J.P. Smol (1993). Rapid

response of treeline vegetation and lakes to past climate warming. *Nature*, **361,** 243–246.

110. M.A. Reasoner, G. Osborn, N.W. Rutter (1994). Age of the Crowfoot advance in the Canadian Rocky Mountains: A glacial event coeval with the Younger Dryas Oscillation. *Geology*, **22**, 439–442.
111. M.A. Reasoner, U.M. Huber (1999). Postglacial palaeoenvironments of the upper Bow Valley, Banff National Park, Alberta, Canada. *Quat. Sci. Rev.*, **18**, 475–492.
112. S.C. Fritz, D.R. Engstrom (1995). Patterns of early lake ontogeny in Glacier Bay as inferred from diatom assemblages. In: D.R. Engtrom (Ed.), *Proceedings of the Third Glacier Bay Symposium*, 1993 (pp. 147–151). National Park Service, Washington, D.C.
113. R. Stolarski. R. Bojkov, L. Bishop, C. Zerefos, J. Staehelin, J. Zawodny (1992). Measured trends in stratospheric ozone. *Science*, **256**, 342–349.
114. M.S.V. Douglas, J.P. Smol (1994). Limnology of high arctic ponds (Cape Herschel, Ellesmere Island, N.W.T.). *Arch. Hydrobiol.*, **131**, 401–434.
115. M.S.V. Douglas, J.P. Smol, W. Blake Jr. (1994). Marked post-18th century environmental change in high arctic ecosystems. *Science*, **266**, 416–419.
116. D.A. Hodgson P.E. Noon, W. Vyverman, C.L. Bryant, D.B. Gore, P. Appleby, M. Gilmour, E. Verleyen, K. Sabbe, V.J. Jones, J.C. Ellis-Evans, P.B. Wood (2001). Were the Larsemann Hills ice free through the Last Glacial Maximum? *Antarc. Sci.*, **13**, 440–454.
117. J.C. Ellis-Evans, J. Laybourn-Parrry, P. Bayliss, S.J. Perriss (1998). Physical, chemical and microbial community characteristics of lakes of the Larsemann Hills, Continental Antarctica. *Arch. Hydrobiol*, **141**, 209–230.
118. W.F. Vincent, A. Quesada (1994). Ultraviolet radiation effects on cyanobacteria: Implications for Antarctic microbial ecosystems. In: C.S. Weiler, P.A. Penhale (Eds), *Ultraviolet Radiation in Antarctica: Measurement and Biological Effects* (Antarc. Res. Ser., Vol. **62**, pp. 111–124). American Geophysical Union, Washington, DC.
119. M. Ehling-Schulz, W. Bilger, S. Scherer (1997). UV-B-induced synthesis of photoprotective pigments and extracellular polysaccharides in the terrestrial cyanobacterium *Nostoc commune*. *J. Bacteriol.*, **179**, 1940–1945.
120. B. Demmig-Adams, W.W. Adams (2000). Photosynthesis – Harvesting sunlight safely. *Nature*, **403**, 371–374.
121. J.A. Wolin, H.C. Duthie (1999). Diatoms as indicators of water level change in freshwater lakes. In: E.F. Stoermer, J.P. Smol (Eds), *The Diatoms: Applications for the Environmental and Earth Sciences* (pp. 183–202). Cambridge University Press, Cambridge.
122. W. Blake Jr. (1992). Holocene emergence at Cape Herschel, east-central Ellesmere Island, Arctic Canada: Implications for ice sheet configuration. *Can. J. Earth Sci.*, **29**, 1958–1980.
123. R.S. Bradley (1990). Holocene palaeoclimatology of the Queen Elizabeth Islands, Canadian High Arctic. *Quat. Sci. Rev.*, **9**, 365–384.
124. S.R. O'Brien, P.A. Mayewski, L.D. Meeker, D.A. Meese, M.S. Twickler, S.I. Whitlow (1995). Complexity of Holocene climate as reconstructed from a Greenland ice core. *Science*, **270**, 1962–1964.
125. D.W. Schindler S.E. Bayley, B.R. Parker, K.G. Beaty, D.R. Cruikshank, E.J. Fee, E.U. Schindler, M.P. Stainton (1996). The effects of climatic warming on the properties of boreal lakes and streams at the Experimental Lakes Area, northwestern Ontario. *Limnol. Oceanogr.*, **41**, 1004–1017.
126. M. Scheffer (2001). Climatic warming causes regime shifts in lake food webs. *Limnol.*

Oceanogr., **46**, 1780–1783.

127. K.R. Laird, S.C. Fritz, K.A. Maasch, B.F. Cumming (1996). Greater drought intensity and frequency before AD1200 in the Northern Great Plains, USA. *Nature*, **384**, 552–554.
128. R. Carignan, P. D'Arcy, S. Lamontagne (2000). Comparative impacts of fire and forest harvesting on water quality in boreal shield lakes. *Can. J. Fish. Aquat. Sci.*, **57**(suppl.2), 105–117.
129. A.S. Dixit, R.I. Hall, P.R. Leavitt, R. Quinlan, J.P. Smol (2000). Effects of sequential depositional basins on lake response to urban and agricultural pollution: A palaeoecological analysis of the Qu'Appelle Valley, Saskatchewan Canada. *Freshwat. Biol.*, **43**, 319–337.
130. A.D. Hasler, O.M. Brynildson, W.T. Helm (1951). Improving conditions for fish in brown-water lakes by alkalization. *J. Wildlife Manag.*, **15**, 347–352.

Chapter 17

UVR effects on aquatic ecosystems: a changing climate perspective

Craig E. Williamson and Horacio E. Zagarese

Table of contents

Abstract

Only recently has the potential for strong interactions between climate change and UV effects on aquatic ecosystems been recognized. Chromophoric dissolved organic matter (CDOM) largely mediates these interactions. CDOM influences water transparency and the attenuation of both visible and UVR in the water column. It also influences thermal gradients, mixing processes, and the timing and extent of ice cover. Here we review existing knowledge of these processes and present some conceptual diagrams to illustrate our current understanding of how the key processes interact. The general conclusion from numerous studies is that, for inland waters, climate change and consequent changes in CDOM are likely to have a greater effect on underwater UV environments than will stratospheric ozone depletion. Recent quantitative modeling efforts are also briefly reviewed. Sensitivity testing with these models indicates that mixing depth and water column attenuation of UV are at least as important as ozone depletion in regulating underwater UV exposures. Ozone depletion may be of greater importance for smaller organisms such as bacteria that are less protected from UV damage. Variations in ozone may also be of relatively greater importance in marine environments where CDOM concentrations are low and UV transparency high.

17.1 Introduction

One of the most challenging questions in environmental biology today is how human-accelerated changes in multiple environmental variables are influencing natural communities and ecosystems. One of the more pervasive of these changes is the increase in UVR related to stratospheric ozone depletion. In recent years it has become increasingly clear that climate change may have at least as great an effect as ozone depletion on future changes in underwater UV environments in mid to high latitude freshwater ecosystems [1–4]. Here we briefly summarize previous studies that deal with interactions between climate change and UV effects on aquatic ecosystems, and discuss some of the quantitative modeling approaches that are being used to look at UV impacts. Photochemical transformations in DOM that play such a key role in climate change–UV interactions are dealt with in more detail in another chapter in this volume (Chapter 6).

17.2 Climate change and ozone depletion: A brief global synopsis

The Antarctic ozone hole was the largest ever in 2000 (>28 million km^2, http://pao.gsfc.nasa.gov/gsfc/EARTH/environ/ozone/ozone.htm). Significant trends of increasing UV-B related to ozone depletion have also been observed at Arctic and north temperate latitudes [5–7]. Modeling efforts estimate that localized losses of column ozone may be as high as two-thirds of the Arctic ozone column in the years 2010–2019. Atmospheric processes related to climate change are also influencing stratospheric ozone depletion. In particular, greenhouse gases are expected to enhance radiative cooling of the lower stratosphere and increase stability of the Arctic polar vortex, creating conditions that will further aggravate ozone depletion [6,8]. Recovery from ozone depletion and elevated UV is likely to be slow over the next half century [9].

At the same time that stratospheric ozone has been depleted, there is evidence that increases in tropospheric ozone and other pollutants in the lower levels of the atmosphere have contributed to decreases in incident UVR in highly populated and industrialized urban areas [10–13].

17.3 CDOM as a mediator of climate–UV interactions

In addition to the effects of climate change on ozone depletion, it is now widely recognized that climate change can influence the UV transparency of inland waters by changing the concentration and characteristics of UV-absorbing compounds. This means that underwater UV environments will continue to change even if ozone depletion is arrested or reversed. Climate change effects on underwater UV are largely mediated by CDOM, the light-absorbing component of the yellow–brown "gelbstoff" that results from the incomplete decomposition of living organisms. CDOM is a useful metric of ecosystem level processes in inland waters [14] and it exhibits strong variation in its chemical composition and spectral slope related to variation in source and prior exposure to microbial or photolytic breakdown [15,16]. This complex nature of CDOM has led to a variety of both optical and chemical methods of quantification. The simplest optical characterization of CDOM involves measuring the absorbance of a water sample at a given wavelength (320 and 440 nm have recently been proposed) in a laboratory spectrophotometer [14,17]. The simplest chemical measurement involves measuring DOC. Recent advances in methods of chemical characterization can also provide useful information on CDOM source [18,19] as discussed further below and in Chapter 6.

The bulk of the CDOM in inland waters is generally derived from allochthonous sources in the surrounding catchment areas or littoral zone [20]. Plant matter from terrestrial ecosystems and wetlands in particular are of primary importance [21–23], along with the type of vegetation and hydrology of the catchment area [24–26]. Seasonal variations in hydrology that may be influenced by climate change are also important in determining CDOM inputs as indicated by strong peaks in CDOM in rivers just preceding the peak of the spring flow [27]. Seasonal increases in CDOM in rivers can lead to an increase in heterotrophic microbial activity even at very low temperatures, and increase bacterial biomass [28]. These spring pulses of CDOM can also depress the pH of river waters and contribute to mobilization of aluminium as well as changes in aluminium speciation [29]. The mechanism of action of CDOM in controlling the UV environment in surface waters is largely through its selective absorption of the shorter, more damaging wavelengths of UV [30,31]. The UV absorptivity and the elemental composition of CDOM change with the source of the DOC and the chemical characteristics of the fulvic acid [28,32]. In particular, the C : N ratio of the fulvic acid may vary from < 10 to > 90, and algal-derived fulvic acid tends to be less aromatic and thus less color-absorbing than terrestrially-derived fulvic acid [28,32]. Recent advances in fluorometric characterization allow these sources of fulvic acids to be distinguished by comparing emission intensity at 450 and 500 nm during excitation with 370 nm [19].

CDOM also influences the underwater UV environment by altering temperature and water mixing depths (see section 17.3.2). In the process it may also alter other important ecosystem processes such as CDOM biolability (lability to microbial processing) and photolability (lability to photolysis), and bacterial biomass and growth efficiency (Figure 1, see also Chapter 6). This central role of CDOM in regulating water transparency in inland waters is in contrast to the open oceans where terrigenous CDOM is at low concentrations, and transparency is regulated largely by biogenic sources in the water column [33,34]. Although coastal and estuarine regions may be highly influenced by terrigenous CDOM from runoff and river plumes that may be altered by climate change, much of the CDOM is precipitated or biodegraded when it reaches the sea water [31,35].

17.3.1 DOC and water transparency

Several studies have clearly demonstrated a strong predictive empirical relation-

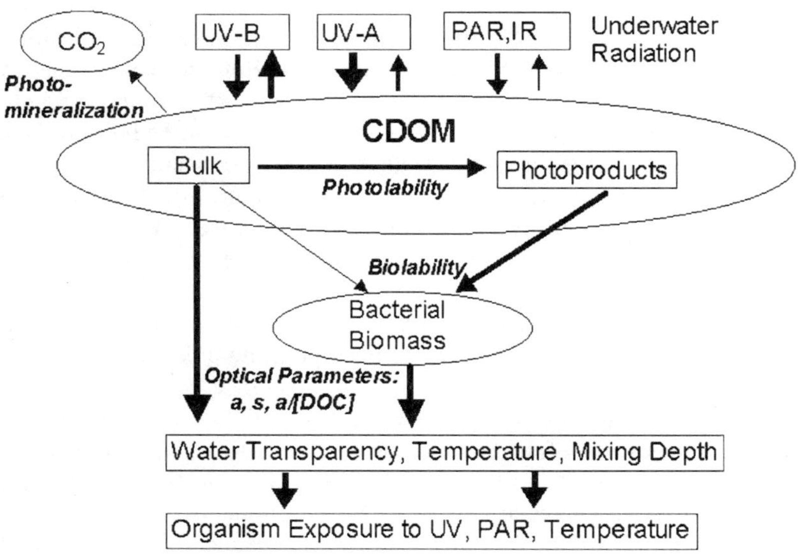

Figure 1. Conceptual diagram showing the role of CDOM in regulating solar radiation and temperature regimes in aquatic ecosystems, as well as the processes by which UV alters CDOM biolability, photolability, and photomineralization. The width of the arrows approximates the strength of the effects.

ship between DOC concentration and underwater UV transparency wherein DOC accounts for approximately 90% of the variation in UV attenuation [15,23,36,37]. The strong relationship between terrestrially derived DOC and UV attenuation is supported by studies along a chronosequence of recently deglaciated lakes in Glacier Bay Alaska [38], and by paleolimnological studies of fossil algal pigments [4,39] (see Chapter 16). UV attenuation depths (1% of surface irradiance at a given wavelength) estimated from DOC concentrations across North America vary greatly among geographic regions (see also Chapter 3). For some regions in the western and northwestern U.S.A. and Newfoundland, 25% of the lakes are estimated to have 320 nm attenuation depths of 4 m or more [2]. In other regions, such as Florida, the upper Midwestern USA, and Nova Scotia, 75% of the lakes have estimated 320 nm attenuation depths less than 0.5 m. Streams and rivers with low CDOM concentrations may be even more vulnerable to climate change than are lakes due to the fact that they are generally much shallower [2].

There is a general consensus that climate change and other factors that influence CDOM concentrations in streams, rivers, lakes, and coastal marine systems will have a greater effect than ozone depletion on future underwater UV environments [2,4,40,41]. However, to date no comprehensive studies have actually documented changes in underwater UV transparency related to climate change. Rather, they examine changes or differences in DOC and infer UV changes from empirical relationships between DOC and UV. However, a space-for-time substitution study was carried out in lakes of different age along a deglaciation chronosequence in Glacier Bay Alaska where UV as well as DOC and biotic response were directly measured. This study revealed a relationship between the timing of changes in terrestrial inputs of DOC from the surrounding watershed, UV transparency of the water, and zooplankton community structure over time (Figure 2) [38]. Very young, recently deglaciated lakes with little terrestrial vegetation in their watersheds have low DOC concentrations and high UV penetration into the water column, with consequent low diversity of zooplankton. As lakes age, terrestrial inputs of CDOM increase, UV transparency of the water decreases, and zooplankton communities become correspondingly more diverse (Figure 2). While the deglaciation that has occurred in Glacier Bay is more related to regional hydrologic balance than to global climate change, similar deglaciation events may be expected in alpine and polar regions if climate warming continues.

A major mechanism thought to link CDOM to climate change is drought, which reduces the export of terrestrially-derived CDOM to aquatic ecosystems [3,43,44]. Drought may also influence water export from wetlands or peatlands that are an important determinant of CDOM inputs to other downstream aquatic ecosystems [21–25]. Acidification can also reduce CDOM, as well as increase toxic metals [3,42,45]. When DOC is low (<1–$2\ \mathrm{mg\,l^{-1}}$) UV transparency increases very rapidly with declines in DOC (Figure 3). This close relationship between DOC and water transparency in low DOC systems suggests that water transparency in these is a particularly sensitive indicator of climate change, acidification, and other anthropogenic stresses in high latitude ecosystems

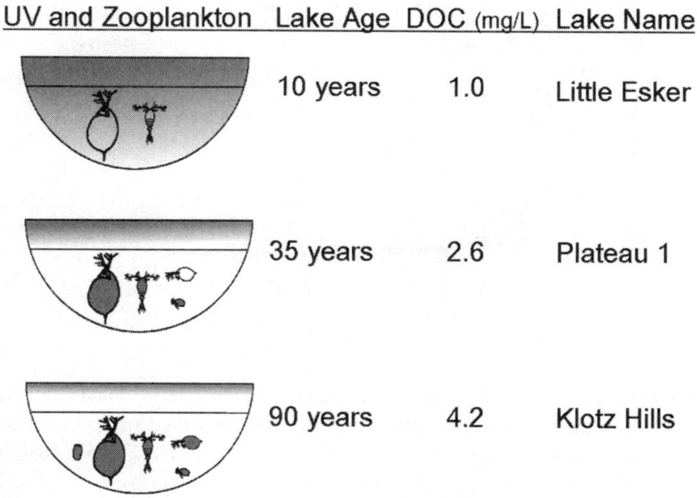

Figure 2. Diagrammatic representation of the UV exposure and low temperature constraints placed on zooplankton at different depths in lakes of different ages in a deglaciation chronosequence of lakes in Glacier Bay, Alaska [38]. The presence of 320 nm UV from the surface to the 1% (of surface irradiance) UV attenuation depth is indicated by shading of the lake basin, while the depth of the thermocline ($<1\,^{\circ}$C change m^{-1}) is indicated by a horizontal line. The presence of the zooplankton species in the lakes is shown diagrammatically where filled organisms represent established populations (>1 l^{-1}), and unfilled organisms represent sparse populations (<1 l^{-1}). All three lakes have thermal gradients that offer a demographic advantage to zooplankton that reside in the surface waters. In the oldest lake, where all five zooplankton species have established populations, potentially damaging UVR is attenuated rapidly, providing a potential refuge from both constraints of cold temperatures and high UV in the warm surface waters. In lakes of intermediate age, zooplankton are faced with either high UV exposure in the warm surface waters, or lower temperatures in the deeper, low-UV strata. In the youngest lake, where only two species exist in very low numbers, substantial UV exposure occurs throughout most of the water column.

[22,46,47].

In high elevation lakes where terrestrial vegetation is often highly reduced, and productivity low due to a combination of low temperatures, low nutrient inputs, and short growing season, terrestrial CDOM inputs are reduced and in-lake concentrations often extremely low [15,48]. Climate change may have different effects on these high elevation lakes depending on their position along the elevation gradient and the response of terrestrial ecosystems and hydrology to climate change. For example, lakes above the vegetation zone that have extremely low CDOM concentrations may actually have reduced transparency if the vegetation zone moves up to higher elevations with climate warming [48,49]. On the other hand, for lakes in high elevation mountain meadows, where temperature keeps evaporation low, soils are water saturated, and CDOM concentrations moderately high, increased temperature may alter the hydrologic balance to reduce soil saturation, decrease CDOM inputs to the lakes, and thus elevate

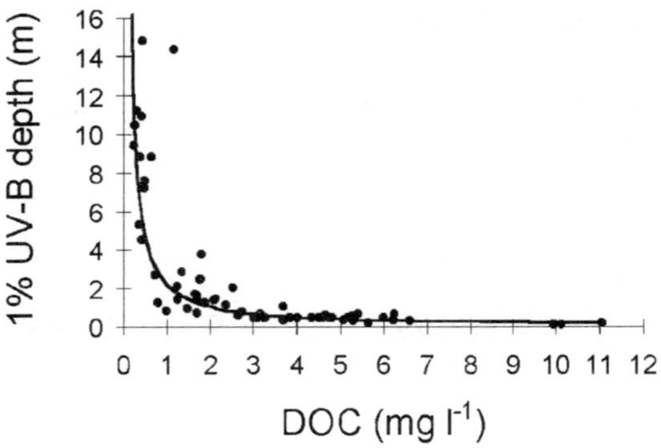

Figure 3. Relationship between the attenuation depth (depth to which 1% of surface 320 nm UV-B radiation penetrates the water column) and dissolved organic carbon (DOC) concentrations in 65 glacial lakes in North and South America [2,15]. Note the rapid increase in the depth of penetration of UV below DOC concentrations of 1–2 mg l^{-1}. [Modified from [2] with permission.]

UV levels. For example, lakes that one of us has sampled (C.E.W.) in the Beartooth Mountains of Wyoming and Montana, USA at about 3000 m, include meadow lakes with DOC concentrations in the 2–4 mg l^{-1} range, and UV attenuation depths at 320 nm of less than 0.2 m.

17.3.2 Transparency, temperature and thermal stratification

A large number of studies have demonstrated that CDOM is also an important mediator of visible water transparency and thermal stratification in lakes [22,23,47,50,51]. This suggests that climate change can alter both the mixing regime in the water column as well as the temperature at which organisms are exposed to UV. Some of the most compelling data have come from studies on Canadian lakes that have the advantage of either long-term data [1,52,53] or large numbers of lakes [22,47]. These studies have demonstrated clear relationships between CDOM, water transparency and mixing depth [22,47], as well as climate variables such as air temperature and precipitation [52]. Increases in air temperature and decreases in precipitation are associated with lower CDOM, greater water transparency, and increased heating of deeper waters. Simultaneous UV measurements in one of these studies indicate that the 1% attenuation depth for UV-B and UV-A may exceed the depth of the thermocline in some lakes [22]. In another study in small (< 500 ha) Canadian Shield lakes, CDOM was found to be the major regulator of water transparency and the depth of the summer mixed layer [54]. Climate change (a $2 \times CO_2$ scenario) was predicted to increase mixing depths of these small lakes by as much as 1–2 m. This is important because variation in vertical mixing is considered one of the most fundamental factors in regulating the exposure of phytoplankton and zooplankton to ambient UV in freshwater and marine systems [40,55,56] (see Chapter 4). Simulation modeling efforts based on 40 years of data from 71 shallow Dutch lakes revealed significant increases in lake temperature as well as increases in the probability and timing of periods of high water transparency, the so-called "clear-water phase" events [57]. While these clear-water phase events have been well demonstrated for visible light in many lakes [58–60], only recently have these seasonal patterns of increasing transparency also been demonstrated for UVR [16,61]. These seasonal changes in transparency are largely mediated by UV photolysis [16] and can result in increases in the 1% (of surface irradiance) attenuation depths for 320 nm UV from 3 m in the spring and autumn to 17 m in the early summer [61].

Climate change is also altering the timing of ice cover in both marine and freshwater systems. Between 1978–1998 the extent of multiyear ice cover in the Arctic Sea has been reduced by about 14%, with simultaneous reductions in the thickness of the ice cover [62,63]. The 20-year time scale of these studies is, however, inadequate to be able to distinguish whether these changes are related to short-term climate fluctuations or longer-term climate trends. Longer-term records are available for ice cover on Northern Hemisphere lakes where ice-out is occurring an average of 6.5 days earlier per 100 years [64]. These changes in the pattern and timing of ice cover may be important for several reasons. Reductions in the extent or duration of ice cover may increase exposure to ambient UV because ice is less transparent to UV than is water [65], particularly when there are many air bubbles in, substantial snow cover on, the ice. In spite of this reduced transparency of ice to UV, in some systems such as Lake Vanda in Antarctica, enough UV to cause inhibition of algal growth can penetrate several metres of ice [66]. In temperate lakes following early ice-out, vertical mixing may be sustained for longer periods in lakes due to lower solar irradiance or greater winds earlier in the year. The cooler temperatures at which increased UV exposure will occur following ice out may also reduce the effectiveness of enzyme-driven, temperature-dependent photorepair of UV-damaged DNA [67,68].

Temperature and UV generally vary along the same environmental gradients, but changes are not parallel over space and time. Seasonal changes in temperature generally lag behind changes in UV in a way that creates an early season peak in the UV : temperature (UV : T) ratio (Figure 4). Inverse relationships between temperature and UV with elevation can cause even wider variation in seasonal UV : T ratios. For example, UV increases and temperature decreases with increasing elevation [69] such that alpine lakes are likely to experience some of the highest UV : T ratios of any aquatic ecosystems.

Seasonal patterns in both temperature and UV are also being altered by regional and global climate changes. Observed increases in UV-B that are related to ozone depletion at north temperate latitudes tend to be most severe during the late winter to early spring. A 1.6°C warming has been reported for North American boreal forests during the period between 1970–1990 [70]. Increases in CDOM photobleaching [71] and declines in CDOM concentrations [1,70] that further elevate underwater UV may accompany such regional climate change. Photobleaching of CDOM can lead to substantial seasonal increases in water transparency to damaging UVR [16]. The timing of the peak

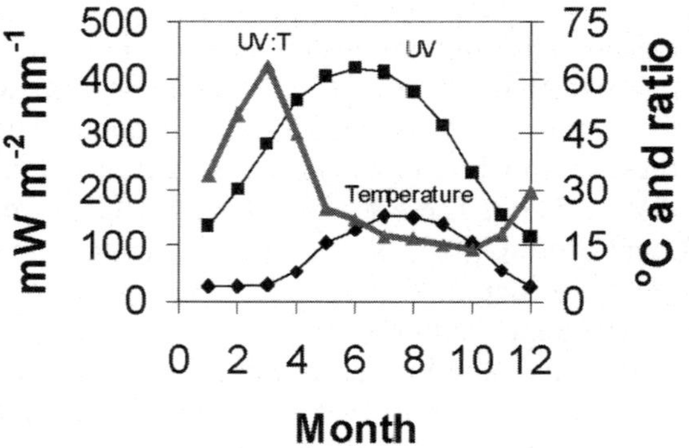

Figure 4. Seasonal variations in ambient UV (squares, 320 nm, from Biospherical Instruments Inc. RT95 radiative transfer model) and surface water temperatures (diamonds, monthly averages from our Lake Giles data 1989–1993) lead to strong variation in seasonal UV : T ratios (heavy line, triangles). These ratios may be critical to aquatic organisms due to the temperature dependence of molecular repair of UV damage. These seasonal UV : T patterns are for the surface of a lake. Underwater UV and temperature exposures of organisms will depend on interactions between CDOM, temperature, and UV at the ecosystem level (Figures 1,6), while the response of organisms to a given exposure level will depend on the interactive effects of temperature and UV at the molecular level (Figure 5). [Modified from [68], with permission.]

UV transparency is close to summer solstice when incident solar radiation is also high. This has important implications for the timing of UV impacts on aquatic organisms that may show strong seasonal peaks in reproduction [61,72]. Such time constraints imposed by seasonal variations in UV and temperature may have severe consequences for reproductive success and subsequent survival of some fish [73,74].

Both seasonal and spatial variations in UV : T ratios are likely to alter UV impacts (Figure 5). At the most fundamental level, organisms may have differences in their temperature optima. In addition, both antioxidants and molecular repair of UV damage are enzyme-driven and thus one would expect them to be temperature dependent. UV damage itself on the other hand is likely to be largely temperature independent. These relationships are not straightforward however. There is evidence that while in some cases UV tolerance does indeed increase at higher temperatures [67,75] in other cases UV tolerance shows either little change with temperature [76], or even decreases at higher temperatures [77,78]. The complexity of these processes is clearly illustrated by the fact that the specific activity of two antioxidant enzymes in some *Daphnia* (catalase and glutathione transferase) increases at higher temperatures, yet survival may be greater at lower temperatures over this same temperature range [78]. A partial explanation may be related, as mentioned above, to variations in the temperature optima of *Daphnia* and other aquatic organisms [79,80], and the importance of temperature-dependent photorepair of UV damaged DNA relative to photoprotection in a given species [68].

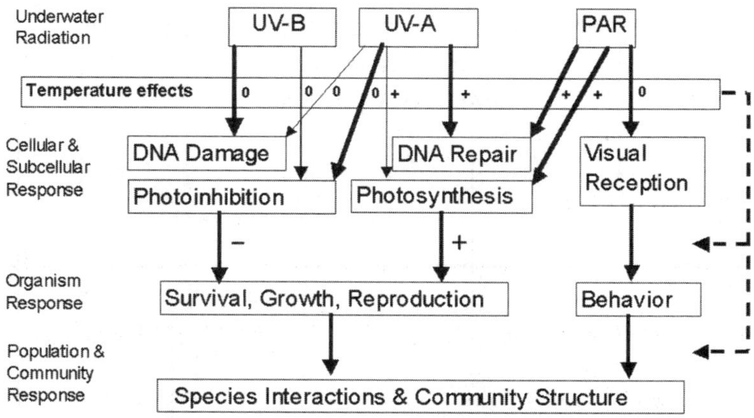

Figure 5. Conceptual model showing how temperature influences the effects of different wavelengths of solar radiation on some components of living organisms (+) but not others (0). The nature of the "+" effects (positive or negative) will depend on the temperature optimum of the enzyme system involved. The width of the arrows approximates the strength of the effects.

The impact of temperature–UV interactions on aquatic organisms may also be modulated by changes in development rates as well as by other environmental factors. For example, lower temperature at higher elevations and higher latitudes may slow development rates and prolong the period of exposure of the juvenile stages of UV-sensitive vertebrates to damaging UV [81,82]. Other environmental factors such as pH may also interact with UV. For example, in some cases pH may act synergistically with UV exposure to reduce survival in amphibians [83,84], while in other cases it may not [85]. The presence of pesticides and other specifically phototoxic compounds may also influence UV–temperature interactions and consequent responses of zooplankton and other aquatic organisms [86–88].

17.4 Modeling climate–UV interactions in aquatic ecosystems

The complex spatial and temporal interactions between climate change and UVR make it difficult to model and predict effects of these processes on aquatic ecosystems. Ozone depletion, as modified by greenhouse gases, will alter the levels of UV incident at the water's surface. In contrast, climate change is more likely to alter the underwater UV environment by influencing watershed and in-lake processes (Figure 6). Removing riparian or littoral vegetation, clear-cutting forests, increasing water usage, and other anthropogenic or climate-induced changes all may alter water transparency, surface water temperatures, and mixing depths. The exposure of aquatic ecosystems to underwater UV involves a complex set of feedback loops between CDOM, UV, PAR, IR, and temperature (Figures 1 and 6). These diagrams (Figures 1 and 6) do not even consider the complex interactions between UV damage and variability in the

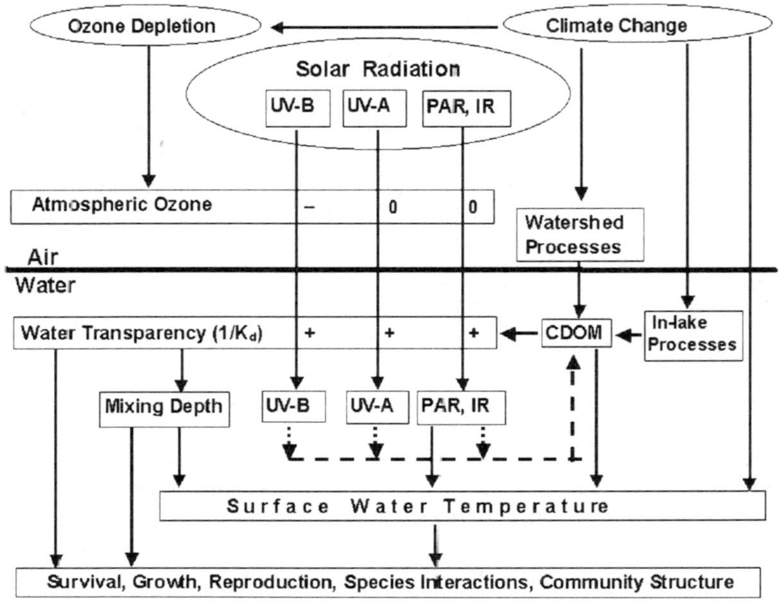

Figure 6. Conceptual model of how the key ecosystem level processes involving CDOM (Figure 1) and the molecular, organism, population, and community level effects of temperature and UV (Figure 5) can be integrated within the context of climate change and ozone depletion. This makes lakes particularly good models for examining the interactive effects of temperature and UVR on natural ecosystems.

responses of organisms from the cellular and subcellular to the population and community levels (Figure 5).

Some progress has been made in recent years toward quantitative modeling of the diverse processes that influence the effects of UV on aquatic ecosystems. At the heart of these models is the biological weighting function (BWF) concept. Biological weighting functions quantify the wavelength-dependence of photon-specific effects of UV, where shorter wavelengths of UV are generally much more damaging than longer wavelengths [89,90]. In order to estimate the biological effectiveness of a given set of UV conditions, the incident UV spectrum must be multiplied by the BWF for the organism of interest (Figure 7). While this quantitative approach has its limitations, it is essential to understanding how climate change-induced variations in the quality and quantity of CDOM alter the incident UV spectrum, and how aquatic organisms and processes will respond. For example, even within a single lake there are strong variations in the wavelength-specific absorptivity of CDOM that alter the spectral slope and thus the selective attenuation of different wavelengths of UV within lakes [16]. Another example is found in endorheic systems where CDOM is exposed to sunlight over a period of many years and consequently this CDOM absorbs very little UV compared to younger CDOM. Climate change is likely to reduce stream flows, create more endorheic systems, and reduce CDOM inputs from

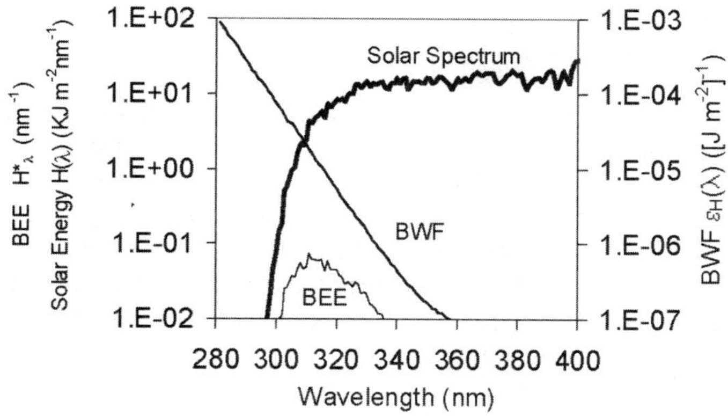

Figure 7. Biologically effective exposure (BEE) for mortality in the cladoceran *Daphnia pulicaria* as estimated from a 7 h solar phototron exposure experiment. BEE is estimated by multiplying the biological weighting function (BWF, an estimate of the wavelength-specific effects of UV), times the cumulative solar energy spectrum (here a 7 h exposure period during midday). [Modified from [59], with permission.]

wetlands. All of these factors will potentially increase the UV exposure of ecosystems. The strong changes in the spectral composition of UV related to changes in quality and quantity of CDOM as well as to changes in water depth must be accounted for when estimating the impact of these increased UV exposures. The responses of both living organisms and chemical processes are also strongly wavelength dependent, and must be accounted for with weighting functions.

To date, BWFs have been developed for a variety of aquatic organisms ranging from marine phytoplankton [89–91], to marine copepods and fish [92,93], freshwater copepods [94], and freshwater cladocerans [95]. The application of BWFs is dependent upon the reciprocity principle, which states that the effect of a given dose is independent of the rate at which the dose is administered [89,96]. In essence, the reciprocity principle is not valid if a high dose rate administered over a short period of time gives a different response than a low dose rate of the same total dose administered over a longer time period.

Some of the first modeling efforts to quantify the effects of ozone depletion on aquatic ecosystems were with phytoplankton photosynthesis in Antarctica where the close proximity of regions of high and low ozone facilitated intercomparisons [97,98]. These studies indicated that inhibition of phytoplankton photosynthesis related to ozone depletion was $\leq 5\%$. In more recent years enough information has accumulated on BWFs for photosynthesis in phytoplankton that comparative summaries are possible. Work on the Rhode River, a sub-estuary of the Chesapeake Bay, has demonstrated short-term variation in the BWFs for phytoplankton, but no significant patterns of seasonal variation [99]. Comparisons among systems in this same study demonstrated striking similarities for the BWFs from the Rhode River and Antarctic phytoplankton.

Three recent modeling efforts have made particularly important progress towards quantifying and integrating the various factors that influence the effects of UV on aquatic organisms. Neale and others [55] developed a model for estimating the effects of ozone depletion and vertical mixing on photosynthesis in Antarctic phytoplankton. This modeling effort included sensitivity analyses of the effects of variations in ozone depletion, phytoplankton sensitivity to UV, vertical mixing, and cloud cover. The results indicated that a 50% reduction in stratospheric ozone could reduce water column integrated photosynthesis by up to 8.5%. The other factors tested were predicted to have a stronger influence than ozone depletion. This included effects of variations in phytoplankton sensitivity to UV (±46%), vertical mixing (±37%), and cloud cover (±15%). Subsequent modeling efforts by Neale include separate terms for repair as well as damage by UV [100].

Huot et al. [40] developed a numerical simulation model for estimating DNA damage in marine bacterioplankton based on spectral scalar irradiance, biological weighting functions for DNA damage and photoenzymatic repair, and vertical mixing. The effects of chlorophyll and CDOM on spectral composition were incorporated into the irradiance portion of the model, and excision repair was included. One of the strengths of this model is the explicit inclusion of photoenzymatic repair (PER) of UV-damaged DNA, based here on the action spectrum for PER in *Escherichia coli*. This is an important step forward because of the dependence of the application of BWFs on the validity of the reciprocity principle [89,96]. In the presence of PER, the reciprocity principle may be invalid [101], and the ability to extrapolate BWFs to different exposure regimes is severely compromised [64]. The results of the Huot et al. model agreed reasonably well with estimates of net DNA damage (cyclobutane pyrimidine dimers) from field measurements and experimental incubations of dosimeters and natural bacterioplankton.

Kuhn et al. [41] have developed a numerical simulation model for looking at UV effects on the embryos of marine copepods and fish that also includes spectral irradiance, BWFs for embryo mortality, and vertical mixing processes. Their model differs from that of Huot's in that it uses downwelling rather than scalar irradiance, and it models mortality (daily survival) rather than net DNA damage. The results of the simulations suggest that cod embryos (*Gadus morhua*) are insensitive to environmental UVR in the St. Lawrence estuary, with the average daily survival rates on the order of 99%. The calanoid copepod *Calanus finmarchicus* was found to be more sensitive to environmental UV with daily survival rates reduced to about 90%. These modeling efforts do not include the potential for more chronic sublethal effects on these organisms and in this sense they may underestimate UV impacts.

Sensitivity analyses were carried out with both of these simulation models to examine which factors were most important in regulating UV damage in aquatic ecosystems. Both models found that changes in UV attenuation by the water column due primarily to mixing processes are of primary importance in determining UV damage [40,41]. Over the range of DOM concentrations for the marine systems tested by Huot et al. (0–0.5 g m^{-3}, much lower than most freshwater systems), variations in DOM were found to be of little importance.

Extrapolating these results to freshwater systems where CDOM concentrations are generally much higher and more variable and vertical gradients in UV much steeper would suggest that variations in the quantity or quality of CDOM is critical to the extent of UV damage. The sensitivity analyses gave contrasting results for the importance of ozone depletion. Net DNA damage was found to be quite sensitive to changes in atmospheric ozone [40], while ozone depletion had only a minimal influence on mortality in *C. finmarchicus* [41].

These kinds of quantitative models are essential if we are to integrate the complex processes involving the effects of climate change on UV effects in aquatic ecosystems. The strong wavelength-selective absorption of UV by CDOM, and the variation in the optical characteristics of this CDOM, make such a quantitative approach essential. For example, CDOM with a higher aromaticity will tend to absorb more short wavelength UV than CDOM with lower aromaticity. If one simply estimates UV exposure from changes in CDOM without taking into account the spectral absorption of the CDOM as well as the spectral sensitivity of the organisms or processes of interest, this could result in misleading interpretations or predictions. On the other hand, collecting quantitative data to validate these models will be a great challenge, particularly for higher trophic levels where reciprocity [101], temperature dependent photoenzymatic repair [68] and active behavioral avoidance [102] may vary greatly among taxa and even within taxa for different life history stages. Other environmental factors such as pH, temperature tolerance, photosensitizers, and toxicants will also have to be taken into account.

Acknowledgements

We thank Don Morris, Patrick Neale, and David Mitchell for their input into the CDOM portions of the conceptual diagrams and figures, Steve Lichak for his help drafting the figures, and Gabriella Grad Dee for her comments on the manuscript. This work was supported in part by NSF grants DEB 9973938 and DEB 0210972, International Foundation for Science (grant A/2325-3) and the Interamerican Institute for Global Change Research (grant CRN-016).

References

1. D.W. Schindler, P.J. Curtis, B.R. Parker, M.P. Stainton (1996). Consequences of climate warming and lake acidification for UV-B penetration in North American boreal lakes. *Nature*, **379**, 705–708.
2. C.E. Williamson, R.S. Stemberger, D.P. Morris, T.M. Frost, S.G. Paulsen (1996). Ultraviolet radiation in North American lakes: attenuation estimates from DOC measurements and implications for plankton communities. *Limnol. Oceanog.*, **41**, 1024–1034.
3. N.D. Yan, W. Keller, N.M. Scully, D.R.S. Lean, P.J. Dillon (1996). Increased UV-B

penetration in a lake owing to drought-induced acidification. *Nature*, **381**, 141–143.
4. R. Pienitz, W.F. Vincent (2000). Effect of climate change relative to ozone depletion on UV exposure in subarctic lakes. *Nature*, **404**, 484–487.
5. J.B. Kerr, C.T. McElroy (1993). Evidence for large upward trends of ultraviolet-B radiation linked to ozone depletion. *Science*, **262**, 1032–1034.
6. M. Rex, N.R.P. Harris, P. von der Gathen, R. Lehmann, G.O. Braathen, E. Reimer, A. Beck, M.P. Chipperfield, R. Alfier, M. Allaart, F. O'Connor, H. Dier, V. Dorokhov, H. Fast, M. Gil, E. Kyro, Z. Litynska, I.S. Mikkelsen, M.G. Molyneux, H. Nakane, J. Notholt, M. Rummukainen, P. Viatte, J. Wenger (1997). Prolonged stratospheric ozone loss in the 1995–96 Arctic winter. *Nature*, **389**, 835–838.
7. J. Herman, R.L. McKenzie (1999). Ultraviolet Radiation at the Earth's surface. In: C.A. Ennis (Ed.), *Scientific Assessment of Ozone Depletion*: 1998, Geneva (pp. 9.1–9.44).
8. D.T. Shindell, D. Rind, P. Lonergan (1998). Increased polar stratospheric ozone losses and delayed eventual recovery owing to increasing greenhouse-gas concentrations. *Nature*, **392**, 589–592.
9. S. Madronich, R.L. McKenzie, L.O. Bjorn, M.M. Caldwell (1998). Changes in biologically active ultraviolet radiation reaching the Earth's surface. *J. Photochem. Photobiol. B: Biol.*, **46**, 5–19.
10. R. Stolarski, R. Bojkov, L. Bishop, J. Zerefos, J. Staehelin, J. Zawodny (1992). Measured trends in stratospheric ozone. *Science*, **256**, 342–349.
11. C. Bruhl, P.J. Crutzen (1989). On the disproportionate role of tropospheric ozone as a filter against solar UV-B radiation. *Geophys. Res. Lett.*, **16**, 703–706.
12. S.C. Liu, S.A. McKeen, S. Madronich (1991). Effect of anthropogenic aerosols on biologically active ultraviolet radiation. *Geophys. Res. Lett.*, **18**, 2265–2268.
13. S. Madronich (1994). Increases in biologically damaging UV-B radiation due to stratospheric ozone reductions: A brief review. *Arch. Hydrobiol. Beih. Ergebn. Limnol.*, **43**, 17–30.
14. C.E. Williamson, D.P. Morris, M.L. Pace, O.G. Olson (1999). Dissolved organic carbon and nutrients as regulators of lake ecosystems: Resurrection of a more integrated paradigm. *Limnol. Oceanogr.*, **44**, 795–803.
15. D.P. Morris, H.E. Zagarese, C.E. Williamson, E.G. Balseiro, B.R. Hargreaves, B. Modenutti, R. Moeller, C. Queimaliños. (1995). The attenuation of solar UV radiation in lakes and the role of dissolved organic carbon. *Limnol. Oceanogr.*, **40**, 1381–1391.
16. D.P. Morris, B.R. Hargreaves (1997). The role of photochemical degradation of dissolved organic carbon in regulating the UV transparency of three lakes on the Pocono Plateau. *Limnol. Oceanogr.*, **42**, 239–249.
17. I.D. Cuthbert, P. del Giorgio (1992). Toward a standard method of measuring color in freshwater. *Limnol. Oceanogr.*, **37**, 1319–1326.
18. D.M. McKnight, R. Harnish, R.L. Wershaw, J.S. Baron, S. Schiff (1997). Chemical characteristics of particulate, colloidal, and dissolved organic material in Loch Vale Watershed, Rocky Mountain National Park. *Biogeochemistry*, **36**, 99–124.
19. D.M. McKnight, E.K. Boyer, P.K. Westerhoff, P.T. Doran, T. Kulbe, D.T. Andersen (2001). Spectrofluorometric characterization of dissolved organic matter for indication of precursor organic material and aromaticity. *Limnol. Oceanogr.*, **46**, 38–48.
20. R.G. Wetzel (1992). Gradient-dominated ecosystems: sources and regulatory functions of dissolved organic matter in freshwater ecosystems. *Hydrobiologia*, **229**, 181–198.
21. D.R. Engstrom (1987). Influence of vegetation and hydrology on the humus budgets

of Labrador lakes. *Can. J. Fish. Aquat. Sci.*, **44**, 1306–1314.
22. J.M. Gunn, E. Snucins, N.D. Yan, M.T. Arts (2001). Use of water clarity to monitor the effects of climate change and other stressors on oligotrophic lakes. *Environ. Monit. Assess.*, **67**, 69–88.
23. P.A. Bukaveckas, M. Robbins-Forbes (2000). Role of dissolved organic carbon in the attenuation of photosynthetically active and ultraviolet radiation in Adirondack lakes. *Freshwat. Biol.*, **43**, 339–354.
24. J.B. Rasmussen, L. Godbout, M. Schallenberg (1989). The humic content of lake water and its relationship to watershed and lake morphometry. *Limnol. Oceanogr.*, **34**, 1336–1343.
25. C.S. Cronan (1990). Patterns of organic acid transport from forested watersheds to aquatic ecosystems. In: E.M. Perdue, E.T. Gjessing (Eds), *Organic Acids in Aquatic Ecosystems* (pp. 245–260). John Wiley & Sons Ltd., New York.
26. C.S. Cronan, G.R. Aiken (1985). Chemistry and transport of soluble humic substances in forested watersheds of the Adirondack Park, New York. *Geochim. Cosmochim. Acta*, **49**, 1697–1705.
27. G.M. Hornberger, K.E. Bencala, D.M. McKnight (1994). Hydrological controls on dissolved organic carbon during snowmelt in the Snake River near Montezuma, Colorado. *Biogeochemistry*, **25**, 147–165.
28. D.M. McKnight, R.L. Smith, R.A. Harnish, C.L. Miller, K.E. Bencala (1993). Seasonal relationships between planktonic microorganisms and dissolved organic material in an alpine stream. *Biogeochemistry*, **21**, 39–59.
29. P.G.C. Campbell, H.J. Hansen, B. Budreuil, W.O. Nelson (1992). Geochemistry of Quebec north shore salmon rivers during snowmelt: Organic acid pulse and aluminum mobilization. *Can. J. Fish. Aquat. Sci.*, **49**, 1938–1952.
30. J.T.O. Kirk (1994). Optics of UV-B radiation in natural waters, *Arch. Hydrobiol. Beih. Ergeb. Limno.*, **43**, 1–16.
31. J.T.O. Kirk (1994). *Light & Photosynthesis in Aquatic Ecosystems.* Cambridge University Press, New York.
32. D.M. McKnight (1994). Aquatic fulvic acids in algal-rich Antarctic ponds. *Limnol. Oceanogr.*, **39**, 1972–1979.
33. R.C. Smith, K.S. Baker (1978). Optical classification of natural waters. *Limnol. Oceanogr.*, **23**, 260–267.
34. K.S. Baker, R.C. Smith (1982). Bio-optical classification and model of natural waters. 2. *Limnol. Oceanogr.*, **27**, 500–509.
35. M.A. Moran, W.M.J. Sheldon, J.E. Sheldon (1999). Biodegradation of riverine dissolved organic carbon in five estuaries of the Southeastern United States. *Estuaries*, **22**, 55–64.
36. N.M. Scully, D.R.S. Lean (1994). The attenuation of ultraviolet radiation in temperate lakes. *Archiv Hydrobiol. Beih. Ergeb. Limnol.*, **43**, 135–144.
37. I. Laurion, W.F. Vincent, D.R.S. Lean (1997). Underwater ultraviolet radiation: development of spectral models for northern high latitude lakes. *Photochem. Photobiol.*, **65**, 107–114.
38. C.E. Williamson, O.G. Olson, S.E. Lott, N.D. Walker, D.R. Engstrom, B.R. Hargreaves (2001). Ultraviolet radiation and zooplankton community structure following deglaciation in Glacier Bay, Alaska. *Ecology*, **82**, 1748–1760.
39. P.R. Leavitt, R.D. Vinebrooke, D.B. Donald, J.P. Smol, D.W. Schindler (1997). Past ultraviolet radiation environments in lakes derived from fossil pigments. *Nature*, **388**, 457–459.
40. Y. Huot, W.F. Jeffrey, R.F. Davis, J.J. Cullen (2000). Damage to DNA in bacterio-

plankton: A model of damage by ultraviolet radiation and its repair as influenced by vertical mixing. *Photochem. Photobiol.*, **72**, 62–74.
41. P.S. Kuhn, H.I. Browman, R.F. Davis, J.J. Cullen, B.L. McArthur (2000). Modeling the effects of ultraviolet radiation on embryos of *Calanus finmarchicus* and Atlantic cod (*Gadus morhua*) in a mixing environment. *Limnol. Oceanogr.*, **45**, 1797–1806.
42. N.D. Yan, W. Keller, N.M. Scully, D.R.S. Lean, P.J. Dillon (1996). Demographic and genetic evidence of the long-term recovery of *Daphnia galeata mendotae* (Crustacea: Daphniidae) in Sudbury lakes following additions of base: the role of metal toxicity. *Can. J. Fish. Aquat. Sci.*, **53**, 1328–1344.
43. P.J. Curtis, D.W. Schindler (1997). Hydrologic control of dissolved organic matter in low-order Precambrian Shield lakes. *Biogeochemistry*, **36**, 125–138.
44. P.J. Curtis (1998). Climatic and hydrologic control of DOM concentration and quality in lakes. In: D.O. Hessen, L. Tranvik (Eds), *Aquatic Humic Substances* (pp. 93–105). Springer-Verlag Berlin.
45. P.A. Bukaveckas, C.T. Driscoll (1991). Effects of whole-lake base addition on the optical properties of three clearwater acidic lakes. *Can. J. Fish. Aquat. Sci.*, **48**, 1030–1040.
46. W. F. Vincent, I. Laurion, R. Pienitz (1998). Arctic and Antarctic lakes as optical indicators of global change. *Ann. Glaciol.*, **27**, 691–696.
47. E. Snucins, J. Gunn (2000). Interannual variation in the thermal structure of clear and colored lakes. *Limnol.Oceanogr.*, **45**, 1639–1646.
48. R. Sommaruga, R. Psenner, E. Schafferer, K.A. Koinig, S. Sommaruga-Wograth (1999). Dissolved organic carbon concentration and phytoplankton biomass in high-mountain lakes of the Austrian Alps: Potential effect of climatic warming on UV underwater attenuation. *Arctic Antarctic Alpine Res.*, **31**, 247–253.
49. H.E. Zagarese, M. Diaz, F. Pedrozo, C. Úbeda (2000). Mountain lakes in North-western Patagonia. *Verh. Int. Verein. Limnol.*, **27**, 533–538.
50. D.W. Schindler (1971). Light, temperature and oxygen regimes of selected lakes in the Experimental Lakes Area, Northwestern Ontario. *J. Fish. Res. Bd. Can.*, **28**, 157–169.
51. S.W. Effler, G.C. Schafran, C.T. Driscoll (1985). Partitioning light attenuation in an acidic lake. *Can. J. Fish. Aquat. Sci.*, **42**, 1707–1711.
52. D.W. Schindler, K.G. Beaty, E.J. Fee, D.R. Cruikshank, E.R. DeBruyn, D.L. Findlay, G.A., Linsey, J.A. Shearer, M.P. Stainton, M.A. Turner (1990). Effects of climatic warming on lakes of the central boreal forest. *Science*, **250**, 967–970.
53. D.W. Schindler (1998). A dim future for boreal waters and landscapes. *BioScience*, **48**, 157–164.
54. E.J. Fee, R.E. Hecky, S.E.M. Kasian, D.R. Cruikshank (1996). Effects of lake size, water clarity, and climatic variability on mixing depths in Canadian Shield lakes. *Limnol. Oceanogr.*, **41**, 912–920.
55. P.J. Neale, R.F. Davis, J.J. Cullen (1998). Interactive effects of ozone depletion and vertical mixing on photosynthesis of Antarctic phytoplankton. *Nature*, **392**, 585–589.
56. H.E. Zagarese, B. Tartarotti, W. Cravero, P. Gonzalez (1998). UV damage in shallow lakes: the implications of water mixing. *J. Plankton Res.*, **20**, 1423–1433.
57. M. Scheffer, D. Straile, E.H. van Nes, H. Hosper (2001). Climatic warming causes regime shifts in lake food webs. *Limnol. Oceanogr.*, **46**, 1780–1783.
58. W. Lampert, W. Fleckner, H. Rai, B.E. Taylor (1986). Phytoplankton control by grazing zooplankton: A study on the spring clear-water phase. *Limnol Oceanogr.*, **31**, 478–490.
59. U. Sommer, Z.M. Gliwicz, W. Lampert, A. Duncan (1986). The PEG-model of seasonal succession of planktonic events in fresh waters. *Archiv. Hydrobiol.*, **106**,

433–471.
60. C. Luecke et al. (1990). Seasonal regulation of *Daphnia* populations by planktivorous fish: implications for the spring clear-water phase. *Limnol. Oceanogr.*, **35**, 1718–1733.
61. C.E. Williamson, B.R. Hargreaves, P.S. Orr, P.A. Lovera (1999). Does UV play a role in changes in predation and zooplankton community structure in acidified lakes? *Limnol. Oceanogr.*, **44**, 774–783.
62. O. M. Johannessen, E.V. Shalina, M.W. Miles (1999). Satellite evidence for an Arctic sea ice cover in transformation. *Science*, **286**, 1937–1939.
63. K.Y. Vinnikov, A. Robock, R.J. Stouffer, J.E. Walsch, C.L. Parkinson, D.J. Cavalieri, J.F.B. Mitchell, D. Garrett, V.F. Zakharov (1999). Global warming and Northern Hemisphere sea ice extent. *Science*, **286**, 1934–1937.
64. J.J. Magnuson, D.M. Robertson, B.J. Benson, R.H. Wynne, D.M. Livingstone, T. Arai, R.A. Assel, R.G. Barry, V. Card, E. Kuusisto, N.G. Granin, T.D. Prowse, K.M. Stewart, V.S. Vuglinski (2000). Historical trends in lake and river ice cover in the Northern Hemisphere. *Science*, **289**, 1743–1746.
65. D.K. Perovich, J.W. Govoni (1991). Absorption coefficients of ice from 250 to 400 nm. *Geophys. Res. Lett.*, **18**, 1233–1235.
66. W.F. Vincent, R. Rae, I. Laurion, C. Howard-Williams, J.C. Priscu (1998). Transparency of Antarctic lakes to solar UV radiation. *Limnol. Oceanogr.*, **43**, 618–624.
67. J.C. Roos, W.F. Vincent (1998). Temperature dependence of UV radiation effects on Antarctic cyanobacteria. *J. Phycol.*, **34**, 118–125.
68. C.E. Williamson, G. Grad, S. Gilroy, H.J. De Lange, D. Karapelou (2002). Temperature dependent UV responses in zooplankton: Implications of climate change. *Limnol. Oceanogr.*, **46**, in press.
69. M. Blumthaler, W. Ambach, R. Ellinger (1997). Increase in solar UV radiation with altitude. *J. Photochem. Photobiol. B: Biol.*, **39**, 130–134.
70. D.W. Schindler, S.E. Bayley, B.R. Parker, K.G. Beaty, D.R. Cruikshank, E.J. Fee, E.U. Shindler, M.P. Stainton (1996). The effects of climatic warming on the properties of boreal lakes and streams at the Experimental Lakes Area, northwestern Ontario. *Limnol. Oceanogr.*, **41**, 1004–1017.
71. P.J. Dillon, L.A. Molot (1997). Dissolved organic and inorganic carbon mass balances in central Ontario lakes. *Biogeochemistry*, **36**, 29–42.
72. C.E. Williamson, S.L. Metzgar, P.A. Lovera, R.E. Moeller (1997). Solar ultraviolet radiation and the spawning habitat of yellow perch, *Perca flavescens*. *Ecol. Applic.*, **7**, 1017–1023.
73. R.L. Eshenroder (1977). Effects of intensified fishing, species changes, and spring water temperatures on yellow perch, *Perca flavescens*, in Saginaw Bay. *J. Fish. Res. Board Can.*, **34**, 1830–1838.
74. D.E. Schindler (1999). Migration strategies of young fishes under temporal constraints: the effect of size-dependent overwinter mortality. *Can J. Fish. Aquat. Sci.*, **56**, 61–70.
75. H. Pakker, R.S.T. Martins, P. Boelen, A.G.J. Buma, O. Nikaido, A.M. Breeman (2000). Effects of temperature on the photoreactivation of ultraviolet-B-induced DNA damage in *Palmaria palmata* (Rhodophyta). *J. Phycol.*, **36**, 334–341.
76. R. Rae, W.F. Vincent (1998). Effects of temperature and ultraviolet radiation on microbial foodweb structure: potential responses to global change. *Freshwat. Biol.*, **40**, 747–758.
77. N.G. Hairston, Jr. (1979). The effect of temperature on carotenoid photoprotection in the copepod *Diaptomus nevadensis*. *Comp. Biochem. Physiol.*, **62A**, 445–448.
78. J. Borgeraas, D.O. Hessen (2000). UV-B induced mortality and antioxidant enzyme

activities in *Daphnia magna* at different oxygen concentrations and temperatures. *J. Plankton Res.*, **22**, 1167–1183.
79. M.V. Moore, C.L. Folt, R.S. Stemberger (1996). Consequences of elevated temperature for zooplankton assemblages in temperate lakes. *Arch. Hydrobiol.*, **135**, 289–319.
80. J.F. Gillooly, S.I. Dodson (2000). Latitudinal patterns in the size distribution and seasonal dynamics of new world, freshwater cladocerans. *Limnol. Oceanogr.*, **45**, 22–30.
81. M. Battini, V. Rocco, M. Lozada, B. Tartarotti, H.E. Zagarese (2000). Effects of ultraviolet radiation on the eggs of landlocked *Galaxias maculatus* (Galaxidae, Pisces) in Northwestern Patagonia. *Freshwat. Biol.*, **44**, 547–552.
82. J. Merilä, M. Pahkala, U. Johanson (2000). Increased ultraviolet-B radiation, climate change and latitudinal adaptation – a frog perspective. *Ann. Zool. Fenn.*, **37**, 129–134.
83. A.C. Hatch, A.R. Blaustein (2000). Combined effects of UV-B, nitrate, and low pH reduce the survival and activity level of larval cascades frogs (*Rana cascadae*), *Arch. Environ. Contam. Toxicol.*, **39**, 494–499.
84. L.E. Long, L.S. Saylor, M.E. Soule (1995). A pH/UV-B synergism in amphibians. *Cons. Biol.*, **9**, 1301–1303.
85. M. Pahkala, A. Laurila, J. Merila (2000). Ambient ultraviolet-B radiation reduces hatchling size in the common frog *Rana temporaria*. *Ecography*, **23**, 531–538.
86. M.V. Moore, C.L. Folt (1993). Zooplankton body size and community structure: effects of thermal and toxicant stress. *Trends Ecol. Evol.*, **8**, 178–183.
87. D.P. Arfsten, R. Davenport, D.J. Schaeffer (1994). UV-A coexposure enhances the toxicity of aromatic hydrocarbons, munitions, and metals to *Photobacterium phosphoreum*. *Biomed. Environ. Sci.*, **7**, 101–108.
88. B.L. Preston, T.W. Snell, R. Kneisel (1999). UV-B exposure increases acute toxicity of pentachlorophenol and mercury to the rotifer *Brachionus calyciflorus*. *Environ. Pollut.*, **106**, 23–31.
89. J.J. Cullen, P.J. Neale (1997). Biological weighting functions for describing the effects of ultraviolet radiation on aquatic systems. In: D.P. Hader (Ed.), *The Effects of Ozone Depletion on Aquatic Ecosystems* (pp. 97–118). Academic Press, San Diego.
90. P.J. Neale (2000). Spectral weighting functions for quantifying the effects of ultraviolet radiation in marine ecosystems. In: S.J. de Mora, S. Demers, M. Vernet (Eds), *The Effects of UV Radiation on Marine Ecosystems* (pp. 73–100). Cambridge University Press, Cambridge.
91. J.J. Cullen, P.J. Neale, M.P. Lesser (1992). Biological weighting function for the inhibition of phytoplankton photosynthesis by ultraviolet radiation. *Science*, **258**, 646–650.
92. J.H.M. Kouwenberg, H.I. Browman, J.J. Cullen, R.F. Davis, J.-F. St-Pierre, J.A. Runge (1999). Biological weighting of ultraviolet (280–400 nm) induced mortality in marine zooplankton and fish. I. Atlantic cod (*Gadus morhua*) eggs. *Mar. Biol.*, **134**, 269–284.
93. J.H.M. Kouwenberg, H.I. Browman, J.A. Runge, J.J. Cullen, R.F. Davis, J.-F. St-Pierre (1999). Biological weighting of ultraviolet (280–400 nm) induced mortality in marine zooplankton and fish. II. *Calanus finmarchicus* (Copepoda) eggs. *Mar. Biol.*, **134**, 285–293.
94. B. Tartarotti, W.E. Cravero, H.E. Zagarese (2000). Biological weighting function for the mortality of *Boeckella gracilipes* (Copepoda, Crustacea) derived from experiments with natural solar radiation. *Photochem. Photobiol.*, **72**, 314–319.
95. C.E. Williamson, P.J. Neale, G. Grad, H.J. de Lange, B.R. Hargreaves (2001). Beneficial and detrimental effects of UV radiation on aquatic organisms: Implications of

variation in spectral composition. *Ecol. Applic.*, **11**, 1843–1857.
96. M. Luckiesh (1930). *Artificial Sunlight, Combining Radiation for Health with Light for Vision.* D. Van Nostrand Co. Inc. New York, N.Y.
97. R.C. Smith, B.B. Prézelin, K.S. Baker, R.R. Bidigare, N.P. Boucher, T. Coley, D. Karentz, S. MacIntyre, H.A. Matlick, D. Menzies, M. Ondrusek, Z. Wan, K.J. Waters (1992). Ozone depletion: ultraviolet radiation and phytoplankton biology in Antarctic waters. *Science*, **255**, 952–959.
98. N.P. Boucher, B.B. Prezelin (1996). Spectral modeling of UV inhibition of in situ Antarctic primary production using a field derived biological weighting function. *Photochem. Photobiol.*, **64**, 407–418.
99. A.T. Banaszak, P.J. Neale (2001). Ultraviolet radiation sensitivity of photosynthesis in phytoplankton from an estuarine environment. *Limnol. Oceanogr.*, **46**, 592–603.
100. P.J. Neale, J.J. Fritz, R.F. Davis (2001). Effects of UV on photosynthesis of Antarctic phytoplankton: Models and applications to coastal and pelagic assemblages. *Rev. Chil. Hist. Nat.*, **74**, 283–292.
101. G. Grad, C.E. Williamson, D.M. Karapelou (2001). Zooplankton survival and reproduction responses to damaging UV radiation: A test of reciprocity and photoenzymatic repair. *Limnol. Oceanogr.*, **46**, 584–591.
102. D.M. Leech, C.E. Williamson (2001). In situ exposure to UV radiation alters the depth distribution of *Daphnia*. *Limnol. Oceanogr.*, **46**, 416–420.

Subject Index